1st International Conference on Fluid Dynamics and Computational Science (FDCS 2022)

AA003786

Journal of Physics: Conference Series Volume 2441

Chengdu, China
12-13 November 2022

ISBN: 978-1-7138-7155-2
ISSN: 1742-6588

Printed from e-media with permission by:

Curran Associates, Inc.
57 Morehouse Lane
Red Hook, NY 12571

Some format issues inherent in the e-media version may also appear in this print version.

This work is licensed under a Creative Commons Attribution 3.0 International Licence.
Licence details: http://creativecommons.org/licenses/by/3.0/.

No changes have been made to the content of these proceedings. There may be changes to pagination and minor adjustments for aesthetics.

Printed with permission by Curran Associates, Inc. (2023)

For permission requests, please contact the Institute of Physics at the address below.

Institute of Physics
Dirac House, Temple Back
Bristol BS1 6BE UK

Phone: 44 1 17 929 7481
Fax: 44 1 17 920 0979

techtracking@iop.org

Additional copies of this publication are available from:

Curran Associates, Inc.
57 Morehouse Lane
Red Hook, NY 12571 USA
Phone: 845-758-0400
Fax: 845-758-2633
Email: curran@proceedings.com
Web: www.proceedings.com

TABLE OF CONTENTS

Preface

1st International Conference on Fluid Dynamics and Computational Science (FDCS 2022) included original and peer-reviewed research papers and held on November 12-13, 2022, in Chengdu, China. This conference was co-organized by Southwest Jiaotong University, Huazhong University of Science and Technology, Tiangong University, sponsored by CCCC Second Harbour Engineering Company Ltd., Asian Union of Information Technology.

FDCS is an annual international conference devoted to the discussion of original work on theoretical, computational, and experimental aspects of the mechanics of fluids, with special regards to the Navier-Stokes equations. FDCS conference covers the field of CFD; heat transfer and thermodynamics; multi-phase flows; fluid-structure interaction; combustion; turbulence; shock waves and explosions; plasma flow.

FDCS aims to provide a knowledge exchange and sharing platform of physical and computational sciences for the aviation industry, automobile industry, shipbuilding industry, etc. We welcome experts in fields related to fluid dynamics and computational science to join us.

This conference was organized in technical association with the Asian Union of Information Technology and included invited keynote talks and oral paper presentations from both academia and industry. The topics covered in this FDCS 2022 are: 1. Computational Fluid Dynamics; 2. Heat Transfer and Thermodynamics; 3. Physical and Computational Sciences; 4. Other Topics in Fluid Dynamics. Each keynote speakers lasted for 30 minutes, including 5 minutes for Q&A. and for other sessions, there was 20 minutes for each presentation, including 5 minutes for Q&A. The book is useful to professionals and scientists involved or interested in fluid mechanics and industrial applications. Hopefully, all participants and other interested readers would benefit scientifically from the proceedings and also find it stimulating in the process.

We would like to thank all the author submitted papers to this FDCS 2022 and thank all the reviewers for their time and effort in reviewing articles. Especially we would like to thank the organizing committee for their valuable advices in the organization and helpful peer review of the papers.

Conference Organizing Committee,
Wuhan, China

Content from this work may be used under the terms of the Creative Commons Attribution 3.0 licence. Any further distribution of this work must maintain attribution to the author(s) and the title of the work, journal citation and DOI.
Published under licence by IOP Publishing Ltd

Journal of Physics: Conference Series **2441** (2023) 011001 doi:10.1088/1742-6596/2441/1/011001

Committees

Chairmen

Prof. Sanbao Hu, Wuhan University of Technology, China
Dr. Bian Gong, Shenzhen Technology University, China

Members

Prof. Dongtao Hu, China University of Geosciences, China
Dr. Xiaohe Deng, Wuhan University of Technology, China
Dr. Wei Wang, Chongqing Jiaotong University, China
Prof. Jianjun Xu, Northeast Petroleum University, China
Prof. Feng Liu, Huazhong University of Science and Technology, China
Dr. Qifei Du, Tiangong University, Tianjin, China
Dr. Jianghao Niu, Civil Aviation University of China, Tianjin, China
Dr. Jianpeng Shi, Xi'an Jiaotong University, China
Prof. Y.D. Shi, Zhejiang University, China
Dr. X.L. Li, Peking University, China
Dr. D.M. Li, Shandong University of Science and Technology, China
Prof. H. Chen, Shanghai University of Engineering Science, China
Associate Prof. S. Zhu, Wuhan University, China
Dr. X.W. Tian, Xi'an University of Science and Technology, China
Dr. Y.H. Dou, Xi'an Shiyou University, China
Dr. Xintian Zhou, Jiangsu Ocean University, China
Dr. Lei Jiang, Nanjing University of Science and Technology, China
Dr. H. Ma, Qingdao University of Science and Technology, China
Dr. R.J. Ma, Tongji University, China
Dr. Yong Wang, Chongqing University, China
Dr. Zhigang Fang, Wuhan University of Technology, China

Journal of Physics: Conference Series **2441** (2023) 011002 doi:10.1088/1742-6596/2441/1/011002

Peer Review Statement

All papers published in this volume have been reviewed through processes administered by the Editors. Reviews were conducted by expert referees to the professional and scientific standards expected of a proceedings journal published by IOP Publishing Publishing.

- **Type of peer review:** Double Anonymous
- **Conference submission management system:** Morressier
- **Number of submissions received:** 121
- **Number of submissions sent for review:** 73
- **Number of submissions accepted:** 73
- **Acceptance Rate (Submissions Accepted / Submissions Received × 100):** 60.3
- **Average number of reviews per paper:** 2.04
- **Total number of reviewers involved:** 13
- **Contact person for queries:**
  Name: Jie Wu
  Email: isfdcs@163.com
  Affiliation: Wuhan University of Technology

Content from this work may be used under the terms of the Creative Commons Attribution 3.0 licence. Any further distribution of this work must maintain attribution to the author(s) and the title of the work, journal citation and DOI.
Published under licence by IOP Publishing Ltd

The analysis on the flow characteristics of supersonic inlet considering structural vibration

Bing Chen, Hongyang Chen, Tao Mo, Shenshen Liu and Hongyin Jia*

Computational Aerodynamics Institute, China Aerodynamics Research and Development Center, Mianyang, Sichuan, China

*Corresponding author e-mail: hongyinjia@foxmail.com

Abstract. Inlet is the key part which can influence the performance of airframe-propulsion integration for the air-breathing vehicle. In supersonic or hypersonic condition, the local vibration of the structure will change the integrated performance greatly. The work is to study the influence of the structural vibration of the compression surface on the inlet's performance, and the vibration factors such as amplitude, wavelength and frequency are evaluated. The influence of the vibration factors on the inlet's performance is studied. The research shows that the structure of the flow field for the inlet is changed. The influence is more obvious when the vibration occurs to be more closed to the lip. Besides, the wavelength has more influence than the amplitude, and the influence of the frequency is smallest.

1. Introduction

Airframe-propulsion integration is a typical layout form of air-breathing vehicle, which could induce seriously coupling problem of inflow and outflow. Therefore, the integration performance of air breathing vehicle is highly sensitive to the change of layout's geometric parameters. The structural vibration will induce the oscillation of inlet's performance, which will have a great impact on the final integration design. Therefore, it is of great practical significance to study the influence of structural vibration on inlet's performance.

Aeroelasticity is one of the important challenges faced by aircraft design. For air breathing hypersonic vehicles, with the in-depth study of integrated coupling mechanism, researchers gradually realize that the elastic deformation and aerothermal problems of fuselage have a significant impact on the airframe-propulsion integration. The effects of aeroelasticity and aerothermoelasticity on the coupling of inflow and outflow are studied. Jack J. McNamara [1] studied the aeroelastic and thermoaeroelastic behaviour in hypersonic flow, and Adam J. Culler [2] studied the coupling between flow field, thermal field and structural field of thermoaeroelastic problem in hypersonic flow. Aiming at the coupling problem between aeroelasticity and airframe-propulsion integration of air breathing vehicle, Andrew D. Clark [3] established the analysis model, and studied the response characteristics of integrated configuration under the influence of aeroelasticity by CFD and engineering estimation. Therefore, the influence of airframe's elastic deformation on integrated performance was clarified. Relevant tools can be used to analyze the dynamic characteristics of airframe-propulsion integration of air-breathing vehicle [4-9].

Content from this work may be used under the terms of the Creative Commons Attribution 3.0 licence. Any further distribution of this work must maintain attribution to the author(s) and the title of the work, journal citation and DOI.
Published under licence by IOP Publishing Ltd

Due to the complex coupling relationship, many scholars have studied the influence of structure on inflow and outflow, which is to clarify the influence degree and explore the influence law. Kline et al. studied the effect of inlet's deformation on the performance of scramjet, and the results show that the deformation of 1.63 μm will lead to a 4.5% thrust change [10]. Ye Kun et al. studied the effects of combustion's plate amplitude, vibration wavelength and vibration frequency on combustion efficiency and total pressure loss for a single scramjet [11], and further clarified the sensitivity of ramjet performance to structural vibration.

Total pressure recovery and flow coefficient are important index parameters to measure inlet performance, and propulsion performance is directly proportional to inlet performance. Therefore, when studying the influence of structure on inflow, it can be equivalent to studying the influence of structural vibration on inlet performance. For the supersonic inlet, the compression surface is the main area of structural vibration induced by aerodynamics. Under the action of aerodynamic force, the structural vibration of the compression surface will show different amplitude, frequency and vibration wavelength, which will have different effects on the performance of the inlet.

The effect of structural vibration of multi-stage external compression surface on the performance of supersonic inlet is studied in the manuscript. The computation method is CFD. The effects of vibration factors such as amplitude, vibration frequency and vibration wavelength on the inlet's flow field are studied, and the oscillation characteristics of inlet's performance induced by structural vibration are clarified.

2. Problem description

A typical model of supersonic inlet is shown in Figure 1. Under the condition of supersonic or hypersonic flight, the dynamic pressure is large, so the plate of the external compression surface will deform and vibrate under aerodynamic induction.

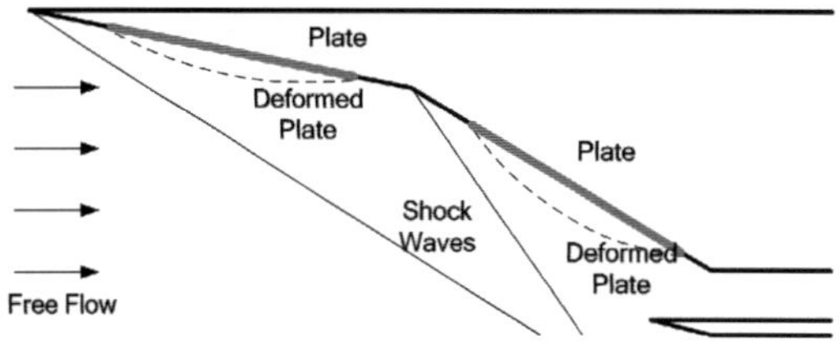

Figure 1. The model of supersonic inlet.

The structural vibration of the inlet compression surface is generally induced by the high dynamic flow after the external compression shock waves, and its impact on the inlet's performance is mainly reflected in two factors.

Firstly, the vibration leads to the disturbance of the flow near the wall, which propagates downstream along the gas flow direction, and then leads to the inlet's performance oscillation.

Secondly, the disturbance caused by structural vibration affects the external compression waves, resulting in the bending of the shock waves, so the flow field after the shock wave is non-uniform, and changes the flow features of the inlet.

Assuming that the plate is sinusoidal oscillation, the main vibration factors include amplitude, frequency and wavelength.

3. Methodology

3.1. Computation method

The inlet's flow characteristics are complex, and the boundary layer effect must be considered in the numerical calculation. The three-dimensional NS equation considering viscosity is used for aerodynamic computation. Its conservative form can be expressed as

$$\frac{\partial U}{\partial t}+\frac{\partial E}{\partial x}+\frac{\partial F}{\partial y}+\frac{\partial G}{\partial z}=\frac{1}{\mathrm{Re}}\left(\frac{\partial E_v}{\partial x}+\frac{\partial F_v}{\partial y}+\frac{\partial G_v}{\partial z}\right) \tag{1}$$

For the plate of inlet's compression surface, the parameters affecting its vibration characteristics mainly include vibration amplitude, vibration wavelength and vibration frequency. The vibration equation can be expressed as

$$y = A\sin\left(2\pi x/\lambda\right)\sin\left(2\pi ft\right) \tag{2}$$

Where, A is amplitude, λ is wavelength, f is frequency.

The flow field is calculated by moving grid technology, and the calculation time step of unsteady flow field is taken as 1 / 50 of vibration period.

3.2. Computation model

In the process of CFD computation, the mesh is continuously adjusted with the structural deformation. In order to reduce the reconstruction time of moving mesh, only moving mesh is used near the structural deformation area, and fixed mesh is still used in other areas. The grid model is shown in Figure 2.

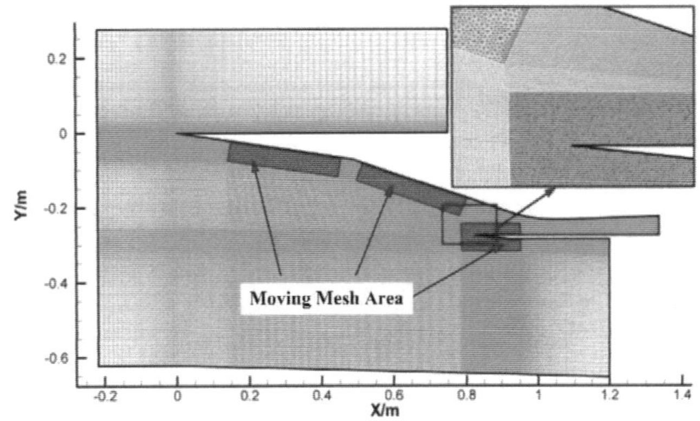

Figure 2. The mesh model.

In the calculation area, the unstructured grid is applied in the moving mesh area, and other areas adopt structural grid, with a total number of 1.04 million grids. The CFD solution adopts the density based second-order implicit solver, the turbulence model is k-omega SST model, and the flux vector splitting method is AUSMDV scheme.

3.3. Method verification

The model in Figure 4 is a typical airframe-propulsion integrated configuration of hypersonic vehicle. The inlet is the core component affecting the inflow and outflow. This paper only studies the flow of the inlet, which is simplified based on this model. The computational accuracy of CFD method without structural vibration is verified by wind tunnel test.

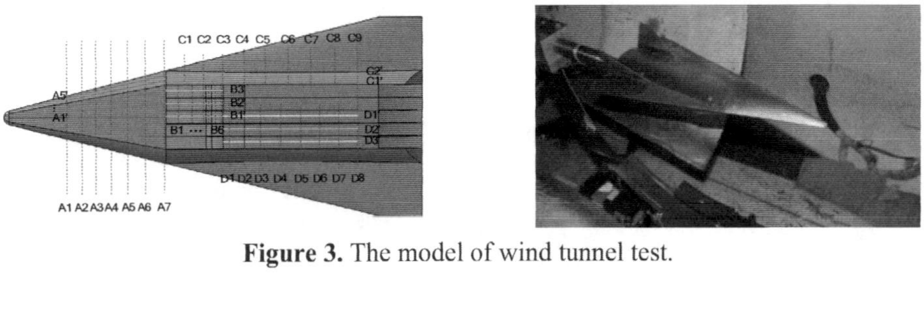

Figure 3. The model of wind tunnel test.

Under the condition of Mach 6 and 0 ° angle of attack, the pressure distribution at the longitudinal section of the windward side of the fuselage is obtained based on wind tunnel test and CFD calculation. The comparison results are shown in Figure 4.

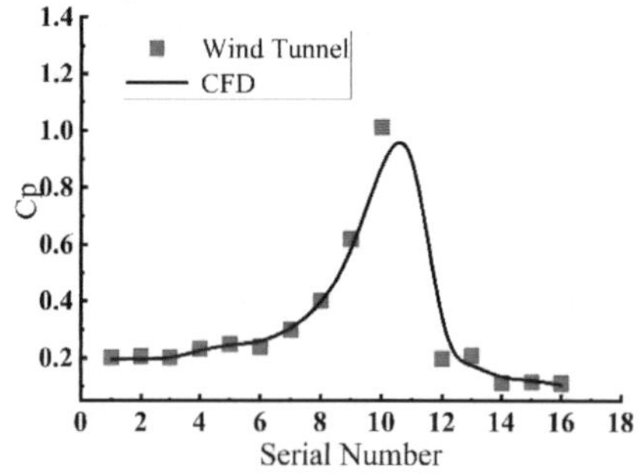

Figure 4. The comparison of CFD and wind tunnel test.

In Figure 4, the front end of the lower surface of the fuselage (before point 10) is the compression surface of the inlet, and the rear end is the outer channel. Through comparison, it can be seen that the CFD results are in good agreement with the experimental results.

4. Results and discussion

The influence of structural vibration on the vehicle's integrated performance is mainly reflected in the influence of structural vibration on the characteristics of flow field. In this paper, only unidirectional coupling is considered, and the influence of flow field on the structure is ignored.

For the two-stage external compression inlet, the effects of the two plates' vibration on the inlet's performance are quite different. In this section, the effects of the vibration of the first and second compression surface on the performance of inflow and outflow are studied respectively, which is to determine the effects of the vibration factors, mainly including vibration amplitude, vibration wavelength and vibration frequency.

The incoming Mach number of CFD calculation is 6, the angle of attack is 0 °, and the altitude is 23km.

4.1. Effect of amplitude

When the vibration frequency is 50Hz and the wavelength is 0.5m, the amplitudes of the first and second compression surface are 1mm, 3mm and 5mm respectively. The vibration period of the plate is 0.02s, the calculation time step of the unsteady flow field is 0.0005s, and the total calculation time is 0.2s, including 10 vibration periods.

Under the action of plate's vibration of the first and second compression surface with different amplitudes, the static pressure contours at t = 0.2s are shown in Figure 5 and Figure 6.

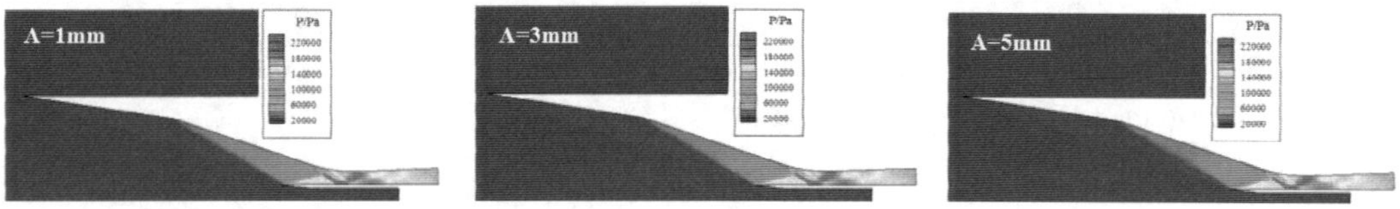

Figure 5. The pressure contours under the action of the first stage compression surface.

Under the action of the vibration of the first compression surface, the shape of the secondary compression shock wave is changed, and makes the shock wave bend. At the same time, the vibration leads to the accumulation of the high-pressure region in the inner compression section. As time goes, the high-pressure region gradually expands downstream. When the amplitude is small, the flow after the first compression shock wave is relatively stable.

With the increase of amplitude, the compression plate's vibration leads to the increase of pressure near the first stage compression surface. Finally, the first stage compression shock wave appears bending under the exclusion of the flow behind the shock wave, lead to the increase of the shock angle, and the inlet's overflow phenomenon is more obvious.

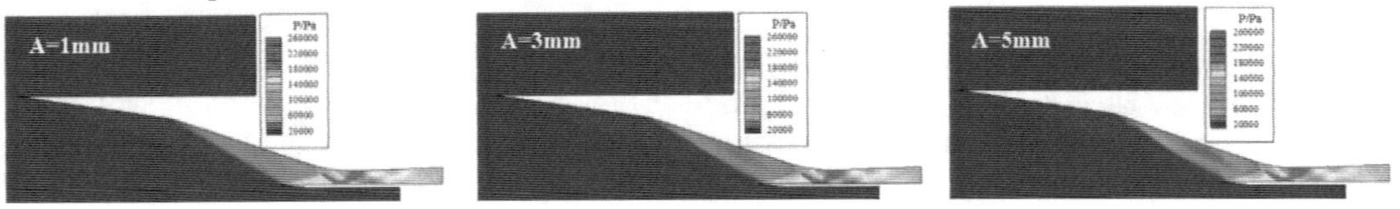

Figure 6. The pressure contours under the action of the second stage compression surface.

Compared with the first stage compression surface, the plate's vibration of the second stage compression surface will also increase the pressure of the inflow, and the high-pressure area gradually expands backward with time. The flow feature is similar to the vibration of the first stage compression surface. However, with the increase of the amplitude, the pressure distribution after the shock wave is uneven.

Under the action of the vibration, the flow after the secondary shock wave produces some weak shock waves and expansion waves, and there are obvious high-pressure area and low-pressure area, which destroys the uniformity of the flow after the wave. At the same time, the high-pressure flow also squeezes the secondary compression shock wave. It leads to the bending of the secondary compression shock wave and increases overflow of the inlet.

The performance curve of the inlet with first and second compression surface vibration is shown in Figure 7.The performance of the inlet is the parameters on inlet's outlet by mass average of CFD results. The parameters mainly contain static pressure, total pressure and Mach number.

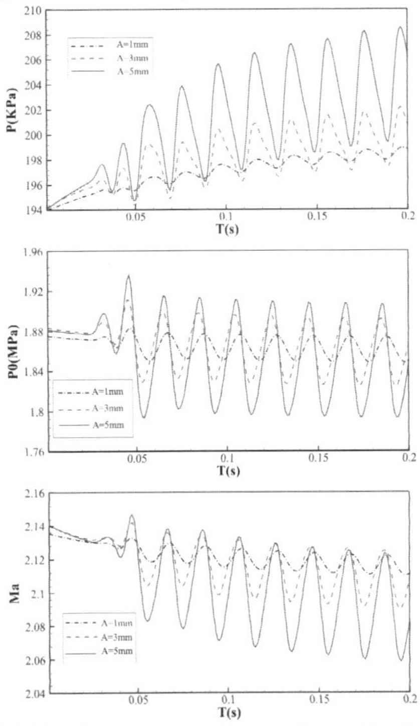

(a) The first compression surface vibration

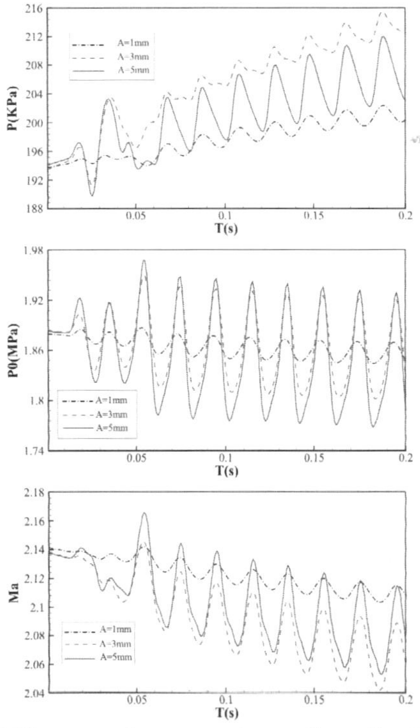

(b) The second compression surface vibration

Figure 7. The flow characteristics on the inlet's outlet.

It can be seen from the figure that the unsteady flow on the inlet's outlet has a time lag relative to the plate's vibration. The lag time of the first compression surface is about 0.028s and that of second compression surface is about 0.015s.

For the first stage compression surface, when the vibration amplitude of the plate is 1mm, 3mm and 5mm respectively, the change amplitude of pressure on the inlet's outlet accounts for 0.5%, 2.3% and 5.2% of the value of the steady flow respectively, and the Mach number accounts for 0.9%, 1.9% and 3.3% respectively. Therefore, the variation of different flow parameters with the vibration amplitude is not a simple linear relationship, and with the passage of the plate's vibration time, the average pressure on the inlet's outlet continues to increase, while the average Mach number continues to decrease. Under different vibration amplitudes, the average pressure increases by about 1%, 2% and 4.6% respectively in 10 vibration cycles, the average Mach number decreases by about 0.7%, 1.4% and 2.3% respectively in 10 vibration cycles.

For the second compression surface, when the vibration amplitude of the plate is 1mm, 3mm and 5mm, the change amplitude of pressure on the inlet's outlet accounts for 1.5%, 2.1% and 4.6% of the steady flow respectively, and the Mach number accounts for 0.9%, 1.9% and 3.3% respectively. With the continuous action of the plate's vibration, the average pressure on the inlet's outlet increases, while the average Mach number decreases. The pressure increases by about 2.6%, 10.3% and 7.2% respectively in 10 vibration cycles, while the Mach number decreases by about 1.4%, 3.3% and 2.8% respectively in 10 vibration cycles.

There is little difference between the first stage compression surface's vibration and the second stage compression surface's vibration on the amplitude of performance parameters, but the second stage compression surface's vibration has a greater impact on the average value of the performance parameters. In the same vibration cycle, the amplitude of average pressure increase and average Mach number decrease is significantly higher than that of the first stage compression surface's vibration.

For the total pressure, the average total pressure loss is about 1.06% in 10 vibration cycles of the first and second compression surface, but the oscillation amplitude of the total pressure caused by the vibration of the first compression surface is less than that of the second compression surface.

Therefore, from the perspective of vibration amplitude, more attention should be paid to the vibration of secondary compression surface.

4.2. *Effect of wavelength*

When the vibration frequency is 50Hz and the amplitude is 3mm, the vibration wavelengths of the first and second compression surface of the inlet are 0.5m, 0.25m and 0.125m respectively. Other settings are the same as Section 4.1. The cloud diagram of analysis results when t = 0.2s is shown in Figure 8 and Figure 9.

The influence of the vibration wavelength of the first compression surface on the flow feature is very obvious. Under the action of the vibrating plate, the uniform flow after the first compression shock wave generates some shock waves and expansion waves, resulting in an obvious high-pressure and low-pressure mixing zone. Under the action of the high and low pressure zone, the first and second stage compression shock waves appear to bend. With the passage of time, the high-pressure region of the inflow gradually accumulates and extends downstream.

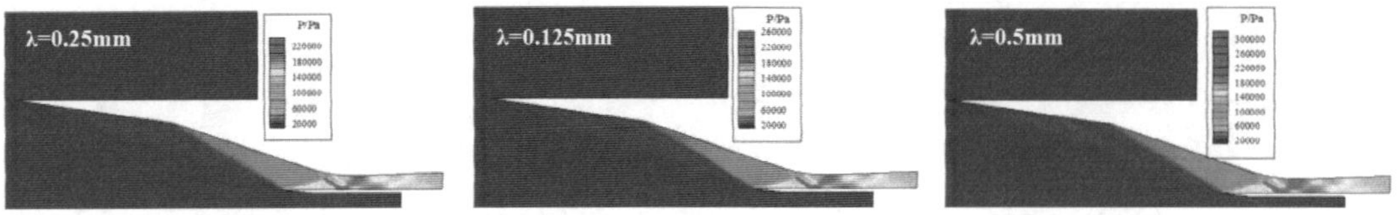

Figure 8. The pressure contours under the action of the first stage compression surface.

The vibration wavelength of the second compression surface has a certain impact on the inflow. With the passage of vibration time, the smaller the vibration wavelength produces a more obvious high-pressure region. The smaller the vibration wavelength will generate the more shock waves in the flow after the second compression shock wave. These shock waves interact with the second compression shock wave. There is obvious shock waves intersection, and even the compression shock wave is blocked by the shock waves generated by the vibrating plate.

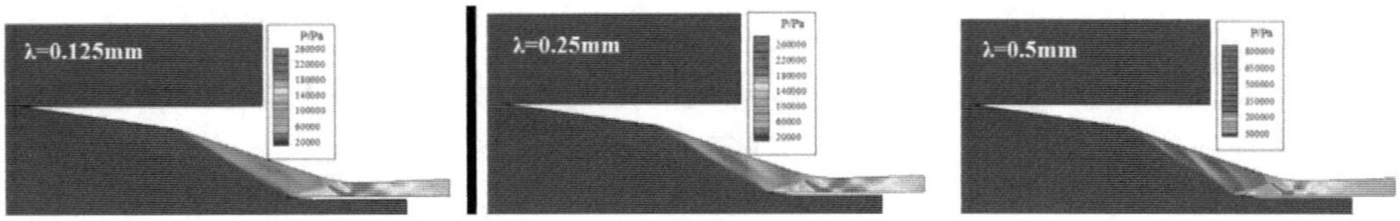

Figure 9. The pressure contours under the action of the second stage compression surface.

The performance curve of the inlet with first and second compression surface vibration is shown in Figure 10.

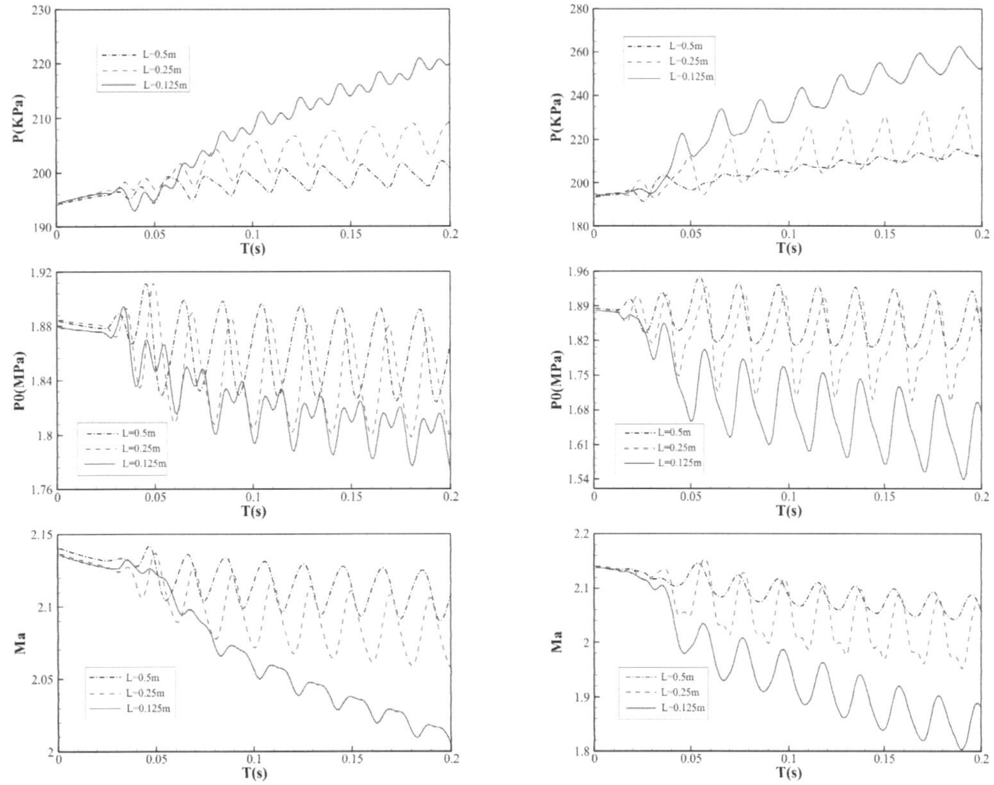

(a) The first compression surface vibration (b) The second compression surface vibration

Figure 10. The flow characteristics on the inlet's outlet.

Under the influence of the plate's vibration wavelength, the average performance parameters change greatly. When the wavelength is small, it even leads to the anharmonic oscillation of the parameters. With the decrease of the wavelength, the average parameters of the inlet's performance decrease sharply, especially for the second compression surface.

The influence of the vibration wavelength on the total pressure increases sharply with the decrease of the wavelength. When the vibration wavelength is 0.125m, the total pressure loss caused by the vibration of the first compression surface is 4.3% in 10 vibration cycles, while the total pressure loss caused by the vibration of the second compression surface is 14.4%.

4.3. *Effect of frequency*

When the vibration amplitude is 3mm and the wavelength is 0.5m, the vibration frequencies of the first and second compression surface are 30Hz, 50Hz and 70Hz respectively. Other settings are the same as Section 4.1. The cloud diagram of analysis results when t = 0.2s is shown in Figure 11 and Figure 12.

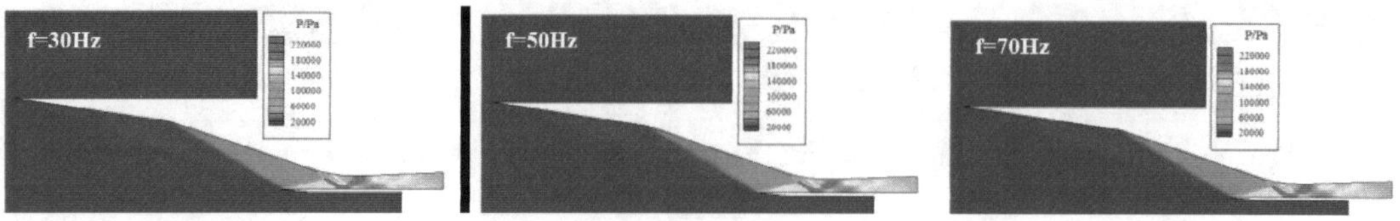

Figure 11. The pressure contours under the action of the first stage compression surface.

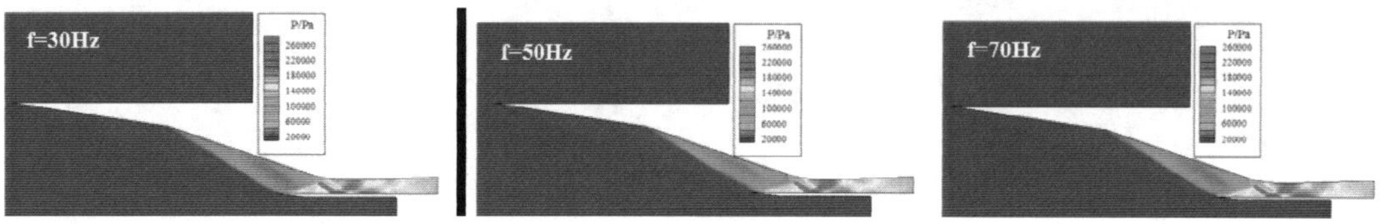

Figure 12. The pressure contours under the action of the first stage compression surface.

For the first and second compression surfaces, the influence of the vibration frequencies on the flow feature is relatively consistent. With the passage of vibration time, the high-pressure region of the flow accumulates and develops downstream, and the higher frequency will produce a more obvious high-pressure region.

The performance curve of the inlet with first and second compression surface vibration is shown in Figure 13.

With the increase of the vibration frequency of the first stage compression surface, the increase of static pressure is 2.06%, 3.09% and 5.2% respectively, and the decrease of Mach number is 1.4%, 1.86% and 2.34% respectively. With the increase of the vibration frequency of the second stage compression surface, the static pressure increases by 3.1%, 7.7% and 11.3% respectively, and the Mach number decreases by 1.86%, 2.8% and 4.7% respectively. The influence of the vibration frequency of the second stage compression surface on the performance parameters is higher than that of the first stage compression surface. The vibration frequency of the first stage compression surface has little influence on the total pressure, but the vibration frequency of the second stage compression surface will reduce the average total pressure. At the vibration frequency of 70Hz, the total pressure loss is about 3.2%.

From the perspective of amplitude, wavelength and frequency, the wavelength of the compression surface affects the final performance, the amplitude affects the oscillation amplitude of performance parameters, and the frequency affects the oscillation frequency of performance parameters of the inlet. From the perspective of the average performance, the influence degree of each vibration factor is: wavelength > amplitude > frequency. The influence of the vibration of the second stage compression surface on the inlet's performance is higher than that of the first stage compression surface, that is, the influence of the vibration on the inlet's performance is more obvious when the location of the vibration is farther.

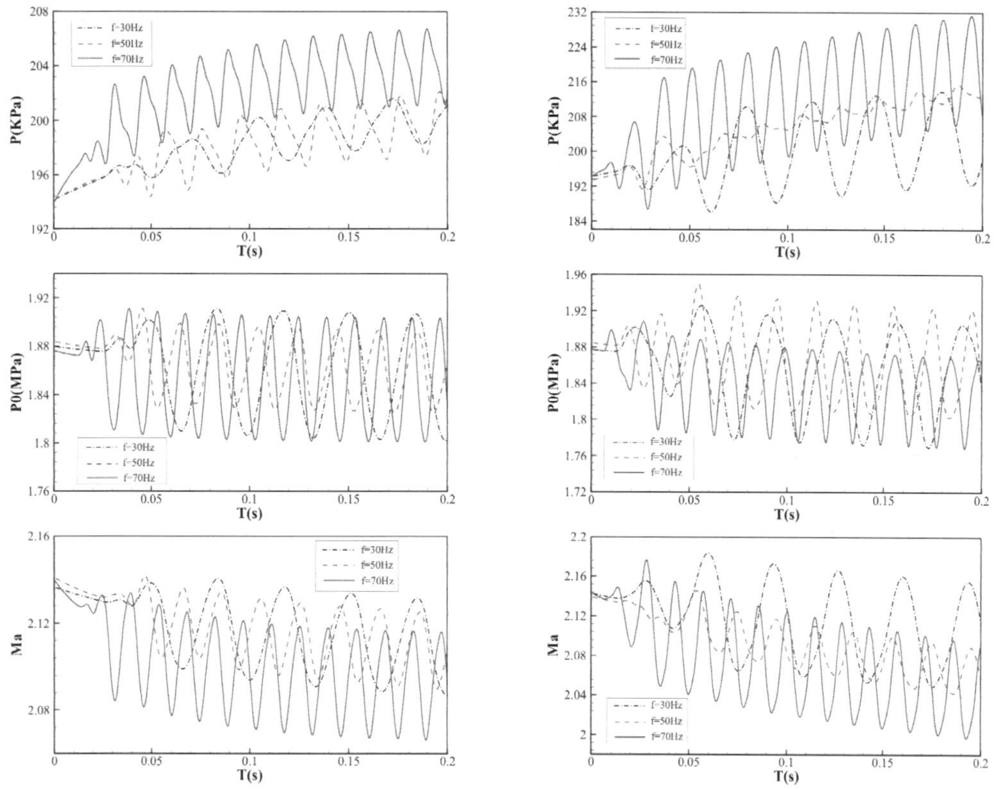

(a) The first compression surface vibration (b) The second compression surface vibration

Figure 13. The flow characteristics on the inlet's outlet.

5. Conclusion

In the paper, the influence of compression surface vibration on inlet's performance is analyzed by CFD, including various vibration factors. Due to the continuous cumulative effect of the flow, when the inlet's feature vibrates harmonically, the performance parameters of the inlet are often not a simple harmonic oscillation. However, the average performance of the inlet changes gradually with the accumulation of vibration time. Relevant research conclusions include:

(1) The influence of the vibration of the compression surface on the performance of the inlet is mainly reflected in the change of the external compression flow field and compression shock waves. The influence is more obvious when the vibration is closer to the inner compression region.

(2) The influence degree of each vibration factor on the inlet's performance is different. Among them, the vibration wavelength has the most obvious influence, followed by the amplitude, and the vibration frequency has the weakest influence on the inlet's performance.

Acknowledgments

This work was financially supported by National Numerical Windtunnel (NNW) Project.

References

[1] Jack J. McNamara etc. Aeroelastic and aerothermoelastic behavior in hypersonic Flow [J]. AIAA Journal, 2008, 46(10):2591-2612.
[2] Adam J. Culler and Jack J. McNamara. Studies on fluid–thermal–structural couplingfor aerothermoelasticity in hypersonic flow [J]. AIAA Journal, 2010, 48(8):1721-1738.
[3] Andrew D. Clark, Maj Dean Mirmirani and etc. An aero-propulsion integrated elastic model of a

generic airbreathing hypersonic vehicle [C]. AIAA Guidance, Navigation and Control Conference and Exhibit, Keyston, Colorado, 2006.

[4] Frank R. Chavez and David K. Schmidt. Analytical aeropropulsive/aeroelastic hypersonic vehicle model with dynamic analysis[J]. Journal of Guidance Control and Dynamics, 1994, 17(6): 1308-1319.

[5] David K. Schmidt. Dynamics and control of hypersonic aeropropulsivel/aeroelastic vehicles [C]. American Institute of Aeronautics and Astronautics Guidance, Navigation and Control Conference, Hilton Head Island, 1992.

[6] Frank R. Chavez and Dr. David K. Schmid. An integrated analytical aeropropulsivel aeroelastic model for the dynamic analysis of hypersonic vehicles[C]. American Institute of Aeronautics and Astronautics Guidance, Navigation and Control Conference, Hilton Head Island, 1992.

[7] Frank R. Chavez and Dr. David K. Schmidt. Dynamics of hypersonic flight vehicles exhibiting significant aeroelastic and aeropropulsive interactions[C]. American Institute of Aeronautics and Astronautics Guidance, Navigation and Control Conference, Monterey, 1993.

[8] Hui MA, Jian-Ping YUAN and Qun FANG. An analytical model of dynamics for generic airbreathing hypersonic vehicles[C]. American Institute of Aeronautics and Astronautics 14th AIAA/AHI Space Planes and Hypersonic Systems and Technologies Conference, Canberra, 2006.

[9] Baris Fidan, Matthew Kuipers, and Petros A. Ioannou. Longitudinal motion control of air-breathing hypersonic vehicles based on time-varying models[C]. American Institute of Aeronautics and Astronautics 14th AIAA/AHI Space Planes and Hypersonic Systems and Technologies Conference, Canberra, 2006.

[10] Kline H, Palacios F, Alonso J J. Sensitivity of the performance of a 3-dimensional hypersonic inlet to shape deformations[C]. 19th AIAA Space Planes and Hypersonic Systems and Technologies Conference, Atlanta, 2014.

[11] YE Kun, YE Zheng-yin, QU Zhan. Effects of panel vibration on performance of combustor for scramjet [J]. Journal of Propulsion Technology, 2017,38(2):386-398.

The impact of inlet mach number and domain shape on subsonic nozzle flow via small disturbance equation

Xuehui Qian*

School of Science and Engineering, the Chinese University of Hong Kong, Shenzhen, 518000, China

*Corresponding author e-mail: XuehuiQian2022@foxmail.com

Abstract. Computational fluid dynamics (CFD) is a cross-disciplinary field that combines mathematics, fluid dynamics, and computer programming to simulate and analyze fluid flow problems using numerical methods. Nowadays, the use of CFD to conduct numerical simulations rather than costly and time-consuming ignition experiments has become an essential aspect of nozzle design. Therefore, it is significant to study the nozzle by numerical simulation. In this paper, a numerical method called small disturbance equation (SDE) is adopted to investigate the impact of inlet Mach number and domain shape on subsonic nozzle flow. Conducting the small disturbance equation via python code, the impact of the inlet Mach number and domain shape is revealed by comparing the different errors of the method, streamlines of the nozzle flow, velocity fields. It is found that with the increase of inlet Mach number, the error of the SDE method grows, and the velocity in x-direction increases while the velocity in y-direction stays the same. Besides, with the shrink of the domain shape, the error of the SDE method gets larger, meanwhile, the maximum of the velocity in each direction increases while the minimum of the velocity decreases. This research would make advantages to identify the optimal designs of a subsonic nozzle.

Keywords: Computational fluid dynamics; Nozzle; Small disturbance equation.

1. Introduction

A nozzle is a flow-accelerating device with a changeable cross sectional area. [1] Research on the nozzle flow via numerical methods has a long tradition. The flow phenomenon in a nozzle can be solved using a variety of numerical methods. In many inviscid flow investigations published in open literatures, solving Euler equations is one of the most extensively used methods. The other widely used way is small disturbance equation method (SDE) [2]. Colona et al [3] utilized Euler equations to study one-dimensional and steady nozzle flow. It is concluded that the simulation using Euler equations worked well but for the transonic region numerical errors and discontinuity occurred. Huang [4] proposed the two numerical methods, one-dimensional Euler equations and two-dimensional SDE methods based simulation to study the nozzle flow. It is found that the SDE method accurately caught the shock bubbles in the nozzle flow, whereas the one-dimensional Euler method was insufficient for two-dimensional nozzle flow modeling due to the appearance of a mingle of subsonic and supersonic flow. Therefore, in this paper, the SDE method is chosen to investigate the nozzle flow. Cheng et al [5] simulated nozzle

Content from this work may be used under the terms of the Creative Commons Attribution 3.0 licence. Any further distribution of this work must maintain attribution to the author(s) and the title of the work, journal citation and DOI.
Published under licence by IOP Publishing Ltd

flow using small disturbance equation and Euler method. In their paper, they compared the solutions under various inlet Mach number and they found out that the velocity distribution of the nozzle various with distinct Mach number. Quintao [6] studied the design optimization of nozzle shape. In her paper, Quintao demonstrated how changes in the parameters that define the shape of a nozzle can affect nozzle exit flow non-uniformities.

This paper mainly focuses on the subsonic nozzle flow. Besides, how the inlet Mach number and domain shape of the nozzle flow would influence the subsonic nozzle flow would be investigated. To study this, the results get from different inlet Mach number and domain shape would be compared and analyzed.

The rest of this paper is organized as follows. In Section 2, the methods utilized to investigate the impact of inlet Mach number and domain shape, including the control equations for the subsonic nozzle flow, the discretization and the computational strategy, are introduced. In Section 3, the results of the impact of inlet Mach number and domain shape are presented in four aspects, the iteration and responding error, the streamline, velocity field in x-direction and velocity field in y-direction. Finally, conclusions are drawn in Section 4.

2. Method

2.1. Control Equations for Nozzle Flow:

In this paper, the small disturbance equation, which is derived from the full potential equation, is used to analyze the subsonic flow in a nozzle. The full potential equation is described as [7]:

$$(1-M_x^2)\varphi_{xx}+(1-M_y^2)\varphi_{yy}+(1-M_z^2)\varphi_{zz}-2M_xM_y\varphi_{xy}-2M_xM_z\varphi_{xz}-2M_zM_y\varphi_{yz}=0 \quad (1)$$

$$M_x=\frac{\varphi_x}{c},M_y=\frac{\varphi_y}{c},M_z=\frac{\varphi_z}{c},M^2=\frac{|\nabla\varphi|^2}{c^2} \quad (2)$$

$$c^2=(\gamma-1)\left[H_0-\frac{1}{2}|\nabla\varphi|^2\right] \quad (3)$$

In this equation, M denotes the Mach number and c denotes the speed of the nozzle flow, while H_0 denotes the stagnation enthalpy. The small disturbance equation is a simplification of the full potential equations in case of small obstacles. In this paper, the equations are restricted to 2-dimension. Since the obstacle is small, the effect on the nozzle flow is small. Therefore, perturbation is considered to uniform flow with velocity magnitude U_∞ in the x-direction. The potential can be represented as:

$$\varphi=U_\infty(x+\emptyset) \quad (4)$$

Besides, the velocities are recovered from the potential using the formulas:

$$u=U_\infty(\mathrm{x}+\emptyset_x) \quad (5)$$

$$v=U_\infty\emptyset_y \quad (6)$$

where u represents the velocity in x-direction and v represents the velocity in y-direction. Therefore, the equation becomes:

$$(1-{M_x}^2)\varphi_{xx}+\varphi_{yy}=0 \quad (7)$$

In this case, the boundary condition becomes:

$$v=(U_\infty+u)f'(x)\approx U_\infty f'(x) \quad (8)$$

where $f(x)$ is the shape of the flow and $f'(x)$ is the first derivative of $f(x)$. In this paper, the subsonic nozzle flow is studied and the $f(x)$ is defined as:

$$f(x)=\begin{cases}\cos\pi(x-1), & 1\ll x\ll 2.\\ 0, & otherwise.\end{cases} \quad (9)$$

This paper is aimed to solve the subsonic small disturbance equation with Dirichlet boundary condition.

2.2. Discretization:

Simultaneous techniques for CFD rely heavily on appropriate discretization. [8] To By applying the finite difference theory, the following discretization could be obtained:

$$\varphi_{j.i} = \frac{\varphi_{j+1,i} + \varphi_{j-1,i} + \left(1 - {M_x}^2\right)\varphi_{j,i+1} + \left(1 - {M_x}^2\right)\varphi_{j,i-1}}{2\left(1 - {M_x}^2\right) + 2} \quad (10)$$

However, it should be noted that eq.10 only applies to the calculation domain's internal grid points. In order to get the discretization for the bottom and top boundaries of the nozzle, further process is required. The initial discretization of φ_{yy} in eq.7 is:

$$\left(\varphi_{j+1,i} - \varphi_{j-1,i}\right) - \left(\varphi_y\right)_{boundary} \cdot \Delta x \quad (11)$$

where Δx is the length of space discretization in x-direction. Supposing the nozzle is x-direction positioned, and a shape function of $f(x)$, where x is the horizontal length, is used to define the upper and lower boundaries. Then the relation between the velocity in y-direction at boundary and nozzle shape can be defied as:

$$\left(\varphi_y\right)_{boundary} = \Delta x \cdot f'(x) \quad (12)$$

The discretization format of the upper and lower border can be generated after this process.

2.3. Computational Strategy:

In order to find a numerical solution to the small disturbance equation, the successive over relaxation method is utilized. Successive over relaxation method (SOR) is an iteration method aimed to solve the linear system of equations. This method is derived by extrapolating the Gauss-Seidel method. The final SOR formula can be written as follows [9]:

$$u_i^{(k)} = (1 - \omega)u_i^{(k-1)} + \frac{\omega}{a_{ii}}[b_i - \sum_{j=1}^{i-1} a_{ij}\, u_j^{(k)} - \sum_{j=i+1}^{n} a_{ij}\, x_j^{(k-1)}] \quad (13)$$

In this case, ω is set as 1.4, where an optimal solution would be get using SOR method.

3. Result and Discussion

In this paper, the impact of the inlet Mach number and domain shape on the subsonic nozzle flow is explored. In order to find out in what extend they would make a difference to the nozzle, respectively, control variates would be introduced. To be more specific, the subsonic nozzle flow is discussed in two conditions. When exploring the impact of the inlet Mach number, the domain shape will remain the same while the inlet Mach number is changing. And when exploring the impact of the domain shape, the inlet Mach number will remain as a constant and only the domain shape will change. The following are the results of the two cases.

3.1. Impact of Mach Number

Since the nozzle flow is a subsonic one, the Mach number should be less than 1. In this case, the domain shape keeps the same and the Mach number is changed into 0.1, 0.2, 0.3 and 0.5, respectively. To represent the impact on the nozzle flow in different aspects better, the results will be showed in four groups, and each group indicates the difference among the four different Mach numbers on different parameters. Here M_{inf} is used to represent the inlet Mach number.

3.1.1. Iteration and Error. Fig. 1 represents the iteration times versus the natural algorithm of error of different input Mach number. It can be observed that as the Mach number increases, as the iteration grows, the error becomes smaller slower, which indicates that the equation converges slower as the Mach number grows larger. What's more, there is a significant drop when the Mach number is taken as 0.5. And in this condition, the error finally converges to a number smaller than other conditions (0.1, 0.2 and 0.3). Therefore, as Mach number increases, the precise also decreases.

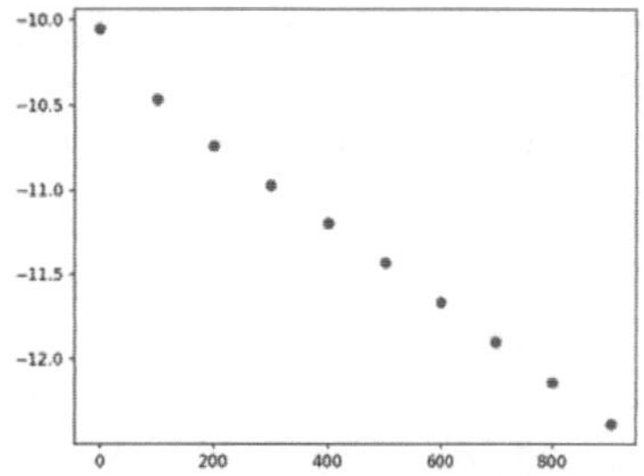

(a) $M_{inf} = 0.1$

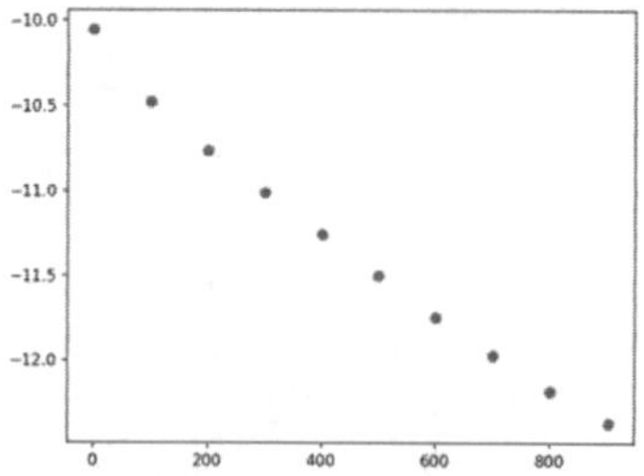

(b) $M_{inf} = 0.2$

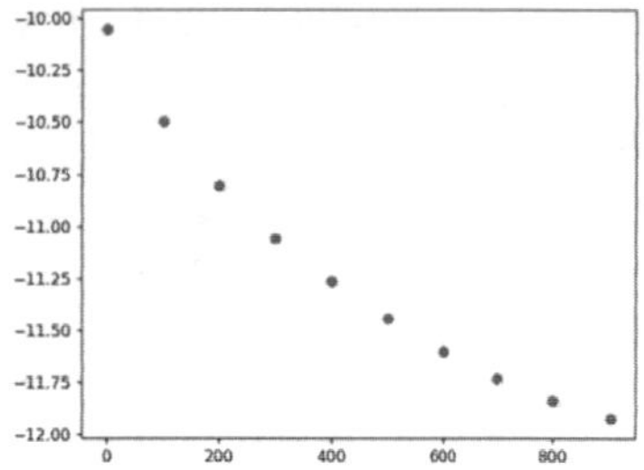

(c) $M_{inf} = 0.3$

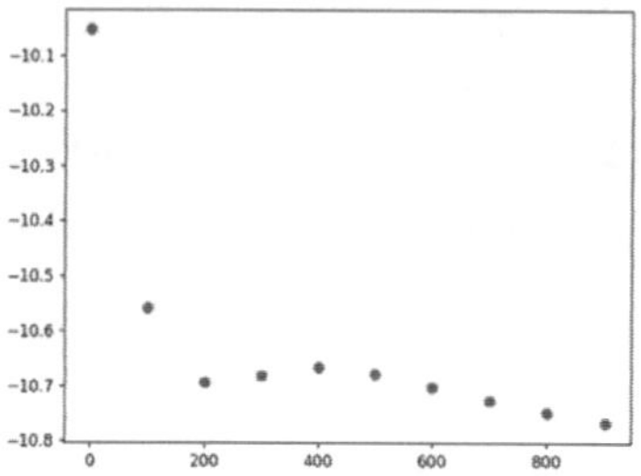

(d) $M_{inf} = 0.5$

Fig. 1 iteration vs. error of different M_{inf}

3.1.2. Streamline. Fig. 2 shows the streamlines among different Mach numbers. As the picture represents, while Mach number increasing, the streamlines become less extreme.

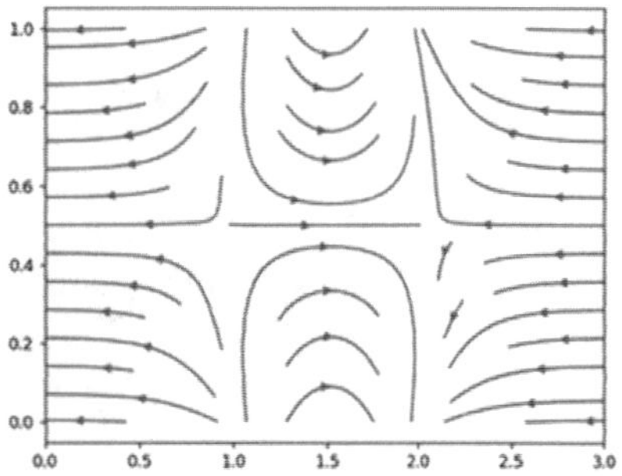

(a) $M_{inf} = 0.1$

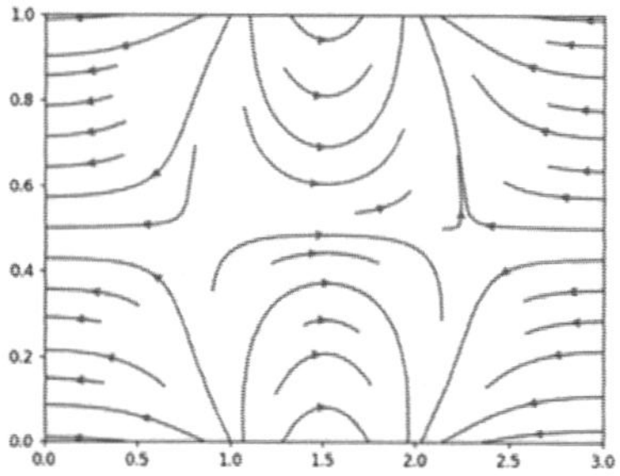

(b) $M_{inf} = 0.2$

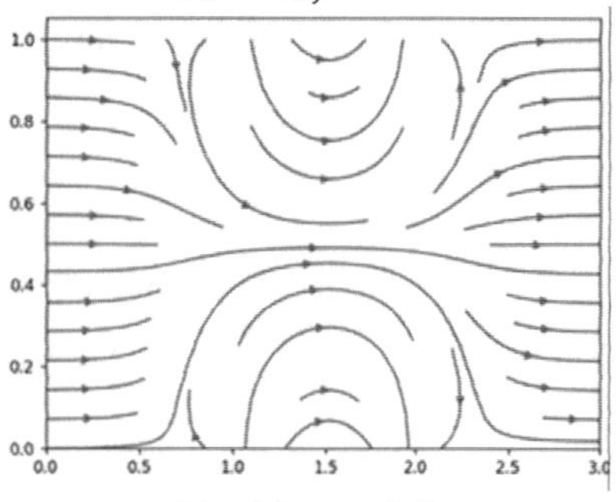

(c) $M_{inf} = 0.3$

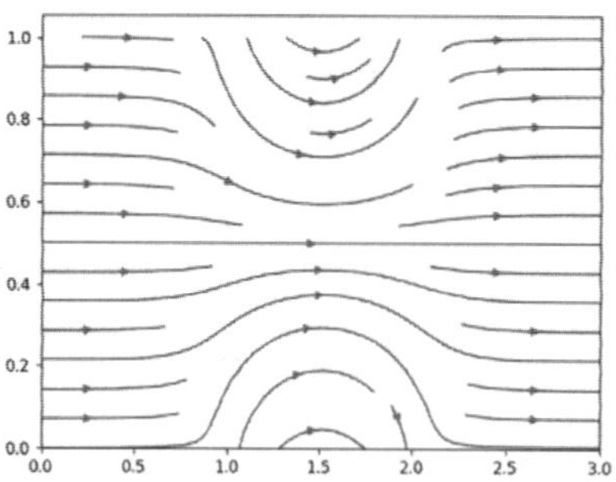

(d) $M_{inf} = 0.5$

Fig. 2 streamlines of different M_{inf}

3.1.3. Velocity in x-direction. Fig. 3 is about the velocity in x-direction of different Mach numbers. It can be easily observe that as the Mach number increases, the minimum of the velocity together with the maximum of the velocity both increase. However, the range of the velocity doesn't vary a lot. The whole graph moves up.

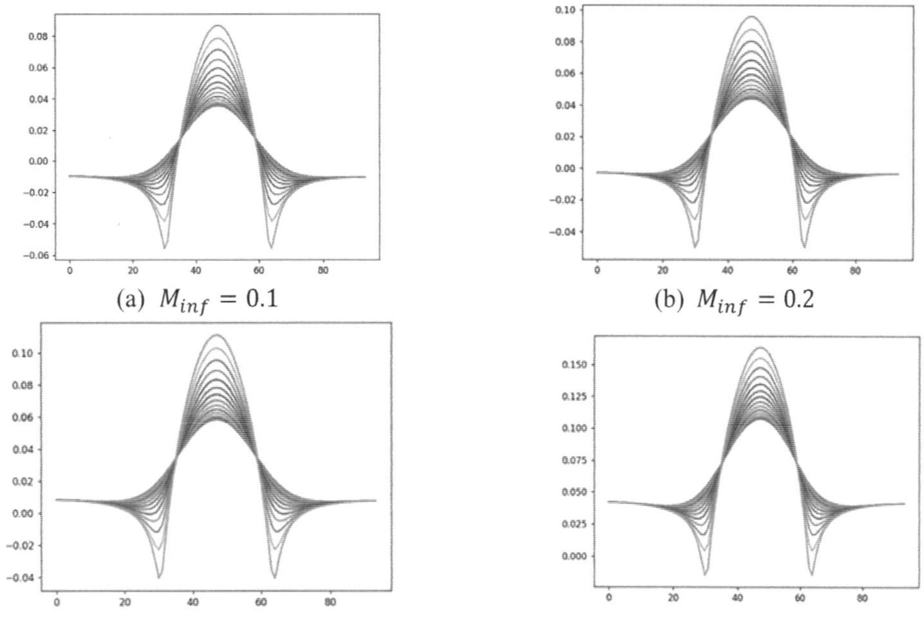

(a) $M_{inf} = 0.1$ (b) $M_{inf} = 0.2$

(c) $M_{inf} = 0.3$ (d) $M_{inf} = 0.5$

Fig. 3 velocity in x-direction of different M_{inf}

3.1.4. Velocity in y-direction. This figure is about the velocity in y-direction of different Mach numbers. However, the velocity in y-direction doesn't vary with Mach numbers.

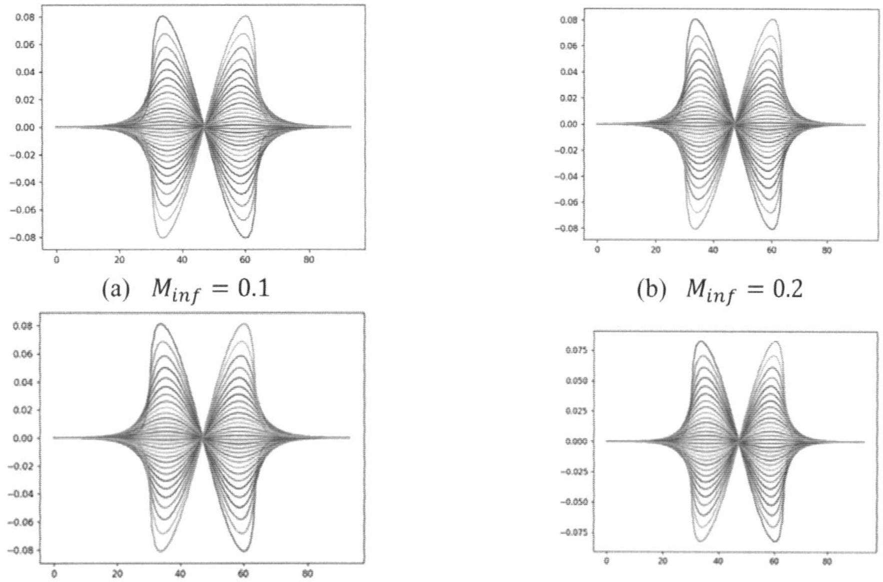

(a) $M_{inf} = 0.1$ (b) $M_{inf} = 0.2$

(c) $M_{inf} = 0.3$ (d) $M_{inf} = 0.5$

Fig. 4 velocity in y-direction of different M_{inf}

3.2. Impact of the Domain Shape

In this case, the Mach number will be kept as a constant. The domain shape is defined as $f(x)$ in eq.9. In order to change the domain shape, a changing parameter β will be added into $f(x)$ as shown below:

$$\beta f(x) = \begin{cases} \beta\cos\pi(x-1), & 1 \ll x \ll 2 \\ 0, & otherwise \end{cases} \tag{14}$$

The same rule as the first case will be conducted. By changing β into 0.1, 0.5, 1 and 1.5, respectively, the impact will be revealed by four different groups of results.

3.2.1. Iteration and Error. Fig. 5 represents the iteration times versus the natural algorithm of error of different domain shape. It can be observed that as β grows larger, the natural algorithm of error also grows larger, which means that the error becomes larger. Therefore, the precise decreases.

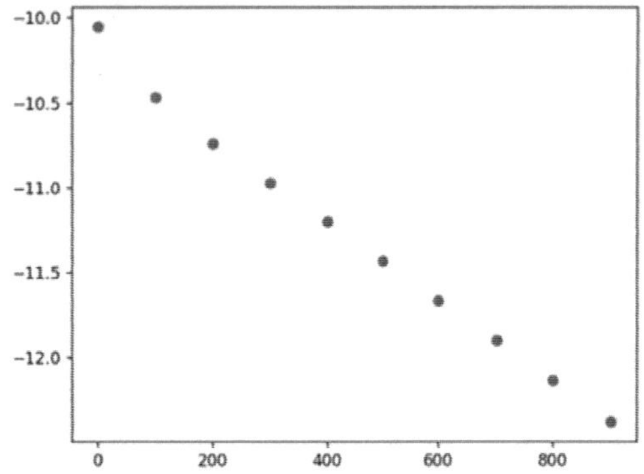

(a) $\beta = 0.1$

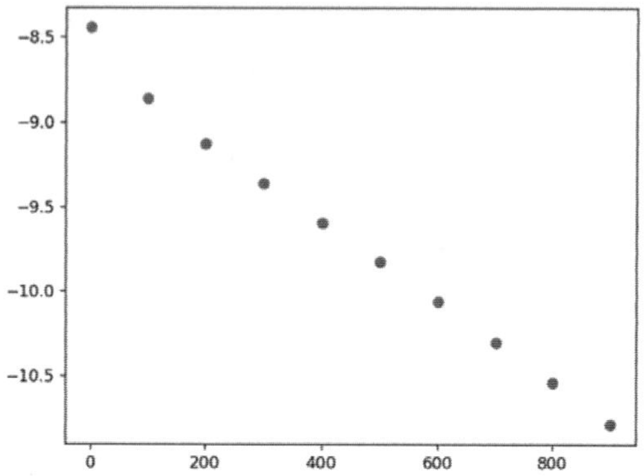

(b) $\beta = 0.5$

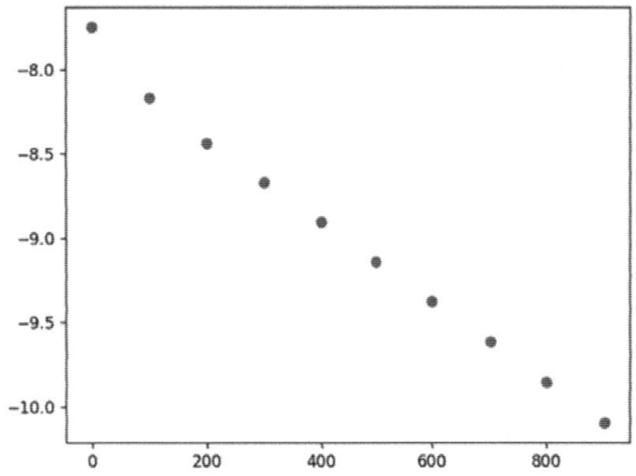

(c) $\beta = 1$

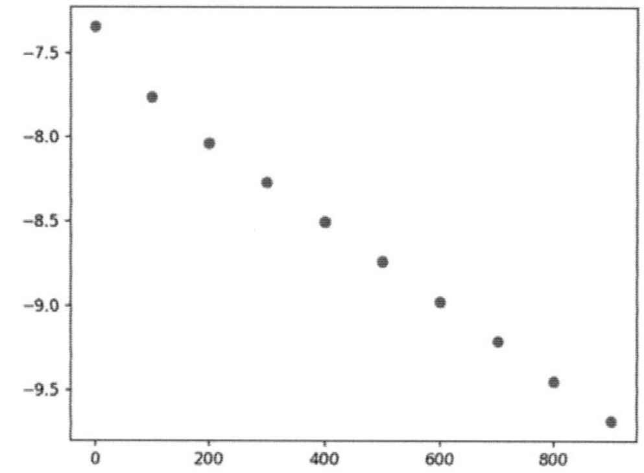

(d) $\beta = 1.5$

Fig. 5 iteration vs. error of different β

3.2.2. Streamline. Fig. 6 shows the streamlines of different domain shape. As the β number increases, the streamline doesn't tend to converge much as the previous one.

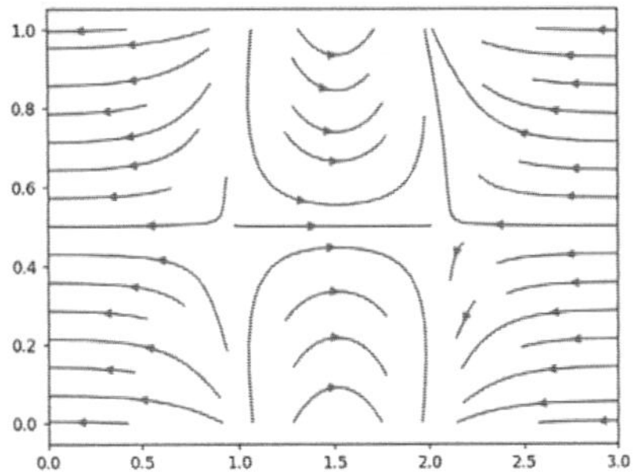

(a) $\beta = 0.1$

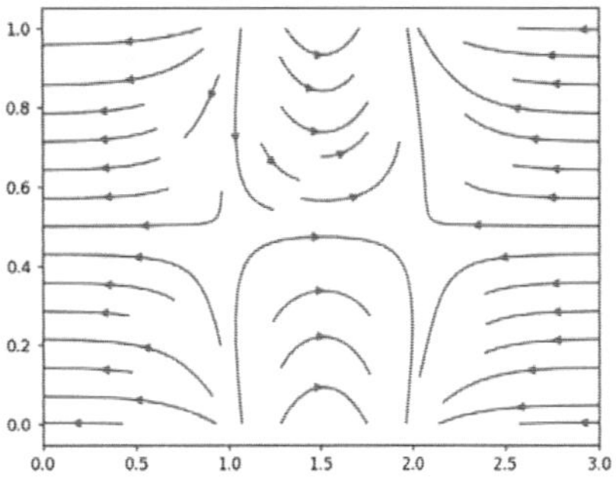

(b) $\beta = 0.5$

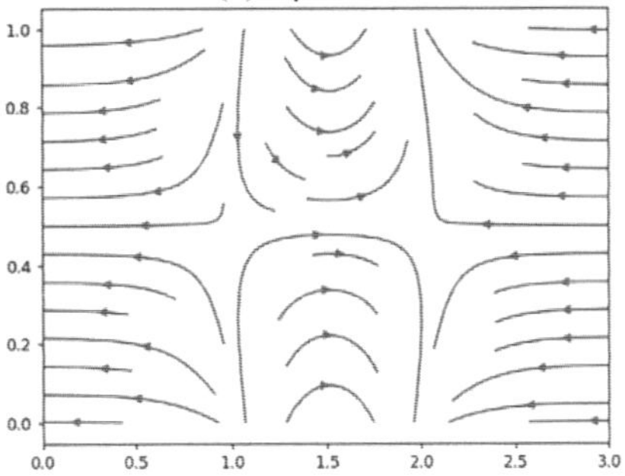

(c) $\beta = 1$

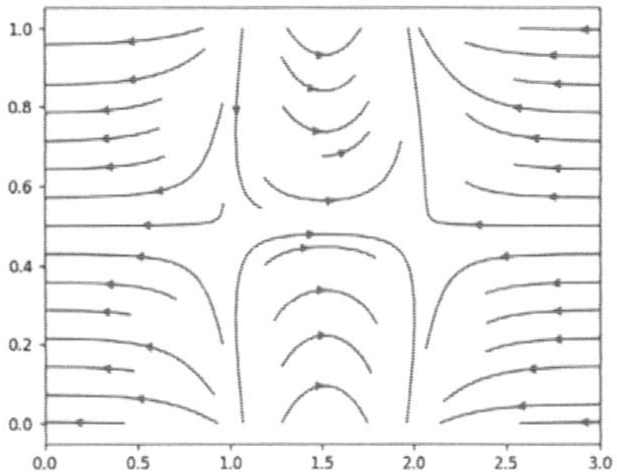

(d) $\beta = 1.5$

Fig. 6 streamlines of different β

3.2.3. Velocity in x-direction. Fig. 7 shows the velocity in x-direction among different domain shape. This figure illustrates that as β becomes larger, the maximum of the velocity increases while the minimum of the velocity decreases. The whole graph stretches.

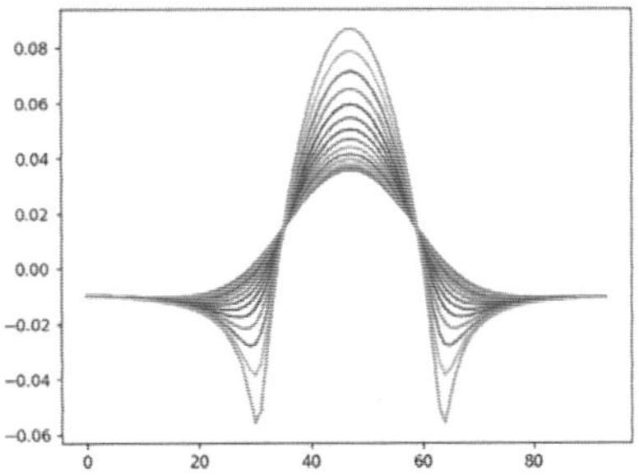

(a) $\beta = 0.1$

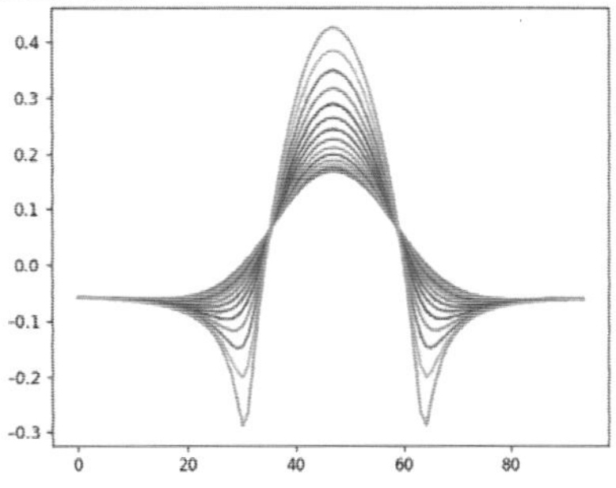

(b) $\beta = 0.5$

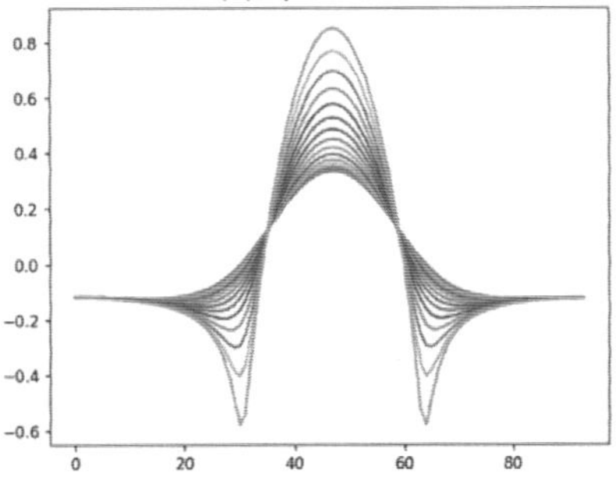

(c) $\beta = 1$

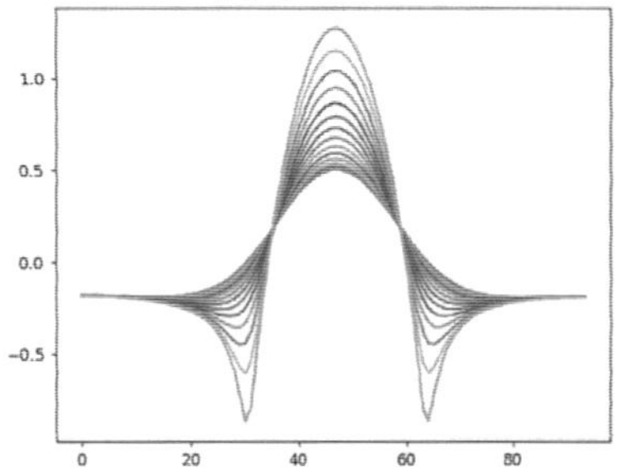

(d) $\beta = 1.5$

Fig. 7 velocity in x-direction of different β

3.2.4. Velocity in y-direction. Fig. 8 shows the velocity in y-direction when the domain shape changes. The minimum and maximum of the velocity of different iteration tends to be a symmetrical pair. Besides, it can be observed that as β increases, the maximum of the velocity also increases. Meanwhile, the minimum of the velocity decreases. What's more, the change of the maximum and minimum tends to be less extreme as β grows larger.

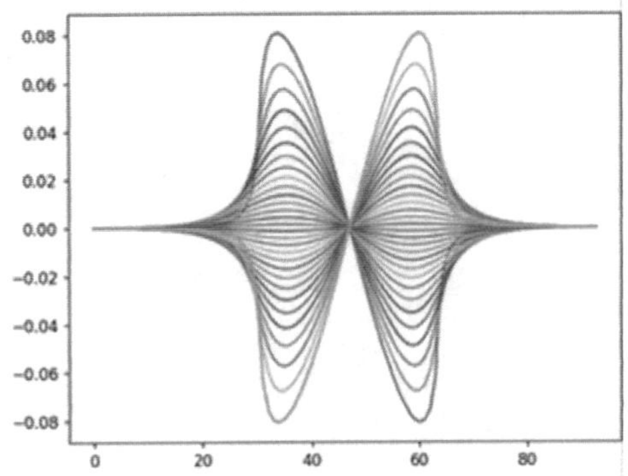

(a) $\beta = 0.1$

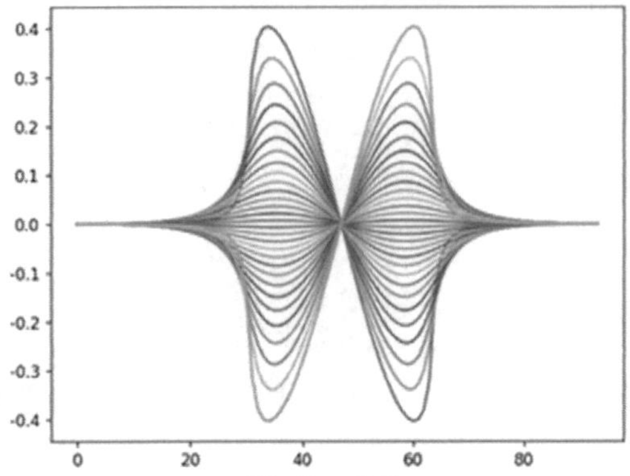

(b) $\beta = 0.5$

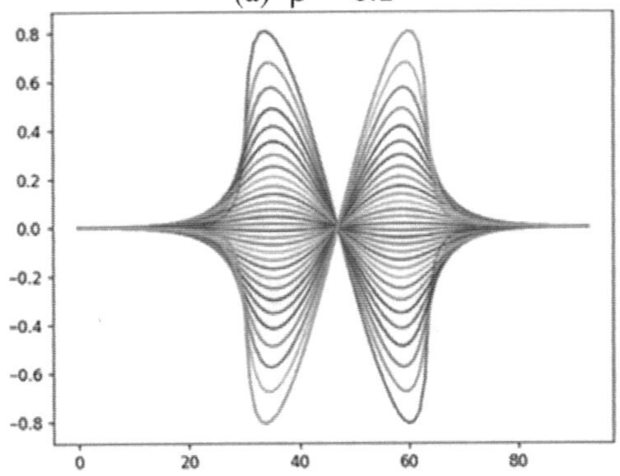

(c) $\beta = 1$

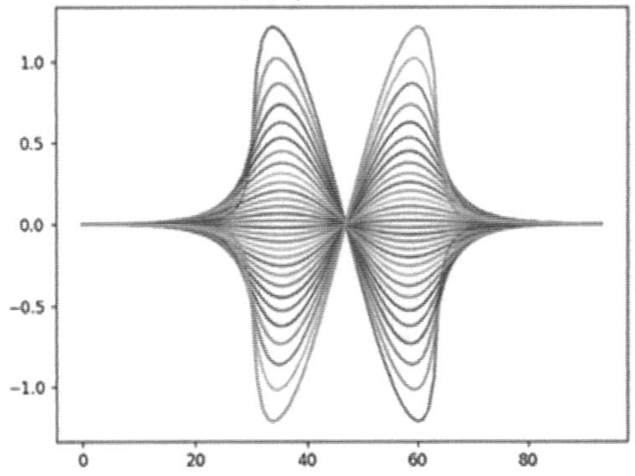

(d) $\beta = 1.5$

Fig. 8 velocity in y-direction of different β

4. Conclusions

This paper explored the impact on the subsonic nozzle flow of different inlet Mach number and domain shape based on small disturbance equation. By comparing the results among different inlet Mach numbers or different domain shapes, the conclusions can be summarized as follows:

(1) When the inlet Mach number grows, the error becomes larger, which means that the equation converges slower and the precise becomes less exact. Besides, as the inlet Mach number increases, the streamline becomes less extreme and the maximum and the minimum of the velocity in x-direction both increases with almost the same amplitude. Meanwhile, the velocity in y-direction doesn't vary with inlet Mach numbers.

(2) As for the domain shape, the change is reflected by a coefficient β. When β increases, the error becomes larger, which means the precise decreases. What's more, as the coefficient β grows, the streamline becomes less convergent. As for the velocity in x-direction, the maximum of it increases while the minimum of it decreases with the increasing coefficient. Besides, it can be concluded that as the coefficient increases, the maximum of the velocity in y-direction increases and the minimum of the velocity in y-direction decreases. And the magnitudes of the maximum and minimum are opposite number.

References

[1] Parmar D., Setiya M., Baloni B., Channiwala S. (2017) Flow Analysis of Subsonic and Supersonic Nozzles. In: Saha A., Das D., Srivastava R., Panigrahi P., Muralidhar K. (eds) Fluid Mechanics and Fluid Power – Contemporary Research. Lecture Notes in Mechanical Engineering. Springer, New Delhi. https://doi.org/10.1007/978-81-322-2743-4_39.

[2] Tannehill, J., Anderson, D., Pletcher, R. (1997) Computational Fluid Mechanics and Heat Transfer. Taylor & Francis, Washington, DC.

[3] Colona, G., Tuttafesta, M., Giordano, D. (2001) Numerical methods to solve Euler equations in one-dimensional steady nozzle flow. Computer Physics Communications.

[4] K.H. Huang. (2021) Euler and Small Disturbance Equation-based Simulation of a nozzle flow. Journal of Physics: Conference Series, 2012. 012048. 10.1088/1742-6596/2012/1/012048.

[5] F.R. Cheng, Y. Xu, W.B. Zhou. (2021) Nozzle Flow Simulation by Small Disturbance Equation and Euler Method. Journal of Physics: Conference Series.

[6] Quintao, Karla K. (2012) Design Optimization of Nozzle Shapes for Maximum Uniformity of Exit Flow, FIU Electronic Theses and Dissertations, 779.

[7] Anderson, J. D. (2002) Modern compressible flow. McGraw-Hill. pp. 358–359.

[8] S. Huang, K.X. Wang, W.F. Chen, J.H. You, X. Chen, J.X. Qian, Z.J. Zhao, Lorenz T.Biegler. (2012) Simplified MFE with power-change adaption strategy for the dynamic optimization of HTR-PM. Computer Aided Chemical Engineering. pp. 1301-1305.

[9] Timothy, S., (2012) Numerical Analysis, Second Edition. Pearson. pp. 108-111.

[10] Batchelor, G.K. (1973) An Introduction to Fluid Dynamics. Cambridge University Press. pp. 71-130.

The effects of geometry parameters on the aerodynamic performance of CFJ airfoil used by a STOL transport

Cheng Chen[1,*], Zhengyao Xie[2], Chengjun He[1], Xian Chen[1] and Jiangtao Huang[1]

[1]Aerospace technology institute, China Aerodynamics Research and Development Center, Mianyang, China
[2]School of aeronautics, Northwestern Polytechnical University, Xi'an, China

cc02005083@126.com

Abstract. With the purpose of investigating the effects of geometry parameters of CFJ (Coflow jet) device on the aerodynamic characteristics of the supercritical bacj airfoil used by a STOL transport, we have carried out a computational analysis of CFJ airfoil with different values of injection slot size and location under the low-speed condition. The results show that the lift coefficient decreases gradually with decreasing stall angle, when the injection slot size is increased. When α=8°, with increasing injection slot size, the velocity of blowing jet decreases with decreasing power consumption and increasing aerodynamic effectiveness. Further, with the injection slot shifted towards the downstream direction, the α value corresponding to the maximum effective lift-drag-ratio increases with increasing maximum lift coefficient.

1. Introduction

The ability of Short Take Off and Landing (STOL) could render the large/medium transport airplanes to take off and land in the airport at the plateau environment and on the gravel runway and leads to their stronger adaptability to the environment, beneficial to rewriting the distribution situation of strategy/livelihood resources and the rapid transport of human and material resources. Thus, the researches on the key techniques for STOL transport airplanes have both strategic significance and social value. One of the largest challenges for STOL transport is the problem of low-speed taking off and landing. Seen from the geographical environment of our country, the abominable climate, natural condition and road condition at the western plateau are the main factors restricting the performance of the transport airplanes. On the one hand, the altitude of the plateau is relatively high with thin air and small dynamic pressure of airflow. On the other hand, the airport equipment at the plateau is relatively crude with abominable runway conditions and it is also impossible to build large modern airports recently. Further, the engine has a thrust loss due to the thin air and may fail to start in the plateau environment. Thus, in order to increase the transport ability of strategy/livelihood resources, it is necessary to solve the lift enhancement problem of STOL transport under the low-speed condition, with the purpose of increasing the lift coefficient and the ability of adapting to the abominable runway condition.

Coflow jet is a popular flow control way in recent years. This technique is previously proposed by Zha *et al.*[1] As shown in Fig.1, CFJ airfoil opens the injection slot AC near leading edge and the suction slot BD near trailing edge on the suction surface of airfoil, and has an inside pump system. A high-speed jet is injected tangentially by the pump near leading edge and meanwhile the same amount of mass flow

Content from this work may be used under the terms of the Creative Commons Attribution 3.0 licence. Any further distribution of this work must maintain attribution to the author(s) and the title of the work, journal citation and DOI.
Published under licence by IOP Publishing Ltd

is sucked in near trailing edge. Thus, it has the advantages of avoiding introducing airflow from the engine additionally and achieving performance enhancement without relying on Coanda effect. The combination of the injection of high-speed airflow into the boundary layer and the jet mixing with the main flow is confirmed to be beneficial to delaying the boundary layer separation and the formation of additional circulation, leading to the delay of stall phenomena and substantial lift enhancement. The experimental and numerical results by Zha *et al.*[1-2] imply that CFJ technique could increase the maximum lifts of NACA2415 airfoil and NACA0025 airfoil by 80% and 220%, respectively.

Nowadays the researches on CFJ technique used as the measures for the enhancement of lift and rudder effect are relatively rare, except that Shi *et al.*[3] applied CFJ technique in the vertical tail to increase the rudder efficiency and Zhi *et al.*[4] used CFJ technique for the airfoil with a trailing-edge rudder to increase the airfoil lift. Considering that the requirement of lift coefficient for taking off and landing in the plateau environment can not be fulfilled solely by the regular lift-enhancement device, it is necessary to investigate the application situation of the active flow control techniques (such as CFJ technique) in the transport airplane and analyze their effect laws on the lift characteristics of the transport. With this in mind, this paper carries out the research on the effects of geometry parameters on the aerodynamic performance of the cross-section airfoil of a STOL transport by using CFD method, laying a good foundation for the design of the CFJ technique scheme of the full airplane.

2. Physical model

The supercritical airfoil studied is the bacj airfoil designed by the Boeing company and has a chord length of c=1m and a maximum thickness of 10%c at the location of 40%c from the leading edge, which is dedicated to the wing design of a STOL transport. For the CFJ airfoil formed from bacj airfoil, there are four important geometry parameters for the injection slot and suction slot, including the size of injection slot (AC), the location of injection slot (the distance between the injection slot and the leading edge), the size of suction slot (BD) and the location of suction slot (the distance between the suction slot and the leading edge).

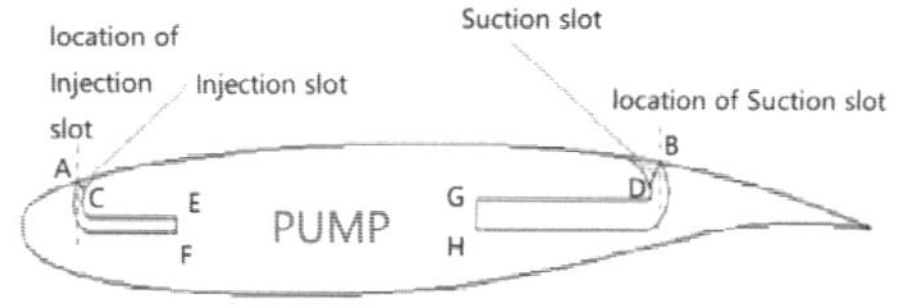

Fig.1 Sketch of CFJ airfoil used by a STOL transport

3. Numerical method

The numerical simulation is carried out using the CFX software, which is employed to solve the compressible Reynolds-averaged Navier-Stokes (RANS) equation using SST turbulence model[5] and unstructured grid. The Roe scheme with a third order MUSCL reconstruction is used to compute the inviscid fluxes and the Lower-Upper Symmetric-Gauss-Seidel (LU-SGS) method is used to carry out pseudo time marching.

The computational domain is [-48c, 79c]×[-48c, 48c], which is shown in Fig. 2 with the setting of boundary conditions. The pressure farfield condition is set at the upper and lower boundaries of the domain. At the left and right boundaries, the inflow and outflow conditions specified are based on the characteristics of the flow.[6] The no-slip condition is imposed at the airfoil surface. While at the inlet of high-pressure cavity EF, the inflow boundary condition is used with the total pressure, total temperature and flow angle fixed and with inflow velocity extrapolated from the interior; at the outlet of low-pressure cavity GH, the outflow boundary condition is employed with the static pressure specified and the density, horizontal velocity and vertical velocity are extrapolated from the interior. In the simulation of pseudo time marching, the total pressure at the inflow boundary EF and the static pressure at the outflow boundary GH are adjusted continuously until the desired jet momentum coefficient C_μ is achieved at both the injection slot AC and suction slot BD.

The mesh distributions for the configurations with CFJ and without CFJ are generated by the multi-block patched grid technique to resolve the locations of free shear layer and separation region. The height of the first layer mesh near the airfoil surface is $10^{-6}c$ and the growth rate is set to be 1.1. The high-pressure cavity and low-pressure cavity are both included in the computation of CFJ airfoil in order to simulate the developments of shear layers at the jet exit and the suction slot better. The Mach number of the freestream is fixed as 0.2.

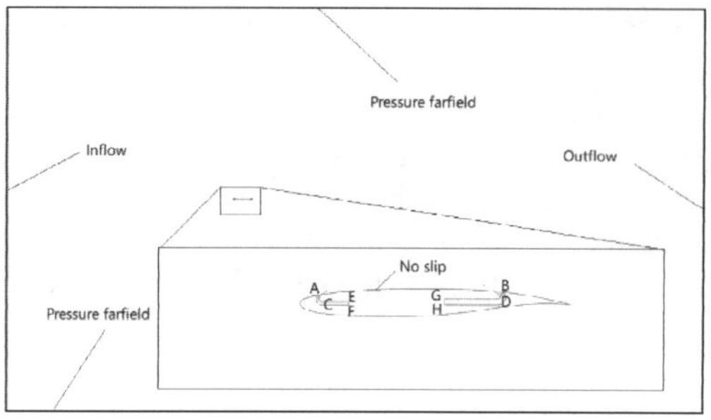

Fig.2 The computational domain and the setting of boundary conditions

For CFJ airfoil, it is usually to use jet momentum coefficient C_μ to reflect the strength of jet, which is defined as

$$C_\mu = \frac{\dot{m}V_j}{0.5\rho_\infty V_\infty^2 S} \tag{1}$$

where $\dot{m}$ is the mass flow rate, V_j is the injection jet velocity, ρ_∞ is the freestream density, V_∞ is the freestream velocity, the reference area S is equal to the airfoil chord c for the present two-dimensional cases. Different from the regular airfoil, due to the work of the pump P, the drag of CFJ airfoil should include the power consumption of the pump and the effective drag as well as the effective drag coefficient with the power consumption considered can be written as

$$D_e = D + P/V_\infty \tag{2}$$

$$C_{D_e} = C_D + P_C \tag{3}$$

where C_D is the drag coefficient for the conventional airfoil. The effective lift-drag-ratio is defined as

$$\left(\frac{L}{D}\right)_e = \frac{L}{D_e} = \frac{L}{D+P/V_\infty} \tag{4}$$

4. Results

This paper studies mainly the effects of the size and location of the injection slot on the aerodynamic characteristics of CFJ airfoil, as the size and location of the suction slot are proven to have a weak effect on the lift and drag features (the relevant results are not given here).

4.1. The effect of injection slot size on the CFJ airfoil

Fig. 3 shows the comparison of aerodynamic forces with different injection slot sizes at C_μ=0.05 for CFJ airfoil, where the 'injectionXX' phrase represents the ratio of the injection slot size to chord c. It can be found that all the lift coefficients of CFJ airfoil are larger than that of the original bacj airfoil and the effective drag coefficients of CFJ airfoil are larger than that of the original airfoil at small α, leading to lower values of the effective lift-drag-ratio than the bacj airfoil. Further, with the increase of injection slot size, the lift coefficient of CFJ airfoil decreases gradually with decreasing stall angle, and the effective drag coefficient decreases before α=8°.

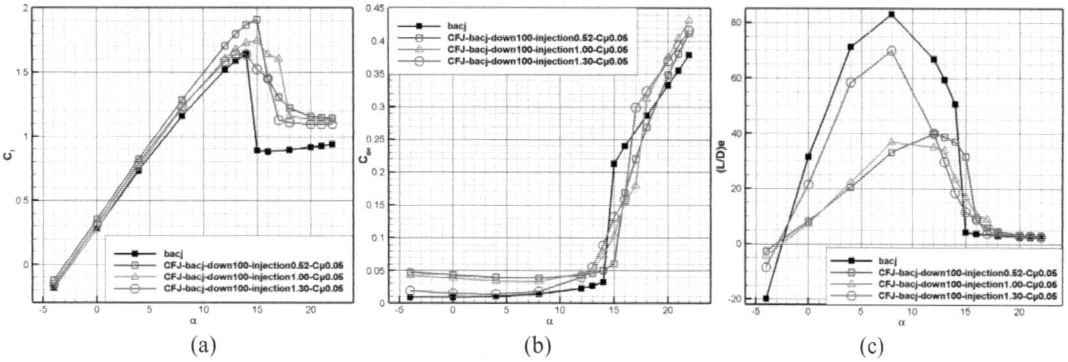

Fig.3 The comparison of aerodynamic forces of CFJ airfoil with different injection slot sizes at C_μ=0.05. (a) the lift coefficient. (b) the effective drag coefficient. (c) the effective lift-drag-ratio.

Fig. 4 and Fig.5 represent the comparison of the plots of Ma contour and streamline with different injection slot sizes for CFJ airfoil and the original airfoil at C_μ=0.05, α=8°, 18°. When α=8°, the bacj airfoil and CFJ airfoils are all at the attached status. With increasing injection slot size, the velocity of blowing jet decreases with decreasing power consumption and increasing aerodynamic effectiveness. While α=18°, the separation vortex is generated for both the bacj airfoil and CFJ airfoils. The velocity of blowing jet decreases with increasing injection slot size. When the injection slot size achieves the value of 1.30%*c*, the range of separation vortex is relatively large compared with the case with a smaller injection slot, and more power consumption is needed to maintain the mass conservation between the injection slot and suction slot, leading to a low aerodynamic efficiency. Thus, a reasonable choose of injection slot size could increase the aerodynamic efficiency of CFJ airfoil.

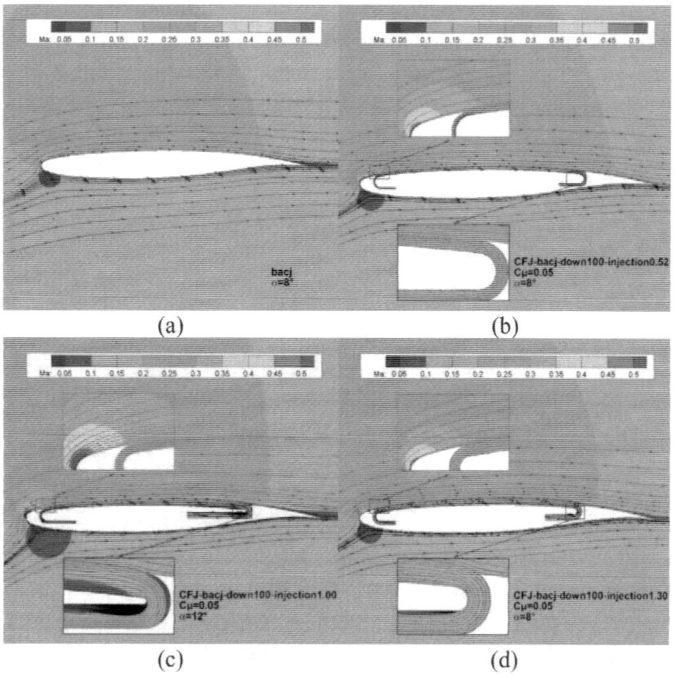

Fig.4 The plots of Ma contour and streamlines for bacj airfoil and CFJ airfoil, α=8°. (a) bacj airfoil. (b) CFJ airfoil with the injection slot size of 0.52%*c*. (c) CFJ airfoil with the injection slot size of 1%*c*. (d) CFJ airfoil with the injection slot size of 1.3%*c*.

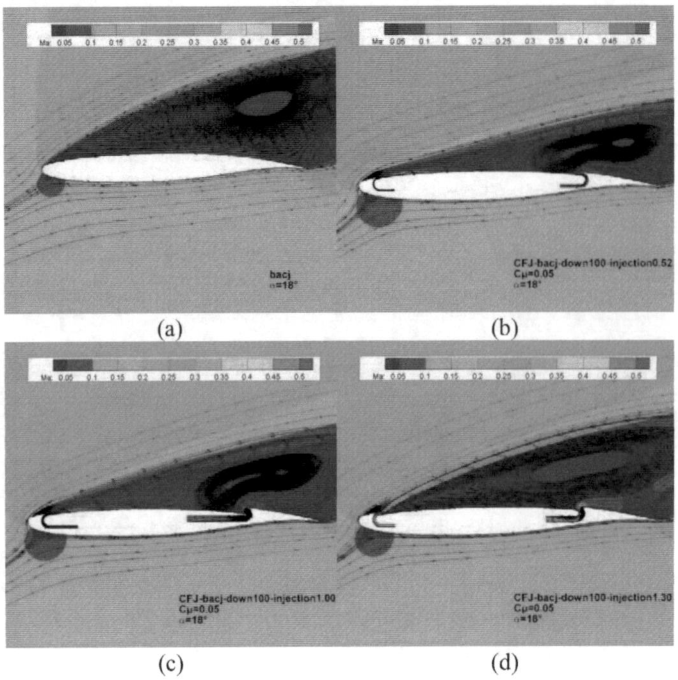

Fig.5 The plots of Ma contour and streamlines for bacj airfoil and CFJ airfoil, α=18°. (a) bacj airfoil. (b) CFJ airfoil with the injection slot size of 0.52%*c*. (c) CFJ airfoil with the injection slot size of 1%*c*. (d) CFJ airfoil with the injection slot size of 1.3%*c*.

4.2. *The effect of injection slot location on the CFJ airfoil*

Fig. 6 gives the comparison of aerodynamic forces with different injection slot locations at C_μ=0.05 for CFJ airfoil, where the phrase 'locxx' denotes the distance between the injection slot and the leading edge. It can be seen that the lift coefficients of CFJ airfoil are larger than that of bacj airfoil and the effective drag coefficients of CFJ airfoil are larger than that of bacj airfoil at small α, leading to lower values of the effective lift-drag-ratio compared with the original airfoil. With the location of injection slot shifted towards the downstream direction, the α value corresponding to the maximum effective lift-drag-ratio increases with increasing maximum lift coefficient. Further, the CFJ airfoil with the location of injection slot at x=4.5%*c* has a larger value of maximum effective lift-drag-ratio.

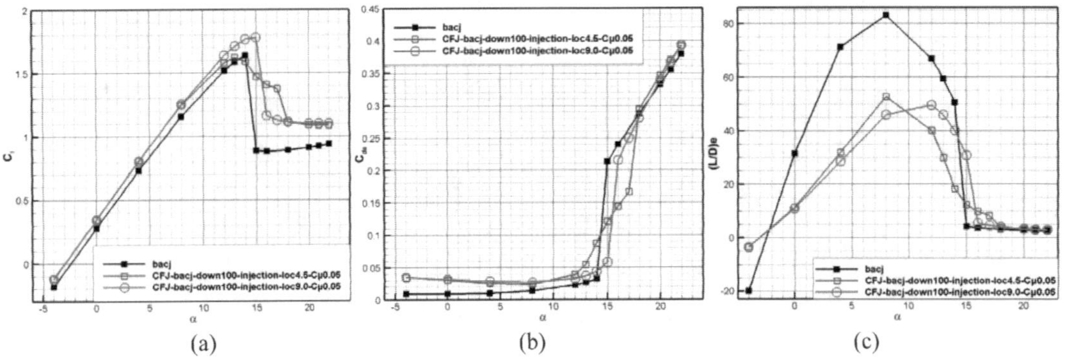

Fig.6 The comparison of aerodynamic forces of CFJ airfoil with different injection slot locations at C_μ=0.05. (a) the lift coefficient. (b) the effective drag coefficient. (c) the effective lift-drag-ratio.

Fig. 7 and Fig. 8 show the comparison of the plots of Ma contour and streamline with different injection slot locations for CFJ airfoil and the original airfoil at C_μ=0.05, α=14°, 18°. When α=14°, the bacj airfoil and the CFJ airfoil with the location of injection slot at *x*=9%*c* are at the attached status. While the CFJ airfoil with the location of injection slot at *x*=4.5%*c* has a separation with a small streamwise range and a stall trend. Thus, the blowing jet could suppress the flow separation better when the location of injection slot is further away from the leading edge. As for the CFJ airfoil with the location of injection slot at *x*=4.5%*c*, it is difficult to suck in the same amount of airflow at the suction slot as the injection slot and the total pressure at the suction slot is decreased compared with that for other locations of blowing jet. To maintain the mass conservation between the injection slot and suction slot, more power consumption is needed, leading to a lower aerodynamic efficiency. When α=18°, the separation vortex with a large streamwise range is generated for both the bacj airfoil and CFJ airfoils. The blowing jet for the CFJ airfoil with the location of injection slot at *x*=9%*c* has no flow separation at the injection slot and there is a separation near the injection slot for the CFJ airfoil with the location of injection slot at *x*=4.5%*c*, making the sucking of airflow at the suction slot more difficult.

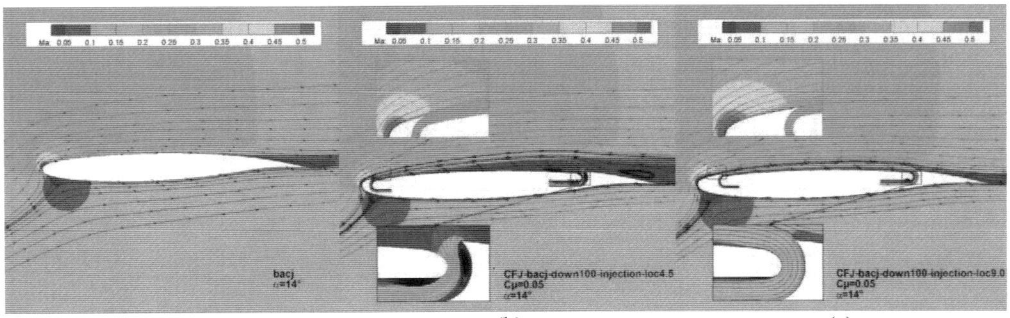

(a) (b) (c)

Fig.7 The plots of Ma contour and streamlines for bacj airfoil and CFJ airfoil, *α*=14°. (a) bacj airfoil. (b) CFJ airfoil with the location of injection slot at *x*=4.5%*c*. (c) CFJ airfoil with the location of injection slot at *x*=9%*c*.

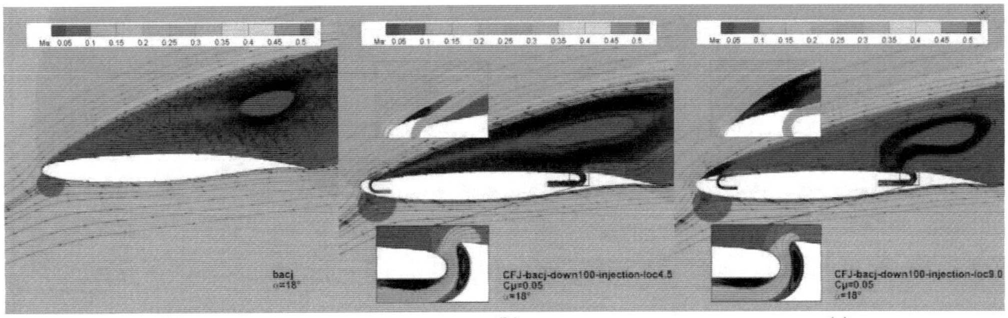

(a) (b) (c)

Fig. 8 The plots of Ma contour and streamlines for bacj airfoil and CFJ airfoil, *α*=18°. (a) bacj airfoil. (b) CFJ airfoil with the location of injection slot at *x*=4.5%*c*. (c) CFJ airfoil with the location of injection slot at *x*=9%*c*.

5. Conclusion

In this paper, we use the CFX software to investigate the effects of the size and location of injection slot on the aerodynamic characteristics of CFJ airfoil. The results show that the lift coefficients of CFJ airfoil are larger than that of the original bacj airfoil and the effective drag coefficients of CFJ airfoil are larger than that of the original airfoil at small α. With the increase of the size of the injection slot, the lift coefficient of CFJ airfoil decreases gradually with decreasing stall angle, implying that the size of 0.52%*c* should be chosen for the design of CFJ airfoil. As seen from the flow fields, the aerodynamic effectiveness is increased with increasing injection slot size, when α=8°. While α=18°, the size of injection slot should not be too large for maintaining a large aerodynamic effectiveness. With the location of injection slot shifted towards the downstream direction, the α value corresponding to the maximum effective lift-drag-ratio increases with increasing maximum lift coefficient, and the blowing jet could suppress the flow separation better.

References

[1] Zha G C, Paxton C. A novel airfoil circulation augment flow control method using co-flow jet[C]// The 2nd AIAA Flow Control Conference, Portland, USA, June 28-July 1, 2004.

[2] Zha G C, Carrol B F, Paxton C D, et al. High performance airfoil using co-flow jet flow control [C] // The 43rd AIAAAerospace Sciences Meeting and Exhibit, Reno, USA, January 10-13, 2005.

[3] Shi Z J, Xu H Y, Guo R J, et al. Application Research of Flow Control Using Co-flow Jet on a Vertical Tail[J/OL]. Advances in Aeronautical Science and Engineering, 2021. (in Chinese)

[4] Zhi H L, Zhu Z H, Lu Y J, et al. Aerodynamic performance enhancement of co-flow jet airfoil with simple high-lift device[J]. Chinese Journal of Aeronautics, 2021, 34(9): 143-155.

[5] Gao Z. Numerical investigation of flow control in the aircraft wing using Co-flow jet[D]. Northwestern Polytechnical University, 2020. (in Chinese)

[6] Xu H Y, Xing S L, Ye Z Y. Numerical study of the S809 airfoil aerodynamic performance using a co-flow jet active control concept[J]. Journal of Renewable and Sustainable Energy, 2015, 7: 023131.

Numerical Simulation of Flow Over Airfoil and Its Optimization

Rick E. Xu [1, *], Zhe Wu [2, a]

[1] William A. Shine Great Neck South High School, Lake Success, NY, USA
[2] Nanjing University of Aeronautics and Astronautics, Nanjing, China

*Corresponding author e-mail: Xuyy04@gmail.com, [a] 252455858@126.com

Abstract. Airfoils are one of the most important factors in determining an airplane's ability to fly. As such, their optimization holds important ramifications for the continual need to improve airplane performance. In this paper, the optimization of airfoil design is explored through numerical computational fluid dynamics. This paper first investigates the effects of Angle of Attack on airfoil lift/drag ratios, which is shown to have a significant influence on the performance of airfoils. Then, the effects from thickness are explored, and it is found that the influence of thickness is highly dependent on the specific shape of airfoil. In addition, in order to further improve airfoil aerodynamic performance, a type of modified airfoil with an additional tip is developed and simulated. The results indicate that airfoils with additional tips can perform better, but only with the proper combination of tip parameter values. For example, the modified airfoil with a tip of 0.03 L length at a 25° tip angle bested that same airfoil without an additional tip. Finally, it is found that the Backpropagation machine-learning algorithm, when trained with prior simulation data, can quickly predict lift and drag coefficients of airfoils. The results in this paper could facilitate the further optimization of airfoil design.

Keywords: Optimization of Airfoils, Computational Fluid Dynamics, Aerodynamic Performance.

1. Introduction

Proper airfoil design is of utmost importance in order to satisfy technical and economic aircraft design requirements. Efficient airfoils can enable aircraft to carry heavier loads, fly at faster speeds, and decrease fuel consumption. One widely-used metric for evaluating an airfoil's performance is the lift/drag ratio, which is calculated by dividing the airfoil's lift coefficient by the drag coefficient. Improving the aerodynamic performance of airfoils is a vital area of research for the design of next-generation airplanes.

In recent years, advances in computing power have enabled the rapid proliferation of computational fluid dynamics (CFD) techniques. CFD has found applications in many research fields, including airfoil design. Not only is CFD quicker than the traditional method of wind-tunnel testing, but it can also produce accurate results. For example, Kanimozshi [1] validated the accuracy of lift and drag coefficient values derived from CFD in comparison to wind tunnel tests. Additionally, significant research has been

Content from this work may be used under the terms of the Creative Commons Attribution 3.0 licence. Any further distribution of this work must maintain attribution to the author(s) and the title of the work, journal citation and DOI.
Published under licence by IOP Publishing Ltd

conducted on the accuracy of specific models of CFD. Saraf [2] compared three k-ε and k-ω models against wind tunnel data and confirmed their ability to map trends and closely follow experimental data. Within the airfoil design field, a focus of research has been evaluating the performance of NACA airfoils through lift/drag ratio calculations. Some focus on determining optimal angles of attack (AOA), such as Gorgulu [3], who aimed to determine the most effective AOA for a NACA 0009 airfoil using CFD. Others compared different NACA models, such as Chumbre [4], who evaluated 5 NACA airfoils based on their lift/drag ratios from CFD simulations. Because CFD problems are defined by boundary conditions, it has allowed researchers to simulate more than just airfoils in isolation. Zaheer [5] simulated the ground effect by setting a moving wall as the bottom boundary. Some research has also focused on the modeling itself, such as Aqilah [6], who determined that mesh quality affected the results of a CFD simulation. Even though the study of airfoils has been well established, the systematic investigation of airfoil design is not complete and there are a variety of avenues for continued exploration.

This paper, focusing on airfoil design optimization, investigates the combination of the Backpropagation (BP) algorithm with CFD methodologies to quickly determine the aerodynamic performance of a specific airfoil. Additionally, it studies the effect of an additional tip on an airfoil's lift/drag ratio. The rest of the paper is organized as follows: Section 2 introduces the methodology and numerical settings, Section 3 details the results and discussions, and finally, the conclusions of the paper are presented in Section 4.

2. Methodology and Numerical Settings

In this section, we present the methodology and the numerical settings of the simulation. The formulation for the simulation includes the continuity equation,

$$\frac{\partial u_i}{\partial x_i} = 0 \tag{1}$$

and the Navier-Stokes equations,

$$\frac{\partial u_i}{\partial t} + u_j \frac{\partial u_i}{\partial x_j} = -\frac{1}{\rho}\frac{\partial \mathrm{p}}{\partial x_i} + (v + v_T)\frac{\partial^2 u_i}{\partial {x_j}^2} \tag{2}$$

Here, u_i represents the velocity components, p is the pressure, v is kinematic viscosity, v_T is the eddy viscosity, which is determined by the k-ε model [7], and ρ is the density of the fluid. The aerodynamic performance of the airfoil is represented by the coefficients for drag and lift. Here, the drag coefficient is defined as [8],

$$C_d = \frac{2F_d}{\rho u^2 A} \tag{3}$$

where F_d is the drag force, ρ is the mass density of the fluid, u is the fluid relative velocity, and A is the area perpendicular to the direction of fluid flow. The lift coefficient follows a similar form, and is defined as,

$$C_l = \frac{2F_l}{\rho u^2 A} \tag{4}$$

with F_l as the lift force. Since all lift force comes at the expense of drag force, the critical lift/drag (C_l/C_d) ratio represents the efficiency of an airfoil at balancing the two forces. This allows the C_l/C_d ratio to be used in evaluating the relative performance of an airfoil.

3. Results and Discussion

3.1. The Effects of Angle of Attack

The performance of airfoils at all thicknesses tested followed a similar trend as their AOA changed. Figure 1 illustrates the data collected from simulations of the airfoils with 100% thickness.

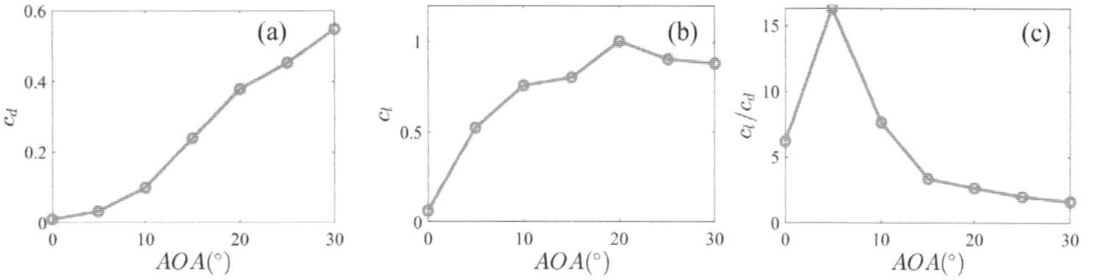

Figure 1. The simulation data at various AOA for airfoils with thicknesses of 100%: (a) Drag coefficient, (b) Lift coefficient, and (c) C_l/C_d ratio.

As shown in (a), the drag coefficients continually increase as AOA increases. In contrast, (b) demonstrates that the lift coefficients increase rapidly at lower AOA but plateau at higher AOA. However, it is not clear what the angle of optimal performance is until both coefficients are combined. (c) shows that for 100% thickness airfoils, 5° is the AOA with the highest C_l/C_d ratio and therefore the best performance. All other airfoil thicknesses tested also achieved a maximum C_l/C_d ratio at 5° except for 150% thickness airfoils, which achieved their maximum at 10°. In order to further understand the mechanisms behind this phenomenon, we will present velocity vector, velocity contour, and pressure gradient figures.

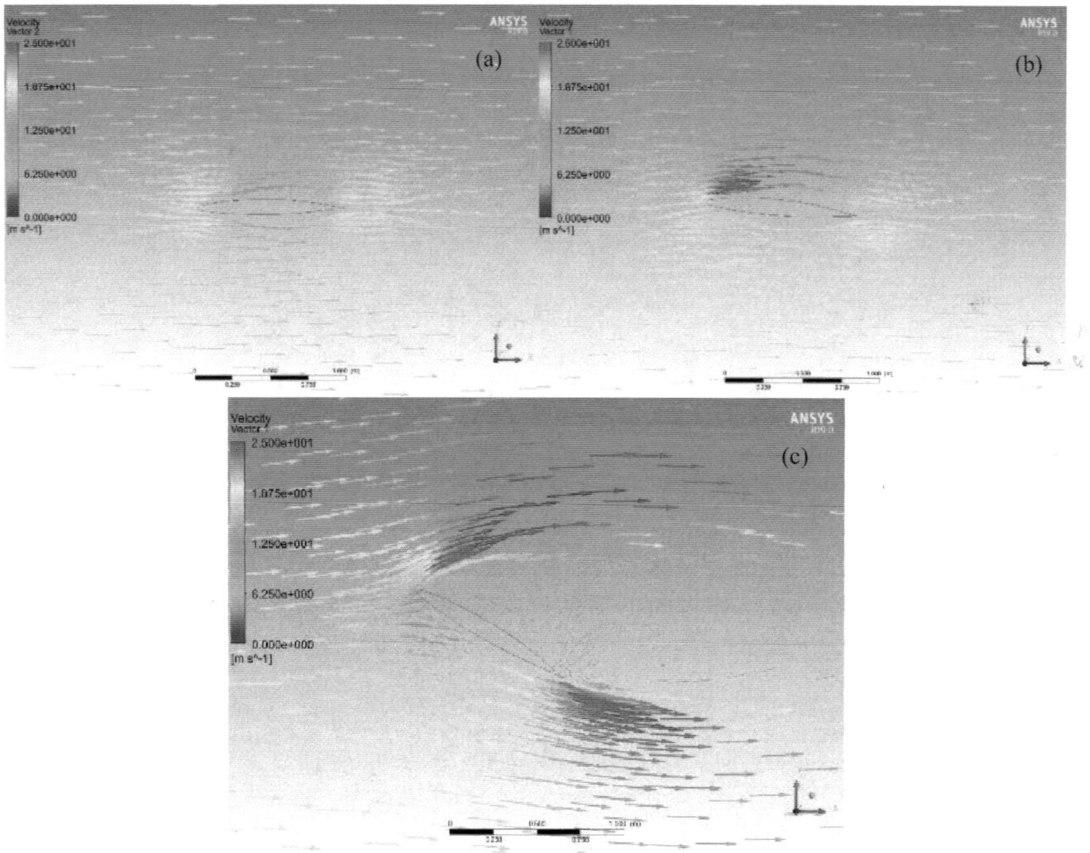

Figure 2. Velocity vectors around the airfoils at various AOA with thicknesses of 100%: (a) 0° AOA, (b) 5° AOA, and (c) 30° AOA.

As the AOA of the airfoil increases in Figure 2, the increasing disruption to homogenous airflow is indicated by the presence of red and blue-colored vectors, designating faster and slower-moving air, respectively. (a) demonstrates the relatively balanced vectors on the top and bottom sides of airfoils at low AOA, resulting in relatively smaller C_l and C_d. However, the air above the higher-AOA airfoil in (b) is significantly faster, increasing the lift coefficient, and a larger surface perpendicular to airflow increases the drag coefficient. Finally, despite there being more fast-moving air present in extreme-AOA airfoils such as (c), its presence on both sides of the airfoil nullifies much of the lift force. Furthermore, the phenomenon of flow separation is visible in (c), as indicated by the large vortices of slow-moving air, which cause significant drag on the airfoil. The velocity contours in Figure 3 will illustrate these features again and reveal more informative insights.

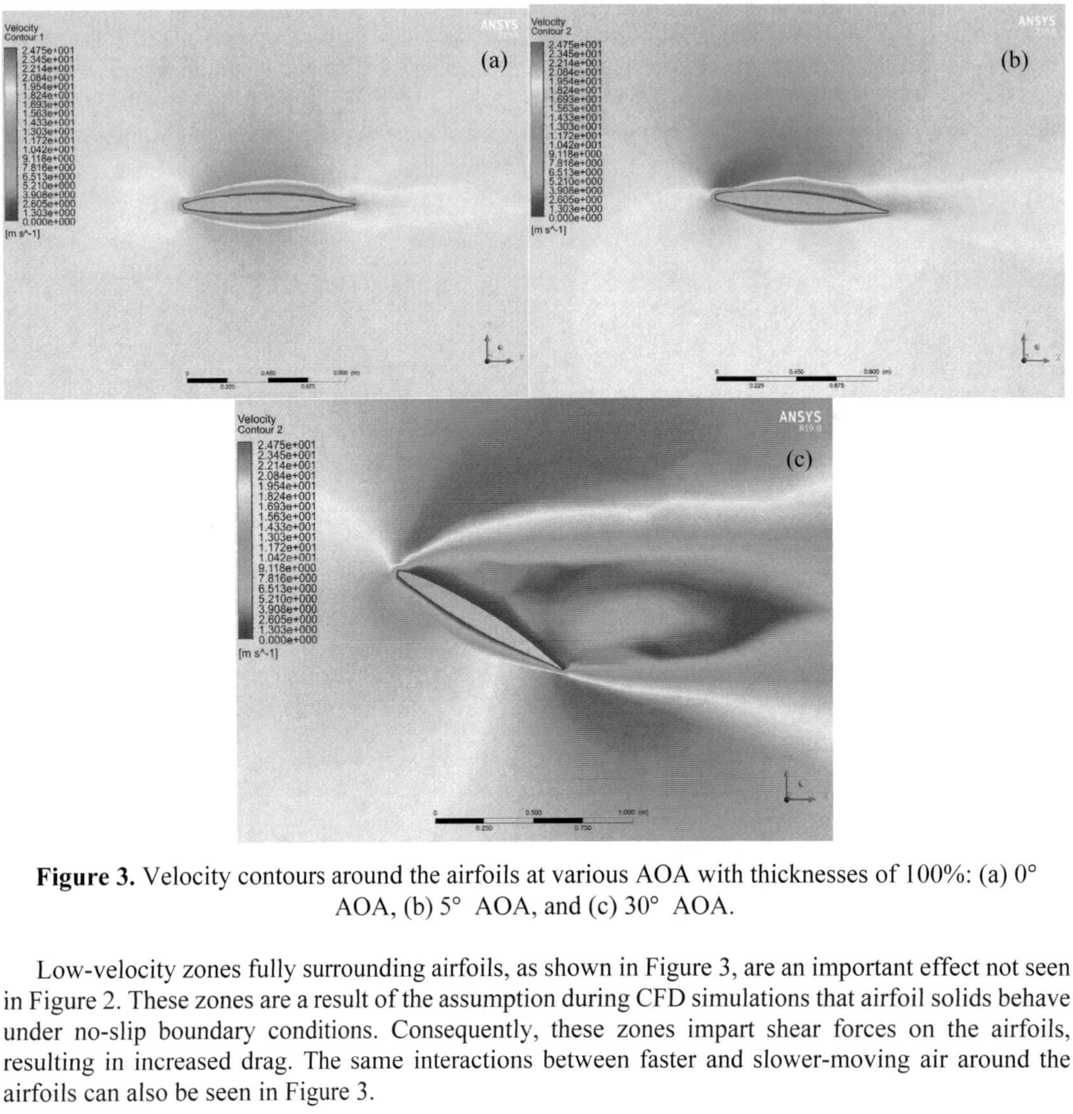

Figure 3. Velocity contours around the airfoils at various AOA with thicknesses of 100%: (a) 0° AOA, (b) 5° AOA, and (c) 30° AOA.

Low-velocity zones fully surrounding airfoils, as shown in Figure 3, are an important effect not seen in Figure 2. These zones are a result of the assumption during CFD simulations that airfoil solids behave under no-slip boundary conditions. Consequently, these zones impart shear forces on the airfoils, resulting in increased drag. The same interactions between faster and slower-moving air around the airfoils can also be seen in Figure 3.

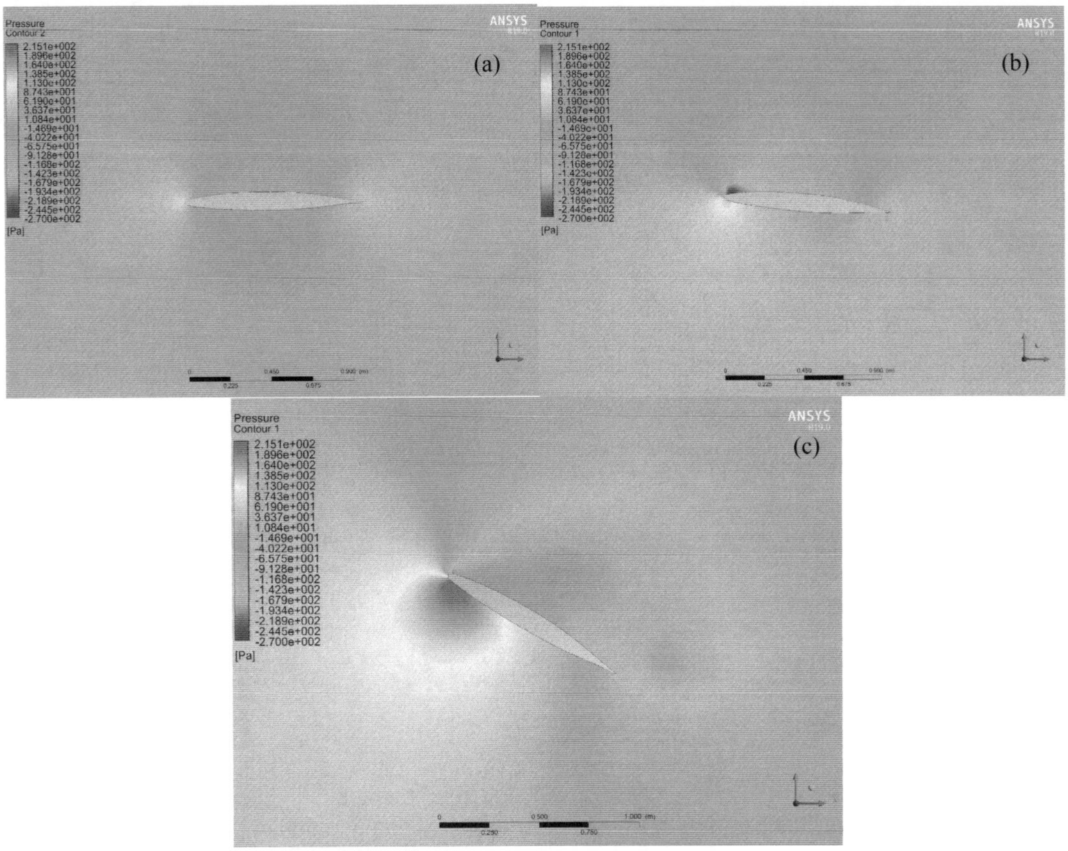

Figure 4. Pressure contours around the airfoils at various AOA with thicknesses of 100%: (a) 0° AOA, (b) 5° AOA, and (c) 30° AOA.

As shown by the red and blue-colored areas, Figure 4 illustrates the phenomenon of higher pressure below and lower pressure above an airfoil, respectively. This phenomenon is a result of the conservation of energy principle as detailed in Bernoulli's equation, which states that the sum of kinetic, potential, and pressure energy must always remain the same. As air faces more resistance and slows down, it loses kinetic energy. Because potential energy remains the same, the pressure increases to hold the extraneous energy. The same applies with lower-pressure and faster-moving air. As a result of this pressure differential, lift force is generated. Then, if the pressure differential increases, the lift coefficient will generally increase as well, which is true of all airfoils we tested.

The interactions of varying velocities and pressures drive the effects of varying the AOA. As AOA increases, the data reveal a consistent pattern of increasing drag coefficients and increasing, then plateauing, lift coefficients. We found that the optimal lift/drag ratio generally occurs around an AOA of 5°.

3.2. The Effects of Thickness

In this section, where we evaluated the effects of thickness on airfoils, it is found that varying the thickness of airfoils presents more complex results than varying the AOA.

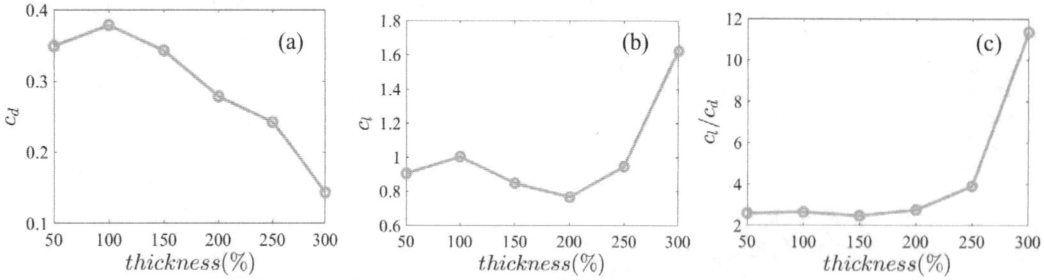

Figure 5. The simulation data at various thicknesses for airfoils with an AOA of 20°: (a) Drag coefficient, (b) Lift coefficient, and (c) C_l/C_d ratio.

The increase of airfoil thicknesses produced inconsistent results in the parameters we collected data for. Despite the clear downward trend in Figure 5 (a), the drag coefficient also trended upwards for airfoils at other AOA. Similarly, the upward trend in (b) was not indicative of any general lift coefficient trend, nor was the C_l/C_d ratio spike at 300% thickness in (c) representative of the data collected from airfoil simulations at other AOA. Our results did not indicate any broadly applicable trends from increasing the thickness of airfoils. However, there are still clear effects of thickness on a case-by-case basis, and studying specific cases will produce useful information.

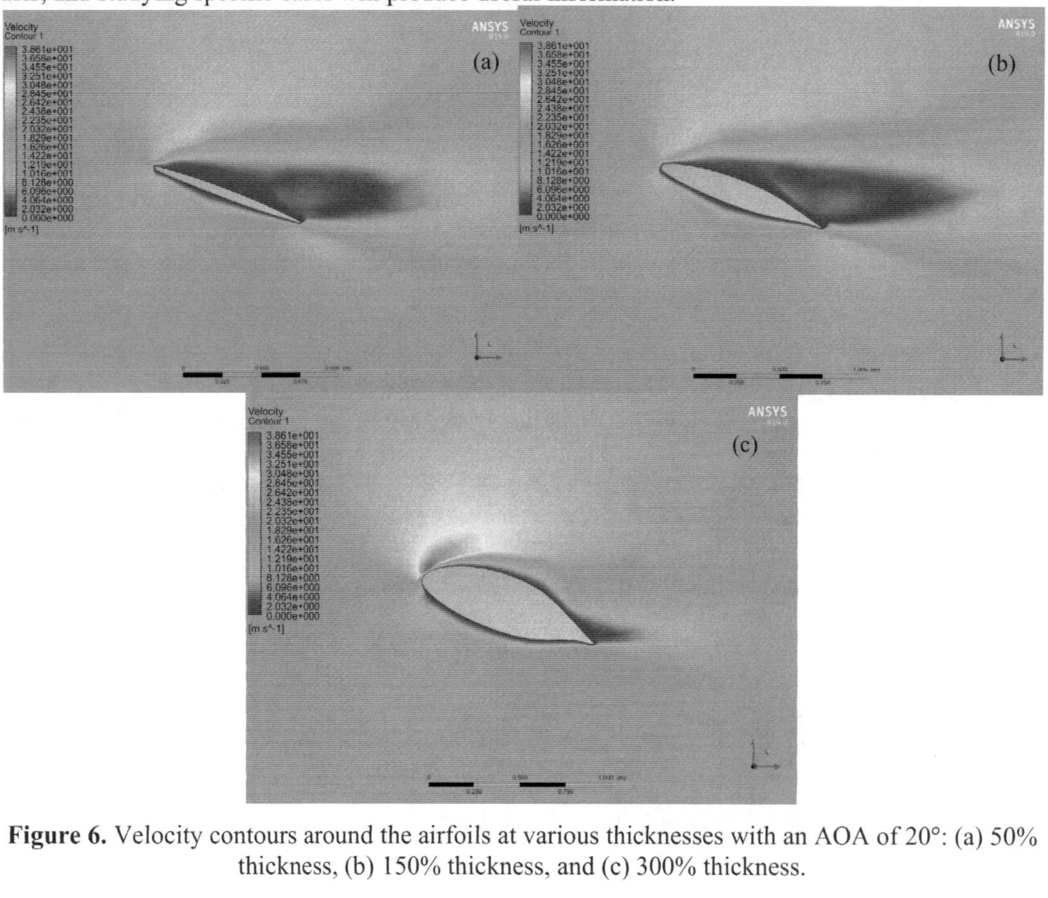

Figure 6. Velocity contours around the airfoils at various thicknesses with an AOA of 20°: (a) 50% thickness, (b) 150% thickness, and (c) 300% thickness.

Figure 6 shows the differences in velocity contours of airfoils with AOA of 20° as their thickness increases. It is found that (a) and (b), which both produce significant vortices, have high drag coefficients, while (c) only produces a relatively minor vortex. One likely effect of a thicker airfoil is reducing the production of vortices by allowing the fast-moving air flowing around the edges to continue traveling on the airfoil for longer. Furthermore, the airfoil in (c) causes far greater maximum velocity than in (a) or (b) because the flatter leading edge compresses the air more than in (a) or (b). Because of this greater difference in air speeds between the top and bottom of the airfoil in (c), there is a correspondingly greater pressure differential and lift coefficient.

3.3. The Effects of an Additional Tip

To further improve the aerodynamic performance of airfoils, we modified an original 100% thickness at 5° AOA airfoil to create 6 new versions each with an additional tip. There were two parameters used to create 8the tips on the modified airfoils: tip length (l_T) and tip angle (β). To examine the effect each additional tip would have on the performance of the original airfoil, the results from the simulations of modified airfoils were compared to the results from an original airfoil, as shown in Figure 7.

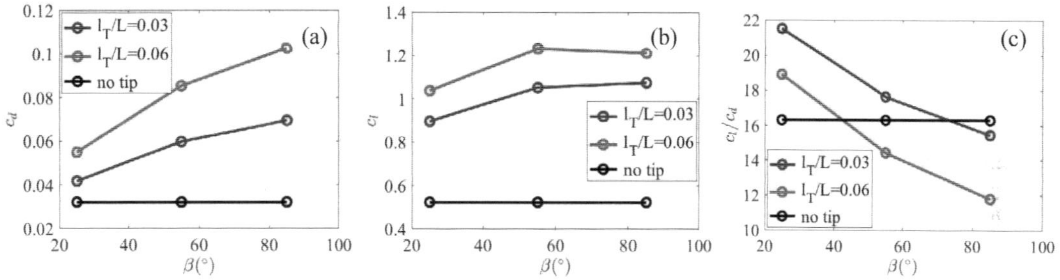

Figure 7. Simulation data for modified airfoils with 0.03 L and 0.06 L tip lengths at varying β compared to data from an original airfoil: (a) Drag coefficient, (b) Lift coefficient, and (c) C_l/C_d ratio.

Table 1. Comparison of C_l/C_d Ratio Between Original and Modified Airfoils

Case #	l_T (L)	β (°)	C_l/C_d	$\frac{(C_l/C_d)-(C_l/C_d)_{ori}}{(C_l/C_d)_{ori}} \times 100\%$ (%)
ori	—	—	16.31	0
1	.03	25	21.52	31.94
2	.03	55	17.64	2.02
3	.03	85	15.46	-5.21
4	.06	25	18.91	15.94
5	.06	55	14.43	-11.53
6	.06	85	11.8	-27.65

The detailed values of the C_l/C_d ratio values for the original and modified airfoils can be found in Table 1. To quantify the relative performance of the modified airfoils compared to the original, Column 5 in Table 1 calculates the percentage difference in the C_l/C_d ratio between the modified and original airfoils. For example, Case 1 showed a 31.94% improvement over the original airfoil. When a Case's percentage difference is negative, then the Case performed worse than the original airfoil. As shown in Figure 7 (a), the modified airfoils experienced larger-than-baseline and continually increasing drag coefficients as the β increased, with the 0.06 L tip airfoils increasing at a greater rate than the 0.03 L tip airfoils. Similarly, the lift coefficient results in (b) were greater for the modified airfoils, although they began to increase at a lesser rate or even decrease as the β increased. With both coefficients increasing as the β increases, and both coefficients remaining consistently above the baseline, it is difficult to identify an optimal combination of tip lengths and angles without calculating C_l/C_d ratios. Figure 7 (c) indicates that out of the variations tested, the additional tip improves the performance at lower β, but the drag coefficient increases too greatly relative to the lift coefficient to continue providing benefit at higher β. Furthermore, (c) also demonstrates the overall superiority of 0.03 L tips compared to 0.06 L tips. In sum, modifying airfoils with an additional tip can improve performance if the proper length and angle are selected.

A closer look at the velocity and pressure contours of the tested airfoils will help us understand these results.

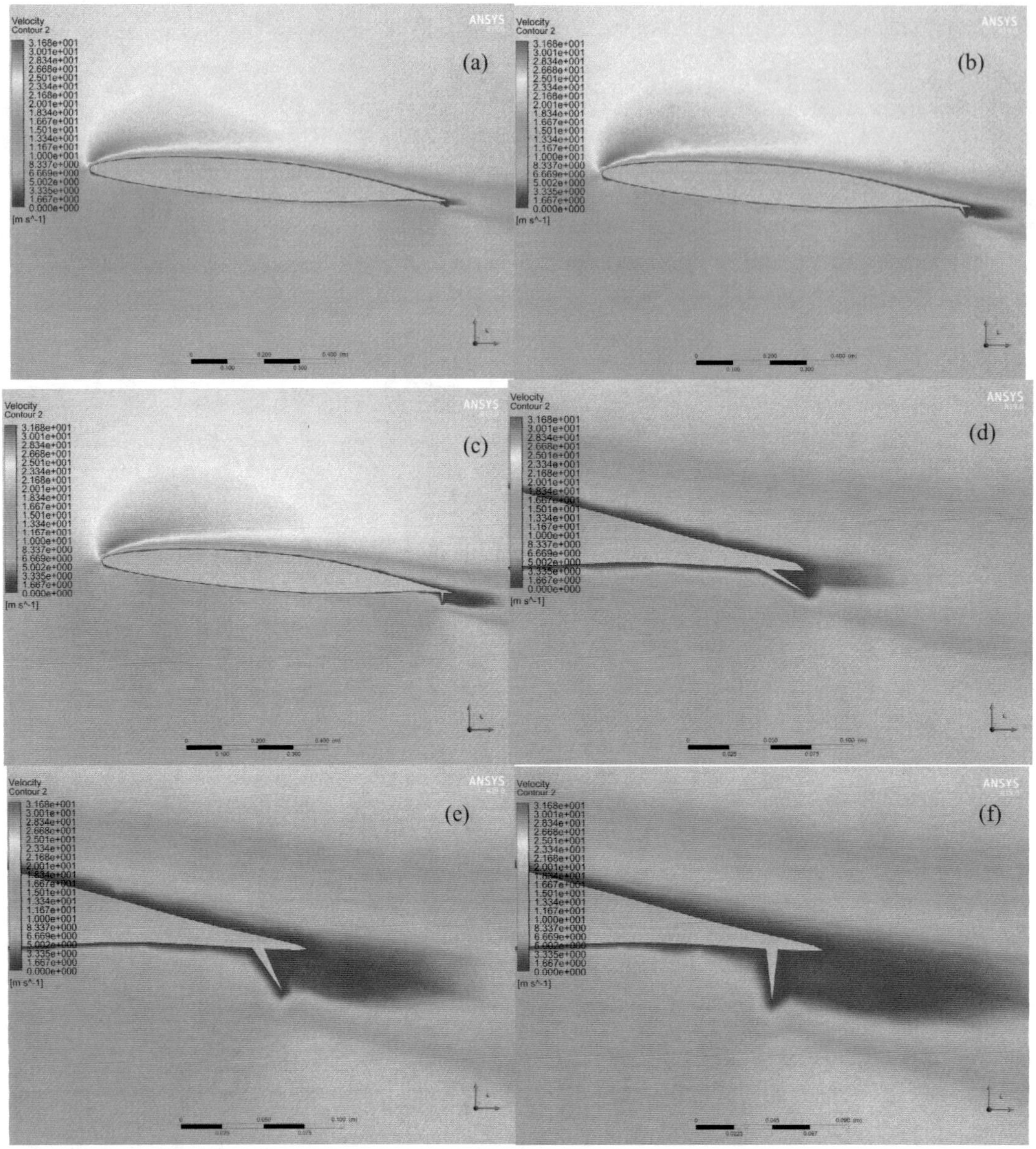

Figure 8. Velocity contours of modified airfoils with 100% thickness at 5° AOA: (a). 0.03 L tip at 25° β, (b) 0.03 L tip at 55° β, (c) 0.03 L tip at 85° β, (d) Closeup of 0.03 L tip at 25° β, (e) Closeup of 0.03 L tip at 55° β, and (f) Closeup up of 0.03 L tip at 85° β.

Figure 8 shows the velocity contours of modified airfoils with 0.03 L tips at various β. In comparison to Figure 3 (b), which is the velocity contour of the original airfoil, the modified airfoils experience a greater maximum velocity of about 32 m/s at the leading edge in comparison to roughly

25 m/s. Since the velocity of air is inversely proportional to the pressure, a greater maximum velocity and a constant minimum velocity of 0 m/s corresponds to a greater pressure differential, which causes the greater coefficients of lift as shown in Figure 1. Another difference is the presence of vortices behind the modified airfoils, with the increasing size of the vortices as the angle of the tip increases explaining much of the increase in the drag coefficient.

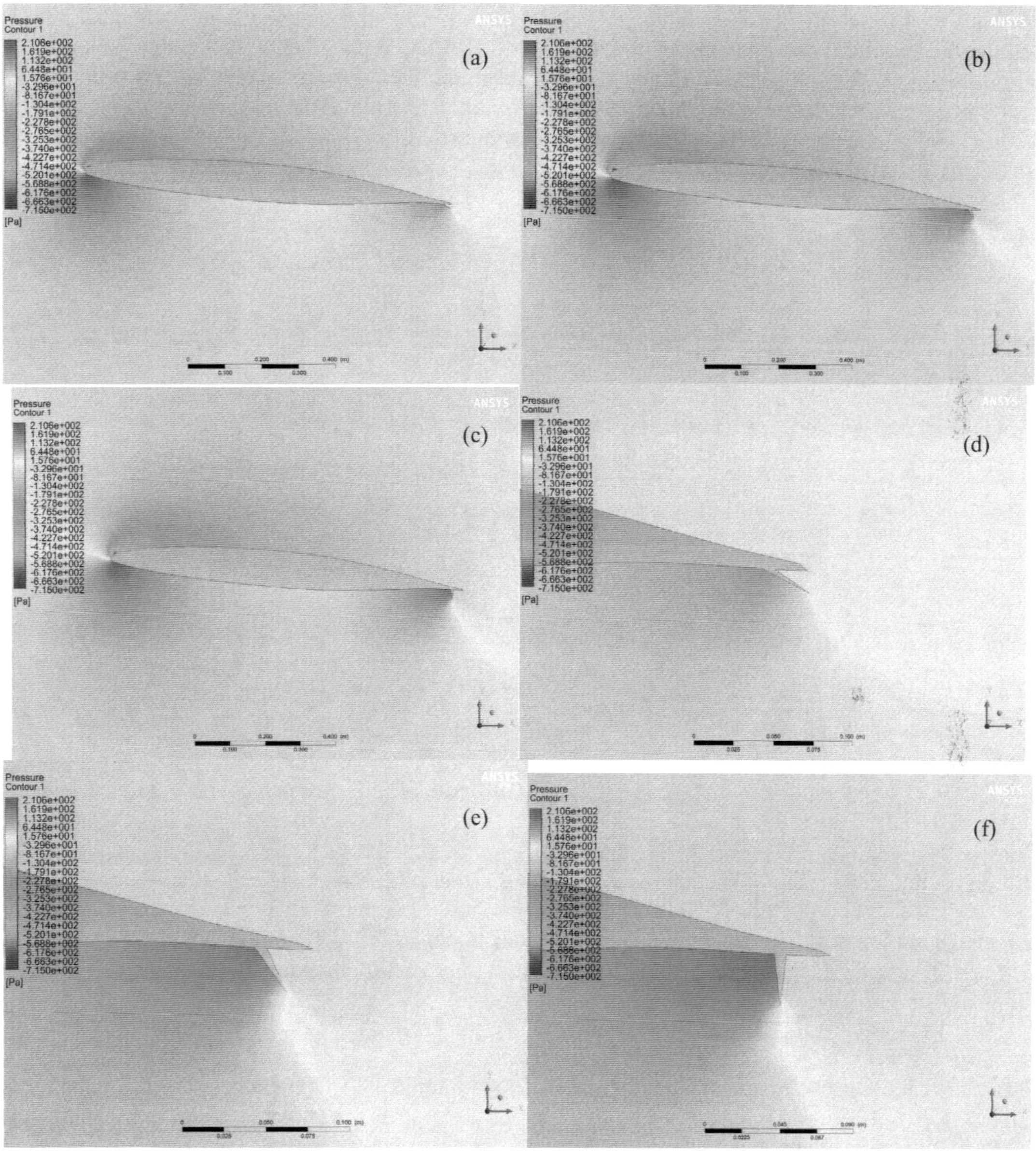

Figure 9. Pressure contours of modified airfoils with 100% thickness at 5° AOA: (a). 0.03 L tip at 25° β, (b) 0.03 L tip at 55° β, (c) 0.03 L tip at 85° β, (d) Closeup of 0.03 L tip at 25° β, (e) Closeup of 0.03 L tip at 55° β, and (f) Closeup up of 0.03 L tip at 85° β.

The pressure differentials of the modified airfoils in Figure 9 are far greater than the pressure differential of the original airfoil in Figure 4 (b), with the greatest difference coming from the lower minimum pressure of about -715 Pa in comparison to roughly -270 Pa. What is also evident from these pressure contours is that the additional tips serve to restrict the dispersal of high pressure underneath the airfoil, with increasing effectiveness as the pitch increases, as corroborated by Figure 7 (b). This restricts the balancing and subsequent pressure differential reduction as seen in the original airfoil.

3.4. Backpropagation Algorithm for Quick Prediction of Airfoil Performance
Although we were able to extract useful lift and drag coefficient data from the analyses that would help inform the selection of an optimal airfoil design, it is difficult to predict the lift and drag coefficients for any arbitrary airfoil without undergoing a time-consuming CFD process. To solve this problem, we introduce a machine-learning algorithm which can quickly determine lift and drag coefficients for any airfoil. Below, Figure 7 displays the actual and predicted C_l/C_d ratio values of seven airfoils with randomly-selected AOA and thicknesses.

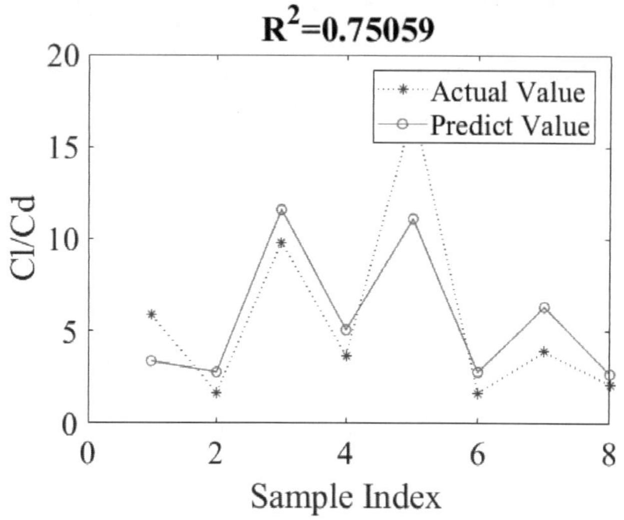

Figure 10. Comparison between predicted and actual C_l/C_d values

The coefficient of determination is R^2=0.75059. Here, R^2 is the percentage of the dependent variable variation that a linear model explains, which is always between 0 and 1. Usually, the larger the R2, the better the regression model fits the observations. In the present case, the R^2 ~1, showing the BP algorithm works well for the current dataset. This can also be observed by the lines in the figure, as the predicted values (red circles) are near the actual values (blue stars). This case demonstrates that the BP algorithm can work well for the aerodynamic prediction of airfoils. By combining the CFD workflow and the BP machine learning algorithm, it is possible to predict the previously-unknown performance of an airfoil with greater speed than traditional methods.

4. Conclusion
This paper focuses on the optimization of airfoil design through CFD approaches. The key parameters affecting airfoil performance, such as the AOA and thickness, are investigated, and it is found that the AOA has significant and consistent effects on the performance of airfoils, whereas the effects of thickness are highly dependent on the specific airfoil. Then, it is demonstrated that airfoils with an additional tip can perform better than unmodified airfoils so long as the tip parameters are set appropriately. In the future, the exploration of a wider array of β, lengths, and designs should be conducted. In addition, it is further shown that the Backpropagation algorithm, trained on data from the previously-run CFD simulations, can be used to quickly determine the performance of an airfoil.

Nomenclature

C_d	drag coefficient	(-)
C_l	lift coefficient	(-)
AOA	angle of attack	(°)
l_T	tip length	(m)
L	chord length	(m)
u	velocity	(m/s)
p	pressure	(Pa)
ρ	density	(kg/m^3)
υ_T	eddy viscosity	(m^2/s)
ν	kinematic viscosity	(m^2/s)
β	tip angle	(°)

References

[1] Kanimozhi, V. G., (2018). Analysis of Airfoil Flow Pattern Using CFD. International Journal of Engineering Research & Technology, Special Issue.

[2] Saraf, A. K., Singh, M. P., & Chouhan, T. S. (2017). Aerodynamic analysis of NACA0012 airfoil using CFD. International Journal of Mechanical and Production Engineering, 5(12), 21-25.

[3] GÖRGÜLÜ, Y. F., ÖZGÜR, M. A., & Ramazan, K. Ö. S. E. (2021). CFD analysis of a NACA 0009 aerofoil at a low reynolds number. Politeknik Dergisi, 1-1.

[4] Chumbre, V., Rushikesh, T., Umatar, S., & Kerur, S. M. (2018). CFD analysis of aerofoil sections. International Research Journal of Engineering and Technology, 5(7), 349-353.

[5] Zaheer, Z., Roy, K. R., Nair, G. S., Ragipathi, V., & Niranjan, U. V. (2019, November). CFD analysis of the performance of different airfoils in ground effect. In Journal of Physics: Conference Series (Vol. 1355, No. 1, p. 012006). IOP Publishing.

[6] Aqilah, F., Islam, M., Juretic, F., Guerrero, J., Wood, D., & Ani, F. N. (2018). Study of Mesh Quality Improvement for CFD Analysis of an Airfoil. IIUM Engineering Journal, 19(2), 203-212.

[7] Wilcox, D. C. (2008). Formulation of the kw turbulence model revisited. AIAA journal, 46(11), 2823-2838.

[8] Munson, B. R., Young, D. F., & Okiishi, T. H. (1995). Fundamentals of fluid mechanics. Oceanographic Literature Review, 10(42), 831.

Research on the pressure distribution under different airfoil types of aircraft

Xinying Chen [1, a, †], Xinyu Cheng [2, b, †] and Junlu Tian [3, *, †]

[1] McGill University faculty of science, Montreal, Quebec, H3A 0G4, Canada
[2] Aeronautics and Astronautics College, Sun Yat-sen University, Shenzhen, Guangdong, 518107, China
[3] Ulsan Ship and Ocean College, Ludong University, Yantai, Shandong, 264001, China

[a] xinying.chen@mail.mcgill.ca; [b] Chengxy37@mail2.sysu.edu.cn;
*Corresponding author's e-mail: 20181902192@m.ldu.edu.cn

[†] These authors contributed equally.

Abstract. Airfoils produce life force for aircraft, which is the reason for planes flying in the sky. The formation of the airfoil depends on its shape, so airfoil designs play an important role in airplane designs. The airfoil design is also prioritized in the process of aircraft design, its speed of creating affecting the progress of the entire project. The goal of airfoil's design is not simply to create "good" wings, because it does not exist. This means that when airfoils are adapting to a certain airflow or flying condition, their performance might not be satisfied due to environmental variations. Therefore, the application of Computational Fluid Dynamics (CFD) techniques and the Small Disturbance Equation, this study uses python to perform numerical analysis to simulate the surface pressure of the ideal wings under certain flying statues, and then applying a series of algorithms to calculate the shape of the target airfoils, which can find the most suitable airfoil shape under the flying circumstances. According to the researches, the best possible wings satisfying which the pressure below be as larger as possible than the pressure above to produce lift force, is f(x) = k*(x-1)^4 for downside and f(x) = k*sin(pi(x-1)) for the top. Besides, after a series of calculations, this paper realized that the smaller the k value can be, the better fit it is to an ideal simulated airfoil shape.

1. Introduction

With the rapid development in technology and computer science, Computational Fluid Dynamics (CFD), which uses electronic computers to simulate fluid motion numerically, has been developed and widely used in aerospace design, automotive design, and other engineering fields. It reduces the cost of aerodynamic design, shortens the computational cycle, and improves the design quality. However, CFD has its limitations, as the current level of computer development and computer performance constraints make the results of CFD simulations often inaccurate with the actual situation [1]. In airfoil design, CFD technology alone is not enough, but a series of effective methods are needed to reshape the base model according to the gap between the aerodynamic performance calculated by CFD and the target

Content from this work may be used under the terms of the Creative Commons Attribution 3.0 licence. Any further distribution of this work must maintain attribution to the author(s) and the title of the work, journal citation and DOI.
Published under licence by IOP Publishing Ltd

aerodynamic performance until the expected requirements are met [2]. Among them, numerical simulation using python is one of the most effective airfoil aerodynamic performance modification methods.

The earliest airfoils were made by imitating the shape of a bird's wing, which the Wright brothers and Santos-Dumont used in their respective aircraft manufacturing processes. This type of airfoil had more lift but also very high drag and relatively low lift resistance. At the beginning of the last century, Göttingen University in Germany researched the Göttingen series of airfoils through repeated design experiments using the theoretical airfoils of the Russian scientist Zhukovsky and the thin wing theory of the German scientist Menke. Since then, all the major aviation countries in the world have carried out their airfoil development. The most famous of them is the NACA series developed by the United States. After the 1930s, some researchers proposed the theory of laminar flow airfoil. The laminar airfoil allows the airflow on the airfoil surface to maintain laminar flow, reducing the airflow disturbance on the airfoil surface when the aircraft speed increases, thus reducing the airflow friction resistance on the airfoil surface and the heat transfer from friction, allowing the aircraft to maintain high speed and smooth flight. The laminar flow airfoil has an important significance in the history of airfoil development, and after continuous improvement by designers, the laminar flow airfoil is widely used in high subsonic and supersonic aircraft. When the airflow over the airfoil exceeds the local speed of sound, it is easy to generate a surge, which causes an increase in frictional drag on the airfoil and a rise in temperature, which adversely affects the flight of the aircraft, so early studies avoided designing airfoils with airflow over the speed of sound [2]. Later, the development of random airfoil design techniques led to the design of transonic airfoils that do not or only produce weak surges at surface airflow velocities above the speed of sound. These airfoils include the "supercritical airfoil" and "spike airfoil" which have been widely used. Nowadays, the number of airfoil types is extremely large, and they are widely used in different types of airplanes and other aircraft and wind turbines for different purposes. Therefore, this research uses Python to numerically simulate the pressure distribution on the airfoil surface under specific flight conditions to explore whether different airfoils can meet the pressure conditions of the target airfoil.

In summary, this paper uses a combination of theory and numerical simulation to study the pressure distribution characteristics of different airfoil types from hydrodynamic analysis, clarify the mechanism of airfoil influence on aircraft design, and provide theoretical and technical support for optimizing the accuracy of airfoil selection.

2. Methods

2.1. Physical model

The wing is the main component of an aircraft used to generate lift force. It is generally divided into two airfoils, left and right, which are arranged symmetrically on either side of the fuselage [4]. The main function of the wing is to generate lift force to support the aircraft in the air. It also plays a stabilizing and maneuvering role. There are a variety of wing planes, including rectangular, trapezoidal, swept-back, triangular, double-delta, arrow, and side-bar wings. Normally, modern aircraft are monoplane, but biplanes (two wings overlapping above and below), triplanes, and multiplanes have also been popular throughout history. Wings have a symmetry plane. The cross-section of the wing, cut parallel to the plane of symmetry, is called the wing profile and often referred to as the airfoil, the geometry of which is one of the basic geometric characteristics of the wing. The aerodynamic characteristics of the airfoil directly affect the aerodynamic characteristics of the wing and the aircraft as a whole, and have an important place in aerodynamic theory and the aircraft. Common airfoil shapes are divided into six types: plano-convex: the lower arc is a straight line. Symmetrical: the upper and lower arcs are both convex and symmetrical. Concave: the lower arc is at the wing chord line, with a large lift coefficient. Biconvex: both upper and lower arcs are convex but asymmetrical. s: the centre arc is a flat lying s. This type of wing is often used for tailless aircraft as the centre of pressure does not change when the angle of attack changes. Special type: Other airfoil types [3].

Air is physically fluid and in physics, there is fluid mechanics to study fluids. As in Figure 1, it is assumed that the existence of an ideal fluid that is free of turbulence, vortices and viscosity. This ideal fluid flows from the left side of a tube to the right side. the ideal fluid flows in from the left side of the tube with a velocity of v_1, a density of ρ_1 and a cross-sectional area of A_1, and out from the right side with a velocity of v_2, a density of ρ_2 and a cross-sectional area of A_2. the inflow and outflow from the two ends of a tube in the same time Δt The masses are $\Delta m_1 = \rho_1 A_1 v_1 \Delta t$ and $\Delta m_2 = \rho_2 A_2 v_2 \Delta t$ respectively. In the absence of source and leakage, since dm/dt is a constant, there is $\rho_1 A_1 v_1 = \rho_2 A_2 v_2$, i.e. $\rho A v = cons\tan t$, and in particular, for incompressible fluids, there is $Av = cons\tan t$, which is the continuity equation. Since the choice of the two positions on the left right is random, at any point within the fluid one has, $p + \rho v_2 + \rho g z = cons\tan t$, which is the Bernoulli equation [5].

This paper focuses on the physical study of the NACA series of airfoils, a series of airfoils proposed by the National Advisory Committee on Aeronautics (abbreviated as NACA, now NASA) after the 1930s, which has an important place in the history of airfoil development. It has evolved to include the NACA four-digit airfoil, the NACA five-digit airfoil and the NACA six-digit airfoil. The range of applications covers light aircraft, large passenger aircraft and even supersonic fighters. The NACA family of airfoils has been used in many of the current airfoil design studies.

In this paper, a set of NACA four-digit airfoils are used as the subject of study and are briefly described here. The NACA four-digit airfoil is one of the earliest members of the NACA family of airfoils and is still widely used today. It uses four digits to describe a single airfoil and prefixes these four digits with NACA, which represents the name of the current airfoil. Figure 1 illustrates the geometric parameters of the NACA2415 [2].

The leading point of the airfoil is called the leading edge point of the airfoil and the trailing edge point of the airfoil. The line between these two points is called the geometric chord of the airfoil. However, for some airfoils where the lower surface is mostly straight, this line is also defined as the geometric chord. The distance between the leading and trailing edge points, or between the projection of the leading and trailing edge on the chord, is called the chord length of the airfoil and is denoted by c.

The upper and lower surface curves of the airfoil are expressed as a function of the relative coordinates of the chord lengths as follows [6-7].

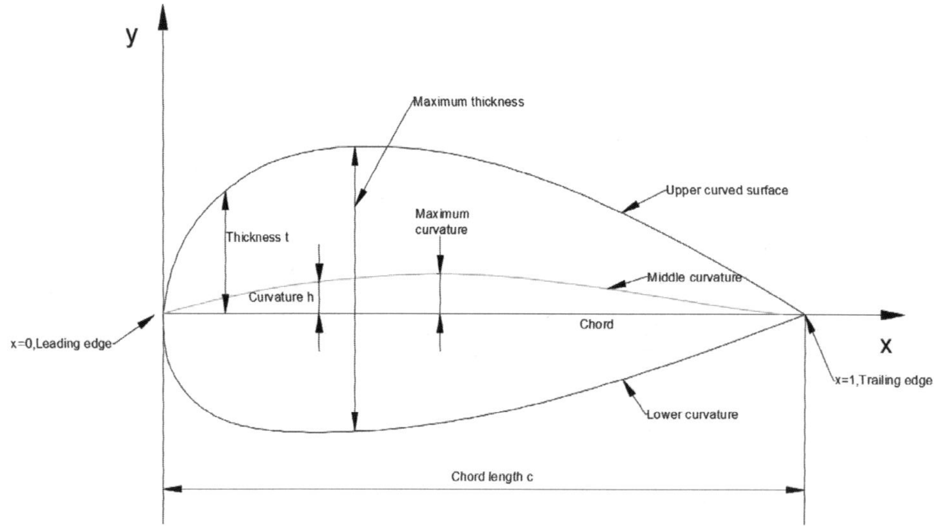

Figure 1. Schematic diagram of the geometric parameters of the NACA2415

The upper surface coordinates $x = \frac{\overline{x}}{c}$, the lower surface coordinates $yd = \frac{\overline{yd}}{c}$, and the horizontal coordinates $x = \frac{\overline{x}}{c}$. The distance between the upper and lower surfaces is called the thickness, $t = yu - yd$ the maximum thickness being $t\,max = max(yu - yd)$. The distance between the central arc and the chord is the curvature $h\,max = max\frac{1}{2}(yu - yd)$, the maximum curvature$h\,max = max\frac{1}{2}(yu - yd)$.

2.2. *Mathematical model*

2.2.1. 1D Euler Equations. This mathematical formula, together with physical principles.
1) Conservation of mass
The Continuity Equation:

$$\frac{\partial \rho}{\partial t} + div(\rho U) = 0 \tag{1}$$

2) Newton's second law
The Momentum Equation:

$$\frac{\partial(\rho U)}{\partial t} + div(\rho U \otimes U + pI) = div\tau \tag{2}$$

3) Conservation of energy
The Energy Equation:

$$\frac{\partial(\rho E)}{\partial t} + div(\rho E U) = div(k\nabla T) + div(\sigma U) \tag{3}$$

lead the Navier-Stokes equations-the general equations of fluid flow.

These equations need to be supplied with a domain, boundary conditions, initial conditions. For stationary problems, this study sets the time derivative to 0.

After a series of operations, the Euler equation can be obtained.

$$\frac{\partial}{\partial t}\begin{pmatrix} \rho \\ \rho u \\ \rho E \end{pmatrix} + \frac{\partial}{\partial x}\begin{pmatrix} \rho u \\ \rho u^2 + p \\ \rho u(E + p/\rho) \end{pmatrix} = 0 \tag{4}$$

$$p = (\gamma - 1)\rho e \tag{5}$$

$$c^2 = \gamma p/\rho \tag{6}$$

If vorticity is zero$\nabla \times v = 0$,
Then $v = \nabla\phi$,
And the continuity equation becomes,

$$\frac{\partial \rho}{\partial t} + div(\rho\nabla\phi) = 0 \tag{7}$$

This study will derive an equation for the density in terms of the potential. For steady flow, the time derivative disappears and the relation shows that

$$\frac{\rho}{\rho_0} = \left(1 - \frac{|\nabla\phi|^2}{2H_0}\right)^{1/(\gamma-1)} \tag{8}$$

This flow model involves a single unknown, the potential which is a scalar function.

This model was used for many decades to design airplane until in the late 80's and 90's replaced with the Euler and the Navier-Stokes equations.

In many applications, this paper focuses on a steady flow and therefore omits all time derivatives,

$$c^2 = (\gamma - 1)\left[H_0 - \frac{1}{2}|\nabla\phi|^2\right] \tag{9}$$

with wall BC

$$v_n = \frac{\partial\phi}{\partial n} = 0 \tag{10}$$

This is the steady full potential equation in non-conservative form. This form reveals more about the equation than its conservative counterpart.

2.2.2. The Small Disturbance Approximation Equation. A simplification of the full potential equations is done in the case of thin obstacles, such as thin airfoils. Since the obstacle is "small", its effect on the flow is small and this research considers perturbation to uniform flow with a velocity of magnitude in the x-direction [8-9]. The potential has a representation

$$\phi = U_\infty(x + \Phi) \tag{11}$$

Where velocities are recovered from the potential using the formulas,

$$u = U_\infty(1 + \Phi_x) \tag{12}$$

$$v = U_\infty\Phi_y \tag{13}$$

The equation becomes $(1 - M_x^2)\Phi_{xx} + \Phi_{yy} = 0$, which can be simplified further to $(1 - M_\infty^2)\Phi_{xx} + \Phi_{yy} = 0$.

The wall BC becomes $v = (U_\infty + u)f'(x) \approx U_\infty f'(x)$where f(x) is the shape of the airfoil.

2.3. Python code analysis

The Python Code developed for analysis is based on the Small Disturbance Equation. The small disturbance equation is initially associated with the classical hypersonic similitude, which is a direct result of the theory. It is extremely useful for investing the two-dimensional and axisymmetric irrotational equation of motions [10]. In figure 1, it shows the model that has been used in the analysis. Simulate a linear model, the small disturbance model

$$\left(1 - M_\infty^2\right)\phi_{xx} + \phi_{yy} = 0 \tag{14}$$

2.4. Boundary condition of compressible flow

Five helper functions are used in the analysis. The first one is the velocity function, velocity_vec, which is for the velocity field, it takes two input variables, phi and h. Phi is the velocity potential [11]. The velocity vector is divided into x-vector and y-vector, which is represented in the code as u and y. By following the process of calculating the difference between the two vectors in two directions, the velocity field of all points is solved. Cal_pressure is the second helper function, which takes 4 input variables. This is the pressure function to solve the pressure field. Adding in more changeable constants and using the velocity field which is solved from velocity_vec to calculate the pressure field. Next, BC_F is the function for boundary conditions. The purpose of this function is for the boundary condition of the field, which also defines the shape of the wing being investigated. It sets up a function and a domain of both x and y. If it's inside the shape, it returns a value, if not, it returns zero. The following one is the error function, error. It takes three input variables. Every a hundred times of the calculation above, this function will calculate the error and residual. Last but not the least, it is the SOR function. This function uses the iteration method to solve the linear algebra equation to get the potential. Using the sort function to get the initial condition and then solve for a new phi. Every 100 solves, it calculates the error. After that, using the set of pi to calculate the velocity and then get the pressure.

3. Result and discussion

3.1. five equations

(1) f(x)=K*(x-1)

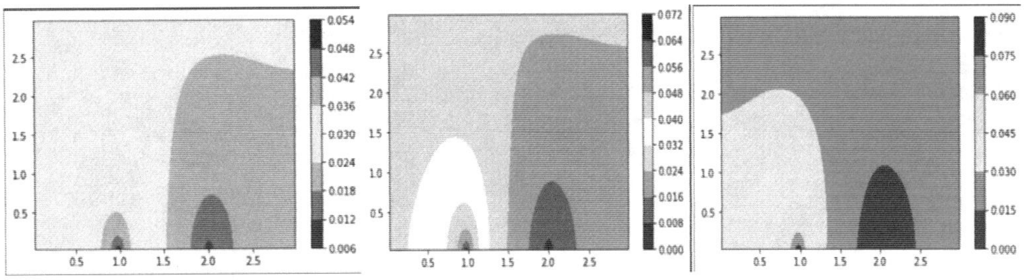

Figure 2. Pressure distribution of f(x)=K*(x-1) (K=0.1 0.15 0.2).

In Figure 2, it can get that when the shapes f (x) = K * (x-1) ^ 3, the pressure value of the wing first increases and then decreases in the range of (1,2), in which the maximum pressure point is located at the coordinate (1,1.5).

(2) f(x)=K*(x-1)^2

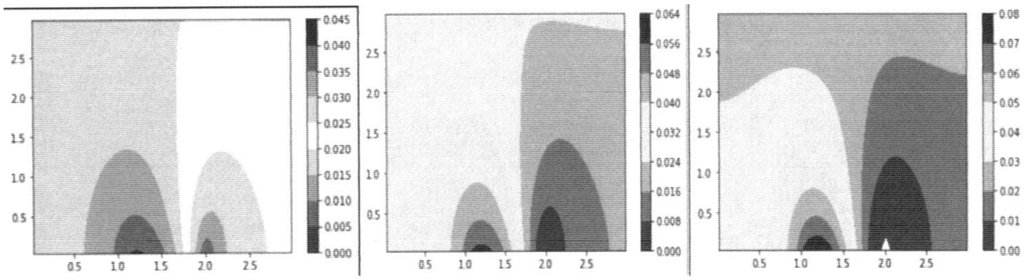

Figure 3. Pressure distribution of f(x)=K*(x-1)^2 (K=0.1, 0.2, 0.3).

In Figure 3, the shape is f (x) = K * (x-1) ^2, and the pressure value of the wing first increases and then decreases between 1 and 2. With the increasing K value, the maximum pressure area gradually decreases.

(3) f(x) = K*(x-1)^3

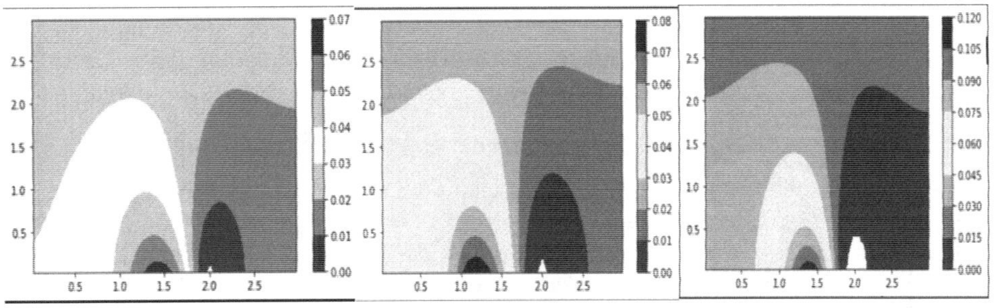

Figure 4. Pressure distribution of f(x) = K*(x-1)^3 (K=0.2, 0.3, 0.4).

As it can be seen from Figure 4, when the wing forms f (x) = K (x-1), the pressure of the wing decreases monotonically at (1,2), in which the maximum pressure point is concentrated at x = 1 and the minimum point is located at x = 2.

(4) f(x)= K*(x-1)^4

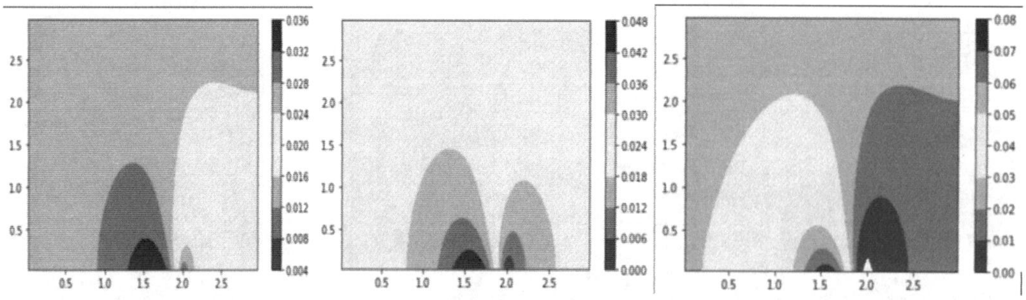

Figure 5. Pressure distribution of f(x)= K*(x-1)^4 (K=0.05, 0.1, 0.2).

It is shown in Figure 5 that with the increase of K value, the pressure area decreases moderately, but the maximum point is concentrated at x = 1.5.

(5) f(x)=K*sin(pi(x-1))

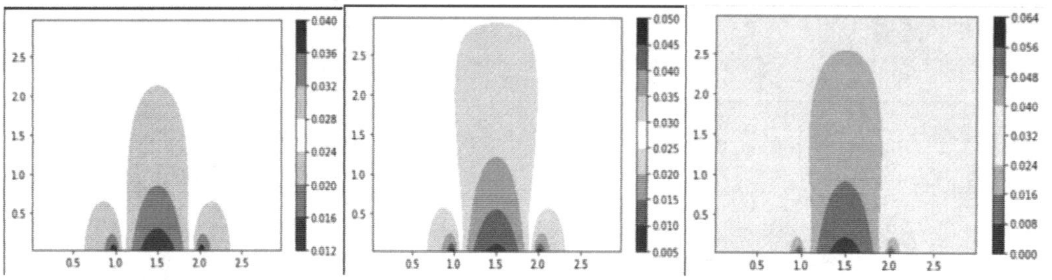

Figure 6. Pressure distribution of f(x)=K*sin(pi(x-1)) (K=0.1, 0.15, 0.2).

From Figure 6, due to the rise of K, the pressure within the study range slowly decreases, and the minimum pressure is concentrated at x = 1.5.

3.2. Comparison among the five equations

A comparison between all results from all shapes that have been got from the simulations above. (plots and description). By comparing the color depth and distribution in the region where y belongs to (0-1) and X belongs to (1-2), it can be found that the pressure in the function f (x) = k * sin (PI (x-1)) (Figure 5) is the smallest and evenly distributed in the region where x belongs to 1-2, especially when the coefficient K is larger, the pressure in the region as a whole is smaller. The function f (x) = k * (x-1) ^ 4 (Figure 4) has the deepest color in X belonging to 1-2, that is, the pressure is the largest and most widely distributed. Especially, when the coefficient K is smaller, the pressure in the region as a whole is larger.

4. Conclusion

With the continuous progress of computing technology and computer technology, the computational fluid force using electronic computers and discrete numerical methods to simulate fluid motion has been greatly developed and has been widely used in aerospace design and other fields. Although the application of CFD can solve some problems, it still has limitations. Therefore, to meet the expected requirements in the design of airfoil and CFD technology, a series of effective methods are needed to modify the basic airfoil. Through the small disturbance equation, this paper first gives the ideal simulated airfoil surface pressure distribution in a specific flight state by python, and then directly calculates the airfoil shape parameters that meet the design objectives by using a series of algorithms. Finally, it gets the conclusion by comparison. The main conclusions are

(1) The most suitable lower side function of the aircraft airfoil is f (x) = k * (x-1) ^ 4, and the upper side function is f (x) = k * sin (PI (x-1)).

(2) The selection principle of the two values of K is that the smaller the better.

References

[1] Zhang, J.P. (2006) Perspectives on Aeronautical Aerodynamics Development. International Aviation, 2: 80–83.
[2] Xu, Y.F. (2012) Rapid design of aircraft airfoil based on CST parameterization method. Nanjing University of Aeronautics and Astronautics, 1: 11–15.
[3] Wang, P.Y. (2018) Study on the wing shape of aircraft and its lift generation. China Science and Technology, 12: 1–3.
[4] Zheng, E. B. (2017) Lifting force from fluid expansion and compression. Science and Technology Wind, 19: 132.
[5] An, y. (2017) Exploring the force of fluids - a brief description of Bernoulli's equation. Scholastic, 11: 121–122.
[6] Cao, T.Z. (2017) Application and simplified derivation of Bernoulli's equation in flying vehicles. Digital Communication World, 02: 71–72.
[7] Lu, D.S. (1998) College physics. Higher Education Press, Beijing.
[8] Usherwood, J.R. (2009) The aerodynamic forces and pressure distribution of a revolving pigeon wing. Experiments in Fluids., 46: 991–1003.
[9] Mohammad, K., Gandomi, F. (2021) The Effects of National Academy of Sports Medicine and Sahrmann Training on Foot Pressure Distribution in Flexed Posture Students. Iranian Rehabilitation Journal., 19: 99–100.
[10] Jendzelovsky, N., Antal, R. (2021) CFD and Experimental Study of Wind Pressure Distribution on the High-Rise Building in the Shape of an Equilateral Acute Triangle. Fluids., 6: 81.
[11] Carpegna, G., Alovisi, M., Paolino, D.S. (2020) Evaluation of Pressure Distribution against Root Canal Walls of NiTi Rotary Instruments by Finite Element Analysis. Applied Sciences., 10: 2981.

Aerodynamic optimization analysis of the front bumper side of the vehicle

Bingheng Wu, Lin Chen, Qingyun Guan, Ling Xiao and Xiayi Yuan *

Guanzhou Automobile Group Co., Ltd Automotive Engineering Institute, Guangzhou, China

*Corresponding author e-mail: wubingheng@gacrnd.com

Abstract. The front bumper is the earliest part of the vehicle that is exposed to the head-on airflow, and the optimization of the front bumper side plays an extremely important role in the development of the air drag of the vehicle. In this paper, the front bumper side of a model is optimized by numerical simulation method, which greatly improves the air drag of the vehicle.

1. Introduction

Aerodynamic performance is one of the most important performance in the vehicle performance development process, which mainly affects vehicle fuel economy and electric consumption, occupant comfort, and overall vehicle thermal management, among which aerodynamic resistance is the main factor affecting fuel economy or electric consumption[1]. At a speed of 60km/h, the aerodynamic drag accounts for about 50% of the overall drag, and increases with speed.[2]. Therefore, in the design and development, aerodynamic resistance is the most important index of aerodynamic performance optimization design.

When the vehicle is moving forward, the front bail is the earliest part of the vehicle to contact the oncoming airflow, and its airflow separation status directly affects the airflow flow through the body and rear of the vehicle, if the airflow separates prematurely when passing through the front bail to produce complex turbulence, the airflow is very likely to be difficult to flow smoothly through the rear of the vehicle, therefore, managing the airflow at the side of the front bail and optimizing the side of the front bail play a vital role in the development of the aerodynamic drag of the vehicle. In this paper, we optimize the front bumper flank of a model by numerical simulation method, and greatly improve the wind resistance performance.

2. Optimisation methods

CFD is short for Computational Fluid Dynamics. CFD simulation uses electronic computers as a tool to apply various discrete mathematical methods for numerical experiments, computer simulations and analytical studies of various problems in fluid dynamics, which can present details of flow, heat transfer, mass transfer, combustion and other processes.

Compared with wind tunnel tests, CFD simulation has the advantages of low cost, the ability to simulate multiple conditions at the same time, and the ability to simulate more complex or ideal processes, which can broaden the scope of experimental research and reduce the costly experimental workload[3]. At present, the mainstream CFD software includes Star-CCM+, Fluent, powerflow and so

Content from this work may be used under the terms of the Creative Commons Attribution 3.0 licence. Any further distribution of this work must maintain attribution to the author(s) and the title of the work, journal citation and DOI.
Published under licence by IOP Publishing Ltd

on. Currently, these CFD software can easily exchange values between each other, and use unified pre- and post-processing tools, the model mesh inheritance is good, save the user in computer methods, programming, pre- and post-processing and other aspects of repetitive, inefficient labour. The CFD software used in this paper is Star-CCM+, and different optimisation scenarios have been simulated and analysed.

3. Optimisation theory

In the boundary layer, the fluid immediately adjacent to the object surface due to the effect of molecular gravity, completely adhered to the object surface, and the relative velocity of the object is zero, the boundary layer flow from the object surface detachment phenomenon called boundary layer separation. There are two cases of boundary layer separation, one is to occur in the smooth surface, the other is to occur in the object surface has a sharp corner or other shape interruption or discontinuity.

On a smooth object surface, the gas within the boundary layer will have a velocity gradient due to viscosity, and the gas near the wall is decelerated, but the gas away from the wall still maintains its large inertia force. The viscous stress is resistance to the fluid in the boundary layer, so as the fluid flows backward along the object, the fluid in the boundary layer will gradually decelerate and increase pressure, and as the flow develops, the deceleration becomes more intense, the pressure gradually increases, and the gas near the wall side will generate negative velocity, forming a reverse pressure gradient ($\partial p/\partial x>o$). When the negative velocity is generated, the gas away from the boundary layer keeps its inertia, and the gas near the boundary layer generates negative velocity and reverses the flow, then the gas starts to separate, and the gas near the wall will generate vortex rotation at the back because of the negative velocity reversal and form turbulence. The location where the gas is detached from the object surface is called the separation point, and the separated boundary layer forms a larger vortex area downstream, but it may also attach to the object surface again somewhere downstream to form a local reflux area or bubble.

In the object surface has a sharp corner or other shape interruption or discontinuity where the boundary layer separation occurs because the fluid flow along the object surface is interrupted, can not continue to flow against the object surface, the outflow velocity near the object surface is very large, the pressure is very small, so there must be a large downstream inverse pressure gradient, in its role, the boundary layer that is from the sharp corner or interruption of the separation.

In the front bailout side region, the pressure of the laminar boundary layer without separation is smaller than that of the turbulent boundary layer with separation, and the front bailout region of the front of the vehicle has less resistance increase when the boundary layer is not separated than when it is separated. For the vehicle, the boundary layer separation leads to the vortex area behind the winding object, and the small air pressure in the vortex area makes the object produce differential pressure resistance, and the range of the vortex area can be reduced by avoiding or postponing the airflow separation, i.e., the low pressure area at the rear of the vehicle is reduced, so we should avoid or postpone the occurrence of gas separation as much as possible when reducing the resistance to the vehicle.The separation of gases can be determined by some of the following technical parameters:

1) ISO surface

The surface formed by connecting the points where the total pressure on the surface of the object is 0 is called ISO surface, and the schematic diagram is shown in Figure 3.1. By viewing the ISO surface, the location of airflow separation and the distribution of vortex volume can be directly determined.

The airflow has a large velocity gradient between the complete surface and the ISO surface, with the airflow appearing turbulent and turbulent; outside the ISO surface the airflow is in a more stable state with a small velocity gradient. On the side of the front bumper, managing the airflow means controlling the separation of the airflow and reducing the volume between the ISO surface and the overall surface. In Fig. 1 the side trim of the front bumper has a bulge that causes airflow separation, resulting in a large volume on the ISO surface in the outer area of the front wheels.

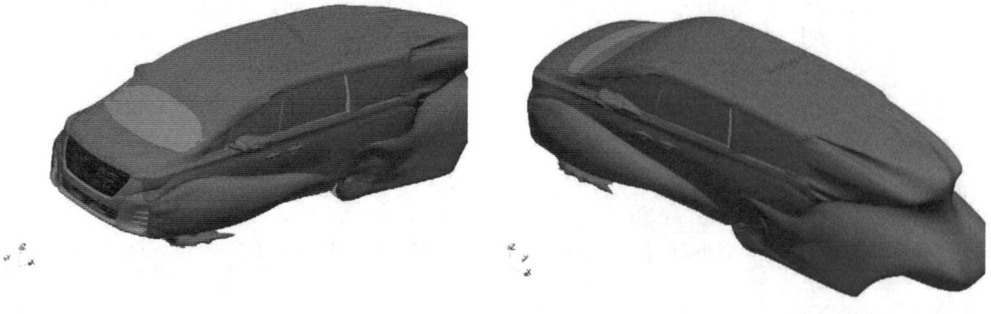

(a) (b)

Figure 1. The ISO surface of a vehicle

2)Total pressure coefficient

The total pressure coefficient can be used as a dimensionless parameter for the study of fluid flow and is defined as:

$$C_{pt} = \frac{p + \frac{1}{2}\rho V^2}{p_\infty + \frac{1}{2}\rho_\infty V_\infty^{\ 2}} \qquad (1)$$

where p is the relative static pressure at the study point; V is the velocity at the study point; p_∞ is the relative static pressure away from any disturbance, defined in this paper as 0; ρ_∞ is the independent fluid density (at sea level atmospheric pressure and temperature 15°C); and V_∞ is the velocity of the independent fluid. In Fig. 2 (b) Total pressure coefficient diagram, there is a clear gradient at the separation interface, the red and blue part of the diagram; the separation zone has low pressure and no clear gradient, the blue region of the diagram.

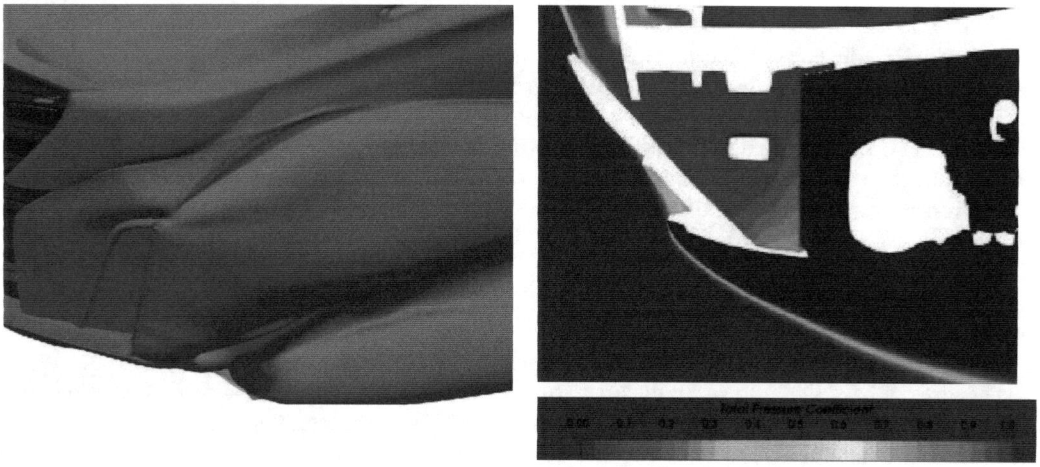

(a) (b)

Figure 2. The ISO surface (a) and the total pressure coefficient diagram (b) of a vehicle

3)Velocity coefficient and velocity direction

The velocity coefficient can be used as a dimensionless parameter for the study of fluid flow and this dimensionless coefficient is defined as

$$C_V = \frac{V}{V_\infty} \tag{2}$$

Where V is the magnitude of the velocity at the point of study, V_∞ is the magnitude of the velocity of the independent fluid.

In Fig. 3 (b): Velocity diagram, the velocity coefficient is represented by the colour, and it can be seen that there is a clear gradient at the separation interface, the red and blue part of the diagram; the velocity in the separation zone is small and there is no clear gradient, the blue region of the diagram.The direction of velocity is the direction of the actual movement of the airflow. In Fig. 3 (b): Velocity diagram, the direction of velocity is represented by the flow line, which shows that the separation interface and the front bail side have a certain angle when the separation of the airflow occurs. Slowing down the separation of the airflow means that the separation point is as far back as possible and the angle between the separation interface and the front bumper side is reduced so that the airflow can flow parallel to the front bumper side towards the rear of the vehicle.

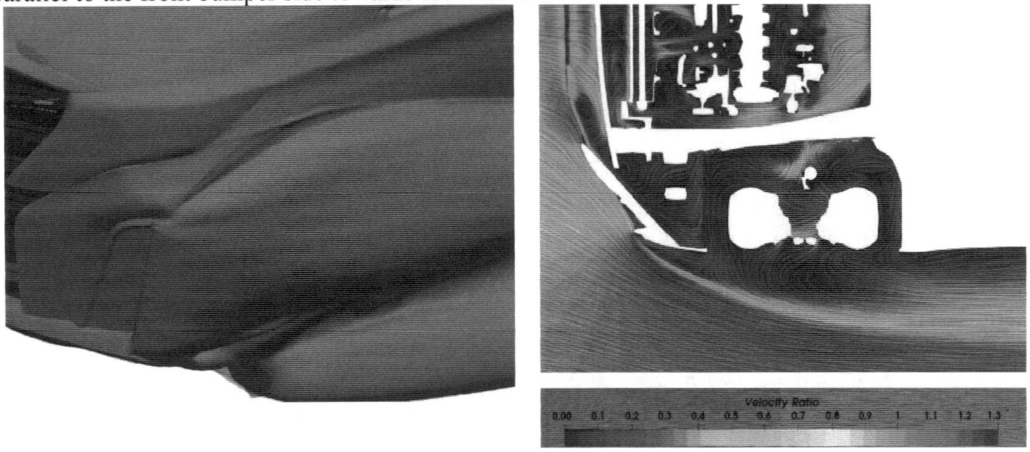

(a) (b)

Figure 3. The ISO surface (a) and the velocity diagram (b) of a vehicle

4. Optimized Solutions

The current shape of the front bumper side of the vehicle can be discussed in three general categories: 1) a full and continuous smooth front bumper side; 2) a full and discontinuous front bumper side with a small camber; and 3) a front bumper side with a concave shape in the corner.

In the first two cases the airflow separation is generally mild and can be achieved by rounding off the camber, in the third case the airflow separation is more severe. Nowadays, some vehicles are designed to be characterised by their exterior design and tend to be designed with concave corners. The air flow from the fog light area through the outer trim to the side is easily separated on the outside of the trim due to the sudden change in curvature, and the air flow no longer flows against the front bumper side towards the rear of the vehicle, as shown in Fig. 4. Optimisation of this situation requires the outermost side of each protrusion to form a rounded profile on the whole to fit the airflow separation surface, allowing the airflow to reattach to the front bumper side as it leaves the outermost side of the front bumper.

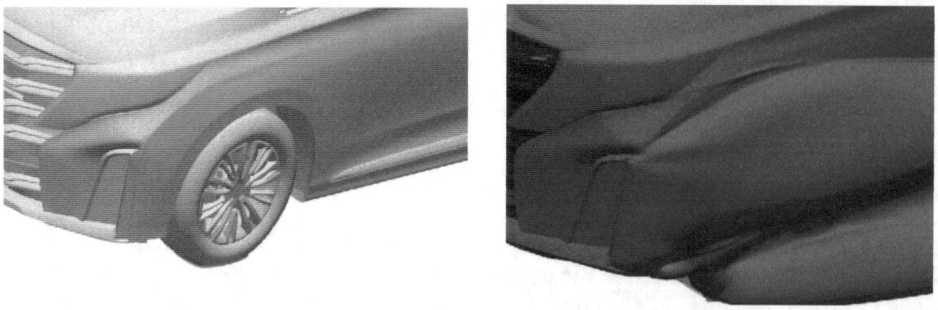

(a) (b)

Figure 4. Outer profile surface (a) and ISO surface (b) of a vehicle

Option 1: Profile optimization

The fog light outer trim is moved back to optimise the airflow to delay separation as it passes over the trim and flows against the side to the rear of the vehicle. Compared to before the optimisation, the volume of the bulge between the ISO surface and the sides is reduced, the range of areas with a total pressure coefficient equal to 1 in the pressure coefficient diagram is greater, the range of areas with a large velocity coefficient in the velocity diagram is greater and the angle between the separation interface and the front bumper side is smaller.

Parameters	Before optimization (a)	Optimised (by moving the trim back) (b)
ISO surface	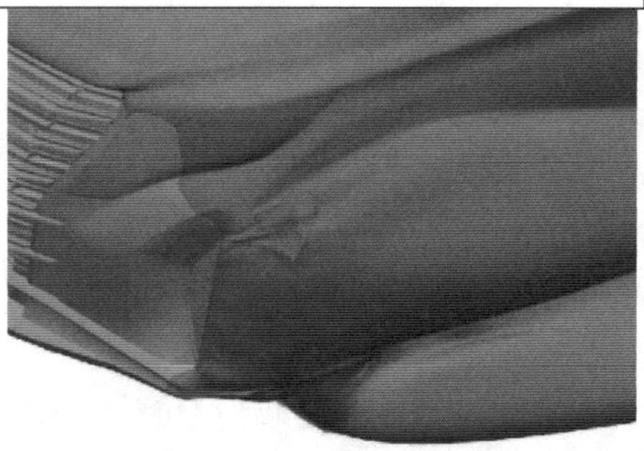	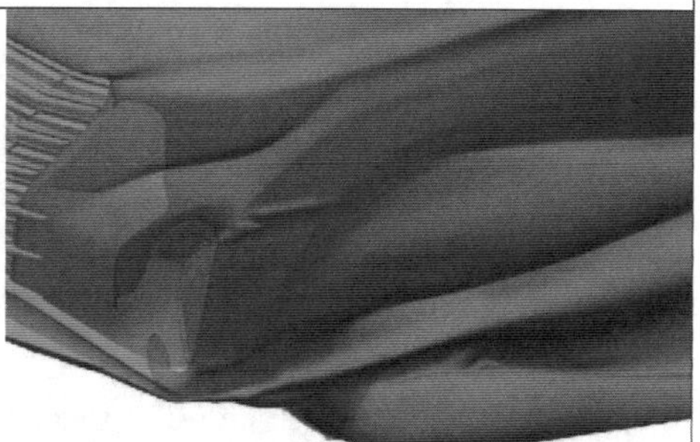
C_{pt}	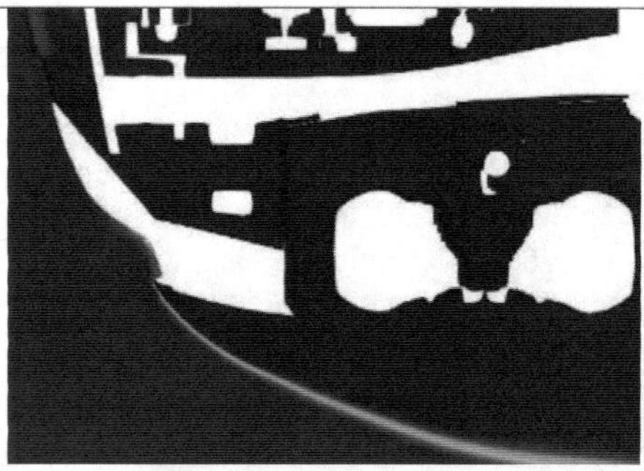	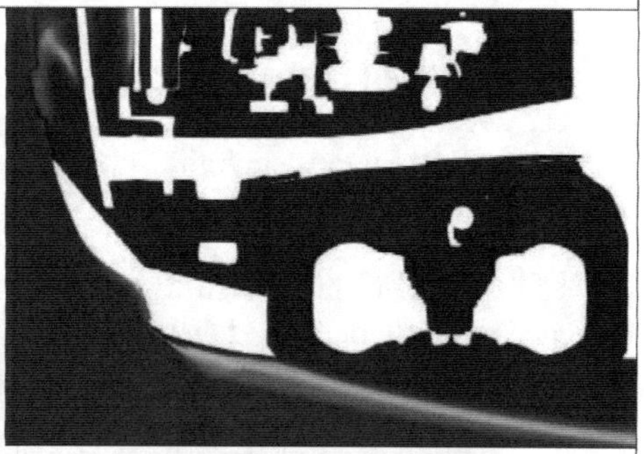

Journal of Physics: Conference Series **2441** (2023) 012006 doi:10.1088/1742-6596/2441/1/012006

Velocity	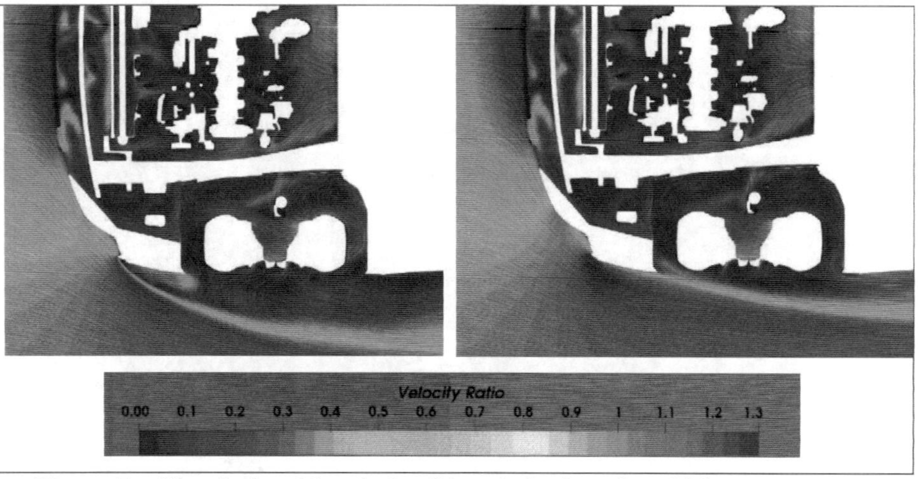

Figure 5. Flow before (a) and after (b) optimisation of a vehicle

Option 2: Move the fog light trim back + open the air curtain

The fog lamp area is concave and the airflow builds up. By opening the air curtain in the fog lamp area, the airflow can be directed to the tyre area and the airflow no longer builds up in the fog lamp area. Compared to before the optimisation, the volume of the bulge between the ISO surface and the side is reduced, the area with a total pressure coefficient equal to 1 in the pressure coefficient diagram is wider, the area with a large velocity coefficient in the velocity diagram is wider and the angle between the separation interface and the front bumper side is smaller.

Parameters	Before optimization (a)	Optimised (Move the fog light trim back + open the air curtain) (b)
ISO surface	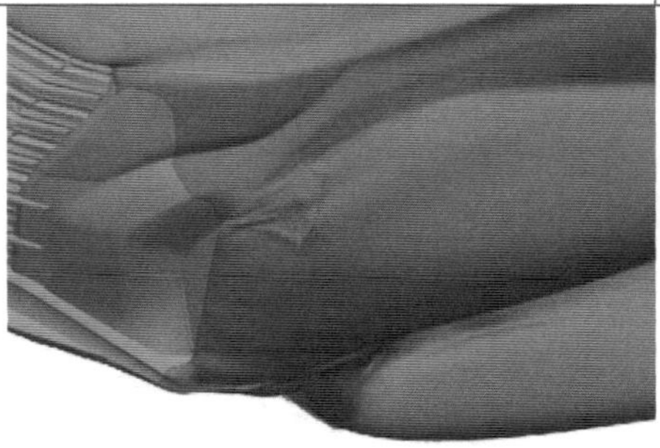	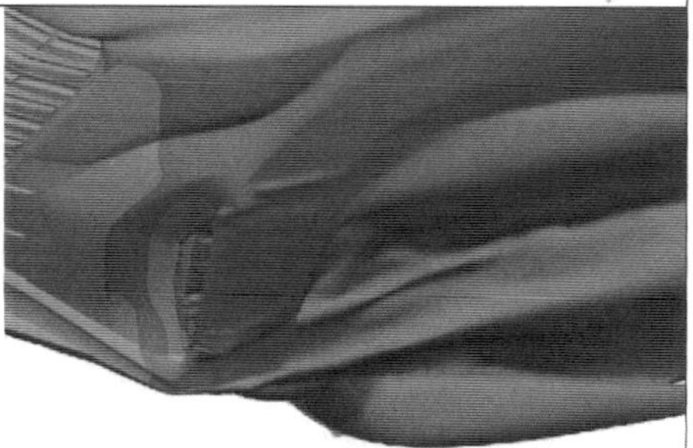

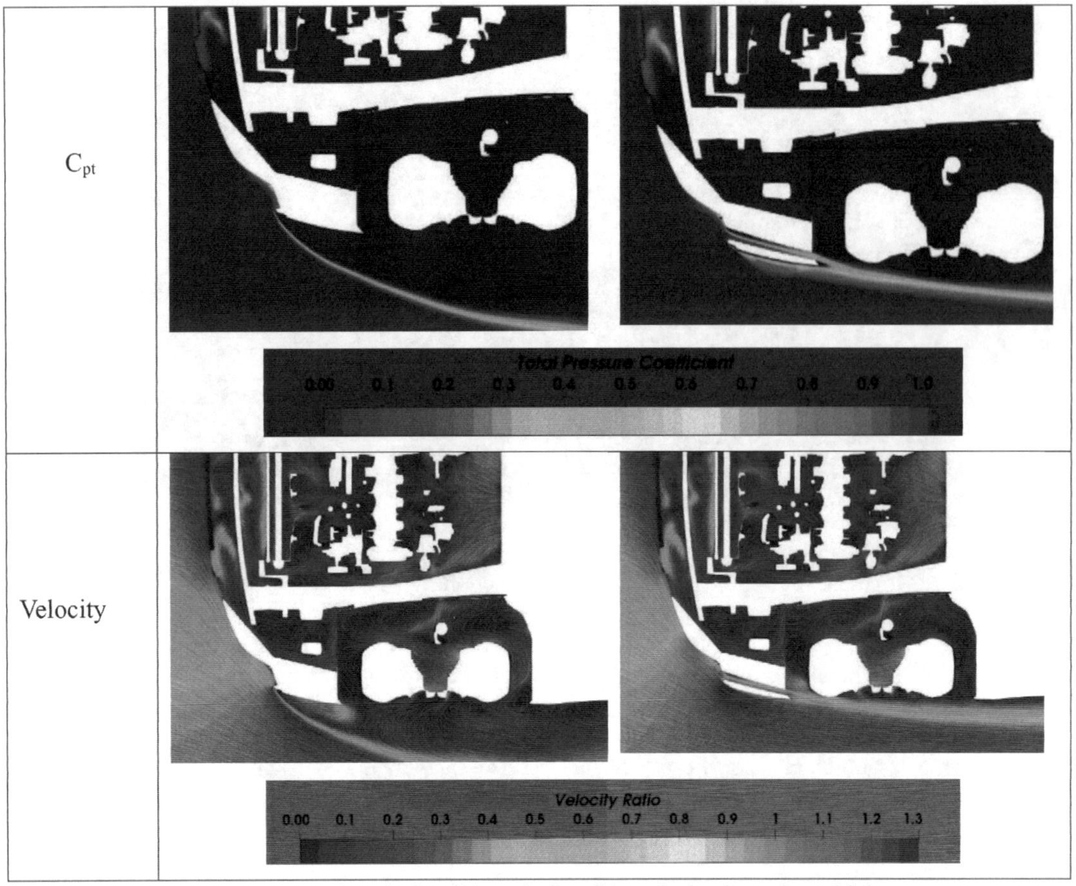

Figure 6. Flow before (a) and after (b) optimisation of a vehicle

By optimising the sides of the vehicle's front bumper, the airflow fits more closely to the front bumper after passing through it and is less likely to separate, and the vehicle aerodynamic drag coefficient has improved considerably, as shown in the Table 1.

Table 1. Three Scheme comparing.

Numble	△Cd(counts)
base	-
Option 1 : Move the fog light trim back	-30
Option 2 : Move the fog light trim back + open the air curtain	-32

5. Conclusion

In this paper, the modeling and simulation analysis of a model is vehicleried out by Star CCM+ software to optimise the front bumper side of a vehicle, resulting in a 32 counts reduction in wind resistance, which greatly improves the fuel economy of the vehicle, and this optimisation method can also provide a reference for the aerodynamic development of subsequent models.

References

[1] J. Howell, C. Sherwin, M. Passmore,et al, Aerodynamic drag of a compact SUV as measured on-road and in the wind tunnel, SAE Paper, 2002, pp.583-590.
[2] Z. Q. Gu, Automotive aerodynamics, People's Traffic Press, Beijing, 2005, pp.80-115.
[3] L. M. Fu, Vehicle aerodynamics, China Machine Press, Beijing, 1988, pp.30-254.

Numerical Simulation of the Two-phase Flow of a Constant-pressure Rear-spray Low Recoil Weapon with Two Chambers

Zijian Li[1*], Ming Qiu[1], Jie Song[1], Liang Yang[1] and Zhenqiang Liao[1,2]

[1]School of Mechanical Engineering, Nanjing University of Science and Technology, Nanjing, 210094, Jiangsu, China
[2]School of Mechanical Electrical Engineering, Global Institute of Software Technology, Suzhou, 215163, Jiangsu, China
*Corresponding author's e-mail: lzjnjust@163.com

Abstract. To study the effect of a constant-pressure rear-spray low recoil weapon with two chambers on weapon launch performance, based on two-phase flow theory, the valve-spring type constant-pressure rear-spray low recoil weapon firing process with two chambers was studied numerically. Considering the movement of the valve-spring system, a mathematical model of the weapon launch process combining the two-phase flow of gas and solid in the front and back chambers, the movement of the projectile and the transient flow of gas in the lateral rear nozzle is established. Using a 30 mm chain gun as an example, numerical simulation was carried out, and the distribution law of the flow field in the two chambers and the lateral rear nozzle during the firing of the weapon was obtained and compared with the internal flow field of the traditional weapon chamber. The results show that by reasonably matching parameters to control the flow of gunpowder gas in the two chambers, the recoil impulse can be reduced by 81.58% under the premise that the initial velocity of the projectile does not decrease, and the purpose of making full use of gunpowder gas energy to greatly reduce arms recoil is realized.
Keywords. Recoil reduction; nozzle; two-phase flow

1. Introduction

Modern warfare requires that the new generation of weapons should not only have greater power, but also put forward higher requirements for their mobility, diversity of carrying platforms, and battlefield adaptability. However, due to its own structure and launch method, the technical way to increase the power of weapon will greatly increase the recoil. The huge recoil makes it necessary for the weapon system to be equipped with complex anti-recoil devices and support fixtures, reducing the weapon's ability to act quickly . It can be seen that how to greatly reduce recoil under the premise of ensuring power is a key bottleneck that restricts the mobility of weapons and the further improvement of the adaptability of the mounted platform. Considering that the gunpowder gas energy discharged from the muzzle of the gun muzzle when the general gun weapon is not used when firing accounts for more than 40% of the total combustion energy of the gunpowder, scholars have done a lot of research on the use of gunpowder gas energy to reduce the recoil of weapons, such as the muzzle brake [1], the rarefaction wave gun [2], the laval nozzle reduction recoil weapon [3], etc. However, these recoil reduction techniques or recoil reduction efficiency are limited, and it is difficult to achieve the purpose of taking into account the basic elimination of recoil and the initial speed of the projectile.

Content from this work may be used under the terms of the Creative Commons Attribution 3.0 licence. Any further distribution of this work must maintain attribution to the author(s) and the title of the work, journal citation and DOI.
Published under licence by IOP Publishing Ltd

Kathe [4] proposed expansion wave artillery using the gunpowder gas in the chamber after spraying balance recoil, cannot affect the initial speed of the projectile without affecting the recoil momentum of the recoil, but because the gunpowder gas directly from the rear of the body tube spray, the rear spray device cannot be reused, so it is difficult to achieve continuous weapon shooting. Qiu [5], Si [6] and others proposed a lateral back-spraying sparse wave reduction recoil control method, which controls the opening of the lateral rear-spraying device by cleverly designing the back-spray trigger device associated with the projectile movement, while achieving the effect of reducing recoil, the initial speed of the projectile is not reduced and continuous shooting is achieved, but the gunpowder gas energy that can be used to reduce recoil is limited, and its recoil reduction efficiency is generally not more than 45%.

In this paper, a method is proposed to control the opening and closing of the exhaust pipe through the valve-spring device, and the gas in the chamber is exported laterally and sprayed to decrease arms recoil. Due to the special structure of the weapon such as two chambers, a valve-spring system and a nozzle, the firing process is quite different from that of traditional weapons. Considering the motion of the valve-spring system, this paper establishes a launch process model that combines the front and back chamber gas-solid two-phase flow, projectile motion and the transient flow of gas in the exhaust pipe and performs numerical simulation. In order to analyse the influence of various parameters on the initial velocity of the projectile, the recoil impulse and the pressure distribution in the chamber, it finally laid a theoretical foundation for the practical engineering application of the weapon.

2. The Recoil Reduction Principle of a Constant-pressure Rear-spray Low Recoil Weapon with Two Chambers

The basic principle of a constant-pressure rear-spray low recoil weapon with two chambers is shown in figure 1. Firstly, the propellant in the back combustion chamber is ignited, the gas of propellant breaks through the sheet, it flows to the front combustion chamber through the fire transmitting orifice and ignites the propellant in the front chamber; at this moment, the pressure of the back chamber is greater than the pressure of the chamber ahead, the gas of propellant from the back chamber continues to flow to front chamber, and after overcoming the projectile extrusion pressure, it pushes the projectile forward with the gunpowder gas in the front chamber, but at this time, back chamber pressure is not higher than the precompression force of the spring, and valve will block the exhaust pipe under the action of the spring. When the back chamber pressure increases to higher than the spring precompression force, the valve begins to move, and the back chamber gunpowder gas is ejected from the divergent nozzle to the back through the exhaust pipe; because the divergent nozzle is directly connected to the atmosphere, the back chamber pressure drops faster than the front chamber. When the pressure at the bottom of the bullet in the back chamber was less than that in the front chamber,front chamber gunpowder gas begins to flow into the back chamber and eventually spews out at high speed to the back along with that from the back chamber. Through the rational design of front and back chamber structure, ignition and flow of gunpowder in the two chambers are controlled, and the following two goals are finally achieved: 1) The sum of the impulse of forward gunpowder gas and the impulse of the projectile is basically equal to the total impulse of backward-flowing gunpowder gas, so as to achieve the purpose of basically eliminating the recoil impulse;2) Reasonably control the timing of the flow of the front chamber into the back chamber, so that the gas flows out of the fire hole at the bottom of the front chamber, and the sparse wave that causes the pressure drop in the chamber to move at the local air velocity to the direction of the chamber just catches up with the bottom of the projectile at the chamber, so as to ensure that the initial speed of the projectile is not reduced.

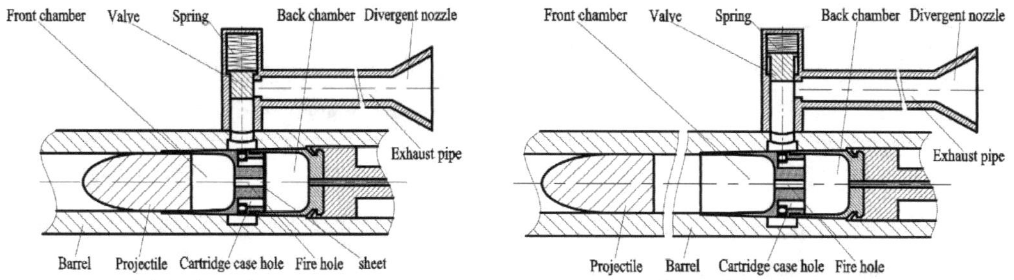

(a) Valve close state (b) Valve open state

Figure 1. Schematic diagram of constant-pressure rear-spray low recoil with two chambers

3. Two-phase Flow Numerical Model

The mass, momentum and energy exchange of propellant gas between the front and back chambers considering two-way flow is reflected by items $\boldsymbol{J}$ and $\boldsymbol{J}'$; The mass, momentum and energy exchange of gas between the back chamber and the valve body are reflected in item $\boldsymbol{J}''$; two-phase flow model is shown as follows:

$$\frac{\partial \boldsymbol{U}_{\mathrm{r}}}{\partial t}+\frac{\partial \boldsymbol{F}_{\mathrm{r}}}{\partial x}=\begin{cases}\boldsymbol{H}_{\mathrm{r}}-\boldsymbol{J} & \text{The gunpowder gas in the back chamber flows into the front chamber}\\ \boldsymbol{H}_{\mathrm{r}}-\boldsymbol{J}-\boldsymbol{J}'' & \text{The gunpowder gas in the back chamber flows into the valve body}\\ \boldsymbol{H}_{\mathrm{r}}+\boldsymbol{J}'-\boldsymbol{J}'' & \text{The gunpowder gas in the front chamber returns to the back chamber}\end{cases} \quad (1)$$

$$\boldsymbol{U}_{\mathrm{r}}=\begin{bmatrix} A\varphi_{\mathrm{r}}\rho_{\mathrm{gr}} \\ A(1-\varphi_{\mathrm{r}})\rho_{\mathrm{p}} \\ A\varphi_{\mathrm{r}}\rho_{\mathrm{gr}}u_{\mathrm{gr}} \\ A(1-\varphi_{\mathrm{r}})\rho_{\mathrm{p}}u_{\mathrm{pr}} \\ A\varphi_{\mathrm{r}}\rho_{\mathrm{gr}}\left(e_{\mathrm{gr}}+\dfrac{u_{\mathrm{gr}}^{2}}{2}\right) \end{bmatrix} \quad \boldsymbol{F}_{\mathrm{r}}=\begin{bmatrix} A\varphi_{\mathrm{r}}\rho_{\mathrm{gr}}u_{\mathrm{gr}} \\ A(1-\varphi_{\mathrm{r}})\rho_{\mathrm{p}}u_{\mathrm{pr}} \\ A\varphi_{\mathrm{r}}\rho_{\mathrm{gr}}\left(u_{\mathrm{gr}}^{2}+\dfrac{p_{\mathrm{a}}}{\rho_{\mathrm{gr}}}\right) \\ A(1-\varphi_{\mathrm{r}})\left(\rho_{\mathrm{p}}u_{\mathrm{pr}}^{2}+p_{\mathrm{r}}+\tau_{\mathrm{p}}\right) \\ A\varphi_{r}\rho_{\mathrm{gr}}u_{\mathrm{gr}}\left(e_{\mathrm{gr}}+\dfrac{u_{\mathrm{gr}}^{2}}{2}+\dfrac{p}{\rho_{\mathrm{gr}}}\right) \end{bmatrix} \quad (2)$$

$$\boldsymbol{H}_{\mathrm{r}}=\begin{bmatrix} A(m_{\mathrm{cr}}+m_{\mathrm{ign}}) \\ -Am_{\mathrm{cr}} \\ A(m_{\mathrm{cr}}u_{\mathrm{pr}}+m_{\mathrm{ign}}u_{\mathrm{ign}}-f_{\mathrm{sr}})+p_{\mathrm{r}}\dfrac{\partial A\varphi_{\mathrm{r}}}{\partial x} \\ A(f_{\mathrm{sr}}-m_{\mathrm{cr}}u_{\mathrm{pr}})+p_{\mathrm{r}}\dfrac{\partial A(1-\varphi_{\mathrm{r}})}{\partial x}+\tau_{\mathrm{p}}(1-\varphi_{\mathrm{r}})\dfrac{\partial A}{\partial x} \\ A(m_{\mathrm{cr}}(e_{\mathrm{p}}+\dfrac{p_{\mathrm{r}}}{\rho_{\mathrm{p}}}+\dfrac{u_{\mathrm{pr}}^{2}}{2})+m_{\mathrm{ign}}H_{\mathrm{ign}}-f_{\mathrm{sr}}u_{\mathrm{pr}}-Q_{\mathrm{s}})-p_{\mathrm{r}}\dfrac{\partial A\varphi_{\mathrm{r}}}{\partial x} \end{bmatrix} \quad (3)$$

Where: ρ_{gr} is the density of gas phase material in the back chamber; φ is the gas phase void fraction of the back chamber; u_{gr} and u_{pr} are the gas phase velocity and solid phase velocity of the back chamber respectively; A is the cross-sectional area of the barrel;m_{cr} and m_{ign} are respectively the generation rate of gunpowder gas in unit volume and the generation rate of ignition gas of various ignition elements; p_r is the gas phase pressure of the back chamber; u_{ign} refers to the flow velocity of gas generated by igniting powder of various ignition elements in the chamber; f_{sr} is the resistance between gas and solid in the unit volume of the back chamber; e_{gr} is the gas relative internal energy of the back chamber; e_{pr} is the chemical energy of solid phase;k is the specific heat ratio of gunpowder and gas; H_{ign} is the stagnation enthalpy of ignition element ignition gas; Q_s is the heat exchange of gas-solid two phases per unit surface area of the back chamber.

The flow model of the front chamber and nozzle is similar to that of the back chamber, which will not be repeated here. Relevant auxiliary equations such as propellant combustion law and interphase resistance can be found in a reference [7].The difference scheme can be referred to references [8] and [9].

4. Analysis of numerical simulation results

4.1. Comparative Analysis of Weapon Chamber Pressure and Projectile Velocity

Based on a 30mm chain gun, the traditional weapon is transformed into a constant-pressure rear-spray low recoil weapon with two chambers according to the principle of figure 1. The known structure and gunpowder parameters are shown in table 1. The whole launching process of the weapon is numerically simulated and compared with the traditional weapon.

Table 1. Calculation parameters

Parameter	Barrel diameter/m	Barrel length/m	Powder solid particle density/(kg·m^{-3})	Gunpowder energy/(J·kg^{-1})	Nozzle diameter/m
Value	0.03	2.457	1600	3600000	0.026

4.2. Comparative Analysis of Weapon Chamber Pressure and Projectile Velocity

As can be seen from Figures 2 and 3, the traditional weapon ignites gunpowder particles by ignition cartridge, while the improved weapon ignites powder particles in the front chamber by firing powder gas in the back chamber, and the front and back chambers are separated by a sheet. Only when the back chamber pressure reaches the set value, the sheet will break, and the gas from the back chamber will flow to the front. Therefore, the time when the chamber bottom pressure of the improved weapon begins to rise relative to the traditional weapon lags slightly. The breech pressure of the traditional weapon reaches the peak value of 330.783MPa when the time is 2.138ms, while the breech pressure of the improved weapon reaches the peak value of 331.712MPa when the time is 2.485ms. The start-up time of the projectile lags slightly. When the internal ballistic time is 1.502ms, the projectile of the improved weapon begins to move. After that, the upward trend of the projectile speed is similar to that of the traditional weapon, but the upward trend of the projectile speed of the traditional weapon is slightly lower than the projectile speed of the improved weapon. When the internal ballistic time is 5.490ms, the projectile exits the muzzle. At this time, the muzzle velocity is 960.303m/s, which is slightly higher than that of the traditional weapon.

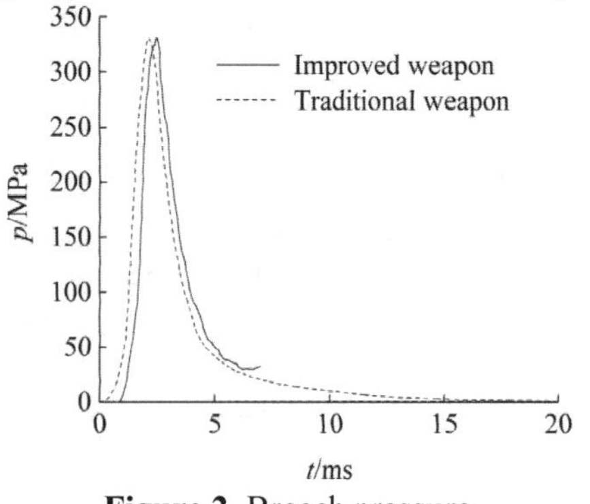

Figure 2. Breech pressure

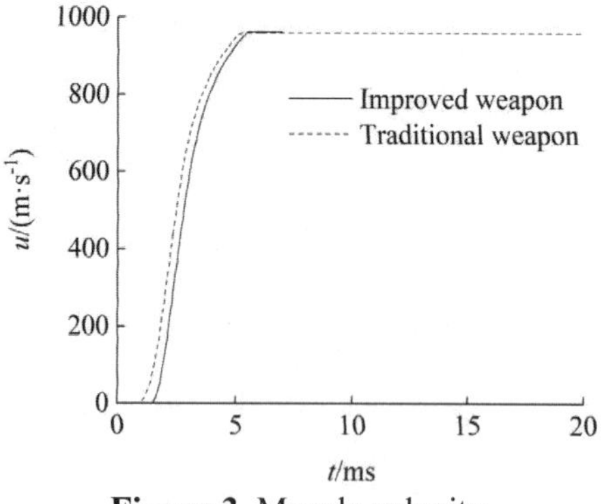

Figure 3. Muzzle velocity

4.3. Front and Back Chamber Pressure Difference and Recoil Impulse

As can be seen from Figures 4 and 5, the recoil impulse of improved weapon is reduced from 552.982N·s to 101.855N·s on the premise that the chamber pressure and projectile muzzle velocity are basically unchanged, and the recoil reduction efficiency is 81.58%. It is proved that the recoil reduction device can realize the function of recoil reduction, and the peak value of the pressure difference between the front and back chamber is 68.864MPa, which is within the safety range, and the safety of the weapon during launch is also guaranteed.

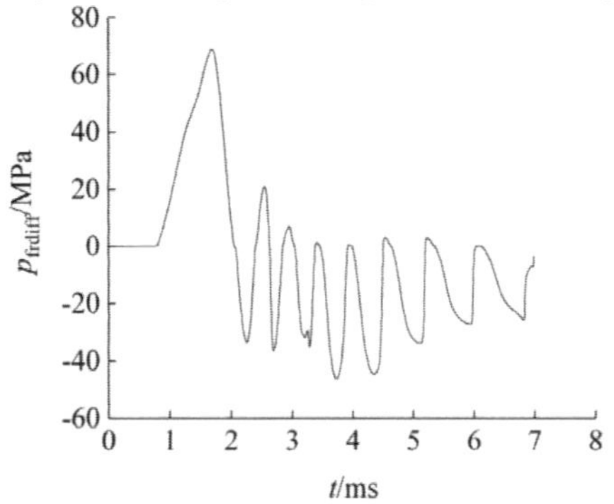

Figure 4. Pressure difference curve

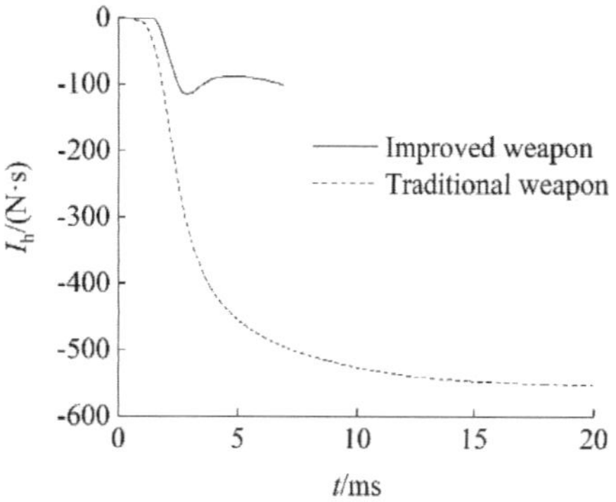

Figure 5. Recoil impulse curve

4.4. Analysis of Flow Field in Nozzle

For the convenience of narration, five representative time points are selected, of which 0.824ms is the time when the gunpowder gas flows into the front chamber from the back chamber, 1.502ms is the time when the bullet begins to move, 2.035ms is the time when the exhaust inlet is fully opened, 2.065ms is the time when the gas from the front chamber returns to the back, and 5.490ms is the time when the projectile exits and the internal ballistic process ends.

As can be seen from Figures 6 and 7, the valve does not move at 0.824ms, the exhaust pipe inlet does not open, the pressure in the nozzle is atmospheric pressure, and the gas phase velocity is 0m/s. shortly after the exhaust pipe inlet is opened at 1.502ms, the gas pressure and velocity at the exhaust pipe inlet jump due to the injection of high-pressure gunpowder gas, but the gunpowder gas has not filled the whole pipe at the moment, so it decays rapidly after a certain distance. After 2.035ms, because the gunpowder gas has filled the whole nozzle and is expanding, the gas pressure along the length of the nozzle shows a downward distribution and the velocity shows an upward distribution; Due to the further expansion and acceleration of the gunpowder gas at the divergent nozzle, the gas velocity increases greatly and the pressure decreases greatly, indicating that the divergent nozzle can obtain a large back injection impulse of the gunpowder gas.

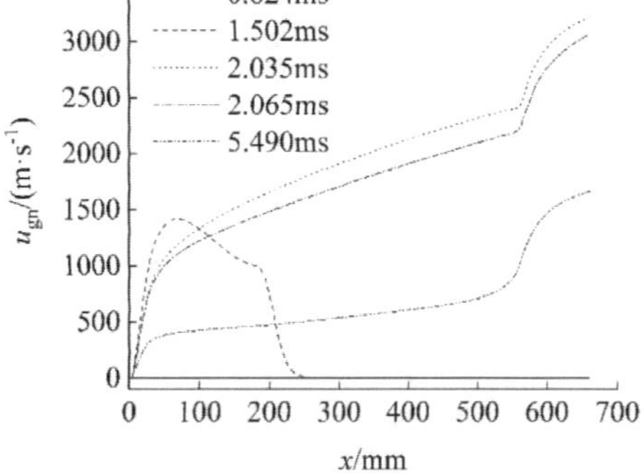

Figure 6. Gas velocity curve

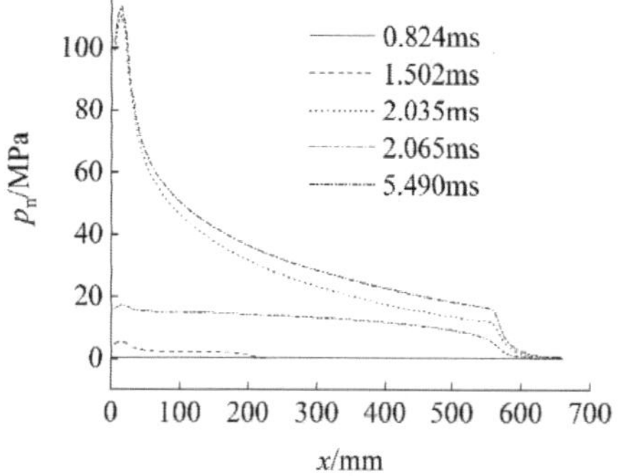

Figure 7. Gas pressure curve

5. Conclusion

In this paper, the launching process of a constant-pressure rear-spray low recoil weapon with two chambers is numerically simulated based on the two-phase flow theory. The changes of flow field in the front and back chambers and nozzle are analysed. The following conclusions are drawn:1) The recoil reduction mechanism of a constant-pressure rear-spray low recoil weapon with two chambers is revealed. At the same time, the backward injection timing of the gunpowder gas side is controlled through the valve movement, and the high-pressure gunpowder gas in the front and back chambers is introduced as much as possible, so as to greatly reduce arms recoil.2) Two-phase flow theory well describes the internal flow field characteristics of the recoil weapon with two chambers.

Acknowledgments
This project was supported by the National Natural Science Foundation of China(Grant No. 12072161 and No. 51676090).

References

[1] Wang G, Cheng C, Zhang X B and Huang X 2021 Numerical simulation and analysis of muzzle flow during a rarefaction wave gun firing *Propellants Explosives Pyrotechnics* **46**(12) 1902-1913.
[2] Li P and Zhang X B 2020 Numerical research on adverse effect of muzzle flow formed by muzzle brake considering secondary combustion *Defence Technology* **17**(4) 1178-1189
[3] Chen Y, Liao Z Q, Liu G and Wang T 2008 Performance comparison of two recoilless structures with laval nozzles *Journal of Ballistics* **20**(4) 88-91.
[4] Kathe J and Gim H 2012 Numerical and experimental studies on the dynamic behaviors of a gun that uses the soft recoil system *Journal of Mechanical Science and Technology* **26**(7) 2167-2170.
[5] Qiu M, Si P, Song J and Liao Z Q 2021 Recoil reduction method of gun with side to rear jet controlled by piston motion *Symmetry Basel* **13**(3) 396-396.
[6] Si P, Qiu M, Liao Z Q, Song J and Ma L X 2021 Numerical simulation of two-phase flow in a side spray gun considering piston reset motion *Explosion and Shock Waves* **41**(8) 125-138.
[7] Jin Z M and Weng C S 2003 *Advanced Interior Ballistics* (Beijing: Higher Education Press).
[8] Liang D F, Falconer R A and Lin B L 2006 Comparison between TVD-MacCormack and ADI-type solvers of the shallow water equations *Advances in Water Resources* **29**(12SI) 1833-1845.
[9] Liang D F, Lin B L and Falconer R A 2007 Solving the depth-integrated solute transport equation with a TVD-MacCormack scheme *Environmental Modelling & Software* **25**(13SI) 1619-1629.

Numerical Simulation of External Flow-Field of a Car Body Based on CFD

Dan Zhao*

School of Management, Tianjin University of Technology, Tianjin, China

*Corresponding author e-mail: andier712@126.com

Abstract. With the rapid development of computer technology in recent years, this paper explores the feasibility of applying computational fluid dynamics (CFD) to study the external flow-field of automobile body. The computer numerical solution provides the pressure and velocity distribution of the car body, such as aerodynamic pressure field and velocity field, in order to obtain the information of aerodynamic resistance, aerodynamic lift, aerodynamic lateral force and related torque, which can be further used to guide the design and development of automobile. The comparison between numerical simulation results and experiments shows that CFD plays a strong guiding role in the further development of automobile shape and structure.

key words: Flow-Field; Car Body; CFD.

1. Introduction

With the rapid development of computer technology, it is possible to apply computational fluid dynamics to predict the external flow-field of automobile body. At present, many general software for solving incompressible or even compressible fluid flow-fields are beginning to enter the market. Theoretically, these softwares can be applied to the aerodynamic study of automobile design process. Especially for automobile designers and developers, getting the aerodynamic parameters of the vehicle at different design stages will help to optimize the vehicle model, improve the speed, improve safety and comfort, and increase fuel economy.

Many foreign automakers, especially Ford (Ford Motor Company), incorporated Computational fluid Dynamics (CFD) into the C3P plan of their "FORD2000" plan as early as 1993. After years of trying, Ford has reduced the development cycle of new models from 36 months to 12-18 months. But this does not mean that the simulation technology of CFD applied to the flow-field of automobile body has been mature. Satyan Chandra et al. used CFD to simulate and analyze the external flow-field of Formula one racing car, improved the tail wing design, adjusted the lower pressure and air resistance of the car, and increased the high-speed stability and maneuverability of the car [1]. Fabritius, Junhyeok and others effectively improved the simulation accuracy from the grid topology connection mode through the research on the generation strategy of tetrahedron, hexahedron and hybrid grid [2,3]. In fact, how to solve the grid establishment and solution model of automobile external flow-field are still two key problems in the application of CFD in automobile aerodynamics.

Nowadays, people from all walks of life have higher and higher requirements for automobiles, so they need to be able to apply a variety of high-tech means to improve the overall application effect of

Content from this work may be used under the terms of the Creative Commons Attribution 3.0 licence. Any further distribution of this work must maintain attribution to the author(s) and the title of the work, journal citation and DOI.
Published under licence by IOP Publishing Ltd

automobiles, ensure their quality up to standard, and provide more diversified services. In this context, the application of CFD technology has strong application advantages, showing stronger reliability and stability. After the application of CFD technology, we can focus on the overall process of redesign, through simulation technology to get a more high-quality external flow-field design, and through some scientific data calculation, related optimization work. It can not only effectively enhance the performance of the car, but also improve the safety of the car.

2. Automobile Mathematical Model and Its CFD Application

Two kinds of equations are usually used to describe the external flow-field of the automobile body, one is the potential flow theory equation, and the other is the Navier-Stokes equation which describes the general phenomenon. In order to make the calculation reflect the external flow-field of the car body as truly as possible, the Navier-Stokes equation is used here. Considering that the speed of the car is usually lower than that of 300km/h, that is, 83.3m/s, even in the case of head-on wind, the relative velocity of air acting on the car is lower than the speed of sound. It can be seen that for ordinary household or commercial vehicles, the Mach number Ma is much less than 1, so it can be simplified to incompressible fluid treatment without considering the compressibility of the gas. The equation is

$$\frac{\partial u_i}{\partial x_i} = 0 \quad (1)$$

$$\frac{\partial}{\partial x_j}(u_j u_i) - \frac{\partial}{\partial x_j}\left(m\frac{\partial u_i}{\partial x_i}\right) + \frac{\partial p}{\partial x_i} - \frac{m}{3}\frac{\partial}{\partial x_i}\left(\frac{\partial u_j}{\partial x_j}\right) = 0 \quad (2)$$

In the form p ——Fluid pressure

m——Dynamic viscosity

u_i, u_j——The component of the fluid velocity in the i and j direction (i, j = 1, 2, 3)

x_i, x_j——i, j coordinates (i, j = 1, 2, 3)

Car movement can be seen as a kind of damage to smooth flow. In fact, due to the interaction between the vehicle shape and the high-speed airflow, the flow-field around the vehicle is very complex, and the direction and velocity of the airflow will change greatly. The turbulence model here adopts the k-$\boldsymbol{\varepsilon}$ model. The above N-S equation and k-$\boldsymbol{\varepsilon}$ model can reflect the aerodynamic status of the car, and also reveal other relevant parameters of the car flow-field. At present, researchers still use the early experimental model method as the main means to study the external flow-field of the car body.

There are many methods for car shape design, and the most widely used is the "clay mold" method. In the initial stage of the car shape design, a 1:5 clay model is established in order to change its detailed structure, and then the model is placed in the model wind tunnel for blowing experiments. Modify the model according to the experimental data, and then put the modified model into the model wind tunnel for blowing experiment. German designers call it "aerodynamic tuning". After repeated for many times to achieve the ideal effect, a full-size model was done according to the modified 1:5 clay model. Due to the change of body size, the aerodynamic parameters are bound to change, so some corrections need to be made. This kind of work is usually repeated dozens or even hundreds of times [4]. With the application of CFD, aerodynamic simulation can be carried out at any stage of automobile design, and the calculation results can also be quickly fed back to the car shape designer, which is convenient for the car shape designer to improve the design, thus greatly shortening the cycle of vehicle design and development.

3. Grid Realization and Numerical Analysis

The general software PHOENICS introduced in this study is used for numerical analysis. Although this software also has some support for the finite element method, most of it is the finite volume method. From a mathematical point of view, the methods of constructing difference equations by finite element method and finite volume method are different. The finite element method is only integrated on a certain element and is not associated with the adjacent elements, so it is also called the unstructured grid method. The finite volume method must refer to the adjacent elements when constructing the difference equation, and considering the computational efficiency, these grids must be regular grids, which brings inconvenience to the division of the computational region. Because the division of the grid is affected not only by the whole flow-field, but also by the position of the car in the flow-field and the shape of the car body. Therefore, at present, the body-fitted coordinate system (BFC) grid is widely used to study the flow around a car-like bluff body with complex outer surface. It mainly constructs the mesh by algebraic transformation or solving differential equations, and ensures that the boundary of the solution region coincides with the coordinate surface. Although the emergence and development of BFC has made a great contribution to the promotion and application of CFD, and BFC is still the most widely used and studied method at present, some researchers still doubt the practicability and application prospect of this method. The main doubtful points are: the applicability to the complex shape area is not very ideal; the grid generation needs to be programmed separately; the grid generation takes a long time, sometimes even more than the computing itself; and BFC is not recommended in the PHEONICS3.3 version. On the basis of previous experience, this paper uses the PARSOL method in the software to calculate. PARSOL retains the characteristics of regular grid generation and simple calculation, and uses the local flow method to improve the calculation accuracy, which has a good adaptability to complex shapes. The basic idea is that the mesh elements completely occupied by the body are completely blocked, that is, there is no flow, while for the elements that can not be fully occupied because of the inclined curved edge of the body itself, the percentage is calculated and flowed according to the percentage, that is, the so-called local flow.

In order to verify the practicability of PARSOL method, we calculate and compare the flow-field around a cylinder and the flow-field around a NACA4412 airfoil. It can be seen from figure 1 that when the Reynolds number is equal to 1.54, the flow is close to the surface of the cylinder, and there is no flow separation and wake vortex, which is consistent with classical fluid mechanics. Figure 2 shows the pressure coefficient of NACA4412 at an angle of attack of 8°. As can be seen from the diagram, the calculated value lies between the theoretical value of the potential flow and the experimental value, and keeps a good trend with the experimental value.

Fig. 1 The velocity-field of a cylinder

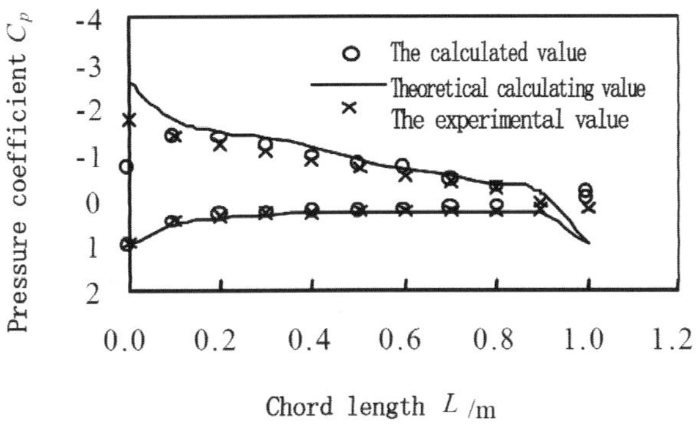

Fig. 2 Pressure coefficients diagram for NACA4412

4. Flow-Field Analysis of Ideal Car Body

The vehicle selected in this paper is an ideal car which is scaled down at 1:2 and supported by four pillars instead of wheels in the Lockheed low-speed wind tunnel [5]. Its size is 2.760m × 0.864m × 0.640m and the height from the car to the ground is 127mm. It takes about 10 hours to use PHOENICS to run on a microcomputer with IntelPentiumCeleron333 CPU and 64MB memory. Compared with the calculated and experimental pressure coefficient distribution, the results are basically satisfactory.

4.1. Flow-Field Analysis of Car Body

Fig. 3, 4 shows the velocity field before and after the car body. Through observation, it can be found that there is no vortex, separation and reattachment between the engine hood and the windshield, and between the rear windshield and the suitcase. Basically, it flows smoothly close to the car body. It is just that there is a more obvious tail vortex at the rear of the car, which is caused by the intersection of the air flow from the upper surface and the air flow from the bottom of the car. It is worth noting that the tail vortex only rotates at the rear of the car and gradually detaches from the rear of the car while driving, preventing the air from curling up and causing dust to fall on the suitcase and the rear windshield, which helps to keep the rear of the car clean. As can be seen from the above phenomenon, this kind of car shape is exactly what the car body stylist pursues, which can be called the ideal car body.

Fig. 3 Velocity vectors for front quarter

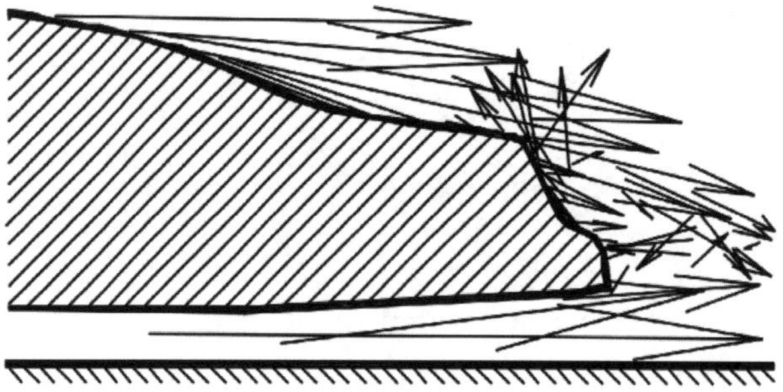

Fig. 4 Velocity vectors for rear quarter

4.2. Comparison Between Numerical Calculation and Experiment

Figure 5 shows the comparison between the pressure coefficient Cp of the ideal upper surface of the car body calculated by PHOENICS and the experimental value. As can be seen from the figure, the calculated values coincide with the experimental values at the front of the car, but there is some deviation at the top of the car. This may be because in the calculation, we set the velocity of the upper boundary of the calculation region as the incoming flow velocity, that is, we think that the disturbance at the upper boundary is zero, while in the experimental wind tunnel, because of the inevitable velocity stagnation on the wall of the wind tunnel, this will inevitably affect the velocity pressure on the upper surface of the car, resulting in the phenomenon that the calculated value is too large. In spite of this, we can see from the pressure coefficient curve that the pressure distribution of the car body is consistent with the experiment, which shows that the numerical calculation still has a certain guiding significance for the car shape design. For the lower surface, that is, the bottom of the car, it can be seen in figure 6 that the calculated values in the front part of the car body are in good agreement with the experimental values[6], but in the middle and rear part, the calculated values form an "arch", which is different from the trend of keeping the "level" in the experiment. But this is understandable because we compare the two-dimensional calculation results with the three-dimensional experiments, and the four pillars are not considered in the calculation (used to replace the wheel in the experiment). Now, we only try to describe this deviation from the Bernoulli equation: in the three-dimensional experiment, due to the restriction of the two pillars on the incoming flow, the flow velocity is bound to increase and the pressure decreases as the air flows through the props, resulting in two low peaks in the bottom pressure coefficient curve at the front and back pillars. In the two-dimensional calculation of PHOENICS, we only consider the case that there is no pillar in the longitudinal section of the car, in addition, the bottom of the calculated car body is an upward curved surface at both ends, which makes the bottom air flow accelerate at first and then decelerate, and the pressure decreases at first and then increases, forming an "arch" pressure coefficient curve. This conjecture will be verified in our next three-dimensional calculation.

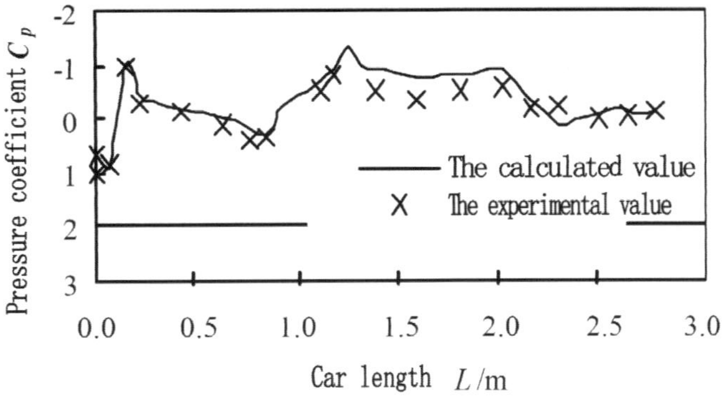

Fig. 5 Pressure coefficient of upper surface

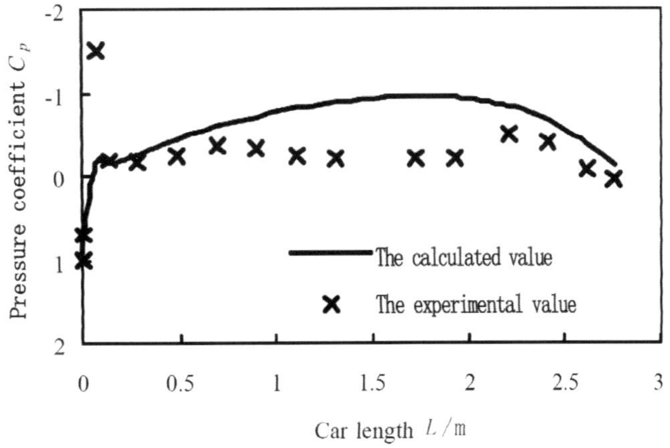

Fig. 6 Pressure coefficient of lower surface

5. Conclusion and Prospect

In this paper, the calculation results of the numerical simulation of automobile flow-field still have the problem of rear and bottom deviation, which will be improved in the next three-dimensional calculation.

However, through the numerical analysis and simulation calculation in this paper, it has been preliminarily shown that the application of CFD can optimize the vehicle shape design, improve the aerodynamic characteristics, improve the vehicle performance and shorten the development cycle. Especially when there is a certain gap in the level of automobile design and manufacture between China and the developed countries, if we can vigorously develop CFD to simulate the external flow-field of the vehicle, which can be used to optimize the vehicle shape, reduce drag and noise, and improve fuel economy, its significance will not be limited to improving the quality and competitiveness of the vehicle, and its economic benefits will also be considerable. In particular, the modification of existing vehicles, such as the increase of wound current, negative lift meter, etc., will significantly reduce the amount of fuel, which is conducive to saving energy and promoting economic development.

6. Prospect

With the development of computer technology, CFD technology is more and more applied to many fields of automobile R & D and design, but there are still some problems that need to be further solved.

(1) In-depth study of the pre-processing technology of CFD. The geometric model in the field of automotive engineering is so complex that due to the limitations of current mesh pre-processing tools, reasonable computational grids can only be generated by simplifying or even ignoring many parts. This distortion affects the accuracy of numerical simulation to a great extent.

(2) Improve the calculation method of CFD. At present, the improvement of calculation speed depends more on the improvement of computer speed, but the shortening effect of calculation time brought by the improvement of calculation method is not obvious. Advanced CFD technology should have higher solution accuracy and solution speed.

The design and manufacture of automobile is a systematic project. China's automobile industry should start from the basic manufacturing capacity and improve its basic manufacturing capacity through continuous accumulation. At the same time, advanced computer technology can be used to design the automobile, provide a theoretical basis for automobile production, reduce the production cost in the later stage, and improve the reliability of the automobile.

References

[1] Satyan Chandra, Allison Lee, Steven Gorrell, et al. CFD analysis of PACE formula-1 Car [J]. Computer–Aided Design and Applications, 2011(1): 1 – 14.

[2] Fabritius B, Tabor G. Improving the quality of finite volume meshes through genetic optimisation [J]. Engineering with Computers, 2016, 32(3): 425-440.

[3] Choi Junhyeok, Kim Harrim, Sastry Shankar Prasad, et al. A Deviation-Based Dynamic Vertex Reordering Technique for 2D Mesh Quality Improvement [J]. Symmetry, 2019, 11(7): 895.

[4] Hucho W. Aerodynamics of road vehicles[J]. An-nual Review of Fluid Mechanics, 1993, 25(3): 45~60.

[5] Williams J. A calibration study of CFD for automotive shapes and CD [R]. SAE940323, 1994, 1~20.

[6] Kim K H, Sukeyuk Kobayashi. Pressure distribution on propeller blade surface using numerical lifting surface theory[A]. Propellor'84, Sympoium, SNAME [C], May, 1984.

Study on the influence of aerodynamic configuration on vortex vibration of steel pipe

Guo Huang [1,*], Xiaojun Li [2], Xiang Xia [3]

[1] China Electric Power Research Institute, Beijing, 100055, China
[2] Lishui Zhengyang Electric Power Design Institute Co., Ltd., Lishui, 323050, China
[3] State Grid Zhejiang Electric Power Co., Ltd., Hangzhou, 310063, China

*Corresponding author e-mail: huangguocn@126.com

Abstract. In this paper, for steel pipe members with a slenderness ratio of 120, spoilers with different numbers and spacings are arranged to test the vibration reduction effect. Wind tunnel test results show that 7 sets of spoilers with a size of 1D×0.5D can basically suppress the vibration of the steel pipe member. When the spoiler spacing is 1D~4D, the reduction of the maximum amplitude is more than 90%.

1. Introduction

Under the action of wind, steel pipe members of steel pipe tower are prone to vibration. The wind will generate alternating vortices behind the circular rods after flowing through that, forming a Karman vortex street. It generates a vortex excitation force that changes the displacement of the rod in the crosswind direction, thereby generating vibration. In engineering, it is called breeze vibration or vortex-induced vibration. If the steel pipes undergo repeated vibrations, it will cause fatigue damage to the rods and affect the safety of the entire transmission line. It is significant to study how to prevent and control the vortex induced vibration of steel pipe.

Simplifying the steel pipe rod into a cylinder, the problem is transformed into the vortex-induced resonance problem of the cylinder. Scholars have carried out many studies on the vortex detachment mode and numerical simulation of the flow around a cylinder [1-3]. In order to suppress the vortex-induced resonance caused by the flow around the cylinder, the shape of the cylinder is usually changed to interfere with the conditions for the occurrence of vortex detachment to weaken the vortex excitation force. Yang [4] conducted wind tunnel tests on the outer spiral of steel pipe members with slenderness ratios of 120 and 160, and recommended the arrangement and parameter design of the outer spiral. Bai [5] set up a deflector at the central axis of the cylindrical tail through numerical simulation and found that it can well control the vortex detachment of the cylindrical surface and suppress vibration. Scholars have also developed many passive control technologies in the past few decades. According to their mechanism, they can be divided into two categories, namely, near wake stabilizers and surface geometry adjustment devices. The former device relies on the near wake and shear layer changes around the vibrating structure to keep the wake stable and suppress the occurrence of vortex-induced resonance, such as fairing [6], wake splitter [7], guide foil [8], Wake C-shaped deflector [9]. The feature of this type of device is that it can rotate smoothly with the change of the airflow direction, and the mechanical design is complicated. The second type of device uses surface changes, such as protrusions and grooves, to manipulate vorticity distribution and separation points of the boundary layer on the vibrating structure

Content from this work may be used under the terms of the Creative Commons Attribution 3.0 licence. Any further distribution of this work must maintain attribution to the author(s) and the title of the work, journal citation and DOI.
Published under licence by IOP Publishing Ltd

to suppress vortex induced resonance. This passive control method is less affected by the wind direction angle and is easier to implement in engineering. Taking into account construction and maintenance, it is a more preferable plan to arrange a certain number of spoilers on the steel pipe members, but the arrangement scheme is usually based on experience, without theoretical guidance. In this paper, according to the common engineering pipe rod with slender-to-length ratio of 120, different spoiler layout scheme is used for wind tunnel test, the spoiler is optimized, and the recommended spoiler layout scheme is given, which provides guidance for the actual engineering.

2. Wind Tunnel Test

2.1. The device

The test wind tunnel is a low-speed wind tunnel with a size of 2.4m×1.8m. The instantaneous velocity was measured using a Cobra probe manufactured by Turbulence Instruments Australia (TFI), as shown in Figure 1. The displacement sensor (IL-300) is used to measure the vibration displacement at a certain point of the model, as shown in Fig. 2. The vibration displacement signal measured by the displacement sensor is collected and output by DH8303 dynamic signal test and analysis system produced by Donghua Testing Company.

Figure 1. Cobra probe

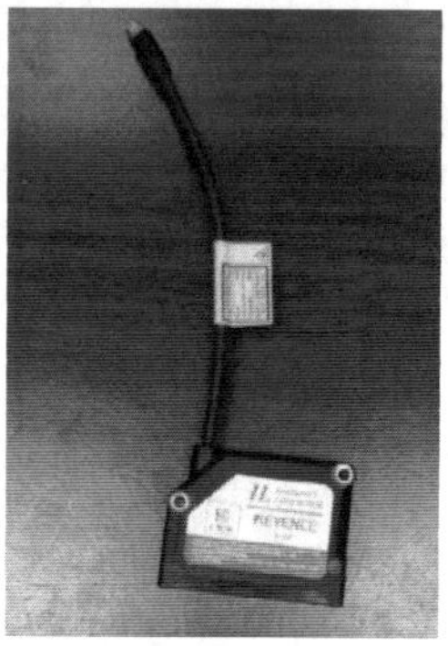

Figure 2. The displacement sensor

2.2. Model scale ratio

The selected steel pipe size is shown in Table 1.

Table 1. Steel tube prototype

Specification	Diameter (mm)	wall thickness (mm)	length (mm)
Q345	187	4	7770
slenderness ratio λ	cross-sectional area (mm^2)	unit length mass (kg)	
120.1	2299.65	18.05	

Due to the limitation of the size of the wind tunnel, the test of the steel pipe is usually carried out using scaled models. According to the prototype steel pipe, considering the size of the wind tunnel. The width of the wind tunnel is 2400mm, and the length of the model is set to 2200mm. According to the geometric scale ratio, various indexes of the test model are finally determined. The prototype is a steel pipe with a density of 7850kg/m^3 and a modulus of elasticity of 2.06×1011N/m^2. Considering the difficulty of processing and the constraints of wind tunnel wind speed, the model material is finally made of aluminum tube. The density of the aluminum tube is 2700kg/m^3, and the elastic modulus is 7.2×1010N/m^2. The basic scale is as follows:

Geometric scale ratio: $\lambda_L = \frac{l_m}{l_p} = \frac{2200}{7770} = \frac{1}{3.53}$. l_m is the model length. l_p is the prototype length.

Wind speed scaling ratio: $\lambda_U = \frac{U_m}{U_p} = 1$. U_m is the model wind speed. U_p is the prototype wind speed.

Density scale ratio: $\frac{\rho_m}{\rho_p} = \frac{2700}{7850} = \frac{1}{2.9}$. ρ_m is the model density. ρ_p is the prototype density. ρ_s is the structure density. ρ is the air density.

The size of the model obtained is shown in Table 2. According to the model size and the cross-sectional size of the wind tunnel test section, the blockage ratio of the flow section of aluminum tube is 2.7%, which is lower than the limit of 5% for the blockage ratio in the wind tunnel test.

Table 2. Test the characteristics of aluminum pipe

Outside diameter (mm)	inside diameter (mm)	wall thickness (mm)
53	47	3
length (mm)	cross-sectional area (mm^2)	unit length mass (kg)
2200	471.239	1.272

2.3. Model making
Set the size of the spoiler as 1D×0.5D (length×width), where D is the outer diameter of the model. The spoiler used in the experiment was made by a laser cutting machine. The spoiler is light in weight and has almost no effect on the structural quality. The material is not easy to deform and bend, and can keep the shape basically unchanged, which meets the test requirements. According to the design of the working conditions, quick-drying glue was used to fix the spoiler on the smooth round tube.

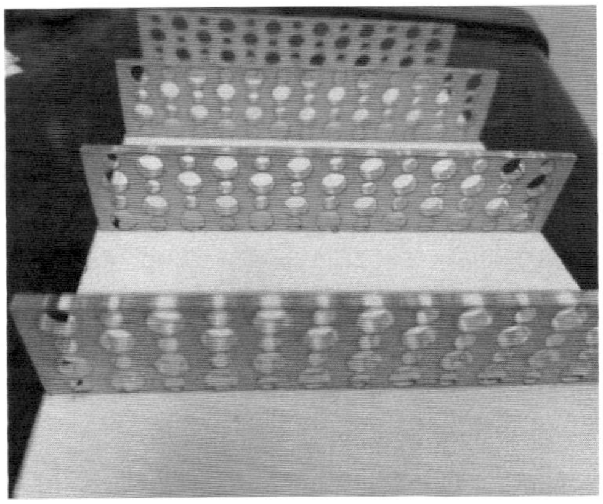

Figure 3. Spoiler specimen

Figure 4. Spoiler installation

2.4. Testing process
A cobra speed probe is used to record real-time wind field information, and a laser displacement meter is used to record the vertical displacement of the apex of the circular tube. The recording time interval is 0.001s. The test uses a low-to-high loading method to increase the wind speed. First, the maximum amplitude of the smooth circular tube is tested. Then the spoilers are arranged at 90° intervals along the circumference of the tube to ensure that the vibration damping effect can still be achieved when the wind direction changes. The 4 spoilers arranged at 90° intervals along the circumference of the tube are called 1 group of spoilers, and 3 to 7 sets of spoilers are arranged along the length of the tube. Adjust the spacing between each group of spoilers to test the maximum amplitude of the round tube. The test conditions are shown in Table 3.

Table 3. Test conditions

Groups	3	4	5	6	7
spacing (mm)	0~6D: 0, 53, 106, 159, 212, 265, 318	0~6D: 0, 53, 106, 159, 212, 265, 318	0~6D: 0, 53, 106, 159, 212, 265, 318	0~6D: 0, 53, 106, 159, 212, 265, 318	0~5D: 0, 53, 106, 159, 212, 265

3. Result analysis

The test results are shown in Figure 5 to Figure 9.

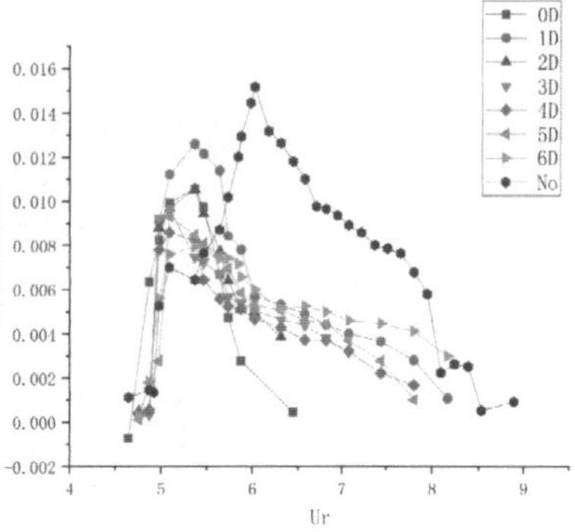

Figure 5. 3 groups

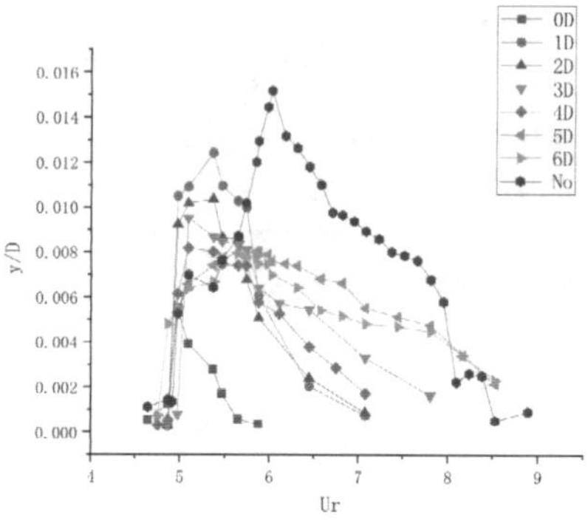

Figure 6. 4 groups

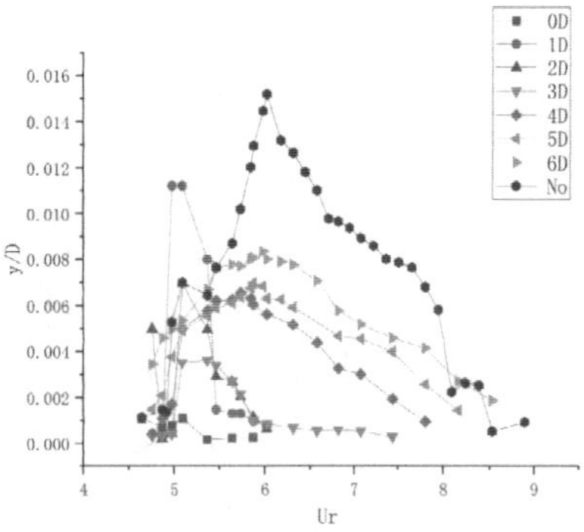

Figure 7. 5 groups

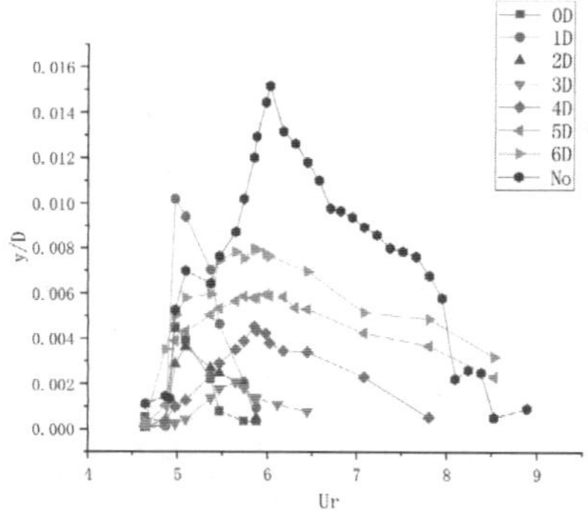

Figure 8. 6 groups

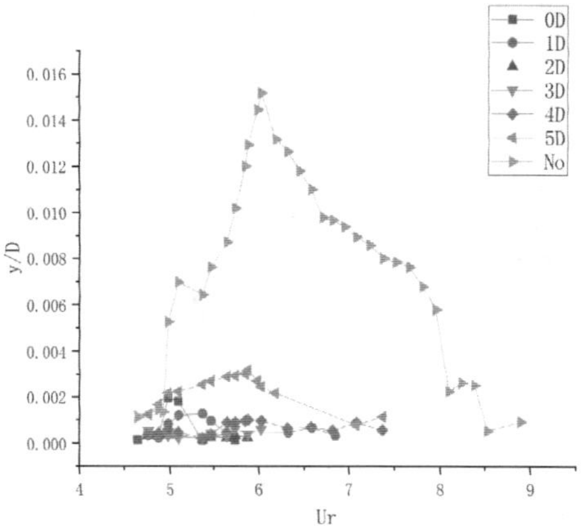

Figure 9. 7 groups

According to the wind tunnel test data of a circular tube with a slenderness ratio of 120, it can be seen that the dimensionless amplitude-converted wind speed curve of the circular tube is a parabola with an approximate opening downward. It can be seen from Figure 5 that the overall vibration reduction effect of the three sets of spoilers is not significant. After the spoilers are added, the peak point of the curve is located on the left side of the light circle curve. Generally, the curve shows a narrowing trend, and the top appears more steep. In the conditions, the apex is the highest when the arrangement distance is 1D, and the vibration reduction effect is the worst. The curve range is the narrowest when the arrangement distance is 0D. After arranging 4 sets of spoilers, it can be seen that all the curves narrow towards low wind speeds. When the arrangement spacing is 0D, the apex of the curve is the lowest and the vibration reduction effect is the best. However, the apex of the curve is the highest when the arrangement spacing is 1D, and the vibration reduction effect is poor. As the layout spacing increases, the curve width increases. Arranging 5 sets of spoilers can also see that when the spacing is 1D, the apex is the highest, and the maximum amplitude reduction is limited. The damping effect is best when the spacing is 0D. Except for this condition, it can be seen that the curve with a spacing of 3D is located below the other curves, which also has a good damping effect. After arranging 6 sets of spoilers, the spacing is the highest at the 1D apex, and the maximum amplitude reduction is limited. At this time, the 3D spacing curve is below all the curves. It can be seen that the vibration reduction effect is the best at this time. It can be seen from Fig. 9 that the damping effect of the 7 groups is better. When the layout spacing is 5D, the curve is at the top, while the data points of the remaining spacing curves are basically near zero, with a high degree of coincidence.

It can be seen from the above that as the number of layout groups increases, the dimensionless amplitude-converted wind speed curve with the same layout spacing shows a trend that the resonance wind speed region narrows towards low wind speeds and the amplitude decreases. Secondly, within the working condition of the same arrangement group number, with the increase of the arrangement spacing, the resonance area shows a trend of widening, which is closer to the wind speed range of the resonance area of the smooth circular pipe.

The highest point (maximum amplitude) of each curve is discussed separately as the research object, as shown in Figure 10. First of all, it can be seen that the influence of the number of layout groups is very significant. The increase in the number of layout groups is closer to the zero point compared with the curve with a low number of groups, which significantly reduces the maximum amplitude. The arrangement of 7 groups of spoilers can basically suppress the vibration of the steel pipe. The maximum amplitude-arrangement spacing curve is a parabola with an opening downward as a whole. That is to say, in the process of the arrangement spacing from small to large, there is an optimal arrangement spacing which makes the vibration damping effect the best. For the arrangement of 7 sets of spoilers, with a spacing of 1D~4D, the maximum amplitude reduction can reach more than 90%.

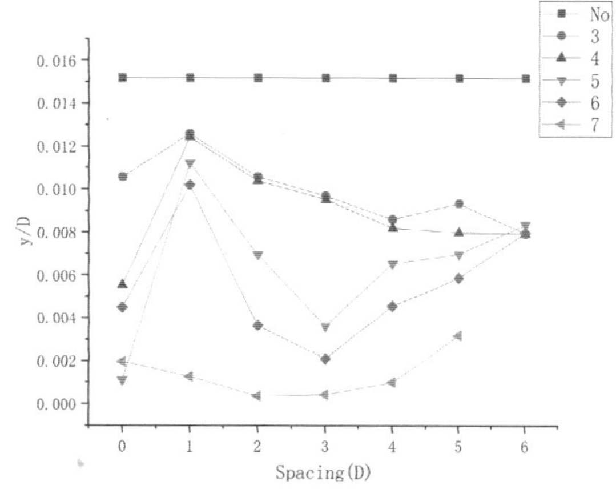

Figure 10. Maximum dimensionless displacement-spacing curve

4. Conclusion

In this paper, for steel pipe members with a slenderness ratio of 120, spoilers with different numbers and spacings are arranged to test the vibration reduction effect. Wind tunnel test results show that 7 sets of spoilers with a size of 1D×0.5D can basically suppress the vibration of the steel pipe member. When the spoiler spacing is 1D~4D, the reduction of the maximum amplitude is more than 90%.

References

[1] Jauvtis N., Williamson C.H.K. The effect of two degrees of freedom on vortex-induced vibration at low mass and damping. Journal of Fluid Mechanics, 2004, 509: 23-62.

[2] Laneville A. On vortex-induced vibrations of cylinders describing X-Y trajectories. Journal of Fluids & Structures, 2006, 22(6): 773-782.

[3] Raghavan K., Bernitsas M. M. Experimental investigation of Reynolds number effect on vortex induced vibration of rigid circular cylinder on elastic supports. Ocean Engineering, 2011, 38(5-6): 719-731.

[4] Yang J.B., Han J.K., Hua X.G. Wind-induced Vibration Suppression of Tubular Towers for 1000 kV Double-circuit Transmission Lines on the Same Tower From Huainan to Shanghai. Proceedings of the CSEE, 2010, 30(7): 104-110.

[5] Bai H., Li J.W., Xia Y. Numerical Simulation and Control Measures of Flow Around nCircular Cylinders at Low Reynolds Number. Journal of Architecture and Civil Engineering, 2010, 27(4): 39-43.

[6] Gustavo R.S. Assi, Peter W. Bearman, Michael A. Tognarelli. On the stability of a free-to-rotate short-tail fairing and a splitter plate as suppressors of vortex-induced vibration, Ocean Engineering, 2014, 92: 234-244.

[7] Assi G.R., Bearman P., Kitney N. Low drag solutions for suppressing vortex-induced vibration of circular cylinders. J. Fluids Struct. 2009, 25(4): 666-675.

[8] Galvao R., Lee E., Farrell D. Flow control in flow–structure interaction. J. Fluids Struct. 2008, 24(8): 1216-1226.

[9] Law Y.Z., Jaiman R.K. Wake stabilization mechanism of low-drag suppression devices for vortex-induced vibration. J. Fluids Struct. 2017, 70: 428-449.

The Effects of Roadside Buildings Roof Shapes and Tree Plantings on the Vehicle Exhaust Gas Pollution Dispersion in the Urban Street

Frances Hsu [1, a], **Ajanae Bennett** [2, b]

[1] The Lawrenceville School, New Jersey, United States of America
[2] University of Michigan, Lawrenceville, USA

[a] Fhsu23@lawrenceville.org, [b] abennett@lawrenceville.org

Abstract. In this study, the effects of roadside building roof shapes and roadside tree plantings on the dispersion of vehicle exhaust gas pollution in the urban street have been carefully explored. Three typical roof shapes, i.e., flat, triangle and round shapes, are considered in the present work. Firstly, the numerical simulations of air pollution transport in the atmospheric turbulent boundary layer with/without the vegetation canopy have been performed using Computational Fluid Dynamics (CFD). Secondly, the mean flow patterns and the distribution of turbulent kinetic energy (TKE) are then examined. Finally, the concentrations of the air pollution have been carefully explored. It is found that the triangle roof could result in a lower concentration of air pollution in the street canyon. The existence of the roadside trees could decrease the air pollution concentration, although its influence is much weaker than the roof shape. It is also found that the concentration of air pollution is quite related to the flow pattern and TKE distribution. This study could provide helpful guidance on the urban planners on the design of the urban street in the future.

Keywords: CFD, Air Pollution, Building Roof, Street Canyon, Vehicle Exhaust.

1. Introduction

1.1. Background

Air quality in cities is important not just for the health of the residents, but also for the cleanliness of the environment. A huge portion of the pollution comes from vehicle exhaust. These vehicles, including cars, trucks, trains, and motorcycles, decrease the quality of air greatly. Even more, throughout the roads in cities, higher levels of pollutant levels have been detected in street canyons. This higher level of pollutant levels is caused mainly by the increased traffic emissions within a tighter space, and the reduced natural ventilation in the tighter space.

With different roadside building designs that form different mass shapes, the difference in the roof will alter the ventilation of the canyon and the extent of which pollutant transportation. Six different geometric models are considered in this study. The different geometric models include different roof shapes and the existence of trees beside the street canyons. Case A would be the regular flat roof without

Content from this work may be used under the terms of the Creative Commons Attribution 3.0 licence. Any further distribution of this work must maintain attribution to the author(s) and the title of the work, journal citation and DOI.
Published under licence by IOP Publishing Ltd

trees. Case B would be a building of the same height as Case A building, but the top is a triangular-shaped roof. Case C is also at the same height but with a rounded roof. Case AT would be in the same roof configuration as Case A with just the exception of added trees on both sides of the canyon between the buildings. Case BT is at the same roof configuration as Case B with the expectation of trees between the buildings. Case CT would be the same as the last 2, with its roof configuration being the same as the configuration of Case C, and with trees in between. The vehicle exhaust gas pollution in the urban street canyons will be investigated in these six different cases through numerical simulations, with focus on the effects of the roadside building roof shape and the addition of trees.

1.2. Literature Review

To decrease air pollution in street canyons, building roof geometry, building height, and wind direction are important factors affecting the ventilation of street canyons. Past researchers have always focused only on one factor at a time. Mohamed F. Yassin1 explored the effect the height and shape of a building roof on the air quality in urban street canyons. He studied various building heights along with various roof shapes to determine if any combination would improve the air quality in the urban street canyons. The research concluded that a higher building roof could decrease the amount of pollutants left in the street canyons. Regarding the roof shapes, trapezoid-shaped and slanted roofs both help decrease the pollutant concentration. Klukova et.al. 2 investigated the combined effects of roof shape, the difference in roof heights, and the position of the ground-level line on to the pollutant dispersion in street canyons. They found that the best-ventilated street canyon, which was the street canyon with the least pollution density, was the pitched roofs with the pitched roofs higher on the windward wall than on the leeward wall.Llaguno-Munitxa3 investigated the effect of the roof the façade geometry on the mean wind flow and turbulence in the street canyon and found that the pitched and round roofs could increase the in-canyon mean and turbulent velocities. Akhter et al.4 recently numerically explored the effects of street canyons on the business street of Dhaka city in Bangladesh. It was found that the upstream building height plays a significant role in dispersing pollutants inside the canyon. Cai 5 considered the effects of the differential wall heating in street canyons on the dispersion of a passive scalar using large-eddy simulation. It was suggested that for the street-released scalar, the resistance between the bottom canyon and the upper canyon is significant compared with the one between the upper canyon and the urban boundary layer.

Although these previous studies revealed a lot of information regarding the importance of building height and roof shape, none of these studies touched upon the question regarding the existence of trees in street canyons, and their relation to different roof shapes. This is an important topic of study as the more we know about which factors help increase street ventilation, urban planners can design buildings that help increase ventilation in street canyons more easily. This study offers a new direction in the study of the ventilation of pollutants in street canyons, and that is including both trees and roof shapes in the calculation.

1.3. Present Work

This study will investigate the vehicle exhaust pollutant dispersion by using Computational Fluid Dynamics (CFD). This study will investigate six different cases regarding different shapes and the involvement of trees. This paper is organized as follows: Section II will start by explaining the methodology used to configure this experiment, the different parameters that were set, and the equations that are used in this study. Section III will continue to explain the different figures and results that were obtained. Finally, section IV will discuss the results and conclusion that can be made.

2. Numerical Methodology

2.1. Governing Equations

For the present numerical simulation, iso-thermal and incompressible fluid flows are considered. The governing equations for the conservations of the mass and momentum are as follows.

The mass Conservation equation:

$$\frac{\partial \bar{U}_i}{\partial x_i} = 0 \tag{1}$$

The momentum equation:

$$\bar{U}_j \frac{\partial \bar{U}_i}{\partial x_j} = -\frac{1}{\rho}\frac{\partial P}{\partial x_i} + \frac{\partial}{\partial x_j}\left(v\frac{\partial \bar{U}_i}{\partial x_i} - \overline{U_i' U_i'}\right) \tag{2}$$

where $\bar{U_i}$ (i=x, y, z) represents for the mean velocity component, P is the pressure, v is the kinetic viscosity and ρ is the air density. The $-\overline{U_i'U_i'}$ is the turbulent Reynolds stress, which is needed to be modelled. Here, the standard $k-\varepsilon$ turbulent model is used, which has been widely used in the flow simulations on the turbulent boundary layer flows.

The vehicle exhaust air pollution is treated as a passive scalar here. And the transport equation of the scalar concentration, C (kg/m3), is applied as follows:

$$\bar{U}_j \frac{\partial C}{\partial x_j} = -\frac{\partial}{\partial x_i}\left[(D_m + D_t)\frac{\partial C}{\partial x_j}\right] + S_c \tag{3}$$

where D_m and D_t are respectively the molecular and turbulent diffusivity of the air pollution. S_c is the source term, and here it represents for the vehicle exhaust release rate at the street canyon, which is set as 1e-5 $kg/m^3/s$.

All the equations are discretized in second-order scheme, and the pressure-velocity coupling is solved using SIMPLE algorithm. The non-dimensional residuals for all variables are set as 10^{-6}.

2.2. Vegetation modeling

The tree planting is simulated as two rectangular boxes. Different from the solid obstacles, the tree planting zone is treated as porous zones, since the air flow could pass through the tree crown. In the present work, we use the porous fluid zones to simulate the tree planting. And the vegetation modeling equations used in the porous fluid zones are given as following:

$$S_{\bar{U}_i} = -\rho C_d LAD \bar{U}_i U \tag{4}$$

$$S_k = \rho C_d LAD(\beta_p U^3 - \beta_d U k) \tag{5}$$

$$S_\varepsilon = \rho C_d LAD \frac{\varepsilon}{k}(C_{\varepsilon 4}\beta_p U^3 - C_{\varepsilon 5}\beta_d U k) \tag{6}$$

where$S_{\bar{U_i}}$, S_k, and S_ε are respectively the additional source terms for the momentum, turbulent kinetic energy and turbulent dissipation transport equations for the porous fluid zones. $\boldsymbol{U}$ is the velocity magnitude. Here, $C_d = 0.2$, LAD = 1 m2/m3, which is the leaf area density.

2.3. Flow domain and Boundary conditions

The flow domain of the street canyon and the boundary conditions are pivotal for the CFD calculation results. With different scales and sizes, the direction of the wind will be inspected when calculating. The current domain stands as a large rectangle, with the two buildings and trees in the middle of this rectangle, as shown in Fig. 1. This sizable domain allows the investigation of the airflow to be more precise and follows exactly what the experiment needs.

For this simulation, it has been established that there are 6 cases that will be simulated, as shown in Fig. 2. Case A is the regular flat roof without trees. Case B is a building of the same height as Case A building, but the top is a triangular-shaped roof. Case C is also at the same height but with a rounded roof. Case AT, BT, and CT are respectively the same geometrical model as their counterparts but with trees on either side of the street canyon. The model scale is designed as 1:1. The whole airflow area is designed with the measurements of 108 m in height and 458 m in width. Both buildings on either side of the street canyons are 18 meters wide and 18 meters tall. The canyon itself is also 18 meters between the houses. The buildings are set 144 meters away from the inlet. For the triangular and round-shaped buildings, the body of the building is set to be 15 meters, while the roof, both triangular and dome-shaped, is at a height of 3 meters. In the cases where a tree is necessary, the tree is 1 meter away from the buildings. The tree itself is 2 meters wide, and its height is 8 meters.

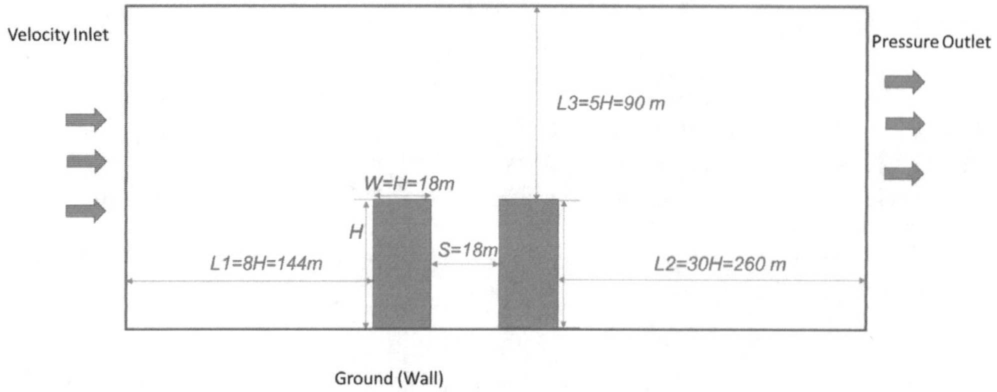

Figure 1. The sketch of the calculation domain and boundary conditions (for Case A as an example)

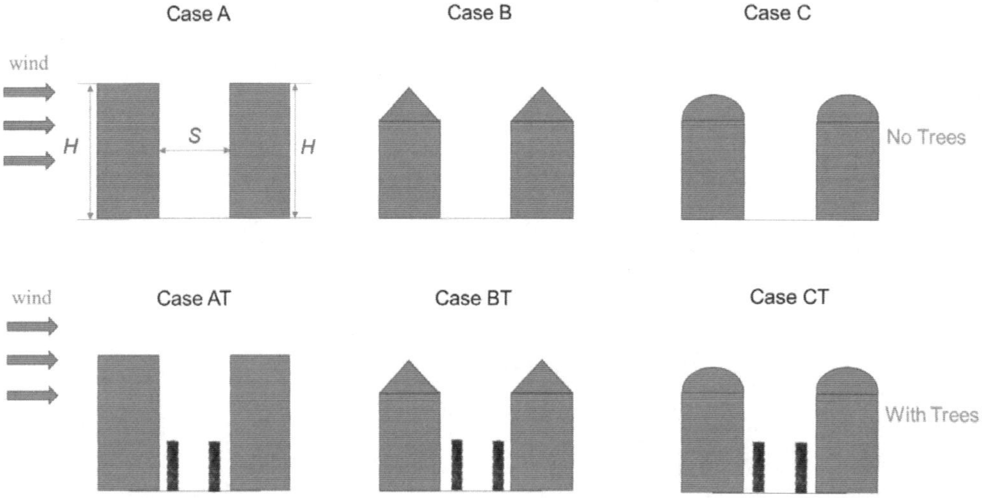

Figure 2. The schematics of the 6 different geometrical models considered in this work

3. Results and Discussion

3.1. Mean flow pattern

The dispersion of the vehicle exhaust pollution in the urban street is quite related to the air flow pattern there. And thus, the mean flow patterns for different cases are first examined. Figure 3 shows the contours of the mean velocity magnitude for different cases. As shown in Fig. 3, all the mean flow patterns for different cases are generally similar as expected. A small recirculation region near the windward side of the front building near the ground is captured. And a recirculation region also exists in the street canyon, and a very long low-velocity region behind the back building which is due to a large recirculating flow appear when the wind flow past the urban street. However, through careful comparisons, one can note that the downstream low-velocity region is a little shorter for Case B/BT and is longest for the Case A/AT. This is mainly due to that the triangle-shaped roof could activate the flow separations near the roof. For the round-shaped roof, the mean flow patterns are very similar to the triangle-shaped-roof cases.

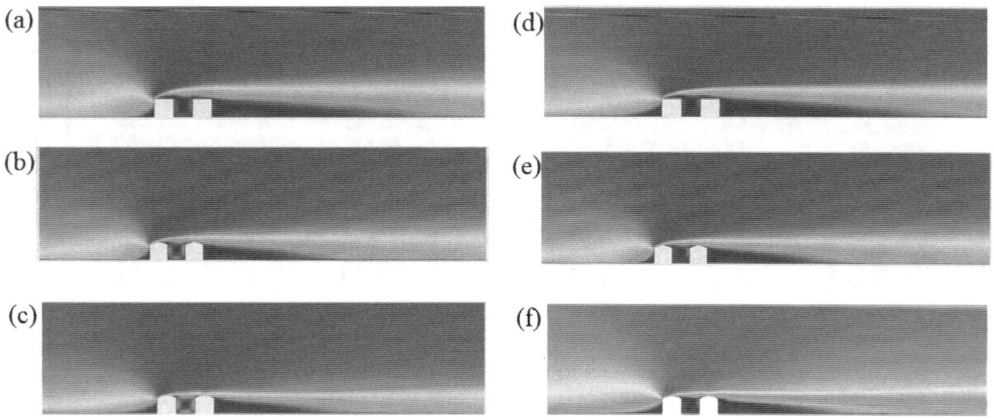

Figure 3. Contours of mean velocity magnitude for different cases: (a) Case A, (b) Case B, (c) Case C, (d) Case AT, (e) Case BT and (d) Case CT.

3.2. Turbulent kinetic energy

The turbulent kinetic energy (TKE) is a significant variable, which represents for the intensity of the turbulent fluctuations. As we know, the turbulent dispersion of the air pollution is related to the turbulent fluctuations. Therefore, it is essential to explore the distributions of the TKE in the urban streets for different configurations. Figure 4 depicts the contours of the TKE for different cases. For the flat-shaped-roof cases, high TKE region exists in the shear layer which is separated from the front side of the roof of the front building. It is to note that the separated shear layer could extend further downstream and touch the back building, as shown in Figs. 4(a) and (d). For the triangle-shaped-roof cases, high TKE region in the shear layers from the roof of the front building is also observed. But compared with the flat-roof-cases, the TKE values in the further downstream behind the back building is higher than that for the flat-roof cases, as shown in Figs. 4(b) and (e). For the round-shaped-roof cases, the high TKE regions at the separated shear layers from the roof of the front building are also observed. It is to note that the TKE values at the shear layer from the roof the back building is also very high. Furthermore, the TKE in the further downstream region behind the urban street is higher than those other two cases. It should be noted that for all these cases, the TKE at the street canyon is very low, which could result in the weak turbulent transport of the air pollution. But through careful examination, one can note that for the TKE becomes slightly higher in the region close to the windward wall of the back building at the street canyon for the triangle-shaped-roof cases.

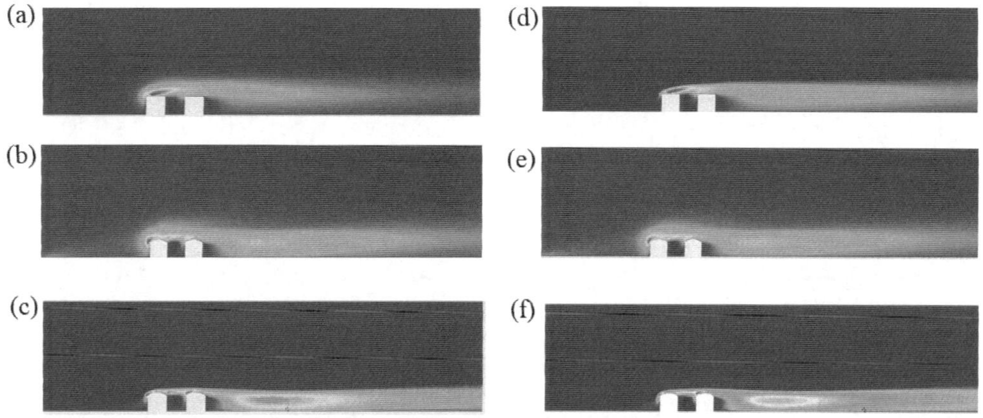

Figure 4. Contours of turbulent kinetic energy (TKE) for different cases: (a) Case A, (b) Case B, (c) Case C, (d) Case AT, (e) Case BT and (d) Case CT.

3.3. Air Pollution Distribution

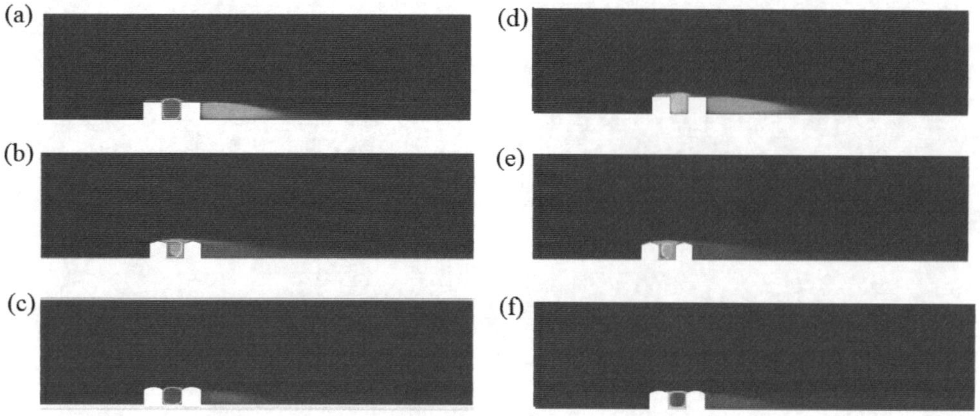

Figure 5. Contours of concentration of the air pollution for different cases: (a) Case A, (b) Case B, (c) Case C, (d) Case AT, (e) Case BT and (d) Case CT.

In this subsection, the contours of the vehicle exhaust pollution are presented in Fig. 5. For the flat-roof Case A, it is found that the air pollution would accumulate in the street canyon and high concentration of the air pollution is observed there, as shown in Fig. 5(a). This is because the very low velocity in the street canyon. It is worth to keep in mind that the vehicle exhaust pollution release is set at the height less than 3 m. It can thus be found that the air pollution due to the vehicle exhaust would occupy the whole street canyon. Actually, it is due to the recirculating flows in the street canyon. In the region behind the back building, there also is the concentration of the air pollution. It can be found that the distribution of the concentration of the is similar to the velocity and TKE contours, indicating that the transport of the air pollution is quite related to the momentum and TKE transport. For the corresponding flat-roof case with tree plantings, i.e. Case AT shown in Fig. 5(d), it can be found that the value of the concentration of the air pollution becomes lower, suggesting that the tree planting could reduce the concentration of the vehicle exhaust.

For the triangle-shaped-roof Case B, it is very interesting to observe that the concentration of air pollution in the street canyon is much lower and the concentration in the further downstream region behind the back building is also much decreased, as shown in Fig. 5(b). Fig. 5 (e) displays the corresponding case with tree plantings. It is found that the tree plantings could further decrease the concentration of air pollution. This could be related to the higher velocity and TKE in the street canyon due to the sharped roof which could activate the flow separation.

For the round-shaped-roof Case C, it is found that high concentration of air pollution is observed in the street canyon, as shown in Fig. 5(c), which is very similar to the flat case. However, it is to note that very low concentration of air pollution is observed in the further downstream region behind the back building. This is associated with the high TKE there as shown in Fig. 4(c). It again proves that the concentration of air pollution is quite related to the distribution of the TKE. High TKE indicate high turbulence there which will enhance the turbulent dispersion of the air pollution and will result in low concentration in that region. The correspondingly configuration with tree plantings is presented in Fig. 5(f). The concentration of air pollution is generally very similar to the Case C. Through careful comparisons, one can note that the concentration of air pollution is a little lower in the regions close to the tree plantings than Case C. It can thus found that the effect of the tree planting is weaker than that of the roof shape.

4. Conclusion

The goal of this project is to gain further insight into the different ways houses and apartments can be built along with the design of street canyons to minimize the flow of pollution in street canyons. In this research, after running 6 different simulations in ANSYS (3 different types of roofs and trees on both sides of the street canyon), the data from the graphs generated strongly suggests that case of BT, the building with the triangle roof and trees on either side of the street canyon, has the least amount of pollution residue left in the street canyons. It is also found that the roof shape could have a significant influence on the ventilation in the street canyon and thus the dispersion of vehicle exhaust pollution. The tree planting could further improve the air quality in the street canyon, although the effects of the tree planting is much weaker than the roof shape. The present work could provide some guidance on the urban planners to design urban streets in the future.

5. References

[1] Mohamed F. Yassin, 2011. Impact of height and shape of building roof on air quality in urban street canyons, Atmospheric Environment, 45, 5220-5229.
[2] Zuzana Klukova, Stepan Nosek, Vladimír Fuka, Zbynek Janour, Hana Chaloupecka, Jana Doubalov, 2021. The combining effect of the roof shape, roof-height non-uniformity and source position on the pollutant transport between a street canyon and 3D urban array, Journal of Wind Engineering and Industrial Aerodynamics, 208, 104468.
[3] Llaguno-Munitxa M, Bou-Zeid E, Hultmark M, 2017. The influence of building geometry on street canyon air flow: Validation of large eddy simulations against wind tunnel experiments. Journal of Wind Engineering and Industrial Aerodynamics,165, 115-30.
[4] Akhter MN, Ali ME, Rahman MM, Hossain MN, Molla MM, 2021. Simulation of air pollution dispersion in Dhaka city street canyon. AIP Advances, 11, 065022.
[5] Cai X, 2012. Effects of differential wall heating in street canyons on dispersion and ventilation characteristics of a passive scalar. Atmospheric Environment, 51, 268-77.

Research on Fast CFD Simulation of Automobile Flow Field Based on Artificial Intelligence

Xiaofeng Cao*, Qiang Wang, Hongyan Li, Hai Ma

Aviation Key Laboratory of Science and Technology on Aerodynamics of high speed and high Reynolds number, AVIC Aerodynamics Research Institute, Shenyang 110136, China

*Corresponding author e-mail: jlcxf_ruc@126.com

Abstract. Fast automotive aerodynamic evaluation can extremely reduce the design cycles of automotive. This paper establishes a rapid simulation model of the automobile flow field based on the artificial intelligence surrogate model. The end-to-end network is used to map the geometric model, incoming flow conditions, and result. The decoder is used to splice high-dimensional and low-dimensional features to achieve feature sharing. The average MAE error of the optimal model is 5.249%. The average calculation time of a single example is 1.2968s, which is about 0.62% of CFD solver. The simulation result demonstrate that the deep learning method can not only accelerate the calculation, but also can improve the design efficiency of automobile aerodynamic profile with high accuracy.

1. Introduction

With the development of the automobile industry, less automobile resistance and better automobile performance have gradually become the goal of automobile design. In the process of driving, overcoming air resistance is the main factor affecting the performance of the vehicle, in addition to energy loss. At present, the acquisition of vehicle shape flow fields mainly depends on wind tunnel tests and computational fluid dynamics simulation [1]. Wind tunnel test requires a lot of time for model iteration. Computational fluid dynamics simulation can effectively reduce the number of wind tunnel tests and improve efficiency. However, there is also some limitations, the cost of calculation time is too long. To reduce time-consuming CFD calculation, some researchers use surrogate models to improve the efficiency of flow field simulation. Flow field simulation based on surrogate models has the advantages of a short calculation cycle and low cost. The surrogate model based on artificial intelligence can quickly obtain the flow field distribution around the vehicle in a short time, shortening the time to obtain the flow field, and facilitating the update and iteration of the vehicle shape.

Kei Akasaka [2] used the machine learning method to predict the air resistance coefficient of automobiles, the mean average relative error of resistance training was 2.4%. Thuerry [3] used the deep learning method to realize the flow field reconstruction of the airfoil under the condition of incompressible flow. Cao [4] discovered the flow field reconstruction technology under the condition of compressible flow and directly carried out fast CFD simulation using the free flow state. The error between the calculated result of the artificial intelligence flow field and the calculation time of a single example was reduced to 1/120 of that of the CFD solver. Huang [5] studied the aerodynamic drag

Content from this work may be used under the terms of the Creative Commons Attribution 3.0 licence. Any further distribution of this work must maintain attribution to the author(s) and the title of the work, journal citation and DOI.
Published under licence by IOP Publishing Ltd

reduction of the high-resistance vehicle model using artificial intelligence and combined excitation and obtained the maximum drag reduction rate of the Ahmed vehicle model under unsteady combined excitation. The problem of automobile drag reduction has gradually become an essential task in automobile design. The mapping relationship between the geometric model and flow field can be quickly established by using artificial intelligence methods to assist automobile shape design to achieve drag reduction.

A surrogate model based on artificial intelligence carried out in paper to establish a whole process method from data pre-processing, deep neural network computing, and post-processing of computing result. In specific fields, such as the design of automobile shapes and automotive aerodynamic performance, flow field rapid evaluation plays a huge role. The flow field result can be obtained quickly by using the deep neural network method and decrease the design cycle of the vehicle for the drag reduction problem of vehicle components.

2. Artificial intelligence model

2.1. Dataset

Ahmed model is the vehicle body model after Ahmed [6] abstracted and simplified the actual vehicle reasonably in 1984. The main body of the model car is divided into three parts: the headstock, body, and tailstock. The arc chamfer semidiameter at the headstock is 100mm, and the maximum length at the headstock is 288mm; The body is a box with a cross-section of 1044mm long and 288mm wide, as shown in Figure 1. The tailstock is composed of an inclined tail window and a vertical rear back, with sharp edges on each edge. The fully developed turbulent boundary layer will separate at the sharp edges of the tailstock and form a recirculating region. The included angle between the model rear window and the horizontal plane is defined as the rear window inclination ϕ, it is used to study the influence of the rear window inclination on the vehicle wake and the aerodynamic forces acting on the vehicle. A large number of studies show that the wake generated by the Ahmed model can well reflect the main characteristics of the actual vehicle wake, so it is widely used in the relevant theoretical research of vehicle wake, aerodynamic characteristics and drag reduction control.

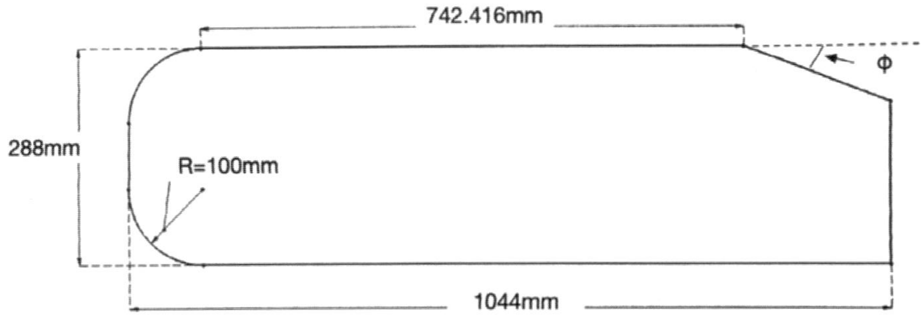

Figure 1. Two-dimensional automobile geometry

This paper selects the Ahmed model for rapid simulation of the vehicle flow field by changing the rear window inclination φ to build the geometric models. The rear window inclination varies from 0 ° to 30 °. Samples are taken every 0.5 ° to get a total of 60 geometric models. The CFD solver is used to build the dataset. The computational grid is a two-dimensional unstructured hybrid grid of triangles and quadrangles. The thickness of the first layer of the object surface is 1 × 10-6m, with about 60000 grid cells and 50000 grid nodes, as shown in Figure 2.

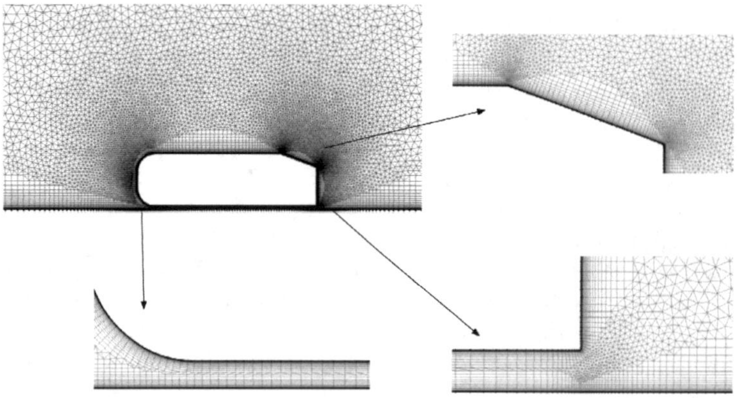

Figure 2. CFD calculating mesh

Using the UNSMB solver for aerodynamic calculation of the flow field, the finite volume method is used to solve the steady flow field, and the Reynolds Average Navier Stokes (RANS) equation is solved. The SA model is used for the turbulence model, the multi-grid technology is used to accelerate the convergence, and the MPI technology is used for parallel calculation and accelerated flow field calculation. The calculation state is that 60 geometric models are randomly sampled at 14m/s-40m/s (50km/h-140km/h) to generate a 6-channel dataset. The first three channels u, p, and m are convolutional neural network inputs, and the last three channels u, v and p are training labels,. A total of 4000 data samples are generated.

2.2. Neural network structure

The convolutional neural network is a kind of feedforward neural network with convolution computation and deep structure, which is one of the representative algorithms of deep learning[7]. The convolution layer uses the convolution kernel to extract features from input data. Features with different complexity can be extracted by changing the kernel size, stride, and padding. The activation function in the convolution layer introduces the nonlinear characteristics into the neural network to help the neural network learn and understand complex nonlinear functions. The pooling layer performs feature selection and information filtering after feature extraction of convolution features to reduce computational complexity. The fully connected layer combines the extracted features nonlinearly to get the output and uses the high-dimensional features at this stage to complete the learning goal. Based on the end-to-end U-Net[8], the fast mapping between the incoming flow conditions and the flow field distribution is realized. The end-to-end network used in this paper is an improved U-Net network, which consists of 7 convolution module encoders and 7 deconvolution module decoders.

Each convolution module in the encoder includes a convolution layer, a normalized layer, and a nonlinear activate layer. The initial geometric shape and physical state are extracted layer by layer through the convolution module to obtain different features of high and low dimensions. The encoder is used to conduct a stride deconvolution of the image so that the network can extract large-scale abstract information from the growing number of feature channels.

The decoder is a feature reducer. Each module contains a deconvolution layer, a normalization layer, and a nonlinear activation layer. U-Net uses the convolution layer to extract spatial information features and introduces a skip connection between the encoder and decoder to ensure that the information in the output layer can be used by the neural network discriminator. All the features extracted by the encoder are connected to the corresponding decoder through the feature merging structure, effectively doubling the number of channels in each decoder. The network structure is shown in Figure 3.

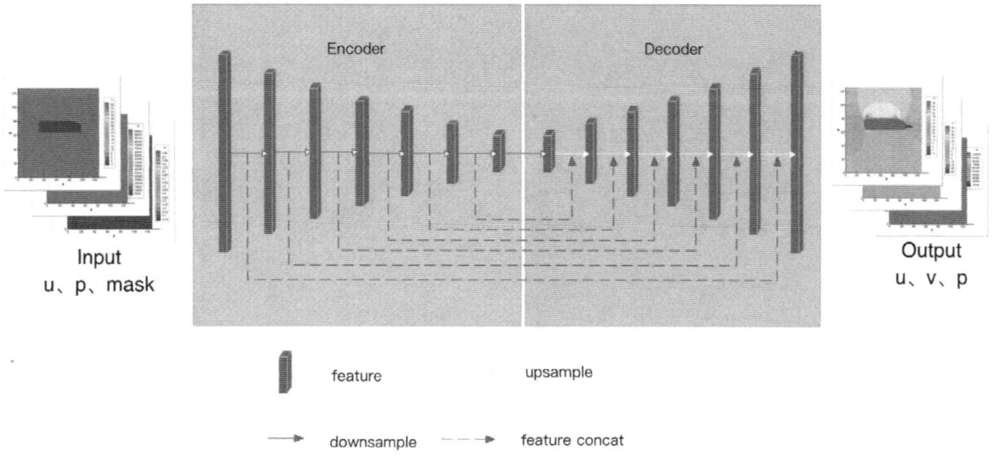

Figure 3. Neural network structure

3. Simulation results and analysis

3.1. Model train and evaluation

Based on the Pytorch, a variant U-Net network was built. During the training process, Adam optimizer and L1 regularization are used for network structure training, the learning rate is 0.0006, the number of iterations is 10000, and the probability of dropping out randomly is 0.2. Deep neural network training is based on a workstation. CPU is Intel Xeon (R) Silver 4210R CPU (2.40GHz, 10 cores) × 2. 64G memory, NVIDIA Quadro RTX 4000 GPU, based on Ubuntu 20.04.3 LTS.

200 samples are used as the testset for the model test set. The mean absolute error (MAE) and the MAE of each physical quantity (velocity, pressure, and density) are used to judge the model performance. The mean absolute error represents the average of the absolute errors between the predicted and observed values. MAE can evaluate the change degree of data, and the calculation formula of MAE is:

$$\text{MAE} = \frac{1}{N}\sum_{t=1}^{N} | O_t - P_t |$$

Where: O_t is the target value, P_t is the predicted value.

3.2. Result Analysis

The surrogate model based on artificial intelligence has achieved good results on the current dataset, and can greatly improve the calculation efficiency meanwhile ensure calculation accuracy. On the testset, the overall average MAE of the model is 5.249%, the MAE of the pressure is 7.016%, the MAE of the velocity component in the x is 0.771%, the MAE of the velocity component in the y is 6.274%, and the MAE of the overall velocity is 1.203%. The fitting effect of the model on the velocity component in x is better than that on the pressure of the velocity in y direction. The model test results are shown in the Table1. Figure 4 is the sample of testset.

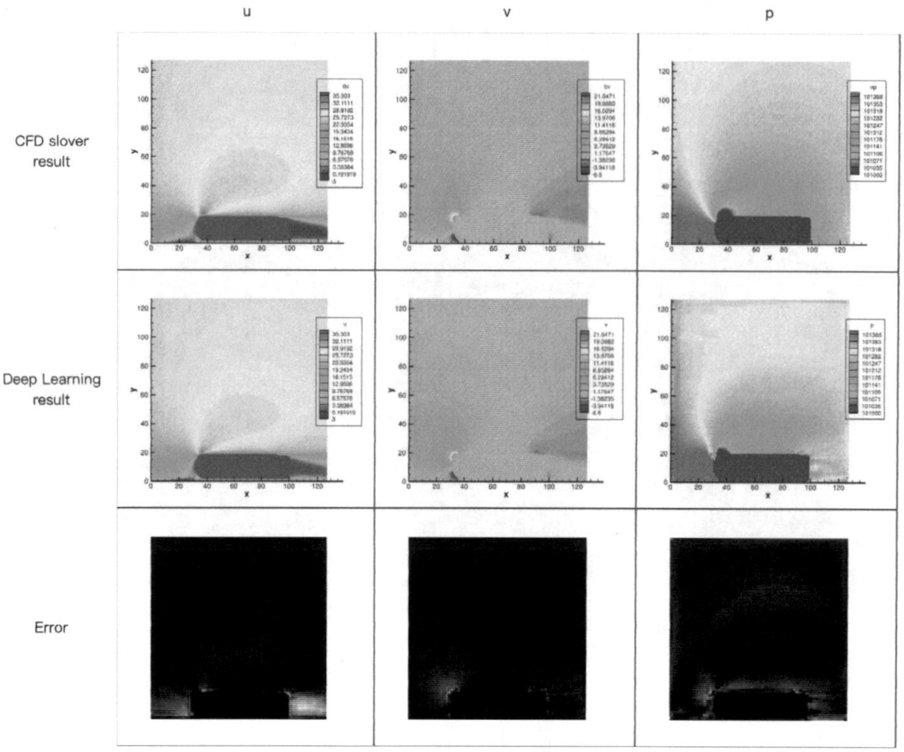

Figure 4. The sample of testset.

(The rear window inclination is 8°, and the speed is 24.95m/s (89.82km/h); In CFD result and Deep Learning result: u is the component of velocity in the x, v is the component of velocity in the y, p is the pressure. In Error, u, v, p indicates the error of CFD and deep learning.)

Table 1. The result of testset

Pressure MAE	V_x MAE	V_y MAE	V MAE	Mean MAE
7.016%	0.771%	6.274%	1.203%	5.249%

Compared with CFD, the calculation time cost of a single case of the AI model is also greatly improved. The calculation time of the AI model is about 1/160 of that of the conventional CFD, and the average calculation time of a single case is 1.2968s. The comparison is shown in Table2.

Table 2. The calculation time of AI and CFD solver

	CFD	Deep Learning
Time(s)	210	1.2968

According to the results of the time-averaged flow field reported in early literature [9] [10], there are three main characteristics in the flow field of the Ahmed model: a separation field on the inclined rear window, a pair of counter-rotating flow direction vortices or C-column vortices along both sides of the rear window, and a pair of backflow on the vertical rear side. In the rapid simulation of the two-dimensional vehicle flow field, deep learning effects a good capture on a pair of backflow bubbles on the vertical back side. As shown in Figure 5, the deep learning prediction result can capture the reflow region well. There are two circumstances, in high speed, a pair of reflow bubbles on the back of the vertical can be captured by deep learning. The centre and morphology of a pair of reflow bubbles is a little inconsistent between CFD result and AI. Another, in low speed, the reflow bubbles can be predicted as well, but in some test results, a pair of reflow bubbles will be coupled into one reflow bubble. The prediction accuracy of reflow bubbles on the vertical rear face needs to be further improved.

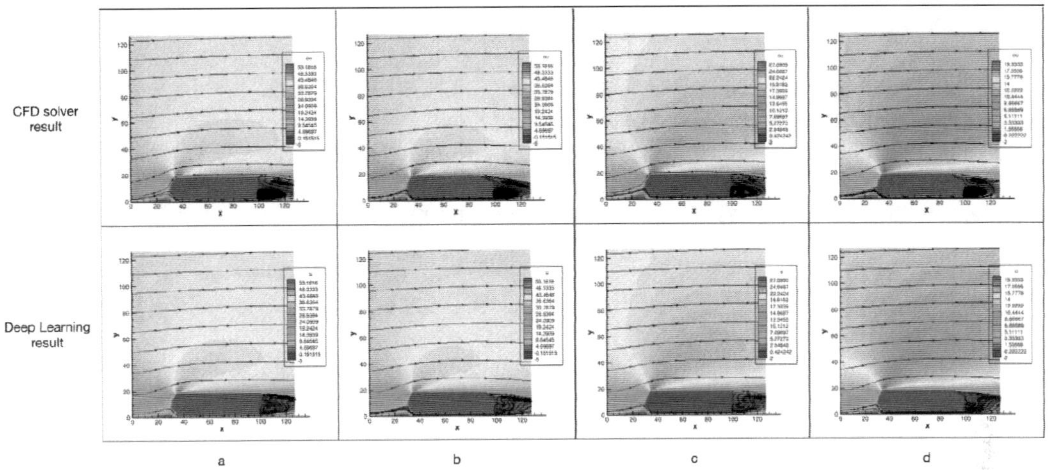

Figure 5. Comparison of AI and CFD solver vertical backflow bubbles

(a: Tail window inclination: 2°, speed:20.53m/s (73.91km/h) b: Tail window inclination:25.5°, speed: 15.32m/s (55.15km/h) c: Tail window inclination:0°, speed:37.64m/s (135.50km/h) d: Tail window inclination: 24°, speed:37.15m/s (133.74km/h))

4. Discussion

In any case, the correlation between the prediction result of the model and CFD calculation result is strong. However, there still some difference exist between the prediction result and CFD result. The error is usually shown in the variation of the flow field behind the vertical wake. The relative error of velocity component Vy is larger than Vx. This phenomenon caused by the lack of latest information which extracted from the dataset, so we can consider introducing the attention mechanism in the neural network to enable it to extract information for specific regions and improve the model performance and generalization ability. Although there is a relative error between the predicted result and the real flow field, from the prediction visualization result, the trend of the flow field predicted in this paper is the same as that of the real flow field.

The rapid simulation method of automobile flow field based on artificial intelligence proposed in this paper can quickly obtain the surrounding flow field according to the current incoming flow conditions, which verifies the application of deep learning algorithm in the field of the aerodynamic and shows the feasibility of deep learning algorithm in fast CFD solution. The deep learning algorithm can effectively improve the efficiency of the flow field solution. In addition, more detailed geometric shape description methods (such as point cloud and SDF) can be carried out to establish the mapping between the flow field and geometric shape; Furthermore, distinguish the near field and far field in the flow field label, establish the near field label, avoid the precision loss caused by interpolation, and establish a more refined mapping relationship.

References

[1] Wang Fujun. Computational fluid dynamics analysis: Principle and application of CFD software[M].Beijing: Tsinghua University Press, 2004.

[2] Kei Akasaka, Fangge Chen, Takchito Teraguchi. Interactive prediction of automotive aerodynamic drag coefficient using machine learning. The 34th Annual Conference of The

Japanese Society for Artificial Intelligence, 2020.
[3] Thuerey N, WeiBenow K, Prantl L, et al. Deep learning methods for Reynolds-averaged Navier-Stokes simulations of airfoil flows[J]. AIAA Journal,2020,58(1):25-36.
[4] Cao Xiaofeng, Li Hongyan, Guo Chengpeng, et al. Research on two-dimensional airfoil flow field reconstruction based on deep learning [J]. Aeronautical Science & Technology, 2022, 33(07): 106-112.
[5] Huang Yajie. Aerodynamic drag reduction of high drag vehicle model based on artificial intelligence and combined actuators. Harbin Institute of Technology, Shenzhen, Master dissertation, 2019.
[6] Ahmed S R, Ramm G, Faltin G. Some salient features of the time-averaged ground vehicle wake[J]. SAE Transactions, 1984, 840300:473-503.
[7] Goodfellow, I., Bengio, Y., Courville, A.. Deep learning (Vol. 1). Cambridge: MIT press, 2016: 326-366.
[8] Ronneberger O, Fischer P, Brox T. U-Net: Convolutional networks for biomedical image segmentation[C]//International Conference on Medical image computing and computer-assisted intervention. Springer, Cham, 2015:234~241.
[9] Lienhart H, Stoots C, Becker S. Flow and Turbulence Structures in the Wake of a Simplified Car Model (Ahmed Modell)[C] // New Results in Numerical and Experimental Fluid Mechanics III. Berlin, Heidelberg : Springer Berlin Heidelberg, 2002 : 323-330.
[10] Lienhart H, Becker S. Flow and Turbulence Structure in the Wake of a Simplified Car Model[J]. SAE Transactions, 2003, 2003-01-0656 : 785-796.

Research on turbulence model for simulation of wind flow in mountain areas of micro-topography

Yuan Yin[1], Yangsen Li[1], Ruizong Lin[1], Zhiwei Liu[2], Benzhao Fu[2], Guo Huang[3,*], Hongjie Zhang[3]

[1] State Gird Fujian Electric Power Co. Ltd, Fuzhou 350003, China
[2] State Grid Fujian Economic Research Institute, Fuzhou 350003, China
[3] China Electric Power Research Institute, Beijing, 100055, China

*Corresponding author e-mail: huangguocn@126.com

Abstract. The accuracy of wind flow simulation in mountain area in micro-topography is related to discrete scheme, turbulence model and wall function. In this paper, the standard hill model is adopt to analyze the calculation accuracy of different turbulence models. Four turbulence models, including RNG k-ε, k-ω, SST k-ω and TSST k-ω, were used in the calculation, and were compared with the result of wind tunnel test. The calculation results with the second-order upwind scheme, the SST or TSST turbulence model and the scalable wall function are basically consistent with results of wind tunnel test.

Keywords: micro-topography; wind flow simulation; turbulence model; hill model.

1. Introduction

At present, in order to consider the acceleration effect of air flow in mountain areas of micro-topography, the design of wind speed for the transmission line in mountain area is increased more 10% than that in the plains. However, there is no relevant regulation in the code for the wind profile. In order to gain the acceleration ratios of wind speed velocity and the coefficients of wind profile in mountain areas, and to improve the safety of transmission lines, it is necessary to obtain wind flow in mountain areas with field measurements, wind tunnel test and CFD method. Due to the high cost of field measurement and wind tunnel test, the CFD method is used to carry out flow analysis, which can reduce costs. The terrain point cloud data from satellite telemetry is used to obtain the geometric configuration of micro-terrain.

Many scholars have done numerical simulations of complex terrain wind flow and got a series of research results. In 1999, Uchida and Ohya used digital elevation data to simulate mountainous terrain and studied the atmospheric boundary layer flow [1]. Kim found that the RNG k-ε turbulence model was better than the standard k-ε turbulence model [2]. Blocken's research showed that the Realizable k-ε model could be used to evaluate complex three-dimensional flows [3]. Petry conducted wind tunnel test and numerical simulation on the atmospheric boundary layer flow, and found that significant changes of turbulence characteristics only occurred in the wake regions of the flow. Compared with the k-ε turbulence model, the simulation result of the SST k-w turbulence model is more consistent with the result of wind tunnel experimental [4]. Risan verified the feasibility and effectiveness of the RANS and LES hybrid model [5]. Some scholars used a lot of turbulence models to carry out CFD numerical

Content from this work may be used under the terms of the Creative Commons Attribution 3.0 licence. Any further distribution of this work must maintain attribution to the author(s) and the title of the work, journal citation and DOI.
Published under licence by IOP Publishing Ltd

simulation of complex mountain wind flow, and found that the RANS turbulence model had better performance on the leeward side. Although the LES model can provide better simulation results, the computational cost of LES method is huge, and it is not realistic at present.

The mountain has obvious bluff body characteristics, and the mountain flow also has obvious bluff body flow characteristics. The accuracy of the simulation results is directly and significantly related to the discrete scheme, turbulence model and wall function. Therefore, it is necessary to research the CFD numerical simulation of mountain micro-topography. In this paper, the standard hill model is used to analyze the turbulence models, so as to provide theoretical guidance for the wind flow simulation of mountain micro-topography.

2. Turbulence Model

The N-S equation is not closed, and turbulence models are required to provide constraints. Turbulence models commonly include standard k-ε, RNG k-ε and SST (shear stress transport) k-ω turbulence model.

2.1. Standard k-ε

The N-S equation can be closed by the standard k-ε turbulence model, and the flow near the wall with low Reynolds number is handled by the standard wall function. The standard k-ε turbulence model is as follows:

$$\frac{\partial(\rho \boldsymbol{k})}{\partial \boldsymbol{t}}+\frac{\partial}{\partial \boldsymbol{x}_i}(\rho \boldsymbol{k}\boldsymbol{u}_i)=\frac{\partial}{\partial \boldsymbol{x}_j}\left[\left(\mu+\frac{\mu_t}{\sigma_k}\right)\frac{\partial \boldsymbol{k}}{\partial \boldsymbol{x}_j}\right]+\boldsymbol{G}_k-\rho\varepsilon \tag{1}$$

$$\frac{\partial(\rho\varepsilon)}{\partial \boldsymbol{t}}+\frac{\partial}{\partial \boldsymbol{x}_i}(\rho\varepsilon \boldsymbol{u}_i)=\frac{\partial}{\partial \boldsymbol{x}_j}\left[\left(\mu+\frac{\mu_t}{\sigma_\varepsilon}\right)\frac{\partial \varepsilon}{\partial \boldsymbol{x}_j}\right]+\boldsymbol{C}_{1\varepsilon}\frac{\varepsilon}{\boldsymbol{k}}\boldsymbol{G}_k-\boldsymbol{C}_{2\varepsilon}\rho\frac{\varepsilon^2}{\boldsymbol{k}} \tag{2}$$

where $\mu_t=\rho C_\mu \frac{k^2}{\varepsilon}$, $G_k=\mu_t S^2$, $S\equiv\sqrt{2S_{ij}S_{ij}}$, $S_{ij}=\frac{1}{2}\left(\frac{\partial u_j}{\partial x_i}+\frac{\partial u_i}{\partial x_j}\right)$. The model constants are as follows: C_μ=0.09, $C_{1\varepsilon}$=1.44, $C_{2\varepsilon}$=1.92, σ_k=1.0, σ_ε=1.3.

2.2. RNG k-ε

The RNG k-ε turbulence model was proposed by Yakhot and Orzag. In this model, the small-scale motions are removed from the governing equations by modifying the viscosity term. The k and ε equations are very similar to the standard k-ε equation:

$$\frac{\partial(\rho k)}{\partial t}+\frac{\partial(\rho k u_\mathrm{i})}{\partial x_\mathrm{i}}=\frac{\partial}{\partial x_\mathrm{j}}\left(\alpha_k u_\mathrm{eff}\frac{\partial k}{\partial x_\mathrm{j}}\right)+G_\mathrm{k}+G_\mathrm{b}-\rho\varepsilon-Y_\mathrm{M}+S_\mathrm{k} \tag{3}$$

$$\frac{\partial}{\partial t}(\rho\varepsilon)+\frac{\partial}{\partial x_\mathrm{i}}(\rho\varepsilon u_\mathrm{i})=\frac{\partial}{\partial x_\mathrm{j}}\left(\alpha_\varepsilon u_\mathrm{eff}\frac{\partial\varepsilon}{\partial x_\mathrm{j}}\right)+C_{1\varepsilon}^*\frac{\varepsilon}{k}\left(G_k+C_{3\mathrm{g}}G_\mathrm{b}\right)-C_{2e}\rho\frac{\varepsilon^2}{k}-R_\varepsilon+S_\mathrm{e} \tag{4}$$

where $u=u+u_\mathrm{i}$, $\alpha_\mathrm{K}=\alpha_\varepsilon$=1.39, $C_{1e}^*=C_{1e}-\frac{\eta(1-\eta/\eta_\mathrm{o})}{1+\beta\eta^3}$, η_o=4.377, β=0.012, $\eta=\left(\alpha E_\mathrm{ij}\ E_\mathrm{ij}\right)^{1/2}\frac{k}{\varepsilon}$, $E_\mathrm{ij}=\frac{1}{2}\left(\frac{\partial u_\mathrm{i}}{\partial x_\mathrm{j}}+\frac{\partial u_\mathrm{i}}{\partial u_\mathrm{i}}\right)$.

Compared with the standard k-ε model, the biggest feature of the RNG k-ε turbulence model is that the mainstream time-averaged strain rate E_{ij} is introduced into the calculation of the coefficient $C_{1\varepsilon}^*$ of the generation term of the ε equation. The RNG k-ε model takes into account the swirling flow, and can solve flows with high strain rates and large streamline curvature better.

2.3. SST k-ω

The SST k-ω turbulence model can be used to close the N-S equation, where the transport equations of turbulent kinetic energy k and specific dissipation rate ω are:

$$\frac{\partial(\rho k)}{\partial t}+\frac{\partial}{x_i}(\rho k u_i)=\frac{\partial}{\partial x_j}\left(\Gamma_k\frac{\partial k}{\partial x_j}\right)+\tilde{G}_k-Y_k \tag{5}$$

$$\frac{\partial(\rho\omega)}{\partial t}+\frac{\partial}{x_i}(\rho\omega u_i)=\frac{\partial}{\partial x_j}\left(\Gamma_\omega\frac{\partial\omega}{\partial x_j}\right)+G_\omega-Y_\omega+D_\omega \tag{6}$$

2.4. Transition SST k-ω

For typical cross-sections of bluff bodies such as cylinders and cylinder-like models, the low-velocity flow around them involves complex boundary layer transitions, separations, reattachments and periodic shedding. The Transition SST k-ω turbulence model can not only simulate the flow separation and reattachment, but also capture the transition flow from the laminar flow to the turbulent flow. The turbulence model consists of five governing equations The transport equations of turbulent kinetic energy k and specific dissipation rate ω are the same as (5)~(6).

The equation for the intermittent factor γ is:

$$\frac{\partial(\rho\gamma)}{\partial t}+\frac{\partial}{x_j}(\rho u_j\gamma)=P_{\gamma1}-E_{\gamma1}+P_{\gamma2}-E_{\gamma2}+\frac{\partial}{\partial x_j}\left((\mu+\mu_t)\frac{\partial\gamma}{\partial x_j}\right) \tag{7}$$

The transition start criterion is:

$$\gamma_{sep}=\min\left(2\max\left[\left(\frac{\mathrm{Re}_v}{3.235\,\mathrm{Re}_{\theta c}}-1,0\right)F_{reattach},2\right]\right)F_{\theta t} \tag{8}$$

The transition governing equation interacts with turbulent kinetic energy in the following way:

$$\frac{\partial(\rho k)}{\partial t}+\frac{\partial}{x_j}(\rho k u_j)=\tilde{P}_k-\tilde{D}_k+\frac{\partial}{\partial x_j}\left((\mu+\sigma_k\mu_t)\frac{\partial k}{\partial x_j}\right) \tag{10}$$

Equations (5) ~ (10) are governing equations of Transition SST k-ω turbulence model.

3. Simulation calculation

The hill model is often adopt to show the accuracy of the algorithm, which is used for the research work. The maximum slope of the model is about 32°, the contour expression is shown in the formula (11), and the model is shown in Figure 1.

$$z(x,y)=h\cos^2\left(\frac{\pi(x^2+y^2)^{1/2}}{2L}\right) \tag{11}$$

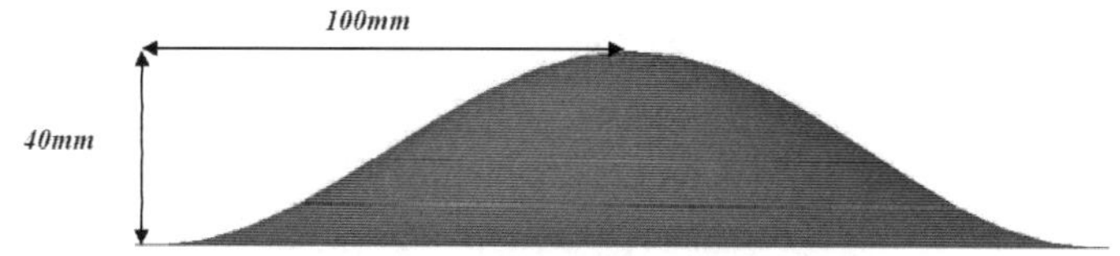

Figure 1. Standard Hill Model

The RNG k-ε, k-ω, SST k-ω, and TSST k-ω turbulence models were used in the calculation. The specific setting conditions are shown in the Table 1. Figure 2 shows the comparison between the simulated and experimental results of wind speed profiles at different positions of the hill model under the condition of the B-type inflow profile. From the comparison results in Table 1 and Figure 2, it can be shown that the numerical simulation result is basically consistent with the wind tunnel test result with the second-order upwind scheme, SST or TSST turbulence model and Scalable Wall Function. Such a combination of parameter settings is used in simulation of the micro-topographic wind flow.

Table 1. Setting conditions of discrete scheme, turbulence model and wall function

No.	Setting Cases			Parameter Description
	Discrete scheme	Turbulence model	Wall function	
1	High-Resolution	RNG	Scalable Wall Function	RNG k-ε and High Resolution
2	2nd upwind	RNG		RNG k-ε and 2nd upwind
3	2nd upwind	KO		k-ω and 2nd upwind
4	2nd upwind	SST		SST k-ω and 2nd upwind
5	2nd upwind	TSST		Transition SST k-ω and 2nd upwind
6	1st upwind	TSST		Transition SST k-ω and 1st upwind

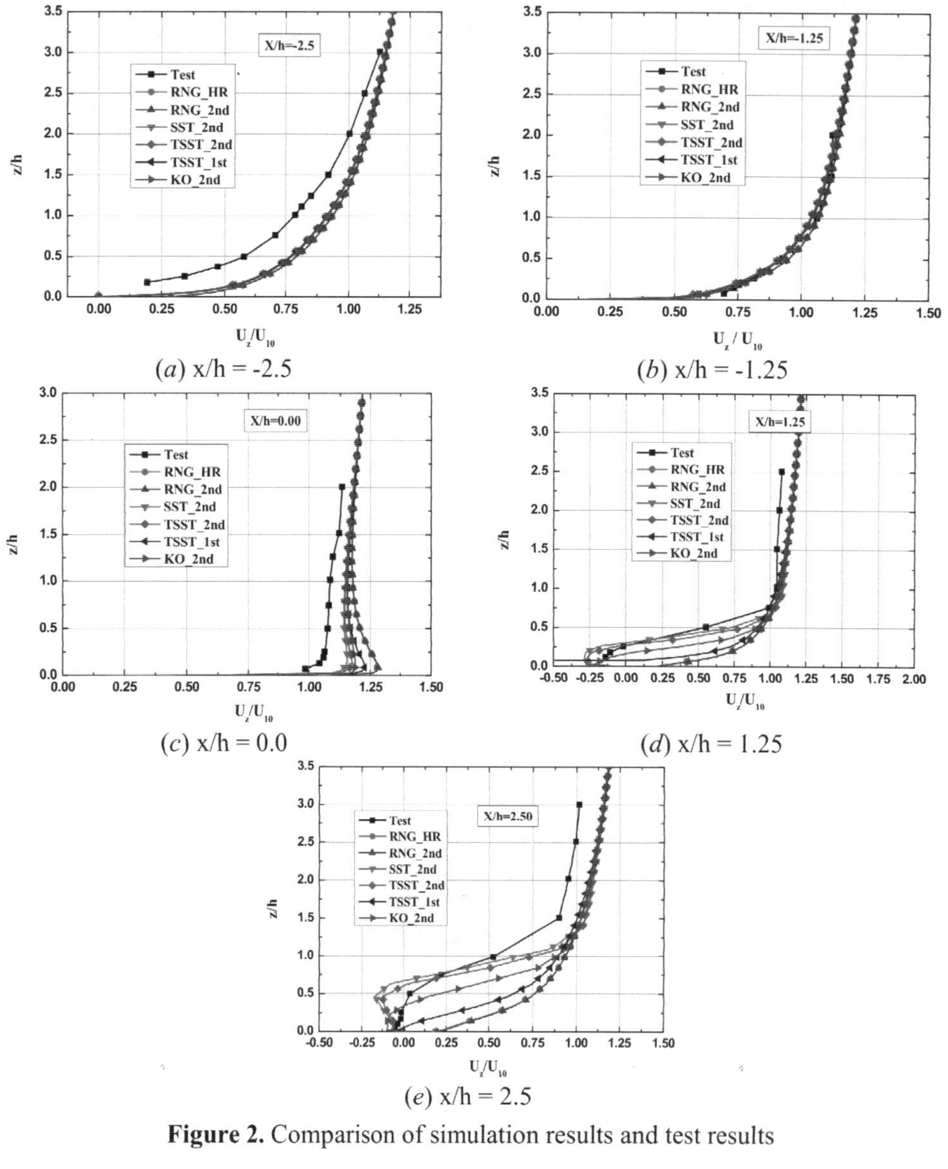

(*a*) x/h = -2.5

(*b*) x/h = -1.25

(*c*) x/h = 0.0

(*d*) x/h = 1.25

(*e*) x/h = 2.5

Figure 2. Comparison of simulation results and test results

4. Conclusion

Under the condition of type-B inflow profile, the air flow around the standard hill model is simulated. The RNG k-ε turbulence model, the k-ω turbulence model, SST k-ω turbulence model, and TSST k-ω turbulence models were used in the calculation, and compared with the results of wind tunnel test. The results of numerical simulation are basically consistent with the results of wind tunnel test, in which the second-order upwind scheme, SST or TSST turbulence model and Scalable Wall Function are adopt.

Acknowledgments

This work was supported by the State Grid Fujian Economic Research Institute. The project code is 52130N210007, with the name of Research on assessment technology of mechanical state of towers based on numerical wind tunnel of atmospheric boundary layer.

References

[1] Uchida T., Ohya Y. Numerical simulation of atmospheric flow over complex terrain[J]. Journal of Wind Engineering an d Industrial Aerodynamics, 1999, 81(1): 283-293.
[2] Kim H. G. Patel V,. Lee C. M. Numerical simulation of wind flow over hilly terrain[J]. Journal of Wind Engineering an d Industrial Aerodynamics, 2000, 87(1): 45-60.
[3] Blocken B., van der Hout A., Dekker J., et al. CFD simulation of wind flow over natural complex terrain: case study with validation by field measurements for Ria de Ferrol, Galicia Spain[J]. Journal of Wind Engineering and Industrial Aerodynamics, 2015, 147: 43-57.
[4] Petry A. P., Loredo-Souza A. M., Filho D. G. R. D. F., et al. Wind tunnel and CFD analysis of wind flow over a complex terrain[R]. German: German Wind Energy Conference-Dewek, 2012.
[5] Risan A., Lund J. A., Chang C.Y., et al. Wind in Complex Terrain-Lidar Measurements for Evaluation of CFD Simulations[J]. Remote Sensing, 2018, 10(1): 59.

Investigation on Shallow Trenched-Hole Film Cooling Effectiveness at Turbine Blade Pressure Side

Jiangjiang Yin [1], Kenan Huang [2, *], Ying Huang [1]

[1] Shanghai Aircraft Design and Research Institute, Shanghai, China
[2] School of Mechanical and Energy Engineering, Zhejiang University of Science and Technology, Zhejiang, China

*Corresponding author e-mail: huangkenan@nuaa.edu.cn

Abstract. Numerical simulations are performed to investigate the effects of shallow trenches on the adiabatic film cooling effectiveness (η_{ad}) of single row film cooling on the pressure side of a turbine guide vane, under the cascade representative conditions. The mainstream inlet Reynolds number is set as 4×10^5. A single row of film cooling holes are located at 30% axial chord length at the pressure side. Three shallow trenches with different widths (w/d=2, 2.5 and 3) but have the same depth (0.75d) and a cylindrical film cooling hole are compared in the current study at M= 2.0. The results show that the shallow trenched-hole on the pressure side significantly suppresses lift-off of the secondary flow, and effectively improves the spanwise uniformity of film coverage. Therefore, compared with the conventional cylindrical hole (CH), the shallow trenched-hole shows a significant enhanced film cooling performance at M=2.0. By comparisons, it is confirmed that the shallow trenches with w/d=2 and 2.5 are more pronounced than the shallow trench with w/d=3. While the shallow trench with w/d=2.5 takes a little advantage over the w/d=2.

Keywords: turbine blade pressure side; film cooling; film cooling effectiveness; shallow trenched-hole; numerical simulation.

1. Introduction

The cooling technology of hot end components is one of the key technologies supporting the development of high-performance aeroengines. In order to solve the outstanding contradiction between the increasing heat load due to the increase of thrust-to-weight ratio and the limitation of the cooling air, more efficient enhanced cooling technology of hot end components must be developed [1].

Film cooling is one of the effective thermal protection measures of hot end components. In practical applications, due to structural constraints, discrete film cooling hole injection is often used. A large number of studies show that the coherent interaction between the jet and the main stream induces a kidney shaped vortex pair, which brings damage to the cooling film coverage. Therefore, suppressing the formation of the key vortex pair is a key mechanism of improving the performance of discrete hole film cooling [2-4]. In this regard, researchers at home and abroad have carried out a lot of research and successively developed such enhanced film cooling structures as shaped holes [5-8], upstream ramp [9-12] and shallow trenched-hole [13-16]. Relatively speaking, it is easier to process the shallow trenched-

Content from this work may be used under the terms of the Creative Commons Attribution 3.0 licence. Any further distribution of this work must maintain attribution to the author(s) and the title of the work, journal citation and DOI.
Published under licence by IOP Publishing Ltd

hole, since it can be prepared during the process of thermal barrier coatings. Therefore, the shallow trenched-hole is considered to be an enhanced film cooling structure with good thermal mechanical properties [17].

The shallow trenched-hole cooling structure consists of a shallow trench and a row of discrete holes embedded in it. Research shows that by embedding the discrete holes in the shallow trench, the jet penetration is reduced due to the sudden expansion of the area. Besides, it forms a spanwise connection in the trench cavity, which is conducive to restraining the kidney vortex pair induced by the interaction between the jet and the mainstream. Waye and Bogard [18] conducted an experimental study on the structure of shallow trenched-hole cooling structure, and pointed out that the vertical structure of the trailing edge is more conducive to the formation of the spread of jet. Zhang Dahai et al. [19] compared and analyzed the ηad of wide trench, narrow trench and inclined trench through numerical research, and pointed out that the inclined trench has more significant improvement on the ηad under small blowing ratio, while the narrow trench has more advantages under large blowing ratio. Harrison and Bogard [20], Lu et al. [21], Dai Ping and Lin Feng [22] studied the effects of trench depth and width numerically and experimentally, respectively. Their results revealed a complex dependence with the blowing ratio. Fan Fangsu et al. [23] numerically investigated the coherent flow structure of the trenched-hole (TH) film cooling on a flat plate by applying LES method. They pointed that the low frequency small-scale vortex dominated the flow field under high blowing ratios. In recent years, researchers have proposed several transverse trench types such as W-shaped or sinusoidal trenches [22-26] to better guide the spanwise expansion of the film jet into adjacent film holes in the trenches.

At present, studies on the shallow trenched-hole film cooling are mostly performed on flat plates. While in actual turbine cascade channel, film cooling is applied in a highly complex condition influenced by multiple factors such as surface curvature, pressure gradient [27-29]. Therefore, the application shallow trenched-hole cooling structure in actual turbine blades needs further investigation. In this study, the numerical simulations of single-row shallow trenched-hole film cooling are performed under a real turbine cascade condition, in order to reveal the influence of the trench on the η_{ad} on the pressure side of the blade.

2. Computational Procedures

2.1. Computational domain

The adopted turbine guide vane is taken from literature [30], as shown in Figure 1 (a). The blade pitch (*P*) is 53.6mm, the blade chord length (*C*) is 74.4mm, the axial chord length (C_x) is 42.3mm, the inlet angle (β) is 90°, and the installation angle (α) is 35.7°. The single row of film cooling holes are located at 30% axial chord length on the pressure side, as shown in Figure 1 (b). The diameter (*d*) of the film hole is 0.6 mm, the angle (θ) between the film hole axis and the normal direction of the local blade profile is 60°, and the pitch between two neighboring injection holes is 3*d*. The injection holes are buried in a transverse shallow trench, as shown in Figure 2 (c). The trench depth is 0.75 *d*, and the trench width *w/d* are taken as 2, 2.5, and 3 respectively.

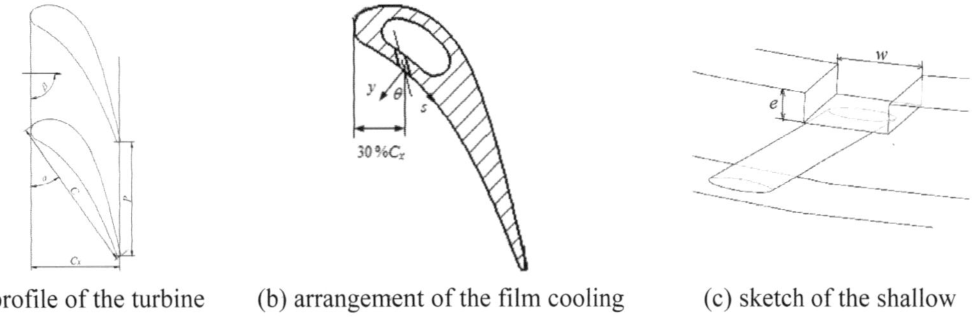

(a) profile of the turbine guide vane (b) arrangement of the film cooling holes on pressure side (c) sketch of the shallow trench

Fig. 1 Schematic of the blade contour and shallow trenched-hole

Based on the periodic characteristics of the cascade, only one cascade pitch and 3 hole-to-hole pitch are adopted in the current simulation. Accordingly, the boundary conditions are defined as follows:

The mainstream inlet is treated as mass flow inlet, with the Reynolds number of $Re_\infty=4\times10^5$ and the total temperature of 2100K. The cooling air is fed from both sides of the plenum, as shown in Figure 2. The inlet of the cooling air is a mass flow inlet, with the total temperature of 900K. The mass flow of the cooling air is determined by the blowing ratio (M). The outlet of the cascade is treated as a pressure outlet with the pressure of 1.3MPa. Symmetrical conditions are applied to both sides of cascade passage. While periodic conditions are applied to the upper and lower parts of the cascade. The rest walls of the computational domain are set as insulated no-slip wall.

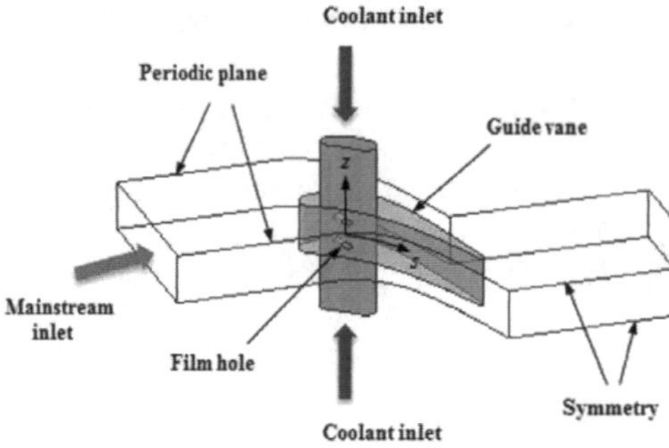

Fig. 2 Schematic of computational zone

2.2. Computational Methodology

In this paper, the commercial software ANSYS Fluent CFD segregated implicit solver is used to solve the time-averaged N-S equation (RANS) of steady-state compressible flow [31]. Both the mainstream and the cooling air are treated as ideal gas. The dependence of its thermal conductivity and viscosity coefficient on temperature is treated by empirical formula (kinetic theory and Sutherland fomula).

The commercial software GAMBIT was used for mesh generation. 2.4 million structured hexahedron cells are used in the computational domain. Figure 3 shows the grids of the domain.

Many previous studies have proved that the Realizable k- ε turbulence model had good accuracy in predicting film cooling performance [33-35]. Therefore, the Realizable k- ε Turbulence model with the enhanced wall treatment is adopted for the turbulence closure in the current simulation.

(a) local grids at film-hole centerline plane (b) local grids in the vicinity of trenched film hole

Fig. 3 Grids of the computational domain

3. Results and Discussions

Figure 4 shows the Mach number distribution of the turbine guide vane without film cooling. It reveals that the η_{ad} on the pressure side is exposed to the mainstream environment where the flow accelerates slowly.

Figure 5 shows the distribution of η_{ad} on the pressure side under M=2.0. It can be seen that the η_{ad} downstream of the cylindrical hole deteriorates significantly at such high blowing ratio. Jet from the CH penetrates deeply into the mainstream, inducing a large-scale kidney vortex pair, as shown in Figure 6 (a). While as shown in Fig. 6 (b), although there also forms a kidney vortex pair downstream of the shallow trench, its scale is significantly smaller than that of the conventional CH. Since the cooling air firstly flows into the shallow trench to form a good lateral expansion before it encountering with the mainstream. A better film coverage is realized in the zone between adjacent film holes. Besides, the intrusion of the high-temperature main flow to the lower part of the film is effectively restrained. As a result, significant improvement in the film cooling performance is presented in Fig. 5 (b), (c) and (d). Comparing Fig. 5 (b), (c) and (d), it can be seen that the film cooling performance of the trench with w/d=2.5 is the best.

Figure 7 shows the distribution of the $\eta_{ad,l\text{-}av}$ on the pressure side. It is clear that comparing with the conventional CH, the shallow trenched-hole significantly improves the downstream film cooling performance. Besides, the $\eta_{ad,l\text{-}av}$ curve shows slight enhancement from w/d=2.0 to w/d=2.5, while it shows an obvious decline as w/d increased from 2.5 to 3.0.

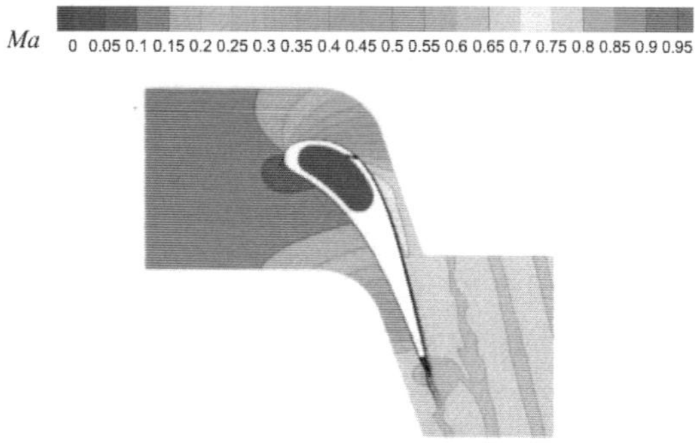

Fig. 4 Mach number distribution

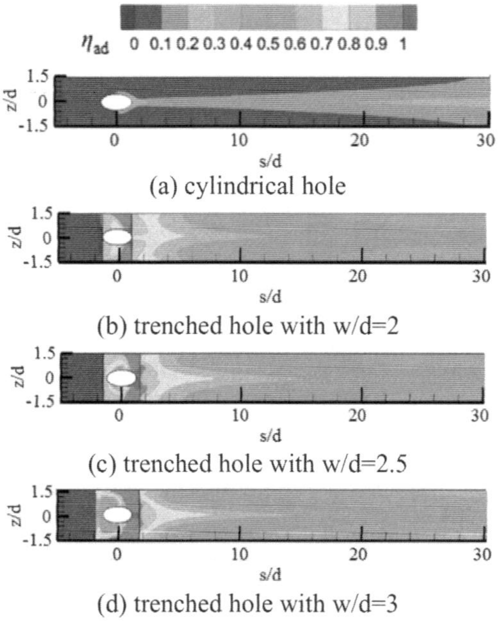

(a) cylindrical hole

(b) trenched hole with w/d=2

(c) trenched hole with w/d=2.5

(d) trenched hole with w/d=3

Fig. 5 Adiabatic film cooling effectiveness distribution

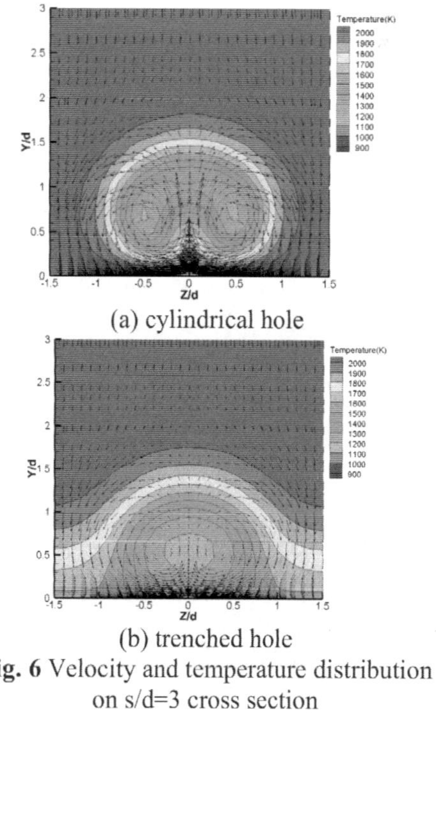

(a) cylindrical hole

(b) trenched hole

Fig. 6 Velocity and temperature distribution on s/d=3 cross section

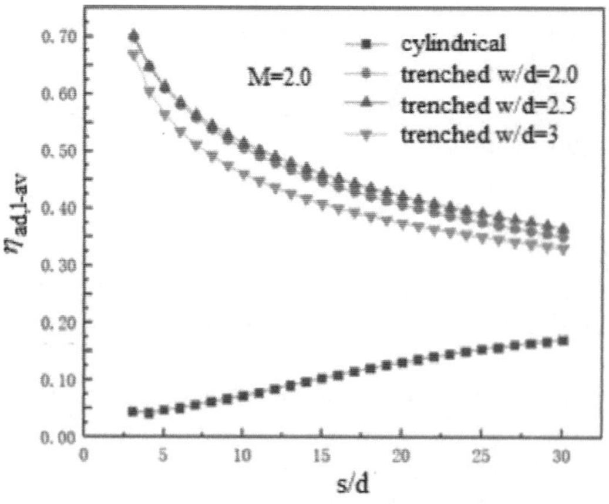

Fig. 7 Laterally-averaged adiabatic film cooling effectiveness at pressure side

4. Conclusions

Numerical simulations are performed to investigate the effects of shallow trenches on the η_{ad} of single row film cooling on the pressure side of a turbine guide vane, under the mainstream inlet Reynolds number of 4×10^5 and blowing ratio of 2.0. The film cooling holes are located at 30% axial chord length at the pressure side. Three trenches with *w/d*=2, 2.5 and 3 and a CH are compared in the current study. The main results are as follows:

(1) The shallow trenched-hole significantly improves the η_{ad} when comparing with the conventional cylindrical hole.

(2) The η_{ad} shows slight enhancement from *w/d*=2.0 to *w/d*=2.5, while it shows an obvious decline as *w/d* increased from 2.5 to 3.0.

References

[1] Bunker R S. Gas turbine heat transfer: ten remaining hot gas path challenges[J]. ASME Journal of Turbomachinery, 2007, 129: 193-210.

[2] Andreopoulos J, Rodi W. Experimental investigation of jets in a crossflow[J]. Journal of Fluid Mechanics, 1984, 138: 93-127.

[3] Haven B A, Kurosaka M. Improved jet coverage through vortex cancellation[J]. AIAA Journal, 1996, 34(11): 2443-2444.

[4] Haven B A, Kurosaka M. Kidney and anti-kidney vortices in crossflow jets[J]. ASME Journal of Fluid Mechanics, 1997, 352: 27-64.

[5] Bunker R. S. A review of turbine shaped film cooling technology[J]. ASME Journal of Heat Transfer, 2005, 127: 441-453.

[6] ZHU Yan-xin, TAN Xiao-ming, GUO Wen, et al. Numerical simulation on effects of different film cooling holes on plate[J]. Journal of Propulsion Technology, 2013, 34(4) : 499-505.

[7] WU A-sai, ZHU Hui-ren, LIU Cun-liang, et al. Visualization experiment on film distribution characteristics of shaped film holes based on thermochromic liquid crystal[J]. Journal of Aerospace Power, 2015, 30(2): 376-383.

[8] ZHU Xing-dan, ZHANG Jing-zhou, TAN Xiao-ming. Effects of rotation on shaped-hole film cooling on a turbine blade[J]. Journal of Propulsion Technology, 2016, 37(9) : 1713-1719.

[9] Na S, Shih T I-P. Increasing adiabatic film-cooling effectiveness by using an upstream ramp[J]. ASME Journal of Heat Transfer, 2007, 129: 464-471.

[10] HE Li-ming, JIANG Yong-jian, KANG Qiang, et al. Numerical investigation on improving film cooling effectiveness with an upstream ramp[J]. Journal of Propulsion Technology, 2009, 30(1) : 9-13.

[11] Zhou W, Hu H. Improvements of film cooling effectiveness by using Barchan dune shaped ramps[J]. International Journal of Heat and Mass Transfer, 2016, 103: 443-456.
[12] Zheng D R, Wang X J, Zhang F, et al. Numerical investigation on the effects of the divided steps on film cooling performance[J]. Applied Thermal Engineering, 2017, 124: 652-662.
[13] Bunker R S. Film cooling effectiveness due to discrete holes within a shallow trench[R]. ASME Paper GT-2002-30178, 2002.
[14] Harrison K L, Bogard D G. CFD Prediction of flim cooling adiabatic effectiveness for cylindrical holes embedded in narrow and wide shallow trenches[R]. ASME Paper GT2007-28005, 2007.
[15] CHEN Xin, HE Li-ming, YU Jin-lu, et al. Experiments on improving film cooling effectiveness with transverse slots[J]. Journal of Aerospace Power, 2010, 25(2) : 291-294.
[16] FENG Hong-ke, WANG Chun-hua, FAN Fang-su, et al. Optimization research of film cooling structures with cratered cylindrical hole[J]. Journal of Propulsion Technology, 2019, 40(5) : 1099-1106.
[17] Davidson F T, KistenMacher D A, Bogard D G. Film cooling with a thermal barrier coating: round holes, craters, and trenches[J]. ASME Journal of Turbomachinery, 2014, 136: 041007.
[18] Waye S K, Bogard D G. High resolution film cooling effectiveness measurements of axial holes embedded in a shallow trench with various trench configurations [J]. Journal of Turbomachinery, 2007, 129(2) : 294-302.
[19] ZHANG Da-hai, CHEN Qiu-yang, ZENG Min, et al. Numerical investigation on the effect of different shallow trench configurations on film cooling effectiveness[J]. Journal of Aerospace Power, 2008, 23(4) : 611-616.
[20] Harrison K L, Bogard D G. CFD Prediction of flim cooling adiabatic effectiveness for cylindrical holes embedded in narrow and wide shallow trenches[R]. ASME Paper GT2007-28005, 2007.
[21] Lu Y P, Dhungel A, Ekkad S V, et al. Effect of trench width and depth on film cooling from cylindrical holes embedded in trenches[J]. ASME Journal of Turbomachinery, 2009, 131: 011003-1-13.
[22] DAI Ping, LIN Feng. Numerical investigation on the influence of transverse slot configurations on film cooling effect[J]. Journal of Propulsion Technology, 2011, 32(2): 253-260.
[23] FAN Fang-su, WANG Chun-hua, FENG Hong-ke, et al. Large eddy simulation of film cooling from a shallow trench-hole[J]. Journal of Propulsion Technology, 2020, 10.13675/j.cnki.tjjs. 190244.
[24] Oguntade H I, Andrews G E, Burns A D, et al. Improved trench film cooling with shaped trench outlets[J]. ASME Journal of Turbomachinery, 2013, 135: 021009.
[25] LI Guang-chao, CHEN Yu-kai, LIU Yong-quan, et al. Mechanism on increasing film cooling effectiveness by W shape slots[J]. Journal of Propulsion Technology, 2016, 37: 520-526.
[26] Wei J S, Zhu H R, Liu C L, et al. Experimental study on the film cooling characteristics of the cylindrical holes embedded in sine-wave shaped trench[R]. ASME Paper GT2016-56856, 2016.
[27] Gao Z H, Narzary D P, Han J C. Film cooling on a gas turbine blade pressure side or suction side with axial shaped holes[J]. International Journal of Heat and Mass Transfer, 2008, 51: 2139-2152.
[28] Liu C L, Zhu H R, Bai J T, et al. Film cooling performance of converging slot-hole rows on a gas turbine blade[J]. International Journal of Heat and Mass Transfer, 2010, 53: 5232-5241.
[29] Zhu X D, Zhang J Z, Tan X M. Numerical assessment of round-to-slot film cooling performances on a turbine blade under engine representative conditions[J]. International Communications in Heat and Mass Transfer, 2019, 100: 98-110
[30] Yao Y, Zhang J Z, Tan X M. Numerical study of film cooling from converging slot-hole on a gas turbine blade suction side[J]. International Communications in Heat and Mass Transfer, 2014, 52: 61-72.
[31] ANSYS Fluent 14.0 User's Guide[M]. ANSYS Inc, Canonsburg, PA, 2012.
[32] Celik I B, Ghia U, Roache P J, et al. Procedure for estimation and reporting of uncertainty due to

discretization in CFD applications[J]. ASME Journal of Fluids Engineering, 2008, 130: 078001.

[33] Harrison K, Bogard D. Comparison of RANS turbulence models for prediction of film cooling performance[R]. ASME Paper GT2008-50366, 2008.

[34] Silieti M, Kassab A J, Divo E. Film cooling effectiveness: comparison of adiabatic and conjugate heat transfer CFD models[J]. International Journal of Thermal Science, 2009, 48: 2237-2248.

[35] Ely M J, Jubran B A. A numerical evaluation on the effect of sister holes on film cooling effectiveness and the surrounding flow field[J]. Heat and Mass Transfer, 2009, 45: 1435-1446.

Numerical Evaluation on Effects of Upstream Sand-Dune-Shaped Ramp and Backward Coolant Injection on Heat Transfer

Yanni Deng [1], Liang Yu [1], Ying Huang [1, *], Shengchang Zhang [2]

[1] Shanghai Aircraft Design and Research Institute, Shanghai, China
[2] College of Energy and Power Engineering, Nanjing University of Aeronautics and Astronautics, Nanjing, China

*Corresponding author email: huangying011@nuaa.edu.cn

Abstract. Numerical simulations are performed for the forward and backward injection film cooling with the presence of upstream ramp under three typical blowing ratios of 0.5, 1.0 and 1.5. Totally nine film cooling configurations are taken into considerations in the current study, where two ramps (straight-wedge-shaped ramp (SWR) and sand-dune-shaped ramp (SDR)), two coolant injection modes (forward injection (FW) and backward injection (BW)), and two film-hole shapes (cylindrical hole (CH) and fan-shaped hole (FH)) are involved. From the current study, the effects of coolant injection scheme and upstream ramp on the film cooling performance are evaluated in the viewing of the heat transfer. Based on the numerical results, it is demonstrated that a wide high net heat flux reduction (NHFR) zone is created in the forward injection cases at M(the blowing ratio)=0.5, but it is shrunk or disappeared rapidly at M=1.5, except the FH-FW-SDR case. While for the backward injection case, two high heat flux reduction zones are generated around the hole exit where the separate bubbles locate, and it remains even at M=1.5. Among the current cases, the FH-FW-SDR is demonstrated to be the best one for achieving the highest NHFR under the all three blowing ratios. With regard to the coolant injection scheme, its influence on heat transfer coefficient is more complicated, tightly associated with the blowing ratio, film-hole shape and the upstream ramp.

Keywords: film cooling; forward and backward injections; cylindrical and shaped holes; sand-dune-shaped ramp; heat transfer and aerodynamic loss.

1. Introduction

In the advancement of high thrust-to-weight ratio aero-engines, the thermal protection of hot section components is a critical challenge such that the working material can bear higher and higher temperature of hot gas without significant deterioration by using more and more limited amount of coolant for cooling purpose [1]. To meet the development requirement of the gas turbine engines, developing highly-efficient cooling scheme is one of the key technological routines.

It is well known that the film cooling plays a dominant role in the thermal protection of turbine-

Content from this work may be used under the terms of the Creative Commons Attribution 3.0 licence. Any further distribution of this work must maintain attribution to the author(s) and the title of the work, journal citation and DOI.
Published under licence by IOP Publishing Ltd

engine hot section components [2]. The upstream ramp is also an effective flow-control means for the jet-in-cross-flow. Barigozzi et al. [3, 4] experimentally studied the effect of the upstream ramp on the aerodynamic losses and adiabatic film cooling effectiveness on a flat plate model, under the blowing ratios ranging between 0.3 and 1.0. It was indicated that there were lager film-hole discharge coefficients in the presence of upstream ramp. But the aerodynamic losses were also prompted as a consequence of the separated flow. Among various upstream ramps, of particular interest was the novel sand-dune-shaped ramp proposed by Zhou and Hu [5, 6], which was a favourable aerodynamic contour to the oncoming airflow. By using the sand-dune-shaped ramp, it was found that the film cooling effectiveness over the test plate would be enhanced significantly.

Over the past decades, film cooling with forward coolant injection was the main focus. Recently, a research group from Lamar University inspired the researches on film cooling with backward coolant injection [7-10]. These studies provided a clear understanding on the backward injection film cooling. In comparison with the forward coolant injection, the backward coolant injection altered the mutual interaction between discrete coolant jets and primary flow, leading to a better jet spreading in the lateral direction but a worse jet spreading along the streamwise direction on a whole.

Both the upstream ramp and the backward coolant injection are demonstrated to be the effective means for controlling the mutual interaction between primary flow and coolant jet. However, some issues remain further investigations, such as the individual effect of contoured upstream ramps, dual effects of upstream ramp/coolant injection and upstream ramp/shaped hole, as well as and the combined effects of upstream ramp, coolant injection and shaped film cooling hole, etc. For addressing more insights into the above issues, numerical simulations are carried out in the current study. Two ramps (SWR and SDR), two coolant injection modes (forward injection and backward injection), and two film-hole shapes (cylindrical hole and fan-shaped hole) are taken into considerations, under three typical M (M=0.5, 1.0 and 1.5). From the current study, the effects of the coolant injection on the film cooling performances of cylindrical and fan-shaped film cooling holes in the situations with or without upstream ramps, as well as the effects of the SDR on the film cooling performances in the situations with forward or backward injections, are analyzed in details. In this study, the heat transfer coefficient and the NHFR are involved.

2. Numerical Modelling

2.1. Physical model and computational domain

Single-row film cooling on a flat plate was concerned in the current study. As shown in Figure 1 and Figure 2, the detailed schematic of computational domain and film cooling configurations has been described in our previous study [14]. The mail geometric parameters of SDR were illustrated in Table 1. For the purpose of comparison, a simple straight-wedge-shaped ramp (SWR) was also adopted, as seen in Figure 2(c), which had the same ramp height as the SDR. Its length was the same as the front-body length of SDR. The distance between the SWR and the film-hole center (L_2) was selected as the sum of Lm and L1 in the SDR situation.

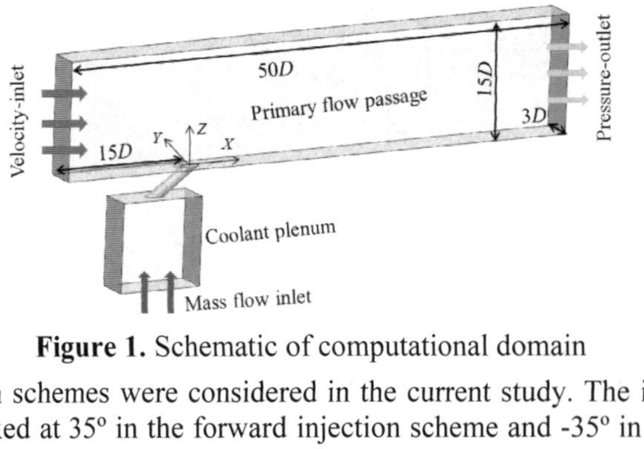

Figure 1. Schematic of computational domain

Two coolant injection schemes were considered in the current study. The inclination angle (α) of film cooling hole was fixed at 35° in the forward injection scheme and -35° in the backward injection scheme respectively, in relative to the primary flow direction. Also, two types of film cooling holes were considered. Both the cylindrical hole and the fan-shaped hole had the same height (H) of 9.2mm or the same length-to-diameter ratio (L/D) of 4.0. For the fan-shaped hole, its diffused portion was orientated at 1.5D downstream of the film-hole inlet and the laterally diffused angle (γ) was set as 7°, as seen in Figure 2(d).

Totally, nine cases were involved in the current study, as summarized in Table 2.

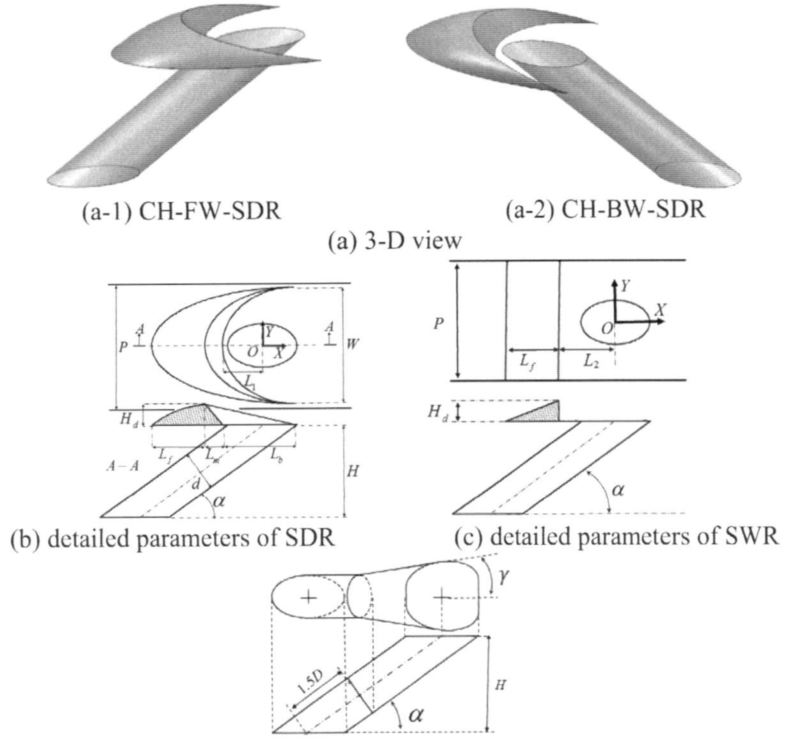

Figure 2. Schematic of film cooling configurations

Table 1. Geometric parameters of SDR

Parameter	Unit/mm	Parameter	Unit/mm
H_d	2.0	L_m	1.8
W	11.4	L_b	7.2
L_f	5.4	L_1	4.0

Table 2. Film cooling structures and short descriptions

Case	Hole shape	Injection mode	Ramp shape	Descriptions
No.1	Cylindrical	Forward	-	CH-FW
No.2	Cylindrical	Backward	-	CH-BW
No.3	Fan-shaped	Forward	-	FH-FW
No.4	Fan-shaped	Backward	-	FH-BW
No.5	Cylindrical	Forward	Straight-wedge	CH-FW-SWR
No.6	Cylindrical	Forward	Sand-dune	CH-FW-SDR
No.7	Cylindrical	Backward	Sand-dune	CH-BW-SDR
No.8	Fan-shaped	Forward	Sand-dune	FH-FW-SDR
No.9	Fan-shaped	Backward	Sand-dune	FH-BW-SDR

2.2. Boundary conditions and Computational scheme

The heat transfer to a film cooled surface is defined as Eq.4, as reported by Goldstein [14]:

$$q = h_{\mathrm{f}}(T_{\mathrm{aw}} - T_{\mathrm{w}}) \tag{1}$$

where T_{w}, h_{f} and q denote the wall temperature, heat transfer coefficient and heat flux with film cooling respectively. Based on the linearity and homogeneity of the simplified boundary differential equations (as reported by Schulz et al. [12]), the heat transfer is also defined as:

$$q = h\ (T_{\infty} - T_{\mathrm{w}}) \tag{2}$$

Dividing Eq.5 with Eq.4, it follows that

$$\frac{h}{h_{\mathrm{f}}} = \frac{T_{\mathrm{aw}} - T_{\mathrm{w}}}{T_{\infty} - T_{\mathrm{w}}} \tag{3}$$

Then, using a dimensionless temperature

$$\Theta_{\mathrm{w}} = \frac{T_{\infty} - T_{c}}{T_{\infty} - T_{\mathrm{w}}} \tag{4}$$

The *NHFR* is presented by Sen et al. [13], which denotes the ratio of reduction in heat transfer to the surface with film cooling to heat transfer without film cooling.

$$NHFR = \frac{q_0 - q}{q_0} = 1 - \frac{h_{\mathrm{f}}}{h_0}(1 - \eta_{\mathrm{ad}}\Theta_{\mathrm{w}}) \tag{5}$$

where q_0 is the heat flux without film cooling. Based on the actual gas turbine blade application, a typical value of 1.6 for Θ_{w} is suggested by Sen et al. After a simple monotonicity analysis for Eq.9, it is easy to found that the *NHFR* increases with increasing of h_{f}/h_0 when $\eta_{\mathrm{ad}}>1/\Theta_{\mathrm{w}}$, but when $\eta_{\mathrm{ad}}<1/\Theta_{\mathrm{w}}$, it decreases with increasing of h_{f}/h_0.

The detailed boundary conditions, computational scheme and validation has been described in our previous study [14].

3. Results and Discussions

3.1. Effect of coolant injection scheme in situations without upstream ramp

Figure 3 presents the local *NHFR* distributions for the forward and backward injection cylindrical hole cases. For the CH-FW case at M=0.5 , a tapered zone with $NHFR>1$ is observed along the hole centerline, which means that the film cooling heat flux (q) is less than zero (q_0-q>q_0 or $q<0$). In this zone, the film layer plays a dual positive role (DPR) in film cooling: isolating the cooled surface from being eroded by the hot primary flow and taking away the heat flux from the target surface. Out of the DPR-zone, a wider zone with the *NHFR* within the range of 0 to 1 ($0< q< q_0$) appears, in which the film layer only plays a single positive role (SPR) in film cooling, i.e. isolation. Except the two zones, there are two small zones, where the *NHFR* is less than zero ($q> q_0$), appear far downstream the hole exit, which increases the heat flux from the mainstream to the wall, playing actually negative role (NR) in film cooling. For simplicity, the three typical zones are denoted as DPR-zone, SPR-zone and NR-zone respectively. Under a moderate to high blowing ratio, the DPR-zone is completely disappeared and the SPR-zone is shrunk to a very narrow strip for the CH-FW case, but the NR-zone is gradually widened. Therefore, the coolant jet is not only wasted, but also brings negative effects on film cooling. For the CH-BW case, the DPR-zone is mainly located around the hole exit where the separate bubble appears, and it is also the region where the η_{ad} and the heat transfer coefficient are very high. Differing from the CH-FW case, the DPR-zone does not disappear and the SPR-zone is widened with increasing of M, but a "C-shaped" strip with negaive *NHFR* appears, surrounding the film hole exit, producing adverse effect on film cooling.

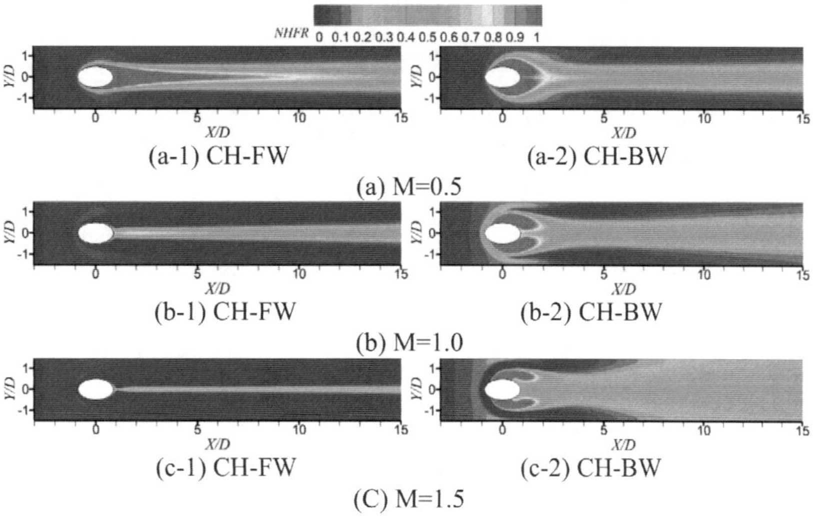

Figure 3. Local NHFR distributions for CH-FW and CH-BW

Figure 4 presents the local *NHFR* distributions for the forward and backward injection fan-shaped hole cases. For the FH-FW case, the DPR-zone is found to be wider than the CH-FW case (as seen in Figure 3), and it remains even the blowing ratio is as high as 1.5, but the detrimental NR-zone also appears when increases the blowing ratio. For the FH-BW case, the DPR-zone and SPR-zone are both wider than the CH-BW case under all *M* (as seen in Figure 3). Under a low to moderate *M*, the FH-FW case produces more favorable *NHFR* distribution than the FH-BW case. While under a high *M*, the *NHFR* distribution of the FH-BW case seems to be better.

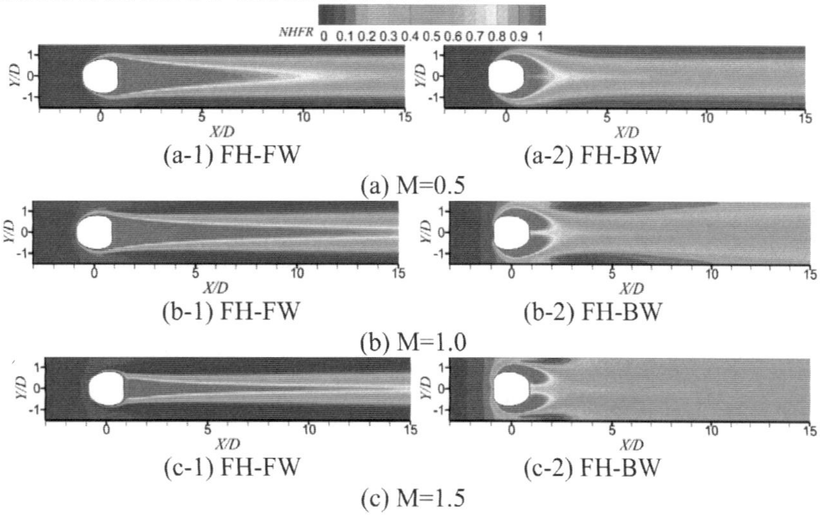

Figure 4. Local NHFR distributions for FH-FW and FH-BW

3.2. Effect of upstream ramp shape on cylindrical-hole film cooling with forward injection

Figure 5 presents the local *NHFR* distributions for the CH-FW-SWR case and the CH-FW-SDR case. Under a low blowing rato of 0.5, both rampes present favorable *NHFR* distributions around and downstream the hole exit, and the SDR creates a larger DPR-zone than the straight-wedge ramp. While under a moderate to high *M*, due to the decrease of η_{ad} and the increase of heat transfer coefficient, two strip-shaped NR-zones are formed beside the SPR-zone, causing the film cooling to be useless.

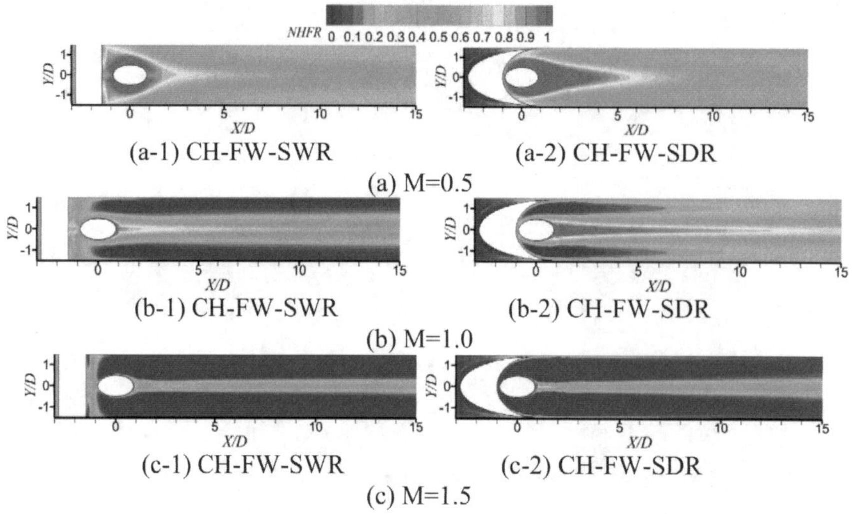

Figure 5. Local NHFR distributions for CH-FW-SWR and CH-FW-SDR

3.3. *Effect of coolant injection scheme in situations with upstream ramp*

Figure 6 presents the local *NHFR* distributions for the FH-FW-SDR case. It is obvious that the FH-FW-SDR performs the best *NHFR* distribution over the other eight film cooling confirgurations involved in this paper under all blowing ratios. Due to the favorable distributions of η_{ad} or film-layer coverage, a wide DPR-zone is created to protect the cooled surface.

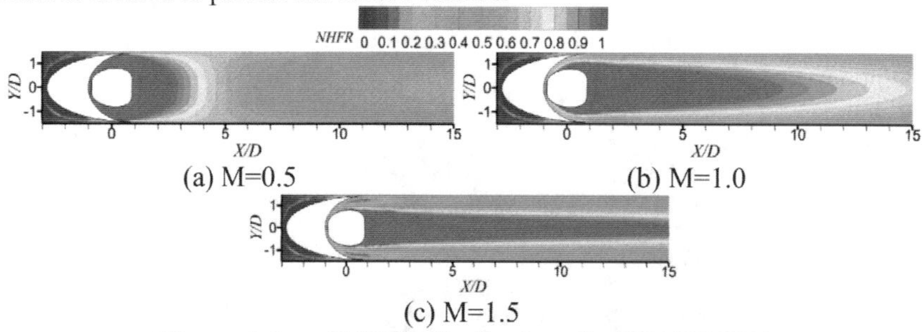

Figure 6. Local NHFR distributions for FH-FW-SDR

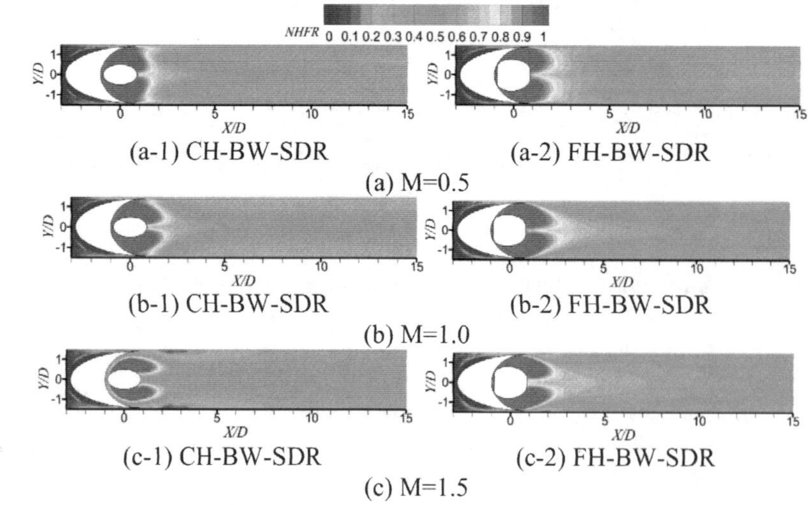

Figure 7. Local NHFR distributions for CH-BW-SDR and FH-BW-SDR

Figure 7 presents the local NHFR distributions for the CH-BW-SDR case and the FH-BW-SDR case. As the separate wake vortices dominating flow pattern is not fundamentally altered by using an upstream SDR or shaped hole, the NHFR distributions for both cases are quite similar, forming two DPR-zones around the hole exit and a wide SPR-zone downstream the hole exit.

3.4. Comprehensive comparison of the laterally-averaged heat transfer coefficient and NHFR

Figure 8 shows a comprehensive comparison of the laterally-averaged heat transfer coefficient ($E_{g,av}$) along streamwise direction. At M=0.5, for the cases without upstream ramps, the backward injection produces relatively higher $E_{g,av}$ than the forward injection; while for the cases with an upstream SDR, the forward injection produces higher $E_{g,av}$ than the backward injection when X/D>3. Among the nine cases, the CH-FW-SWR shows the lowest $E_{g,av}$. At M=1.0, the heat transfer coefficient of the FH-FW-SDR decreases slightly, making it present the lowest heat transfer coefficient when X/D>3. While for the CH-FW-SDR case, its heat transfer coefficient increases rapidly and exceeds the other eight cases when X/D<6. At M=1.5, except the FH-FW-SDR case, all other forward injection cases show relatively higher $E_{g,av}$ over the corresponding cases with backward injection at far downstream the hole exit, which is confirmed to be harmful to cooled surface, because the η_{ad} in this region is commonly less than $1/\Theta_w$, and the increase of heat transfer coefficient will undoubtedly increase the heat flux to the wall. For the FH-FW-SDR case, it maintains a relatively low heat transfer coefficient distribution at far downstream of the hole exit, which is favorable to the film cooling.

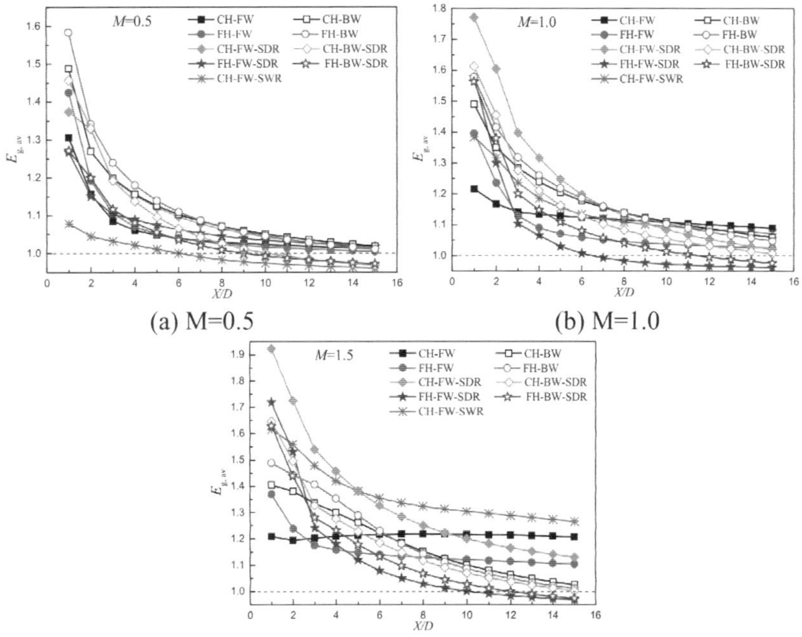

(a) M=0.5 (b) M=1.0

(c) M=1.5

Figure 8. Laterally-averaged heat transfer coefficient along streamwise direction

Figure 9 presents a comprehensive comparison of the laterally-averaged *NHFR* along streamwise direction. Under a low blowing ratio of 0.5, it is seen that all forward injection cases produce relatively higher laterally-averaged *NHFR* than the corresponding backward injection cases, and the laterally-averaged *NHFR* of the FH-FW-SDR is the highest among the involved film cooling configurations. At M=1.0, the adverse effects of the forward injection begin to appear in the CH-FW case and the CH-FW-SDR case, where the CH-FW shows the poorest laterally-averaged *NHFR* distribution. At M=1.5, it is obvious that the backward injection cases perform relative higher laterally-averaged *NHFR* than the corresponding forward injection cases except the FH-BW-SDR case. For the CH-FW case, the CH-FW-

SWR case and the CH-FW-SDR case, negative laterally-averaged *NHFR* appears in a large area along the streamwise direction, which means that more heat flux is transmitted to the cooled surface. In general, the FH-FW-SDR is suggested to be the best film cooling configurations among the involved structures.

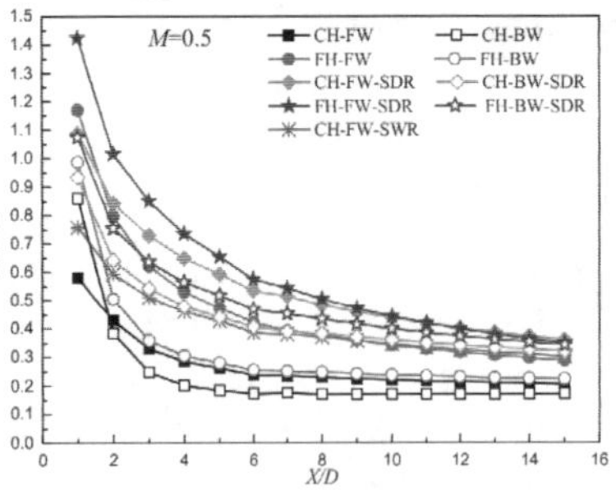

(a) M=0.5

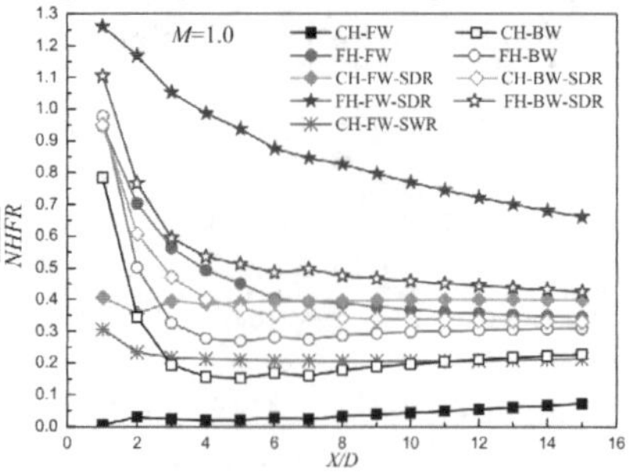

(b) M=1.0

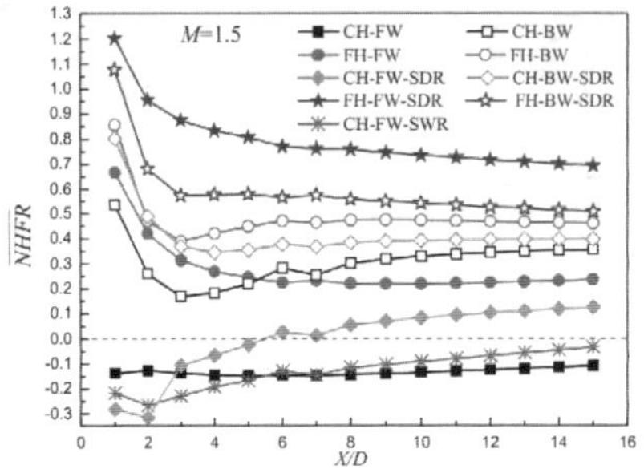

(c) M=1.5

Figure 9. Laterally-averaged NHFR distribution along streamwise direction

For further comparison, the variation of area-averaged *NHFR* with blowing ratios is shown in Figure 10, where the *NHFR* is averaged within $0<X/D<15$ and $-1.5<Y/D<1.5$. It is seen that the area-averaged NHFR decreases with increasing of blowing ratio for forward injection cases except the FH-FW-SDR; while for the backward injection cases, the area-averaged NHFR increases with increasing of *M*. Evidetly, for the FH-FW-SDR case, the area-averaged NHFR increases first and then decreases with the increase of *M*, and it keeps the highest values over the other eight cases for all three blowing ratios.

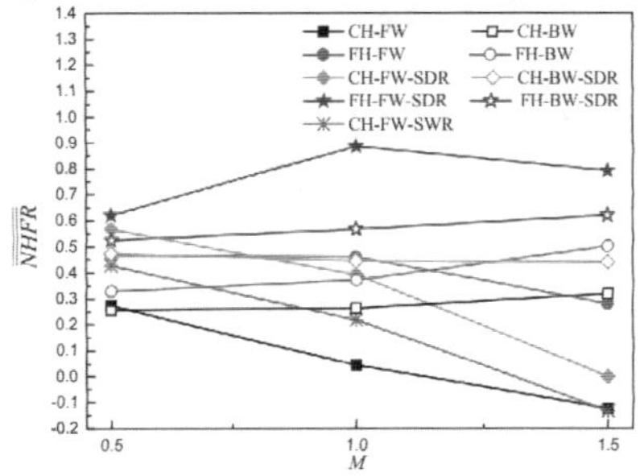

Figure 10. The variation of area-averaged NHFR with blowing ratios

4. Conclusions

Totally nine film cooling configurations are taken into considerations in the current study, where two ramps (SWR and SDR), two coolant injection modes (forward injection and backward injection), and two film-hole shapes (cylindrical hole and fan-shaped hole) are involved. From a series of numerical simulations under three typical *M* (0.5, 1.0 and 1.5), the effects of coolant injection and upstream ramp on the heat transfer on a flat surface are illustrated. The main conclusions are summarized as the follows.

(1) At M=0.5, for the cases without upstream ramps, the backward injection produces relatively higher $E_{g,av}$ than the forward injection; while for the cases with an upstream SDR, the forward injection arises higher heat transfer coefficient than the backward injection when X/D>3. With the blowing increases, the variation of $E_{g,av}$ becomes quite complicated, but it is illustrated that, except the FH-FW-SDR case, all other forward injection cases show relatively higher $E_{g,av}$ over the corresponding cases with backward injection at far downstream the hole exit under M=1.5.

(2) In the viewing of *NHFR*, a wide duel-positive-role zone (DPR-zone) is created in the forward injection cases under M=0.5, but it is shrunk or disappeared rapidly with the M increases to 1.5, except the FH-FW-SDR case. While for the backward injection case, two separate DPR-zones are generated around the hole exit where the separate bubbles locate, and it remains even the blowing ratio is as high as 1.5. In addition, all forward injection cases produce relatively higher laterally-averaged *NHFR* than the corresponding backward injection cases under M=0.5, but under a high M of 1.5, the backward injection is demonstrated to be more favorable, except the FH-BW-SDR case.

(3) Among the current cases, the FH-FW-SDR is demonstrated to be the best one for achieving the highest *NHFR* under the all three M.

References

[1] R.S. Bunker, Gas turbine heat transfer: ten remaining hot gas path challenges, ASME J. Turbomach. 129 (2007) 193-210.
[2] D.G. Bogard, K.A. Thole, Gas turbine film cooling, AIAA J. Propul. Power, 22 (2006) 249-269.
[3] G. Barigozzi, G. Franchini, A. Perdichizzi, The effect of an upstream ramp on cylindrical and fan-shaped hole film cooling: part I-aerodynamic results, ASME Paper GT2007-27077, 2007.
[4] G. Barigozzi, G. Franchini, A. Perdichizzi, The effect of an upstream ramp on cylindrical and fan-shaped hole film cooling: part II-adiabatic effectiveness results, ASME Paper GT2007-27079, 2007.
[5] W. Zhou, H. Hu, Improvements of film cooling effectiveness by using barchan dune shaped ramps, Int. J. Heat Mass Transfer 103 (2016) 443-456.
[6] W. Zhou, H. Hu, A novel sand-dune-inspired design for improved film cooling performance, Int. J. Heat Mass Transfer 110 (2017) 908-920.
[7] S. Shetty, X.C. Li, G. Subbuswamy, Numerical simulation on gas turbine film cooling of curved surface with backward injection, ASME Paper HT2012-59469, 2012.
[8] X.C. Li, G. Subbuswamy, J. Zhou, Performance of gas turbine film cooling with backward injection, Energy Power Eng. 5 (2013) 132-137.
[9] G. Subbuswamy, X.C. Li, K. Gharat, Numerical simulation of backward film cooling with fan-shaped holes, ASME Paper HT2013-17801, 2013.
[10] G. Subbuswamy, X.C. Li, K. Gharat, Numerical study of aerodynamic performance of film cooling with backward injection holes, ASME Paper HT2013-17803, 2013.
[11] Goldstein R J, Film Cooling, Advances in Heat Transfer, 7 (1971) 321-379.
[12] Gritsch M, Schulz A, Wittig S, Heat Transfer Coefficient Measurements of Film-Cooling Holes With Expanded Exits, ASME 1998 International Gas Turbine and Aeroengine Congress and Exhibition. American Society of Mechanical Engineers, 1998.
[13] Sen BB, Schmidt DL, Bogard DG, Film Cooling With Compound Angle Holes: Heat Transfer, ASME J. Turbomach. 118 (1996) 800-806.
[14] Zhang S, Zhang J , Tan X . Improvement on shaped-hole film cooling effectiveness by integrating upstream sand-dune-shaped ramps, Chinese Journal of Aeronautics, 34 (2021) 42 - 55.

Numerical study of different micro-channel structures on fluid flow and heat transfer performance

Zijian Chen*

College of Chemical Engineering, Beijing University of Chemical Technology, Beijing, 100029, China

*Corresponding author's e-mail: zjchen@buct.edu.cn

Abstract. This research mainly proposed three new types of micro-channel heat sink structures. Numerical simulations and simulation calculations were carried out on the heat sink structure. The three groups of inlet flow rates were used as the main variables to analyse the three new types of heat sinks, model flow characteristics and heat transfer performance. In addition, it was compared with the traditional rectangular micro-channel heat sink (RT-MCHS). Based on and compared with RT-MCHS, three new microchannel heat sinks, reverse concave microchannel heat sink (RC-MCHS), co-directional concave microchannel heat sink (CC-MCHS) and convex reverse concave microchannel heat sink (CRC-MCHS), were designed and studied numerically. The comparative analysis of MCHS design performance includes fluid flow characteristics, fluid heat transfer performance and heat transfer efficiency. The effects of inlet velocity on the Darcy friction coefficient, pressure drop, Nusselt number, average outlet temperatures and efficiency factors were studied numerically as inlet velocity from 1 m·s-1 to 5 m·s-1. The results show that the optimal strategy in the new design is the CRC-MCHS, which can hinder the flow of fluid and enable sufficient heat exchange between the liquid and the substrate. In addition, at the optimal inlet flow rate of 3 m·s-1, the outlet water temperature of CC-MCHS is the lowest, indicating the heat transfer performance of the electronic components is the best.

1. Introduction

With the continuous development of microelectronics and micromechanical systems, many electronic products begin to develop in a compact and highly integrated direction, thus causing more heat when the electronic components are working. These microelectronic components devices reduce productivity and life due to high temperatures during use and are easily damaged at high temperatures, and affect the normal operation of the connected work equipment. Therefore, the microchannel heat sink has become a hot spot problem.

In order to solve the cooling problem of high-temperature electronic equipment, in 1981, the concept of the microchannel heat sink (MCHS) was firstly proposed by Tuckerman and Pease [1, 2]. A cheaper microchannel heat sink prototype, which was much cheap than pure silicon thin film, was designed in 1985 by Swift, Migliori and Wheatley [3].

With the development of research on microchannels and the uprising demand for high efficiency electronic cooling systems, the advantages of micro-channel heat sinks have already emerged. The first

Content from this work may be used under the terms of the Creative Commons Attribution 3.0 licence. Any further distribution of this work must maintain attribution to the author(s) and the title of the work, journal citation and DOI.
Published under licence by IOP Publishing Ltd

trial application of micro-channel heat sink on cooling electronic devices was in the 1980s, and the micro-channel heat sink system has some successful application on automotive air conditioning systems at present [4].

Compared to conventional heat exchangers, the micro-channel heat sink has different flow and heat transfer characteristics for differences in structure. Some new phenomena appearing in the microchannel can be attributed to the scale effect. An enhanced fluid compression effect could be attributed to the reduced size [5]. An increase in the drag coefficient could be attributed to the increase in roughness [6]. The area-related forces in the microchannel, including surface force and viscous force, will increase as the ratio of the surface area to volume increases [7], which will enhance the influence of the axial heat conduction [8]. The two-phase flow characteristics of R-134a as a refrigerant in a 0.5 mm channel and the relationship between mass flow and steam quality under four different flow patterns have been confirmed in the study of Revellin [9]. With small Knudsen number and Mach number, Hetsroni [10] analysed the characteristics of the fluid in different shape microchannels with different hydrodynamic diameters based on a continuous model under the conditions of small Knudsen number and Mach number. Morini's [11] research pointed out that in some cases, the experimental data, including the friction factor and Nusselt number, in the microchannel are inconsistent with the traditional theory, but they also seem to be inconsistent. Among the various theories that have been used to explain the abnormal behaviour of the transmission mechanism through the microchannel, the thinness and compressibility effects, the viscous dissipation effect, the electroosmotic effect (EDL), the property change effect, the channel surface condition (mainly refers to relative roughness) and experimental uncertainty are more credible theoretical explanations. The study of Harirchian and Garimella [12], where the effect of size and mass on the fluid boiling state of the microchannel was studied through the high-speed photography channel, showed that different from the flow pattern in smaller microchannels, the flow pattern in microchannels with a width of 400 μm and larger is similar and the heat transfer coefficient is almost unchanged. Wang and Peng's [13] research showed that in the single-phase forced fluid flowing through rectangular cross-section microchannels, the transformation and laminar heat transfer behaviour in the microchannels are very unusual and complicated is significantly affected by the fluid temperature, velocity and size of the microchannel. Wu and Cheng [14] conducted experimental studies on the heat transfer and pressure drop of water in different silicon microchannels. Alfaryjat [15] studied the influence of the geometric parameters of the microchannels on the water flow and heat transfer characteristics in MCHS with various cross-sectional channel shapes. Qu and Mudawar's [16] research showed that the temperature rise of the solid and fluid regions in the rectangular MCHS along the flow direction could be approximately linear, and the highest temperature appears on the heating base surface of the radiator above the channel outlet.

Based on previous research, flow interruption technologies, such as convex and concave, can enhance the heat exchange performance of MCHS at the expense of high pressure drop. However, as far as the author knows, at present, the relationship between the distribution of convex and concave and the thermal and flow performance of the MCHS is still unclear. Therefore, the purpose of this research is to propose a new design that uses the convex-concave structure to enhance the thermal performance of MCHS.

In this study, a total of four kinds of MCHS structure designs will be proposed, and Solidworks software will be used to model the new heat sink structure in three dimensions. Using Fluent software to carry out the conjugate three-dimensional calculation, the flow performance and heat transfer performance of the MCHS with the convex-concave structure will be studied, and the outlet temperature, Nusselt number, friction coefficient and pressure drop will be obtained. Meanwhile, the influence of the groove geometry and flow rate on the microchannel fluid and heat transfer characteristics will be analysed.

2. Method

2.1. 3D model

The microchannel heat sink (MCHS) consists of 10 microchannels. For saving the computational cost, only half of MCHS, which is symmetrical, is adopted in the present simulation. All 4 different microchannel structures are based on the same sink plate, the MCHS plate. The CAD design of the MCHS plate structure is illustrated in Fig.1, and the main parameters are illustrated in Table 1. As shown in Fig.2, the length of microchannels *L2*, the total width of microchannel area *C2* are 10 mm and 1.9 mm, respectively. The depth of microchannels and microchannel's wall, which separates two microchannels, *h* is 0.35 mm. With the same height of microchannels and walls and the top plate sealing, flow in every channel will be independent to another. A chamber under the inlet hole (left in the experiment) is designed to divide flow to every microchannel evenly.

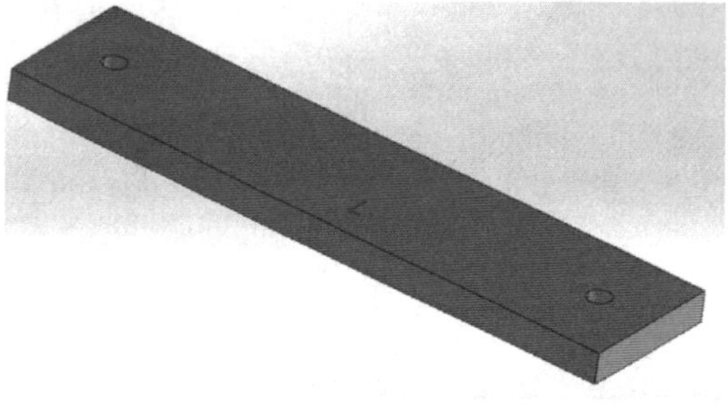

Figure 1. Overview of MCHS

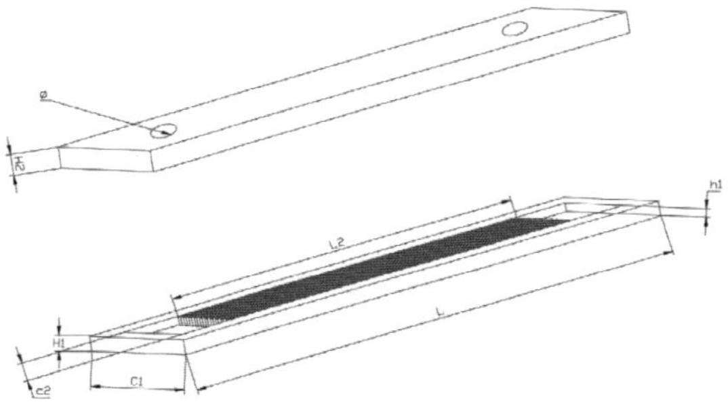

Figure 2. Structure of MCHS plate.

Table 1. Geometrical parameters of MCHS plate

L /mm	L2 /mm	H1 /mm	H2 /mm	h /mm	Φ /mm	C1 /mm	C2 /mm
14	10	0.35	0.5	0.2	0.5	3.1	1.9

First, a rectangular microchannel heat sink (RT-MCHS) was designed as a control group, and its structure was illustrated in Fig.3, where the width of microchannel and wall, L1 and L2, are 0.1 mm and 0.09 mm, respectively. Then 3 different structures were designed based on RT-MCHS: reverse concave microchannel heat sink (RC-MCHS), co-directional concave microchannel heat sink (CC-MCHS) and convex reverse concave microchannel heat sink (CRC-MCHS) were shown in Fig.4, Fig.5 and Fig.6 respectively. The geometrical parameters of these structures are shown in Table 2.

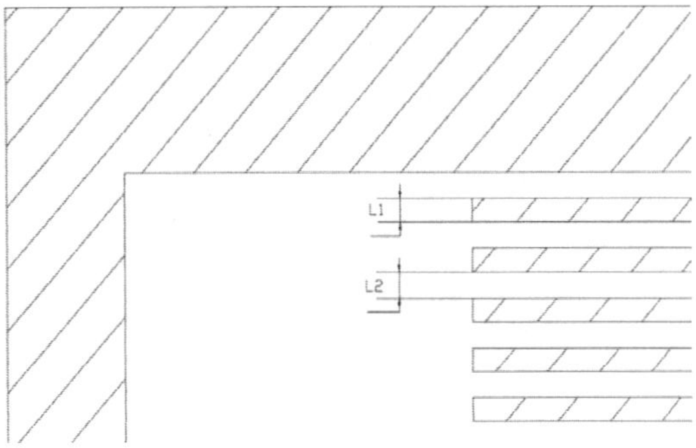

Figure 3. Structure of rectangular microchannel heat sink (RT-MCHS).

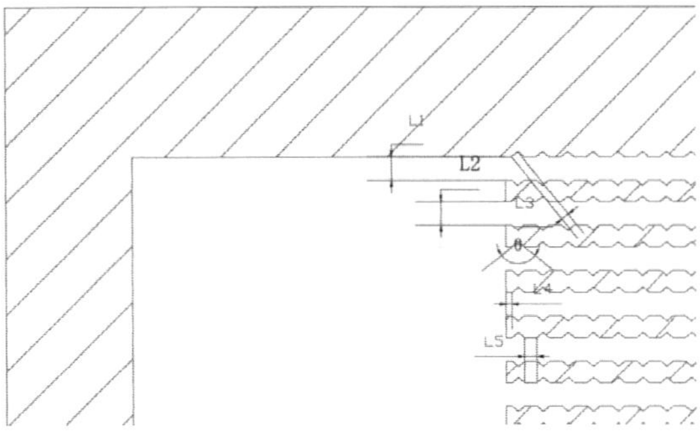

Figure 4. Structure of reverse concave microchannel heat sink (RC-MCHS).

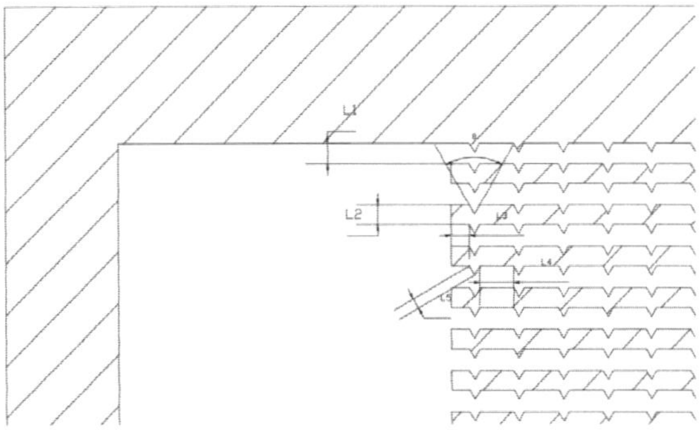

Figure 5. Structure of co-directional concave microchannel heat sink (CC-MCHS).

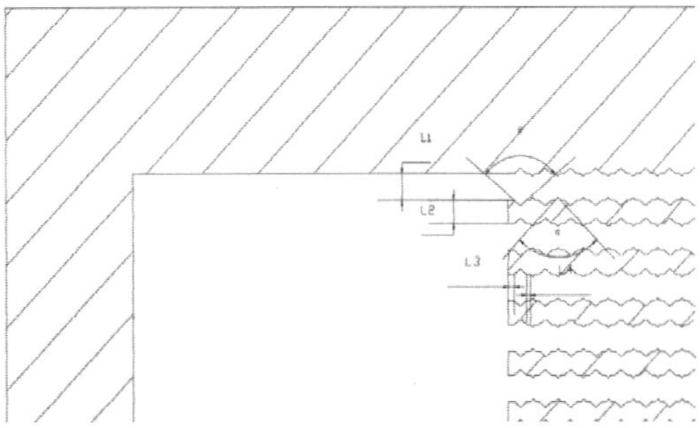

Figure 6. Structure of convex reverse concave microchannel heat sink (CRC-MCHS).

Table 2. Geometrical parameters of 3 microchannel heat sink structures

Structure	Geometrical parameters					
RC-MCHS	L1 /mm	L2 /mm	L3 /mm	L4 /mm	L5 /mm	θ /°
	0.1	0.09	0.05	0.03	0.05	60
CC-MCHS	L1 /mm	L2 /mm	L3 /mm	L4 /mm	L5 /mm	θ /°
	0.1	0.09	0.08	0.15	0.05	30
CRC-MCHS	L1 /mm	L2 /mm	L3 /mm	L4 /mm	θ /°	α /°
	0.1	0.09	0.03	0.02	60	60

2.2. Governing equations

The three-dimensional heat transfer and fluid flow equations are solved by the computational fluid dynamic (CFD) software FLUENT. The following assumptions are set in the CFD process:

(1) Since the Knudsen number (Kn) is less than 10^{-3}, the continuum approach based on the Navier-Stokes equations are used in this research.

(2) The viscous dissipation, radiation heat transfer and gravitational force are neglected.

(3) Thermo-physical properties are set as constant.

(4) State of both fluid flow and heat transfer are set as steady-state.

(5) The flow in all structures is Newtonian, incompressible and turbulence fluid.

As shown below, based on the previous assumptions, the governing equations for mass, momentum, and energy are listed.

Continuity equation:

$$\frac{\partial u}{\partial x}+\frac{\partial v}{\partial y}+\frac{\partial w}{\partial z}=0 \tag{1}$$

where u, v and w are the velocity components, in x, y and z directions, respectively.

Momentum equations:

$$u\frac{\partial u}{\partial x}+v\frac{\partial u}{\partial y}+w\frac{\partial u}{\partial z}=-\frac{1}{\rho_f}\frac{\partial P}{\partial x}+\frac{\mu_f}{\rho_f}\left(\frac{\partial^2 u}{\partial x^2}+\frac{\partial^2 u}{\partial y^2}+\frac{\partial^2 u}{\partial z^2}\right) \tag{2}$$

$$u\frac{\partial v}{\partial x}+v\frac{\partial v}{\partial y}+w\frac{\partial v}{\partial z}=-\frac{1}{\rho_f}\frac{\partial P}{\partial y}+\frac{\mu_f}{\rho_f}\left(\frac{\partial^2 v}{\partial x^2}+\frac{\partial^2 v}{\partial y^2}+\frac{\partial^2 v}{\partial z^2}\right) \tag{3}$$

$$u\frac{\partial w}{\partial x}+v\frac{\partial w}{\partial y}+w\frac{\partial w}{\partial z}=-\frac{1}{\rho_f}\frac{\partial P}{\partial z}+\frac{\mu_f}{\rho_f}\left(\frac{\partial^2 w}{\partial x^2}+\frac{\partial^2 w}{\partial y^2}+\frac{\partial^2 w}{\partial z^2}\right) \tag{4}$$

where ρ_f , μ_f, and p are the density, viscosity pressure of the coolant fluid, respectively.

The energy equation for the coolant:

$$u\frac{\partial T_f}{\partial x}+v\frac{\partial T_f}{\partial y}+w\frac{\partial T_f}{\partial z}=\frac{k_f}{\rho_f C_{pf}}\left(\frac{\partial^2 T_f}{\partial x^2}+\frac{\partial^2 T_f}{\partial y^2}+\frac{\partial^2 T_f}{\partial z^2}\right) \tag{5}$$

where T_f , C_{pf} and k_f are temperatures, specific heat and thermal conductivity of the coolant fluid, respectively.

The energy equation for the solid region:

$$k_s\left(\frac{\partial^2 T_s}{\partial x^2}+\frac{\partial^2 T_s}{\partial y^2}+\frac{\partial^2 T_s}{\partial z^2}\right)=0 \tag{6}$$

where T_s and k_s are solid temperature and thermal conductivity, respectively.

Boundary condition set as follow:

At the inlet of MCHS, the total flow rate of the fluid inlet is inputted as a constant number. The fluid outlet is set as a pressure outlet. Set the room temperature to 293 K, and the initial temperature of the medium fluid to be introduced to 300 K. The surface of MCHS in contact with the electronic components is provided with a uniform and constant heat flow. The heat flow density released by the electronic components to MCHS is 106 W·m-2. The convective heat transfer coefficient between the wall surface of MCHS and the air except for the heat load loading surface is 10-5 W·mm-2·°C. The material of MCHS is aluminium, and the medium introduced is liquid water.

2.3. Grid independence and CFD simulation

In Fluent, four types of microchannel heat sink models are grided. For all four models, the grid quality is set to medium, and the maximum surface size is set to 0.002 mm. The total number of grids and nodes generated for different models are shown in Table.3. After the model meshing is completed, the quality of the grid needs to be checked. If the quality of the grid is 0.7 or above, the quality of the grid is good. If it is lower than 0.7, the result of the grid division will have a significant impact on the calculation, and the model needs to be re-meshed. The. cas and .dat files obtained by FLUENT are processed by the post-processing software TECPLOT to obtain the pressure, temperature and flow rate diagrams of all models.

Table 3. Grid and numbers for 4 MCHS models.

model	grid	node
RT-MCHS	2477187	458845
RC-MCHS	4903493	963048
CC-MCHS	5283847	1023837
CRC-MCHS	6039843	2938474

2.4. Data reduction

The equations below are used to calculate the characteristics of fluid flow and heat transfer in 4 types of MCHS models.

The pressure drop (Δp):

$$\Delta p = \rho_f g h_f \tag{7}$$

where ρ_f represents the density of the fluid, g represents the acceleration of gravity, h_f represents the flow resistance in the channel caused by the influence of the viscosity of the liquid, and its expression is as follows:

$$h_f = f\frac{L}{D_h}\frac{u}{2g} \tag{8}$$

where D_h, L and u represent the hydraulic diameter, length of the microchannel, mean velocity of the fluid in the microchannel,

Darcy friction factor (f):

$$f = \frac{2\Delta P D_h}{\rho_f L u^2} \tag{9}$$

Nusselt number (Nu):

$$Nu = \frac{hD_h}{\lambda_f} \tag{10}$$

where λ_f and h represent the thermal conductivity and convective heat transfer coefficient of the fluid.

The convective heat transfer coefficient (h):

$$h = \frac{qA_b}{A_{tot}(T_w - T_f)} \tag{11}$$

where A_b, A_{tot}, T_w and T_f represent the heating area, heat exchange area, temperature of the heat exchange surface and liquid temperature,

The efficiency factor (E) characterizes the comprehensive performance of the microchannel heat transfer, and its expression is as follows:

$$E = \frac{Nu}{f^{\frac{1}{3}}} \tag{12}$$

Journal of Physics: Conference Series **2441** (2023) 012015 doi:10.1088/1742-6596/2441/1/012015

3. Results and discussion

3.1. Fluid flow characteristics

Taking the inlet velocity of 3 m·s^{-1} as the boundary condition, using Fluent calculations and using the Tecplot post-processing software, the fluid velocity cloud diagram along the *x* axis in the *x-z* plane at the microchannel *y* = 0.2 mm is obtained. Velocity distribution and pressure distribution contours for different structures on *x-z* plane are illustrated in Fig.7 and Fig.8, respectively. The flow velocity of the fluid near the concave and the internal convex has been significantly reduced. The pressure value of the fluid near the tip of the cone increases slightly. The fluid is obstructed by the tip of the cone. As a result, the rule of fluid is broken flow along the straight channel. The flow velocity is reduced under the obstruction of the cone, and the dynamic pressure is converted into static pressure. In the same-directional cavity heat sink model, the decrease of the flow velocity is the most obvious. From this, it can be judged that the CC-MCHS model has the greatest resistance to fluid flow.

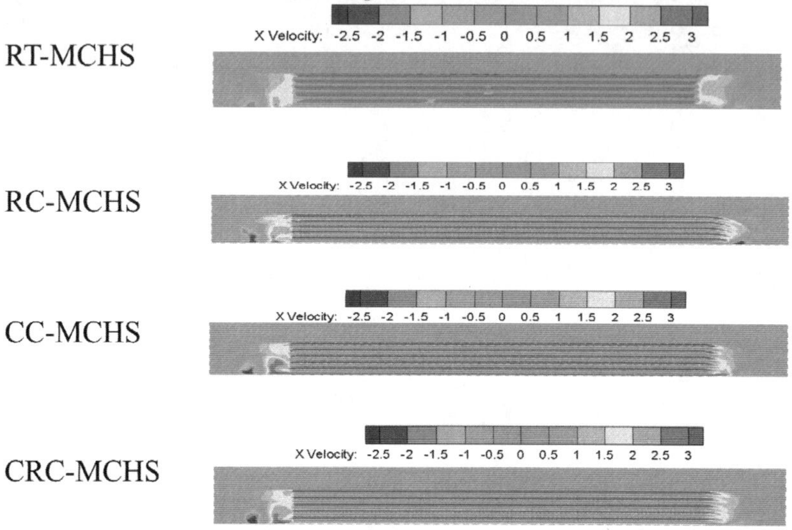

Figure 7. Velocity distribution for different MCHS structures on x-z plane (y = 0.2 mm) with inlet velocity as 3 m·s^{-1}.

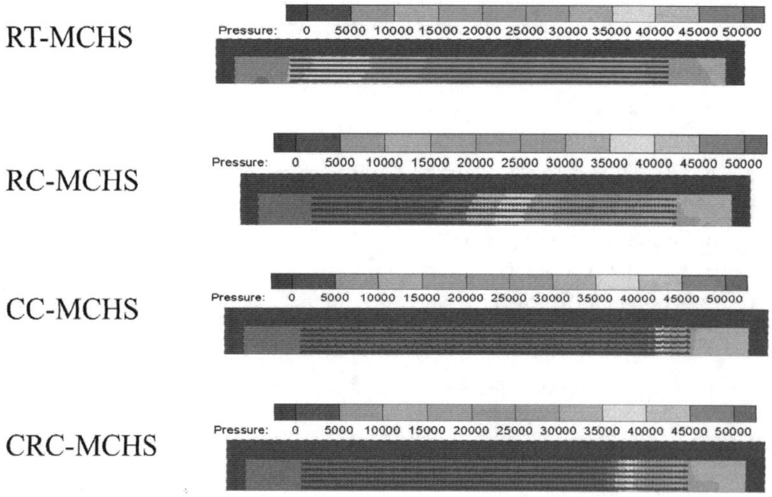

Figure 8. Pressure distribution contours for different MCHS structures on x-z plane (y = 0.2 mm) with inlet velocity as 3 m·s^{-1}.

The variation of Darcy's friction coefficient with velocity is shown in Fig.9. With the gradual increase of velocity, the Darcy friction factor of the four MCHS models is continuously decreasing, which means that the higher the flow velocity, the better the fluid flow characteristics of the heat sink model. At most flow rates, the Darcy friction factor of the CRC-MCHS is the largest, so its flow characteristics are worse than the other three heat sink models. The Darcy friction factor of the RC-MCHS is the smallest, so its flow characteristics are the best. The pressure drop of the four types of microchannel heat sinks varies the same with the inlet flow rate, as shown in Fig.10. When the flow rate gradually increases, the pressure drop of the microchannel heat sink model gradually increases. At the same flow rate, the pressure drop of the RC-MCHS is the smallest, indicating that its fluid flow performance is better. The larger pressure drop of the CRC-MCHS indicates that its flow performance is slightly worse than others.

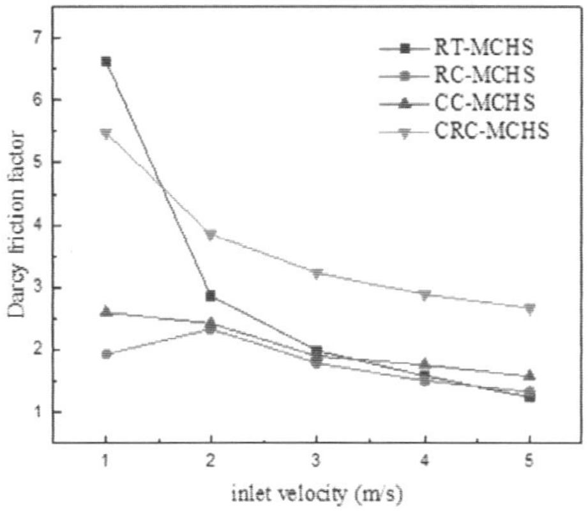

Figure 9. Darcy friction coefficient at different inlet velocities.

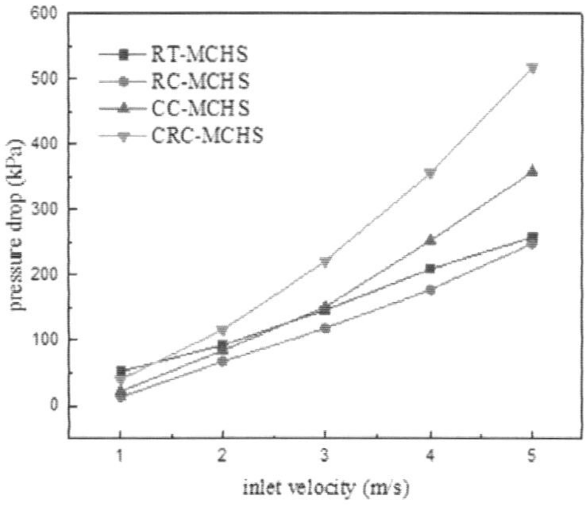

Figure 10. Pressure drops at different inlet velocities.

3.2. Fluid heat transfer performance

Taking the inlet velocity of 3 $m \cdot s^{-1}$ as the boundary condition, using Fluent calculations and using the Tecplot post-processing software, the fluid velocity cloud diagram along the x axis in the x-z plane at the microchannel y = 0.2 mm is obtained. Temperature distribution contours for different structures on x-z plane is illustrated in Fig.11. The temperature of the liquid and solid at the water inlet is the lowest, and the temperature of the liquid and solid at the water outlet is the highest. This is because the low-temperature liquid water passes through the microchannel to change the temperature of the solid area. At the water outlet end, the heat of the water inlet end is taken away by the flow of the liquid, which increases its own heat and decreases the heat exchange performance, resulting in the solid area temperature at the water outlet end being higher than the temperature of the water inlet. Among the four models, the CC-MCHS has the largest increase in fluid temperature and the highest heat transfer efficiency.

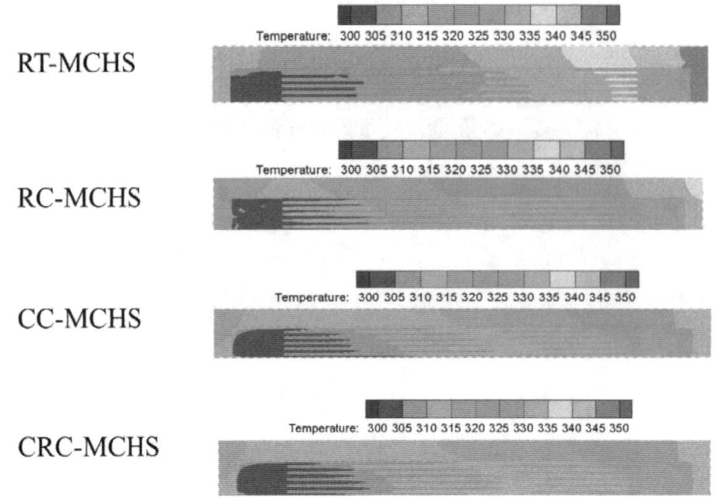

Figure 11. Temperature distribution contours for different MCHS structures on x-z plane (y = 0.2 mm) with inlet velocity as 3 $m \cdot s^{-1}$

Nusselt number and average outlet temperature at different inlet flow rates are shown in Fig.12 and Fig.13, respectively. With increasing flow velocity, the Nusselt number of each model increases, indicating that the Nusselt number increases with increasing velocity. The Nusselt number of the CC-MCHS is the largest at the same flow velocity, which indicates that the CC-MCHS has better heat transfer performance than the other three MCHS models at different flow rates. As the flow rate increases, the average outlet temperature of each model gradually decreases. Among the four MCHS models, the maximum outlet temperature of the CC-MCHS indicates its better heat transfer performance, while the average outlet temperatures of the other three MCHS models are not much different.

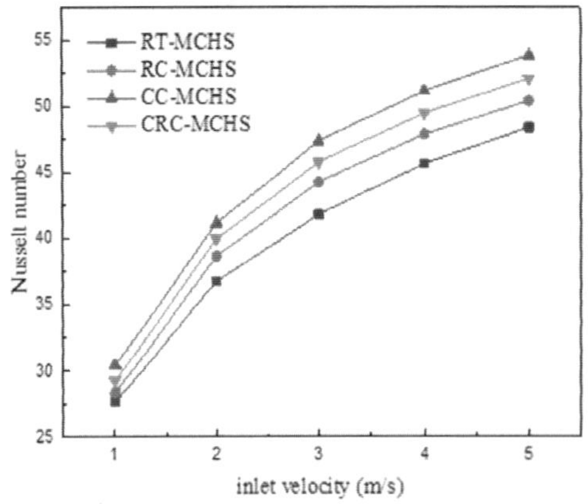

Figure 12. Nusselt numbers at different inlet velocities.

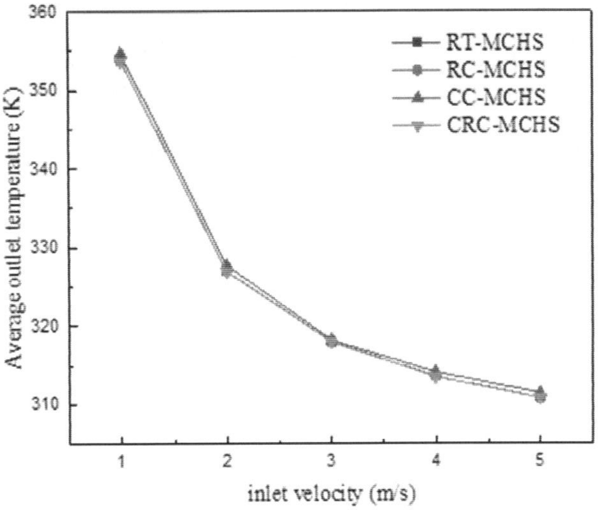

Figure 13. Average outlet temperatures at different inlet velocities.

3.3. Heat transfer efficiency

Efficiency factors at different inlet flow rates are shown in Fig.14. As the flow rate gradually increases, the efficiency factor also increases. The efficiency factor of the CC-MCHS is the largest among the four models at different flow rates. The heat transfer performance of the CC-MCHS is the best, followed by the RC-MCHS, and the heat transfer efficiency of the CRC-MCHS is lower.

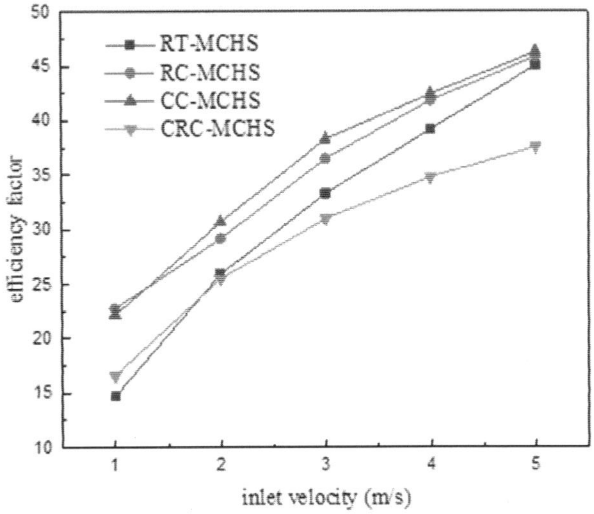

Figure 14. Efficiency factors at different inlet velocities

4. Conclusion

This research mainly proposed three new types of MCHS structures. Numerical simulations and simulation calculations have been carried out on MCHSs. The three groups of inlet flow rates have been used as the main variables to analyse the three new types of MCHSs, model flow characteristics and heat transfer performance. In addition, it has been compared with the rectangular microchannel heat sink (RT-MCHS). The following conclusions have been drawn through research:

(1) Compared with the heat exchange temperature distribution of the traditional rectangular straight microchannels, the concave and the internal convex auxiliary structure can hinder the flow of the fluid so that the liquid and the substrate can conduct sufficient heat exchange.

(2) The fluid pressure in the heat sink microchannel gradually decreases with the flow of the fluid. The fluid pressure at the inlet of the heat sink is the largest, and the outlet pressure at the outlet is the lowest. When the fluid flows near the concave and the internal convex, it is hindered by the cavity and the internal aid, and the dynamic pressure is converted into static pressure.

(3) With the gradual increase of the flow rate, the heat transfer performance of the microchannel gradually increases. The co-directional concave microchannel heat sink (CC-MCHS) has the highest comprehensive heat transfer efficiency among the four heat sink models, while the RT-MCHS has the lowest comprehensive heat transfer efficiency. The three new MCHS models proposed in this paper have higher heat transfer efficiency than the traditional RT-MCHS.

(4) The concave structure increases the heat transfer area in MCHS, improves the heat transfer performance of MCHS, and makes the cone column corresponding to the contact surface of the heat sink and the electronic component have a clear low temperature area. The area of the low temperature area at the cone column is large, and the temperature change is concentrated in the upper half of the heat sink.

(5) Under the same conditions with the water inlet flow rate of 3 m·s-1, the temperature of the water outlet of the CC-MCHS is the lowest, indicating the best heat transfer performance for electronic components and the highest pressure and flow rate. The overall performance of the traditional RT-MCHS is the worst, while the overall performance of the reverse concave microchannel heat sink (RC-MCHS) and the convex reverse concave microchannel heat sink (CRC-MCHS) is between them.

References

[1] Tuckerman, D.B., Pease, R.F. (1981) High-performance heat sinking for VLSI. IEEE Electr. Device L., 2 (5): 126–129.

[2] Tuckerman, D.B., Pease, R.F. (1983) Microcapillary Thermal Interface Technology for VLSI Packaging. In: Symposium on VLSI Technology. Digest of Technical Papers. Hawaii. pp. 60–61.

[3] Swift, G., Migliori, A., Wheatley, J. (1985) Construction of and measurements with an extremely compact cross-flow heat exchanger. J Heat Transfer Eng. 6 (2): 39–47.

[4] Han, Y.H., Liu, Y., Li, M. Huang, J. (2012) A review of development of micro-channel heat exchanger applied in air-conditioning system. Energy Procedia, 14: 148–153.

[5] Tang, G.H., Li, Z., He, Y.L., Tao, W.Q. (2007) Experimental study of compressibility, roughness and rarefaction influences on microchannel flow, Int. J. Heat. Mass. Tran., 50: 2282–2295.

[6] Zhang, C., Chen, Y., Shi, M. (2010) Effects of roughness elements on laminar flow and heat transfer in microchannels. Chem. Eng. Process., 49: 1188–1192.

[7] Guo, Z. (2000) Characteristics of microscale fluid flow and heat transfer I MEMS. In: Proc. Heat Transfer and Transport Phenomena in Microscale. Banff. Canada. pp.24–31.

[8] Li, Z.X., Wang, W., Guo, Z.Y. (2003) Effects of Axial Heat Conduction in the Wall on Convection in Microtubes. ASME 2003 1st International Conference on Microchannels and Minichannels. pp.327–333.

[9] Revellin, R., Dupont, V., Ursenbacher, T., Thome, J.R., Zun, I. (2006) Characterization of diabatic two-phase flows in microchannels: Flow parameter results for R-134a in a 0.5mm channel. Int. J. Multiphas. Flow., 32: 755–774.

[10] Hetsroni, G., Mosyak, A., Pogrebnyak, E., Yarin, L.P. (2005) Fluid flow in micro-channels. Int. J. Heat Mass Tran., 48: 1982–1998.

[11] Morini, G.L. (2004) Single-phase convective heat transfer in microchannels: a review of experimental results. Int. J. Therm. Sci., 43: 631–651.

[12] Harirchian, T., Garimella, S.V. (2009) Effects of channel dimension, heat flux, and mass flux on flow boiling regimes in microchannels. Int. J. Multiphas. Flow., 35: 349–362.

[13] Wang, B.X., Peng, X.F. (1994) Experimental investigation on liquid forced-convection heat transfer through microchannels. Int. J. Heat Mass Tran., 37: 73–82.
[14] Wu, H.Y., Cheng, P. (2003) An experimental study of convective heat transfer in silicon microchannels with different surface conditions. Int. J. Heat Mass Tran., 46: 2547–2556.
[15] Alfaryjat, A.A., Mohammed, H.A., Adam, N.M., Ariffin, M.K.A., Najafabadi, M.I. (2014) Influence of geometrical parameters of hexagonal, circular, and rhombus microchannel heat sinks on the thermohydraulic characteristics. Int. Commun. Heat Mass., 52: 121–131.
[16] Qu, W., Mudawar, I. (2002) Analysis of three-dimensional heat transfer in micro-channel heat sinks. Int. J. Heat Mass Tran., 45: 3973–3985.

Studying the effect of geometry of micro-cavity on pool boiling by LBM

Qiang Huang [1, 2, 3], **Jingzhi Zhou** [1, 3, *], **Xiulan Huai** [1, 2, 3], **Feng Zhou** [1, 2, 3]

[1] Institute of Engineering Thermophysics, Chinese Academy of Sciences, Beijing 100190, China
[2] School of engineering science, University of Chinese Academy of Science, Beijing 100049, China
[3] Nanjing Institute of Future Energy System, Nanjing, 211135, China

*Corresponding author e-mail: huangqzstu@163.com

Abstract. Pool boiling heat transfer from structured surface is simulated numerically with a 3D OTRT (optimal two relaxation time) Lattice Boltzmann method. This work shows in detail that multiple bubbles grow and coalesce over hydrophilic ($\theta = 50^o$) surfaces in 3D view, heat flux transfers from heater surface to fluid, and temperature of heater surface varies with time. Specifically, four types of geometry in microstructure surfaces are examined and compared. It is found that heat flux of S5 is the highest and its wall temperature is the lowest. When compared with S0, the heat flux of S5~S7 increased a lot due to the easier formation of vapor blanket, and cubic shaped cavity alleviates largely this deterioration. Next, the large quantity of heat from the three phase contact line in the vicinity of side wall accounts for bubble nucleation and energy from the evaporation of thin liquid between bubbles and superheated wall helps bubbles to grow further. Besides, it is found that S5 increases nucleate site density, reduces boiling incipience superheat, and enhances remarkably heat transfer performance.

Keywords: Pool boiling; 3D LBM; Micro-cavity.

1. Introduction

Pool boiling is widely used in engineering systems [1, 2] because of its high heat transfer performance. Nucleation boiling stage has been found to play important roles on boiling heat transfer performances. In nucleation boiling, the geometry of nucleation sites due to micro-structure entirely effect on bubble dynamics. Recently, some numerical simulations [3-5] and experiments [6-11] have been performed to investigate boiling mechanism of artificial nucleation site by lattice Boltzmann method (LBM) [12-21]. Most of simulations only focus on nucleation sites. Ma [22] studied firstly dry area on micro-structure hydrophilic heaters by 3D LBM. Li [23] researched modified surface to by 3D LBM. In addition, some experiments have been made to enhance nucleate boiling, mainly focusing on micro- or nano- structured surfaces because of the rapid growth of surface modification and fabrication technologies. Jiang [24] investigated cylindrical shaped micro-pillar arrays with various sizes. Liang [25] showed that micro-pillar could decrease boiling starting superheat, and enhanced boiling heat transfer performance. Vishal V [26] revealed that the tunnel structure improved liquid-vapor two phase fluid in phase change region.

Content from this work may be used under the terms of the Creative Commons Attribution 3.0 licence. Any further distribution of this work must maintain attribution to the author(s) and the title of the work, journal citation and DOI.
Published under licence by IOP Publishing Ltd

However, relatively few 3D numerical papers studied the effect of different geometries of micro-cavity on pool boiling.

In this work, 3D LBM was used to research on the heat transfer performance of microstructure under hydrophilic circumstance. First, the effect of four types of micro-cavity geometry under hydrophilic circumstance on heat transfer was investigated. Second, the influence of geometry of micro-cavity structured surface on bubble dynamics was examined.

2. Model

2.1. Dimensionless parameters

The boiling heat flux q' is defined as

$$q' = \frac{q}{q_0} = \frac{1}{S_{heater}\Delta t}\int_{t_1}^{t_1+\Delta t}\left[\iint_0^{S_{heater}} k_s\left(\frac{\partial T}{\partial y}\right)_{y=0}\frac{\Delta y}{T_{cr}}\,dxdz\right]dt\Big/q_0 \tag{1}$$

Where S_{heater} is region of heater. The based boiling heat flux was set by

$$q_0 = \frac{\mu_l h_{lv}}{l_0} \tag{2}$$

In practice, micro-structure plays an essential role in boiling phenomena. In order to accurately illustrate the underlying characteristic behind boiling, it is important to study pure micro-structure effects by simulations.

2.2. Profile of micro cavity

Fig. **1** shows 3D views of the profile of the four kinds of structured surfaces with micro-cavity array.

(i) Cubic shaped(S5) (ii) Cylindrical shaped(S6) (iii) Spherical shaped(S7) (iv) Conical shaped(S8)

Fig. 1 Schematic of surfaces with micro-cavity array

3. Results and discussions

3.1. The effect of geometry of micro-cavity on boiling heat transfer

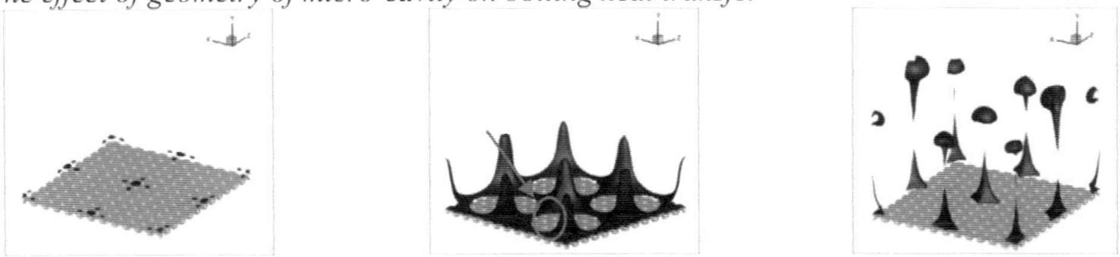

(i) t*=84.8 (ii) t*=118.8 (iii) t*=135.8

(a) S5

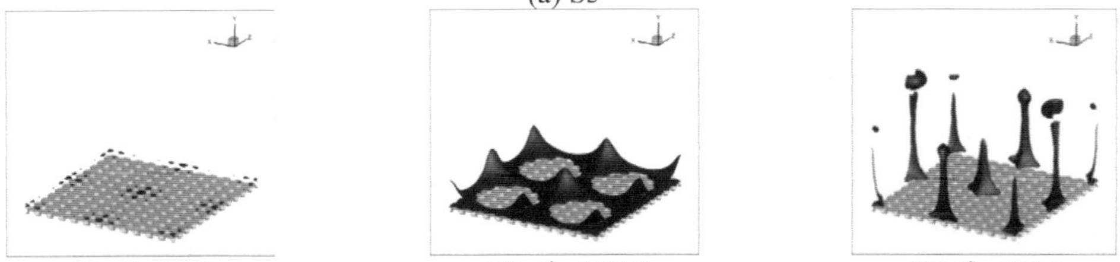

(i) t*=84.8 (ii) t*=118.8 (iii) t*=135.8

(b) S6

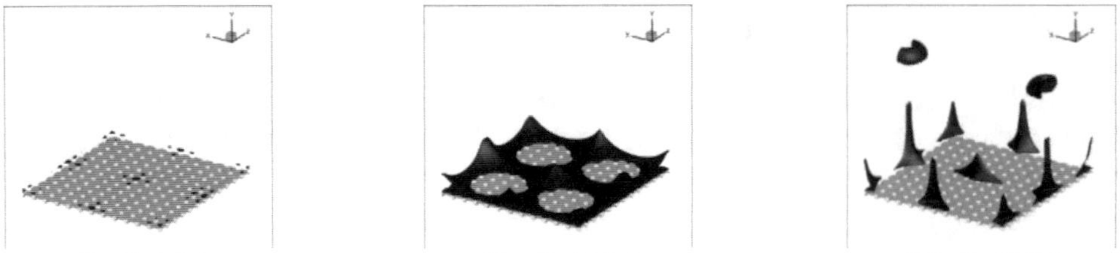

(i) t*=84.8 (ii) t*=118.8 (iii) t*=135.8

(c) S7

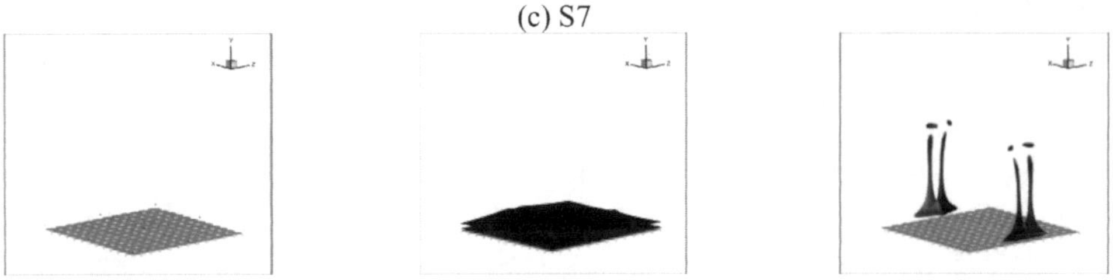

(i) t*=84.8 (ii) t*=118.8 (iii) t*=135.8

(d) S8

Fig. 2 Vapor distribution on micro-cavities hydrophilic surfaces (at z = Hb) (Θ = 50°) at Ja = 0.15.

From the data of Fig. **2**, the side wall of cavities forces the bubble to only expand in the vertical direction before bubble reaches the cavity mouth. Once the height of bubble exceeds that of cavity mouth, bubble usually grow along upper hemisphere direction. This expansion promotes the rapid merging of bubbles in those adjacent cavities. When the place where the bubbles contact with heated surface becomes large, remarkable bubble blanket forms over S8, which result in deterioration of heat transfer. However, bubbles coalescent has been suppressed steep side wall of cavities over S5~S7. The red arrow points out the resistance of S5 for vapor flow is biggest.

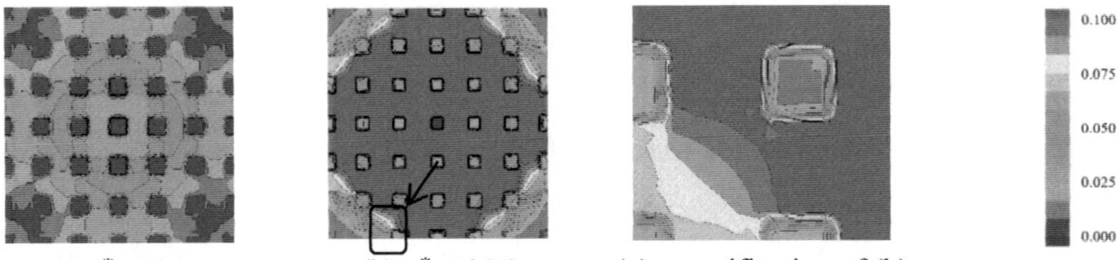

(a) t*=84.8 (b) t*=135.8 (c) magnification of (b)

Fig. 3 Cross sectional view for heat flux of S5 (at y = Hb) at Ja = 0.15.

Once bubbles nucleate, they then grow through evaporation of liquid at the solid-liquid-vapor phase contact line [27], as also shown in Fig. **3**(a). However, a large number of heat transferring to liquid is from the side wall of micro-cavity rather than the bottom of micro-cavity. That could be explained as the extended meniscus in the vicinity of the contact line. Next, Fig. **3**(b) displays in detail the heat flux contour. The interesting thing is a majority of heat is from superheated side wall rather than cavities to fluid zone. Fig. **3**(c) is the magnification of zone labeled by the black square symbol in Fig. **3**(b). Fig. **4** show that growing bubbles could trap a thin layer of liquid between itself and the superheated side wall, and evaporation of this trapped liquid accounts for bubble further growth.

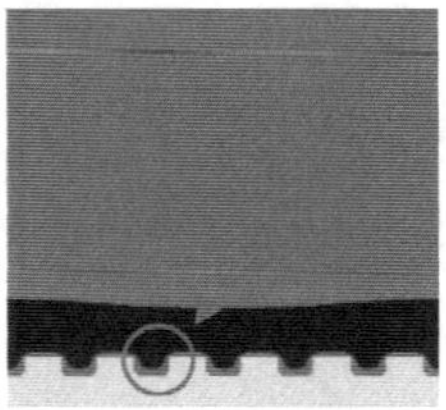
(a) the thin layer of liquid beneath bubble

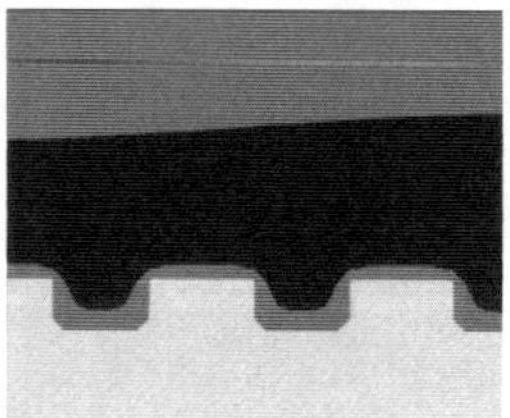
(b) the magnification of (a)

Fig. 4 Cross sectional view for density of S5 (at z = 0.5 * Lz) at Ja = 0.15. Color scheme: vapor phase (light purple), solid phase (dark yellow), liquid phase (dark cyan).

Fig. **5** display that q' rises remarkably for a long time after bubbles nucleate, while Tw decreases slightly. The increment indicates that due to bubble nucleation, the liquid contacting with solid wall (at z = Hb) tended to absorb a lot of heat to meet the latent heat of vaporization. Fig. **5**(a) shows that q' is 0.078 for S5, 0.073 for S6, 0.054 for S7, 0.029 for S8 and 0.026 for S0. Besides, the trend of q' of S5~S7 is basically coincide.

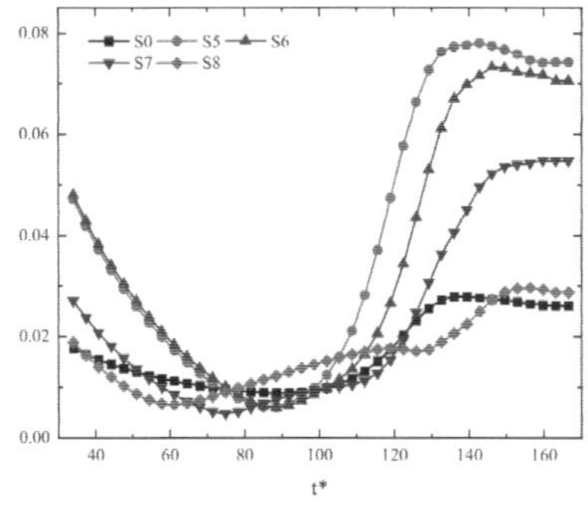

(a) transient heat flux

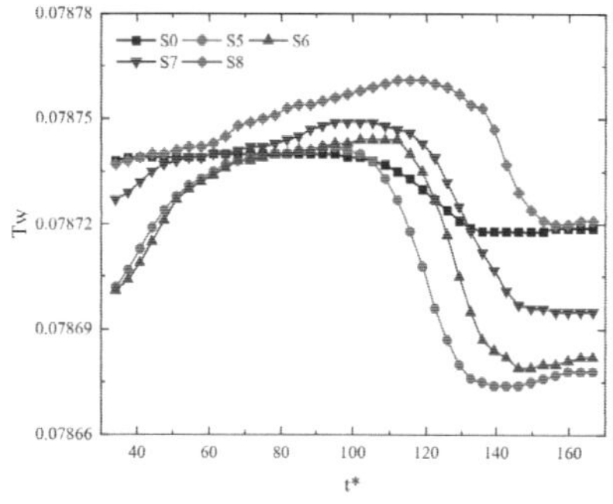

(b) transient wall temperature

Fig. 5 Geometric effects on simulated heat transfer curves for infinite micro-cavity structured hydrophilic surfaces at Ja = 0.15.

Compared with S0 from Fig. **5**, it could be found that: firstly, the q' of S5, S6, S7 increases a lot while S8 increases a little. This may result from that the nucleate sites of micro-cavity structure surface are the place where micro-cavity structure surface is just the small bottom of cavities. Fig. **6**(b) shows that the nucleate sites of micro-cavity structure surface are then more than that of S0, as shown in Fig. **6**. And, the bubbles nucleating in cavity tends to flow out cavity and coalesce over substrate, which results in forming vapor blanket and blocking fresh liquid from flowing the cavity. Whereas, it is interesting that the trapped vapor in the cavities of S5 and S6, Deep and narrow cavities of them prohibit vapor from flowing out and forming vapor blanket over substrate. Therefore, q' of S5 and S6 is still better than S7 and S8. As for q' of S7, curved side wall of cavity of S7 doesn't trap vapor as easy as S5 and S6 and helps to replenish liquid. Secondly, their nucleate time of S5 and S6 are delayed while S7 and S8 nucleates ahead, which may result from small nucleate zone, namely cavity. Thirdly, from the data of Fig. **5**(b), the Tw of S5~S8 move downward due to short distance between surface of heating wall and nucleation zone.

3.2. The effect of surface with micro-cavity array on bubble dynamic

This work turns its attention to boiling heat transfer trends under hydrophobic surface. It examines micro-cavity structure surface to assess the effect of geometry on bubble dynamics.

Fig. **6** depicts in detail the effect of geometry of micro cavity on bubble dynamics. There are three important things needed to know from Fig. **6**. Firstly, it is the much more nucleation sites that result in the huge enhancement of heat transfer for S5~S7. Secondly, the number vapor phase of S5 in bubble growth stage is remarkable less than other surface due to its high flow resistance. Thirdly, many bubble seeds after bubbles departure promote heat transfer.

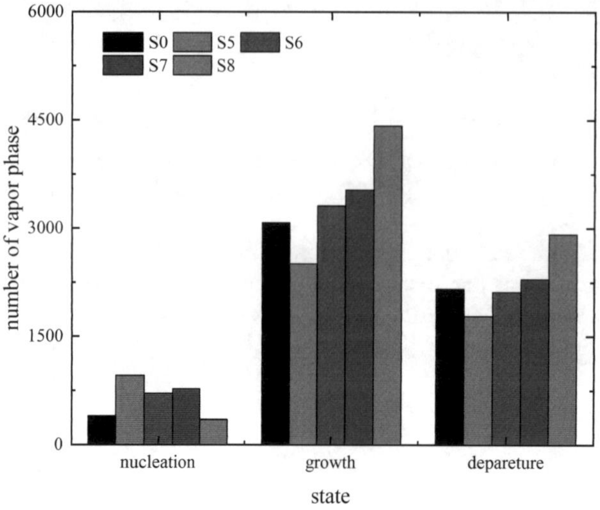

Fig. 6 Number of vapor phase over surface with micro-cavity(Θ = 50°) at Ja = 0.15.

4. Conclusion

In this paper, simulations of nucleation boiling heat transfer are conducted based on the pseudo-potential LBM [28, 29] with OTRT. Four kinds of nucleation sites are investigated, respectively. The following conclusions can be drawn from this paper:

1. As micro-cavity serves as artificial nucleation sites, boiling incipience takes place preferably at the cavities, indicating that the micro-cavity structured surface can reduce the wall superheat at onset of nucleate boiling, increase the nucleation sites density, and enhance largely heat transfer performance. The heat transfer of S5 performs best among all structured surfaces.
2. At the early stage of bubble growth, bubbles acquire heat in large quantities from three phase contact line. At the later stage of bubble growth, bubbles acquire heat in large quantities from thin liquid between bubble and the superheated wall.

This work only pays its attention on pool boiling of regular micro-cavity surfaces. It is worthwhile to simulate pool boiling of random porous micro-structure surfaces.

Acknowledgment

This research work was sponsored by National Natural Science Foundation of China (Grant No. 52006218).

References

[1] X. Chen, H. Ye, X. Fan, T. Ren, G. Zhang, A review of small heat pipes for electronics, Appl. Therm. Eng. 96 (2016) 1–17.
[2] Attinger, D., Frankiewicz, C., Betz, A.R.,Schutzius, T.M.etc. Surface engineering for phase change heat transfer:a review. MRS Energy Sustain. 1 (2014), 1–40.
[3] X.T. Chang, H.B. Huang, Y.P. Cheng, X.Y. Lu.Lattice Boltzmann study of pool boiling heat transfer enhancement on structured surfaces. Int. J. Heat Mass Transf., 139 (2019), 588-599.
[4] X. Ma, P. Cheng. Dry spot dynamics and wet area fractions in pool boiling on micro-pillar and micro-cavity hydrophilic heaters: a 3D lattice boltzmann phase-change study. Int. J. Heat Mass Transf., 141 (2019), 407-418.
[5] S. Gong, P. Cheng.Two-dimensional mesoscale simulations of saturated pool boiling from rough

surfaces. Part II: bubble interactions above multi-cavities.Int J Heat Mass Tran, 100 (2016), 938-948.
[6] B. Shen, T. Hamazaki, W. Ma, N. Iwata, S. Hidaka, A. Takahara, K. Takahashi, Y. Takata. Enhanced pool boiling of ethanol on wettability-patterned surfaces.Appl. Therm. Eng., 149 (2019), 325-331,
[7] T.P. Koukoravas, G. Damoulakis, C.M. Megaridis. Experimental investigation of a vapor chamber featuring wettability-patterned surfaces.Appl. Therm. Eng., 178 (2020), 115522.
[8] H. Jo, H.S. Park, M.H. Kim. Single bubble dynamics on hydrophobic–hydrophilic mixed surfaces.Int. J. Heat Mass Transf., 93 (2016), 554-565.
[9] X. Kong, Y.H. Zhang, J.J. Wei.Experimental study of pool boiling heat transfer on novel bistructured surfaces based on micro-pin-finned structure.Exp. Therm. Fluid Sci., 91 (2018), 9-19.
[10] Y. Sun, G. Chen, S. Zhang, Y. Tang, J. Zeng, W. Yuan. Pool boiling performance and bubble dynamics on microgrooved surfaces with reentrant cavities. Appl. Therm. Eng., 125 (2017), 432-442.
[11] Jiang Y. Y., Osada H., Inagaki M. et al. Wall thermal conductivity effects on nucleation site interaction during boiling: an experimental study [C]. 14th International Heat Transfer Conference, American Society of Mechanical Engineers, Washington DC, USA, 2010, 637–646.;
[12] Cui, J., Yang, T., Yang, F. et al. An Improved Enthalpy-Based Lattice Boltzmann 3D Model with Added Probability Function for the Growth Characteristics of Frost Layer. J. Therm. Sci.
[13] Taher, M.A., Kim, H.D. & Lee, Y.W. Study of thermal and hydraulic performances of circular and square ribbed rough microchannels using LBM. J. Therm. Sci. 24, 549–556 (2015).
[14] Kan, A., Mao, S., Wang, N. et al. Simulation and Experimental Study on Thermal Conductivity of Nano-Granule Porous Material Based on Lattice-Boltzmann Method. J. Therm. Sci. 30, 248–256 (2021).
[15] Yang, B., Wu, W., Li, M. et al. A Multiple-Relaxation-Time Lattice Boltzmann Model for Natural Convection in a Hydrodynamically and Thermally Anisotropic Porous Medium under Local Thermal Non-Equilibrium Conditions. J. Therm. Sci. 29, 609–622 (2020).
[16] Xin, F., Li, XF., Xu, M. et al. Simulation of gas exothermic chemical reaction in porous media reactor with lattice Boltzmann method. J. Therm. Sci. 22, 42–47 (2013).
[17] Y.T. Mu, L. Chen, Y.L. He, Q.J. Kang, W.Q. Tao Nucleate boiling performance evaluation of cavities at mesoscale level Int. J. Heat Mass Transf., 106 (2017), 708-719.
[18] S. Ahmad, J.T. Chen, C. Eze, J.Y. Zhao.Lattice Boltzmann study of nucleation site interaction and nucleate boiling heat transfer on a hybrid surface with multiple cavity-pillar structures.Int. J. Therm. Sci., 163 (2021).
[19] P. Zhou, Z. Liu, W. Liu, X. Duan. LBM simulates the effect of sole nucleate site geometry on pool boiling.Appl. Therm. Eng., 160 (2019).
[20] P. Zhou, W. Liu, Z. Liu Lattice Boltzmann simulation of nucleate boiling in micro-pillar structured surface Int. J. Heat Mass Transf., 131 (2019) 1-10.
[21] Li, M., et al., Nucleate Boiling Simulation Using Interface Tracking Method, Nuclear Engineering and Design, 369 (2020), 110813, 1-10.
[22] X. Ma, P. Cheng, 3D simulations of pool boiling above smooth horizontal heated surfaces by a phase-change lattice Boltzmann method, Int. J. Heat Mass Transf. 131 (2019) 1095–1108.
[23] W. X. Li, Q. Li, Y. Yu, and K. H. Luo, Nucleate boiling enhancement by structured surfaces with distributed wettability-modified regions: A lattice Boltzmann study, Appl. Therm. Eng. 194, 117-130 (2021).
[24] H.T. Jiang N.Xu et. Experimental investigation of the effect of cylindrical array structure on heat transfer performance during nucleate boiling. Int J Heat Mass Tran, 174 (2021), 121319.
[25] G. Liang, Y. Chen, H. Yang, D. Li, S. Shen. Nucleate boiling heat transfer and critical heat flux (CHF) from micro-pit surfaces Int J Heat Mass Transf, 152 (2020).

[26] Vishal V,.Nirgude,Santosh K,Sahu.Enhancement of nucleate boiling heat transfer using structured surfaces.Chemical Engineering and Processing: Process Intensification, 122(2017), 222-234.
[27] Stephan, P., Hammer, J. A new model for nucleate boiling heat transfer.Warme und Stoffubertragung 30, (1994)119–125.
[28] S. Gong, P. Cheng, A lattice Boltzmann method for simulation of liquid–vapor phasechange heat transfer, Int. J. Heat Mass Tran. 55 (17) (2012) 4923–4927.
[29] C. Zhang, P. Cheng, Mesoscale simulations of boiling curves and boiling hysteresis under constant wall temperature and constant heat flux conditions, Int. J. Heat Mass Transf. 110(2017) 319–329.

Study of low-level pool boiling in vapor chamber of wickless micro-heat pipe based on Lattice Boltzmann pseudopotential method—Effects of superheat and wall wettability

Shalong Xiong [1], **Yanping Du** [1,*], **Rui Wu** [2]

[1] China-UK Low Carbon College, Shanghai Jiao Tong University, Shanghai, 200240, China
[2] School of Mechanical Engineering, Shanghai Jiao Tong University, Shanghai, 20240, China

*Corresponding author e-mail: Yanping.du@sjtu.edu.com

Abstract. The lattice Boltzmann method, a mesoscopic numerical method with a solid kinetic basis, is used to establish a low-level pool boiling model to exhibit the transport phenomena inside an evaporation chamber of a wickless micro heat pipe using DI water as the working fluid. The effects of wall superheat and wall wettability on bubbles' nucleation, growth, and detachment in the evaporation chamber were compared and analysed. Moreover, the influence of the wall superheat on the thermal performance of the micro heat pipe was specifically examined. It is found that the low-level pool boiling occurs with a wall superheat that is higher than $0.35T_c$, indicating a high heat flux for heat removal. Interestingly, when the relatively hydrophobic chamber wall is superheated by $0.46T_c$, there are more nucleation sites, and the complete bubbles cycle containing nucleation, growth, and detachment is longer. In contrast, on a hydrophilic surface under the same and sufficient superheat condition, the number of bubble nucleation sites is less than that on the hydrophobic surface, and the complete cycle of bubbles is shorter. It is worth mentioning that the model has values for predictions of homogeneous nucleation of bubbles and thermal transport on hydrophobic or hydrophilic surfaces.

1. Introduction

Boiling is an effective heat transfer process, widely applied in nuclear reactors, heavy vehicle engines, computer chips, microelectronic devices, and other high-tech equipment and systems. At the same time, boiling is a complicated and elusive process, because physical processes are closely interrelated, such as the nucleation, growth, separation, and coalescence of bubbles, the transport of latent heat, the instability of liquid-vapor interfaces, etc.

As a typical rapid passive cooling technology utilizing boiling process, heat pipe has been widely used in various scenarios with high heat flux. Considering that the heat pipe is becoming thinner and thinner, the heat and mass transfer of phase change process inside the heat pipe becomes more difficult to be observed and studied by experimental methods, it is necessary to use numerical methods to analyze the boiling process in the evaporation chamber of a typical heat pipe. Among them, the lattice Boltzmann method, as a numerical method that owns the advantages of clearer microscopic physical significance

Content from this work may be used under the terms of the Creative Commons Attribution 3.0 licence. Any further distribution of this work must maintain attribution to the author(s) and the title of the work, journal citation and DOI.
Published under licence by IOP Publishing Ltd

and inherent parallelism property, has become an increasingly popular numerical method in the field of multiphase flow.

The lattice Boltzmann pseudo-potential phase change model was firstly proposed by Shan and Chen in 1993 and was perfected in 1994 [1, 2]. The essential idea of it is to introduce a special form of the interaction potential between adjacent fluid micro-clusters and define the interaction force based on this potential energy. Compared with the traditional VOF method and Level-set method, both of which are based on continuum hypothesis, the natural parallelism of the lattice Boltzmann method makes it much faster in those from the perspective of computational efficiency. Furthermore, by introducing the real equation of state for a specific gas, the vapor-liquid separation process can be realized directly without the need to define and capture the phase interface accurately. After a series of development, the lattice Boltzmann pseudopotential model has become an important numerical model in the field of boiling heat transfer and multiphase flow.

In this paper, the lattice Boltzmann pseudopotential model with the exact difference method proposed by Gong and Cheng in 2012 will be used to numerically simulate the process of the low-level pool boiling in the evaporation chamber of the micro heat pipe [3]. Through this simulation, the theory of bubble nucleation is verified and the effects of wall wettability on bubbles' dynamic behaviors are qualitatively studied.

2. Mathematical model

2.1. Momentum equation

The density distribution function evolution equation of the pseudopotential LBM model can be written as [4].

$$f_i\left(x+e_i\delta_t,t+\delta\right)-f_i\left(x,t\right)=-\frac{1}{\tau}\left(f_i\left(x,t\right)-f_i^{eq}\left(x,t\right)\right)+F_i\left(x,t\right) \tag{1}$$

where $f_i\left(x,t\right)$ is the particle distribution with the velocity e_i at position x and time t, τ is the relaxation time. $f_i^{eq}\left(x,t\right)$ is the corresponding equilibrium distribution function given by

$$f_i^{eq}=\omega_i\rho\left[1+\frac{e_i\cdot u}{c_s^2}+\frac{\left(e_i\cdot u\right)^2}{c_s^4}-\frac{u^2}{2c_s^2}\right] \tag{2}$$

with ω_i being the weighting coefficients, ρ being the density and c_s the lattice sound speed. $F_i\left(x,t\right)$ is the body force term and given by the exact difference method [3].

$$\Delta f_i\left(x,t\right)=f_i^{eq}(u+\Delta u,\rho)-f_i^{eq}(u,\rho) \tag{3}$$

where $\Delta u=\dfrac{F\cdot\delta t}{\rho}$ is the velocity change under the effect of body force during time step δ_t and F is given by,

$$F=F_{\text{int}}\left(x\right)+F_s\left(x\right)+F_g\left(x\right) \tag{4}$$

where $F_{\text{int}}\left(x\right)$ is the interparticle interaction force which is in charge of phase separation, $F_s\left(x\right)$ is the interaction force between solid and fluid and $F_g\left(x\right)$ is the gravitational force. And $F_{\text{int}}\left(x\right)$ is given by

$$F_{\text{int}}\left(x\right)=c_0\psi\left(x\right)G\nabla\psi\left(x\right) \tag{5}$$

where $\psi\left(x\right)$ is the "effective mass", which is a function of the local density and determined by the equation of state. It is given by

$$\psi(x)=\sqrt{\frac{2\left(p-\rho c_s^2\right)}{c_0 G}} \tag{6}$$

where c_0 is 6.0 and G is -1. p is using the Peng-Robinson equation of state in this model

$$p=\frac{\rho RT}{1-b\rho}-\frac{a\rho^2\alpha(T)}{1+2b\rho-b^2\rho^2} \tag{7}$$

$$\alpha(T)=\left[1+(0.37464+1.5422\omega-0.2699\omega^2)\right]\left(1-\sqrt{\frac{T}{T_{cr}}}\right)^2 \tag{8}$$

where $a=2/49, b=2/21, R=1$ and ω is chosen as 0.344, which is the acentric factor for water [3].

The mathematical discretization form of equation (5) is as follow [5],

$$F_{\text{int}}(x)=-\psi(x)G\sum_{x'}\omega_i\psi(x')\cdot(x'-x) \tag{9}$$

The wettability of the heated surface is characterized by the force between the fluid and the solid wall and is introduced by following

$$F_s(x)=-\psi(x)G\sum\omega_i\psi(x')\cdot(x'-x) \tag{10}$$

The gravity is calculated by

$$F_g(x)=\left(\rho(x)-\rho_{ave}\right)g \tag{11}$$

where g is gravity acceleration, ρ_{ave} is the average density of the calculated region within each time step. The lattice model we adopt is D2Q9, with corresponding weighting factors in nine velocity directions is $\omega_i=4/9(i=0)$、$1/9(i=1\sim4)$、$1/36(i=5\sim8)$ and their velocity vector is

$$e_i=\begin{cases}(0,0), & i=0\\ (\pm1,0)c,(0,\pm1)c, & i=1\sim4\\ (\pm1,\pm1)c, & i=5\sim8\end{cases} \tag{12}$$

where $c=\delta_x/\delta_t$ is the lattice speed, δ_x is lattice length and δ_t is time step. And in D2Q9, $c_s^2=c^2/3$.

The kinematic viscosity is given by

$$\upsilon=c_s^2(\tau-\frac{1}{2})\delta_t \tag{13}$$

The density and velocity in the above equations are calculated by

$$\rho=\sum f_i \tag{14}$$

$$\rho u=\sum e_i f_i \tag{15}$$

What should be carefully noticed is that the velocity $\boldsymbol{u}$ is not the real fluid velocity and should be corrected by the external force on the above, and its value is transferred to the calculation of distribution function in the energy equation in the next subsection

$$\rho v=\sum e_i f_i+\frac{\delta_t}{2}F \tag{16}$$

2.2. Energy equation

The energy equation is solved by the temperature distribution function as follows [6],

$$g_i\left(x+e_i\delta_t,t+\delta\right)-g_i(x,t)=-\frac{1}{\tau_g}\left(g_i(x,t)-g_i^{eq}(x,t)\right)+\phi \tag{17}$$

where the source term ϕ can be found in [7] and we introduce it here for convenience,

$$\phi = C_i + \left(Q_i + \frac{1}{2}\frac{\partial Q_i}{\partial t} \right) \tag{18}$$

$$C_i = \left(1 - \frac{1}{2\tau_g} \right) \frac{\omega_i e_i \cdot \dfrac{\partial (Tv)}{\partial t}}{c_s^2} \tag{19}$$

$$Q_i = \omega_i T \left[1 - \frac{1}{\rho c_v} \left(\frac{\partial p}{\partial T} \right)_\rho \right] \nabla \cdot v + \frac{\nabla \cdot (k \nabla T)}{\rho c_v} - \nabla \cdot (\alpha \nabla T) \tag{20}$$

where α is the thermal diffusivity and its relation with thermal conductivity is $\alpha = k/\rho c_v$, in which c_p is the specific constant pressure heat capacity. τ_g is the relaxation time for the temperature distribution function. $g_i^{eq}(x,t)$ is the equilibrium distribution function and is given by following

$$g_i^{eq}(x,t) = \omega_i T \left[1 + \frac{e_i \cdot v}{c_s^2} + \frac{(e_i \cdot v)^2}{c_s^4} - \frac{v^2}{2c_s^2} \right] \tag{21}$$

Where temperature T is calculated by summing all temperature distribution values g_i

$$T = \sum g_i \tag{22}$$

For the energy equation, we also adopt D2Q9 discretization scheme. By the following equation, the relaxation time for temperature distribution evolution can be attained

$$\alpha = c_s^2 \left(\tau_g - \frac{1}{2} \right) \delta_t \tag{23}$$

3. Simulation setup

To simulate the heat and mass transfer process in the evaporation chamber of a wickless and micro heat pipe, a two-dimensional rectangular low-level pool boiling geometric model as shown in Figure 1 is established in this paper. The physical length and width of the evaporation chamber are Lx = 800× 10^-6 m and Ly = 300× 10^-6 m respectively, with characteristic length taken as 100×10^-6 m. The lattice length and width are 600 and 225 respectively.

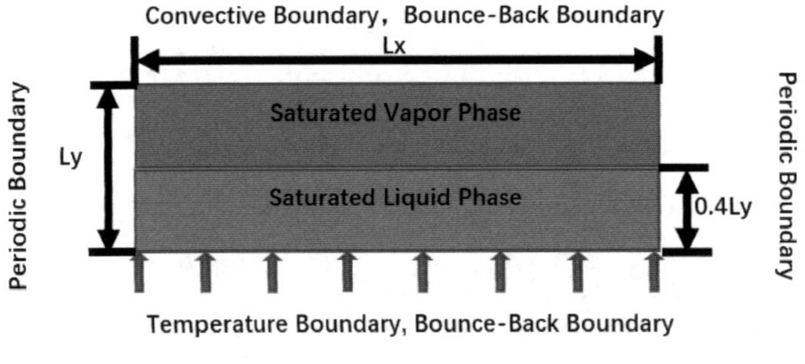

Figure 1. Schematic diagram of evaporation chamber of a typical micro heat pipe (computational domain of the low-level pool boiling model)

The left and right boundaries of the simulation domain are using the periodic boundary for both the density distribution function (momentum equation) and the temperature distribution function (energy equation). At the upper boundary of the simulation domain, the bounce-back boundary using the saturated vapor density is employed for the density distribution function, and the convective boundary is applied for the temperature distribution function. The convective boundary scheme is used to eliminate the possible influence on the internal flow field. At the lower wall, the bound-back boundary reflecting the wettability of the wall is used for the density distribution function, and the temperature boundary condition is used for the temperature distribution function.

Initially, the lower part (40%) and the upper part (60%) of the simulation domain are occupied by saturated vapor and saturated liquid at 388K respectively. Through Maxwell Construction, the corresponding saturated liquid and saturated vapor densities can be obtained. What should be pointed out is that all of the simulation parameters in LBM are usually expressed in lattice units. The real physical properties are obtained by the dimensionless method. To ensure numerical stability, the wall density slowly increases to a density value that reflects the wettability of the wall.

4. Simulation results and discussions

4.1. Contact angle simulation

Surface wettability is usually characterized by the contact angle of a liquid droplet or vapor bubble on a surface. It essentially reflects the interaction intensity between vapor, liquid, and solid. There are currently two popular methods for simulating contact angles: one is to treat solid nodes as fluid nodes, giving them fixed density values, and use the same potential function, which should describe the gas, to calculate the "effective mass" of the wall [8]; This method has two advantages: the force between solid and liquid can be described uniformly with the formula describing the interaction force between fluids; at the same time, the temperature of the wall follows the evolution equation of the temperature distribution function of the fluid. So that programming is greatly facilitated. Although this assumption is defective: the relaxation times of interactions between solid and liquid particles are obviously not consistent with those of interactions between fluid particles. The first method is as follows

$$F_s(\mathbf{x}) = G\psi(\mathbf{x}) \sum w_i \psi(\mathbf{x} + \boldsymbol{e}_i)\ \boldsymbol{e}_i \quad (24)$$

The other method is to define a fixed density to the wall, and the interaction with the fluid node is adjusted by the interaction strength Gw [9].

$$F_s(\mathbf{x}) = G_w \rho(\mathbf{x}) \sum w_i s(\mathbf{x} + \boldsymbol{e}_i)\ \boldsymbol{e}_i$$

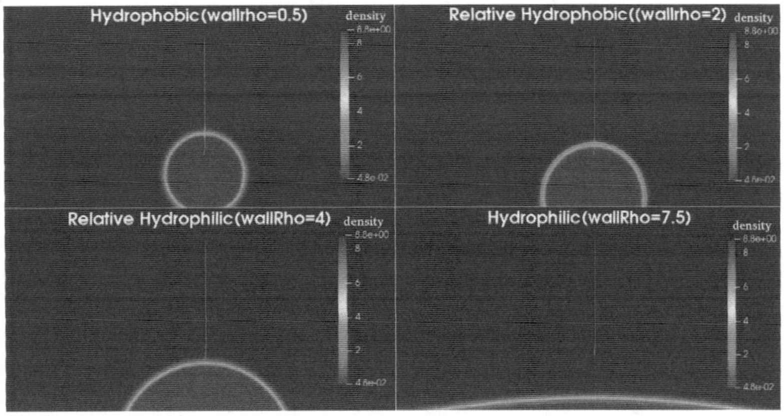

Figure 2. Wall wettability comparison of different value

where s (x + $\boldsymbol{e}_i$) is a switch function. If the fluid wall nodes collide with the solid nodes during streaming, the value is 1, indicating that there is a fluid-solid force; if it is judged not to be a solid node, the value is 0, indicating that there is only the interaction force between fluids. The second method is commonly used in the simulation of porous media by the lattice Boltzmann method. Since the first method is easier to program and implement, it can also achieve simulating the fluid-solid effect.

Therefore, we simulate the contact angle for several different situations according to the first method (shown in Figure 2). It is found that when the set density value of the wall surface is small (0.5 or 2), the wall exhibits hydrophobic properties, which means the contact angle of the droplet is greater than 90°; when the set density value is large (4 or 7.5), the wall features hydrophilic (contact angle of the droplet is less than 90°).

4.2. Effects of wall superheat on start-up of low-level pool boiling

Superheat is the primary reason for inducing phase transition [10], which is defined as the temperature difference between wall and liquid. When the superheat formed by the wall and the liquid phase in the simulation area is less than $0.35T_c$, that is to say when the liquid phase temperature is in the metastable region, the vapor-liquid phase transition may not occur. When the superheat is greater than $0.35T_c$, which is when the liquid phase is in the unstable region, the vapor-liquid phase transition process is prone to occur [10].

In order to exclude the interference of other factors such as wall roughness, only the case of a smooth and horizontal surface is taken into consideration. The homogeneous nucleation situation in the phase transition theory is to be tested. According to this theory, homogeneous nucleation can only occur at the expense of providing a higher degree of superheating. Therefore, the following simulations are performed under five superheat conditions: $0.2T_c$, $0.3T_c$, $0.4T_c$, $0.46T_c$, and $0.51T_c$. The wall density is set to 4, indicating a relatively hydrophilic wall.

4.2.1. Superheat $0.2T_c$. When the superheat of the smooth surface is set to be $0.2T_c$, the iteration reaches the maximum time set by the simulation, and no bubbles appear in the simulation domain (as shown in Figure 3). This shows that under a smooth surface, a superheat of $0.2T_c$ is insufficient to provide a large thermodynamic "perturbation" to make homogeneous nucleation occur in this computational region.

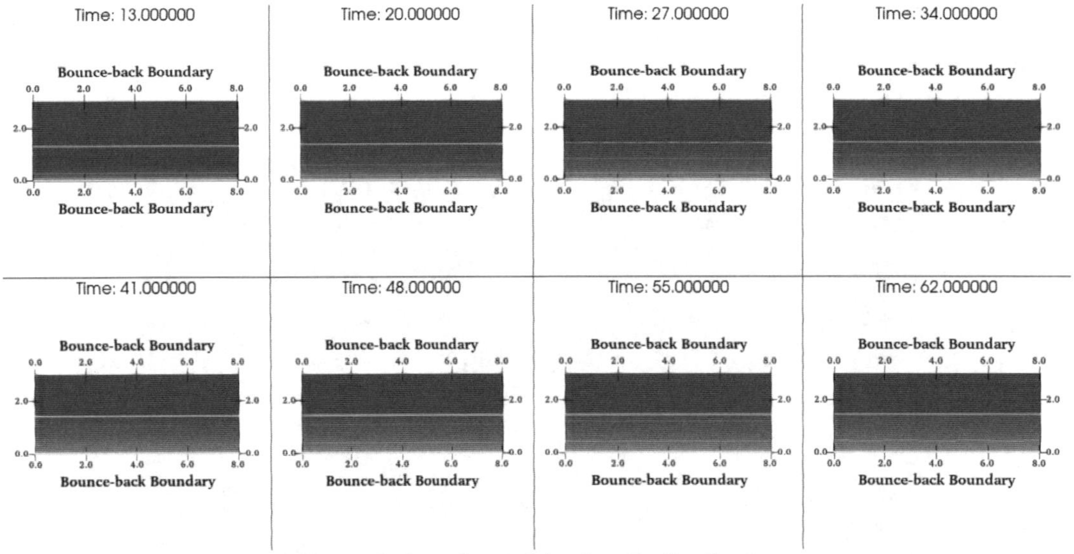

Figure 3. Snapshots of the density distribution (wall density $\rho = 4$, superheat $\Delta T = 0.2T_c$)

4.2.2. Superheat $0.3T_c$. When the superheat of the smooth surface is $0.3T_c$, it is found that there are several density prominences on the heated bottom wall when the iteration step reaches $620000\delta t$, but no bubbles are successfully generated (as shown in Figure 4). This shows that even if the liquid phase in this computational region is given the size of critical state thermodynamic perturbation, homogeneous nucleation is still not easy to occur.

4.2.3. Superheat 0.4T_c. Under this circumstance, there is no bubble nucleation until t = 610000δt, nearly running out of the maximum simulation time in our model (as shown in Figure 5). Despite this value is slightly larger than 0.35T_c, which is approximated when the inter-particle potential is taken as the Lenard-Jones potential [11], Gong's model again demonstrates the correctness and accuracy of the lattice Boltzmann pseudopotential method in the field of multiphase flow. If 0.35T_c is applied, it can be estimated and forecasted that bubbles could be generated in this computational domain, but time-consuming. In Figure 5, there are three bubbles being identified in three locations, two of which are symmetric about the computational area, and the third bubble is right in the middle.

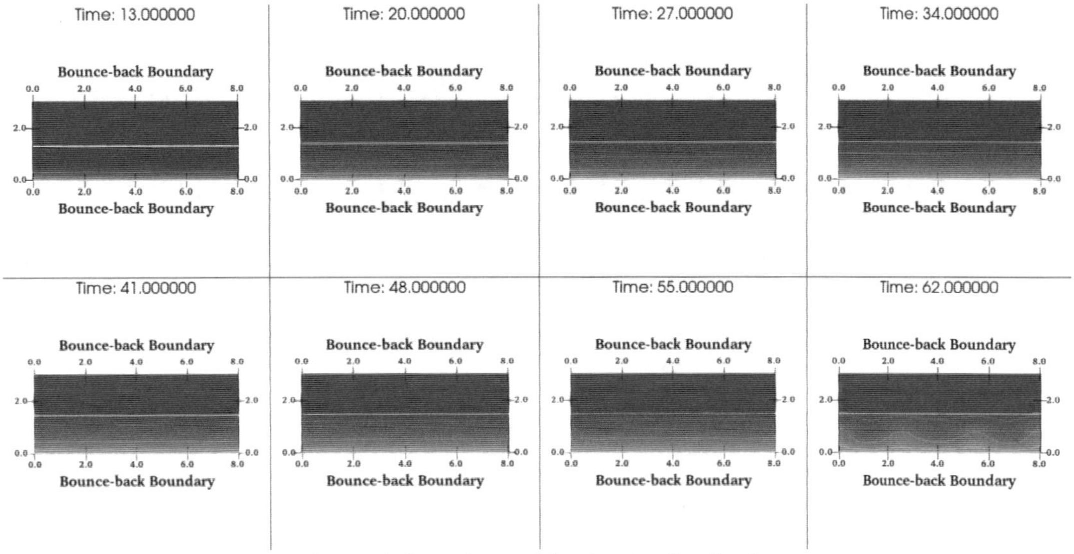

Figure 4. Snapshots of the density distribution (wall density $\rho = 4$, superheat $\Delta T = 0.3T_c$)

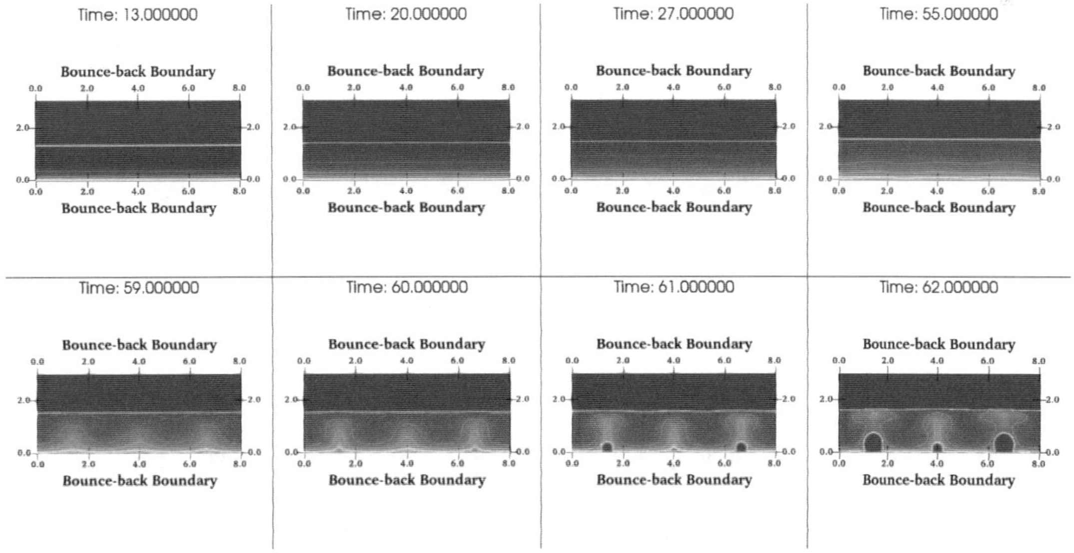

Figure 5. Snapshots of the density distribution (wall density $\rho = 4$, superheat $\Delta T = 0.4T_c$)

4.2.4. Superheat $0.46T_c$. When increasing the superheat of the smooth surface up to $0.46T_c$, the first appearance of complete bubbles is observed when the iteration step reaches $260000\delta t$, much earlier than the $0.4T_c$ condition. The bubbles including 6 symmetrically growing bubbles near lateral sides of the computational area and a vapor film. (as shown in Figure 6)

4.2.5. Superheat $0.51T_c$. In order to further verify the phase change theory, the wall temperature is promoted to $1.21T_c$, six bubbles firstly appear while iterating to $180000\delta t$ (as shown in Figure 7), much earlier than when wall is heated by $1.16T_c$. This means that the slight increase of superheat in the unstable region of the phase diagram does enhance bubble generation. It is worth noting that, the positions where the generated bubbles are also symmetric about the computational domain but the vapor film is obviously much thicker.

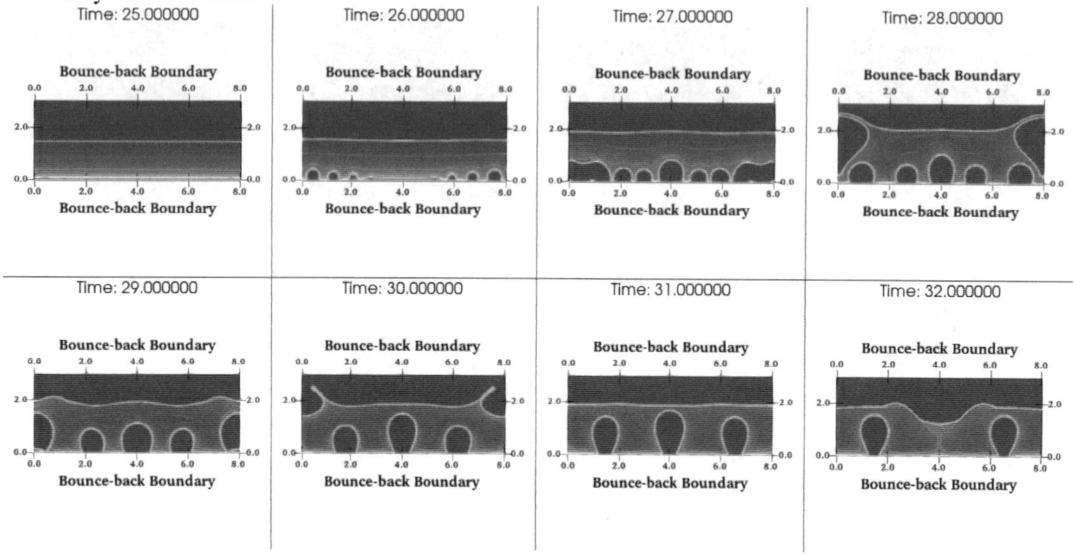

Figure 6. Snapshots of the density distribution (wall density $\rho = 4$, superheat $\Delta T = 0.46T_c$)

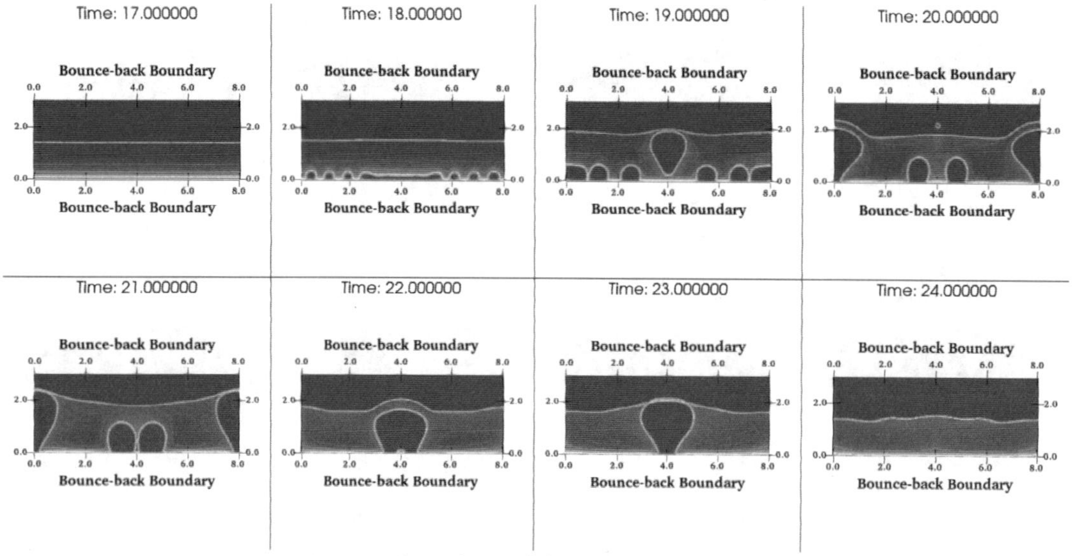

Figure 7. Snapshots of the density distribution (wall density $\rho = 4$, superheat $\Delta T = 0.51T_c$)

4.3. Effects of wall wettability on low-level pool boiling

This section discusses the effects of wall wettability on the low-level pool boiling on horizontal smooth surfaces. Four kinds of surfaces are simulated: purely hydrophobic (wallRho=0.5), purely hydrophilic (wallRho=7.5), relatively hydrophilic (wallRho=4), and relatively hydrophobic (wallRho=3). Although the purely hydrophobic surface is rare in reality, it is worthwhile to look into this extreme case to exhaust all possibilities so as to make a thorough analysis.

4.3.1. Purely hydrophobic wall. According to the previous subsection, it is supposed to generate bubbles with wall applied by a temperature as large as $1.16T_c$(superheat $0.46T_c$) when t = $260000\delta t$. However, as shown in Figure 8, on the purely hydrophobic surface (wallRho=0.5), a vapor film instead of a bubble appears on the bottom wall when iterating to $270000\delta t$ before when it completely covers the bottom wall, and it becomes unstable in the following time steps. When t = $170000\delta t$, two vapor prominences on the lateral sides of the computational area appear to burst into the upper vapor region, behaving similar to bubble growth. If simulating a larger area such as high-level pool boiling in most authentic situations, this could be a film boiling regime. This phenomenon implies that the hydrophobic wall is helpful in bubble coalescence since six bubbles are independently growing when the wall is relatively hydrophilic shown in subsection 4.2.4, and no comparable thickness of vapor film forms under this circumstance.

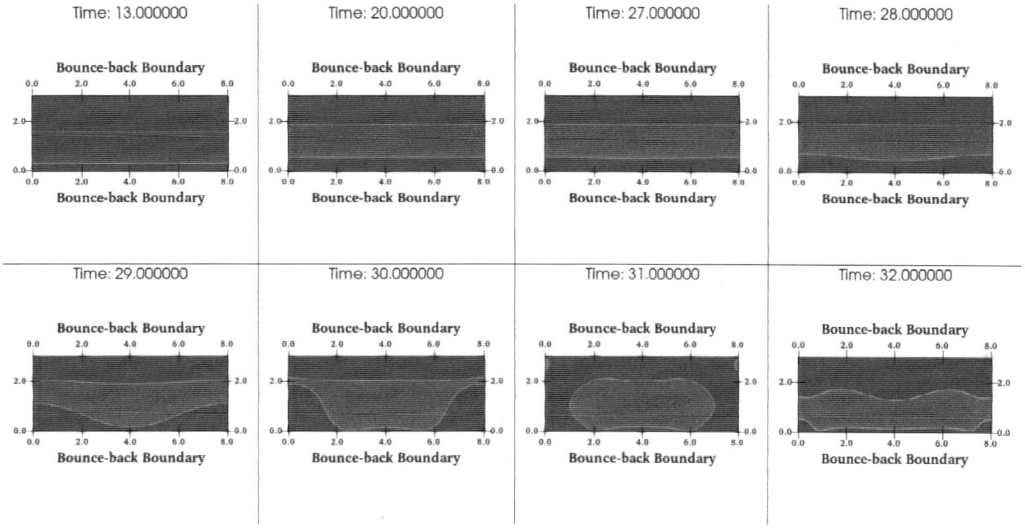

Figure 8. Snapshots of the density distribution (wall density $\rho = 0.5$, superheat $\Delta T = 0.46T_c$)

4.3.2. Purely hydrophilic wall. On a completely hydrophilic surface (wallRho=7.5), bubbles do not nucleate until the iteration step reaches $350000\delta t$. As shown in Figure 9, only two bubbles are generated at a time, and they locate close to lateral sides of the bottom wall as well. After another $2000\delta t$ iterations, two more bubbles are formed subsequently and start to grow with the previous two bubbles independently. Nevertheless, these four bubbles detach the bottom wall as soon as possible, in contrast with the longstanding attachment of bubbles on the relatively hydrophilic wall (wallRho=4) shown in Figure 6, which declares that the wall hydrophilicity is beneficial to bubbles' growth and thus shortens the bubble period. During bubbles growth, it can be seen that the bubbles are always rounded, i.e. the contact angle is close to 180°. After bubbles' detachment, two "small bubbles" are left at the detachment position (shown by the red box in Figure 9). However, the simulation found that the residual "small bubbles" could not re-nucleate and carry out the next bubble cycle under the current conditions. This indicates that hydrophilicity effect plays an important role in surface tension enhancement, an indicator of bubbles' radius-strong surface tension to make a bubble or droplet more rounded.

4.3.3. Relatively hydrophilic wall. As has been simulated in subsection 4.2.3, on a relatively hydrophilic surface (wallRho=4), six symmetrically distributed bubbles in the computational region and a vapor film are observed for the first time when the iteration step reaches $260000\delta t$ (as shown in Figure 6). Compared with the purely hydrophilic surface (wallRho=7.5), the initial nucleation time is delayed. When the iteration step reaches $270000\delta t$, two bubbles merge into larger bubbles and then leave the wall at t = $280000\delta t$. On the other hand, the vapor film in the middle is stabilized as a bubble under the combined action of surface tension and wall interaction, and then it continues to grow and burst into the vapor region (between t = $280000\delta t$ and t = $290000\delta t$ as shown in Figure 6). Furthermore, under this circumstance, only two positions in the computational area are able to become continuous bubble nucleation sites (shown by the red box in Figure 10). Within the maximum iteration step ($620000\delta t$), three complete bubble cycles, including nucleation, growth, and detachment, occurs at these two locations.

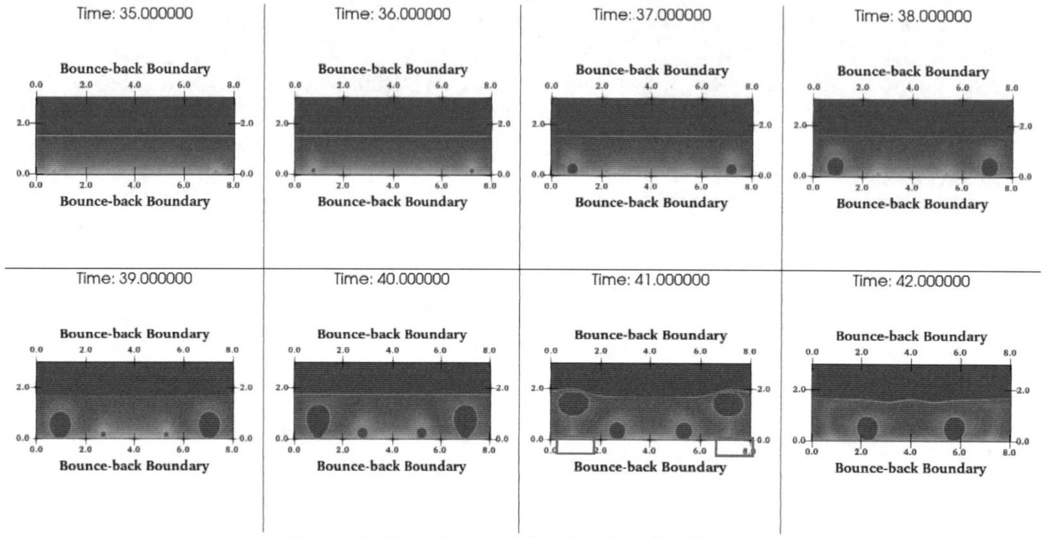

Figure 9. Snapshots of the density distribution (wall density $\rho = 7.5$, superheat $\Delta T = 0.46T_c$)

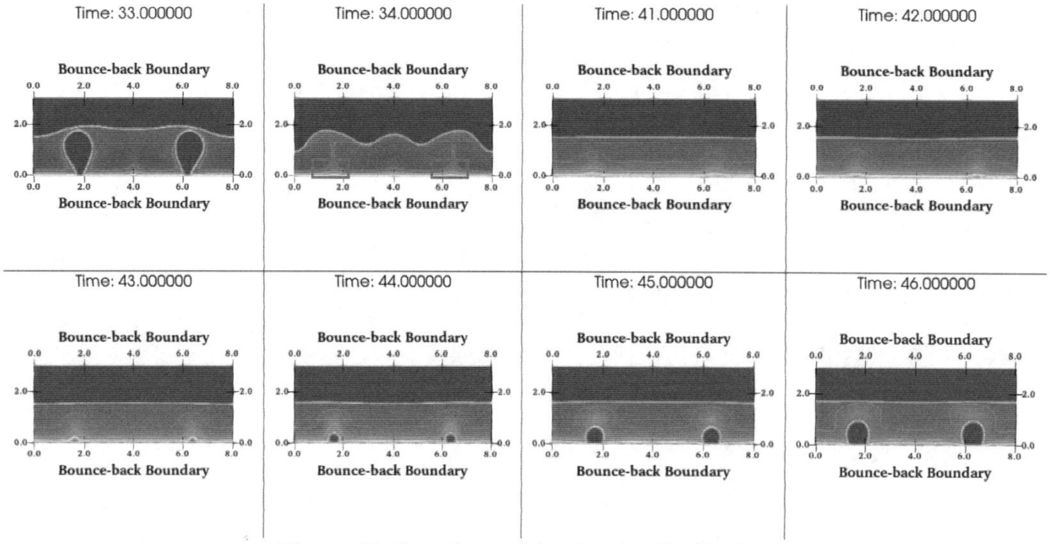

Figure 10. Snapshots of the density distribution (wall density $\rho = 4$, superheat $\Delta T = 0.46T_c$)

4.3.4. Relatively hydrophobic wall. On a relatively hydrophobic surface (wallRho=3), two symmetrically growing bubbles and a vapor film are firstly identified when $t = 170000\delta t$ (as shown in Figure 11). Compared with the relatively hydrophilic condition (wallRho=4), the firstly generated bubbles on the relatively hydrophobic surface are larger and the vapor film is much thicker. After 10000 times iteration, the thick vapor film forms an inverted trapezoidal vapor under the combined action of wall hydrophobic force and surface tension, instead of a regular semicircle, while the two growing bubbles are closely attached to the bottom wall, extending its three-phase bottom line as long as possible ($t = 190000\delta t$, $t = 200000\delta t$, $t = 210000\delta t$, $t = 220000\delta t$. Therefore, when these two bubbles are about to burst into the vapor region, their three-phase contact line is much longer than that of bubbles on the relatively hydrophilic surface (wallRho=4) (as shown in Figure 11 and Figure 12).

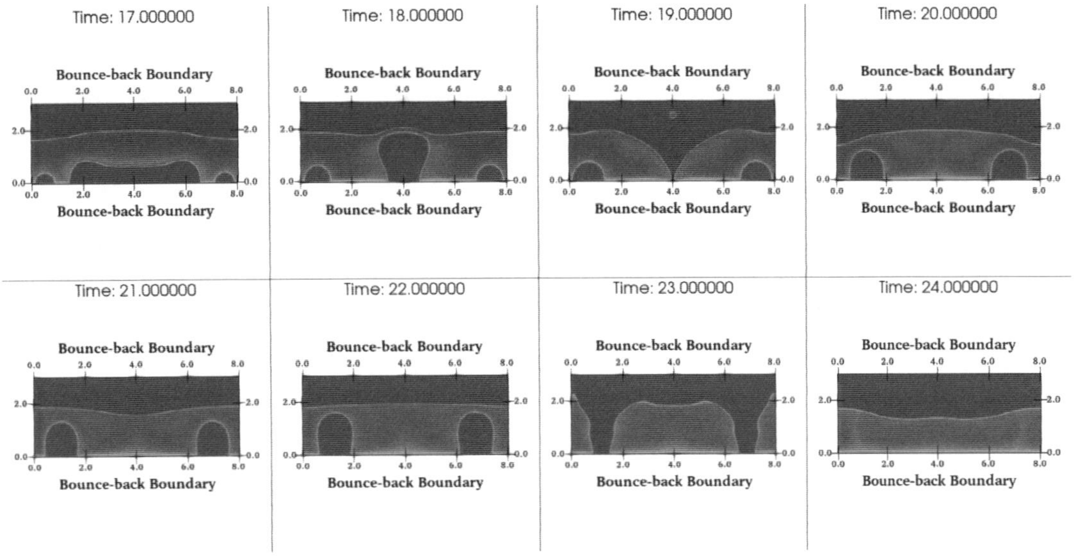

Figure 11. Snapshots of the density distribution (wall density $\rho = 3$, superheat $\Delta T = 0.46T_c$)

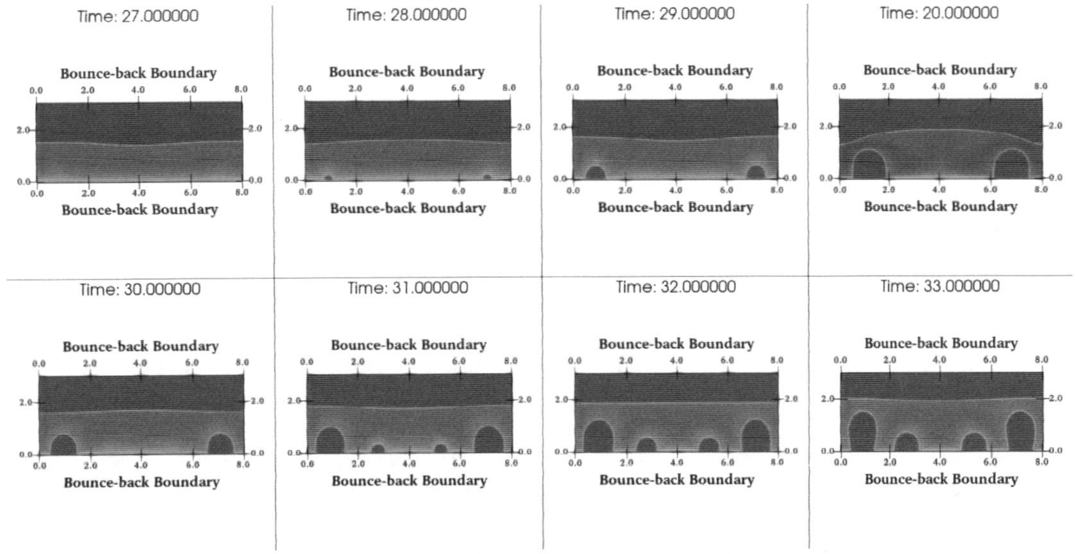

Figure 12. Snapshots of the density distribution (wall density $\rho = 3$, superheat $\Delta T = 0.46T_c$)

In summary, when providing a degree of superheat that enables low-level pool boiling in the computational region, the completely hydrophobic and hydrophilic surface are disadvantage to bubble formation neither. But mechanisms are different: the former one is detrimental to bubble's nucleation by stretching the three-phase contact line and the latter one does harm to bubbles' continuous generation by leaving less rounded nucleation sites. Apart from these two extreme cases, when the wall surface is relatively hydrophobic, the generated bubble is more likely to form a "neck" at the moment of detachment, thus becoming a residual small bubble and serving as the continuous nucleation sites to prepare for the next bubble cycle. When the hydrophilicity of the wall surface increases, the bubbles are more likely to nucleate and grow in the form of a complete circle, so it is not easy to form a "neck" at the moment of detachment. In other words, the hydrophobic wall can possibly leave more nucleation sites and promote re-nucleation, but it also could lead to film boiling which inevitably undermines boiling efficiency because of a long bubble period; the detachment frequency of bubble on a relative hydrophilic is higher, meaning that the energy absorbed by bubbles can be transferred away from the surface more frequently. Finally, the mentioned bubbles' dynamics during the low-level pool boiling are in good agreement with the experimental results, most of which focus on high-level pool boiling, indicating the correctness of Gong's model [12,13].

5. Conclusion

In this paper, based on the lattice Boltzmann pseudopotential phase change method, a low-level pool boiling model imitating the evaporation chamber of a wickless micro-heat pipe is constructed. The model clearly shows the bubble dynamics including nucleation, coalescence, detachment, etc. The homogeneous nucleation theory is examined and the effect of the wall wettability on bubble dynamics is observed. In particular, the model shows that the relatively hydrophobic wall can generate more nucleation sites and the relatively hydrophilic surface provides fewer nucleation sites. We attribute those phenomena to the following reasons--the hydrophilic surface could facilitate surface tension so as to promote bubble departure and shorten the bubble period while the hydrophobic surface undermines surface tension, making bubbles attach to the wall for a longer time and thus lengthen bubble period. Future work will be conducted to quantitatively verify the lattice Boltzmann pseudopotential model in the field of multiphase and optimize the geometry and wettability of the evaporation chamber of a heat pipe.

References

[1] Shan, X. and H. Chen, Lattice Boltzmann model for simulating flows with multiple phases and components. Physical review E, 1993. 47(3): p. 1815.

[2] Shan, X. and H. Chen, Simulation of nonideal gases and liquid-gas phase transitions by the lattice Boltzmann equation. Phys Rev E Stat Phys Plasmas Fluids Relat Interdiscip Topics, 1994. 49(4): p. 2941-2948.

[3] Gong, Shuai, and Ping Cheng. "Numerical investigation of droplet motion and coalescence by an improved lattice Boltzmann model for phase transitions and multiphase flows." Computers & Fluids 53 (2012): 93-104.

[4] Guo, Zhaoli, Chuguang Zheng, and Baochang Shi. "Discrete lattice effects on the forcing term in the lattice Boltzmann method." Physical review E 65.4 (2002): 046308.

[5] Shan, Xiaowen. "Analysis and reduction of the spurious current in a class of multiphase lattice Boltzmann models." Physical Review E 73.4 (2006): 047701.

[6] Gong, S. and P. Cheng, A lattice Boltzmann method for simulation of liquid–vapor phase-change heat transfer. International Journal of Heat and Mass Transfer, 2012. 55(17-18): p. 4923-4927.

[7] Li, Q., P. Zhou, and H.J. Yan, Improved thermal lattice Boltzmann model for simulation of liquid-vapor phase change. Phys Rev E, 2017. 96(6-1): p. 063303.

[8] Mazloomi, M.A., S.S. Chikatamarla, and I.V. Karlin, Entropic lattice Boltzmann method for multiphase flows: Fluid-solid interfaces. Phys Rev E Stat Nonlin Soft Matter Phys, 2015. 92(2): p. 023308.

[9] Benzi, R., et al., Mesoscopic modeling of a two-phase flow in the presence of boundaries: The contact angle. Phys Rev E Stat Nonlin Soft Matter Phys, 2006. 74(2 Pt 1): p. 021509.
[10] Baidakov, V.G., S.P. Protsenko, and Z.R. Kozlova, Thermal and caloric equations of state for stable and metastable Lennard-Jones fluids: I. Molecular-dynamics simulations. Fluid Phase Equilibria, 2008. 263(1): p. 55-63
[11] Nicolas, J.J., et al., Equation of state for the Lennard-Jones fluid. Molecular Physics, 2006. 37(5): p. 1429-1454.
[12] Bourdon, B., et al., Enhancing the onset of pool boiling by wettability modification on nanometrically smooth surfaces. International Communications in Heat and Mass Transfer, 2013. 45: p. 11-15.
[13] Jo, H., et al., Heterogeneous bubble nucleation on ideally-smooth horizontal heated surface. International Journal of Heat and Mass Transfer, 2014. 71: p. 149-157.

Multi-objective Topology Optimization of the Temperature Field Considering the Material Nonlinearities

Guyang Li [1,] *, Lin Li [2], Zhigang Peng [2], Sanbao Hu [3]

[1] China Shipbuilding Corporation Engineering Management Center, Beijing, 100020, China
[2] Wuhan Second Ship Design & Research Institute, Wuhan 430064, China
[3] School of Automotive Engineering, Wuhan University of Technology, Wuhan 430070, China

* Corresponding author email: yzl5927565442@163.com

Abstract. This paper presents a multi-objective topology optimization method for the temperature field considering the material nonlinearities. Although a large number of papers have studied the topology optimizations of the heat transfer problems, none of them considered the influence of the material nonlinearity properties. In this paper, the material nonlinearity is considered. Based on the nonlinear assumption, the mathematical model of the corresponding topology optimization is presented first. Then, the sensitivities of the objectives and the constraints respect to the design variables are derived. And the optimization routines are discussed. After establishing the theories and solving scheme, a software is developed by MATLAB for the topology design of the nonlinear heat transfer. Several nonlinear numerical examples are presented. And the corresponding optimal results generated under the linear assuming are also collected for comparisons. Furthermore, this paper also simply discusses the preventing approaches of instabilities and the implementation methods of some other unusual engineering constraints. These true engineering considerations are very helpful for the real industry design experiences.

Keywords: topology optimization; multiphysics; multi-objective optimization; finite element method; manufacturing constraints.

1. Introduction

Since Bendsøe and Kikuchi proposed the homogenization method of topology optimization, topology optimization methods have been widely used to solve structural design problems such as stiffness design [1], strength design [2] and maximum eigenfrequency design [3] of linear elastic problems. For more than 20 years, topology optimization has always been a research hotspot in the field of structural optimization. With the further development of related research, the topology optimization method has been gradually extended to the conceptual design of multi-physical field problems, and has become a multi-disciplinary design tool. Topology optimization design of temperature field is one of the important research directions. Diaz and Benard [4] studied the topology optimization design of two-dimensional thermal barrier structure; Li et al. [5] studied the topology optimization design of two-dimensional

Content from this work may be used under the terms of the Creative Commons Attribution 3.0 licence. Any further distribution of this work must maintain attribution to the author(s) and the title of the work, journal citation and DOI.
Published under licence by IOP Publishing Ltd

temperature field by using the evolutionary structure method; Gersborg Hansen et al. [6] used the finite difference method to study the topology optimization of two-dimensional temperature field; Sigmund [7-8] studied the topology optimization of two-dimensional multi physical coupled fields including temperature field problems.

For detailed introduction of topology optimization of temperature field problems, please refer to the monograph published by Bendsøe and Sigmund [1] in 2003. However, the above studies all assume that the thermal conductivity of the material is constant, without considering the influence of temperature change on the heat transfer characteristics of the material. However, in engineering materials, materials with constant thermal conductivity do not exist. Although in some specific cases, the heat transfer parameters of the material can be considered as constant approximately because the temperature changes little or the thermal conductivity of the material does not change significantly with the temperature or based on other simplified models. However, when the temperature span is large and the thermal conductivity of materials varies greatly with temperature, the nonlinear thermal conductivity of materials must be considered. Based on the above reasons, the topology optimization problem of temperature field where the thermal conductivity of material is a nonlinear function of temperature is discussed in this, and the influence of material nonlinearity on the optimal topology of structure is fully considered.

2. Governing equation of heat conduction

We take the two-dimensional heat conduction problem as an example to derive the mathematical model of the multi-objective topology optimization problem of the nonlinear heat conduction structure. According to the heat transfer theory [9], the governing equation of steady-state heat conduction in orthotropic solids is:

$$\begin{aligned} -\frac{\partial}{\partial x}\left(k_x\frac{\partial T}{\partial x}\right)-\frac{\partial}{\partial y}\left(k_y\frac{\partial T}{\partial y}\right)&=q_v \text{ in } \Omega(1.\text{a}) \\ T\big|_{\Gamma_2}&=T_1 \text{ on } \Gamma_1(1.\text{b}) \\ -\frac{\partial T}{\partial \mathbf{n}}\Big|_{\Gamma_1}&=q_2 \text{ on } \Gamma_2(1.\text{c}) \\ -k\frac{\partial T}{\partial \mathbf{n}}\Big|_{\Gamma_3}=h\left(T-T_f\right)\big|_{\Gamma_3}& \text{ on } \Gamma_3(1.\text{d}) \end{aligned} \tag{1}$$

Where $\Omega \in R^2$ is the solution domain, Γ is the boundary of Ω, T is the temperature of the object, q_v is the heat source density in the material. Equation (1. a) is the equilibrium equation; Equation (1. b) represents the first kind of boundary conditions, Γ_1 is the boundary of the first type of boundary condition; Equation (1. C) represents the second kind of boundary conditions, Γ_2 is the boundary of the second kind of boundary conditions; Equation (1. d) represents the third kind of boundary conditions, Γ_3 is the boundary of the third type of boundary conditions (convective boundary conditions), where A represents the temperature of the external fluid in contact with B. k_x and k_y are the thermal conductivities of the material in the two main axis directions. Assuming that the material is isotropic and considering its nonlinearity with respect to temperature, then:

$$k_x = k_y = k(T) \tag{2}$$

According to the principle of minimum entropy generation, the "weak form" of steady two-dimensional heat conduction problem can be expressed as:

$$-\int_{\Omega}\left[k\nabla^2 T+q_v\right]\delta T dV+\int_{\Gamma_1} k\frac{\partial T}{\partial \mathbf{n}}\delta T dS$$
$$+\int_{\Gamma_2}\left(k\frac{\partial T}{\partial \mathbf{n}}+q_2\right)\delta T dS \tag{3}$$
$$+\int_{\Gamma_3}\left(k\frac{\partial T}{\partial \mathbf{n}}+h\left(T-T_f\right)\right)\delta T dS=0$$

Where $\mathbf{n}$ represents the outer normal vector of the corresponding boundary. Next, the finite element equation of the nonlinear steady-state temperature field problem is derived from equation (3). The domain Ω is discretized into n finite elements, and the temperature distribution in the element is:

$$T^e=[N]\{T\}^e \tag{4}$$

Where, T^e represents the cell temperature, $[N]$ is the cell interpolation function (shape function), $\{T\}^e$ is the temperature of the node, then the temperature in the domain can be expressed as:

$$T=\sum_{e=1}^{n}[N]\{T\}^e \tag{5}$$

From formula (4):

$$\frac{\partial T^e}{\partial x_i}=\begin{bmatrix}\dfrac{\partial N}{\partial x}\\ \dfrac{\partial N}{\partial y}\end{bmatrix}\{T\}^e=[B]\{T\}^e \tag{6}$$

where

$$[B]=\begin{bmatrix}\dfrac{\partial N}{\partial x} & \dfrac{\partial N}{\partial y}\end{bmatrix}^T \tag{7}$$

Bring equations (5), (6), (7) into (3) to sort out the finite element equation

$$\mathbf{K_T T}=\mathbf{F} \tag{8}$$

where, $\mathbf{K_T}=\mathbf{K}+\mathbf{K_3}$, $\mathbf{F}=\mathbf{F_v}-\mathbf{F_2}+\mathbf{F_3}$, and $\mathbf{K}$, $\mathbf{K_3}$, $\mathbf{F_v}$, $\mathbf{F_2}$, $\mathbf{F_3}$ are respectively assembled from $[K]^e$, $[F_3]^e$, $\{F_V\}^e$, $\{F_2\}^e$ and $[F_3]^e$.

$$[K]^e=\int_{V_e}[B]^T k[B]dV$$
$$\{F_V\}^e=\int_{V_e}q_v[N]^T dV$$
$$\{F_2\}^e=\int_{\Gamma_e}q_2[N]^T dS \tag{9}$$
$$[K_3]^e=\int_{\Gamma_e}[N]^T h[N]dS$$
$$[F_3]^e=\int_{\Gamma_e}hT_f[N]^T dS$$

$[K]^e$ is the stiffness matrix of the element, Its integral domain V_e represents the domain of the element; $\{F_V\}^e$ is the unit heat source vector generated by the heat source q_v; $\{F_2\}^e$ is the heat flow vector generated by the known heat flow on the second kind of boundary conditions; $[K_3]^e$ is the element stiffness matrix caused by the third kind of boundary conditions; $[F_3]^e$ is the heat flow vector on the boundary where the third kind of boundary conditions are known. Please refer to the relevant finite element monograph [10] for the assembly of the global stiffness matrix and the heat flow vector and the specific derivation process of equation (8), which will not be repeated here. Since the thermal conductivity $k(T)$ is a nonlinear function of temperature, $[K]^e$ and $\mathbf{K}_\mathrm{T}$ are also a function of temperature, the resulting finite element equation (8) is a nonlinear algebraic equation group, which can be solved by iterative method.

3. Mathematical model and solution method of multi-objective topology optimization

The mathematical model of multi-objective topology optimization can be generally expressed as follows:

$$\begin{aligned} \min \mathbf{\Phi} &= [\Phi_1 , \Phi_2 ,\ldots \Phi_m] \\ \text{s.t.} : \mathrm{g}_i(T,\ \rho) &\le 0 \quad (i=1,2,\ldots p) \\ \mathrm{h}_j(T,\ \rho) &= 0 \quad (j=1,2,\ldots q) \end{aligned} \tag{10}$$

Wherein the objective function $\mathbf{\Phi}$ is a vector composed of m functions, g_i and h_j represent inequality constraint and equality constraint functions respectively, ρ is the normalized material density. In this paper, SIMP [11] material interpolation method is used to relax the topology optimization problem of 0-1 programming into a continuous variable optimization problem. SIMP method expresses the thermal conductivity k of the material as the following function of density ρ:

$$k_e = k^{\min} + \left(k^0 - k^{\min}\right)\rho_e^{pl} \tag{11}$$

In which, k_e is the thermal conductivity of the unit, ρ_e is the regularization density of the corresponding element. k^0 is the thermal conductivity when the material density is 1, $k^{\min}$ is the thermal conductivity when the material density is $\rho^{\min}$ (considered as a hole, no material). To avoid solving singularities, we usually take $k^{\min} = 0.001\, k^0 , \rho^{\min} = 0.001$, pl is a penalty factor introduced by SIMP method, which makes the relaxed continuous variable problem return to 0-1 problem by implicit penalty method.

The new optimization problem constructed by the weighting method can be expressed as:

$$\begin{aligned} \min \mathbf{\Phi} &= \omega_1\bar{\Phi}_1 + \omega_2\bar{\Phi}_2 + \ldots + \omega_m\bar{\Phi}_m \\ \text{s.t.} : \mathrm{g}_i(T,\ \rho) &\le 0 \quad (i=1,2,\ldots p) \\ \mathrm{h}_j(T,\ \rho) &= 0 \quad (j=1,2,\ldots q) \\ \rho^{\min} &\le \rho \le 1 \end{aligned} \tag{12}$$

wherein, $\bar{\Phi}_m$ represents the mth target of normalization, ω_m is its corresponding weight coefficient, and $\sum_{i=1}^{m}\omega_i = 1$.

a) Sensitivity analysis

For the optimization problem (12), mathematical programming methods, such as sequential linear programming (SLP) and moving asymptote method (MMA), can usually be used to solve it. SLP and MMA are both gradient based methods. When these methods are used, one step (sensitivity) of all objective functions and constraint functions with respect to design variable ρ must be provided. Here, we use the adjoint method to derive the first-order analytical sensitivity of the objective function. Consider the objective function

$$\Phi_m = \int_\Omega A(T,\rho)\,dV \tag{13}$$

In order to derive the sensitivity of the objective function Φ_m, we introduce the Lagrange term to construct a new function Φ_m^*:

$$\Phi_m^* = \Phi_m + \boldsymbol{\lambda}^T \left(\mathbf{K_T T} - \mathbf{F} \right) \tag{14}$$

Take equation (8) into equation (14) and find the first derivative of the design variable ρ from equation (14)

$$\begin{aligned} \frac{d\Phi_m}{d\rho} &= \frac{\partial \Phi_m}{\partial \rho} + \frac{\partial \Phi_m}{\partial T}\frac{dT}{d\rho} \\ &+ \lambda^T K_T \frac{dT}{d\rho} + \lambda^T \frac{\partial K_T}{\partial \rho} T \\ &= \frac{\partial \Phi_m}{\partial \rho} + \frac{\partial \Phi_m}{\partial T}\frac{dT}{d\rho} \\ &+ \lambda^T K_T \frac{dT}{d\rho} + \lambda^T \frac{\partial K_T}{\partial \rho} T \\ &= \frac{\partial \Phi_m}{\partial \rho} + \left(\frac{\partial \Phi_m}{\partial T} + \lambda^T K_T \right) \frac{dT}{d\rho} \\ &+ \lambda^T \frac{\partial K_T}{\partial \rho} T \\ &= \frac{\partial \Phi_m}{\partial \rho} + \left(\frac{\partial \Phi_m}{\partial T} + \lambda^T K_T \right) \frac{dT}{d\rho} \\ &+ \left(k^0 - k^{\min} \right) pl \; \rho^{(pl-1)} \lambda^T K_T T \end{aligned} \tag{15}$$

In order to eliminate the implicit term $\frac{dT}{d\rho}$, we introduce the adjoint problem

$$\mathbf{K_T} \boldsymbol{\lambda} = -\frac{\partial \Phi_m}{\partial T} \tag{16}$$

wherein, $\mathbf{K_T}$ is the overall stiffness matrix in equation (8). It can be seen from equation (16) that as long as the load vector $\mathbf{F}$ in equation (8) is replaced by a new vector $-\frac{\partial A(T,\rho)}{\partial T}$, the value of the Lagrange multiplier vector $\boldsymbol{\lambda}$ can be obtained, and the sensitivity of the objective function Φ_m can be obtained by bringing it into (15):

$$\begin{aligned} \frac{d\Phi_m}{d\rho} &= \frac{\partial \Phi_m}{\partial \rho} \\ &+ \left(k^0 - k^{\min} \right) pl \; \rho^{(pl-1)} \lambda^T K_T T \end{aligned} \tag{17}$$

For volume constraint functions:

$$\Phi_V = \frac{1}{|\Omega|} \int_\Omega \rho d\Omega \tag{18}$$

Wherein, $|\Omega|$ represents the volume (surface) product of the design domain Ω. Although its sensitivity function can also be derived by formula (17), since there is no explicit T term in Φ_V, it can be directly obtained by variational method:

$$\frac{d\Phi_V}{d\rho}=\frac{1}{|\Omega|} \tag{19}$$

Other objective functions, such as centroid position

$$C_{cog}=\left(x_i^*-\frac{1}{V^*}\int_\Omega x_i\rho d\Omega\right)^2-\varepsilon \tag{20}$$

The sensitivity function can also be obtained directly by variational method

$$\frac{dC_{cog}}{d\rho^e}=2\left(x_i^*-\frac{1}{V^*}\int_\Omega x_i\rho d\Omega\right)\left(-\frac{x_i^e}{V^*}\right) \tag{21}$$

Where, C_{cog} represents the position of the center of mass, x_i represents the coordinate of the center of mass on the i th coordinate axis, x_i^* is the desired coordinate of the center of mass on the i th coordinate axis, V^* is the volume of the optimization result, x_i^e is the coordinate of the center of cell e on the i th coordinate axis, and ε is a very small positive number.

b) Mathematical programming

After repeated iterations, Until the convergence requirements are met. The final design point is the optimal solution of the original problem. MMA transforms the implicit optimization problem into a series of explicit and more simply strictly convex approximation subproblems by introducing a moving asymptote. The approximation function is determined by the pre-determined left and right asymptotic points and the derivatives of the original objective function and constraint function at each point. MMA method is the most effective mathematical programming method to solve multi-objective topology optimization [7]. For the detailed introduction of MMA, please refer to the articles of Svanberg [12 - 13] and Bruyneel et al. [14]. The examples given in this paper all adopt the MMA method.

c) Numerical instability prevention

Topology optimization methods based on SIMP model usually led to checkerboard patterns and mesh dependency. Checkerboard phenomenon refers to that in the optimization result, the local cell density values alternate between 0 and 1 to form a checkerboard like grid structure. The appearance of checkerboard phenomenon is not because this material arrangement has the best stiffness, but because the approximate accuracy of finite element is not high, and the local stiffness is larger than the actual value. When a checkerboard appears, the optimization result is difficult to manufacture. Therefore, it must be overcome in the optimization process. Common methods to eliminate checkerboard include: using high-order element, sensitivity filtering method, perimeter control method, non-conforming element, etc. [1]. Grid dependency refers to the phenomenon that the denser the grid division in the solution domain, the more detailed the topology optimization results will be. This phenomenon is fundamentally caused by the local optimality of optimization methods. The sensitivity filtering method is an effective method to eliminate the grid dependence, and it can ensure that the optimization results meet the requirements of the minimum size constraint at the same time. For a detailed introduction of sensitivity filtering, please refer to the article of Sigmund [15]. In this paper, sensitivity filtering method is used to eliminate grid dependence and checkerboard.

4. Numerical examples

This part demonstrates the topology optimization of temperature field with nonlinear thermal conductivity by several typical numerical examples. The examples in this paper are obtained by the program independently developed based on MATLAB language, including single objective and multi-objective problems. Finally, we also briefly discuss the influence of common engineering constraints, such as symmetric constraints on the optimization results. In all calculation examples, the thermal conductivity of the material is measured by experiments, and the specific values are shown in Table 1.

Table 1. Relationship between thermal conductivity and temperature

Temperature (°C)	Thermal conductivity **(W/(m·K)**)
18	14
20	14
22	36
24	63
26	47
28	43
30	37
32	36
34	48
36	52
38	51
40	48
42	43
44	37
46	36
48	48
50	52
52	51
54	43
56	37
58	36
60	48

In the calculation process, the thermal conductivity of the material at a certain temperature is obtained by cubic spline interpolation. The relationship between the thermal conductivity and temperature of the material is shown in Fig. 1, which has strong nonlinear characteristics.

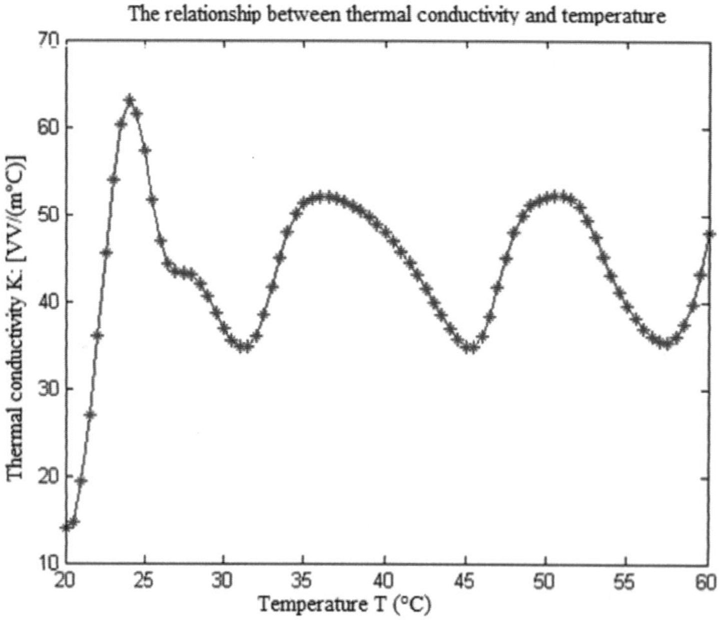

Fig. 1 Relationship between thermal conductivity and temperature of materials

4.1. Single objective problem

As shown in Fig. 2, in the square design domain with side length $1\times 1\text{m}^2$, $q_v = 100\text{W/m}^2$, the upper and left sides are the temperature boundary conditions of T = 40 ℃ and T = 20 ℃ respectively, and the other two sides are the adiabatic conditions. First, we discuss the following single objective problem

$$\begin{aligned} \min \quad & \Phi_1 = \int_{\Omega} T^T K_T T dV \\ s.t. \quad & \Phi_v = \frac{1}{|\Omega|}\int_{\Omega} \rho d\Omega \le 0.4 \end{aligned} \tag{22}$$

The sensitivities of the objective function Φ_1 and the constraint function Φ_v are obtained by equations (17) and (18), respectively. Discretize the design domain into 80 × 80 quadrilateral linear elements. At the same time, the sensitivity filter function proposed by Sigmund [15] is used to eliminate the checkerboard and grid dependence, take $r_{\min} = \frac{1}{80}\times 6 = 0.075m$ and the penalty factor is taken $pl = 3$. For the comparison of the results of the nonlinear design, we first assume that the material is a linear material with a thermal conductivity of 40. The optimization results are shown in Fig. 3 (a), and the temperature distribution under the corresponding optimization results is shown in Fig. 4 (a). If the thermal conductivity parameter of the material is a nonlinear parameter given in Fig. 1, the topology optimization result is shown in Fig. 3 (b), and the temperature distribution under the corresponding optimization result is shown in Fig. 4 (b). It can be seen from Fig. 3 (a) and Fig. 3 (b) that the material is mainly distributed in the area connected with the low-temperature boundary t = 20 ℃ to facilitate heat dissipation, which is consistent with engineering common sense. Although the two optimization results are the same, it is not difficult to find that the two results are very different in the position and angle of material distribution when comparing the fine structure. Therefore, it is necessary to consider the nonlinear thermal conductivity of materials.

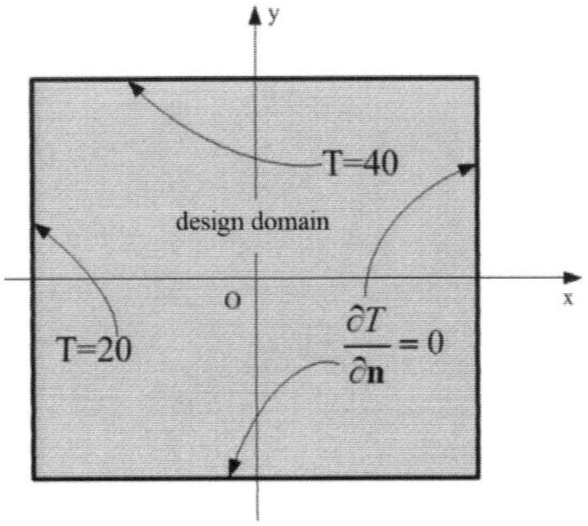

Fig. 2 Schematic diagram of design domain and boundary conditions

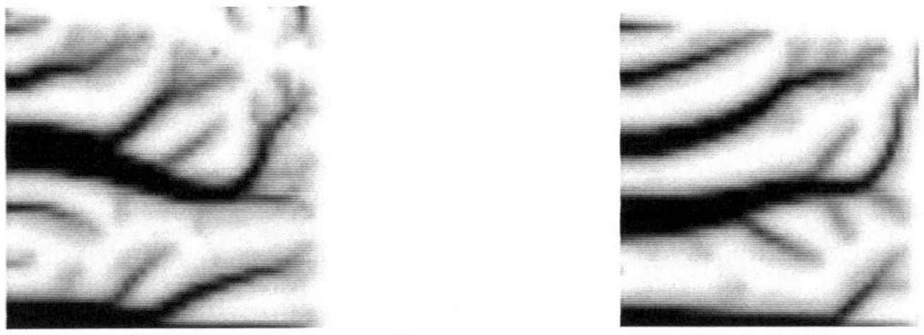

(a) Optimal topology under linear assumption; (b) Optimal topology under nonlinear assumption
Fig. 3 Shows the results of topology optimization.

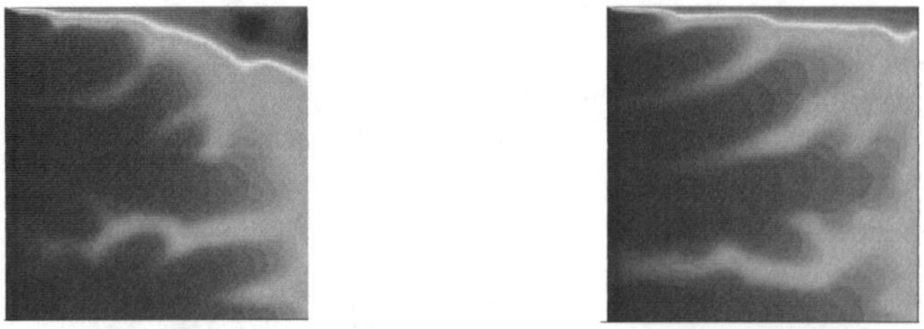

(a) Temperature distribution under linear assumption; (b) Temperature distribution under nonlinear assumption
Fig. 4 Temperature distribution corresponding to two topology optimization results

4.2. Multi objective problem

Now, let's consider a multi-objective topology optimization problem. Similarly, as shown in Figure 2, we use nonlinear materials to discuss the following multi-objective problems:

$$\begin{aligned} \min \quad & \Phi_1 = \int_\Omega T^T K_T T dV \\ & \Phi_2 = \left(0.1 - \frac{1}{V^*}\int_\Omega y\rho d\Omega\right)^2 \\ s.t. \quad & \Phi_v = \frac{1}{|\Omega|}\int_\Omega \rho d\Omega \le 0.4 \end{aligned} \tag{23}$$

In fact, Φ_2 is a target representing a centroid coordinate $y = 0.1$, and its sensitivity is obtained by formula (21). The coordinate origin is located in the center of the square, and the x axis and y axis are perpendicular to the left and right sides and the upper and lower sides, as shown in Fig. 2. Φ_1 and Φ_v is the objective function and constraint function discussed in the previous single objective problem. Similarly, the design domain is discretized into 80 × 80 quadrilateral linear elements. At the same time, the sensitivity filter function proposed by Sigmund [15] is used to eliminate the checkerboard and grid dependence, take $r_{\min} = \frac{1}{80} \times 6 = 0.075m$ and the penalty factor $pl = 3$ and weight coefficient $\omega_1 = \omega_2 = 0.5$ are taken. The nonlinear material parameters shown in Fig. 1 are used. When the centroid target is not considered, the result shown in Fig. 3 (b) is obtained, and the centroid position is $x_{cog} = -0.0851m,\ y_{cog} = -0.0516m$. When considering the centroid target, the optimization results are shown in Fig. 5 (a). The centroid position $x'_{cog} = -0.0726m,\ y'_{cog} = 0.0969m \approx 0.1m$,and the corresponding temperature distribution are shown in Fig. 5 (b).

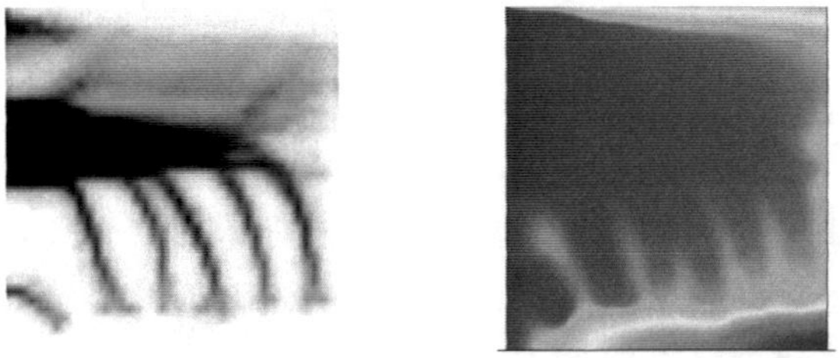

(a) Topology optimization results; (b) Corresponding temperature distribution
Fig. 5 The results of multi-objective topology optimization.

4.3. *Engineering constraints*

Engineering constraints are introduced into topology optimization problems to consider the performance of manufacturing cost, process, product appearance and so on. It is a very important and challenging research direction in topology optimization problems. Asymmetry is one of the most common constraints. Axisymmetric constraints are often considered from the requirements of product appearance and manufacturing (Segmented Manufacturing). Here, in the nonlinear single objective problem (22) shown in Fig. 2, we introduce the axisymmetric constraint, and use the same parameters as before to redesign the topology optimization. So, how to realize axisymmetric constraint? This paper adopts the following methods: first, the design domain Ω is divided into two identical subdomains Ω_1 and Ω_2 according to the symmetry axis, and the density of each element of Ω_1 is selected as the design variable; Then, the density of each element of Ω_2 is matched with the design variable of Ω_1 through the symmetry relationship. If the center of the element is just on the symmetry axis, the density of the element on the symmetry axis should also be taken as the design variable. In this way, the original optimization problem with n design variables is transformed into an optimization problem with design variables close to or equal to n/2, which reduces the scale of the optimization problem while achieving engineering constraints. Figure 6 (a) shows a topology optimization result symmetrical about the straight line $y=0$. Figure 7 (a) shows a topology optimization result symmetrical about the straight line $y=-x$.

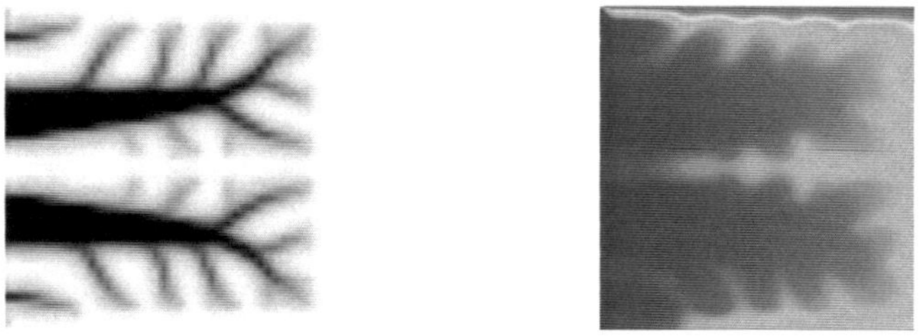

(a) Topology optimization results; (b) Corresponding temperature distribution
Fig. 6 The topology optimization results under the linear symmetry constraint.

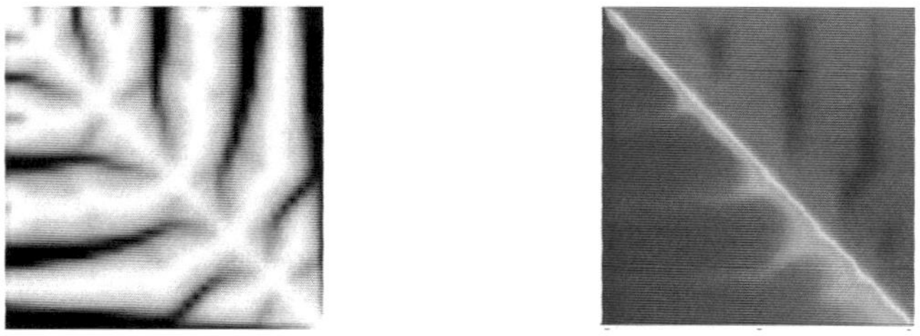

(a) Topology optimization results; (b) Corresponding temperature distribution
Fig. 7 The topology optimization results under the linear symmetry constraint.

5. Conclusion

The multi-objective topology optimization problem of temperature field with thermal conductivity as a nonlinear function of temperature is derived and realized in this paper. Through several typical examples, we can draw the following conclusion: in some cases, considering the nonlinear properties of materials has an important impact on the results of topology optimization. At the same time, it is proved that it is feasible to consider more design objectives and Engineering (Manufacturing) constraints in the conceptual design stage of structures by introducing centroid position objectives and axisymmetric constraints into topology optimization. Although only two-dimensional problems are discussed, the method described in this paper can be extended to three-dimensional problems without any obvious obstacles.

References

[1] Bendsøe MP, Sigmund O. Topology Optimization: Theory, Methods and Applications. Springer-Verlag (Berlin Heidelberg New York). 2003.
[2] Cheng Gengdong, Zhang Dongxu Topology optimization of plane elastic body constrained by stress Journal of Dalian University of technology, 1995, 35 (1): 317-322
[3] Peterson NL. Maximization of eigenvalues using topology optimization. Structural and Multidisciplinary Optimization. 2000, 20:2-11.
[4] Diaz AR, Benard A. Topology optimization of heat-resistant structures. Proc ASME Des Eng Tech Conf. 2003, 2A:633-639.
[5] Li Q, Steven GP, Xie YM. Evolutionary topology optimization for temperature reduction of heat conducting fields. Internation Journal for Heat and Mass Transfer. 2004, 47:5071-50839.
[6] Gersborg-Hansen A, Bendsøe MP, Sigmund O. Topology optimization of heat conduction problems using finite volume method. Structural and Multidisciplinary Optimization. 2006, DOI: 10.1007/s00158-005-0584-3.
[7] Sigmund O. Design of multiphysics actuators using topology optimization - Part I: One-material structures, Computer Methods in Applied Mechanics and Engineering. 2001, 190:6577-6604.
[8] Sigmund O. Design of multiphysics actuators using topology optimization – Part II: Two-material structures, Computer Methods in Applied Mechanics and Engineering. 2001, 190:6605-6627.
[9] Kong Xiangqian Application of finite element method in heat transfer, Third Edition Science Press, Beijing 1998.
[10] Wang Xucheng Finite element method Tsinghua University Press, Beijing,2003.
[11] Bendsøe MP. Optimal shape design as a material distribution problem, Structural Optimization, 1989, 1(4):193-202.
[12] Svanberg K. The method of moving asymptotes - a new method for structural optimization. International Journal for Numerical Methods in Engineering. 1987, 24:359-373.
[13] Svanberg K. A globally convergent version of MMA without linesearch. Proceedings of the First World Congress of Structural and Multidisciplinary Optimization (Goslar, Germany) 1995, 9-16.
[14] Bruyneel M, Duysinx P, Fleury C. A family of MMA approximations for structural optimization. Structural and Multidisciplinary Optimization. 2002, 24:263-276.
[15] Sigmund O, A 99 line topology optimization code written in Matlab, Structural and Multidisciplinary Optimization. 2001, 21:120-127.

The Application of Boundary Element Method for The Solution of The Inverse Heat Conduction

Jin Wen [1, *], Ziyan Xiang [1, a], Longlong Li [2, b], Yulu Zhang [3, c], Huili Yan [3, d] and Yi Zhu [4, e]

[1] College of Mathematics and Statistics, Northwest Normal University, Lanzhou, China.
[2] Computer Science and Engineering College, Northwest Normal University, Lanzhou, China
[3] College of Physics and Electronic Engineering, Northwest Normal University, Lanzhou, China.
[4] College of Business, Northwest Normal University, Lanzhou, China.

*Corresponding author e-mail: wenj@nwnu.edu.cn; [a] xiangziyan1006@163.com; [b]201971010122@nwnu.edu.cn; [c] zyldew@163.com; [d] 2219714872@foxmail.com; [e]2694195574@foxmail.com

All authors are first authors

Abstract. In this paper, the general and supplementary conditions are used to determine the inverse heat conduction in the parabolic equation. The measured conditions ensure that the inverse problem has a unique solution, but the solution is not stable, so the problem is ill-posed. Using the boundary element method to discretize the original problem, it is used to solve the problem. On this basis, the direct method, the least square method and the regular method are tried. Finally, the minimal energy method is used to solve the system. Numerical experiments are carried out and the results are discussed.

Keywords: boundary element method; minimal energy method; heat sourse; inverse problems.

1. Introduction

Different methods can be used to solve the inverse heat conduction, and the impor- tance of solving such problems is explained through the application in engineering and aerospace. And it is proposed that the internal state of the solid can be known by deter- mining the surface heat flux or the internal temperature of the boundary, and the problem can be solved by using the boundary conditions. [1] The uniqueness of the existence of the solution of the inverse problem should also be considered, there is a solution to be found, and the requirements for the solution effect of the problem are put forward, which theoret- ically explains the solvability of such problems as the inverse problem. [2] The minimal energy method and its application are introduced. It can be seen from the text that for ill-posed problems, the optimal solution can be obtained by the minimum energy method derived from the physical principle. In this way, after using the boundary element method, the problem can be obtained. [3] The function of

Content from this work may be used under the terms of the Creative Commons Attribution 3.0 licence. Any further distribution of this work must maintain attribution to the author(s) and the title of the work, journal citation and DOI.
Published under licence by IOP Publishing Ltd

the program e04ucf in the NAG library library, the method of use and the programming principle are introduced. This method can stably solve the numerical solution. Boundary element analysis is used to solve a kind of two-dimensional heat transfer equation, which provides the possibility for further expansion of boundary element method. [4] The conditions are very important for the solvability of the problem. Whether the conditions given by the heat source are suitable can be used to know whether the problem can be solved. [5] Some papers have compared the direct method, the least square method and the boundary element method to solve the heat con- duction equation, and verified the correctness of the theory with numerical experimental analysis. [6] The finite element method is used for analysis, and the method of expanding the solution space is also used, which is an idea to solve such problems. [7] The boundary element method is used for parabolic equations as well as elliptic equations, and relatively good results are obtained. [8] The solution of the integral equation has been studied, and the partial diff erential can be deformed by using Green's formula, and then the deformed integral equation can be solved. [9] Stability of numerical solution was studied, and the importance of the program on the quality of numerical solution was explained through diff erent programs. [10] Problem solved for diff erent thermal conductivity. [11]

The rest of this paper is arranged as follows. The second part explains the problem to be solved, the third part uses the boundary element method to analyze the problem, and the partial diff erential equation is changed to the integral form for discretization. The fourth part explores the four diff erent solutions - direct, least squares, regularization and our final choice - minimal energy method, part 5 conducts numerical experiments on two examples and part 6 concludes this paper.

2. Formulation of the problem

The one-dimensional time dependent inverse heat conduction problem in which the sourse depend on time only and satisfies:

$$(\partial/\partial t)u(x,t) = a^2(\partial^2/\partial x^2)u(x,t) + f(t),\ 0 < x < 1,\ 0 < t \le t_{max}\ ,$$
$$u(x,0) = u_0(x),\ 0 \le x \le 1,$$
$$u(0,t) = 0,\ 0 \le t \le t_{max,}$$
$$u(1,t) = g(t),\ 0 \le t \le t_{max,} \tag{1}$$

with the investigated condition:

$$u(x_0,t) = h(t), 0 \le t \le t_{max,} \tag{2}$$

where $x0 \in (0, 1)$ is the interior location of a thermocouple recording the temperature mea- surement, $u0$ (x), $g(t)$ and $h(t)$ are given functions satisfying the compatibility conditions:

$$u0\ (0) = 0,\ u0\ (1) = g(0),\ u0\ (x0\ ) = h(0), \tag{3}$$

and the function $u(x, t)$ and $f(t)$ are unknown. It is known that if $u(x, t)$ and $f(t)$ is suf- ficientlly and smooth and the initial condition are given then the problem have a unique and stable solution, however we now just have condition (2.1)-(2.5), so it is a ill-posed problem. In other word, the difficulty is that the solution of the backward heat conduction problem dose not continuously depend on the given condition accurately. It is required to find the temperature distribution $u(x, t)$, before that $f(t)$ should also be investigated.

3. The boundary element method

One important step for solving the problem by BEM is transfer the governing differntial equation to an integral form to an integral form and it is achieved by applying a funda- mental solution of the heat conduction equation. [12] It is well known that the fundamental solution of (4) is of the form, see,

$$F(x,t,\xi,\tau) = \frac{H(t-\tau)}{\sqrt{4\pi(t-\tau)}} exp\left(-\frac{(x-\xi)^2}{4(t-\tau)}\right) \tag{4}$$

where $H(t - \tau)$ is the Heaviside function, when $\tau > t$, the function is identically zero. [13]

Using the fundamental solution the equation (2.1), it can be transformed into the intagral form below,

$$(3.1)\eta(x)u(x,t) = \int_0^1 [F(x,t,\xi,\tau)(\partial/\partial n)u(\xi,\tau) - u(\xi,\tau)(\partial/\partial n)F(x,t,\xi,\tau)]\,|_0^1 d\tau$$
$$+\int_0^1 F(x,t,\xi,0)u(\xi,0)dy + \int_0^1\int_0^{t_{max}} F(x,t,\xi,0)f(\tau)d\tau d\xi, \tag{5}$$

where $(x, t) \in [0, 1] \times [0, T]$, and

$$\eta(x) = \begin{cases} 1, & 0 < x < 1, \\ \frac{1}{2}, & x = 0,1. \end{cases} \tag{6}$$

In practice, the analytical solution of the integral equation is almost impossible to be ob- tained, and it is necessary to try other methods to obtain numerical approximate solutions. Considering that the integral equation can be discretized, we first discretize the boundary. So we now discretize the boundaries $0 \times [0, tmax]$ and $1 \times [0, tmax]$ into N time step, the space interval in $N0$ elements, if we take x on the boundary, $x = 0$ or $x = 1$, then the integral function are shown as,

$$\eta(x)u(x,t) = \sum_{j=1}^{N}\int_{t_{j-1}}^{t_j} F(x,t,0,\tau)d\tau(\partial u/\partial n)(0,\widetilde{t_\iota})$$
$$+\int_{t_{j-1}}^{t_j} F(x,t,1,\tau)d\tau(\partial u/\partial n)(1,\widetilde{t_\iota}) - \int_{t_{j-1}}^{t_j} (\partial/\partial n)u(x,t,0,\tau)g(\widetilde{t_\iota})$$
$$+\sum_{k=1}^{N_0} F(x,t,\xi,0)d\xi u_0(x,\widetilde{x_k}) + \sum_{j=1}^{N}\int_{t_{j-1}}^{t_j}\int_0^1 F(x,t,\xi,\tau)d\xi d\tau g(\widetilde{t_\iota}) \tag{7}$$

where $\tilde{t}i$ is the midpoint of $[tj\text{-}1 , tj]$, the definition of the boundary influence matricces are shown as,

$$G_{ij} = \int_{t_{j-1}}^{t_j} F(x_i,\widetilde{t_\iota},\widetilde{\xi_j},\tau)d\tau, i = 1,2,\cdots,2N \tag{8}$$

$$E_{ij} = \int_{t_{j-1}}^{t_j} (\partial/\partial n)F(x_i,\widetilde{t_\iota},\widetilde{\xi_j},\tau)d\tau + 0.5\delta_{ij},\ i = 1,2,\cdots,2N \tag{9}$$

$$F_{ij} = \int_{x_{k-1}}^{x_k} F(x_i,\widetilde{t_\iota},\xi,\tau)d\,\xi,\ k = 1,2,\cdots,N_0, \tag{10}$$

where δij is the Kronecher delta symbol. Then, by applying the integral equation at the point $(0, \tilde{\ } i)$ and $(1, \tilde{t} i)$, we can acquire $2N$ equations.

$$Aq + Bh + Cu_0 = d$$

$$A_{i,j} = \int_{t_{j-1}}^{t_j} F(0,\widetilde{t_\iota},0,\tau)d\tau,$$

$$A_{i,j+N} = \int_{t_{j-1}}^{t_j} F(0,\widetilde{t_\iota},1,\tau)d\tau, i,j = 1,2,\cdots,N$$

$$A_{i+N,j} = \int_{t_{j-1}}^{t_j} F(1,\widetilde{t_\iota},0,\tau)d\tau,$$

$$A_{i+N,j+N} = \int_{t_{j-1}}^{t_j} F(1,\widetilde{t_\iota},1,\tau)d\tau, i,j = 1,2,\cdots,N$$

$$B_{i,j} = \int_{t_{j-1}}^{t_j} (\partial/\partial n)F(0,\widetilde{t_\iota},0,\tau)\,d\tau + 0.5\delta_{ij},$$

$$B_{i,j+N} = \int_{t_{j-1}}^{t_j} (\partial/\partial n)F(0,\widetilde{t_\iota},1,\tau)\,d\tau,\ i,j = 1,2,\cdots,N$$

$$B_{i+N,j} = \int_{t_{j-1}}^{t_j} (\partial/\partial n)F(1,\widetilde{t_\iota},0,\tau)\,d\tau,$$

$$B_{i+N,j+N} = \int_{t_{j-1}}^{t_j} (\partial/\partial n)F(1,\widetilde{t_\iota},1,\tau)\,d\tau + 0.5\delta_{ij},$$

$$C_{i,k} = \int_{x_{k-1}}^{x_k} F(0, \widetilde{t_\iota}, \xi, 0),$$

$$C_{i+N,k} = \int_{x_{k-1}}^{x_k} F(1, \widetilde{t_\iota}, \xi, 0),\ k = 1,2,\cdots, N_0,\ i = 1,2,\cdots, N$$

$$d_i = -\sum_{j=1}^{N} \int_{t_{j-1}}^{t_j} \int_0^1 F(0, \widetilde{t_\iota}, \xi, \tau) d\xi d\tau g(\widetilde{t_\iota}),$$

$$d_{i+N} = -\sum_{j=1}^{N} \int_{t_{j-1}}^{t_j} \int_0^1 F(1, \widetilde{t_\iota}, \xi, \tau) d\xi d\tau g(\widetilde{t_\iota}),$$

q j = (∂/∂n)u(0, ˜ t j), q j+N = (∂/∂n)u(1, ˜ t j), j = 1,2,··· ,N,

s j = 0, s j+N = g(˜ t j), j = 1,2,··· ,N, (11)

we now use the investgated condition (11) at the point (*x*0 , *t*˜*i*) for *i* = 1, 2, · · · , *n*, and we can obtain N equations.

Then, constrain (11) shouid also be considered by applying the integral equation at the point (0, ˜*i*) and (1, *t*˜*i*) , we can acquire 2*N* equations.

$$h = Aq + Bh + Cu_0 - d$$

$$A_{i,j} = \int_{t_{j-1}}^{t_j} F(x_0, \widetilde{t_0}, 0, \tau) d\tau,$$

$$A_{i,j+N} = \int_{t_{j-1}}^{t_j} F(x_0, \widetilde{t_0}, 1, \tau) d\tau,\ j = 1,2,\cdots, N$$

$$h_i = h(\widetilde{t_\iota}),$$

$$B_{i,j} = \int_{t_{j-1}}^{t_j} (\partial/\partial n) F(x_0, \widetilde{t_\iota}, 1, \tau) d\tau + 0.5\delta_{ij},$$

$$B_{i,j+N} = \int_{t_{j-1}}^{t_j} (\partial/\partial n) F(x_0, \widetilde{t_0}, 1, \tau) d\tau,\ i,j = 1,2,\cdots, N$$

$$C_{i,k} = \int_{x_{k-1}}^{x_k} F(x_0, \widetilde{t_\iota}, \xi, 0) d\xi,\ i = 1,2,\cdots, N$$

$$d_i = -\sum_{j=1}^{N} \int_{t_{j-1}}^{t_j} \int_0^1 F(x_0, \widetilde{t_\iota}, \xi, \tau) d\xi d\tau g(\overline{t_i}), k = 1,2,\cdots, N_0,\ i = 1,2,\cdots, N$$

h i = h(˜ t i), i = 1,2,··· ,N, (12)

where *hi* take the value of *h*(*t*) at the mid point of [*y j*-1, *y j*]. Base on above notations we can obtain,

(d˜ f -1 - ˜AA-1d f -1) f + (˜AA-1 B - ˜B)s + (˜C - ˜AA-1C)u0 = h, (13)

In order to solve the above problems, we have transformed the form of the equations. But, the syetem is ill-conditioned due to the ill-posed inverse problems. We are requested to find a stable and accurate method to solve the problem.

4. Methods of solution

The application of various methods can be considered for the solution of the system. We have reduced the IHCP to the linear system of equation *Z f* =*x*, different methods will be used to solve this equation and a relatively good method will be chosen. Then, we transform the equation above into a simple form to learn the problem more easily, we now suppose that,

x = h + ($\widetilde{B}$-AA^{-1}B)s + ($\widetilde{AA}^{-1}$C-$\widetilde{C}$)u_0 , Z = ($\widetilde{d}$f $^{-1}$-AA^{-1}df $^{-1}$)f, and the problem transform to a *N* × *N* system of linear equation,

Z f = x. (14)

Now, we can clearly know that the equation have the unique solution. But, the syetem is ill-conditioned due to the ill-posed inverse problems. We are requested to find a stable and accurate method to solve the problem.

4.1. Direct Mehtod.

When we use the direct solution method, N linear equations have N unknowns. If the matrix Z is invertible, we can get $f = Z$-1 x. but, the problem is inverse heat conduction, is ill-posed, can not be solved by direct method.It is almost impossible to get accurate results.

4.2. Least Sqares Method.

Attempt to solve such a problem using least squares [14], but we still can't get an accurate solution, because there are no constraints added to the solu- tion to the problem and there is not enough information to determine the direction of the solution, then try to use the regularization method. In the following experiments, we used the least squares method and found that the least squares method has low accuracy and high randomness in solving the problem.

4.3. Regularzation Method.

Many regularization methods are applied to solve these ill- posed problems,

$$min\{ \|Z\lambda_\alpha - \tilde{x}\|^2 + \alpha^2\|\lambda_\alpha\|^2\} \tag{15}$$

where $|\cdot|$ presents the Euclidean norm, λ_α for $i = 1, 2, \cdots, N$ are the regular parameter. And the solution may be given through the equation,

$$Q(\lambda_\alpha) = (Z^{tr} Z + \alpha^2 I)^{-1}(Z^{tr}x, \tag{16}$$

where I is a identity matrix. It can be known that the solution of -(17) depends on the regularization parameter. The selection of parameters is very strict. When the parameter selection is too large, the numerical solution will have a large gap with the accurate solution due to quasi-smoothing. However, if the parameter selection is too small, when $\lambda\alpha$ tends to the limit, the solution will become random. To remove the difficulty of choosing regularization parameters and measuring the stability of numerical solutions, we will utilize the minimal energy method.

4.4. Minimization of energy method.

To get a more accurate solution, we change the form of condition (17) to the form of inequality

$$|u(x_0, t) - h(t)| \leq \varepsilon, t \in (0, T) \tag{17}$$

where ε is a preset quantity, clearly the problem has many solutions and the solution of the problen (13) - (17) is included in the solution set of the problem. By discretizing equation (18) combined with the results in the third part, we get an $N \times N$ inequality system, namely,

$$|Zf - x| \leq \varepsilon \tag{18}$$

The difficulty now is that we need a method to find that solution in the solution set in a stable way. Let H^1 [0, 1] and H-1 [0, 1] represent the usual Sobolev spaces, and a Hilbert spaces φ is defined, such that we got,

$u \in L^2 (0, T; H^1 [0, 1]), (\partial/\partial t) \in L2 (0, T; H\text{-}1 [0,1])$ where $u \in \Pi$ and its norm is

$$\|u(x,t)\| = (\int_0^T \|u(\cdot,t)\|_1^2 + \int_0^T \|(\partial/\partial t)u(\cdot,t)\|_{-1}^2 dt)^{\frac{1}{2}}, \tag{19}$$

where $//\cdot//1$ denotes the form of H^1 [0, 1] and $//\cdot//$-1 denotes the form of the space H-1 [0, 1]. We now consider the original problem, for ang $\zeta(x) \in L^2$ [0, 1] there is a unique weak solution such that u(x,t)∈. Π Hence, operator has been defined,

$$\Pi: L\,2\,[0,1] \rightarrow \varphi$$

$$\Pi\,\zeta\,(t) = f(t) \tag{20}$$

if there is a $\zeta \in L\,2\,[0,1]$ such tha is small enough and $u(x, t)$ depends on a classical equation with variational priciple involves the minimazation of the kinetic energy functional,

$$K(u) = \frac{1}{2}\int\int \Omega|\nabla u|^2 dxdy \tag{21}$$

And the relation can be presented in integral form by application of Green's formula,

$$KE(T) = \int_0^t \int_0^1 |(\partial/\,\partial t)u(u,t)|^2 dxd\tau \tag{22}$$

To determine ζ and to satisfy a physical interpretation the energy equation is derived, the governing equation is multiplied by $u(x, t)$ and the domian $[0, 1] \times [0, T]$ is integrated over. Finally, the resulting equation is acquired

$$KE(t) = a^2 \int_0^{t_0} u(1,t)(\partial/\,\partial x)u(1,t)dt - a^2 \int_0^{t_0} u(0,t)(\partial/\,\partial x)u(0,t)dt\ \frac{1}{2}\int_0^1 u^2(x,t_0)dx - \frac{1}{2}\int_0^1 u^2(x,0)dx \tag{23}$$

We consider the energy functional as $H(u)$, $M = \{v\text{: } v = ⪊\zeta \in \Psi, |v - h| \leq \varepsilon\}$ is a convex set. Then, the constrained minimal problem is transformed as,

$$H(u) = \inf H(v),\ v \in M \tag{24}$$

And if we let $K = \{\zeta\text{: } \zeta \in L2, |⪊\zeta - h| \leq \varepsilon\}$, the expressiom can be writtten as,

$$H(⪊\zeta) = \inf H(⪊\rho),\ \rho \in K \tag{25}$$

which expressed in discrete form as,

$$H(u) = \frac{1}{2}\sum_{i=1}^{N_0} \int_{x_{i-1}}^{x_i} \zeta^2(x)dx \tag{26}$$

And the equation maybe rewritten as,

$\left|\sum_{j=1}^{2N}(EI_{ij}EI_{ij}^{-1}F_{ij})u_0(x_i,\tilde{t_j}) + \sum_{k=1}^{N_0}(GI_{ij} - EI_{ij}E_{ij}^{-1}GI_{ij})\,\zeta_j - h_i\right| \leq \varepsilon$ Then, use the notation $\sum_{j=1}^{2N}(EI_{ij}EI_{ij}^{-1}F_{ij})u_0(x_i,\tilde{t_j}) - h_j = \tilde{h}_j$and $GI_{ij} - EI_{ij}E_{ij}^{-1}GI_{ij} = W_{ij}$

We get,

$$\left|\sum_{k=1}^{N_0} W_{ij} - \tilde{h}_i\right| \leq \varepsilon \tag{27}$$

The problem is solved by NAG routine e04ucf, and the routine is intended to minimize an arbitrary continuous function such that it satisfies certain conditions, including constraints on variables, nonlinear smoothing conditions.

5. Numerical Results

In the foregoing, four methods for solving the problem sought are given. Next, nu- merical experiments will be carried out, using the boundary element method to find the numerical solution of the examples, and comparing the numerical solution of the exact so- lution with the solutions obtained by other methods. The first examples assume that $a = 1$, $x0 = 0.5$. Example 1 detecting the effect of different N on the accuracy of the results while Example 2 extending the application range by formal transformation.

Example 1.We set up two simple functions $u(x,t) = e^{-t}\,sin(x) + 3tx^2 + \frac{1}{5}x^5$, $f(x) = -7t$.It is easy to know that $u_0 = e^{-t}\,sin(x) + \frac{1}{5}x^5$, $g(t) = e^{-t}\,sin(1) + 3t + 0.2$,

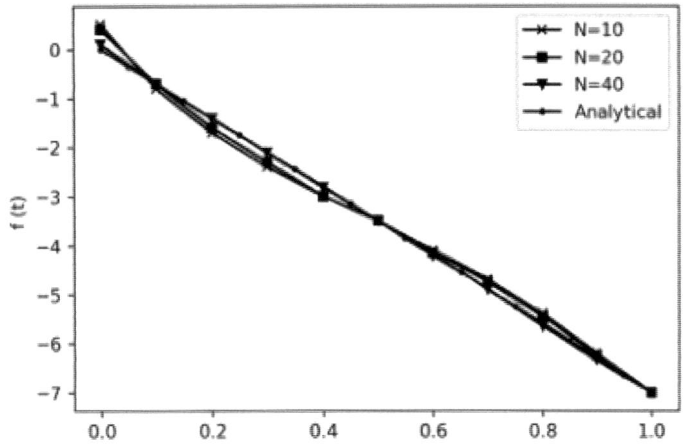

Figure 1. The numerical solutions and analytic solution of f (x) for Ex- ample 1, where ▲ denote the analytical solution, × for N = 10, ■ for N = 20 and • for N = 40.

Example 1 illustrates the different eff ects of *N*, Figure 1 shows the numerical solution and the exact solution obtained by the minimal energy method, where *N*, *N*0 and ε is taken, it can be seen that the numerical solution is close to the exact solution, and when *N* increases, the fitness of the exact solution to the numerical solution also increases. We

Table 1. the numerical solution of f (t) for Example 1

Mesh points(t)	Least-squares method	mininal energy method	analytical solution
0.025	-23433	-0.16209	-0.175
0.075	-70892	-0.53486	-0.525
0.125	-90787	-0.88137	-0.875
0.175	-128778	-1.2266	-1.225
0.225	-148989	-1.57578	-1.575
0.275	-139823	-1.92626	-1.925
0.325	-179302	-2.27654	-2.275
0.375	-218992	-2.62662	-2.625
0.425	-234455	-2.97663	-2.975
0.475	-258904	-3.325	-3.325

also compared the results of diff erent methods, and it is shown in Table 1. We used the direct method, the least squares method and the minimum energy respectively. We found that the difference between the results of the direct method and the least squares method was relatively large. It shows that the minimal energy method can lead a relatively better solution. Since the equation is ill-posed, we tested diff erent point value. And when *N* increases, the solution obtained by the minimal energy method gradually approaches the accurate solution, while the solution obtained by the least square method changes greatly. In addition, when *x*0 takes experimental parameters at diff erent times, it also has a certain impact on the results. After exploring the different eff ects on the results from Table 2 when *N* takes diff erent values, and when using diff erent methods to obtain $f(t)$ by solving $u(x, t)$, we also noticed whether the selection of *x*0 will aff ect the final As a result, different *x*0 values were then chosen and compared with the exact solution.It can be seen that $x0 = 0.25$ is more accurate when *x*0 takes a small value, and when $x0 = 0.5$, it is closer to the analytical solution in the second half.

Table 2. the numerical solution of f (x) for T0 = 0.25, T0 = 0.5, T0 =0.75 and analytical solution for Example 1

mesh points(t)	T_0 = 0.25	T_0 = 0.5	T_0 = 0.75	analytical solution
0.025	-0.16235	-0.16209	-0.16237	-0.175
0.075	-0.53468	-0.53486	-0.53471	-0.525
0.125	-0.88124	-0.88137	-0.88123	-0.875
0.175	-1.22657	-1.2266	-1.22659	-1.225
0.225	-1.57576	-1.57578	-1.57581	-1.575
0.275	-1.92623	-1.92626	-1.92626	-1.925
0.325	-2.27651	-2.27654	-2.27651	-2.275
0.375	-2.62659	-2.62662	-2.62662	-2.625
0.425	-2.9766	-2.97663	-2.97665	-2.975
0.475	-3.32601	-3.325	-332,666	-3.325

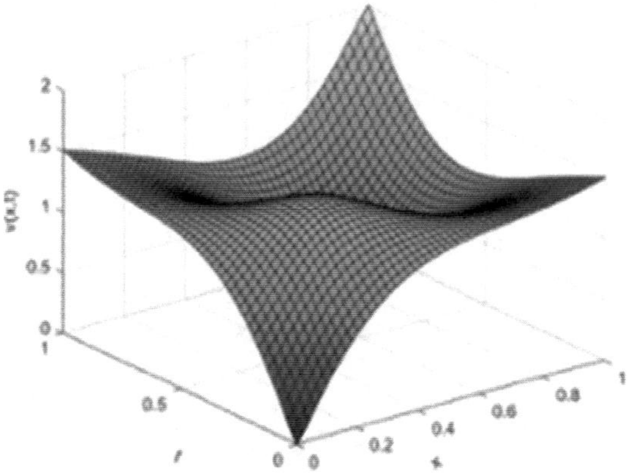

(a)

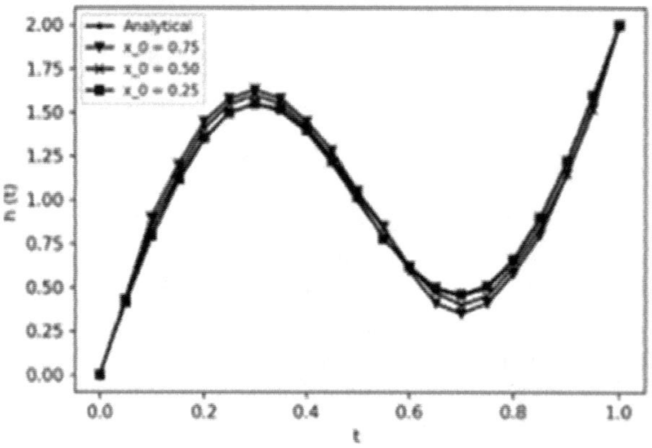

(b)

Figure 2. Figure (a) is the v(x, t) of Example 2, and Figure (b) is h(t) obtained by taking different x0 values as the investigated condition for Example 2

Example 2. In order to expand the application of the equations, we will explore the solvability of the inverse problems of other forms of the heat conduction equation. For example, when $u(0, t) \neq 0$. The $u(x, t)$ and $f(t)$ are represented below,

$$u(x, t) = x^2 + 2t + sin(2\pi t),$$
$$f(t) = 2\pi cos(2\pi t). \quad (28)$$

In this case, it is easy to know that $u(0, t) = 2t + sin(2\pi t)$, which obviously does not meet the conditions of this article. Considering that the change of form may make the problem satisfy the condition, we try to transform the u to $u = \tilde{u} + v$, and then the condition of v satisfy (2.1)-(2.5) and the equation converts to,

$$(\partial/\partial t)v(x, t) - (\partial^2/\partial x^2)v(x, t) = 0,\ 0 < x < 1,\ 0 < t \leq T,$$
$$v(x, 0) = 0,\ 0 \leq x \leq 1,$$
$$v(0, t) = 2t + sin(2\pi t),\ 0 \leq t \leq T,$$
$$v(1, t) = 0,\ 0 \leq t \leq T,$$
$$v(0.5, t) = 0.25 + 2t + sin(2\pi t), \quad 0 \leq t \leq T. \quad (29)$$

Using the combination of boundary element method and minimal energy method, the an- swer to this question can be obtained, We plotted the approximate solution of f (x) using the numerical solution of u(x, t). We illustrated the final solution through a picture, which includes v(x, t) and h(t). As shown in Figure 2, the solution of v(x, t) for Example 2 and

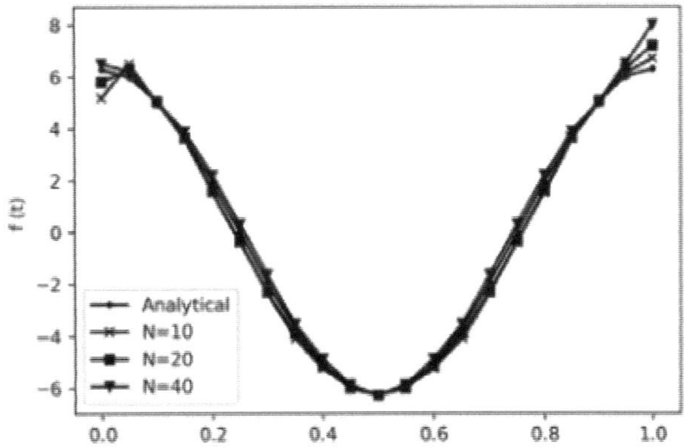

Figure 3. The numerical solutions and analytic solution f (t) for Exam- ple 2, where ▲ denote the analytical solution, × for N = 10, ■ for N = 20 and • for N = 40.

the solution of $h(t)$, it can be seen that when $x0$ takes different values, the results also fluc- tuate, and near the value of $x0$, the accuracy of the numerical solution better. The values of the energy function during the experiment are written in the Table 3.

Table 3. the solution of energy function H(u) for Example 2

N	minimal energy method	analytical value
10	-0.02394	0
20	-0.00379	0

6. Conclusions

In this paper, we use the boundary element method, which is different from the direct method, the least square method and the canonical method, to obtain the optimal solution of the ill-posed problem in the problem compared with other methods by means of the minimal energy method, and carry out numerical experiments. The application range of the solution method is expanded through the second example. In addition, although the calculation result of this method is more accurate, the computer storage required for the calculation process is relatively large. In addition, the method can also be generalized on 2D or 3D problems. proposed regularization method is stable and efficient.

References

[1] D. Sharma, A. K. Parwani, and C. Shrivastava, "A review on solution of inverse heat conduction problems," in International Journal of Engineering Research & Technology, 2014.

[2] A. I. Prilepko and I. V. Tikhonov, "Uniqueness of a solution of the inverse problem for the evolution equation and application to the transport equation," Mathematical Notes, vol. 51, no. 2, pp. 158–165, 1992.

[3] M. Nikolova, "Energy minimization methods," Springer New York, 2015.

[4] F. Wang, C. Wen, and G. Yan, "Boundary element analysis of inverse heat conduction problems in 2d thin- walled structures," International Journal of Heat and Mass Transfer, vol. 91, no. DEC., pp. 1001–1009, 2015.

[5] SAVATEEV and G. E., "On problems of determining the source function in a parabolic equation," Journal of Inverse and Ill-Posed Problems, vol. 3, no. 1, 1995.

[6] D. B. Ingham, Y. Yuan, and H. Han, "Boundary-element method for an improperly posed problem," IMA Journal of Applied Mathematics, no. 1, pp. 61–79, 1991.

[7] H. Han, "The finite element method in a family of improperly posed problems," Mathematics of Computa- tion, vol. 38, no. 157, pp. 55–65, 1982.

[8] D. B. Ingham and Y. Yuan, "The solution of a nonlinear inverse problem in heat transfer," IMA Journal of Applied Mathematics, no. 2, pp. 113–132, 1993.

[9] Phillips and L. David, "A technique for the numerical solution of certain integral equations of the first kind," Journal of the Acm, vol. 9, no. 1, pp. 84–97, 1962.

[10] A. Fatullayev and E. Can, "Numerical procedures for determining unknown source parameter in parabolic equations," Mathematics & Computers in Simulation, vol. 54, no. 1-3, pp. 159–167, 2000.

[11] A. I. Kozhanov, "Solvability of the inverse problem of finding thermal conductivity," Siberian Mathematical Journal, vol. 46, no. 5, pp. 841–856, 2005.

[12] C. A. Brebbia, J. Telles, and L. C. Wrobel, Boundary element techniques:. Boundary element techniques:, 1987.

[13] J. V. Beck, "Inverse heat conduction, ill-posed problems," A Wiley-Interscience Publication, 1985.

[14] X. Zhang, X. F. Pan, and M. W. Lu, "Meshless weighted least-square method," Acta Mechanica Solida Sinica, vol. 17, no. 3, 2003.

Numerical simulation of two-phase evaporation and condensation

Wei Chang *, Fangju Zhang, Kun Fang, Gongli Tan and Pengjie Fan

The 723 research institute of CSIC, Yangzhou, China

*Corresponding author e-mail: 849063259@csic.cn

Abstract. The evaporation and condensation mechanism of two-phase flow is studied. The Euler-Euler pattern is used to express the evaporation and condensation properties of two-phase flow. The research shows that the tracking program can accurately track the interface. The thermal resistance on the confluence of the condensation part and insulation part becomes larger. The working fluids have significant influence of two-phase.

1. Introduction

The common classification method of multi-phase flow is based on the number of phases. The two-phase flow is the most studied by researchers, such as water and steam, acetone and acetone vapor. The phases flow can be sorted on the basis of working fluid properties. There are four main types: liquid-solid phases flow, which implies solid particles flow with the liquid; Gas-solid phases flow, which implies solid particles flow with gas; Liquid-liquid phases flow, which implies two kinds of liquids that cannot be uniformly mixed flow together; Gas-liquid phases flow, which implies gas and liquid flow together [1, 2]. In this article, gas-liquid of two-phase flow is studied.

Continuous medium patterns in two-phase flow include Euler-Lagrange pattern and Euler-Euler pattern. Euler-Lagrangian pattern is used for discrete patterns, one phase is discrete phase and the other phase is continuous phase. Euler-Euler pattern treats all phase as continuous media. The volume of fluid pattern, mixture pattern and euler pattern belong to Euler-Euler pattern. The volume of fluid pattern treats two-phase flow as single mixture. The two phase share the same velocity field and temperature field. It only needs to solve a set of mixed phase momentum equations and mixed phase energy equations.

This article can use the volume of fluid pattern. The volume of fluid pattern is applicable to stratified and free surface flow. The volume of fluid pattern pattern can use the geometric reconstruction pattern to track the gas-liquid interface, which has with the highest accuracy. Hirt and Nichols first used the volume of fluid method to simulate dam break, Rayleigh-Taylor instability and moving phase interface problems in 1981 [3].

2. Mathematical pattern

2.1. The Geometric pattern

In this article, the geometric pattern is shown in Figure 1. The pattern is separated into three parts: condensation part, insulation part and evaporation part from left to right [4].

Content from this work may be used under the terms of the Creative Commons Attribution 3.0 licence. Any further distribution of this work must maintain attribution to the author(s) and the title of the work, journal citation and DOI.
Published under licence by IOP Publishing Ltd

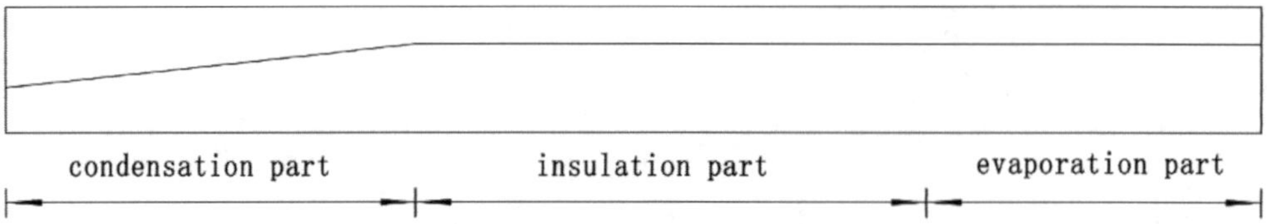

Figure 1. The geometric pattern.

2.2. The grid dividing

The Fluent software is uesd to divide grid. In this pape, the grid uses the quadrilateral structured. On the boundary between the fluid part and the solid part, the grid is concentrated on the fluid part to improve the calculating exactness, as shown in Figure 2.

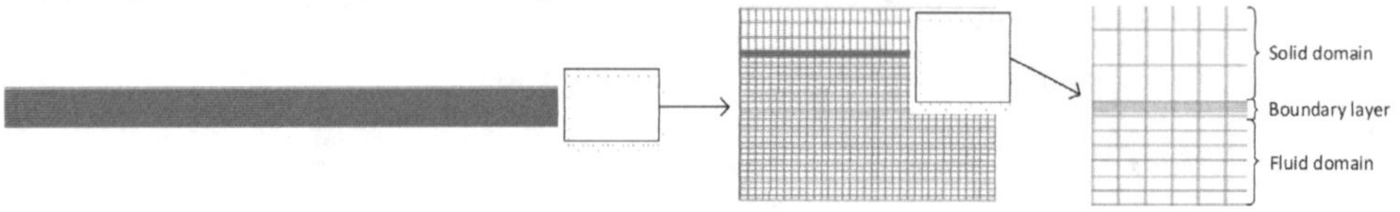

Figure 2. The grid dividing.

2.3. The equations of state

In gas-fluid flow, the confluence between phases is the continuous function by seeking the fraction in the gas phase [5]. The continuous functions are as follows in gas-fluid flow:

$$\frac{\partial \rho_1 \alpha_1}{\partial t}+\frac{\partial \rho_1 \alpha_1 u}{\partial x}+\frac{\partial \rho_1 \alpha_1 v}{\partial y}=S \tag{1}$$

$$\frac{\partial \rho_2 \alpha_2}{\partial t}+\frac{\partial \rho_2 \alpha_2 u}{\partial x}+\frac{\partial \rho_2 \alpha_2 v}{\partial y}=-S \tag{2}$$

There is mass exchange in gas-liquid flow. The gas-fluid flow can occur evaporation and condensation. When the whole parts solve the continuous functions, it only need to solve the fraction of the second phase. In this article, the first phase is gas. It not needs to calculate the continuous function of the gas. The continuous function of the second phase is calculated to gain the volume fraction of the second phase. It can need to calculate by equation (3) in the volume fraction of the second phase.

$$\alpha_1+\alpha_2=1 \tag{3}$$

The momentum equation of gas-fluid in the whole parts is shared [6]. The velocity fields of gas-fluid need to be calculated. The velocity fields in the X and Y axes are built:

$$F_x-\frac{\partial p}{\partial x}+\mu\left(\frac{\partial^2 u}{\partial x^2}+\frac{\partial^2 u}{\partial y^2}\right)=\rho\left(\frac{\partial u}{\partial t}+u\frac{\partial u}{\partial x}+v\frac{\partial u}{\partial y}\right) \tag{4}$$

$$F_y-\frac{\partial p}{\partial y}+\mu\left(\frac{\partial^2 v}{\partial x^2}+\frac{\partial^2 v}{\partial y^2}\right)=\rho\left(\frac{\partial v}{\partial t}+u\frac{\partial v}{\partial x}+v\frac{\partial v}{\partial y}\right) \tag{5}$$

The energy equation in the whole parts is one and the same. The temperature field is built:

$$\rho c\left(\frac{\partial T}{\partial t}+u\frac{\partial T}{\partial x}+v\frac{\partial T}{\partial y}\right)=\lambda\left(\frac{\partial^2 T}{\partial x^2}+\frac{\partial^2 T}{\partial y^2}\right)+S_e \tag{6}$$

In the equations, ρ_1 means density of the first phase. ρ_2 means density of the second phase. α_1 means volume fraction of the first phase. α_2 means volume fraction of the second phase. S means mass term. ρ means density of gas-fluid phase. c means specific heat of gas-fluid phase. S_e means energy term.

3. Numerical Simulation

In order to study the evaporator and condenser on the confluence of phases, it is necessary to find the gas-liquid confluence. Based on the found gas-liquid interface, the energy source term and mass source term are applied to the phase interface [7].

Although the cloud image function of FLUENT software can display the gas-liquid interface, the user cannot directly obtain the relevant information of the gas-liquid interface, and must write a program to track the phase interface to obtain the phase interface position and other information [8].

3.1. The gas-liquid interface

It can be seen from Figure 3 and Figure 4. This figures show the cloud image of gas-liquid on the evaporator part and condenser part. There is a layer of liquid film on the evaporator part and condenser part. The red line is the gas-liquid interface tracked by tracking program. It can be clearly seen that the program accurately tracks the gas-liquid interface. It lays a foundation for the mass term and energy term on the gas-liquid interface.

Figure 3. Interface on evaporation section.

Figure 4. Interface on condensation section.

3.2. Temperature distribution on the wall

The fall of the temperature is produced on the confluence of the condensation part and insulation part, as shown in Figure 5. Due to a slope on condensation part, the water film gathers on the confluence of the condensation part and insulation part. The thickness of the second phase is much thicker on the condensation part. The thermal resistance becomes larger. Because it is the constant heat flow boundary condition, the temperature fall also becomes larger.

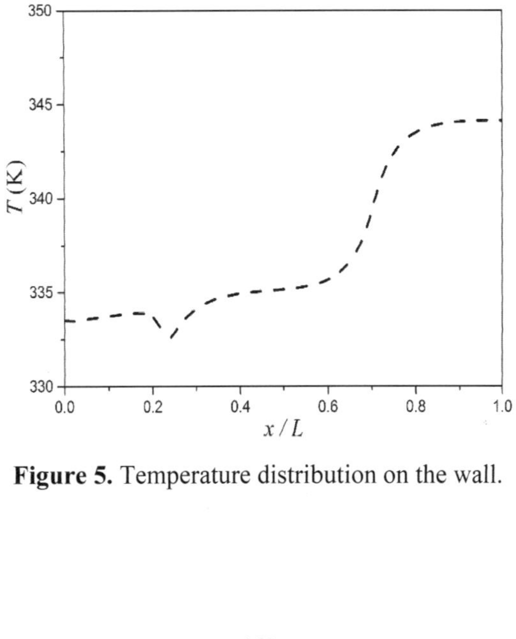

Figure 5. Temperature distribution on the wall.

3.3. Influence of different working fluids

In a similar case, the temperature fall by acetone are larger than water on the condensation part and evaporation part. It becauses that acetone has smaller thermal conductivity, as shown in Figure 6. The working fluids have significant influence of two-phase.

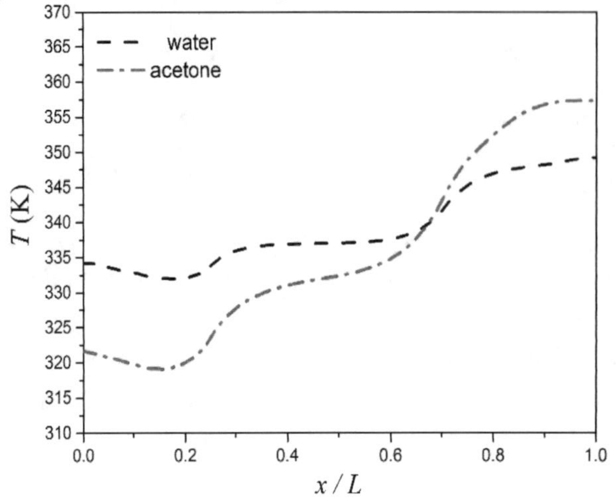

Figure 6. Temperature distribution of different working fluids.

4. Conclusion

A tracking program is used to track the phase interface to obtain the phase interface position and other information. The volume of fluid pattern is used in this article. The volume of fluid pattern is suitable for two-phase condensation and evaporation.

The research shows that the tracking program can accurately track the interface. The thermal resistance on the confluence of the condensation part and insulation part becomes larger. The working fluids have significant influence of two-phase. The results of numerical simulation can accurately describe two-phase condensation and evaporation. This paper lays the foundation for the study of two-phase condensation and evaporation, and it is expected to be applied to experimental research.

References

[1] D.F. Che, H.X. Li, Multiphase flow and its application, Xi'an Jiaotong University Press, 2007.
[2] K. Kerrigan, H. Jouhara, A naturally aspirated convector for domestic heating application with low water temperature sources, Energy and Buildings, 2013, pp. 187-194.
[3] C.W. Hirt, B.D. Nichols, Volume of fluid method for the dynamics of free boundaries, Journal of Computational Physics, 1981, pp. 201-225.
[4] Wei Chang, Yanlong Li, Kun Fang, Fangju Zhang, Numerical simulation of two-phase flow and heat transfer in an axial rotating heat pipe, 2020 IEEE 5th Information Technology and Mechatronics Engineering Conference, 2020, pp. 1332-1336.
[5] L.Asmaie, M.Haghshenasfard, Thermal performance analysis of nanofluids in a thermosyphon heat pipe using CFD patterning, Heat and Mass Transfer, 2013, pp. 667-678.
[6] W.H. Lee, Pressure iteration scheme for two-phase flow patterning, Multiphase Transport Fundamentals, 1980, pp. 407-432.
[7] D. Schepper, G.J. Heynderickx, G.B. Marin, CFD patterning of all gas-liquid and vapor-liquid flow regimes predicted by the baker chart, Chemical Engineering Journal, 2008, pp.349-357.
[8] D.L. Sun, J. Xu, Q. Chen, patterning of the evaporation and condensation phase-change problems with FLUENT, Numerical Heat Transfer, 2014, pp. 326-342.

Examination of Star Formation Rate enhancement in galaxy pairs selected based on data from Illstris TNG

Yiru Chen*

Department of Physics, Tsinghua University, Beijing, China

*Corresponding author's e-mail: chen-yr18@mails.tsinghua.edu.cn

Abstract. Star Formation Rate (SFR) is one of the most important properties of galaxies, which offers a good insight of galaxy evolution. Generally, interactions with the small and large scale environment may affect the SFR of the galaxy. Therefore, an examination of SFR of galaxy pairs is carried out in this paper to investigate the partner galaxy impacts on the evolution process of one galaxy. Compared with isolated galaxies, the SFR of galaxy pairs is enhanced due to interactions with the large scale environment (e.g., the tidal force). This finding confirms that interactions with the large scale environment will affect the galaxy evolution process and further work can be done to examine the impacts of other interactions. These results shed light on the deeper understanding of galaxy evolution process and call on more investigation of large scale environment of the galaxy.

1. Introduction

Understanding the origin of the galaxy formation and evolution is one of the most challenging goals in modern cosmology. Star Formation Rate (SFR) is one key property in the galaxy evolution process, both simulation [1] and observation [2] have seen the enhancement of it in galaxy pairs. However, the problem is that the simulation has to be matched with the observation since many simulations don't consider much about the environment of galaxy pairs [3]. To solve this problem, the data from Illstris TNG will make a huge difference since it simulates the bigger structure. Investigation of some basic ideas of galaxy formation and evolution may be helpful before a deeper examination of SFR.

The basic idea of galaxy formation can be understood by observing the picture in Fig. 1. Though invisible, dark matter plays the core and the leading role in the process of galaxy formation. The original density distribution of the dark matter density field collapses into dark matter halos. The first bound objects emerge in this way. Then these dark matter halos act as seeds for galaxy formation. They accrete neutral hydrogen gas from their surrounding environment. Due to cooling and condensation, the formation of galaxies happens at the centers.

The theory for galaxy evolution is more complicated compared with galaxy formation process, different mechanisms can vary a lot, but the primary mechanism responsible for the growth of a galaxy is the gas accretion from the intergalactic medium (IGM). Apparently, there are plenty of mechanisms can play a role, for example, the subsequent formation of a supermassive black hole at the center of the galaxy and an efficient accretion onto it can trigger AGN (active galactic nucleus) activity. Besides, the jet from the AGN, starburst winds and supernovae explosions can drive gas outflows from the galaxy. The gas ejected from the galaxy eventually cools down and reaccretes again to continue the cycle.

Content from this work may be used under the terms of the Creative Commons Attribution 3.0 licence. Any further distribution of this work must maintain attribution to the author(s) and the title of the work, journal citation and DOI.
Published under licence by IOP Publishing Ltd

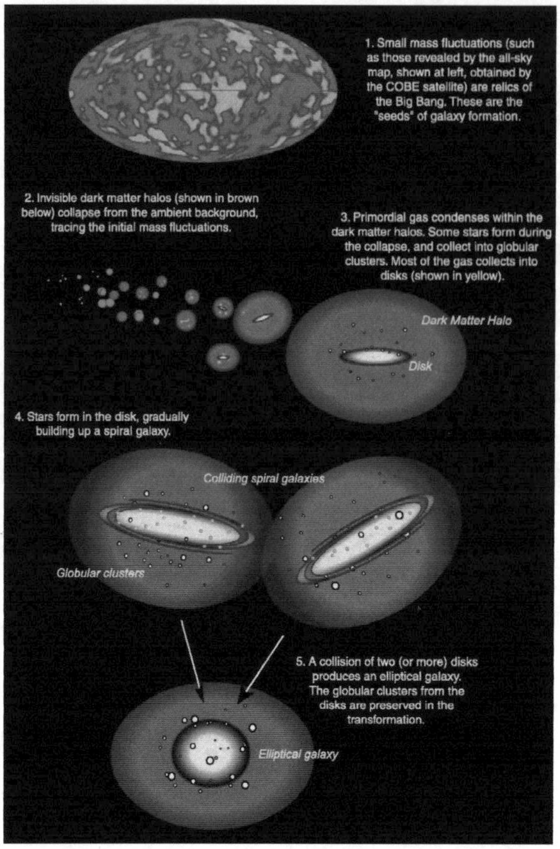

Figure 1. Hierarchical galaxy formation [2].

However, the evolutionary history of a galaxy is not entirely determined by these processes alone. It should be known that the galaxies don't evolve in isolation. Interactions with the small and large scale environment will serve as an important role in the formation and evolution of galaxies. Thus, the partner galaxy will affect the formation and evolution for galaxy pairs. The purpose of this work is to examine differences between the basic properties of galaxy pairs and isolated galaxies.

There are many basic properties of galaxies, an easy and fundamental one is SFR, corresponding to the total mass of stars formed per year. The SFR and the Star Formation History (SFH) are two of the most important parameters describing the galaxy evolution. Both SFR and the SFH can help us have a better understanding of the history of galaxies. A model of the star formation rate as a function of time can be seen in Fig. 2, which can be understood how the Star Formation Rate (SFR) evolve over time.

Besides, the star formation rate has many applications. For example, relationship among Stellar Birth Rate$\mathrm{B(M,t)}$, Star Formation Rate $\psi(t)$, and the Initial Mass Function $\xi(M)$ is shown below:

$$\mathrm{B(M,t)dMdt} = \psi(t)\,\xi(M)\mathrm{dMdt} \qquad (1)$$

Based on the equation above, the SFR is relevant to many basic properties of galaxies, i.e., it has an examination of star formation rate of galaxy pairs is essential to understanding the evolution process of galaxy pairs. On this basis, the plan is to examine the star formation rate of galaxy pairs and compare our findings with isolated galaxies.

The rest part of the paper is organized as follow: Section 1 will talk about some basic ideas of galaxy formation and evolution; Section 2 will show how to examine Star Formation Rate enhancement in galaxy pairs; Section 3 will show our final results; Section 4 will show some drawbacks of this work and the topic of future works; Section 5 will make a summary of this work.

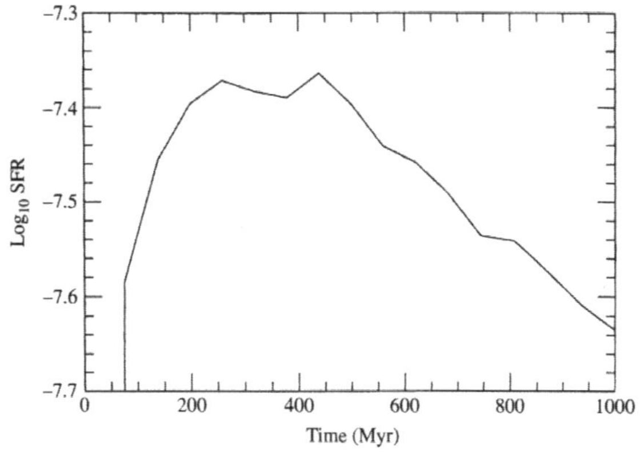

Figure 2. A model of the star formation rate (in units of Msolarpc−2 yr −1)as a function of time [4]

2. METHODOLOGY

In order to achieve the goal described in Section 1, several basic features of data of Illustris TNG need to be understood first to avoid unnecessary mistakes.

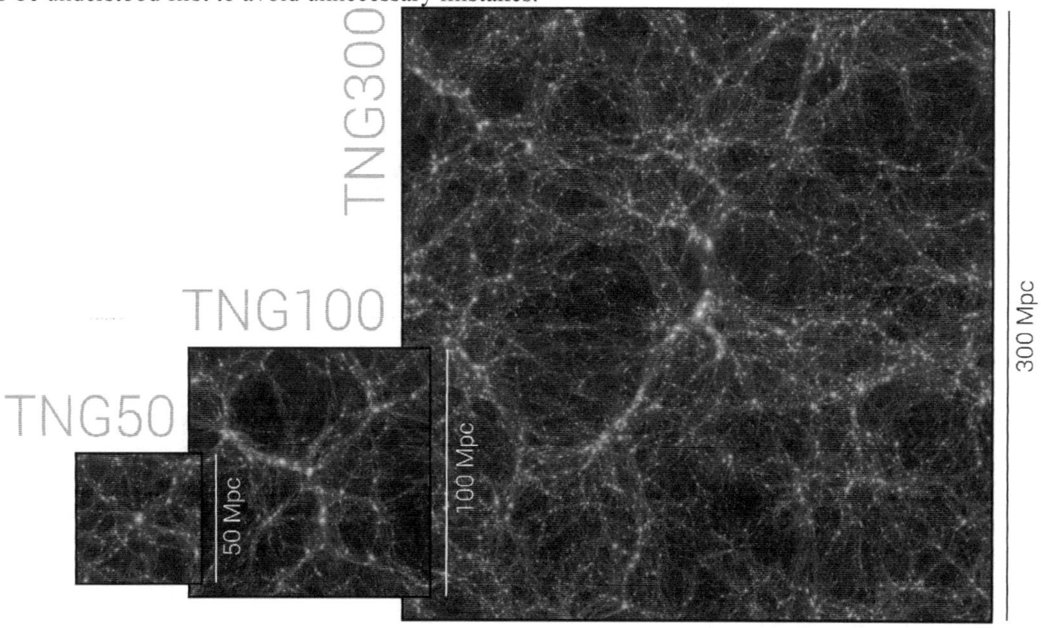

Figure 3. Comparison of the three TNG simulations: TNG50, TNG100, and TNG300. For each projected dark matter density is shown. The three volumes are roughly though not exactly to scale [5].

2.1. Box size

The Illustris TNG project contains three physical simulation box sizes. Hence, the proper size should be chosen for this study. As illustrated in Fig. 3, projected dark matter density of these three physical simulation box sizes is shown. Obviously, different physical simulation box sizes will make a huge difference. None of the simulations can cover everything, the appropriate size and resolution need to be found to better suit the researching goal.

TNG 50 has the best resolution level, which enables the scientists to see some details, e.g., the chemo-dynamical evolution of galaxies. Since TNG 300 has the largest size, it enables the scientists to examine the large structure, e.g., the properties of galaxy clustering. In order to give consideration to both resolution and size, TNG 100 is chosen in this work to study the properties of galaxy pairs.

2.2. Election of galaxy pairs

Galaxy pairs can be distinguished from isolated galaxies based on the distances. If two galaxies are close enough, they are thought to be galaxy pairs. Based on a similar study about the analysis of galaxy pairs from the SDSS [2], it can be supposed that if the distance of two galaxies is smaller than 500 kpc, they are thought to be galaxy pairs.

Nevertheless, with this method galaxy groups will also be selected in this method. Galaxy groups need to be removed from the selection. Fig. 4 exhibits a real example of a galaxy group, which technically is not a galaxy pair. As a consequence, these galaxy groups need to be removed from the selection group. If galaxy pairs are distinguished from isolated galaxies based on the distances, the unit of Illustris TNG needs to be paid attention to. The unit of coordinates of the galaxies in TNG is $ckpc/h$. The 'c' means 'comoving', as opposed to physical. Thus, the comoving distance needs to be multiplied by the scale factor a, which depends on different time of the universe, where h is defined by:

$$H_0 = 100h\ kms^{-1}Mpc^{-1} \quad (2)$$

To simplify, it can be assumed that $h = 0.7$.`

Figure 4. Four of the seven members of galaxy group HCG 16 [6].

2.3. Selection of control group

In order to see how the partner galaxy will influence the formation and evolution process, isolated galaxies need to be selected for com- parison. Whereas, galaxies of different masses normally have different star formation rate. In this case, isolated galaxies need to be selected based on the mass distribution of the galaxy pair samples.

3. Results

3.1. SFR enhancement

Ignoring those galaxies whose SFR is too low (ignoring those galaxies whose star forming process has come to an end), the residual result is illustrated in Fig. 5. The dotted line in Fig. 5 represents the average star formation rate of isolated galaxies. Apparently, the star formation rate of galaxy pairs is higher and closer the pairs are to each other, the star formation rate tends to be higher.

In Section 2.3, it has been known that Galaxies of different masses normally have different SFR, indicating that heavier galaxies and lighter galaxies are distinguished in the galaxy pairs. The blue one represents the star formation rate of the heavier galaxies in the galaxy pairs. The orange one represents the star formation rate of the lighter galaxies in the galaxy pairs. Obviously, the heavier galaxies in the galaxy pairs have higher SFR. Neglecting those galaxies whose SFR is too low, one can find 145 isolated galaxies and 16 galaxy pairs. Thus, the galaxy pairs are not so rare. In this case, having a better understanding of the properties of galaxy pairs is crucial.

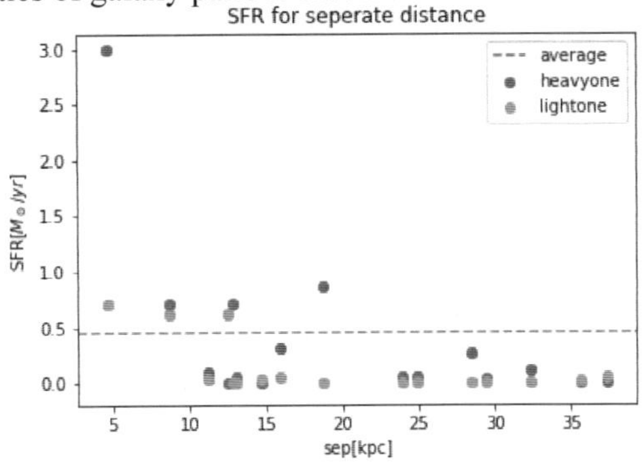

Figure 5. Star formation rate of galaxy pairs for different separate distance. The dotted line represents the average SFR of isolated galaxies. The blue dots represent the SFR of the heavier galaxies in the galaxy pairs. The orange dots represent the SFR of the lighter galaxies in the galaxy pairs.

3.2. *Tidal force*

As for the causality of the star formation rate enhancement, it can be attributed to tidal force, which is the interactions between two galaxies can convert the shape of the galaxy disks and produce starbursts. A starburst galaxy is a galaxy undergoing an exceptionally high rate of star formation. Therefore, the SFR of galaxy pairs may be different from isolated galaxies.

To analyze the tidal force will affect the SFR of galaxy pairs quantitatively, the tidal perturbation parameter f [7, 8] can be helpful follows:

$$f = log(\frac{F_{prim}}{F_{sec}}) = 3log(\frac{R}{b}) + 0.4(m_{sec} - m_{prim}) \tag{3}$$

Where f means the tidal force, R means the radius of the smaller galaxy, b means the distance between two galaxies, m means the magnitude. Obviously, the closer two galaxies are, the bigger the difference of the mass of two galaxies are, the bigger the tidal force are. The bigger the f is, bigger the tidal force is. To have a closer look of the star formation rate enhancement, a SFR enhancement parameter Q can be defined as below:

$$Q = \frac{SFR_{heavier}}{SFR_{lighter}} \tag{4}$$

where $SFR_{heavier}$ means star formation rate of the heavier galaxies in the galaxy pairs, $SFR_{lighter}$ means SFR of the lighter galaxies in the galaxy pairs. The Comparison results of Q with f are presented in Fig. 6. The Q is increasing with the boosting of f, which proves the point that tidal force will affect the star formation rate of galaxy pairs.

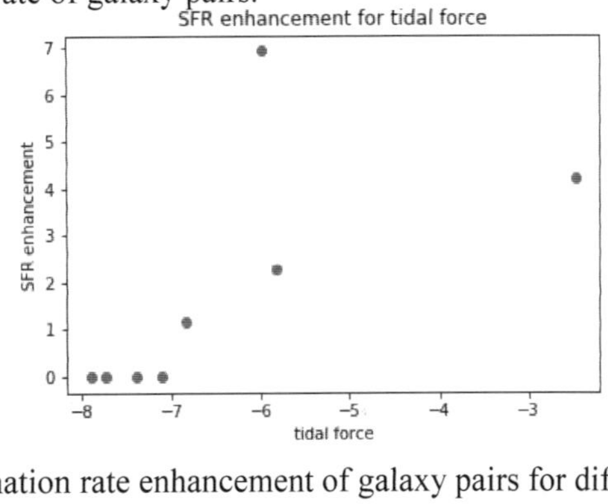

Figure 6. Star formation rate enhancement of galaxy pairs for different tidal force.

3.3. Dark matter

Dark matter may also influence the SFR enhancement of galaxy pairs. To find out the relationship of dark matter and SFR enhancement quantitatively, the dark matter parameter r_{DM} can be helpful, which is defined as:

$$r_{DM} = \frac{r}{R_{lighter}+R_{heavier}} \quad (5)$$

where r means the distance between two galaxies, $R_{lighter}$ means the comoving radius containing half of the total mass of the lighter galaxies in the galaxy pairs, $R_{heavier}$ means the comoving radius containing half of the total mass of the heavier galaxies in the galaxy pairs. Since most mass of the galaxy is made of dark matter, the comoving radius containing half of the total mass of the galaxies can reflect the radius of dark matter. The smaller the r_{DM} is, the bigger difference dark matter makes between two galaxies. The Q as a function of r_{DM} is depicted in Fig. 7. Obviously, the r_{DM} is negatively correlated with Q, which verifies the point that dark matter will affect the star formation rate of galaxy pairs.

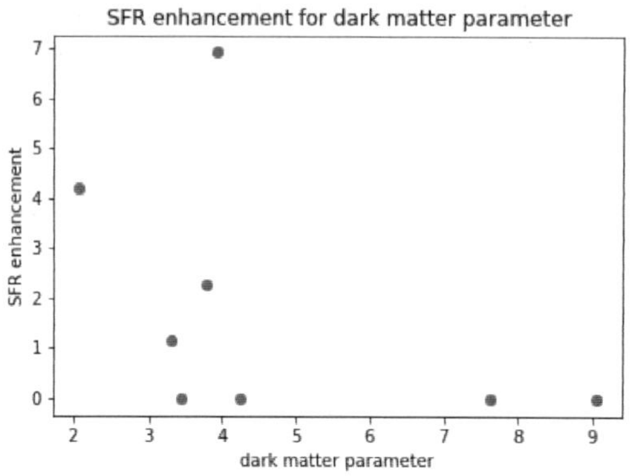

Figure 7. Star formation rate enhancement of galaxy pairs for different dark matter parameter.

4. Discussion

It has been proven that the SFR of galaxy pairs is normally higher than isolated galaxies. In fact, many further studies can be carried out to improve this work. For example, there may be other explanations for star formation rate enhancement in galaxy pairs.

4.1. Other explanations for SFR enhancement in galaxy pairs

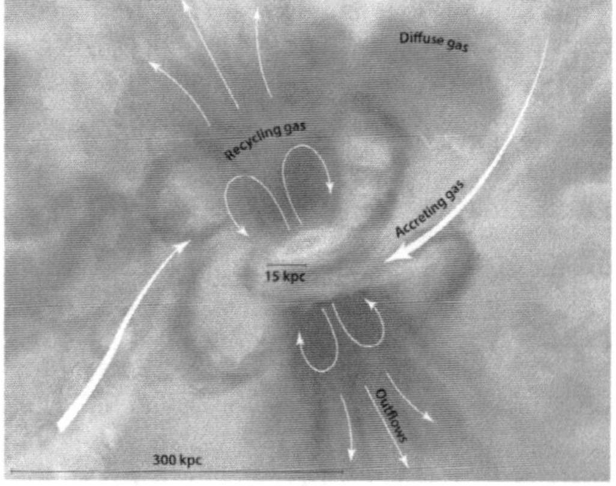

Figure 8. A cartoon view of the CGM [9].

Besides tidal force, the rotation of galaxy pairs may also affect the SFR. The Circumgalactic Medium (CGM), home to large-scale flows of gas that drive galaxy evolution, ought to be considered. These gas flows provide fuel for star formation. If the rotation of galaxy pairs is bigger, it will be harder for the gas to fall into the galaxies, henec the star formation rate may be lower. Fig. 8 exhibits a schematical view of the CGM. The idea that the rotation of the environment of galaxies seems to be proven by a recent work [7]. Fig. 9 proves that the bigger the rotation of the environment of galaxy, the SFR of the galaxy tends to be higher.

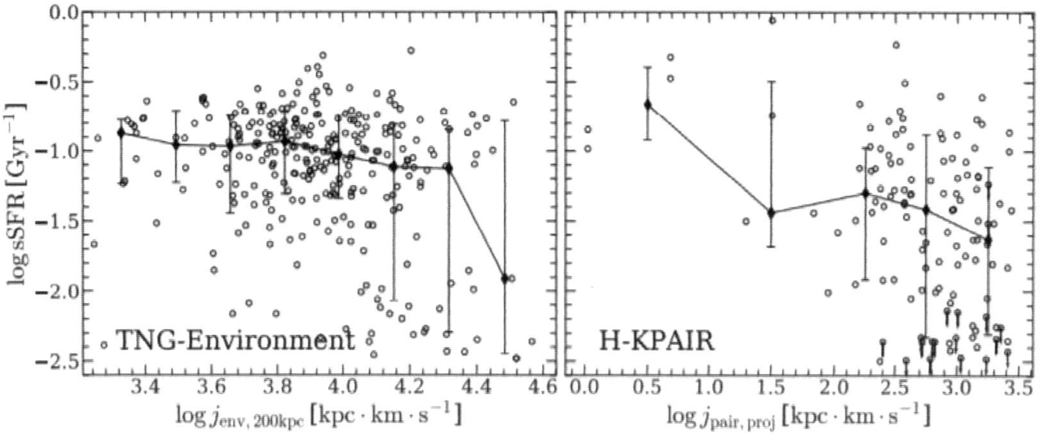

Figure 9. Correlation between the specific SFR (log sSFR) and the environmental angular momentum of the TNG galaxy samples at z≤0.1 (left panel) as well as the SFR and the angular momenta (right panel) [10].

4.2. Further research direction

This study has several drawbacks, e.g., different box sizes of illustris TNG are not examined. Different box sizes and resolution may affect the final results. Besides, the results are compared with observation qualitatively instead of quantitatively. Thereby, further study may find some differences between observation and simulations. More details need to be examined in future work, e.g., the AGN feedback and the gas accretion from the intergalactic medium (IGM) impacts on SFR in galaxy pairs.

5. Conclusions

In summary, SFR and SFH are two of the most important parameters describing the galaxy evolution. Both SFR and the SFH can help us have a better understanding of the history of galaxies. Based on data from Illstris TNG 100, it has been verified that the SFE of galaxy pairs is normally higher than isolated galaxies. The tidal force and the rotation of the environment of galaxies may account for the mechanisms of SFR enhancement in galaxy pairs. However, further studies are still needed to investigate some details in this phenomenon. Overall, these results offer a guideline for galaxy evolution and pave a path for the further direction on the research of large scale environment of galaxy.

References

[1] Moreno J., et al., 2019, Monthly Notices of the Royal Astronomical Society, 485, 1320–1338.
[2] Das A., Pandey B., Sarkar S., Dutta A., 2021, Galaxy interactions in different environments: An analysis of galaxy pairs from the SDSS (arXiv:2108.05874).
[3] Patton D. R., et al., 2020, Monthly Notices of the Royal Astronomical Society, 494, 4969–4985.
[4] Burkert A., Truran J. W., Hensler G., 1992, ApJ, 391, 651.
[5] Nelson D., et al., 2021, The IllustrisTNG Simulations: Public Data Release (arXiv:1812.05609).
[6] https://esahubble.org/news/heic1514/.
[7] Morales-Vargas A., et al., 2020, Monthly Notices of the Royal Astronomical Society, 499, 4370–

4393.
[8] Byrd, G. G., Valtonen, M. J., et al., 1986, A&A 166, 75.
[9] Tumlinson J., Peeples M. S., Werk J. K., 2017, Annual Review of Astronomy and Astrophysics, 55, 389–432.
[10] Lu S., et al., 2021, From large-scale environment to CGM angular momentum to star forming activities – II: quenched galaxies (arXiv:2109.06197).

Comparison of different models for gravitational and electromagnetic radiation of binary neutron star mergers

Boyang Sun *

School of Physics, Nankai University, Tianjin 300070, China

*Corresponding author e-mail: guanghua.ren@gecacademy.cn

Abstract. The neutron star binary merger is one of the most energetic phenomena in our Universe. Based on the calculation of gravitational radiation (GR) with the separate consideration of revolution and stellar rotation, and electromagnetic radiation (ER) with unipolar induction DC circuit model and magnetic dipole model, the results are compared to analyse the extent to which the stellar rotation will affect the total gravitational radiation power. Besides, the relationships between radiation power and the related parameters (e.g., orbital radius and stellar mass) are investigated. Furthermore, the feasibility of different types of binary star merger as the progenitor of fast radio bursts and gamma-ray bursts were studied based on the two models of ER, obtaining that binary neutron star and neutron star-white dwarf systems are among the possible progenitors. Finally, the radiation power of both GR and ER are compared under the same conditions. According to the results, the energy dissipation of the system is dominated by gravitational radiation. These results shed lights on further studies on the radiation processes of binary neutron star mergers and binary star systems.

1. Introduction

1.1. Overview of Neutron Stars

A neutron star is a special celestial body with extreme physical conditions in the Universe. Specifically, the interior of it is almost entirely composed of degenerate neutrons, and it has extremely high density with strong gravitational and magnetic fields around. These extreme physical conditions make the neutron star natural laboratories to investigate the laws of physics under extreme conditions, which can provide a lot of valuable information for theoretical research, i.e., neutron stars have attracted much attention of the scientific community.

The study of neutron stars can be traced back to the 1930s. Less than two years after Chadwick discovered neutrons, Baade and Zwicky proposed that supernova explosions might produce an extreme celestial body composed of dense neutrons and named them neutron stars [1]. Then in 1967, Hewish and Bell detected periodic radio pulse signals from PSR B1919+21, which was the first time that a neutron star was observed [2]. This periodic radiation is originated from the azimuth of the beam-shaped radio radiation at its poles, changing with the rapid rotation of neutron star.

It is generally believed that a star with an initial mass greater than eight times the mass of the sun ($8M_{\odot}$) may collapse into a neutron star after its nuclear fuel is exhausted. Specifically, if the core mass exceeds the Chandrasekhar limit during the collapse, the electronic degeneracy pressure is not enough

Content from this work may be used under the terms of the Creative Commons Attribution 3.0 licence. Any further distribution of this work must maintain attribution to the author(s) and the title of the work, journal citation and DOI.
Published under licence by IOP Publishing Ltd

to support the core, i.e., the core will collapse further. Electrons and protons will react as $p + e^- \rightarrow n + v$ to become neutrons and neutrinos that will completely escape later. Furthermore, the combined effect of strong interaction and neutron degeneracy pressure stops the collapse process and forms a stable star [3].

1.2. Neutron star binary mergers

A binary star system that meets certain conditions can form a binary neutron star system after a complex and long evolution, as schematically shown in Figure 1. The evolution of a binary star system is divided into three stages: inspiral, merger, and ringdown. This is a rather complicated process, partly because four basic interactions are involved. The binary neutron star will produce a variety of energy radiation during the merge phase.

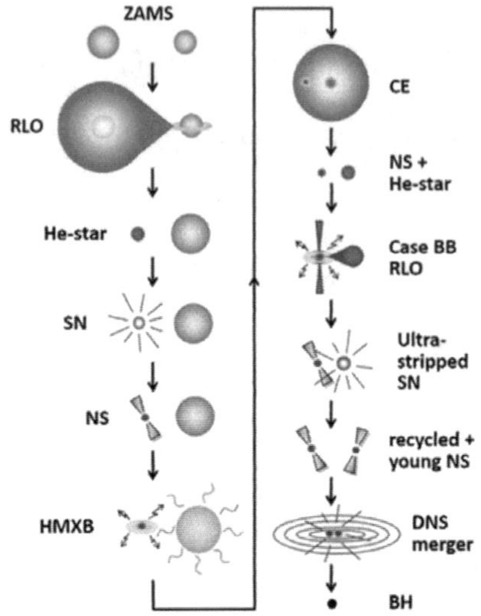

Figure 1. Illustration of the formation of a double neutron star system, from reference [4]

Firstly, according to General Relativity, a pair of objects rotating with each other will cause regular oscillations in the surrounding space-time. This oscillation carries energy and can propagate infinitely, i.e., a gravitational wave. Due to the large mass of the binary neutron star system, the generated gravitational wave signal is extremely strong. Therefore, in 2017, LIGO and Virgo have detected the gravitational wave generated by the binary neutron star merger, GW170817 [5].

Secondly, the binary neutron stars merger will produce a variety of electromagnetic radiation, the most well-known types include gamma-ray bursts (GRB) and afterglow radiation. GRB was accidentally discovered by the Vela satellite of the US in 1967 [6]. It is the highest-energy electromagnetic event known to mankind, which is a typical gamma storm releases energy in a few seconds equivalent to the total energy released by the sun during its entire life cycle [7]. Afterglow radiation are caused by the synchrotron radiation of shock waves accelerated electrons in terms of the interaction of GRB jets with the interstellar medium. In addition, in 2007, Lorimer discovered a special kind of high-energy electromagnetic radiation and named it as Fast Ray Bursts (FRB) [8]. Currently, there is no final conclusion about the mechanism of GRB and FRB, nor the progenitors of these phenomena.

Thirdly, the Kilonova radiation, the mechanism of which is that a large amount of neutron-rich material will be ejected when neutron stars merge. In addition, heavy elements are synthesized in these ejections through fast neutron capture reactions. Subsequently, they undergo radioactive decay to heat the ejection, thus generating radiation [9]. The product of the binary neutron star merger may form supermassive neutron stars (SMNS), black holes, or some theoretical celestial bodies (e.g., quark stars and millisecond magnetars) owing to the different initial masses of the binary stars.

1.3. Arrangement for this article

This paper focuses on electromagnetic and gravitational radiation of neutron star binary mergers. The former includes gamma-ray bursts and fast radio bursts, and the latter involves gravitational wave radiation. The rest part of the paper is organized as following. The Sec. 2 will discuss all four models used in detail with corresponding calculation formula. In Secs. 3 and 4, the calculation results of different models of GR and ER are compared separately, and a number of relevant issues as well as the limitations of the models are discussed. Finally, Sec. 5 gives a brief summary of this paper and offers the outlook for the future work.

2. Analytical models

2.1. Gravitational radiation (GR)

The pulsar binary stars will radiate gravitational energy in the form of gravitational waves during inspiral. This subsection introduces two models for investigating gravitational radiation, where one considers the energy dissipation caused by the revolution of binary stars while the other considers the gravitational radiation caused by the rotation of neutron stars on the basis of the former. On this basis, the latter model can be regarded as an extension and supplement of the former one.

2.1.1. Gravitational radiation powered by binary star revolution. The motion of the binary star satisfies the Einstein field equation

$$R^{\mu v} - \frac{1}{2} g^{\mu v} R = k T^{\mu v} \tag{1}$$

where $R^{\mu v}$ is the Rich tensor, $g^{\mu v}$ is the metric tensor, and $T^{\mu v}$ is the energy - momentum tensor of the matter field. Taking the weak field linear approximation to the field equation (i.e., ignoring the nonlinear effect at the field source) and substituting the mass quadrupole $Q_{kj} = \int \mu x_k x_j dV$ (μ is the particle mass), the gravitational radiation power can be derived as [2]

$$P_{gra} = -\frac{dE}{dt} = \frac{G}{5c^5}\left(\dddot{Q}_{ij}^{\,2} - \frac{1}{3}\dddot{Q}_{kk}^{\,2}\right) \tag{2}$$

Subsequently, Newtonian dynamics will be used to solve the motion of the binary star. This approximation is reasonable since the strong field at the field source is not considered. Supposing that the binary stars revolve around each other in the x-y plane, and ignoring the effect of their rotation and internal structure and shape, the binary stars can be regarded as two mass points with masses m1 and m2, respectively. Taking the centroid coordinate system, one obtains $d_1 = \frac{m_2}{m_1+m_2} d$, $d_2 = \frac{m_1}{m_1+m_2} d$, where $d = d_1 + d_2$ is the binary distance. Therefore, according to Newton's law of gravitation, the orbital equation of the binary star motion can be easily obtained as: $d = \frac{a(1-e^2)}{1+ecos\varphi}$. Apparently, it is a typical ellipse equation, where a is the semi-major axis of the ellipse, e is the eccentricity, and the angular velocity of the revolution of binary star

$$\dot{\varphi} = \frac{[\mathrm{G}(m_1 + m_2)\mathrm{a}(1 - \mathrm{e}^2)]^{1/2}}{\mathrm{d}^2} \tag{3}$$

Since the motion of the double star is limited to the plane, the mass quadrupole moment of the system has only four non-zero components, which can be expressed as:

$$Q_{ij} = \begin{pmatrix} \mu d^2 cos^2\varphi & \mu d^2 sin\varphi cos\varphi \\ \mu d^2 sin\varphi cos\varphi & \mu d^2 sin^2\varphi \end{pmatrix} \tag{4}$$

in which the converted mass has been taken in the formula $\mu = \frac{m_1 m_2}{m_1+m_2}$. Then, the third derivative of the mass quadrupole is calculated and substituted into the angular velocity formular

$$\dddot{Q}_{ij} = \begin{pmatrix} \beta A(\varphi) & -\beta B(\varphi) \\ -\beta B(\varphi) & -\beta C(\varphi) \end{pmatrix} \tag{5}$$

where $\beta = 2Gm_1m_2(\frac{m_1+m_2}{a^5(1-e^2)^5})^{1/2}$, substituting into the radiant power expression, one derives

$$P_{rev} = \frac{8G^4}{15c^5}\frac{m_1^2 m_2^2(m_1+m_2)}{a^5(1-e^2)^5}(1+ecos\varphi)^4[12(1+ecos\varphi)^2+e^2sin^2\varphi] \tag{6}$$

In order to find the average radiated power of a binary star in a revolution period, one ought to first obtain the revolution period according to Kepler's third law

$$\mathrm{T} = \sqrt{\frac{4\pi^2 a^3}{G(m_1+m_2)}} \tag{7}$$

and one knows

$$< P_{rev} > = -\int\frac{dE}{T} = \frac{(1-e^2)^{2/3}}{2\pi}\int_0^{2\pi}\frac{P}{(1+ecos\varphi)^2}d\varphi \tag{8}$$

Substitute Eq. (6), the expression of P is simplified as

$$< P_{rev} > = \frac{32G^4}{5c^5}\frac{m_1^2 m_2^2(m_1+m_2)}{a^5}f(e) \tag{9}$$

where $f(e) = \frac{1+\frac{73}{24}e^2+\frac{37}{96}e^4}{(1-e^2)^{\frac{7}{2}}}$ is the correction factor caused by orbital eccentricity. The binary star system will radiate gravitational waves according to this formula. Obviously, the total energy of the system will gradually decrease over time. This directly leads to the reduction of the distance between the binary stars and the shortening of the revolution period, which is the main reason for the inspiral. This part will be studied in detail in Secs. 3 and 4.

2.1.2. Gravitational radiation powered by stellar rotation. In model one, the rotation of the neutron star is neglect because the model ignores the structure and shape of the neutron star and treats it as a mass point. However, in fact, neutron stars are not true spheres, newborn neutron stars generally rotate rapidly attributed to their extremely small radius and conservation of angular momentum, which also radiates gravitational waves. This model mainly takes this part of energy radiation into consideration.

Since the interior of a neutron star is very dense and the difficulty for external conditions to change its shape, it is reasonable to regard it as a rigid body. Supposing the circular frequency of the neutron star's rotation is ω and introduction the following coordinate system x'^{μ} that rotates with the neutron star, its relationship with the laboratory coordinate system x^{μ} is

$$x^{\mu} = \begin{pmatrix} cos\omega t & -sin\omega t & 0 & 0 \\ sin\omega t & cos\omega t & 0 & 0 \\ 0 & 0 & 1 & 0 \\ 0 & 0 & 0 & 1 \end{pmatrix} x'^{\mu} \tag{10}$$

Taking the ellipsoid of inertia in x'^{μ} system, the moment of inertia can be expressed in the form of a diagonal matrix diag(I_{11}, I_{22}, I_{33}). Set x_1', x_2', x_3' for the three principal axes of inertia ellipsoid, the mass quadrupole moment

$$Q_{11}(t) = \left(\int \mu x_1'^2 dV'\right)cos^2\omega t + \left(\int \mu x_2'^2 dV'\right)sin^2\omega t = I_{11}cos^2\omega t + I_{22}sin^2\omega t \tag{11}$$

Similarly, the other components of the quadrupole can be obtained, and the matrix of the mass quadrupole moment is

$$Q_{ij}(t) = \begin{pmatrix} \frac{1}{2}[(I_{11}+I_{22})+(I_{11}-I_{22})cos2\omega t] & \frac{1}{2}(I_{11}-I_{22})sin2\omega t & 0 \\ \frac{1}{2}(I_{11}-I_{22})sin2\omega t & \frac{1}{2}[(I_{11}+I_{22})+(I_{11}-I_{22})cos2\omega t] & 0 \\ 0 & 0 & I_{33} \end{pmatrix} \tag{12}$$

Introducing the equatorial ellipticity $\delta = \frac{I_{11}-I_{22}}{I}$, $I = I_{11} + I_{22}$, the energy radiation power produced by the rotation of the neutron star can be obtained from the Eq. (2) and (12) as

$$P_{rot} = \frac{G}{5c^5}\left(\dddot{Q}_{ij}^{\,2} - \frac{1}{3}\dddot{Q}_{kk}^{\,2}\right) = \frac{32}{5}\frac{G}{c^5}\omega^6 I^2\delta^2 \tag{13}$$

Adding the subscripts 1 and 2 to the two neutron stars respectively, one derives the total power of the gravitational radiation generated by binary star system due to their rotation

$$P = \sum P_{rot} = \frac{32}{5}\frac{G}{c^5}\sum_{i=1}^{2} \omega_i^6 I_i^2 \delta_i^2 \tag{14}$$

2.2. Electromagnetic radiation (ER)

2.2.1. Unipolar induction model. The Unipolar induction DC circuit model was originally proposed by Goldreich and others during the study of the electromagnetic radiation of Jupiter-Io system. Since the model is concise with clear physical images, it is widely used in astrophysics researches. Figure 2 shows the fundamental ideas of this model.

The premise of applying this model is that there must be a magnetosphere with a strong magnetic field around the binary star. Since all existing observations and theoretical studies have shown that there is a very strong magnetic field on the surface of a neutron star (10^4~10^{11} T), the premise is naturally satisfied. In this way, when the neutron star revolves in the magnetosphere, it will cut the lines of magnetic force and generate the induced electromotive force (EMF). Setting the magnetic field strength in the magnetosphere as B and in terms of Faraday's law of electromagnetic, one obtains

$$E = \int (v \times B) \cdot dl\, P_{rot} \cong 2vBR_c \tag{15}$$

It is easy to know that the speed of a star's motion is $v = (\omega - \omega_s) \times \vec{a} = \Delta\omega \times \vec{a} \cong a\Delta\omega$, where ω is the revolution angular velocity, ω_s is the rotation angular velocity, and the direction of $\vec{a}$ is from the host star to the companion star. This electromotive force will generate a DC circuit loop in the magnetosphere. Substituting the current expression based on Ohm's law into the above equation, one derives

$$I = \frac{E}{R_{tot}} \cong \frac{2aBR_c}{R_{tot}}\Delta\omega \tag{16}$$

where R_{tot} is the total resistance of the circuit, the magnetic resistor R_{meg} and resistor of neutron star R_{NS}. Nevertheless, neutron star, as a compact neutron object, its resistor should be much less than the resistance of the magnetic fluid resistance. Hence, it can be considered as $R_{tot} = 2R_{meg}$, where factor '2' is caused by the DC circuit of both the upper and lower layer.

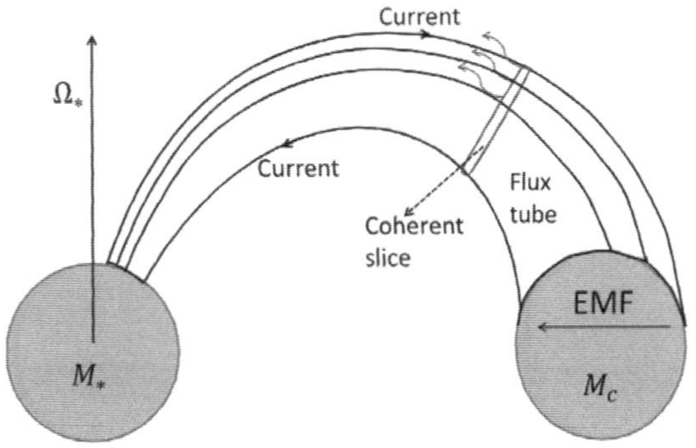

Figure 2. Schematic picture of unipolar inductor model, from reference [10]

In this way, it is obvious that the power lost by the circuit due to internal energy dissipation is

$$P_H = -\frac{dE}{dt} = 2I^2R_{tot} = \frac{8a^2B^2R_c^2}{R_{tot}}\Delta\omega^2 \tag{17}$$

If the contribution of the companion star's magnetic field to the magnetosphere is neglected, the main star's magnetic field can be regarded as the magnetic field produced by the magnetic dipole. Since a>>R*, the magnetic field strength of the magnetosphere at the companion star is approximately $B = B_* \frac{R_*^3}{a^3}$.

Assuming that the magnetospheric resistance is equal to the impedance of free space, i.e., $R_{meg} = \mu_0 c$, the electromagnetic radiation power of the binary star system is

$$P_H = \frac{v^2}{\mu_0 c}\frac{B_*^2 R_*^6 R_c^2}{a^6} \tag{18}$$

In fact, the actual situation of this process is much more complicated than the above calculation, partly because the rotation of the neutron star will produce a toroidal field. As a consequence, it will interfere with the above-mentioned polar field and make the loop unstable. Lai did a detailed study on this and suggested using a quasi-cyclic circuit to describe this process. He proposed that the increase of the toroidal field will lead the circuit to be cut off, and magnetic reconnection will occur to release the magnetic energy, and then the circuit will be reconnected. Magnetic energy is continuously released through this cycle. Based on the above considerations, Lai obtains the energy dissipation power as [11]

$$\dot{E}_{diss} = -T\Delta\Omega = \zeta_\varphi \Delta\Omega \frac{\mu^2 R_c^2}{2a^5} = \left(\frac{v_{rel}}{c}\right)^2 \frac{B_*^2 R_*^6 R_c^2}{\pi a^6} \tag{19}$$

After converting its unit system to UI, the result is numerically equal to Eq. (18). Furthermore, if it is assumed that the dissipated energy is uniformly radiated in all directions in the form of electromagnetic waves, i.e., spherical waves are radiated by the field source, the Poynting vector at the position R from the field source is

$$\bar{S} = \frac{P_H}{4\pi R^2} = 1.4 \times 10^{43}\left(\frac{B_*}{10^{13}GR}\right)^2\left(\frac{a}{30km}\right)^{-7} \vec{n}\ ergs^{-1}m^{-2} \tag{20}$$

This is the electromagnetic radiation energy that can be received per unit area at the position R from the field source.

2.2.2. Magnetic dipole model. In this model, it is assumed that binary neutron stars carry a certain amount of charge. In other words, when the binary stars rotate rapidly, a current loop will be generated, which is a standard magnetic dipole model as illustrated in Figure 3. Therefore, the equation of ER of the system can be derived from the results of magnetic dipole radiation.

The current generated from revolution of two neutron starts with charge of q_1, q_2 and binary revolution period T is $I = \frac{q_1+q_2}{T}$. Substituting Eq. (7) into the formula, one obtains

$$I = \frac{\sqrt{G(m_1+m_2)}}{2\pi}\frac{q_1+q_2}{a^{3/2}} \tag{21}$$

The current loop is approximately regarded as a circular ring, and the radius is equal to the semi-major axis a of the ellipse. Therefore, the magnetic moment corresponding to the current loop is

$$\vec{m} = \pi a^2 I = \frac{\sqrt{G(m_1+m_2)}}{2}(q_1+q_2)a^{1/2}\ \vec{n} \tag{22}$$

The electric field intensity E and the magnetic field intensity B generated by the magnetic dipole oscillation at the distance from the field point are [12]:

$$\begin{cases} \vec{E} = \dfrac{1}{4\pi\varepsilon_0 c^3 r}\hat{e}_r \times [\ddot{m}] \\ \vec{B} = \dfrac{\mu_0}{4\pi c^2 r}\hat{e}_r \times (\hat{e}_r \times [\ddot{m}]) \end{cases} \tag{23}$$

Substituting Eq. (22) into the above two formulas, one obtains the current loop Poynting vector of the electromagnetic radiation at the distance of r:

$$\bar{S} = \frac{1}{2\mu_0}Re\left(\vec{E}\times\vec{B}^*\right) = \frac{\mu_0^2|\ddot{m}|^2}{32\pi^2 c^3 R^2} \tag{24}$$

Since the distance of the binary stars will be reduced due to energy radiation, i.e., the radius of the current loop a is a function of t:

$$\bar{S} = \frac{\mu_0^2 G(m_1+m_2)}{128\pi^2 c^3 R^2}\frac{q_1+q_2}{4}a^{-1}\left(-\frac{1}{2}a^{-1}\frac{da}{dt}+\frac{d^2a}{dt^2}\right)^2 \tag{25}$$

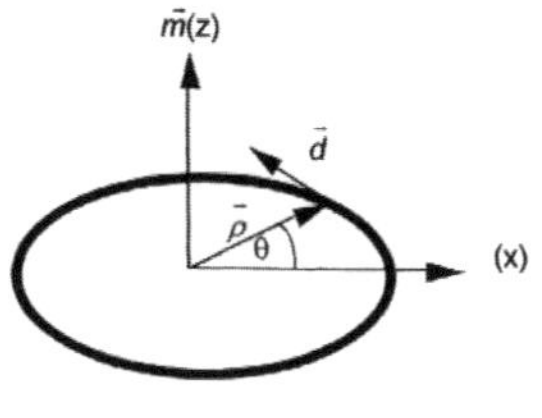

Figure 3. Schematic picture of magnetic dipole model, from reference [13]

It can be seen that in order to obtain the analytical solution of the Poynting vector, one must find the specific form of the function a(t) that exhibits how the distance of binary star evolves with time, which will be discussed in the next section.

3. Comparison of models for GR

3.1. Radiated power of two models

From Eq. (9), the gravitational radiation power generated by the revolution of the double neutron star is $< P_{rev} > = \frac{32G^4}{5c^5}\frac{m_1^2 m_2^2(m_1+m_2)}{a^5}f(e)$. Following four binary neutron star systems are selected and listed in Table 1.

Table 1. Some properties of four binary neutron star systems

Name	M_m ($M_\odot$)	M_c ($M_\odot$)	D(kpc)	e
GW170817	1.46	1.28	40000	≤0.024
GW190425	2.02	1.35	15900	≤0.048
PSR B1913+16	1.441	1.387	6.4	0.617
PSR J0737−3039	1.337	1.250	1.15	0.088

* The data is derived from LIGO Scientific Collaboration

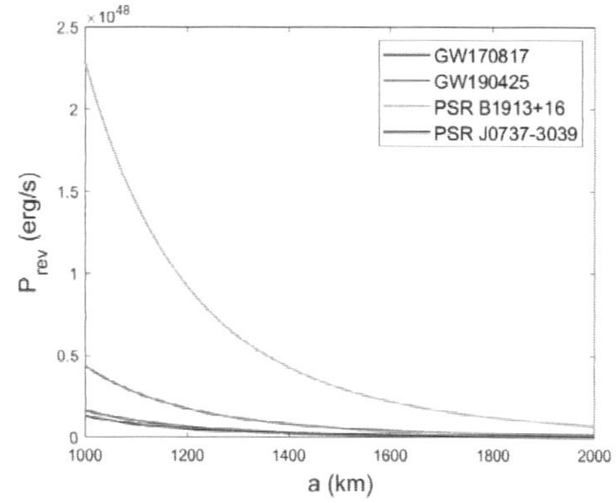

Figure 4. Relation between radiated power of GR and radius of orbit

Figure 4 depicts the relation graph of P_{rev} and a when a ranges from 1000km to 2000km. It can be seen that the radiated energy flow is very high when the binary stars revolve, and the radiant power increases rapidly when the binary stars approach. In order to obtain the relationship between the radiant power of the neutron star's rotation and its mass m, radius r and other parameters, the expression of the moment of inertia needs to be obtained. Although Model 2 assumes that the neutron star is not a sphere, it can be approximately applied to the sphere's moment of inertia formula when calculating the moment of inertia, which is $I = 0.4mr^2$, thus formula (13) turns

$$P_{rot} = \frac{128}{125}\frac{G}{c^5}\omega^6 m^2 r^4 \delta^2 \tag{26}$$

Similarly, four pulsar systems are selected and listed in Table 2.

Table 2. Some properties of four different pulsars

Name	M ($M_\odot$)	R (km)	T_{rot} (ms)	D(kpc)
PSR B1257+12	1.4	10.4	6.22	0.71
PSR J0348+0432	2.01	13	39.1	2.1
PSR J1614-2230	1.908	13	3.15	1.2
PSR J0737-3039A	1.337	~25	22.7	1.15

* The data is derived from Centre de Données astronomiques de Strasbourg (CDS)

From the rotation period of pulsars, the circular frequency can be obtained. Among the above four pulsars, the maximum circular frequency is of order $\omega_{max} = \frac{2\pi}{T_{rot}} = \frac{2\pi}{6.22ms} \sim 10^3$, and the minimum circular frequency is of order $\omega_{min} = \frac{2\pi}{39.1ms} \sim 10^2$. Therefore, the circular frequency is set in the range of 100 s^{-1} to 1000 s^{-1}. Figure 5 presents the relation of radiated power and circular frequency.

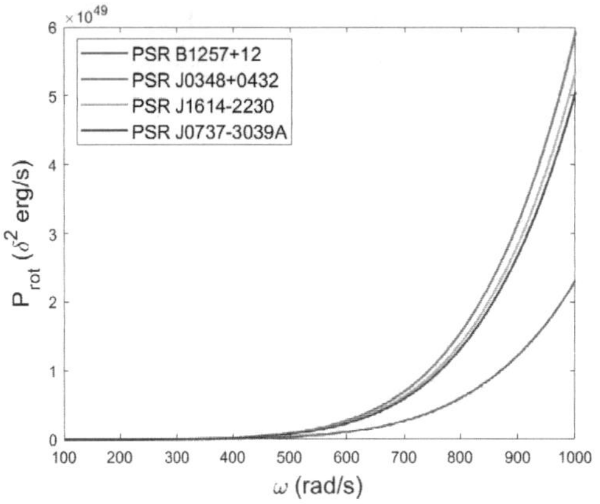

Figure 5. Relation between radiated power of GR and frequency.

Thus, the gravitational radiation caused by the rotation of a neutron star is also quite considerable. Even if the ellipticity of the equator is only 10^{-2}. When the rotation frequency is of order 10^3, the radiation power can still reach the order of 10^{45}. In contrast, for the four selected systems, when the orbit radius of the revolution is 10^3 km, the gravitational radiation caused by the revolution of the binary star is on the order of 10^{46} to 10^{48}. However, the energy will decrease sharply with the shortening of the orbital radius, reaching 10^{56} erg/s when the radius is on the order of 10^2; while the neutron star will reduce its rotation frequency due to energy dissipation after it is formed. The pulsar PSR J1748-2446ad with the shortest rotation period has a circular frequency of only 10^3 s^{-1}. Therefore, for the middle and late stages of the binary star precession, that is, when the orbital radius is less than 10^3 km, the neutron star's rotation relative to the gravitational force generated by the revolution of the binary star radiation is generally extremely small and can be ignored. To further quantify this effect, the PSR J0737-3039 binary star system is chosen as an example. The relevant data is listed in Table 3, and the relationships are shown in Figure 6 (assuming the equatorial ellipticity δ=0.1).

Table 3. Some properties of the PSR J0737-3039 binary star system

Name	M_m ($M_\odot$)	M_c ($M_\odot$)	R (km)	T_{rot} (ms)	e
PSR J0737−3039	1.337	1.250	—	—	0.088
PSR J0737-3039A	1.337	—	~25 [14]	22.7	—
PSR J0737-3039B	1.250	—	~25*	2770	—

* The data is derived from Centre de Données astronomiques de Strasbourg (CDS)

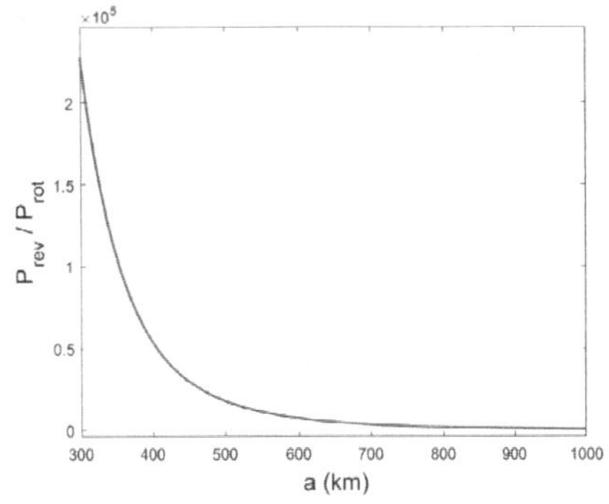

(a)

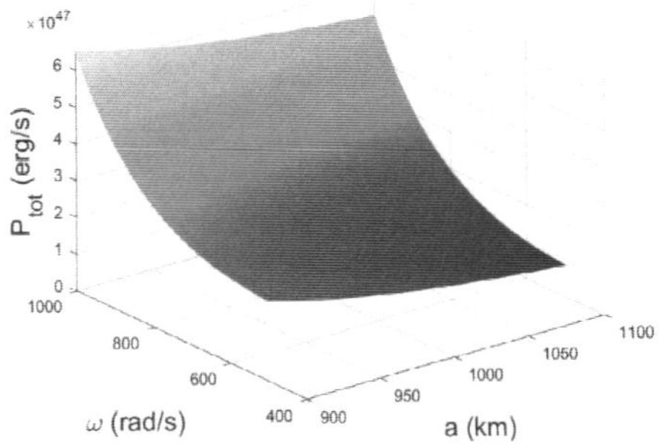

(b)

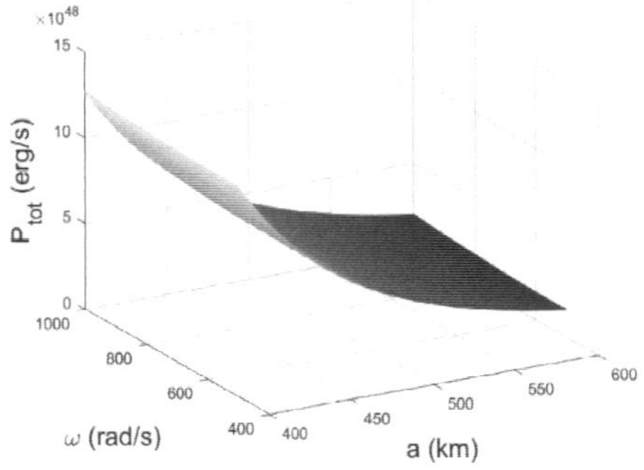

(c)

Figure 6. Figure (a) shows the change of ratio, P_{rev} / P_{rot} with the orbit radius a. Figures (b) and (c) show how the total radiated power is affected by the radius a and the rotation frequency ω. The value of a in Figure (b) is (900km, 1100km), and for Figure (c), a is (400km, 600km). Since the rotation frequency of PSR J0737-3039B is quite small, which has little effect on total radiated power, its rotation frequency change is not considered.

In the early stage of the inspiral (900km<a<1100km), and the rotation frequency of the star is of order 10^3. Their rotation frequency will have a certain impact on the gravitational radiation power. When 400km<a<600km, the gravitational radiation power is already dominated by the orbital radius.

3.2. Evolution of orbit and angular frequency

Next, the precession phenomena caused by energy dissipation in the binary star system are analyzed. According to classical mechanics, the total energy of a binary star system is

$$E = -\frac{Gm_1m_2}{2a} + \frac{1}{5}\sum_{i=1}^{2} m_i\omega_i^2r_i^2 \tag{27}$$

where the first term on the right side of the equation is the total energy of revolution, and the second term is the rotation energy of the binary star system. The evolution of the orbit is only related to the energy of the revolution. To prove this, it can be assumed that the energy loss caused by the rotation of the neutron star has a proportion of η that causes the orbit radius to change, then

$$\frac{da}{dt} = \frac{2a^2}{Gm_1m_2}\left(\frac{dE}{dt} - \frac{2}{5}\eta\sum_{i=1}^{2} m_ir_i^2\omega_i\frac{d\omega_i}{dt}\right) \tag{28}$$

In order to find the relationship between the radius of the orbit and time, the angle term should not appear in the equation. Thus, the average radiant power in one revolution period is used. Hence, substituting the equation (9) into equation (28), one derives

$$< \frac{da}{dt} > = -\frac{64G^3}{5c^5}\frac{m_1 m_2(m_1+m_2)}{a^3}f(e) - \frac{4a^2}{5G}\eta\left(\frac{\sum_{i=1}^{2} m_i r_i^2 \omega_i \frac{d\omega_i}{dt}}{m_1 m_2}\right) \tag{29}$$

The first term on the right side of the formula is the radiation caused by the revolution of a binary star, denoted as $< \frac{da}{dt} >_{rev}$; the second term is the radiation caused by the rotation of neutron star, denoted as $< \frac{da}{dt} >_{rot}$. The first term can be directly integrated

$$< a >_{rev} = \{a_0^4 - 4[\frac{G^3}{5c^5}m_1 m_2(m_1+m_2)f(e)t]\}^{1/4} \tag{30}$$

where 't' is an integer multiple of the period T, and 'a' is the radius of the orbit of the binary star after n cycles. According to the above assumptions, the frequency of the neutron star's rotation circle satisfies

$$P_{rot} = I\omega\frac{d\omega}{dt} = -\frac{32}{5}\frac{G}{c^5}(1-\eta)\omega^6 I^2\delta^2 \tag{31}$$

The corresponding solution is

$$\omega(t) = \omega_0[1 + \frac{256}{25}\frac{G}{c^5}(1-\eta)\omega_0^4 mr^2\delta^2 t]^{-1/4} = \omega_0[1 + 2(1-\eta)f(m,r)\omega_0^4 t]^{-1/4} \tag{32}$$

where $f(m,r) = \frac{128}{25}\frac{G}{c^5}mr^2\delta^2$, is the function related to the moment of inertia of the neutron star, and ω0 is the frequency of the rotation circle corresponding to t=0. It can be solved by the above formula $< a >_{rot} \sim t^{-1/4}$, this formula diverges at t=0, i.e., it has no physical meaning. The only solution is to order $\eta = 0$, which proves that the energy lost by the rotation of double stars will not affect the length of the revolution orbit. Therefore, the expression of the rotation angle frequency is

$$\omega(t) = \omega_0[1 + 2f(m,r)\omega_0^4 t]^{-1/4} \tag{33}$$

4. Comparison of models for ER

4.1. Radiated power of two models

According to the unipolar induction model, the power of electromagnetic waves radiated by the binary star system is

$$P_H = \frac{v^2}{\mu_0 c}\frac{B_*^2 R_*^6 R_c^2}{a^6} = \frac{(2\pi)^2}{\mu_0 c T^2}\frac{B_*^2 R_*^6 R_c^2}{a^4} \tag{34}$$

Substituting Eq. (7) into this equation, one finds

$$P_H = \frac{G(m_1+m_2)}{\mu_0 c}\frac{B_*^2 R_*^6 R_c^2}{a^7} \tag{35}$$

Since the unipolar induction model requires the magnetic field of the host star to be much stronger than the magnetic field of the companion star, it is not suitable for the analysis of general binary neutron star systems. However, since the magnetic field of white dwarfs is generally much smaller than that of neutron stars, this model should be suitable for analyzing the neutron star - white dwarf system. The parameters of four binary star system are listed in Table 4.

Table 4. Some properties of different binary star system

Name	M_c ($M_\odot$)	Rc ($R_\odot$)	T_{rev} (s)
4U 1820-303	0.0648	0.0314	685
4U 0513-40	0.0435	0.0358	1020
M15 X-2	0.0325	0.0395	1356
XTE J1751-305	0.0171	0.0489	2544

* The data is derived from Reference [15], and the radius of companion star is calculated according to the radius-mass relations of white dwarfs $R \sim M^{-1/3}$, as shown in Figure 7.

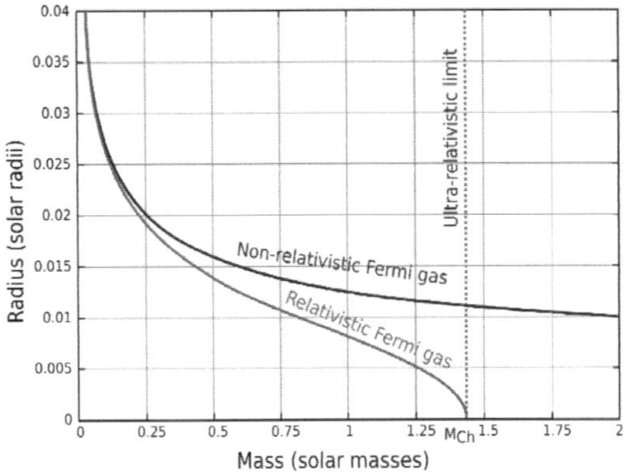

Figure 7. The relationship between radius and mass for a white dwarf star consisting of a cold Fermi gas in two cases whether relativistic effect is considered, plotted by AllenMcC [16].

Assuming that the radius of the neutron star is 10 km and the magnetic field is 106 T, which are reasonable for general neutron stars. Then, the relation graph of the radiation power with the radius of the orbit is exhibited in Figure 8.

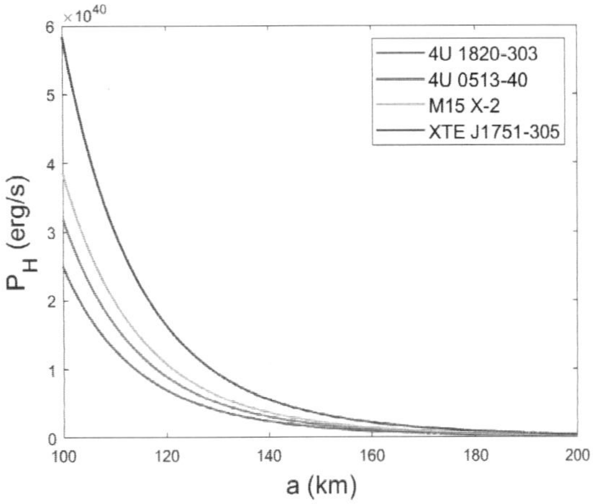

Figure 8. Relation between PH and radius of orbit a based on unipolar induction model.

Therefore, the radiation power in this model is more sensitive to the change of a. When a=100km, the above systems are all of order 10^{40}. Substituting the formula (29) for the evolution of the orbital radius 'a' with 't' into formula (25), the Poynting vector of the binary star system radiating electromagnetic waves is

$$\bar{S} = \frac{\mu_0^2 G(m_1 + m_2)}{128\pi^2 c^3 R^2} \frac{q_1 + q_2}{4} f(m,e)^2 [a^{-9} - 8f(m,e)a^{-12} + 16f(m,e)^2 a^{-15}] \quad (36)$$

where $f(m,e) = \frac{64G^3}{5c^5} m_1 m_2 (m_1 + m_2) f(e)$. Assuming that the radiation energy in all directions at the position R away from the field source is uniform, the total power of electromagnetic waves radiated by the binary stars under this model is

$$P_M = \frac{\mu_0^2 G(m_1 + m_2)}{128\pi c^3} (q_1 + q_2) f(m,e)^2 [a^{-9} - 8f(m,e)a^{-12} + 16f(m,e)^2 a^{-15}] \quad (37)$$

For model two, three binary neutron star systems GW170817, GW190425, PSR J0737-3039 are selected, and the neutron star - white dwarf system NGC1851 and the binary white dwarf system WD J0651+2844 are selected for comparison. The parameters of these five systems are listed in Table 5.

Table 5. Parameters of five different binary star systems

Name	M_m ($M_\odot$)	M_c ($M_\odot$)	D(kpc)	e
GW170817	1.46	1.28	40000	≤0.024
GW190425	2.02	1.35	15900	≤0.048
PSR J0737−3039	1.337	1.250	1.15	0.088
4U 0513-40	1.4	0.0435	12	0.71
WD J0651+2844	0.50	0.26	0.66	0.383

According to the magnetic dipole model, the relation graph of electromagnetic waves radiated by the above five systems are exhibited in Figure 9.

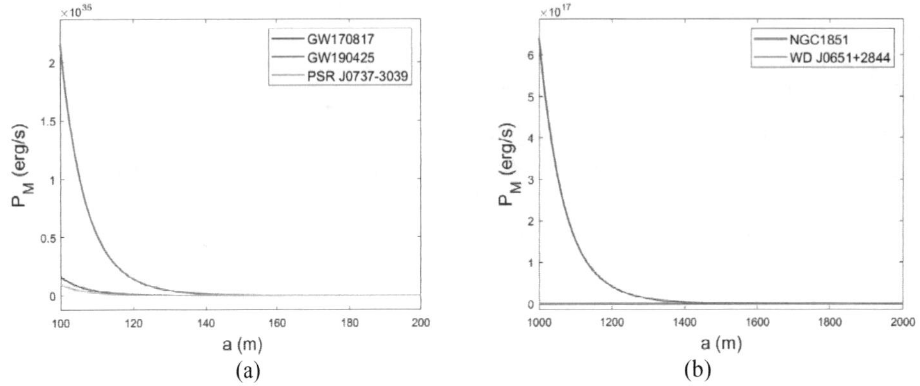

Figure 9. The two Figures shows the relation between P_M and radius of orbit based on magnetic dipole model. Binary neutron stars are compared in Figure (a), with other two systems shown in Figure (b).

It can be seen that the radiation power in this model is very sensitive to the change of a. The radiation power increases sharply due to the inspiral. When a=100 m, the radiation power of the dual neutron star system can reach the order of 10^{34}; in addition, it is also combined with the system. There is a big relationship. When a=100 m, the radiant power of the neutron star - white dwarf system is only on the order of 10^{32}, and the binary white dwarf system is only on the order of 10^{28}.

4.2. The production of FRB and GRB

The currently known astrophysical processes involving the highest energy radiation include fast radio bursts (FRB) and gamma-ray bursts (GRB). This section discusses the feasibility of applying the above two models to explain FRB and GRB.

For a typical FRB, its duration is on the order of milliseconds $t \sim 1\ ms$, and the radiated power at the field source is estimated to be of order 10^{40} erg/s. For the unipolar induction model, the above simulation has given that when the radius of the neutron star is 10 km and the surface magnetic field intensity is 10^6 T, the distance between the double stars is on the order of 10^2 km to meet this energy index. Therefore, the neutron star-white dwarf star system has the possibility of being the progenitor of FRB. Furthermore, for the binary neutron star system, although the model cannot give accurate predictions, in principle, the binary neutron star will also produce a magnetosphere. Moreover, the magnetic field strength will not be lower than that of the neutron star-white dwarf system, i.e., it can also be used as progenitors of FRB in theory.

For the magnetic dipole model, when the dipole radius is 100 m, if the system charge value can reach the order of 10^2 C, the energy index required by FRB can be achieved. Since the radiation power in this model is extremely sensitive to changes in a, the total charge value of the system only needs to be no less than 10^{-6} C to meet the requirements when a=20 m.

Compared to FRB, GRB has higher radiation power. Generally, the duration of short gamma-ray bursts is less than 2 s, but the average radiation power of field sources can reach the order of 10^{50} erg/s. The unipolar induction model is different from the magnetic dipole model, since it requires that the distance between the double stars cannot be too small. Otherwise, the DC loop will be unstable and the current loop cannot be formed. Assuming that the limit of the distance between the binary stars is the radius of the neutron star r~10 km, in order to make the radiation power reach the magnitude that satisfies the GRB, the magnetic field strength of the host star needs to be no less than 10^8 T, which has reached the magnetic field of the magnetar. Thus, if the neutron star-white dwarf system can be progenitors of GRB, the neutron star should be a magnetar. Since white dwarfs cannot generate such high magnetic fields, the binary white dwarf system cannot be a progenitor of GRB.

There are many conjectures about the mechanism of GRB, among which hot fireball model is one of the mainstream views. The model suggests that a high-temperature and compact photosphere will be formed at the end of supernova explosions and binary star mergers, and then the photosphere will expand to a critical radius, leading to the energy dissipated as thermal radiation. In addition, high-energy gamma rays will be emitted during this process. The thermal radiation power satisfies [17]

$$L_{th} = 4\pi R_{ph}^2 \sigma_B T'^4 \Gamma(R_{ph})^2 \sim 4\pi R_0^2 \sigma_B {T_0}^4 \tag{38}$$

where Γ is the Lorentz factor. According to the magnetic dipole model, if the fireball is to meet the radiation power requirements, the total charge of the binary neutron star needs to be no less than 10^{-4} C when the dipole radius is 10 m, which is easy to meet for most neutron stars. For the neutron star-white dwarf system, the total charge requirement is not less than 0.1 C, and for the double white dwarf system, it is not less than 10^3 C, which is impossible for general white dwarf stars. Therefore, the model indicates that binary neutron stars and neutron star - white dwarf systems may be candidates for the progenitors of GRB, excluding binary white dwarf systems.

Finally, two models in describing the electromagnetic radiation have its own limitations: unipolar induction model cannot accurately describe the magnetic field intensity near the binary system, which greatly limits its applications, it only simulates neutron star - white dwarf system herein as an example. Nevertheless, one can still obtain valuable information based on the simulation results of the model considering the actual physical conditions. The limitation of the magnetic dipole model is that the radiation power is too sensitive to changes in the orbit radius ($P_M \sim a^{-15}$), hence the model is unable to make a reasonable prediction of the electromagnetic radiation law during inspiral. Thereby, the model can only be used to explain the high-energy radiation generated during the binary star fusion stage, e.g., FRB and GRB. On the other hand, it cannot explain the intensity of electromagnetic radiation released during the inspiral of binary neutron star systems.

4.3. Comparison of GR and ER

To quantitatively compare the magnitude of electromagnetic radiation and gravitational radiation, a comparison factor is defined

$$\gamma = \frac{P_{GR}}{P_{ER}} \tag{39}$$

It can be seen from Section 3.2 that when a<500 km or the neutron star's rotation period is long, the gravitational radiation caused by the rotation of the neutron star can be ignored compared to the radiation intensity of the revolution of the binary star. On this basis, only the radiation power generated by the revolution for P_{GR} is considered here. Two models of P_H and P_M were taken to get P_{ER}, of which four systems in Table 4 are still used for unipolar induction model, and their orbital eccentricity are set as 0.1. Besides, other three systems are selected for the magnetic dipole model, as listed in table 6 and Figure 10 demonstrates the result of comparison.

Table 6. Parameters of three different binary star systems

Name	M_m ($M_\odot$)	M_c ($M_\odot$)	D(kpc)	e
GW170817	1.46	1.28	40000	≤0.024
GW190425	2.02	1.35	15900	≤0.048
PSR J0737−3039	1.337	1.250	1.15	0.088

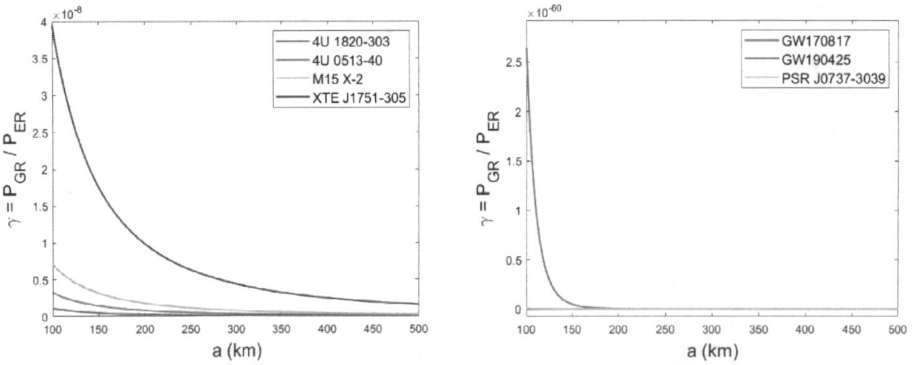

(a) (b)

Figure 10. Figure (a) is the comparison of radiated power of GR and ER for unipolar induction model, and the relation between γ and a (km) for magnetic dipole model is shown in Figure (b).

It can be seen that the energy dissipated by gravitational radiation in the two models is far greater than the energy dissipated by electromagnetic radiation, i.e., the latter can be completely ignored in the precession stage. This is the reason for the calculation of the binary star orbit evolution with time only considering the gravitational radiation.

5. Conclusion

In summary, this paper investigates the laws of gravitational radiation and electromagnetic radiation in the process of binary star merging obtained by different models, as well as analyzes and compares these results. Based on the analysis, the following conclusions are obtained. Firstly, the gravitational radiation in the process of binary star precession includes the contributions of the revolution and rotation of the binary star. The gravitational radiation generated by the revolution is related to the mass and radius of the binary star, and is mainly affected by the radius ($P_{rev} \propto a^{-5}$, $P_{rev} \propto m^{3/2}$). Besides, the gravitational radiation produced by the rotation is mainly affected by the rotation frequency ($P_{rot} \propto \omega^6$). In the early stage of inspiral, the gravitational radiation generated by the rotation may have a greater contribution to the total radiant energy. When a<500 km, the gravitational radiation basically comes from the revolution of the binary star.

Additionally, both the monopole induction model and the magnetic dipole model can explain the electromagnetic radiation of a binary star to a certain extent. With regard to high-energy radiation, both models show that the law of electromagnetic radiation will increase rapidly with the decrease of a. Nevertheless, the electromagnetic radiation power in the unipolar induction model is also proportional to the square of the magnetic field strength of the host star. Whereas, in the magnetic dipole model, it is also proportional to the total charge of the binary star.

Moreover, two models describing electromagnetic radiation give the following predictions for high-energy radiation. The dual neutron star system is a possible candidate for the progenitor of FRB and GRB. Besides, the neutron star - white dwarf system may also be the progenitor of the two, but the conditions are more stringent, i.e., the neutron star is required to belong to a magnetar. In addition, the double white dwarf system is unlikely to be the progenitor of the two.

Finally, the electromagnetic radiation power of the binary star system during the precession phase is much smaller than the gravitational radiation power. Therefore, the precession of the binary star orbit can be regarded as completely dominated by the gravitational radiation.

This article has presented the analytical models for the four models describing gravitational radiation and electromagnetic radiation. Therefore, the accuracy and precision of the model can be verified by further calculation and fitting with the data measured in the experiment. In addition, the unipolar model and the magnetic dipole model only give calculation methods from a phenomenological perspective, but cannot answer the generation mechanism of FRB and GRB. It is necessary for understanding these magnificent phenomena in the universe-in short, which are worthwhile exploring in depth in the future. Overall, these results offer a guideline for future study on the gravitational and electromagnetic radiation processes of binary neutron star merger and binary star systems in certain conditions.

References

[1] W. Baade & F. Zwicky. Remarks on Super-Novae and Cosmic Rays [J]. Physical Review, 1934, 46 (1): 76–77.
[2] A. Hewish, S. J. Bell, et al. Observation of a Rapidly Pulsating Radio Source [J]. Nature, 1968, 217 (5130): 709–713.
[3] S. Weinberg: Gravitation and Cosmology, Principles and Application of the General Theory of Relativity [M]. Wiley-VCH, 1972.
[4] T. M. Tauris, M. Kramer, P. C. C. Freire, et al. Formation of Double Neutron Star Systems [J]. The Astrophysical Journal, 2017, 846(2): 170.
[5] B. P. Abbott, R. Abbott, et al. GW170817: Observation of Gravitational Waves from a Binary Neutron Star Inspiral [J]. Physical Review Letters, 2017, 119(16): 161101.
[6] R.W. Klebesadel, I.B. Strong, R.A. Olson. Observations of Gamma-Ray Bursts of Cosmic Origin [J]. Astrophysical Journal Letters, 1973, 182: L85.
[7] Arizona State University. Massive Star's Dying Blast Caught by Rapid-response Telescopes [N]. PhysOrg. 26, 2017.
[8] D. R. Lorimer, M. Bailes, M. A. McLaughlin, et al. A Bright Millisecond Radio Burst of Extragalactic Origin [J]. Science, 2007,318: 777.
[9] D. Xiao, J.J. Geng, Z.G. Dai. Binary Neutron Star Mergers: Gravitational Wave Emission and Electromagnetic Counterparts [J]. Physics, 2019.
[10] J. S. Wang, Y. P. Yang, X. F. Wu, et al. Fast Radio Bursts from the Inspiral of Double Neutron Stars [J]. ApJL, 2016, 822: L7.
[11] D. Lai. DC Circuit Powered by Orbital Motion: Magnetic Interactions in Compact Object Binaries and Exoplanetary Systems [J]. The Astrophysical Journal Letters, 2012.
[12] S.S. Cai, Y. Zhu, J.J. Xu: Electrodynamics [M]. Higher Education Press, 2002.
[13] K.W. Yung, P.B. Landecker, D.D. Villani (1998). An Analytic Solution for the Force between Two Magnetic Dipoles [J]. Magnetic and Electrical Separation, 2012, Vol. 9, pp. 39-52.
[14] H. H. Zhao and L. M. Song. The Radius Estimation of Double Pulsar PSR J0737-3039A [J]. International Astronomical Union, 2013.
[15] Yu S H, Lu Y J. The Gravitational Wave Radiation from Ultracompact X-ray Binaries (in Chinese) [J]. Sci Sin-Phys Mech Astron, 2019, 49: 019501.
[16] Image from https://commons.wikimedia.org/wiki/File:ChandrasekharLimitGraph.svg [OL]
[17] P. Kumar, B. Zhang. The Physics of Gamma-Ray Bursts & Relativistic Jets [J]. Physics Reports, 2015.

Approximation properties of some Bernstein-Schurer type operators

Boyong Lian

Department of Mathematics, Yang-En University, Quanzhou, Fujian, China

lianboyong@163.com

Abstract. This paper mainly studies the approximation properties of Bernstein-Schurer operators based on a non-negative real parameter λ. By analysis techniques and the method to make the kernel estimation, the author studies the convergence order of this Schurer operators for the function class $\theta BV[0,1+p]$. Our conclusion expands the research results of Çetin.

Keyword: Bernstein-Schurer operataors, Absolutely continuous functions, Bounded variation.

1. Introduction

Let $\varphi(x) \in C[0,1+p]$ and $0 \leq \lambda \leq 1$, Çetin [1] constructed a generalized Bernstein-Schurer defined by

$$V_{m,p,\lambda}(f,x) = \sum_{i=0}^{m+p} \varphi(\frac{i}{m}) q_{m,i}^{(\lambda)}(x), \tag{1}$$

where$q_{m,0}^{(\lambda)}(x) = 1 - x, q_{m,1}^{(\lambda)}(x) = x$, and

$$q_{m,i}^{(\lambda)}(x)=\left[\binom{m+p-2}{i}(1-\lambda)x+\binom{m+p-2}{i-2}(1-\lambda)(1-x)+\binom{m+p}{i}\lambda x(1-x)\right]x^{i}(1-x)^{m+p-i-1}.$$

If $\lambda = 1$, the operators $V_{m,p,\lambda}$ become to the Bernstein-Schurer operators. If $p = 0$, the operators $V_{m,p,\lambda}$ become to the λ -Bernstein operators discussed by Chen [2]. If $\lambda = 1$ and $p = 0$, the operators $V_{m,p,\lambda}$ become to Bernstein polynomial.

In [1], using the famous Korovkin theorem, the author studied a uniform approximation result. He also gave the rate of convergence by another modulus of smoothness. Lastly, the author obtained two kinds of voronovskaya type results.

The approximation of function classes with bounded derivatives is a hot topic. Bojanic-Cheng [3] and other scholars [4-5] have done a lot of work in this direction. In recent decades, there has been a lot of work in this area, so we will not carry out it one by one.

Inspired by the above works, this paper studies the approximation of operators $V_{m,p,\lambda}$ for the function class $\theta BV[0,1+p]$ which is defined by definition 1.1.

Content from this work may be used under the terms of the Creative Commons Attribution 3.0 licence. Any further distribution of this work must maintain attribution to the author(s) and the title of the work, journal citation and DOI.
Published under licence by IOP Publishing Ltd

Definition 1.1

$$\theta BV[0,1+p]=\{\varphi|\varphi(x)=\varphi(0)+\int_0^x \omega(t)dt\},$$

where $x\in[0,1]$ and $\omega(t)$ is bounded variation.

Definition 1.2

$$\xi_{m,\lambda,p}(x,t)=\begin{cases}\sum_{k\le mt} q_{n,k}^{(\lambda)}(x), & 0<t\le 1+p;\\ 0, & t=0.\end{cases},$$

Thus, it's easily get

$$V_{m,p,\lambda}(\varphi,x)=\int_0^{1+p}\varphi(t)\mathrm{d}_t\xi_{m,\lambda,p}(x,t). \quad (2)$$

In order to prove the theorem 3.1 convenience, we first give several conclusions.

2. Some lemmas

Lemma 2.1(see [1]) Let $x\in[0,1+p]$, $e_l=t^l, l=0,1,2$, we get

$$V_{m,p,\lambda}(e_0,x)=1,$$

$$V_{m,p,\lambda}(e_1,x)=(1+\frac{p}{m})x,$$

$$V_{m,p,\lambda}(e_2,x)=(1+\frac{p}{m})^2x^2+\frac{[m+p+2(1-\lambda)]}{m^2}x(1-x).$$

Lemma 2.2(see [1]) Let $x\in[0,1+p]$, we get

$$V_{m,p,\lambda}(t-x,x)=\frac{p}{m}x,$$

$$V_{m,p,\lambda}((t-x)^2,x)=\frac{p^2}{m^2}x^2+\frac{[m+p+2(1-\lambda)]}{m^2}x(1-x)\triangleq\mu_{m,p,\lambda}^2(x).$$

Lemma 2.3 Let $x\in[0,1+p]$, we get

$$V_{m,p,\lambda}(|t-x|,x)\le\mu_{m,p,\lambda}(x). \quad (3)$$

Proof. Using Cauchy-Schwarz inequality, we can get

$$V_{m,p,\lambda}(|t-x|,x)\le\sqrt{V_{m,p,\lambda}((t-x)^2,x)}\cdot\sqrt{V_{m,p,\lambda}(1,x)}=\mu_{m,p,\lambda}(x).$$

Lemma 2.4 (i) Let $0\le t<x<1+p$, we can get

$$\xi_{m,\lambda,p}(x,t)\le\frac{1}{(x-t)^2}\mu_{m,\lambda,p}^2(x).$$

(ii) Let $0<x<v\le 1+p$, we can get

$$1-\xi_{m,\lambda,p}(x,v)\le\frac{1}{(x-v)^2}\mu_{m,\lambda,p}^2(x)..$$

Proof. (i) By Lemma 2.1 and Lemma 2.2, we get

$$\xi_{m,\lambda,p}(x,t)=\int_0^t d_y\,\xi_{m,\lambda,p}(x,y)\le\int_0^t(\frac{x-y}{x-t})^2d_y\,\xi_{m,\lambda,p}(x,y)\le\frac{1}{(x-t)^2}\int_0^{1+p}(y-x)^2d_y\,\xi_{m,\lambda,p}(x,y)$$

$$=\frac{1}{(x-t)^2}V_{m,\lambda,p}((y-x)^2,x)=\frac{1}{(x-t)^2}\mu_{m,\lambda,p}^2(x)$$

(ii) This proof is similar to (i).

3. Theorem

Theorem3.1 Let $\varphi\in\theta BV[0,1+p]$. If $\omega(x+),\omega(x-)$ exist, then we can get

$$\left|V_{m,\lambda,p}(\varphi,x)-\varphi(x)+\frac{px[\omega(x+)+\omega(x-)]}{2m}\right|$$

$$\leq \frac{|\omega(x+)-\omega(x-)|}{2}\mu_{m,\lambda,p}(x) + \frac{2\mu^2_{m,\lambda,p}(x)}{x(1-x)}\sum_{j=1}^{[\sqrt{m}]} \overset{(xj+1+p-x)/j}{\underset{x(j-1)/j}{V}} (\sigma_x) + \frac{1}{4\sqrt{m}} \overset{(x\sqrt{m}+1+p-x)/\sqrt{m}}{\underset{x(\sqrt{m}-1)/\sqrt{m}}{V}} (\sigma_x),$$

where

$$\sigma_x(t) = \begin{cases} \omega(t)-\omega(x+), x<t\leq 1+p; \\ 0, \quad t=x; \\ \omega(t)-\omega(x-), 0\leq t<x. \end{cases}$$

Proof. We know $\varphi \in \theta BV[0,1+p]$, so

$$\varphi(t)-\varphi(x) = \int_x^t \omega(s)\,ds. \tag{4}$$

Also

$$\omega(s) = \frac{\omega(x+)+\omega(x-)}{2} + \sigma_x(s) + \frac{\omega(x+)-\omega(x-)}{2} sgn(s-x) + \tau_x(s)\left[\omega(x) - \frac{\omega(x+)+\omega(x-)}{2}\right], \tag{5}$$

where

$$\tau_x(s) = \begin{cases} 1, & s=x \\ 0, & s\neq x \end{cases}, \quad sgn(s) = \begin{cases} 1, & s>0; \\ 0, & s=0; \\ -1, & s<0. \end{cases}$$

Because $\int_x^t sgn(s-x)\,ds = |t-x|$, $\int_x^t \tau_x(s)\,ds = 0$, so

$$V_{m,\lambda,p}(\varphi,x)-\varphi(x) = V_{m,\lambda,p}(\varphi(t)-\varphi(x),x) = V_{m,\lambda,p}\left(\int_x^t \omega(s)\,ds, x\right)$$

$$= \frac{\omega(x+)+\omega(x-)}{2} V_{m,\lambda,p}(t-x,x) + \frac{\omega(x+)-\omega(x-)}{2} V_{m,\lambda,p}(|t-x|,x) + V_{m,\lambda,p}\left(\int_x^t \sigma_x(s)\,ds, x\right).$$

Thus

$$\left|V_{m,\lambda,p}(\varphi,x)-\varphi(x)+\frac{px[\omega(x+)+\omega(x-)]}{2m}\right|$$

$$\leq \frac{|\omega(x+)-\omega(x-)|}{2} V_{m,\lambda,p}(|t-x|,x) + \left|V_{m,\lambda,p}\left(\int_x^t \sigma_x(s)\,ds, x\right)\right|$$

$$\leq \frac{|\omega(x+)-\omega(x-)|}{2}\mu_{m,\lambda,p}(x) + \left|V_{m,\lambda,p}\left(\int_x^t \sigma_x(s)\,ds, x\right)\right|. \tag{6}$$

Now we estimate another term $V_{m,\lambda,p}(\int_x^t \sigma_x(s)\,ds, x)$.

From (2), the term $V_{m,\lambda,p}(\int_x^t \sigma_x(s)\,ds, x)$ can be stated as

$$V_{m,\lambda,p}\left(\int_x^t \sigma_x(s)\,ds, x\right) = \int_0^{1+p}\left(\int_x^t \sigma_x(s)ds\right) d_t\xi_{m,\lambda,p}(x,t)$$

$$= \int_0^x\left(\int_x^t \sigma_x(s)ds\right) d_t\xi_{m,\lambda,p}(x,t) + \int_x^{1+p}\left(\int_x^t \sigma_x(s)ds\right) d_t\xi_{m,\lambda,p}(x,t) \triangleq \chi + \eta, \tag{7}$$

where

$$\chi = \int_0^x\left(\int_x^t \sigma_x(s)ds\right) d_t\xi_{m,\lambda,p}(x,t), \quad \eta = \int_x^{1+p}\left(\int_x^t \sigma_x(s)ds\right) d_t\xi_{m,\lambda,p}(x,t).$$

Thus, we have

$$\chi = \xi_{m,\lambda,p}(x,t)\int_x^t \sigma_x(s)ds\Big|_0^x - \int_0^x \xi_{m,\lambda,p}(x,t)\sigma_x(s)ds = -\int_0^x \xi_{m,\lambda,p}(x,t)\sigma_x(s)ds$$

$$= -\left(\int_0^{x(\sqrt{m}-1)/\sqrt{m}} + \int_{x(\sqrt{m}-1)/\sqrt{m}}^x\right)\xi_{m,\lambda,p}(x,t)\sigma_x(s)ds.$$

Thus, it follows that

$$|\chi| \le \int_0^{x(\sqrt{m}-1)/\sqrt{m}} \xi_{m,\lambda,p}(x,t) \bigvee_x^t(\sigma_x)dt + \int_{x(\sqrt{m}-1)/\sqrt{m}}^{x} \xi_{m,\lambda,p}(x,t) \bigvee_x^t(\sigma_x)dt.$$

Noting $0 \le \xi_{m,\lambda,p}(x,t) \le 1$, we can get

$$|\chi| \le \mu^2_{m,p,\lambda}(x) \int_0^{x(\sqrt{m}-1)/\sqrt{m}} \frac{\bigvee_x^t(\sigma_x)}{(x-t)^2}dt + \frac{x}{\sqrt{m}} \bigvee_x^{x(\sqrt{m}-1)/\sqrt{m}}(\sigma_x) \tag{8}$$

Let $t = x(p-1)/p$, we get

$$\int_0^{x(\sqrt{m}-1)/\sqrt{m}} \frac{\bigvee_x^t(\sigma_x)}{(x-t)^2}dt = \frac{1}{x}\int_1^{\sqrt{m}} \bigvee_x^{x(p-1)/p}(\sigma_x)\,dp \le \frac{2}{x}\sum_{j=1}^{[\sqrt{m}]} \bigvee_x^{x(j-1)/j}(\sigma_x) \tag{9}$$

It can be deduced from (8) and (9),

$$|\chi| \le \frac{2\mu^2_{m,p,\lambda}(x)}{x}\sum_{j=1}^{[\sqrt{m}]} \bigvee_x^{x(j-1)/j}(\sigma_x) + \frac{x}{\sqrt{m}} \bigvee_x^{x(\sqrt{m}-1)/\sqrt{m}}(\sigma_x) \tag{10}$$

Using the same method, we get

$$|\eta| \le \frac{2\mu^2_{m,p,\lambda}(x)}{1-x}\sum_{j=1}^{[\sqrt{m}]} \bigvee_x^{(xj+1+p-x)/j}(\sigma_x) + \frac{1-x}{\sqrt{m}} \bigvee_x^{(x\sqrt{m}+1+p-x)/\sqrt{m}}(\sigma_x) \tag{11}$$

The proof of Theorem 3.1 is over.

4. Conclusion

By using the classical Bojanic and Cheng's method, the author studies the convergence order of this Schurer operators for the function class $\theta BV[0,1+p]$. In the process of proof, the author use the analysis techniques to make the kernel estimation. Late, another important estimate is obtained. The conclusion expands the research results of Çetin.

Acknowledgments

The author thanks the reviewers for their comments to improve the quality of this paper. This work is supported by the Discipline leader training programs of Yang-en University.

References

[1] Çetin, Approximation by α -Bernstein-Schurer operator, Hacet. J. Math. Stat. Vol. 50(3), 732-743, 2021.

[2] X. Chen, J. Tan, Z. Liu, Approximation of functions by a new family of generalized Bernstein operators, J. Math. Anal. Appl. 450, 244-261, 2017.

[3] R. Bojanic , F. Cheng, Rate of convergence of Bernstein polynomials for functions with derivatives of bounded variation, J. Math. Anal. Appl. 141(1), 1989.

[4] X M. Zeng, F. Cheng, On the rate of approximation of Bernstein type operator, J. Approx. Theory. Vol. 109 (2), 242-256, 2001.

[5] B Y. Lian, Rate of approximation of bounded variation functions by the Bézier variant of Chlodowsky operators, J. Math. Inequal. 7 (4), 647-657, 2013.

Journal of Physics: Conference Series **2441** (2023) 012024 doi:10.1088/1742-6596/2441/1/012024

Acoustic Field Prediction Method and Simulation Research Based on Acoustic Field Geometric Average Theory

Xianwen He [1], Zhizhong Li [1, *], Junping Sun [2]

[1] Naval Submarine Academy, Navigation and Communication Department, Qingdao China

[2] Chinese Academy of Sciences, Institute of Acoustics, Qingdao, China

*Corresponding author e-mail: hxw_minipaper@163.com

Abstract. In order to meet the capability requirements of submarine radiated noise acoustic field prediction in free field and improve the detection and anti-detection capabilities, this paper proposes an acoustic field prediction method based on the geometric average theory of acoustic field. According to the measured self-noise information, a global search method is used. By calculating the average energy of the spherical, the extended compensation for different point distances is achieved, so as to predict the radiation noise at different distances. Through simulation experiments, it is found that the method proposed in this paper has a fast calculation speed, and the acoustic field prediction accuracy is high.

1. Introduction

The improvement of the submarine's acoustic stealth ability not only affects the submarine's concealment ability as a detected object, but also affects the submarine's ability to detect and discover other targets, reflecting the comprehensive underwater detection and anti-detection capabilities [1]. The radiated noise acoustic field prediction of submarines can better improve the detection and anti-detection capabilities [2]. At present, the methods of acoustic field analysis and prediction mainly include numerical methods and experimental methods. The former mainly include finite element method, boundary element method, infinite element method and statistical energy analysis method. The experimental methods mainly include the average square sound pressure method, the sound intensity method and the acoustic holography method [3].

In order to find a simple algorithm for solving the sound radiation problem or scattering problem, the paper [4] proposed the wave superposition method. The superposition of the acoustic field is suitable for analyzing the sound radiation of objects of any shape, and foreign scholars have conducted in-depth research on this. In China, the paper [5] proposed a robust full-wavenumber spatial acoustic field reconstruction technique, which is suitable for analyzing the acoustic radiation of objects of any shape. The paper [6] used the equivalent source method to study the reconstruction of multi-source acoustic field and semi-free acoustic field, and verified the accuracy of the method through speaker experiments. The influence of various factors are analyzed on the reconstruction accuracy of the wave superposition method by numerical simulation in [7].

Geometric acoustics has been used for closed acoustic field simulation and simulation calculation [8]. On this basis, a complex body source acoustic field prediction method based on the geometric

Content from this work may be used under the terms of the Creative Commons Attribution 3.0 licence. Any further distribution of this work must maintain attribution to the author(s) and the title of the work, journal citation and DOI.
Published under licence by IOP Publishing Ltd

average theory of acoustic field is proposed. According to the measured self-noise information, a global search method is used. By calculating the average energy of the spherical, the extended compensation for different point distances is achieved, so as to predict the radiated noise at different distances. Moreover, we expand from a single sound source acoustic field to a complex sound source acoustic field for simulation calculation. The method proposed in this paper has the advantages of fast calculation speed, and the acoustic field prediction accuracy is high, and the effect is better in the case of plane layout.

2. System Model

Fig. 1 denotes the self-noise receiving model. $y(t)$ denotes the output of the system, $s(t)$ denote the noise sources of the system, $x(t)$ denote the measured noise source which is called self-noise. And the DFT (Discrete Fourier Transformation) of $y(t)$ can be expressed as

$$Y(f,T)=\sum_{i=1}^{n}V_i(f,T)+N(f,T) \tag{1}$$

When $V_i(f,T)=H_{iy}(f)S_i(f,T)$, $Y(f,T)$ $V_i(f,T)$, $N(f,T)$ and $S_i(f,T)$ is DFT of $y(t)$, $v_i(t)$, $n(t)$ and $s_i(t)$ during the T period. And the conjugate multiplication can be expressed as:

$$Y^*Y=\sum_{i=1}^{n}\sum_{j=1}^{n}H_{iy}^*H_{jy}G_{iy}^*G_{jy}+N^*N \tag{2}$$

in(2), assumption that $n(t)$ and $s_i(t)$ are uncorrelated, the average is 0. Moreover, the self-correlation spectrum of $y(t)$ can be expressed as:

$$G_{yy}=\lim_{T\to\infty}\left[\frac{E(Y^*Y)}{T}\right] \tag{3}$$

when $s_i(t)\neq x_i(t)$, $H_{nn}(f)\neq 1$. Assumption that $m_i(t)=0$, the cross-correlation spectrum of $x_i(t)$ can be expressed as:

$$G_{x_ix_j}=H_{ii}^*H_{jj}G_{s_is_j} \tag{4}$$

Moreover, the cross-correlation spectrum between input $x_i(t)$ and output $y(t)$ can be expressed as:

$$G_{x_iy}=H_{ii}^*S_i^*Y=\sum_{j=1}^{n}H_{ii}^*S_i^*H_{jy}S_j \tag{5}$$

In the noise measurement, there is correlation between different $x_i(t)$. Because of this , the fig.2 can be simplified to the case that with two independent and unrelated sources $s_1(t)$ and $s_2(t)$, which is seen as Fig.2.

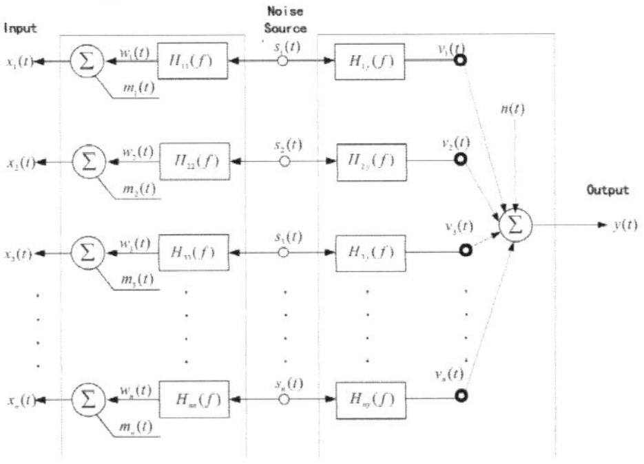

Fig. 1 Self-noise receiving model

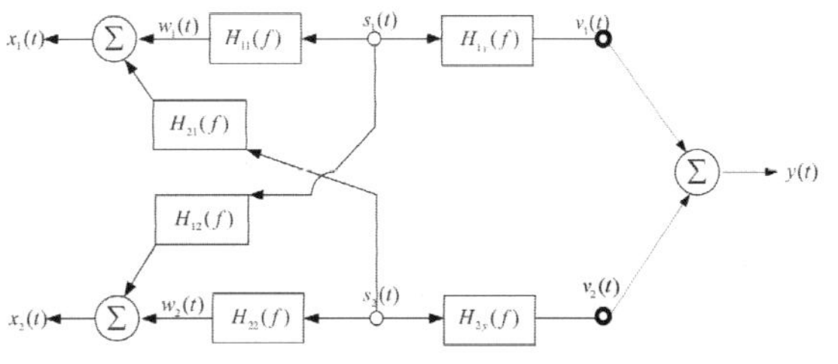

Fig. 2 Self-noise simplified receiving model

$$Y = V_1 + V_2 = H_{1y}S_1 + H_{2y}S_2 \tag{6}$$

$$X_1 = W_1 + H_{21}S_2 = H_{11}S_1 + H_{21}S_2 \tag{7}$$

$$X_2 = W_2 + H_{12}S_1 = H_{12}S_1 + H_{22}S_2 \tag{8}$$

In (6)-(8), Y , V_i , W_i and S_i are DFT of $y(t)$, $v_i(t)$, $w_i(t)$ and $s_i(t)$ during the T period.

$$\begin{aligned} G_{yy} &= \left|H_{1y}\right|^2 G_{s_1s_1} + \left|H_{2y}\right|^2 G_{s_2s_2} \qquad & G_{x_1x_1} &= \left|H_{11}\right|^2 G_{s_1s_1} + \left|H_{21}\right|^2 G_{s_2s_2} \\ G_{x_2x_2} &= \left|H_{12}\right|^2 G_{s_1s_1} + \left|H_{22}\right|^2 G_{s_2s_2} \qquad & G_{x_1x_2} &= H_{11}^* H_{12} G_{s_1s_1} + H_{21}^* H_{22} G_{s_2s_2} \\ G_{x_1y} &= H_{11}^* H_{1y} G_{s_1s_1} + H_{21}^* H_{2y} G_{s_2s_2} \qquad & G_{x_2y} &= H_{12}^* H_{1y} G_{s_1s_1} + H_{22}^* H_{2y} G_{s_2s_2} \end{aligned} \tag{9}$$

Since the self-noise measurement of target radiated noise in water is equivalent to a multi-input and multi-output system, the following methods will be used to process the measured self-noise signal to obtain the equivalent noise source intensity and provide effective sound for the prediction of the sound field in ocean waveguides source.

3. Acoustic field geometric average method

Assumption that, the measured self-noise position coordinates are(x_i , y_i) and the geometric sound source center is(x, y), Then the distance of the sound source from the noise measurement position is

$$r_i(x, y) = \sqrt{(x - x_i)^2 + (y - y_i)^2} \tag{10}$$

When the coordinate (x, y) satisfy the optimal solution of equation in (25), the geometric center of the sound source is established.

$$r_{ALL} = \min(\sum_i r_i(x, y)) \tag{11}$$

The establishment of the geometric center can be obtained by the global search method or the discrete point weighted calculation method, which can reduce the search volume to a certain extent and improve the calculation speed.

a. The self-noise position obtained from the measurement is combined to obtain the center point as

$$x_{ij} = \frac{x_i + x_i}{2}, \quad y_{ij} = \frac{y_i + y_i}{2};$$

b. Repeat the step a with new coodinate (x_{ij} , y_{ij})

c. In the process of step a and b, when an odd number occurs, do the calculating to get new coordinates according the following formula:

$$x_{ijk} = \frac{2x_{ij} + x_k}{2}, \quad y_{ijk} = \frac{y_{ij} + y_k}{2}$$

When the geometric center is established, the equivalent noise source intensity is

$$\left|S(\omega)\right|^2 = (\sum_{ij} r_{ij}^2 \left|P_{ij}(r, \omega)\right|^2 \Delta\theta) / 2\pi \tag{12}$$

When r_{ij} is the distance between measure point and the optimal geometric center, and $\Delta\theta$ is the open angle of the line connecting the location of the measurement point to the geometric center. The specific process of the proposed method can be seen as Fig.3.

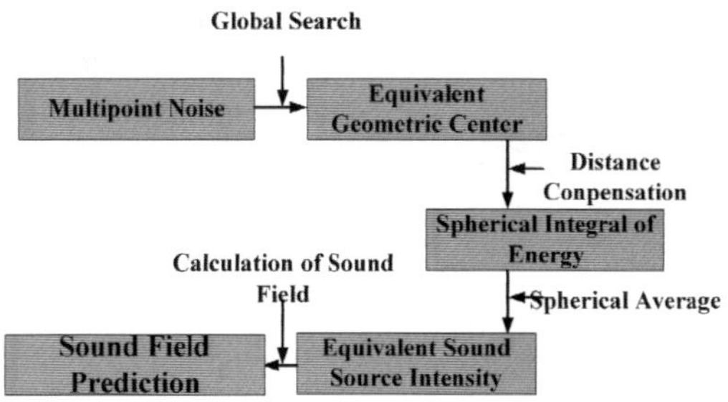

Fig. 3 Calculation process of acoustic field geometric average method

4. Simulation Analysis for Acoustic Field Prediction of Single Sound Source

Assuming that a certain point source is located at the coordinate center position (0, 0, 0), the sound pressure intensity at 1 meter is 1 Pa, as shown in Fig.4. When the number of receiving points is determined, the corresponding number of receiving positions are randomly generated, as shown in Fig.5. The x, y, and z coordinates represent the distance in meter. In Fig.5, the red asterisk represents the position of the sound source, and the blue asterisk represents the position of the receiving point, and its size indicates the intensity of the acoustic field at the location.

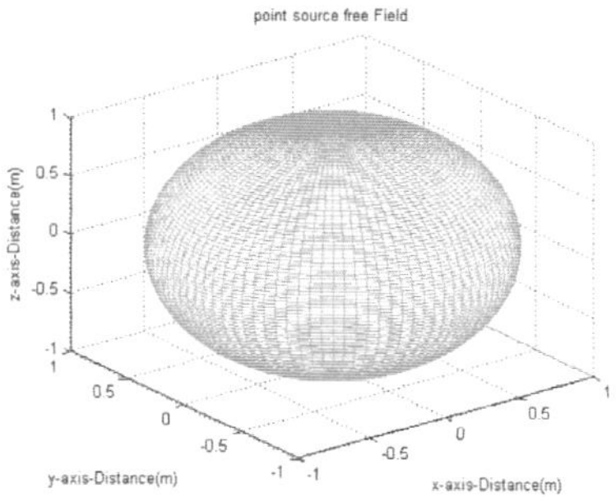

Fig. 4 Equi-phase surface of acoustic field formed by point source

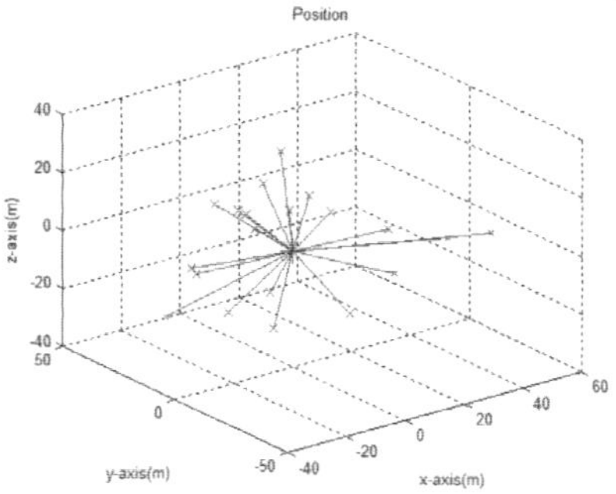

Fig. 5 Schematic diagram of the location of the receiving point

Fig.6 and Fig. 7 show the comparison between the acoustic field of 2000Hz and 100Hz frequency point source and the free field predicted by the acoustic field geometric average method. In the left figure (a), the abscissa is the distance, the ordinate is the intensity, and the blue line is the sound field formed by the real point source, the red line is the sound field predicted by the geometric mean, the right picture (b) is the corresponding position information, the red * is the real point source position, the blue x number is the receiving point location, the cyan + sign is the location of the geometric mean virtual source, (b) The number shown in the title of the figure is the distance between the virtual source and the real source. As shown in the figure, when the real point source is effectively surrounded by the receiving point, after the forecast distance exceeds 10m, the forecast trend of the geometric mean method is basically consistent with the change trend of the real source sound field. Well, there is no obvious frequency-dependent variation in the forecast results.

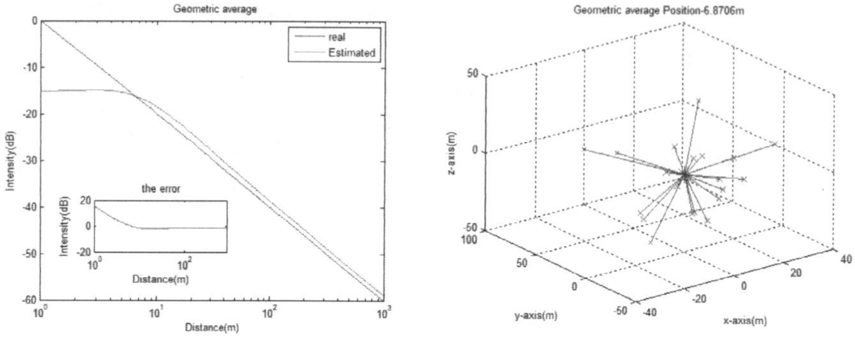

(a) (b)

Fig.6 2000Hz acoustic field prediction

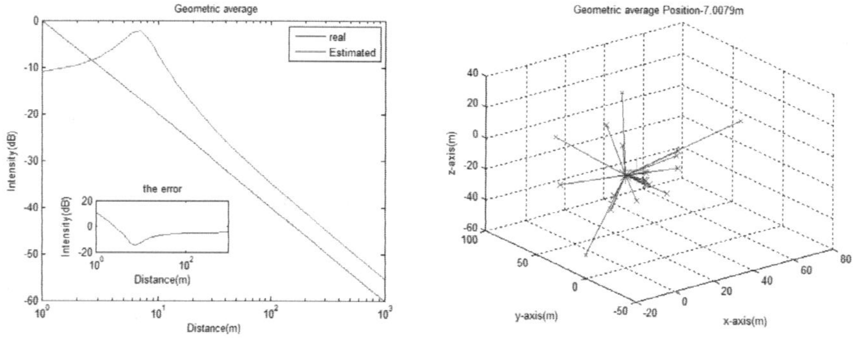

(a) (b)

Fig.7 100Hz acoustic field prediction

The above is the prediction of the acoustic field of a single point source through multiple distributed receiving point sound fields in the near-field space. The number of receiving points is 20. In the case of a single point source surrounded by point sources, the geometric mean method predicts the trend of the sound field after 10m. The real source is similar, the change is monotonous, the disturbance phenomenon is not obvious, and the forecast error is relatively small.

5. Simulation analysis for acoustic field prediction of complex body source

It is assumed that there are 30 point sources randomly distributed in a volume of 1 m^3, forming a complex and unknown volume source, and 20 receiving points are randomly distributed at a certain distance in the three-dimensional space outside the volume source. Fig.8 is a randomly generated 1000Hz sound field stereo pseudo-color image.

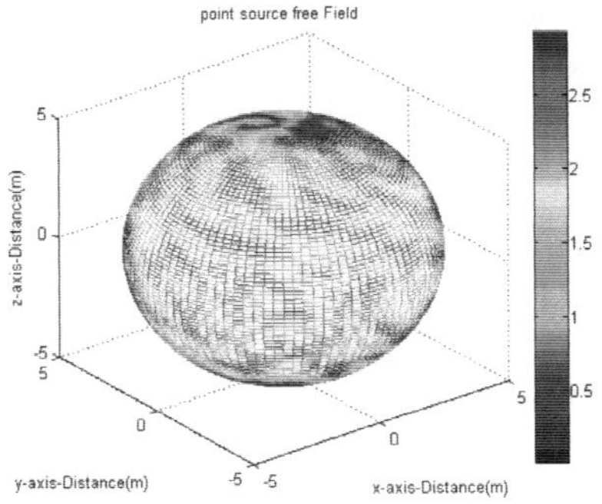

Fig. 8 Acoustic Field Distribution Characteristics of Complex Volume Sources

Fig.9 and Fig.10 are the comparison of the sound field predicted by the geometric mean of the sound field and the real free field of the body source. In the left figure (a), the abscissa is the distance, the ordinate is the intensity, and the blue line is the formation of the real point source. Sound field, the red line is the sound field predicted by the geometric mean, the right picture (b) is the corresponding source and receiving point location information, red * is the real point source location, 30 point sources are randomly distributed in a 1m3 stereo space, blue x The sign is the location of the receiving point, and the cyan + sign is the location of the geometric mean virtual source. Through the analysis of the figure below, when the real point source is effectively surrounded by the receiving point, after the forecast distance exceeds 10m, the prediction trend of the geometric mean method is basically consistent with the change trend of the real source sound field. The error at 1000Hz is relatively large, and the prediction error at 100Hz frequency is relatively smaller.

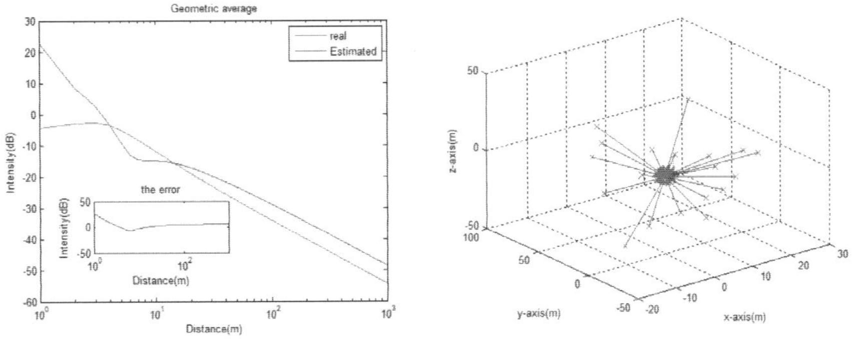

(a) (b)

Fig. 9 1000Hz acoustic field prediction

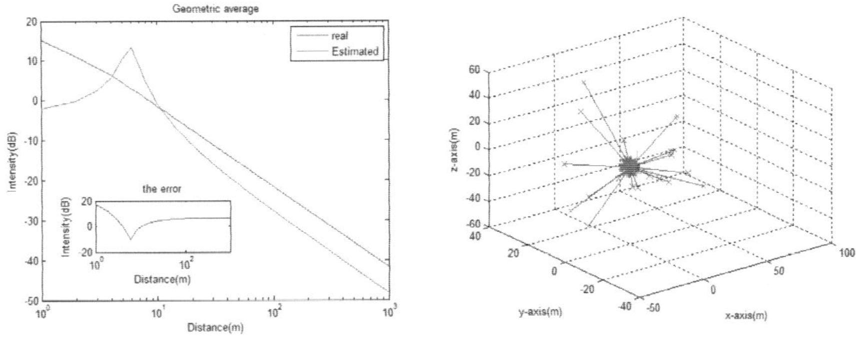

(a) (b)

Fig. 10 100Hz acoustic field prediction

Assuming that there is no prior knowledge of the measured sound source, by randomly generating different numbers of receiving points to study the forecast errors of different methods at a distance of 500 meters in the free field, to explore the variation trend of forecast accuracy with the number of receiver points, and to study the forecast based on different forecast methods with the number of receiver points. Fig.11 shows the prediction errors of different methods at 2000 Hz. The abscissa is the number of receiving points, and the ordinate is the prediction error. The red line in the figure represents the near-field acoustic holography/Huygens superposition method (holography), the blue line represents the sound field geometric average method (ga), which is proposed by this paper and the black line represents the partial coherence/adaptive filtering method (pa). It can be seen from the prediction results that with the increase of the number of randomly distributed receiving points in the whole space, the prediction performance of the acoustic holography/Huygens superposition and the sound field geometric average method is improved, while the prediction accuracy of the partial coherence/adaptive filtering method has a better correlation with the number of receiving points.

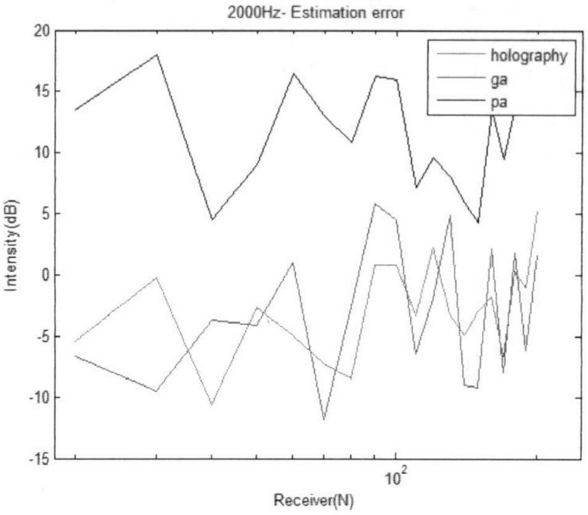

Fig. 11 Forecast Error Analysis

6. Conclusion

In this paper, a complex body source acoustic field prediction method based on the geometric average theory is proposed. According to the measured self-noise information, a global search method is used. By calculating the average energy of the spherical, the extended compensation for different point distances is achieved, so as to predict the radiated noise at different distances. Moreover, we expand from a single sound source acoustic field to a complex sound source acoustic field for simulation calculation. Through simulation calculation, the effects of different methods are analyzed and compared. The geometric average method of acoustic field has a fast calculation speed and the acoustic field prediction accuracy is high. It can predict the radiated noise acoustic field of submarines, which improve capabilities of detection and anti-detection.

References

[1] SHANG De-jiang, HE Zuo yong, The numerical analysis of sound and vibration from a ring-stiffened cylindrical double--shell by FEM and BEM[J]. Acta Acustic, 2001, 26(3): 193—201.

[2] LIU Zhong wen, LI Ttian yun, LUO Dong • ping. The statistical energy analysis for self noise of submarine sonar[J], Ship Science and Technology, 2000, 28(3): 61—64Reference to a chapter in an edited book:

[3] HE Yua nan, HE Zuo yong, JIANG Jun. Near field acoustic holography based on intensity measurement and application in radiating analysis of underwater sound sources. [J]. ActaAcustic, 1996, 21(4): 297—305.

[4] GAO Yon, LI Bing, DAI Jiang. Investigation on underwaterradiation sound power measurements

from cylindrical shells model[J], Journal of Ship Mechanics, 2010, 14(10): 1189—1194.

[5] HE Yu nan, YANG De sen. Experiment study of predicting method for radiming noise by submerged vibrated structure[J]. Journal of Ship Mechanics, 1999, 3(2): 58—62.

[6] JIN Guangwen, HE Lin. Calculation of noises from submerged cylindrical structure by radiator factorf [J]. Acta Acusfic, 2010, 35(4): 427—433.

[7] LONG Jun—bo, WANG Hai—bin, ZHA Dai-feng. Fractional low—order adaptive time—frequency distribution based on stable distribution noise[J]. Computer Engineering, 201 1, 37(18): 81—83.

[8] Ma X, Nikias C L. Joint estimation of time delay and frequency delay in impulsive noise using fractional lower order statistics[J]. IEEE Transactions on Signal Processing, 1996, 44(11) 1: 2669—2687.

The impact of the dark matter on galaxy formation

Junwen Diao [1, †], ShibiaoWei [2, †], Zherui Wei [3, †] and Chang Liu [4, *, †]

[1] Fairfield College Preparatory School, Fairfield, United Sates
[2] Xian Gao Xing NO.1 high school, Xian, China
[3] Western academy of Beijing, Beijing, China
[4] Department of geology China university of mine and technology, Xuzhou, China

*Corresponding author's e-mail: 05191622@cumt.edu.cn

[†] These authors contributed equally.

Abstract. Contemporarily, dark matter exerts dominant impacts on galaxy formation, which is the spine bone of the galaxy. Herein, we investigate the attraction effect of initial dark matter distribution on the normal matter (hydrogen and other gases), which will form the dark matter halos. Subsequently, an introduction to the properties of dark matter halos and some disputes about it are demonstrated. Besides, the cold dark matter (CDM) is chosen to fit the current observation and interpret the influences of dark matter on galaxy formation. Overall, this paper offers a clear procession of the impact of dark matter on galaxy formation, which will be guideline for future development of both galaxy formation theory as well as dark matter detection.

Keywords: Dark matter, dark matter halos, galaxy formation, current observation.

1. Introduction

With the deeper and deeper exploration of universal, plenty of astrophysics and cosmology issues are addressed. However, dark matter remains an unresolved issue, whose properties and components remain unclear. Galaxy formation still has lots of problems needed to be solved. Although there are plenty of scholars that propose various approaches to find out its procession, the formation of galactic disks remains unclear. Therefore, we will give a clear calculation of how dark matter attracts gas to increase its mass based on reviewing articles to delineate the effect of dark matter on galaxy formation [1]. Besides, Millennium N-body simulations will be chosen as the investigation target [2]. In this case, cosmological simulations are carried out to present the distribution of dark matter.

Subsequently, we will compare cold dark matter (CDM) with warm dark matter (WDM) and state the reason for choosing CDM [3]. A major issue with cold dark matter galaxy formation is increasingly severe, which remains an unsolved problem.

The rest part of the paper is organized as follow: Section 2 will introduce the analytical model for galaxy formation and star formation; Section 3 will present some state-of-art experiment results to demonstrate the properties of dark matter for different models; Section 4 will give a brief summary and future guideline eventually.

Content from this work may be used under the terms of the Creative Commons Attribution 3.0 licence. Any further distribution of this work must maintain attribution to the author(s) and the title of the work, journal citation and DOI.
Published under licence by IOP Publishing Ltd

2. Analytical model

2.1. The mechanisms of dark matter attraction of gas and normal matter

Primarily, we establish a spherical coordinate system where the origin is the singularity of big bang, $M(\vec{r})$ is initial dark matter distribution $\vec{r}$ is a vector. Dark matter attracts the distant matter due to gravity

$$\frac{GM}{\vec{r}-\vec{r_1}} = \frac{d^2\vec{r}}{dt^2} = \ddot{\vec{r_1}} \quad (1)$$

Since the dark matter has a certain distribution, we assume that the displacement of distant matter is $\vec{r_1}$

$$\sum \frac{GM}{\vec{r}-\vec{r}_1} = \overrightarrow{r_1} \quad (2)$$

We obtain the acceleration formula of specific position which can in turn extend to all positions in the university

$$\sum \frac{GM}{\vec{r}-\vec{r_1}} = \overrightarrow{r_1}(r) \quad (3)$$

Owing to large mass of dark matter, the other effects on this matter can be neglected. On this basis, one can assume the matter has its certain speed $\vec{v}$, then we derive the acceleration formula that can find out the time required for the matter fuse with dark matter

$$\int \overrightarrow{r_1}(\vec{r})dt = \overrightarrow{r_1}(\vec{r}) + c \quad (4)$$

One can derive the constant c easily because of the specific $\vec{v}$

$$\int_0^z \overrightarrow{r_1}(\vec{r}) + cdt = \vec{r} - \overrightarrow{r_1} \quad (5)$$

where z is the time to get then we can obtain the time of all positions fused with dark matter

$$z(\vec{r}) = t \quad (6)$$

Dark matter exhibits flesh sheet in large scale but we can assume that it will be spherical symmetric in small scale. Because dark matter has its certain diameters, i.e., we assume its radius is r_d.Therefore, we obtain the mass increases per unit of time

$$\int_{r_d}^{r_d+\Delta r} \frac{4\pi r^2 m_1}{\Delta t} dr = \frac{4\pi m_1}{3}\left[\frac{(r_d+\Delta r)^3 - r_d^3}{\Delta t}\right] \quad (7)$$

$$\frac{4\pi m_1}{3}\left[\frac{3\Delta r r_d^2 + \Delta r^2 r_d + \Delta r^3}{\Delta t}\right] = 4\pi m_1 r_d^2 v_d \quad (8)$$

where $m_1$is the matter distribution of matter per unit area. Then, we obtain the formula of increasing mass per unit time because the dark matter does not involve in electromagnetic interaction. In this case, one can assume the radius of dark matter remain unchanged for long time, i.e., we derive relation of $m_1$and t:

$$m_{1end} = (1 + 4\pi r_d^2 v_d t)m_1 \quad (9)$$

However, this formular only makes sense at short time because when the mass increases the radius will shrink.

2.2. An introduction to dark matter

The backbone for the development of galaxies is dark matter, anticipated to develop at the cores of dark matter over densities, known as halos.

2.2.1. The properties of dark matter. Theoretically, CDM or WDM must be invoked to clarify relatively small-scale features in the observable Universe. Apart from this, hot dark matter is introduced, which is no longer possible to use hot dark matter as the only explanation for cosmic galaxy formation. As a result, the production of greater mass aggregates that surround entire galaxy clusters should correspond to the formation of hot dark matter. However, evidence from the cosmic background explorer (COBE) satellite shows that the cosmic microwave background radiation is remarkably uniform.

The energy scale of hot dark matter is about 30 eV, and the dark halo scale is about the size of a large galaxy cloud; the energy scale of warm dark matter is about 2 keV, and the dark halo scale is comparable to that of small galaxy; the energy scale of cold dark matter is 100 GeV, and the dark halo scale is comparable to that of Earth. Since then, the hot dark matter theory has been abandoned, even though it is completely incompatible with experimental findings. The cold dark matter model and the warm dark matter model are now two rival ideas. Cold dark matter models: in the more prevalent cold dark matter model, first form scale and the earth's dark matter halo, then produce more and bigger dark matter halo through an accretion process. Thus, according to the cold dark matter concept, the universe is full of dark matter halo spanning 21 orders of magnitude from Earth to the size of clusters. Because dark matter is substantially more than ordinary baryonic matter, the universe's large-scale structures are all generated by dark matter gravity.

2.2.2. The gravity of dark matter. Cosmological simulations are frequently run with periodic boundary conditions to simulate the large-scale homogeneity and isotropy of the Universe's matter distribution, i.e., the cosmological principle Collisionless Bolzman equations:

$$\frac{df}{dt} = \frac{\partial f}{\partial t} + \vec{v}\frac{\partial f}{\partial \vec{r}} - \frac{\partial \phi}{\partial \vec{r}}\frac{\partial f}{\partial \vec{v}} = 0 \tag{10}$$

and Poisson's equation:

$$\nabla^2 \phi = 4\pi G \int f\, dv \tag{11}$$

2.3. The impacts of dark matter on galaxy formation

Baryons are supposed to discover the dark matter dissemination on scales since they are originally distributed nearly uniformly. As shown in Fig. 1. We predict baryons to gravitate towards the great potential wells of dark matter halos in particular. As a result, these should be the places where galaxies develop.

The cosmos began to cool a little less than 14 billion years after the big bang, and hydrogen helium began to form. Since hydrogen is the most basic element, it is made up of only one proton and one electron. Then there was dark matter, which was the initial prompt since dark matter is the galaxy's spine bone. It will pull the first gas to agglomerate. Then, it will attract an increasing amount of matter. During this time, there will be some high density areas where the star will form. The system will form as a result of a large number of stars forming. The creation of numerous systems will eventually result in the formation of a galaxy. It's known as a supernova. Globular clusters make up the burnt-out materials. The rotating disk will draw more and more gas and dust as it rotates. The reason one should be aware of is that galaxies will have an impact on neighbouring galaxies. The various shapes of the galaxy will merge into a single integral. The irregular galaxy then merges.

Figure 1. The initial bone of dark matter [4].

2.3.1. The start formation. Usually, a large number of stars with varying masses arise from a collapsing cloud of gas. It is natural to believe that stars with low mass are more likely to develop since they require less material. The initial mass function (IMF) is a relative power-law function that describes the relationship between a star's mass and the number of stars with that mass. Its formula is [5]:

$$\xi(M) = (\frac{M}{M_S})^{-\beta} \tag{12}$$

$$\xi(M) = (\frac{M}{M_S})^{-2.35} \tag{13}$$

Expressing the number of stars as a multiple of the number of stars with a mass of one solar mass The exponent, according to modern theories and observations, is not constant, but varies with the mass of the stars. The IMF's formula is still a hot issue of discussion today.

$$\xi(M) = B(\frac{M}{M_S})^{-\beta(M)} \tag{14}$$

With B as a pre-factor, the pre-factor does not have a unit since the mass is given in multiples of the solar mass. For masses less than 0.08 sun masses3, the exponent (M) is 0.3; as for masses between 0.08 and 0.5 solar masses, it is 1.3; and it is 2.3 with regard to masses more than 0.08 solar masses. The pre-factors B are computed using the constraint $\xi_{up}{}^{1}$ (M) = $\xi_{low}{}^{2}$ (M) to obtain a continuous function, hence the values of the two portions of the function must be the same at the limiting mass.

$$\xi(M) = 2(\frac{M}{M_S})^{-\beta(M)} \tag{15}$$

2.3.2. Some feedback to star formation. We concluded that star formation could not take place in all dark matter halos with 100% efficiency. There is evidence for this in a number of observable phenomena, but the two most important are demonstrated in [6, 7].

This second argument may be emphasized further: even if all dark matter halos converted a consistent proportion of their mass to stars, there would still be an excess of dim galaxies in comparison to brilliant galaxies. A plethora of additional observable constraints suggest that the efficiency of galaxy formation is greatly influenced by halo mass. Apparently, a mechanism that selectively inhibits the development of stars in smaller mass dark matter halos is required. Energy/momentum input from supernovae explosions, maybe boosted by star winds, is the most common explanation for this process. In the most large dark matter halos, a similar issue arises. Even while cooling in such halos is inefficient, they can nonetheless cool a large amount of gas over cosmic time. If left unchecked, which will result in the development of galaxies that are far more bright than any that have been observed. The problem's energy needs imply that AGN might be a viable solution.

2.3.3. Galaxy evolution. Galactic development is shifting from hierarchical clustering-dominated early Universe to internal secular processes-dominated future. These are the results of interactions with collective phenomena including bars, oval disks, spiral structure, and triaxial dark halos. As shown in Fig. 2, gas has accumulated form an outside ring, an inner ring around the bar's end, and a dense core concentration after 7 bar revolutions. ESO 426-2 [8] and NGC 3081[9] are barred galaxies with similar properties, which diagram is based on Ref. [10].

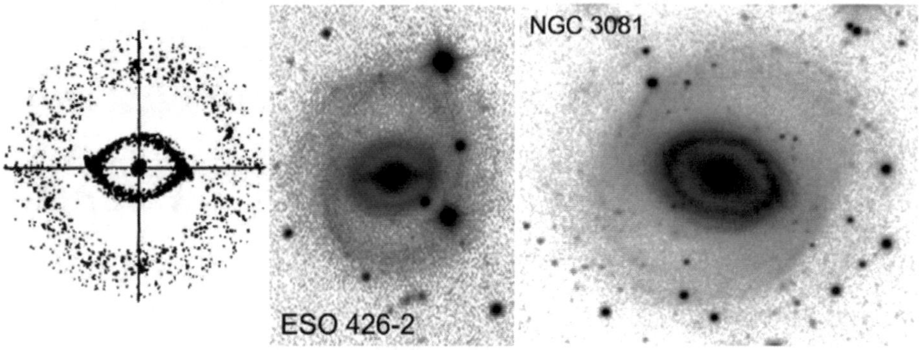

Figure 2. Secular evolution's products include: (left) Gas particle dispersion at the end of a sticky-particle simulation of gas growth in a horizontal but not depicted spinning bar potential.

2.3.4. The discussion of galaxy disks. Numerical simulations are emerging as one of the most significant theoretical methods for investigating the difficult subject of galaxy formation. More advanced treatments of supernova feedback, chemical enrichment models, and approaches to "zoom in" on galaxies of interest within cosmological simulations were used in current studies.

While these research were only partially successful in producing rotationally supported disks, the simulated galaxies seldom matched their observed counterparts. Furthermore, the models' star formation was extremely efficient, transforming an excessively large proportion of the gas into stars by the present day. Plenty of the early modelling flaws may be traced back to the simulations' coarse resolution and the way in which feedback was handled. Furthermore, if star formation in low-mass halos is particularly efficient at high redshifts, a significant number of dense, compact galaxies would develop. If the angular momentum of these baryonic clumps is lost to black halos, the resultant objects will be less in size than if the baryons were accreted smoothly. As a consequence of the early collapse of baryons, a route for gas to lose angular momentum opens, resulting in galaxies that are excessively compact, lack angular momentum, and create stars more efficiently than seen. Following that, greater types of feedback and changes to the cosmological framework were used to try to address these flaws. The above-mentioned issues were aggravated in certain circumstances. Photoheating by a diffuse UV backdrop, for example, was shown to lower the angular momentum content of the simulated galaxies even further. The impact of preheating and gas blowout from tiny halos as well as the development of galaxies in a WDM cosmology were also investigated. Direct comparisons with observation can be realized based on incorporating spectral synthesis techniques into the modelling and employing the Tully-Fisher relation to restrict the hierarchical origin of galaxies. The galaxies in these simulations, however, were still excessively dense. Recent research on disk creation has provided some more encouraging results, which created disk galaxies with angular momentum deficits of less than an order of magnitude. It provides in-depth assessments of simulated galaxies with kinematic and photometric features that are comparable to those seen in observed Sab galaxies. Although different simulations of disk creation have yielded some amazing results, the fundamental physics underpinning this process is still unknown. It also suggests that the low angular momentum of previously simulated galaxies was due, at least in part, to insufficient resolution. Moreover, other scholars argue that implementing feedback that can govern star formation is a critical component for building appropriate disks. [11]

2.3.5. Disputes on dark matter. The major features of dark matter haloes as observed from observations are discussed. More than a thousand galaxies now have detailed rotation curves, demonstrating that they are not so flat in the outer sections, but rise for late-types and decline for early-types according to a well-established finding. More than a thousand galaxies now have comprehensive rotation curves, revealing that they are not as flat in the outer parts as previously thought, but rather increase for late-types and fall for early-types. According to a well-established result, dark matter does not dominate most luminous galaxies' optical disks. LSB are dark matter-dominated galaxies that can be dwarfs or giant galaxies with a lot of HI gas. Their rotation curves can also be used to restrict dark matter theories.

Table 1. The Millennium N-body simulations' selected parameters.

Ωm	Ωb	n_{spec}	h	σs	L_{box} h^{-1} Mpc	Np	Mp h-1 M_0	Mh $h^{-1}M_0$	Label
0.25	0.0455	1.0	0.73	0.9	500	2160^3	8.61×10^8	1.72×10^{10}	MSI
0.25	0.0455	1.0	0.73	0.9	100	2160^3	6.88×10^6	1.37×10^8	MSII
0.272	0.0455	0.97	0.70	0.81	500	2160^3	9.34×10^8	1.87×10^{10}	WM7
0.307	0.04825	0.96	0.68	0.83	542	5040^3	1.06×10^8	$2.12^*\times10^9$	PMILL

3. Experiment measurement

3.1. The Millennium N-body simulations

Cosmological simulations of galaxy formation have been critical in increasing our knowledge of the structure and creation of galaxies in the Universe during the last few decades. These simulations explore the nonlinear history of galaxies across a wide range of sizes, representing a variety of physical processes. Improved numerical approaches, increasing computer capacity, and a better knowledge of the physics important to galaxies have resulted in simulations, which can recreate a significant number of observable galaxy features. Millennium N-body simulations will be used in this article. Some of the simulations results are presented in Table 1 and Figure 3.

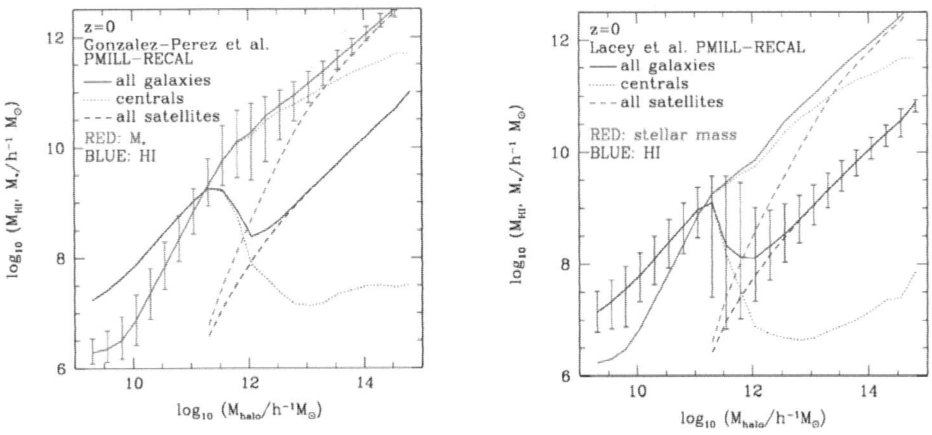

Figure 3. The median stellar mass (red) and HI (blue) contents of dark matter halos in the models as a function of halo mass, showing contributions from all galaxies (solid), centrals (dotted), and satellites within a halo (satellites). (It's worth noting that the specific value of the median mass for satellite galaxies depends on the resolution.) The bars represent the distribution's 10-90 percentile range, and are only given for the total HI content of the haloes in the left panel and the total stellar mass in the right panel for clarity. The predictions of the recalibrated version of the models are shown in the left panel, while the recalibrated model is shown in the right panel.

Journal of Physics: Conference Series **2441** (2023) 012025 doi:10.1088/1742-6596/2441/1/012025

3.2. Parallel N-body, smoothed particle hydrodynamics

We explain star-forming gas using a "multiphase model". Our technique, when implemented in a sub-resolution way, allows for numerically convergent findings. This is especially relevant in CDM cosmology simulations, where early generations of galaxies might be difficult to resolve.

To avoid the difficulties, the parallel N-body, smoothed particle hydrodynamics (SPH) algorithm GADGET2 is utilized in its "conservative entropy" formulation. The initial displacement field is then generated using a typical zooming approach, which involves correctly adding small-scale disturbances in the high-resolution area. It is worth noting that the high-resolution zone doesn't focus on a specific object, with the goal of avoiding bias caused by picking halos that are fundamentally favourable for disk galaxy formation. The high-resolution particles can be divided into two categories: dark matter (HRDM) and gas. As shown in Fig. 4, the volume is adequate for our needs, given our current research does not involve large-scale galaxies or the global mass function, for example. The interesting points are in solitary, galactic-scale objects here, which are mostly unaffected by simulation box size. We include a UV backdrop and radiative cooling and heating, as well as star formation, supernova feedback, and metal enrichment. To characterize the star-forming gas, we use the multiphase model. Without explicitly spatially resolving the distinct phases, our method accounts for some of the important elements of the ISM's multiphase structure. Two fundamental difficulties in understanding star formation on cosmic scales drive our technique. To begin with, there is no underlying explanation for this procedure. Second, large-scale models lack the resolution needed to accurately define the ISM. [11]

Table 2. List of some necessary Parameters

Parameter	Symbol	Value
Supernova evaporation	A_0	1000
Mass Fraction of stars >8 M	β	0.1
Cold cloud temperature (K)	T_{clouds}	1000
Effective supernova temperature (K)	Tsn	10^8

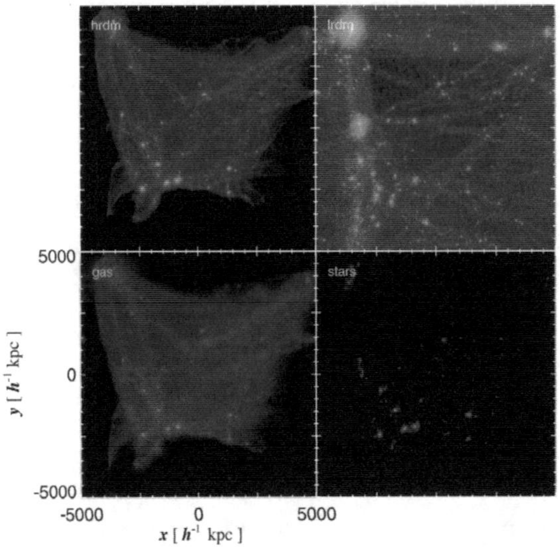

Figure 4. The whole simulation volume projected onto a 10 h^{-1} Mpc square region at z = 0, illustrating the simulation's final state. The HRDM and LRDM particles are shown in the top left and top right panels, respectively. The gas and star particles are depicted in the bottom left and right panels, respectively. The low- and high-resolution particles have mixed at the high-resolution zone's boundary due to gravitational influences, but the LRDM particles have not infiltrated the inner high-resolution region. In areas where the gas density has beyond the threshold density for star formation, star particles have spawned. [11]

4. Conclusion

In summary, we derive the formulae for the attraction force between dark matter and normal matter and choose the CDM model to demonstrate the impact of dark matter on galaxy formation. As a matter of fact, the real properties of dark matter are still unobserved since the solution is obtained based on CDM model, which is under plenty of assumptions and approximations. In the future, it is pursued to construct and propose more accurate model that can be verified based on state-of-art detectors. Overall, these results pave a path for dark matter detection and development of analytical models for dark matter.

References

[1] Baugh, C. M., Gonzalez-Perez, V., Lagos, C. D., Lacey, C. G., Helly, J. C., Jenkins, A., ... & Cole, S. (2019). Galaxy formation in the Planck Millennium: the atomic hydrogen content of dark matter haloes. Monthly Notices of the Royal Astronomical Society, 483(4), 4922-4937.

[2] Vogelsberger, M., Marinacci, F., Torrey, P., & Puchwein, E. (2020). Cosmological simulations of galaxy formation. Nature Reviews Physics, 2(1), 42-66.Secular evolution in disk galaxies: the growth of pseudobulges and problems of cold dark matter galaxy information

[3] Kormendy, J., & Fisher, D. B. (2005). Secular Evolution in Disk Galaxies: The Growth of Pseudobulges and Problems for Cold Dark Matter Galaxy Formation. arXiv preprint astro-ph/0507525.

[4] Wechsler, R. H., & Tinker, J. L. (2018). The connection between galaxies and their dark matter halos. Annual Review of Astronomy and Astrophysics, 56, 435-487.

[5] Hohmann, S. (2020). Calculating the distributions of number, mass and luminosity of the stars with the help of MS Excel®. European Journal of Physics, 42(1), 015601.

[6] Cole, S., Norberg, P., Baugh, C. M., Frenk, C. S., Bland-Hawthorn, J., Bridges, T., ... & Taylor, K. (2001). The 2dF galaxy redshift survey: near-infrared galaxy luminosity functions. Monthly Notices of the Royal Astronomical Society, 326(1), 255-273.

[7] Komatsu, E., Dunkley, J., Nolta, M. R., Bennett, C. L., Gold, B., Hinshaw, G., ... & Wright, E. L. (2009). Five-year wilkinson microwave anisotropy probe* observations: cosmological interpretation. The Astrophysical Journal Supplement Series, 180(2), 330.

[8] Buta, R., & Crocker, D. A. (1991). The outer Lindblad resonance and the morphology of early type disk galaxies. The Astronomical Journal, 102, 1715-1723.

[9] Buta, R., Corwin Jr, H. G., & Odewahn, S. C. (2002, December). The de Vaucouleurs atlas of galaxies. In Disks of Galaxies: Kinematics, Dynamics and Peturbations (Vol. 275, p. 102).

[10] Kormendy, J., & Kennicutt Jr, R. C. (2004). Secular evolution and the formation of pseudobulges in disk galaxies. Annu. Rev. Astron. Astrophys., 42, 603-683.

[11] Robertson B, Yoshida N, Springel V, et al. Disk galaxy formation in a Λ cold dark matter Universe[J]. The Astrophysical Journal, 2004, 606(1): 32.

Mass Function and Density Profile of Dark Matter Halos in Illustris-TNG

Haichen Fan [1, *, †], Lu Lu [2, †], Jianyuan Luo [3, †], Jiaxuan Wang [4, †]

[1] School of Physics and Astronomy, Sun Yat-Sen University, Zhuhai, China.
[2] Department of Physics, The University of Texas at Austin, Austin, Texas, USA.
[3] School of Physics and Technology, Wuhan University, Wuhan, China.
[4] Department of Physics and Astronomy, The University of Toronto Scarborough, Toronto, Ontario, Canada.

*Corresponding author's e-mail: fanhch@mail2.sysu.edu.cn

[†] These authors contributed equally

1. Introduction

A dark matter halo is a hypothetical region predicted by the ΛCDM model [1]. A single dark matter halo may contain multiple virialized clumps of dark matter bound together by gravity, known as subhalos [2]. Both halos and subhalos may contain galaxies [3]. The properties of the dark matter halo are important for investigating the evolution of galaxies and searching for the dark matter, as well as studying the evolution of the universe.

The dark matter halo has two major properties: the Press-Schechter mass function and the density profile. The Press-Schechter mass function, also known as the Press-Schechter formalism, is a mathematical model that can predict the number of objects of a certain mass that can form within a given volume in the universe, e.g., galaxies, galaxy clusters and dark matter halos [4]. It was presented by William H. Press and Paul Schechter in 1974 [5]. According to the theorem of Press and Schechter, the massive objects in the universe (e.g., galaxies, galaxy clusters, etc.) actually come from the collapse of the perturbations imprinted on the universe at very early times.

The halo mass function describes the number of dark matter halos per unit mass, while the density profile describes the mass distribution within a single halo. In this paper, we also examine two density profiles of dark matter halos. One is the Navarro–Frenk–White (NFW) profile [6], which focuses on the spatial mass distribution of dark matter halos and uses N-body simulations to identify the density profile. It was developed by Julio Navarro, Carlos Frenk and Simon White in 1996 [7]. The other one is the Einasto profile, which is a mathematical function commonly used to describe how the density of a spherical stellar system varies in distance. It is developed by Jaan Einasto in 1965 [1]. We collect data from the IllustrisTNG project to examine two density profiles. We also use the Jupyterlab of the IllustrisTNG project to examine the mass evolution of components in subhalos, e.g., stars, galaxies, dark matter, and black holes. The IllustrisTNG is a large galaxy-formation simulations project aiming at predicting the formation and evolution of galaxies and the large-scale structure of the universe.

The rest part of the paper is organized as following: Sec. 2 describes analytical models of mass function and density profile; Sec. 3 shows the results of density profile and mass evolution; Sec. 4 presents the discussion of the results; and Sec. 5 gives a brief summary of what we find.

Content from this work may be used under the terms of the Creative Commons Attribution 3.0 licence. Any further distribution of this work must maintain attribution to the author(s) and the title of the work, journal citation and DOI.
Published under licence by IOP Publishing Ltd

2. Analytic models

2.1. Mass function

It is assumed that the density fluctuations lead to the structure formation. However, it is impossible for us to observe the formation process directly. Instead, we can only see the outcome of this process, i.e., the large-scale structures.

Basically, one can observe their distributions and bring up theoretical models to fit them. The mass function is an example. It predicts the number of structures of a certain mass within a given volume. In this paper, we discuss two kinds of mass functions, Press-Schechter mass function and Sheth-Tormen mass function.

Press and Schechter assumed that the density fluctuation field have a Gaussian probability distribution. Based on the assumption, the probability that the density fluctuation δ exceeds a critical value δ_c can be calculated. Specifically, $\delta_c = 1.686$ is the threshold value which corresponds to virialized structure. In other word, when δ exceeds 1.686, collision will end within a range where the virial theorem is appliable. In addition, Press and Schechter assumed the collision to be spherical [8]. They also treated the under-dense region in the same way as the over-dense region, simply by multiplying a factor 2 in the calculation. The Press-Schechter mass function can be derived through computation.

Although the Press-Schechter mass function agrees well with numerical results, there are some quantitative deviations at low mass and high mass ends. The Sheth-Tormen function offers a modification of the Press-Schechter mass function by changing the spherical collapse model into the ellipsoidal collapse model, which is more realistic. Consequently, it fits numerical results better [9].

2.2. Density profiles

According to NFW profile, the density profiles of different dark matter halos have the same shape, initial density fluctuation spectrum, and values of the cosmological parameters regardless the mass of the halos. Over a broad range of halo mass and redshift, the NFW profile approximates the equilibrium configuration of dark matter halos produced in simulations of collisionless dark matter particles.

Besides, the NFW profile provides a good prediction for about 70% percent of dark matter halos, especially for low-mass halos [10]. If we use gravitational lensing to measure the dark matter density profiles of massive galaxy clusters, one can clearly observe that the data agrees well with the prediction of NFW profiles. However, the NFW dark matter profile is not consistent with observations of the inner regions of low surface brightness galaxies, which have less central mass than predicted. This is known as the cusp-core problem.

In the NFW profile, the density of dark matter halos as a function of radius is given by:

$$\rho(r) = \rho_s \frac{\delta}{\frac{r}{R_s}(1 + \frac{r}{R_s})^2} \tag{1}$$

where $\rho(r)$ is the density of the halo; ρ_s is the critical density; δ is the characteristic overdensity, which is closely related to the formation time and the mass of the halos, the value is about 10^5 for cold dark matter halos with low mass and will decrease as the mass increasing [6]; r is the viral radius of the halo; R_s is the scale radius.

To identify the mass function of the dark matter halo, we introduce the concentration parameter, which is defined as:

$$c = \frac{R_{vir}}{R_s} \tag{2}$$

Here, R_{vir} is the viral radius. According to simulations, the value of the concentration parameter c is about 10 for low-mass dark matter halos (approximate 10 times the mass of the sun) and becomes smaller as the mass of halos increasing [10]. Therefore, the prediction of mass by the NFW profile is:

$$M = \int_0^{R_{vir}} 4\pi r^2 \rho(r) dr \tag{3}$$

Combining Eqs. (2) and (3), we obtain

$$M = 4\pi {R_s}^3 \left[\ln(1+c) - \frac{c}{1+c}\right] \quad (4)$$

The density of dark matter halos by NFW profile can be converted to

$$\frac{\rho(r)}{\rho_s} = \frac{\delta}{c(1+c)^2} \quad (5)$$

Now let's start the prediction of density by the Einasto profile. The Einasto profile possesses a power law logarithmic slope for ρ and r of the form:

$$\frac{d\ln\rho(r)}{d\ln r} = \left(\frac{r}{R_s}\right)^{\alpha} \quad (6)$$

or it can be rearranged as:

$$\frac{\rho(r)}{\rho_s} = e^{\left\{-\frac{2}{\alpha}[(c)^{\alpha}-1]\right\}} \quad (7)$$

where α is the parameter that controls the degree of curvature of the profile.

3. Results

3.1. Density profiles

Figure 1 shows the comparison of the NFW and Einasto profiles of the density of a dark matter halo, where the value of α for the Einasto profile is 0.15. According to the results, the density of the halo is decreasing with the radius increasing. Besides, the prediction of the NFW profile matches well with the Einasto profile when α=0.15.

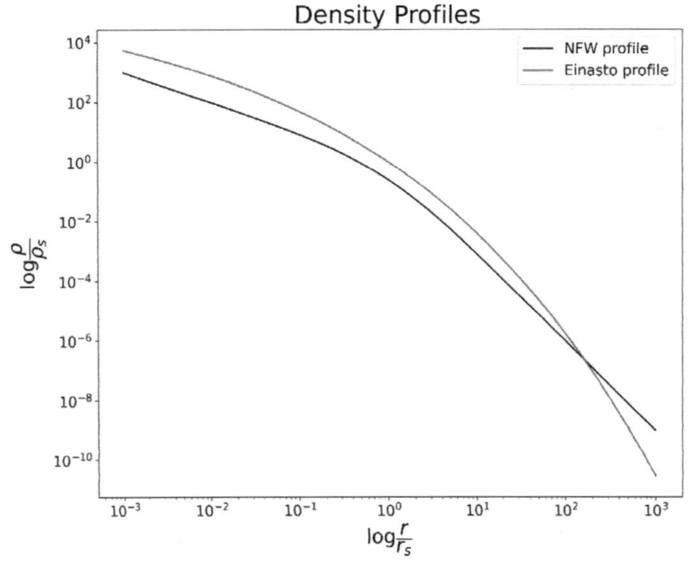

Figure 1. The comparison of NFW and Einasto density profiles with α=0.15.

3.2. Mass evolution

Based on the data from IllustrisTNG, 5 diagrams are exhibited in Fig. 2 to show the mass evolutions of four components of subhalos. The horizontal axis is the snapshot number, each of which corresponds to a specific redshift. Here the snapshot number starts from 68, which corresponds to z=2, and ends at 135, which corresponds to z=0 (i.e., present day). In other words, the horizontal axis represents the redshift that is analogous to time. The vertical axis represents mass. Hence, these 5 graphs show how the masses of four different components of the universe change over time in a specific subhalo.

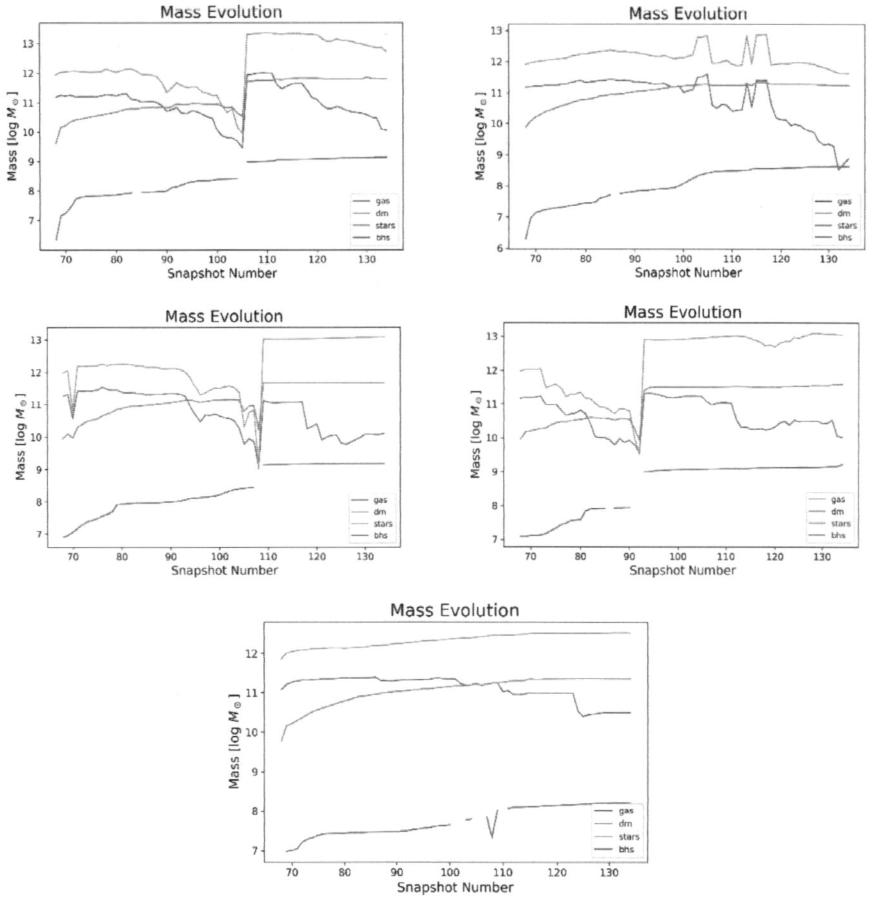

Figure 2. The diagrams of mass evolutions of 5 different subhalos. Their IDs at z=2 are 109974 (upper left), 123175 (upper right), 95711 (middle left), 107743 (middle right), 110822 (lower), respectively. The snapshot number's range is from 68 to 135.

4. Discussion

As illustrated in Fig. 2, the blue, yellow, green, and red lines correspond to gas, dark matter, stars, and black holes, respectively. In the upper left graph with ID 109974, the subhalo falls into a much more massive subhalo around snapshot 105. Notice that in the upper left, middle left, and middle right graphs, where a violent mass evolution occurs, the gas and dark matter's mass experience the same rises and falls before this evolution occurs. However, dark matter lines and gas lines stop to resemble each other after the evolution. Instead, dark matter lines tend to be flat and peaceful, which really resemble stars line, while gas lines continue to undergo fluctuations. In the upper right and lower graphs, with ID 123175 and 110822, in contrast, there is no such violent mass change. In the upper right graph with ID 123175, the dark matter line resembles the gas line with the same fluctuation. In the lower graph with ID 110822, however, nothing interesting happens, and the dark matter line follows the stars line peacefully.

Another thing that needs attention is that in all the 5 graphs, black holes lines at the bottom have some abrupt changes without continuity, which might be attributed to the lack of information in certain snapshot numbers. Besides, one also sees these abrupt changes occur in the upper left, middle left and middle right graphs during the violent mass evolution. It is reasonable to assume the connection between them.

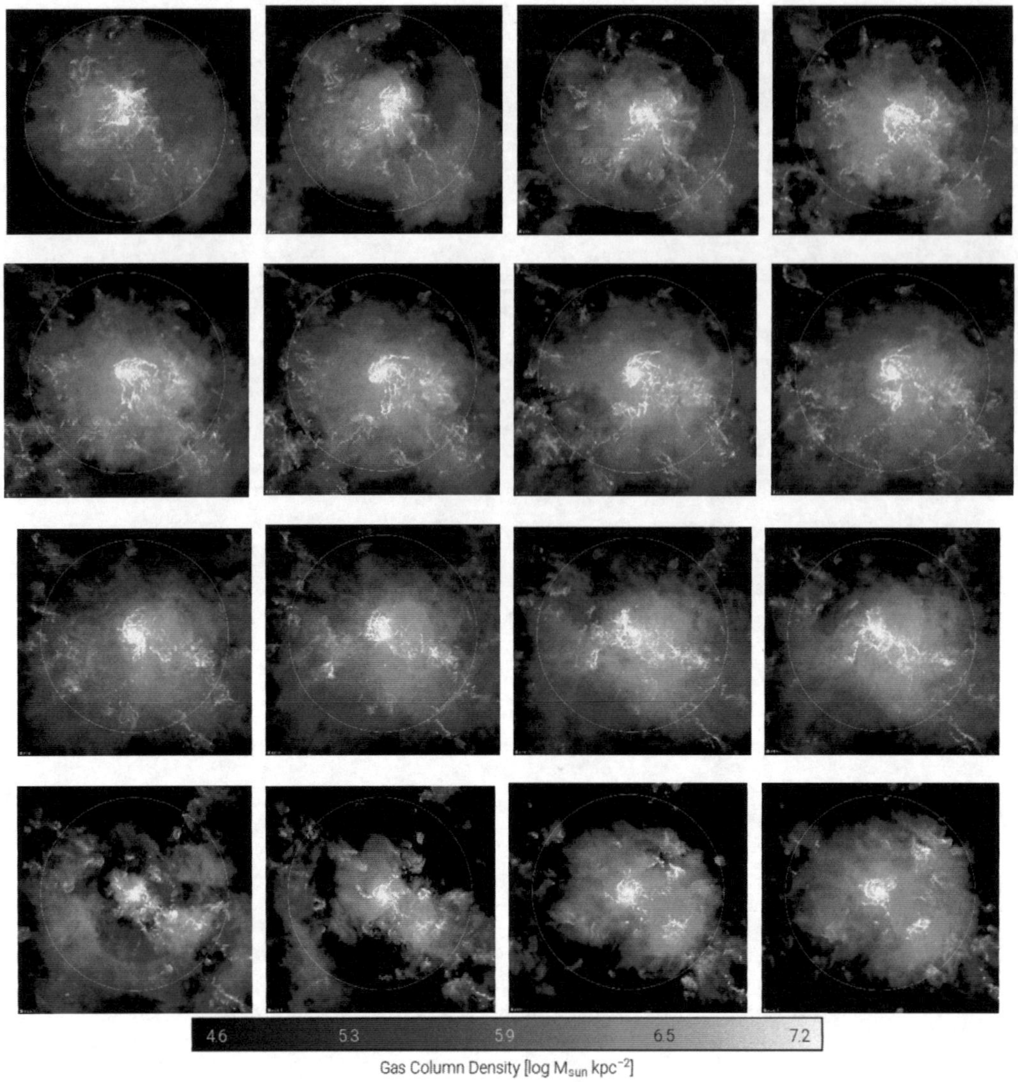

Figure 3. Pictures plotted from the *www.tng-project.org*. The data is from the simulation Illustris-1, the snapshot numbers are from 100 to 115, and the corresponding redshifts are z=0.58, 0.55, 0.52, 0.5, 0.48, 0.46, 0.44, 0.42, 0.4, 0.38, 0.36, 0.35, 0.33, 0.31, 0.3, and 0.27.

All the results are obtained by python API offered by the IllustrisTNG. We determine certain subhalos with their unique IDs at z=2, or snapshot number 68, and track them from z=2 to z=0, using "descendant" command constructed inside the IllustrisTNG package. We also apply the technique called "append" to perform iterations. Apart from that, we examine both the information of all subhalos included in the simulation at a certain redshift and the information of a single subhalo with its specific ID and redshift. This is certainly helpful if one wants to investigate how the mass evolution differs between various subhalos. However, except for the comparison, we didn't find any powerful evidence to support the trends shown in the mass evolution graphs. It is necessary for us to find a cleverer way to address that.

After failing to find powerful evidence from the subhalo information to account for the difference in mass evolution, we further tried to seek some visualization data from IllustrisTNG in order to find what happened at specific snapshot of the same subhalo. The data is obtained by searching the IDs. However, the IDs are different for the same subhalo at different snapshot numbers. To trace the evolution of a subhalo, the sublink of the main descendent branch of one halo is used to obtain the hdf5 file and the different IDs corresponding to different snapshot numbers. Here, the results are displayed in Fig. 3. They describe the gas column density in a specific subhalo whose ID at z=2 is 109974 and that corresponds to the upper left diagram in Fig. 2. The snapshot number ranges from 100 to 115, i.e., redshift from 0.58 to 0.27. As shown in Fig. 2, this subhalo undergoes an abrupt change in mass when the snapshot number is approximately 106, i.e., z=0.44.

5. Conclusion

In this paper, we investigate the density of the dark matter halos in our universe predicted by the NFW profile based on the data from the IllustrisTNG project. According to the analysis, it matches well with that of the Einasto profile for curvature parameter $\alpha = 0.15$. However, unknown conditions that affect the value of α are strict limitations that makes the density profile unpredictable in different redshifts.

In addition, we plot the mass evolution of 5 different subhalos and study their trends based on Illustris TNG. A specific halo is selected (ID 109974 at snapshot number 68), and is observed that it falls into a much bigger subhalo at around snapshot number 106. To test the accuracy of our mass evolution plot, a series of images of what happens during this drastic evolution from snapshot number 100 to 115 are given. According to the results, the subhalo undergoes an obvious change in shape. On this basis, our mass evolution plot is accurate.

Although we accomplish plenty of results, there are still some limitations within our investigation. These results are unable to tell the specific reason why the subhalo evolves into a bigger subhalo at a certain time. We take a look at the data provided by the simulation but cannot derive reasonable explanations from it. Additionally, the data from Illustris TNG is not enough to calculate mass function and density profile of dark matter halo. Therefore, we cannot examine how these two analytical models agree with the simulation. In the future, we may tell why subhalos evolves and predicts their mass evolution based on their information like mass distribution and velocity.

In summary, we dive into two properties of the dark matter halo, the mass function and the density profile, use Illustris TNG to examine the mass evolution of 5 different subhalos, and choose one subhalo to study its change in shape during a violent growth in mass. Our investigation gives us a deeper understanding of the dark matter halo, presents us with the trends of mass evolution in subhalos, and shows us the alteration in appearance of a subhalo under a radical evolution.

References

[1] A Einasto, J.,1965. Kinematics and dynamics of stellar systems, Trudy Inst. Astrofiz. Alma-Ata 5, 87.
[2] Wechsler, Risa; Tinker, Jeremy (September 2018). "The Connection between Galaxies and their Dark Matter Halos". Annual Review of Astronomy and Astrophysics. 56: 435–487.
[3] Mo, Houjun; van den Bosch, Frank; White, Simon (2010). Galaxy Formation and Evolution. Cambridge University Press. pp. 97–98. ISBN 978-0-521-85793-2.
[4] Dark Matter Halos, Mass Functions, and Cosmology: a Theorist's View
[5] Press, W.H., Schechte, P., 1974. Formation of Galaxies and Clusters of Galaxies by Self-Similar Gravitational Condensation. ApJ 187, 425.
[6] Navarro, J., Frenk, C., White, S., 1997. A universal density profile from hierarchical clustering. ApJ 490, 493.
[7] Navarro, J., Frenk, C., White, S., 1996. The Structure of Cold Dark Matter Halos. ApJ 462, 563.
[8] Reed, D., Bower, R., Frenk, C., Jenkins, A.,Theuns, T., 2007. The halo mass function from the dark ages through the present day. MNRAS 374, 2.
[9] Bagla, J. S., Khandai, N., Kulkarni, G., 2009. Mass function of haloes: scale invariant models. arXiv e-prints 0908.2702.
[10] Jing Y., 2000. The density profile of equilibrium and nonequilibrium dark matter halos. ApJ 535, 30.

Differences among IllustrisTNG series and a Brief Comparison with the Illustris Project

Zhengyu Dai [1, †], Jinao Dong [2, †], Runyu Liao [3, *, †] and Qiwei Yue [4, †]

[1] China International University In Israel, Tel Aviv, Israel
[2] Menaul School Qingdao, Qingdao City, China
[3] The School of Physics, Shandong University, Jinan City, China
[4] High School Affiliated to Shanghai Jiao Tong University, Shanghai City, China

* Corresponding author's e-mail: 201700100045@mail.sdu.edu.cn

[†] These authors contributed equally.

Abstract. The IllustrisTNG project is one of the most advanced cosmological gravity and magnetohydrodynamical models that simulates the formation and evolution of galaxies. It focuses on figuring out the physical processes involved based on the moving-mesh code AREPO. This project consists of three series (TNG50, TNG100, TNG300) with different box sizes, resolution, and physical parameters while each series also contains sub-simulations with lower resolution (TNG50-2, TNG50-3, TNG100-2, TNG100-3, TNG300-2, TNG300-3). The purpose of this paper is to illuminate the major differences and connections among different series of the IllustrisTNG project by analysing the data and carrying out corresponding visualization. Moreover, since the IllustrisTNG project is built upon the previous Illustris project, the improvements of the new model have been compared to the former one. Based on the analysis, different IllustrisTNG series differ in physical and numerical parameters, research areas, quantity of sample and so on. In the meantime, there are a lot of connections especially among the sub-simulations and we draw several figures, which provide a clear insight. The situation between the new model and the old one is very similar while the former introduces new methods and includes new physics which make the simulation process much more accurate. These results help to understand the ongoing problems and offer a guideline to enable the establishment of a future simulation model.

1. Introduction

The Next Generation Illustris project IllustrisTNG [1] aims to provide a clear insight into the cosmological process of galaxy formation and evolution in terms of large scale running the moving-mesh code AREPO [2]. Specifically, it is based on magnetohydrodynamics (MHD) [3] within the Λ-cold dark matter (ΛCDM) cosmological paradigm from redshift z=127 to z=0 with cosmological parameters $\Omega_m=\Omega_{dm}+\Omega_b=0.3089$ (matter density), $\Omega_b=0.0486$ (baryonic density), $\Omega_\Lambda=0.6911$ (cosmological constant), $H_0=100h$ km s^{-1} Mpc^{-1} with h=0.6774 (Hubble constant), $\sigma_8=0.8159$ (normalization) and $n_s=0.9667$ (spectral index) [4]. This project is divided into three major categories with different physical simulation box sizes (TNG50 with side length $35h^{-1}\approx50$ Mpc, TNG100 with

Content from this work may be used under the terms of the Creative Commons Attribution 3.0 licence. Any further distribution of this work must maintain attribution to the author(s) and the title of the work, journal citation and DOI.
Published under licence by IOP Publishing Ltd

sidelength $75h^{-1} \approx 100$Mpc and TNG300 with side length $205h^{-} \approx {}^{1}300$Mpc). The three series of simulations focus on different aspects of galaxy formation. So far, scholars have used these models to do several researches: clustering of galaxies and matter (TNG100 and TNG300) [5], stellar mass content of galaxy groups and clusters (TNG100 and TNG300) [6], chemical evolution and enrichment of stars (TNG100) [7], colour bimodality distribution of galaxies (TNG100 and TNG300) [8], magnetic fields and diffuse radio emission in galaxy clusters (TNG100 and TNG300) [9], galactic outflows driven by supernovae and supermassive black hole feedback (TNG50) [10], structure and kinematical evolution of star-forming galaxies (TNG50) [11].

All of these series play different roles in the project. TNG300 with the largest simulation box provides the largest galaxy sample and enables the study of galaxy clustering while TNG50 with the smallest volume and a much higher mass resolution thus offers a more detailed look. As for TNG100, it is performed with the same initial conditions and resolution to the previous Illustris simulation [6]. Therefore, it could be used to make clear comparisons between the new model and the old one (the IllustrisTNG project is the successor of the original Illustris project with parameters Ω_m=0.2726, Ω_b=0.0456, Ω_Λ=0.7274, H_0=100h km s^{-1} Mpc^{-1} with h=0.704, σ_8=0.809 and n_s=0.963 [12]). Moreover, both the IllustrisTNG series and the previous Illustris series contain corresponding sub-simulations with 1 being the highest and 3 the lowest resolution (TNG50-1, TNG50-2, TNG50-3, TNG100-1, TNG100-2, TNG100-3, TNG300-1, TNG300-2, TNG300-3, Illustris-1, Illustris-2, Illustris-3), which could be used to investigtate numerical convergence [5] and to control for resolution effects [11].

In this paper, we focus on investigating the major differences and connections among different IllustrisTNG series and visualize them in order to offer a clear picture of the whole project. We also make comparison between the IllustrisTNG project with the original Illustris project to analyse the advantages and improvement ideas the new model has, which are of great significance for future simulation models. We choose the data from the TNG website [1] and use Jupyterlab to run the python code to realize visualization. Specifically, the specific star formation rate with respect to the galaxy stellar mass, the stellar mass function and the total subhalo mass with respect to the snapshot number are illustrated. For the sake of simplicity, the redshift value is set to be 0 to draw these figures.

The structure of this paper is as follows. In Section 2, we concentrate on the numerical differences and connections among different simulation models. Respective research areas are illustrated in Section 3. Section 4 demonstrates the visualization results while Section 5 explains the differences and connections between the IllustrisTNG project and the Illustris project. Finally, in Section 6, a brief conclusion is given.

Table 1. Physical and numerical parameters for TNG50, 100, 300 and their respective sub-simulations.

Run Name		TNG50-1	TNG50-2	TNG50-3	TNG100-1	TNG100-2	TNG100-3	TNG300-1	TNG300-2	TNG300-3
Volume	[cMpc^3]	51.7^3	51.7^3	51.7^3	110.7^3	110.7^3	110.7^3	302.6^3	302.6^3	302.6^3
L_{box}	[cMpc/h]	35	35	35	75	75	75	205	205	205
N_{GAS}	-	2160^3	1080^3	540^3	1820^3	910^3	455^3	2500^3	1250^3	625^3
N_{DM}	-	2160^3	1080^3	540^3	1820^3	910^3	455^3	2500^3	1250^3	625^3
N_{TRACER}	-	2×2160^3	2×1080^3	2×540^3	2×1820^3	2×910^3	2×455^3	1×2500^3	1×1250^3	1×625^3
m_{baryon}	[M$_\odot$/h]	5.7×10^4	4.6×10^5	3.7×10^6	9.4×10^5	7.6×10^6	6.0×10^7	7.6×10^6	5.9×10^7	4.8×10^8
m_{DM}	[M$_\odot$/h]	3.1×10^5	2.5×10^6	2.0×10^7	5.1×10^6	4.0×10^7	3.2×10^8	4.0×10^7	3.2×10^8	2.5×10^9
m_{baryon}	[10^6M$_\odot$]	0.085	0.68	5.42	1.4	11.2	89.2	11	88	703
m_{DM}	[10^6M$_\odot$]	0.45	3.63	29.04	7.5	59.7	477.7	59	470	3764
$\varepsilon^{z=0}_{DM,stars}$	[kpc]	0.288	0.576	1.152	0.74	1.48	2.95	1.48	2.95	5.90
$\varepsilon_{DM,stars}$	[ckpc/h]	0.39→0.195	0.78→0.39	1.56→0.78	1.0→0.5	2.0→1.0	4.0→2.0	2.0→1.0	4.0→2.0	8.0→4.0
$\varepsilon_{BH,max}$	[ckpc/h]	2.0	4.0	4.60	5.0	5.84	6.84	5.84	6.84	13.7
$\varepsilon_{gas,min}$	[ckpc/h]	0.05	0.1	0.2	0.125	0.25	0.5	0.25	0.5	1.0

2. Numerical differences and connections

Tables 1 and 3 summarize some basic parameters obtained from the TNG website [1] for different simulations while Table 2 and 4 show two parameters which are related to the figures presented in Section 4 subsequently. The parameters given in the table are: the volume and side-length of the simulation boxes. The initial number of gas cells, dark matter particles and Monte Carlo tracer particles. The mass of the target baryons and dark matter particles. The Plummer equivalent gravitational softening of the collisionless component. The maximum softening corresponds to black holes and the minimum gravitational softening corresponds to adaptive gas. The quantity of total subhalos and snapshots. The differences and connections of the parameters are given as following:

- The value and side-length of these simulation boxes differ in different series, but remain unchanged in their respective sub-simulations.
- These value of $N_{\rm gas}$, $N_{\rm DM}$, $N_{\rm TRACER}$, $m_{\rm baryon}$, $m_{\rm dm}$, $\varepsilon^{z=0}_{\rm DM,stars}$, $\varepsilon_{\rm gas,min}$ in sub-simulations of a certain series increase or decrease by an approximately 8-fold or 2-fold relationship.
- TNG100-2 has the same resolution as TNG300-1 while TNG100-3 is in line with TNG300-2.
- TNG100 series use the same initial conditions and resolution as the Illustris project (there does exist a small shift in some cosmological parameters).
- TNG50-2 has a comparable but somewhat better resolution (a factor of 2) to TNG100-1 while TNG50-3 corresponds to TNG300-1 [11].

Table 2. Numerical parameters for TNG50, 100, 300 and their respective sub-simulations.

Run Name	TNG50-1	TNG50-2	TNG50-3	TNG100-1	TNG100-2	TNG100-3	TNG300-1	TNG300-2	TNG300-3
$N^{z=0}_{\rm Subfind}$	5688113	859077	134779	4371211	698336	118820	14485709	2471022	391144
$N_{\rm snap}$	100	100	100	100	100	100	100	100	100

Table 3. Chosen parameters for Illustris-(1, 2, 3).

Run Name		Illustris-1	Illustris-2	Illustris-3
Volume	[cMpc^3]	106.5^3	106.5^3	106.5^3
$L_{\rm box}$	[cMpc/h]	75	75	75
$N_{\rm GAS}$	-	1820^3	910^3	455^3
$N_{\rm DM}$	-	1820^3	910^3	455^3
$m_{\rm baryon}$	[10^6M$_\odot$]	1.26	10.07	80.52
$m_{\rm DM}$	[10^6M$_\odot$]	6.26	50.1	400.82

Table 4. Numerical parameters for Illustris-(1, 2, 3).

Run Name	Illustris-1	Illustris-2	Illustris-3
$N^{z=0}_{\rm Subfind}$	4366546	689785	121209
$N_{\rm snap}$	134	136	136

3. Respective research areas

Some research areas for different IllustrisTNG series are listed in Table 5. As the highest resolution realization of the three series, TNG50 combines large volume with high resolution successfully. It concentrates on small scale properties and structure of galaxies. TNG300 corresponds to the largest volume of the simulation box and provides the largest galaxy sample. It focuses on galaxy clusters and some other rare objects. As for TNG100, it could not only be used to make clear comparison with the previous Illustris project, but is also often used to investigate the properties of galaxies in many aspects.

Journal of Physics: Conference Series **2441** (2023) 012027 doi:10.1088/1742-6596/2441/1/012027

Table 5. Some research areas for TNG50, 100, 300.

TNG50	TNG100	TNG300
Morphology of stellar distribution in galaxies [13]	Optical morphologies of galaxies [18]	Thermodynamical property of hot gas in galaxy clusters [23]
Abundance of satellite galaxies [14]	Properties of the circumgalactic medium around galaxies [19]	Galaxy Cluster X-ray Masses [24]
Cool gas in the circumgalactic medium around galaxies [15]	Properties of strongly barred disc galaxies [20]	Evolution of star formation rate-density relation [25]
Specific angular momentum of low-mass galaxies [16]	Formation of ultradiffuse galaxies in clusters [21]	Turnaround radius of galaxy clusters [26]
Star-forming main sequence of galaxies [17]	Black hole-host galaxy correlations [22]	Stellar property statistics of haloes [27]

4. Visualization

JupyterLab is a web-based user interface suitable for scientific data analysis and the programming language python is supported. Visualization is achieved based on python API offer by the official website and python library Matplotlib.

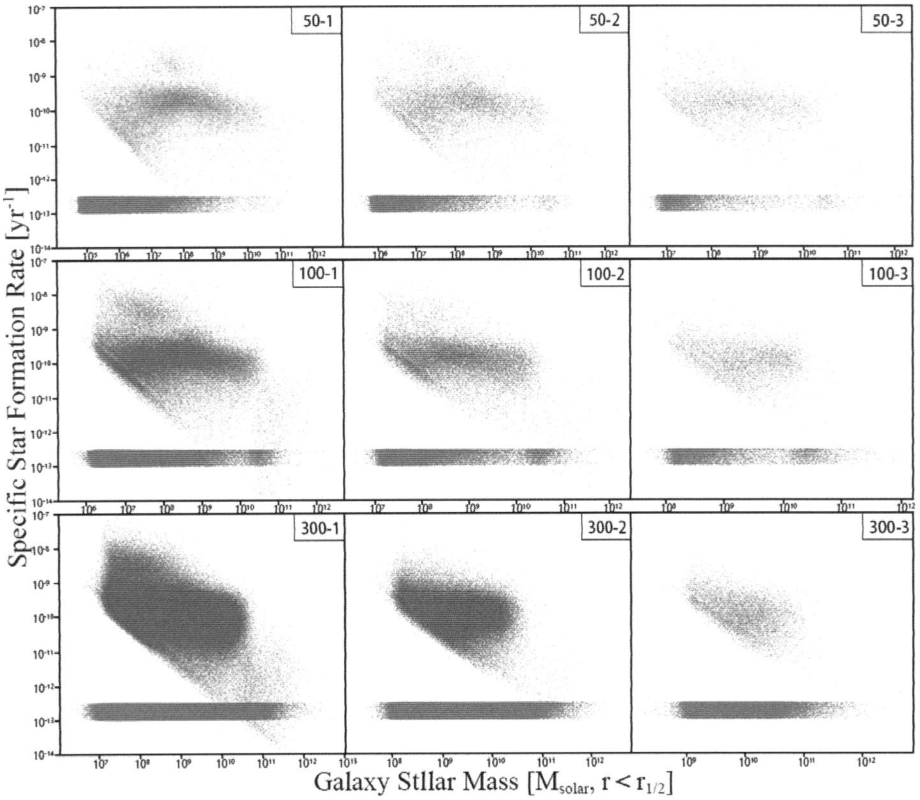

Figure 1. A plot of the specific star formation rate versus galaxy stellar mass within twice the stellar half mass radius for three series of the IllustrisTNG project at z=0 (the upper panel for TNG50, middle panel for TNG100 while the last panel for TNG300). Since it's a log-log plot, systems with 0 star formation rate is not included, we add them back with red dots.

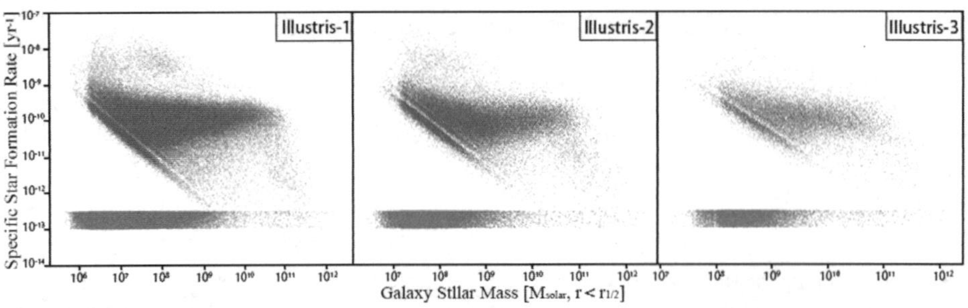

Figure 2. A plot of the specific star formation rate versus galaxy stellar mass within twice the stellar half mass radius for the Illustris project at z=0. Since it's a log-log plot, systems with 0 star formation rate is not included, we add them back with red dots.

The star formation rate (SFR) is of great importance to investigate the formation and evolution process of galaxies. It stands for the total mass of stars formed per year, often given as solar masses per year. As for the specific star formation rate (SSFR), it represents the ratio of SFR to the galaxy stellar mass. From Figures 1 and 2, one draws the following conclusions:

- The SSFR decreases as the galaxy stellar mass increases which means the star formation plays a bigger role to the growth of low-mass galaxies than high-mass galaxies [28].
- As the galaxy stellar mass changes, SSFR mainly revolves around a certain value (10^{-10} yr^{-1} for TNG50-1), which means the SFR is in proportion to the galaxy stellar mass in this range. This phenomenon is in line with the well-known main sequence theory.
- The number of total subhalos for different IllustrisTNG series is listed in Table. 2 and one can have a clear insight from Figure 1. For a certain series, as the resolution decreases, the quantity of sample decreases while TNG300-1 provides the largest galaxy sample. From Table 4, one knows that the number of total subhalos between TNG100 series and the Illustris project is similar to the results demonstrated in Figure 2.
- The galaxy stellar mass and SFR increase with better resolution [6]. As illustrated in Figure 1, for a certain series, as the resolution increases, the quantity of dots in the high-mass range and high-SFR range increase.

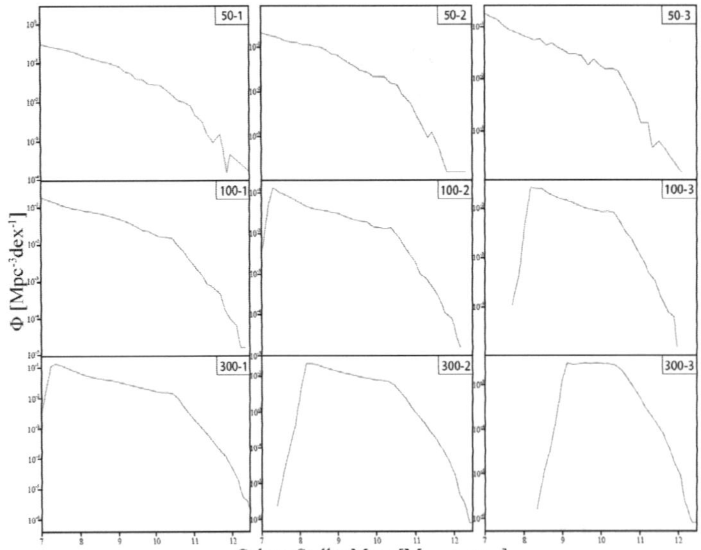

Figure 3. The plot of the stellar mass function for three series of the IllustrisTNG project at z=0 (the upper panel for TNG50, middle panel for TNG100 while the last panel for TNG300).

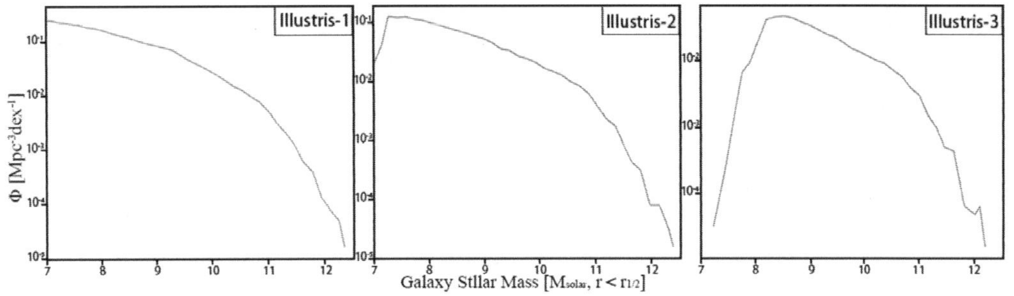

Figure 4. The plot of the stellar mass function for the Illustris project at z=0.

Stellar mass functions are about the number density of galaxies with respect to the galaxy stellar mass. They help to investigate the nature of galaxy population and track the assembly of galaxy stellar masses and evolution of SFR through cosmic time [29]. As illustrated in Figures 3 and 4, one draws the following conclusions:

- The curves of TNG100-2 and TNG300-1 are approximately the same and the curves of TNG100-3 and TNG300-2 are approximately the same, too. That is because TNG100-2 and TNG300-1 share the same spatial and mass resolution, as well as TNG100-3 and TNG300-2 (seen from Table 1).
- As the resolution changes (decreases or increases), the change in curves between TNG100 series and the Illustris project are roughly the same since they have the same initial conditions and resolution.
- In some models with higher resolution, the number density decreases as the galaxy stellar mass increases while in other models with lower resolution the number density first increases and then decreases as the galaxy stellar mass increases.
-

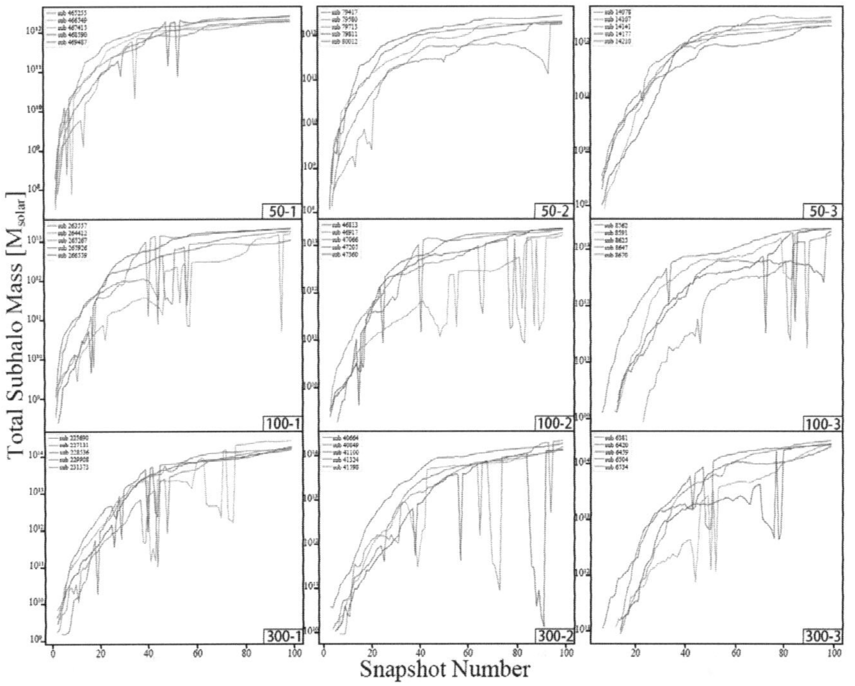

Figure 5. The plot of the total subhalo mass versus snapshot number for three series of the IllustrisTNG project at z=0 (the upper panel for TNG50, middle panel for TNG100 while the last panel for TNG300). Different colours represent the 100th through 104th most massive primaries.

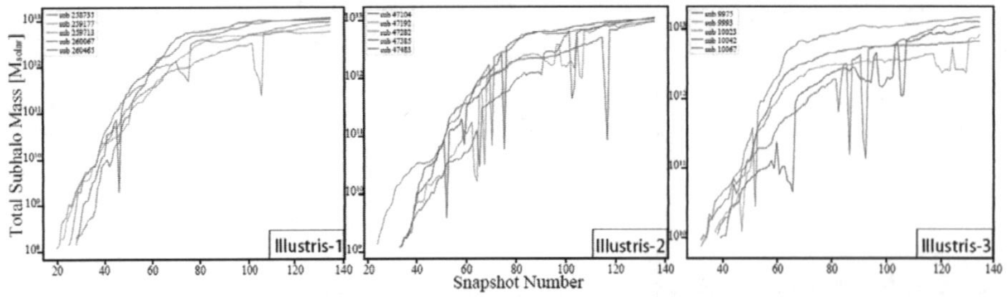

Figure 6. The plot of the total subhalo mass versus snapshot number for the Illustris project at z=0. Different colours represent the 136th through 140th most massive primaries.

Figures 5 and 6 depict the time evolution of five subhalos for different simulations, it offers a clear insight of the mass histories. According to Table 2, the number of snapshots in all IllustrisTNG simulations is 100 while in Table 4 the number of that for Illustris-1 is 134 and 136 for Illustris-2 and Illustris-3. The curvature of the curves is relatively large in the early stage and tends to 0 in the end, which means the majority of masses mainly form in early times. Moreover, there exists a sudden downward trend in some curves, which is associated with the so called 'subhalo switching problem' [30]. It can be noticed that this phenomenon is particularly obvious in TNG100 and TNG300 series.

5. A brief comparsion with the Illustris project

Similarities and differences between the Illustris project and the IllustrisTNG project are demonstrated here:

- Both of them are designed to simulate the formation of galaxies using the moving-mesh code AREPO within the ΛCDM cosmological paradigm, the redshift starts from z=127 and all the simulations have been evolved to z=0 [31]. They have different cosmological parameters while the Illustris project uses the data that is consistent with the Wilkinson Microwave Anisotropy Probe measurements [32] and the IllustrisTNG project uses the data that is consistent with the Planck 2015 results [4].
- The Illustris project is a set of large-scale N-body/hydrodynamical simulations [33] while the IllustrisTNG project is a suite of large-scale magnetohydrodynamical simulations.
- Both of them contain sub-simulations with lower resolution and all the simulations have 'dark matter only' runs (Illustris-1-Dark, TNG50-1-Dark) which are used to explore how galaxies evolve in the universe where only dark matter exists.

The existence of shortcomings [3] of the Illustris project prompted the birth of the IllustrisTNG project and the latter introduces a range of improvements [3] in many aspects. One of the improvements to the numerical techniques is the involvement of magnetic fields. Magnetic fields are of great importance in Astrophysics and play significant roles in many ways, ranging from compact objects to galaxy clusters [34]. Contemporarily, plenty of state-of-art cosmological simulation models (IllustrisTNG, Illustris, Horizon-AGN [35], EAGLE [36-37], etc.) have been successfully constructed, all of them include new physics and advanced methods, however, many issues do exist. For cosmology, TNG50 with the highest resolution is not high enough while TNG300 with the largest volume is not large enough, which provide ideas for the establishment of future simulation models.

6. Conclusion

In summary, different IllustrisTNG series differ in physical and numerical parameters, quantity of sample, quantity of snapshot and research areas. Connections exist among sub-simulations from a certain series and from different series. While TNG100 series and the Illustris project have a lot in common, differences and connections also exist between the IllustrisTNG project and the Illustris project. Overall, these results pave a path for future simulation model establishment and shed light on the detailed information and necessary ideas involved.

References

[1] IllustrisTNG-Main. Avaliable at: https://www.tng-project.org/.
[2] Volker Springel. E pur si muove: Galilean-invariant cosmological hydrodynamical simulations on a moving mesh[J]. Monthly Notices of the Royal Astronomical Society, 2010, 401 (2): 791-851.
[3] Annalisa Pillepich et al. Simulating galaxy formation with the IllustrisTNG model[J]. Monthly Notices of the Royal Astronomical Society, 2018, 473 (3): 4077-4106.
[4] Planck Collaboration. Planck 2015 results-XI. CMB power spectra, likelihoods, and robustness of parameters[J]. Astronomy & Astrophysics, 2016, 594, A11.
[5] Volker Springel et al. First results from the IllustrisTNG simulations: matter and galaxy clustering[J]. Monthly Notices of the Royal Astronomical Society, 2018, 475 (1) : 676-698.
[6] Annalisa Pillepich et al. First results from the IllustrisTNG simulations: the stellar mass content of groups and clusters of galaxies[J]. Monthly Notices of the Royal Astronomical Society, 2018, 475 (1): 648-675.
[7] Jill P Naiman et al. First results from the IllustrisTNG simulations: a tale of two elements-chemical evolution of magnesium and europium[J]. Monthly Notices of the Royal Astronomical Society, 2018, 477 (1): 1206-1224.
[8] Dylan Nelson et al. First results from the IllustrisTNG simulations: the galaxy colour bimodality[J]. Monthly Notices of the Royal Astronomical Society, 2018, 475 (1): 624-647.
[9] Federico Marinacci et al. First results from the IllustrisTNG simulations: radio haloes and magnetic fields[J]. Monthly Notices of the Royal Astronomical Society, 2018, 480 (4): 5113-5139.
[10] Dylan Nelson et al. First results from the TNG50 simulation: galactic outflows driven by supernovae and black hole feedback[J]. Monthly Notices of the Royal Astronomical Society, 2019, 490 (3): 3234-3261.
[11] Annalisa Pillepich et al. First results from the TNG50 simulation: the evolution of stellar and gaseous discs across cosmic time[J]. Monthly Notices of the Royal Astronomical Society, 2019, 490 (3): 3196-3233.
[12] Mark Vogelsberger et al. Properties of galaxies reproduced by a hydrodynamic simulation[J]. Nature, 2014, 509: 177-182.
[13] Razieh Emami et al. Inferring the morphology of stellar distribution in TNG50: twisted and twisted-stretched shapes[J]. The Astrophysical Journal, 2021, 918, 7.
[14] Christoph Engler et al. The abundance of satellites around Milky Way-and M31-like galaxies with the TNG50 simulation: a matter of diversity[J]. Monthly Notices of the Royal Astronomical Society, 2021, 507 (3): 4211-4240.
[15] Dylan Nelson et al. The cold circumgalactic medium in emission: MgII haloes in TNG50[J]. Monthly Notices of the Royal Astronomical Society, 2021, 507 (3): 4445-4463.
[16] Nicolas F. Bouché et al. The MUSE Hubble Ultra Deep Field survey XVI. The angular momentum of low-mass star-forming galaxies: a cautionary tale and insights from TNG50[J]. Astronomy & Astrophysics, 2021, 654, A49.
[17] Erica J Nelson et al. Spatially resolved star formation and inside-out quenching in the TNG50 simulation and 3D-HST observations[J]. Monthly Notices of the Royal Astronomical Society, 2021, 508 (1): 219-235.
[18] Vicente Rodriguez-Gomez et al. The optical morphologies of galaxies in the IllustrisTNG simulation: a comparison to Pan-STARRS observations[J]. Monthly Notices of the Royal Astronomical Society, 2019, 483 (3): 4140-4159.
[19] Guinevere Kauffmann et al. The morphology and kinematics of the gaseous circumgalactic medium of Milky Way mass galaxies-II. Comparison of IllustrisTNG and Illustris simulation results[J]. Monthly Notices of the Royal Astronomical Society, 2019, 486 (4): 4686-4700.
[20] Yetli Rosas-Guevara et al. The buildup of strongly barred galaxies in the TNG100 simulation[J]. Monthly Notices of the Royal Astronomical Society, 2020, 491 (2): 2547-2564.

[21] Laura V Sales et al. The formation of ultradiffuse galaxies in clusters[J]. Monthly Notices of the Royal Astronomical Society, 2020, 494 (2): 1848-1858.
[22] Yuan Li et al. Correlations between black holes and host galaxies in the Illustris and IllustrisTNG simulations[J]. The Astrophysical Journal, 2020, 895, 102.
[23] David J Barnes et al. A census of cool-core galaxy clusters in IllustrisTNG[J]. Monthly Notices of the Royal Astronomical Society, 2018, 481 (2): 1809-1831.
[24] M. Ntampaka et al. A deep learning approach to galaxy cluster X-ray masses[J]. The Astrophysical Journal, 2019, 876, 82.
[25] Ho Seong Hwang et al. Evolution of star fomation rate-density relation over cosmic time in a simulated universe: the observed reversal reproduced[J]. Monthly Notices of the Royal Astronomical Society, 2019, 489 (1): 339-348.
[26] Giorgos Korkidis et al. Turnaround radius of galaxy clusters in N-body simulations[J]. Astronomy & Astrophysics, 2020, 639, A122.
[27] Dhayaa Anbajagane et al. Stellar property statistics of massive haloes from cosmological hydrodynamics simulations: common kernel shapes[J]. Monthly Notices of the Royal Astronomical Society, 2020, 495 (1): 686-704.
[28] Bauer A.E., Drory N., Hill G.J. (2007) SPECIFIC STAR FORMATION RATES. In: DE JONG R. (eds) ISLAND UNIVERSES. Astrophysics and Space Science Proceedings. Springer, Dordrecht. https://link.springer.com/chapter/10.1007/978-1-4020-5573-7_84.
[29] Anna K. Weigel et al. Stellar mass functions: methods, systematics and results for the local Unverse [J]. Monthly Notices of the Royal Astronomical Society, 2016, 459 (2): 2150-2187.
[30] Vicente Rodriguez-Gomez et al. The merger rate of galaxies in the Illustris simulation: a comparison with observations and semi-empirical models[J]. Monthly Notices of the Royal Astronomical Society, 2015, 449 (1): 49-64.
[31] Debora Sijacki et al. The Illustris simulation: the evolving population of black holes across cosmic time[J]. Monthly Notices of the Royal Astronomical Society, 2015, 452 (1): 575-596.
[32] G. Hinshaw et al. Nine-year Wilkinson Microwave Anisotropy Probe observations: cosmological parameter results[J]. The Astrophysical Journal Supplement Series, 2013, 208, 19.
[33] Shy Genel et al. Introducing the Illustris Project: the evolution of galaxy populations across cosmic time[J]. Monthly Notices of the Royal Astronomical Society, 2014, 445 (1): 175-200.
[34] Ruediger Pakmor et al. Magnetohydrodynamics on an unstructured moving grid[J]. Monthly Notices of the Royal Astronomical Society, 2011, 418 (2): 1392-1401.
[35] Y. Dubois et al. Dancing in the dark: galactic properties trace spin swings along the cosmic web[J]. Monthly Notices of the Royal Astronomical Society, 2014, 444 (2): 1453-1468.
[36] Joop Schaye et al. The EAGLE project: simulating the evolution and assembly of galaxies and their environments[J]. Monthly Notices of the Royal Astronomical Society, 2015, 446 (1):521-554.
[37] Robert A. Crain et al. The EAGLE simulations of galaxy formation: calibration of subgrid physics and model variations[J]. Monthly Notices of the Royal Astronomical Society, 2015, 450 (2): 1937-1961.

Investigation of neutral hydrogen distribution and neutral hydrogen halo mass relation based on IllustrisTNG50

Yanlin Wu*

Department of Physics, University of Wisconsin Madison, Madison, WI, USA

*Corresponding author's e-mail: wu437@wisc.edu

Abstract. For the sake of investigating the observational data of the universe mass distribution using neutral hydrogen as a mass tracer, it is necessary to investigate the simulation prediction of the distribution of the neutral hydrogen. In this paper, the TNG50 is selected, which is the latest released part of IllustrisTNG project, as simulation data to study the distribution of gas and neutral hydrogen as well as the relation between the halo mass and the corresponding neutral hydrogen mass contained in each halo. The results show a tendency that most of neutral hydrogen are distributed within the halo and sub-halo structures. The neutral hydrogen halo mass relation is clearly shown and the function can be will produced by the conventional HI halo mass parameterization function. Based on the analysis, this paper fills the blank of other related works with most recent TNG50 data. These results offer a guideline for the following study about the halo model and pave a path to deeper understanding and investigating the hydrogen distribution in the cosmos.

1. Introduction

Generally speaking, the well-known ΛCDM paradigm has been widely proved to be suitable at matching various observations e.g., Cosmic Microwave Background's temperature and polarization. This model contains several free parameters that are related to unsolved cosmological questions. These parameters can describe the geometric structure of the cosmos, the features of dark matter as well as dark energy, and neutrino relevant phenomenon. The main goal of modern astrophysics is to investigate and constrain these physics inherits as accurate as possible. Based on investigating the matter distribution, one obtains large amount of information which can constrain the value of these parameters. One of the ways to achieve this is studying the statistical properties of matter tracers (e.g., neutral hydrogen). A new probe used to map out the matter distribution is the 21 intensity mapping [1-10].

Although this method is facing lots of challenges (e.g., large foreground noises and other technical issues), it is still an efficient method to make the large scale 3D maps of the hydrogen (not ionized) in the universe human beings aware of. The method consider the effect that the spin flip of the electron in neutral hydrogen will release 21 cm wavelength emission. This method is better than other approaches in many ways such as the large volume it can cover efficiently and the fact that the signal amplitude only related to the abundance and clustering of the neutral hydrogen [11]. There are plenty of surveys ongoing about this 21 cm intensity mapping. However, before acquiring the information from real survey data, it is important to make analytical evaluation which will be utilized later to make a

Content from this work may be used under the terms of the Creative Commons Attribution 3.0 licence. Any further distribution of this work must maintain attribution to the author(s) and the title of the work, journal citation and DOI.
Published under licence by IOP Publishing Ltd

comparison with the real survey results. In general, the presented halo model is a rather relatively precise analytic paradigm to predict the accumulation ratios of HI.

In this paper, it will be proved that halos contain most of the neutral hydrogen, which satisfies the initial description and principles of the halo model. There are two types of declarations for the halo model: the first is the mass for the HI accumulated in halo m (as a function of HI halo mass), while the second is the spatial distribution of HI within halos [11]. The IllustrisTNG50 simulation data is applied to analyze the distribution of gas and neutral hydrogen of the universe. The HI halo mass function is produced by plotting HI mass and halo mass of each halos in the simulation box.

This rest part of the paper is organized as follows. In Section 2, it is introduced that the IllustrisTNG data used and the method that is implemented to calculate the mass distribution of the hydrogen for snapshot data and the hydrogen mass for each halos in group catalog data. In Section 3, the plots from the simulation data and the HI halo mass function are studied and comparisons are made with plots from other studies. In Section 4, the results are summarized and future research questions originated from the comparison are brought into discussion.

2. Data &Method

2.1. Data

The simulation data used to analyse the HI halo mass relation function is from the IllustrisTNG Project [12-15]. The TNG 50 cosmological box with 35 Mpc/h comoving side length is used in this paper and this is the latest, the final and most computational demanding simulation part for the IllustrisTNG project [16, 17]. Compared to TNG 100 which is mainly used in Villaescusa-Navarro's paper, TNG 50 is more than one order of magnitude better in the resolution of the properties of mass and 2.5 times better in spatial parameters range [15, 16]. The moving-mesh Arepo code is used to produce this simulation and the code calculates the self-gravity as well as the equations of ideal continuum magnetohydrodynamics MHD [18-20]. It uses Tree-PM approach to calculate the self-gravity and Godunov type finite volume method to calculate the fluid dynamics [16]. In this paper, both the snapshot data and the group catalog data are used. For the group catalog, the standard Friends-to-Friends algorithm (FoF) is applied to identify each halo and give the group fields with linking length b=0.2 [21].

2.2. Method

The major goal is to calculate the fraction of Hydrogen in the gas and the division between neutral and ionized parts. The hydrogen number density can be calculated from the equation: $n_H = \rho X_H / m_p$, where the X_H is the hydrogen mass fraction ~0.75; ρ is the comoving total mass density of the gas cell; m_p is the mass of the proton. The output neutral hydrogen abundance is given by the IllustrisTNG snapshot data and is calculated in the IllustrisTNG. However, it is tricky for star forming gas since the calculation is based on the 'effective temperature of the equation of state instead of physical temperature. Therefore, the neutral hydrogen fraction can be underestimated. Knowing the neutral hydrogen abundance, the neutral hydrogen number density can be then calculated: $n_{H_0} = A \times n_H$ where A represents the neutral hydrogen abundance. However, it is important to know that the neutral hydrogen H_0 contains both H_2 and HI. Since H_2 is only a very small portion of the total neutral hydrogen, the precise measurement of H_2 is unnecessary and would affect little to the final results [11]. This can be a limitation for the investigation of this paper. The H_2 inside the halos and galaxies can cause the inaccuracy of this analysis.

3. Results& Discussion

In order to investigate the clustering properties of the neutral hydrogen, it is important to learn the how many HI is inside the halos. Therefore, the fraction of that is one of the ingredients for the halo model [22-26]. Standard halo model makes the assumption that all neutral hydrogen are resides within the halos. The first step of this paper is to testify the assumption and find the limitations. The map of the gas distribution and neutral hydrogen distribution is made using data from IllustrisTNG50-4. Figure1 shows the distribution comparison for both z=0 and z=5. The neutral hydrogen distribution clearly shows the fibre-like structure of halos and galaxies. From the colour bar on the right side, one sees that the mass distribution of neutral hydrogen is also concentrated inside halos and galaxies. The comparison between distribution maps with different redshifts shows that the fraction of neutral hydrogen located outside of halos and galaxies is larger at higher redshift. These results can be explained by gas in the Interstellar galactic medium IGM can host larger fraction of neutral hydrogen at higher redshift since gas in IGM is denser and UV background amplitude is lower [11]. After analysing the results, the assumption of halo model is testified that most of the neutral hydrogen is resides inside halos and galaxies at low redshift. The assumption may encounter limitations at higher redshift since the portion for neutral hydrogen outside of the halos becomes larger and cannot be ignored. Learning further about this limitation is beyond this paper's scope. Future research can dive more into the effects that the higher fraction of neutral hydrogen outside the halos can bring at high redshift and therefore drive more accurate halo model for high redshift.

The assumption that all neutral hydrogen is inside halos and galaxies is proved to be reasonable for low redshift in the previous part of this paper. Next step is to study the relation with the halo mass $M_{HI}(M,z)$.

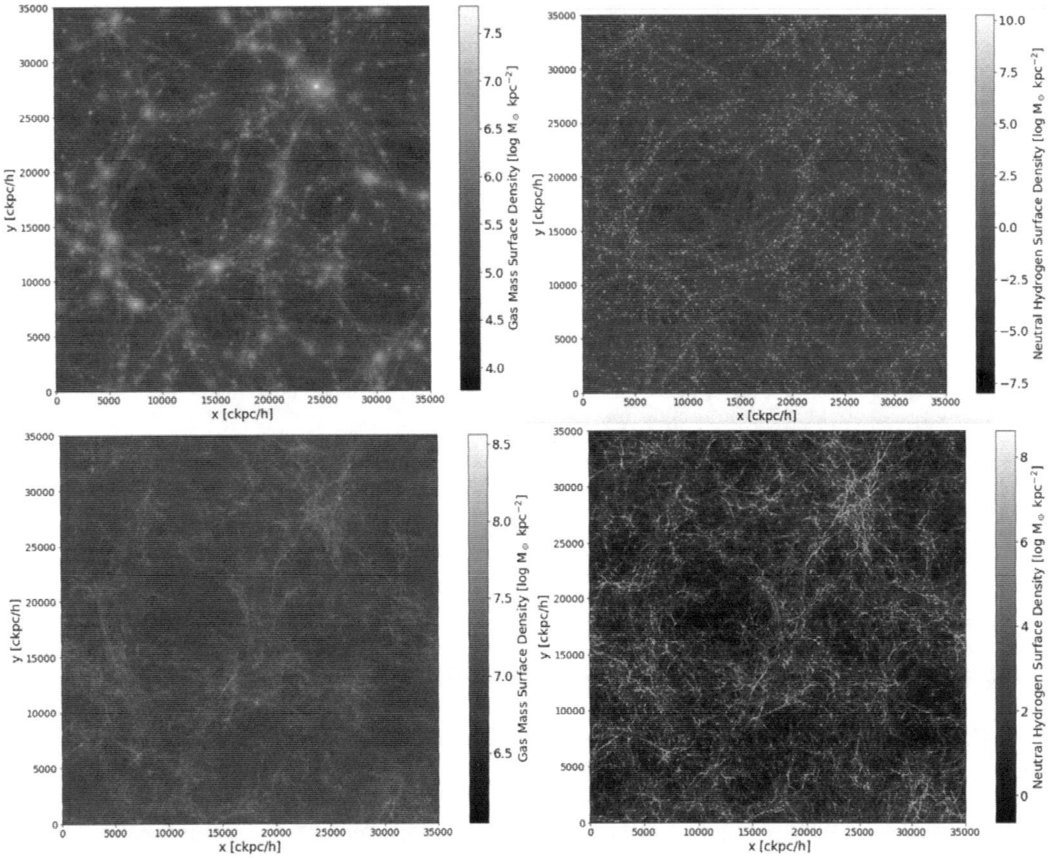

Figure 1. Spatial distribution of neutral hydrogen (right) and gas (left) for the entire TNG50-4 simulation volume. The upper panels are for redshift z=0, while the bottom panels are for redshift z=5.

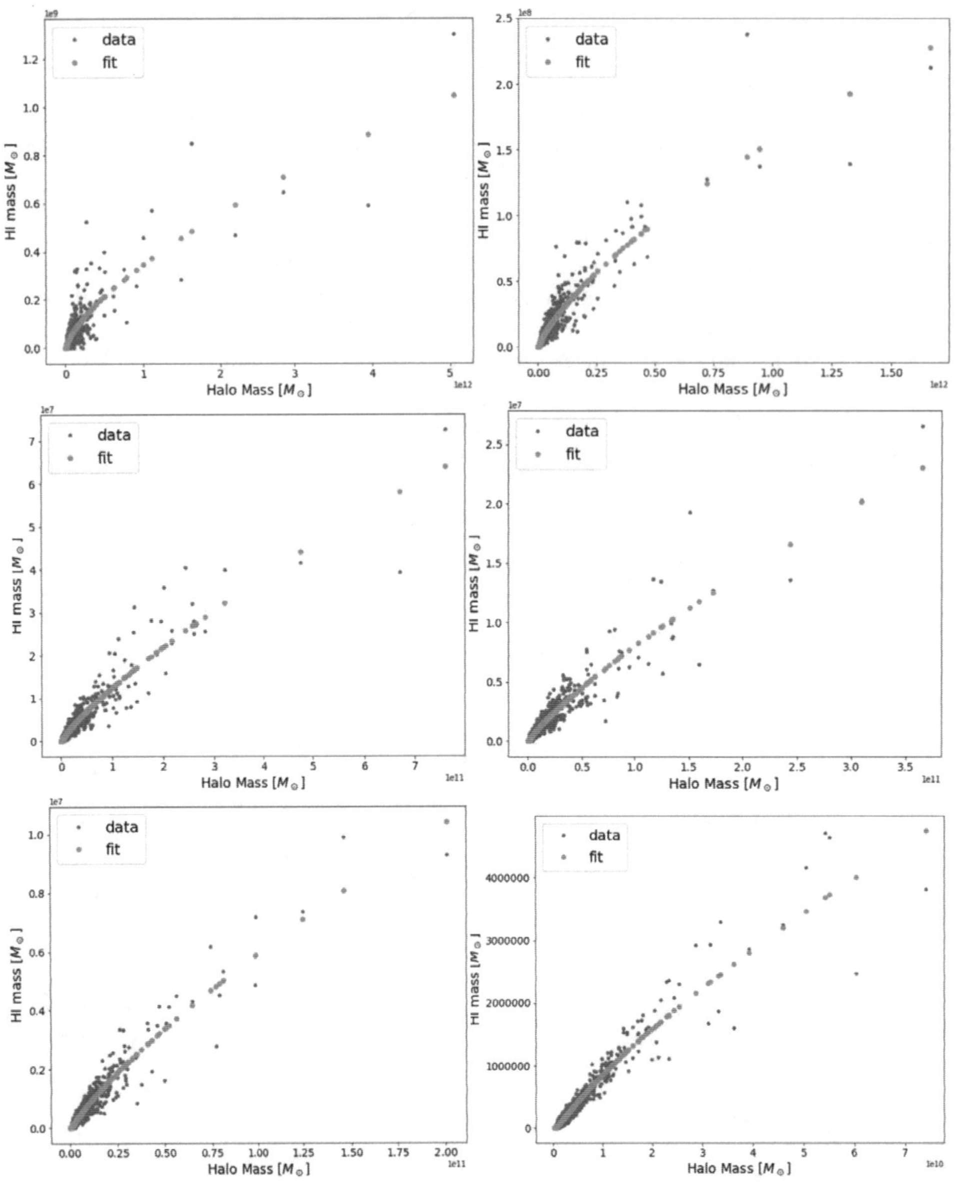

Figure. 2. plots of neutral hydrogen mass of each halo vs halo mass for both z=1 (left) and z=5(right).

For each FoF halo, the neutral hydrogen mass is calculated as well as the halo mass. The halo sample selection is based on its halo id. Halos with halo id from 0 to 300000 due to the limited running storage on Jupyter lab. Halo id is not assigned to each halo based on the mass of the halo or the neutral hydrogen abundance of the halo. Therefore, the selection is random for this analysis and should produce no bias on the final results, The plots are displayed in Figure. 2. The HI mass contained in each halo is plotted versus its halo mass for redshift z=1and z=5. HI mass and Halo mass is positively related. HI mass is monotonically increasing with the increase of halo mass and the slope is higher for low mass halos. There are two kinds of fitting functions, one is the conventional fit function for HI halo mass relation:

$$M_{HI}(M,z) = M_0 \left(\frac{M}{M_{min}}\right)^{\alpha} \exp\left(-\frac{M_{min}}{M}\right) \quad (1)$$

where M_0 indicates the overall normalization; M_{min} is the cutoff mass in this function; and α represents for the slope of the function for high mass end. Another one is mentioned in Villaescusa-Navarro's paper and it changes a little from the original function [11]:

$$M_{HI}(M,z) = M_0(\frac{M}{M_{min}})^{\alpha}\exp(-(\frac{M_{min}}{M})^{0.35}) \quad (2)$$

These two functions are both used to fit the plots for TNG50 data and the more conventional fit function works better. The final function of HI halo mass will be explicitly produced in terms of conventional parameterization function. Each of the free parameters are calculated for different redshifts based on Eq. (1). Table 1 summarizes the results of all parameters for different redshift from z=1 to z=6.

Table 1. Fitting results for HI Halo mass function.

redshift	M_0	M_{min}	α
z=1	9.30765e6	4.916576e9	0.68142
z=2	3.35369e6	4.853738e9	0.72326
z=3	5.78373e5	2.052914e9	0.79659
z=4	2.28401e5	1,221852e9	0.81020
z=5	1.84507e5	0.796866e9	0.79687
z=6	1.73344e5	1.189131e9	0.80510

These results can be used to compare with other studies using the same conventional parameterization [5, 11, 25, 27, 28]. One limitation of this investigation will be the insufficient halo number used in the HI halo mass function. Further study can calculate HI mass and halo mass of larger amount of halos to improve the precision of the results.

4. Conclusion

In summary, the halo model is a promising framework to predict the abundance and clustering of the HI in our universe. The pre hypothesis about the halo model is that almost the whole part of HI is covered within halos and galaxies. In order to prove the assumption, maps of the gas distribution and neutral hydrogen distribution from IllustrisTNG simulation box are obtained to exhibit that the HI is mainly located inside the halos and galaxies. Under that assumption, the halo model can be further studied. One of the ingredients of constructing a more precise halo model is the investigation of the HI halo mass function. The HI mass contained in each halo is being calculated and used to make the plots of HI mass versus halo mass. By fitting these HI halo mass plots for different redshifts from z=1 to z=6, free parameters in HI halo mass function can be determined. Data is collected from the latest released simulation IllustrisTNG 50 and the results can be compared with other studies using different simulation data. This study fills the gap of other studies using TNG100 and 300. The results of this paper will help to improve the halo model and learn about the neutral state hydrogen gas distribution of the cosmos. These parameters of HI halo model function can be used as input data to construct more accuracy analytical halo model.

References

[1] Battye, Richard A., Rod D. Davies, and Jochen Weller. "Neutral hydrogen surveys for high-redshift galaxy clusters and protoclusters." Monthly Notices of the Royal Astronomical Society 355.4 (2004): 1339-1347.

[2] McQuinn, Matthew, et al. "Cosmological parameter estimation using 21 cm radiation from the epoch of reionization." The Astrophysical Journal 653.2 (2006): 815.

[3] Chang, Tzu-Ching, et al. "Baryon acoustic oscillation intensity mapping of dark energy." Physical Review Letters 100.9 (2008): 091303.

[4] Loeb, Abraham, and J. Stuart B. Wyithe. "Possibility of precise measurement of the cosmological power spectrum with a dedicated survey of 21 cm emission after reionization." Physical

Review Letters 100.16 (2008): 161301.
[5] Bagla, J. S., Nishikanta Khandai, and Kanan K. Datta. "H i as a probe of the large-scale structure in the post-reionization universe." Monthly Notices of the Royal Astronomical Society 407.1 (2010): 567-580.
[6] Battye, R. A. , et al. "HI intensity mapping : a single dish approach." Monthly Notices of the Royal Astronomical Society 2(2013):1239-1256.
[7] Switzer, E. R., et al. "Determination of z~ 0.8 neutral hydrogen fluctuations using the 21 cm intensity mapping autocorrelation." Monthly Notices of the Royal Astronomical Society: Letters 434.1 (2013): L46-L50.
[8] Bull, Philip, et al. "Late-time cosmology with 21 cm intensity mapping experiments." The Astrophysical Journal 803.1 (2015): 21.
[9] Anderson, C. J., et al. "Low-amplitude clustering in low-redshift 21-cm intensity maps cross-correlated with 2dF galaxy densities." Monthly Notices of the Royal Astronomical Society 476.3 (2018): 3382-3392.
[10] Villaescusa-Navarro, Francisco, Philip Bull, and Matteo Viel. "Weighing neutrinos with cosmic neutral hydrogen." The Astrophysical Journal 814.2 (2015): 146.
[11] Villaescusa-Navarro, Francisco, et al. "Ingredients for 21 cm intensity mapping." The Astrophysical Journal 866.2 (2018): 135.
[12] Marinacci, Federico, et al. "Non-ideal magnetohydrodynamics on a moving mesh." Monthly Notices of the Royal Astronomical Society 476.2 (2018): 2476-2492.
[13] Naiman, Jill P., et al. "First results from the IllustrisTNG simulations: A tale of two elements–chemical evolution of magnesium and europium." Monthly Notices of the Royal Astronomical Society 477.1 (2018): 1206-1224.
[14] Pillepich, Annalisa, et al. "First results from the IllustrisTNG simulations: the stellar mass content of groups and clusters of galaxies." Monthly Notices of the Royal Astronomical Society 475.1 (2018): 648-675.
[15] Nelson, Dylan, et al. "First results from the IllustrisTNG simulations: the galaxy colour bimodality." Monthly Notices of the Royal Astronomical Society 475.1 (2018): 624-647.
[16] Nelson, Dylan, et al. "First results from the IllustrisTNG simulations: the galaxy colour bimodality." Monthly Notices of the Royal Astronomical Society 475.1 (2018): 624-647.
[17] Pakmor, Ruediger, Andreas Bauer, and Volker Springel. "Magnetohydrodynamics on an unstructured moving grid." Monthly Notices of the Royal Astronomical Society 418.2 (2011): 1392-1401.
[18] Pakmor, Rüdiger, and Volker Springel. "Simulations of magnetic fields in isolated disc galaxies." Monthly Notices of the Royal Astronomical Society 432.1 (2013): 176-193.
[19] Springel, Volker. "E pur si muove: Galilean-invariant cosmological hydrodynamical simulations on a moving mesh." Monthly Notices of the Royal Astronomical Society 401.2 (2010): 791-851.
[20] Davis, Marc, et al. "The evolution of large-scale structure in a universe dominated by cold dark matter." The Astrophysical Journal 292 (1985): 371-394.
[21] Cooray, Asantha, and Ravi Sheth. "Halo models of large scale structure." Physics reports 372.1 (2002): 1-129.
[22] Barnes, Luke A., and Martin G. Haehnelt. "The bias of DLAs at z~ 2.3: evidence for very strong stellar feedback in shallow potential wells." Monthly Notices of the Royal Astronomical Society 440.3 (2014): 2313-2321.
[23] Villaescusa-Navarro, Francisco, et al. "Modeling the neutral hydrogen distribution in the post-reionization Universe: intensity mapping." Journal of Cosmology and Astroparticle Physics 2014.09 (2014): 050.
[24] Padmanabhan, Hamsa, Alexandre Refregier, and Adam Amara. "A halo model for cosmological neutral hydrogen: abundances and clustering H i abundances and clustering." Monthly Notices of the Royal Astronomical Society 469.2 (2017): 2323-2334.

[25] Padmanabhan, Hamsa, and Alexandre Refregier. "Constraining a halo model for cosmological neutral hydrogen." Monthly Notices of the Royal Astronomical Society (2016): stw2706.
[26] Obuljen, Andrej, et al. "High-redshift post-reionization cosmology with 21cm intensity mapping." Journal of Cosmology and Astroparticle Physics 2018.05 (2018): 004.
[27] Pénin, Aurélie, Obinna Umeh, and Mario G. Santos. "A scale-dependent bias on linear scales: the case for H I intensity mapping at z= 1." Monthly Notices of the Royal Astronomical Society 473.4 (2018): 4297-4305.

Main Sequence and Quenching of Star Formation based on Simulations from IllustrisTNG

Xinen Zhou*

University of California, Irvine, USA

*Corresponding author's email: xinenz@uci.edu

Abstract. Star formation plays a crucial role in the field of astrophysics and cosmology. In this paper, the simulation IllustrisTNG is used to recreate the main sequence of star-formation, which explored the relation between the mass of as well as the star-formation rate (SFR). Graphs of different redshifts were investigated based on ordinary least square linear regression fitting, where the trend of star formation main sequence is seen. The slopes and y-intercepts will be discussed regarding the main sequence of star formation, which has already been proved by different researchers in the past few years. However, unlike the main sequence of star formation, the quenching process still doesn't have a proper explanation. Nevertheless, in this paper, star formation quenching was shown in the images and a possible solution to star formation quenching is proposed. Overall, these results offer a new insight toward star formation quenching and offer a guideline for future investigation in terms of star formation.

1. Introduction

Star formation is a process fundamental to galaxy evolution, which corresponds to the collapse phenomenon of warm dense matter cumulated regions containing abundant molecular clouds in interstellar space (also known as "stellar nurseries"/ "star-forming regions"). According to the context of the ΛCDM theories, the galaxies located in dark matter halos are usually considered as generated in the early stage of universe, which are created in layer as gravity boosts perturbations. The inherit structure and distribution of the dark matter halos including the surrounding cosmetic rays have been investigated mostly based on particle simulations. As for the analytical model of simulations, the dark matter is modeled as a special kind of particles with the feature that only interact with each other via the force of gravity instead of other types. Contemporarily, these simulations results have brought up a overall consensus with regard to the dark matter massively distribution. In the canonical theory of galaxy formation, the star formation from galaxies begins when the hot matter (usually the gas) emits rays/mass in terms of radiation (i.e., cools and pressure decays). In addition, the evolution of this kind of procedure is strongly affected by the accumulation of particles (which can be described as "gas") and dark matter, as well as other feedback phenomenon (e.g., the stellar winds, supernovae and active galactic nuclei). Most of the understanding of those galaxies has been established on the relation between different variables, e.g., luminosity, star formation rate, and redshift [1].

Over the last ten years, scholars have found some new relations that are germane to star formation rate (SFR). One of the studies found that star formation rate could be dictated by the stellar mass within a galaxy [2]. This relation has been verified in several observations as well as simulations, which is

Content from this work may be used under the terms of the Creative Commons Attribution 3.0 licence. Any further distribution of this work must maintain attribution to the author(s) and the title of the work, journal citation and DOI.
Published under licence by IOP Publishing Ltd

known as "star formation main sequence" (SFMS). The relationship appears for both low [3, 4] and high (z >= 1 [5]) redshift [1]. The result is also obtained up to redshift equals to 5 by spectroscopic survey [6]. Up to now, it's certain that the SFMS can be found from redshift 0 to 4 [2]. According to previous literatures, it is suggested that the SFR as a function of stellar mass (M∗) satisfies the power law saclings [7, 8]. Specifically, it has an inherit scatter parameter of about 0.2-0.3 dex for stellar galaxies of lower to medium masses [7, 8]. The observed relation indicates that before the accomplishment of star formation, the evolutions galaxy star formation are predominantly regular and smoothly declining towards the end as the mass increases to a high value. The gradual declining trend of the star formation at high stellar mass leads to the so-called "star formation quenching", which is still a puzzle that needs to be resolved.

The objective of the article is to discuss the relation between the star formation rate and the stellar mass based on the result obtained from the simulation IllustrisTNG. Later in the article, "star formation quenching" will also be discussed even though it's still an unsolved problem. The rest of the paper is organized as follows: section 2 will introduce the IllustrisTNG project and demonstrate its way of arranging data; section 3 will present the analysis of the data and image creation; eventually, a brief summary is given in Sec. 4.

2. Data & Method

The data used in this research is collected from the well-known IllustrisTNG project, which is a series of simulations with large calculation parameters (space frame and time steps), multi-cosmological magnetohydrodynamical (MHD) models of galaxy formation. Nowadays, the MHD galaxy formation simulations mainly have sub-resolution models for star formation, radiative cooling processes, stellar evolution and chemical enrichment procedures based on stellar synthesis calculations, and feedback from various sources (e.g., stellar and active galactic nuclei) [9]. The whole TNG program plans to demonstrate the physical processes that explain the galaxy formation. Specifically, it is devoted to understanding the role of galaxies evolve in the formation observations as well as present predictions for state-of-art detection approaches and offer a guideline for future.

In general, the whole IllustrisTNG project are the combination of 18 simulations. The simulations are categorized base on their physical simulation parameter ranges, i.e., the size of the window, mass resolution, and complexity of analytical models. There are three boxes called TNG50, TNG100, and TNG300, respectively. The numbers stand for the side length of the simulation box, the unit of which is Mpc. The three boxes complement each other by advocating attention to all fields of galaxy formation. The largest simulation box with side length of 300 Mpc is capable for the investigation of galaxy clustering, the analysis of rare objects (e.g., galaxy clusters), and offers the sample list of the galaxy., the amounts of rare objects in the smallest simulation volume is limited. However, its advantage goes to its mass resolution, 2 orders of magnitude lager than the larger volume TNG300 simulation. Therefore, the TNG50 volume is more appropriate as for the case of the structural features of galaxies and surrounding gas types with the convergence of multiple physical models. The middle volume simulation of the TNG program, i.e., TNG100, shows balance between these two limits.

The data in the simulation is divided into 5 sets: snapshots, group catalogs, additional files, merger trees, and supplementary data catalogs. There're 100 snapshots stored for every run. Each one corresponds to different redshifts or scale factors. Overall, a whole group catalog will be associated with each snapshot results, where the group files are split into a small number of sub-files. Following catalog will be included for each HDF5 file: Header, Group, Subhalo. However, it should be noted that the IDs of the member particles for each group/subgroup are not saved in the group catalog files. Instead, particles/cells in the snapshot files are in order based on group membership. The group used here is Subhalo, which are obtained follows the Subfind algorithm adding revised baryonic features to the subhalo entry.

In order to exhibit the relation between star formation rates and stellar mass, the field that represents these two variables is selected. On the y-axis, the field "SubhaloSFRinRad" is input, which is the sum of the individual star formation rates of all gas cells in one subhalo restricted to cells within twice the stellar half mass radius. On the x-axis, as mentioned before, should be the stellar mass. Because in the y-axis, the star formation rate is within twice the stealer half mass radius. Similarly, the stellar mass being chosen is the sum of masses of all particles/cells within twice the stellar half mass radius. In this way, one would expect to see the trend of the change of star formation rate.

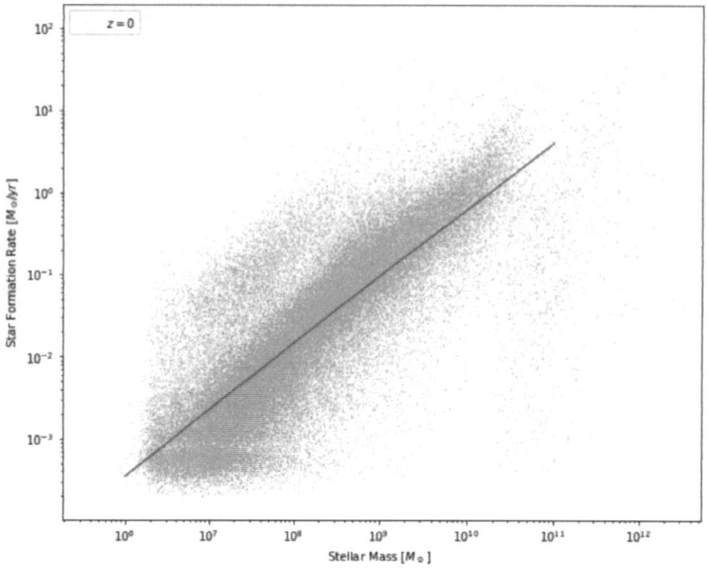

Figure 1. The star formation rate as a function of stellar mass for z = 0. The scatter points denote for the simulation results while the blue lines represent for the linear fitting results.

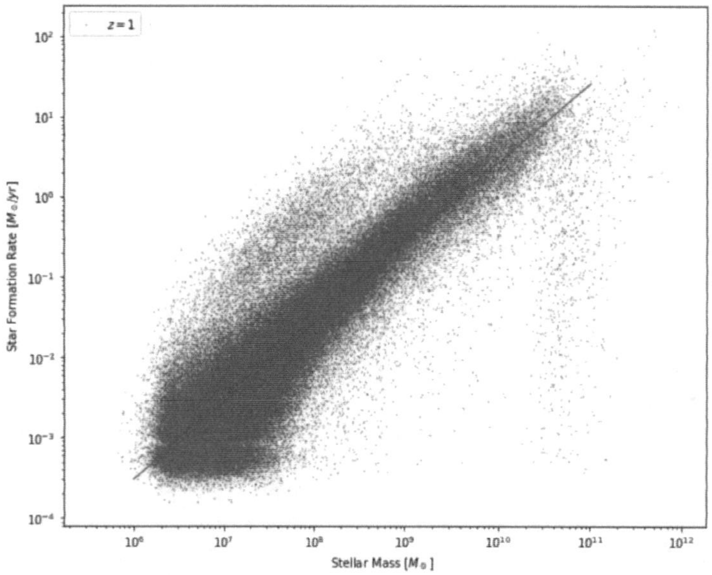

Figure 2. The star formation rate as a function of stellar mass for z = 1. The scatter points denote for the simulation results while the blue lines represent for the linear fitting results.

3. Results & Discussions

The star formation main sequence (SFMS), which shows the relation between the star formation rate and stellar mass, are illustrated in Figs. 1-4 produced based on the data of IllustrisTNG. There are some trends that can be seen from the figures. First, as the redshifts increase, the shape of the dots is more shaped. Especially when z = 4, it can be seen that the angle formed on the top right corner. As the redshift goes to z = 0 and 1, the angle becomes blurred. Additionally, the images of smaller redshift have more dispersive dots on top whereas the image with greater redshift has a shape that almost looks like a solid triangle. Second, the maximum star formation rate is larger when the redshift is greater. As the stellar mass increases to a very large number, the star formation gradually comes to an end. In the image of z = 4, the maximum value is above 100. However, when z = 0, the maximum value is less than 100. Research has found that galaxies at z = 0 are generally less star - forming and more extended than galaxies at z = 3 [10], which corresponds to the image produced here. Only a small fraction of the galaxies at z = 3 are in the 'com-pact' region of the plane, while at z = 0 none are. Similarly, in the figure created, it can also be seen that the star formation is more extended when z = 0 than z = 4. Generally, the trend indicates that the smaller the redshift is, the more extended the star formation will be. Nevertheless, the SFMS may be significantly lower than the observational constraints at intermediate redshifts of z = 1 and z = 2.

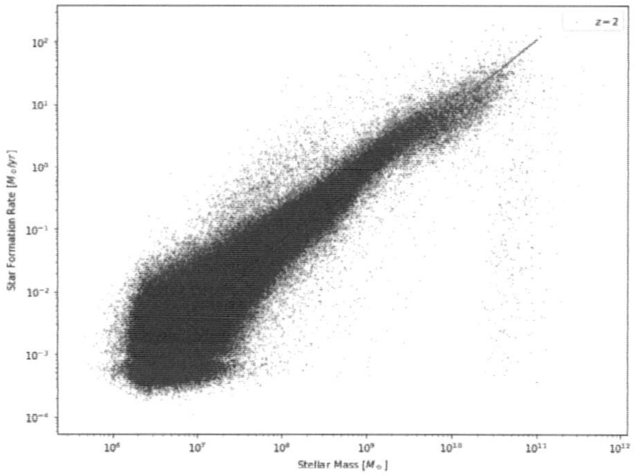

Figure 3. The star formation rate as a function of stellar mass for z = 2. The scatter points denote for the simulation results while the blue lines represent for the linear fitting results.

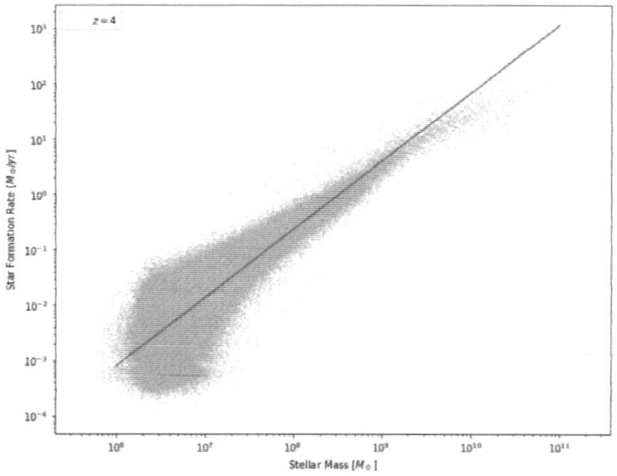

Figure 4. The star formation rate as a function of stellar mass for z = 3. The scatter points denote for the simulation results while the blue lines represent for the linear fitting results.

Moreover, in order to study the trend of the images more clearly, linear regression is applied with all the images. The python function I used is numpy.polyfit. As shown in Table.1, when the redshift z is equal or larger than z = 2, the slope is larger than 1. However, when z = 0, the slope becomes 0.8 which is smaller than 1. It again corroborates the facts that galaxies are less star - forming when z = 0. However, the y-intercept of the image is larger when the redshift is smaller. It's reasonable since the star - formation rate is flatter when the redshift is smaller.

Another thing that is worth noticing might be the star formation quenching shown in Figures. 1-4. Star formation quenching is marked by the phenomenon that the formation of stars gradually disappears as the stellar mass increases to a very high value. From the images, one can see the galaxy reach the star formation quenching earlier if the redshift is smaller. Whereas, the causality of star formation quenching remains unclearly to the contemporary astrophysics community. Recently, an international team led by Professor Faber at UC Santa Cruz has introduced a novel mechanism to explain the observations from lots of detection approaches regarding to galaxy property, supermassive black holes, and the quenching of star formation. In their model, the formation of stars relates to how spread out they are within a galaxy. The reason entails the central blackhole in the galaxy. Since the central blackhole is growing over time, a large galaxy with a great number of stars must grow fast enough so that the central black-hole wouldn't quench the star formation that early. This actually corresponds to the fact that galaxies with smaller radii quench at lower masses (11. Zhu Chen et al 2020).

Table 1. The the slopes and y-intercepts of Figures 1-4

	z = 0	z = 1	z = 2	z = 4
slope	0.81	0.99	1.08	1.23
y-intercept	-19.15	-21.72	-22.73	-24.20

4. Conclusion

In summary, star formation main sequence and star formation quenching are recovered in IllustrisTNG simulation in this paper. These two phenomena could lead to dark matter and blackholes, which are two of the most intriguing problems in modern physics. Based on the analysis, these results corroborate with some of the ongoing research that investigates the relation between stellar mass, star-formation rate, and redshifts. However, the explanation of star formation main sequence is rather obvious while star formation quenching, despite its obvious existence, still remains a puzzle to the whole community. Nevertheless, these results still shed light on star formation investigation and pave a path for further exploration of the attribution of star formation quenching.

References

[1] M. Sparre and O. E. Hartoog 2014 The Astrophysical Journal, Volume 785, Issue 2, article id. 150, 12 pp.
[2] P.Popesso and L.Morselli, Monthly notices of the Royal Astronomical Society: 3213-3226.
[3] J. Brinchmann and S. Charlot 2004 Monthly notices of the Royal Astronomical Society: arXiv:astro-ph/0311060.
[4] Salim, Samir and Rich, R. Michael 2007 The Astrophysical Journal Supplement Series, Vol-ume 173, Issue 2, pp. 267-292.
[5] E. Daddi and M. Dickinson 2007 Astrophysics.J.670:156-172.
[6] L.A.M. Tasca and O. Le Fevre 2014 A&A 581, A54.
[7] K. Whitaker and M. Franx 2015 Astrophysical Journal Letters.
[8] J. S. Speagle 1,2, C. L. Steinhardt 2014 arXiv:1405.2041.
[9] M.Vogelsberger and S.Genel 2014 Monthly Notices of the Royal Astronomical Society, Volume 444, Issue 2, Pages 1518–1547.
[10] S.Genel and M.Vogelsberger 2014 MNRAS 2014 445 (2): 175-200.

Extrasolar Planet Candidates Identified by Single Transit from TESS

Zhixin Wang*

Princeton International School of Mathematics and Science, Princeton, USA

*Corresponding author email: andy.wa2@protonmail.com

Abstract. The discovery of planets around stars other than Sun was a grand milestone to understand origin and evolution of life in Universe. 2019 Nobel Prize for Physics was awarded to the first discovery of exoplanet around Sun-like stars, Peg 51b, in 1995. With development of 25 years, the current number of confirmed exoplanets have reached 4,000. The population of exoplanets presents great diversity and complexity beyond the planets in Solar system, severely challenging our understanding on the origin and evolution of planets. Moreover, clues of extrasolar life have been partially disclosed by detection of exoplanets in habitable zone with temporal surface temperature. In recent years, several promising surveys of exoplanets have been carried out to search for more exoplanets, compiling more complete sample of exoplanets and detecting Earth-like exoplanets. Transiting Exoplanet Survey Satellite (TESS) launched in 2018, is the most import facility for this purpose. It monitors the brightness of nearby stars in all sky with high cadences, searching for exoplanetary transit events. The survey is sensitive to detect terrestrial exoplanets. In this work, we reanalyzed the high quality light curves of TESS with algorithm developed by ourselves and visual examination. Our work was designed complementary to the automatic pipeline developed by TESS science team. TESS pipeline is not optimal for long-period exoplanets, for which only one transit event occurring in the observation sectors of 27.4 days, appearing as single transit single. Those long-period planets are more likely to be habitable with liquid water on their surface in relative large orbits. We took efforts to recover exoplanets with single transit in TESS. First, we identified candidates with significant narrow dimming in light curves uniformly by our automatic program. The singles were further examined visually and fitted by transit profile, selecting transit events. More precise modeling was carried out by MCMC analysis for transit exoplanet candidates. As a preliminary attempt, we only processed the light curves in sector 1 only. Full TESS dataset in all 26 sectors will be studied in future. In summary, we detected 25 exoplanet candidates in publicly accessible light curves in TESS sector 1. Among them, 14 candidates are known exoplanets, and 6 are candidates identified by TESS pipeline in catalog of Tess Objects of Interest (TOI). 5 exoplanet candidates with single transit signal are newly identified by their profiles, and the properties of these system were inferred in this work. The orbital parameters and physical properties derived by us are consistent with those in literature for all the 14 known exoplanets. 2 among 6 of exoplanet candidates in TOI catalog are single transit objects, our results are consistent with values reported in TOI catalog. For the new exoplanet candidates, we report their inferred orbits and physical

Content from this work may be used under the terms of the Creative Commons Attribution 3.0 licence. Any further distribution of this work must maintain attribution to the author(s) and the title of the work, journal citation and DOI.
Published under licence by IOP Publishing Ltd

sizes. We plan to implement our analysis for all light curves in 26 sectors of TESS, Over 100 exoplanet candidates are expected to be recovered, significantly contributing to the current population of 4,000 known exoplanets.

1. Introduction

1.1. Discovery of exoplanets

Founded on the Copernican Principle, that Earth and humans do not occupy a privileged position in the Universe, it has long been hypothesized that lives outside of Earth not only exist, but are also widespread - due to the vast size and relatively long lifespan of the Universe. Some astronomers searched for planets outside the solar system, or exoplanets, that could sustain life. The direct observation of those planets were extremely difficult due to their low brightness and small angular separation from the host star. [12] Therefore, alternative methods that detect the perturbations of the host star's luminosity or motion were suggested. The first exoplanet orbiting around a solar-type star, 51 Pegasi b, was discovered in 1995 by Michel Mayor and Didier Queloz using the radial velocity method, and this research was awarded with the nobel prize for physics in 2019. [7, 10] In 1999, two research teams discovered an exoplanet orbiting HD 209458, a G0V star in constellation Pegasus, using the radial velocity method. Shortly after that, another team used photometry to measure the mass and size of the planet by analyzing the light curve when the planet transits the host star, and the star HD 209458 becomes the first discovered transiting exoplanet. Until today, more than 4800 exoplanets have been confirmed, and the two main methods - transit method and radial velocity method - accounts for more than 90% of the discoveries. It is estimated that at least 5–10% of solar-type stars in the solar neighbourhood harbour massive planets, which, if generalized to the stellar population of our galaxy, would result in billions of those planets. The physical(mass, radius. . .) and orbital(eccentricity, period. . .) properties of exoplanets exhibit great diversity that far exceeds what we observe in the eight planets in the solar system. In fact, solar system planets appear to be outliers in the mass-period distribution. [13]

1.2. Habitable Planet

Extrapolating from the conditions on Earth, where research has shown that water is essential for the carbon-based life emerged on it, having environments in which liquid water is stable is considered necessary for habitable planets. Given sufficient atmospheric pressure, the state of water on the plantatery surface is determined by its temperature, that is, the amount of energy absorbed and emitted by the planet. The habitable zone around a star is thus defined as the range of orbits with suitable temperature for water to remain in liquid state. The surface temperature can be estimated by calculating the exoplanet's equilibrium temperature - the temperature where the amount of stellar flux received balances with the amount of energy radiated and reemitted by the planet. Derivations of the equilibrium temperature and habitable zone are given in section 2.3. The Planetary Habitability Laboratory created the Habitable Exoplanets Catalog(http://phl.upr.edu/projects/habitable-exoplanets-catalog) using three criterias: i) the definition of habitable zones in [5] ii) having a host star with spectral type F, G, K M, and iii) planet radius between 0.5 to 2.5 Earth radii or a minimum mass between 0.1 to 10 Earth masses. The catalog now includes 60 potentially habitable exoplanets, among which 36 are Super-Earths or Mini-Neptunes, 26 are Earth-size, and 1 is Mars-size.

1.3. Method of Detection

Radial velocity and transit are two main methods used in exoplanet surveys, both of which rely on observing the planet's effects on its host star.

1.3.1. Radial Velocity Method. In a two-body system of a star and a planet, both objects orbit around the barycenter of the system. This produces a periodic pattern in the star's radial velocity, where the amplitude corresponds to the star's velocity around the barycenter and the period corresponds to the period of the star's motion. The radial velocity of a star can be accurately measured by using high-resolution spectroscopy to detect its doppler redshifts, since that redshift gives the star's projected motion along the line of sight.

$$z = r(t)\sin(i)\sin(\omega + v) \tag{1}$$

The star's position along the line of sight can be derived using trigonometry, where r(t) is the distance between the star and the barycenter(i.e. A focus of the elliptical orbit), i is the inclination angle, ω is the argument of pericentre, and v is the true anomaly. In polar coordinates, r is given by

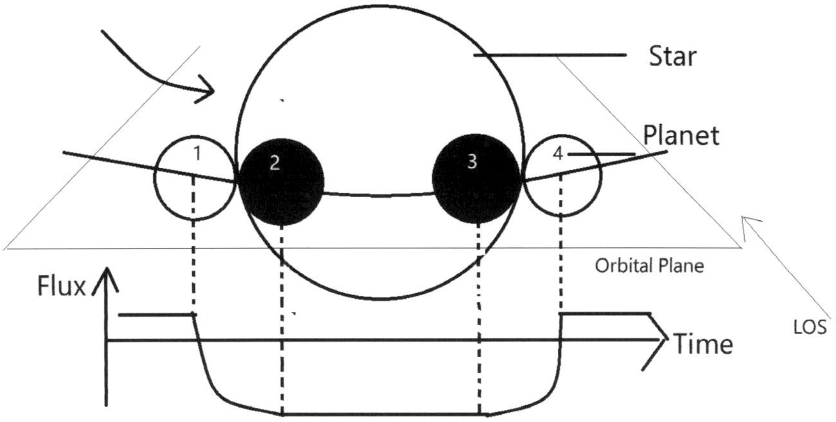

Figure 1. A disc-crossing transit of a planet. The stellar flux drops from 1 to 2, then remains a constant from 2 to 3, and finally returns to normal flux level from 3 to 4.

$$r = \frac{a(1-e^2)}{1+e\cos v} \tag{2}$$

Taking the derivative of z, the radial velocity of the star is

$$v = \frac{dz}{dt} = \sin i(\frac{dr}{dt}\sin(w+v) + \frac{dv}{dt} r\cos(w+v)) \tag{3}$$

Substitute r and $\frac{dr}{dt}$, we have

$$v = \frac{a(1-e^2)\sin i}{(1+e\cos v)^2}\frac{dv}{dt}(e\cos w + \cos(w+v)) \tag{4}$$

Some algebraic substitutions lead to

$$v = \frac{2\pi}{P}\frac{a_* \sin i}{\sqrt{1-e^2}}(e\cos w + \cos(w+v)) \tag{5}$$

where a_* is the semi-major axis of the star's orbit around the barycenter. [13]

1.3.2. Transit Method. As shown in Figure 1, if the line of sight aligns with the orbital plane of an exoplanet, an eclipse will occur when the planet passes in front of the star and obscuring part of it, which leads to a drop in the star's brightness. Therefore, by continuously monitoring the brightness of the candidate stars, transit events can be extracted from the light curves and the planet's radius and orbits can be estimated. When combined with the mass measurements from the radial velocity method, the planet's density, and thus composition, can be estimated.

Shown in Figure 2 is a simple geometric model of a transit event. The star and planet were modeled as two overlapping flat disks, as seen from the observer's perspective. If we assume that the star has uniform luminosity, then the proportion of obstructed flux equals the area obstructed by the eclipsing planet. In the figure, r_* is the stellar radius, r_p is the planetary radius, d is the planet-star separation, p and z are these quantities in the unit of stellar radius. [6] The law of cosine shows that

$$\begin{cases} K_1 = \cos^{-1}\left(\frac{z^2+1-p^2}{2z}\right) \\ K_0 = \cos^{-1}\left(\frac{z^2+p^2-1}{2pz}\right) \end{cases} \tag{6}$$

Then, with simple geometry we find that the percentage of obstructed flux is

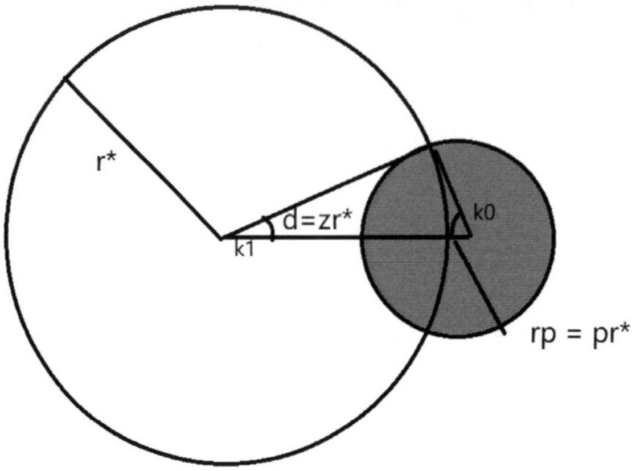

Figure 2. Geometric model of transit

$$\lambda(p,z) = \frac{1}{\pi}(p^2K_0 + K_1) - \sqrt{\frac{4z^2-(1+z^2-p^2)^2}{4}} \tag{7}$$

After considering the three other possible scenarios with different planet-star separation, we have

$$\lambda\,(p,z) = \begin{cases} 0, & (1+p<z) \\ \frac{1}{\pi}(p^2K_0 + K_1) - \sqrt{\frac{4z^2-(1+z^2-p^2)}{4}}, & \text{(intersected)} \\ p^2, & (z \le 1-p). \\ 1, & (z \le p-1) \end{cases} \tag{8}$$

In the transit model used in our research, we also took limb darkening - the variation in luminosity across the star's disk - into account by applying the non-linear limb darkening law from [3].

1.4. TESS and Kepler Survey

TESS and Kepler were two space-based missions dedicated to exoplanet detection via the transit method. The Kepler Space Telescope was launched on 2009 March 6 into an Earth-trailing heliocentric orbit. During its prime mission, Kepler focuses on one single 115 square degrees field of view (FOV), monitoring around 150,000 stars with brightness down to 16 magnitude. This mission is expected to find Earth-size and smaller planets at the habitable zone of solar-like stars. During its 9-years service, Kepler observed 530,506 stars and discovered 2,662 exoplanets, 34 of which were potentially habitable. [1] Following the legacies of Kepler, NASA launched the new Explorer Program, Transiting Exoplanet Survey Satellite (TESS), in 2018. TESS performs an all-sky survey of transiting exoplanets. The TESS cameras cover a FOV of 2300 sq degrees, twenty times larger than Kepler. It also follows a different observation strategy: instead of constantly monitoring a small sky region, TESS observes the entire sky in partially overlapping sectors, each for two orbits (or 27.4 days). Each hemisphere is tiled by 13 overlapping sectors, and 350 days of coverage are available at the Continuous Viewing Zones near the ecliptic

poles. TESS focuses on stars that were brighter and closer than Kepler targets, with $T_{mag} < 12$ or $T_{mag} < 13$ and $T_{eff} < 5500$. Therefore, follow-up observations such as radial-velocity measurements were more feasible for TESS targets. Until today, TESS has confirmed 154 exoplanets and 4471 candidates.

The observing strategy of TESS, as shown in Figure 3, where the blue regions only undergo single-sector observation, means that most long-period planets would be detected as single transits, if not completely absent, in the 27.4 days observation.

Baseline (days)	Sky Coverage (deg²)	Sky Coverage (%)
0	6023	14.6
27.4	25989	63.0
54.8	6270	15.2
82.2	1238	3.0
109.6	231	0.56
137.0 to 301.4	578	1.4 (total)
328.8	215	0.52
356.2	701	1.7

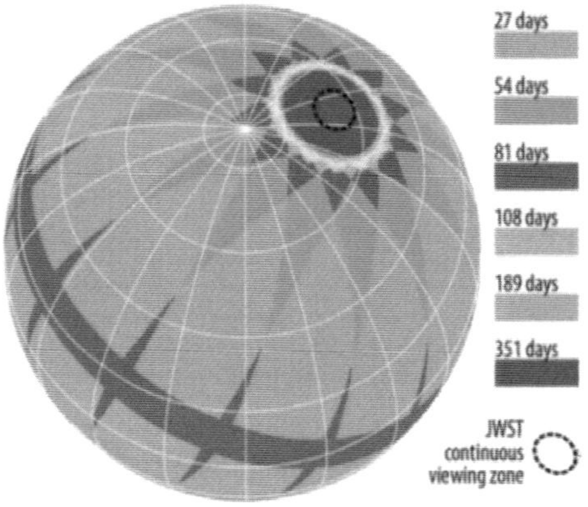

Figure 3. TESS observation baselines and sky coverage

2. Basic Theories

2.1. Keplerian orbits in space

The motion of one planet around its host star can be approximated by a two-body motion, where two point masses move only under gravitational force. Two point masses m_1 and m_2 were located at positions $\mathbf{r}_1$ and $\mathbf{r}_2$. Vector

$$\mathbf{r} = \mathbf{r}2 - \mathbf{r}1 \tag{9}$$

Is the relative position of m2 with respect to m1, while r is the length of r. Newton's second law and the law of universal gravitation gives

$$\mathbf{F}_1 = m_1\ddot{\mathbf{r}}_1 = G\frac{m_1 m_2}{r^3}\mathbf{r} \tag{10}$$

$$\mathbf{F}_2 = m_2\ddot{\mathbf{r}}_1 = -G\frac{m_1 m_2}{r^3}\mathbf{r} \tag{11}$$

Next, let's consider the relative motion of m2 with respect to m1

$$\ddot{\mathbf{r}} = \ddot{\mathbf{r}}_1 - \ddot{\mathbf{r}}_2 = -(G\frac{m_1 m_2}{r^3})\mathbf{r} \tag{12}$$

By solving **r** in this equation of relative motion, we can find the relative motion of m_2 with respect to m_1 Taking the vector product of r with both sides of equation x,

$$\mathbf{r} \times \ddot{\mathbf{r}} = 0 \tag{13}$$

Since that vector product also follows the product rule, integrating both sides of Equation 13 gets,

$$\mathbf{r} \times \dot{\mathbf{r}} = \mathrm{h} \tag{14}$$

where **h** is a constant vector. Therefore, the position and velocity of the relative motion were always on a plane perpendicular to **h**, known as the orbit plane. Next, we set up a polar coordinate in the orbit plane with its origin centered on A and a reference line of $\theta = 0$. Let $\hat{\mathbf{r}}$ and $\hat{\boldsymbol{\theta}}$ be unit vectors parallel and perpendicular to **r**

Then, in polar coordinates,

$$\mathbf{r} = r\hat{\mathbf{r}} \tag{15}$$

$$\dot{\mathbf{r}} = \dot{r}\hat{\mathbf{r}} + r\dot{\theta}\hat{\boldsymbol{\theta}} \tag{16}$$

Also,

$$\dot{\mathbf{r}} = \dot{\theta}\hat{\boldsymbol{\theta}} \tag{17}$$

Therefore,

$$\ddot{\mathbf{r}} = (\ddot{r} - r\dot{\theta}^2)\hat{\mathbf{r}} + \frac{1}{r}\frac{d}{dt}(r^2\dot{\theta})\hat{\boldsymbol{\theta}} \tag{18}$$

Substituting $\dot{\mathbf{r}}$ into Equation 14,

$$\mathbf{h} = \mathbf{r} \times \dot{r}\hat{\mathbf{r}} + r\dot{\theta}\hat{\boldsymbol{\theta}} = \hat{\mathbf{r}} + \hat{\boldsymbol{\theta}}(r^2\dot{\theta}) \tag{19}$$

Then, $h = r^2\dot{\theta}$ is a constant, and h points perpendicular to the orbit plane. Consider the area swept by the segment connecting m1 and m2 in a δt

$$\delta A \approx \frac{1}{2}r(r + \delta r)\sin\delta\theta \approx \frac{1}{2}r^2\delta\theta \tag{20}$$

And dividing both sides by δt,

$$\frac{\mathrm{dA}}{\mathrm{d}t} \approx \frac{1}{2}r^2\dot{\theta} = \frac{1}{2}h \tag{21}$$

Therefore, the segment sweeps out equal area in equal time, which is Kepler's second law of planetary motion.

Next, we want to derive the shape of the planet's orbit, that is, r as a function of θ in the polar coordinate. To convert Equation 18 to a scalar equation of r, we substitute the expression for $\ddot{r}$.

$$(\ddot{r} - r\dot{\theta}^2)\ \hat{\mathbf{r}} + \frac{1}{r}\frac{d}{dt}(r^2\dot{\theta})\hat{\theta} = -(G\frac{m_1 + m_2}{r^3})\hat{\mathbf{r}} \tag{22}$$

Since that $\hat{\mathbf{r}}$ and $\hat{\theta}$ are perpendicular, we only take the $\hat{\mathbf{r}}$ components,

$$\ddot{r} - r\dot{\theta}^2 = -G\frac{m_1 + m_2}{r^2} \tag{23}$$

The solution to this differential equation is

$$r = \frac{p}{1 + e\cos(\theta - \varpi)}, \tag{24}$$

Where $p = \frac{h^2}{G(m_1+m_2)}$ In polar coordinates, this is the equation of a conic section. The orbit of two-body motion is closed, and thus it is elliptical, or in some special cases, circular. (Kepler's first law of planetary motion) In an ellipse, we have

$$r = \frac{a(1-e^2)}{1 + e\cos(\theta - \varpi)}, \tag{25}$$

where a is the semi-major axis, e is the eccentricity, and ϖ is the longitude of pericentre. Additionally, consider the area swept by the two body motion during one orbital period T:

$$\pi ab = \frac{dA}{dt}T = \frac{1}{2}hT \tag{26}$$

$$\pi^2 a^2 b^2 = \pi^2 a^4 (1-e^2) = \frac{1}{4}h^2 T^2 = \frac{1}{4}a(1-e^2)G(m_1+m_2)T^2 \tag{27}$$

$$T^2 = \frac{4\pi^2}{G(m_1+m_2)}a^3 \tag{28}$$

$\frac{4\pi^2}{G(m_1+m_2)}$ is a constant, and thus T^2 is proportional to a^3.(Kepler's third law of planetary motion)

2.2. Transit Probability

Assuming that planets are randomly oriented and the orbit is circular, the probability that a planet on a circular orbit to be favourably aligned for a transit is

$$P_t \approx \frac{R_*}{a} = 0.005(\frac{R_*}{Rsun})(\frac{a}{1au})^{-1}, \tag{29}$$

in solar radius and Earth orbit. The transit probability is equivalent to the proportion of area on the celestial sphere centered on the star that could be under the planet's shadow during one orbital period, since those were and were the only regions where transits can be observed. This area has a ribbon shape on the celestial sphere, as shown in fig x. The transit probability is [2]

$$p = \frac{2\pi(a+b)m}{4\pi(a+b)^2} = \frac{m}{2(a+b)} \tag{30}$$

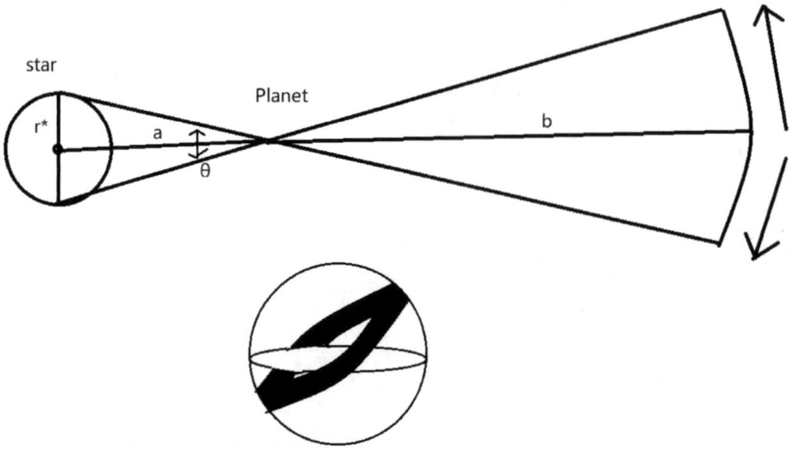

Figure 4. A schematic of the regions on the celestial sphere (centered at the planet) where transits are detectable

Similarity gives

$$p = \frac{2r_* b / a}{2(a+b)} \tag{31}$$

As b approaches ∞,

$$P \approx \frac{r_*}{a} \tag{32}$$

2.3. Habitable Zone

As explained in section 1.2, planets that lie at a suitable star-planet separation have a moderate temperature, and thus could be conductive to life. Based on the conditions on Earth, we want to search for planets with a temperate environment where liquid water exists. The surface temperature of planets is determined by several factors, including the luminosity of the host star, the star-planet separation, and the albedo of the planet. Assuming that the star is a black body that emits radiation isotropically, from Stefan-Bolzmann law, the total power radiated by the star is

$$L = 4\pi R_s^2 \sigma T^4 \tag{33}$$

The flux per area received by the planet is

$$F_{rec} = \frac{L}{4\pi a^2} \tag{34}$$

When the planet is at an equilibrium temperature, the total flux absorbed is equal to the total flux emitted

$$F_{abs} = F_{emt} \tag{35}$$

A fraction of the solar radiance is reflected by the planet, according to bond albedo A_B

$$F_{abs} = (1 - A_B) F_{rec} = (1 - A_B) \frac{L}{4\pi a^2} \tag{36}$$

Also assuming that the planet is a blackbody, the power emitted is

$$F_{emt} = \sigma T_{eq}^4 \tag{37}$$

Solving the equation, we have

$$T_{eq} = T\sqrt{\frac{R}{2a}}(1 - A_B)^{\frac{1}{4}} \tag{38}$$

The equilibrium temperature may deviate from the true surface temperature due to several causes. For planets without an atmosphere, poor heat transport and retention can lead to significant temperature variation across the surface that deviates from the mean equilibrium temperature. Also, certain internal heating mechanisms, such as tidal heating or geothermal energy, could heat up the planet above the equilibrium temperature.

3. Simulation for Single and Multiple Transits

3.1. Simulation for Occurrence Rate

The chance of observing one or at least one transit event within the one-sector period (27.4 days) in TESS for a given star can be expressed as the product of three probabilities: $P_p(x)$, the occurrence rate of type x planet, $P_t(x)$, the occurrence rate of a transit event for randomly oriented planet x, and P_d, the detection rate of such transit event within the time window. Using data from the Kepler Mission, [4] estimated the percentage of stars that have one or more of a certain type of planet with a certain orbital period. Specifically, the planets are classified into Giants, Large Neptunes, Small Neptunes, Super earths, and Earths; the periods are divided into 11 different bins.

Assuming that the transit epoch for a planet with period P is random, the probability that a single transit event is detected within the 27.4 days long sector is

$$P_d = \begin{cases} 0, & (P < 0.5t) \\ P/t, & (0.5t < P < t) \\ t/p, & (P > t) \end{cases} \tag{39}$$

And the probability that two or more transit events are detected is

$$P_d = \begin{cases} 1, & (P < 0.5t) \\ 1 - P/t, & (0.5t < P < t) \\ 0, & (P > t) \end{cases} \tag{40}$$

Figure 5 shows the results of our simulation, the probability that transit events of any type of planet can be detected in one TESS sector observation assuming that the host star is similar to the sun. The full result for each type of planet can be found in the appendix. We found that for any planet type, those with longer orbital periods (¿17 days bin) tend to be single-transit planets.

3.2. Importance of Single Transit Candidates from TESS

As we have discussed in section 2.4, one necessary condition for planet habitability is adequate temperature such that water exists in liquid form. Under 1 atm (1.01325e+5 Pa) atmospheric pressure, the corresponding temperature range is 273.15 K - 373.15 K. Consider a two-body system of a planet and its host star, where the star has one solar radius and mass. Then, as shown in Figure 6 is the possible circumstellar habitable zone for different stellar temperatures (FGK), and this distance ranges from 0.2 au for the coolest K stars to 1.7 au for the hottest F-stars, which corresponds to a period range of 32.6 to 809.0 days. This long period indicates that most habitable planets would appear as single-transit targets during the 27.4 days observation of one TESS sector. Therefore, inferring orbital periods and other planetary parameters from single-transit lightcurve data is crucial for discovering habitable planets in sky regions with TESS single sector coverage.

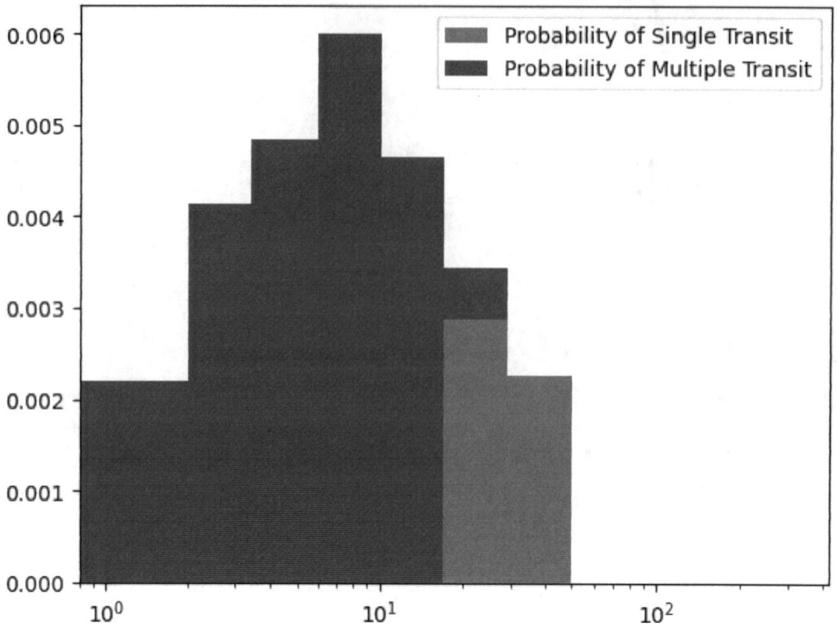

Figure 5. Probability of single and multiple transit events being detected for any type of planet; x-axis: Period, y-axis: Probability.

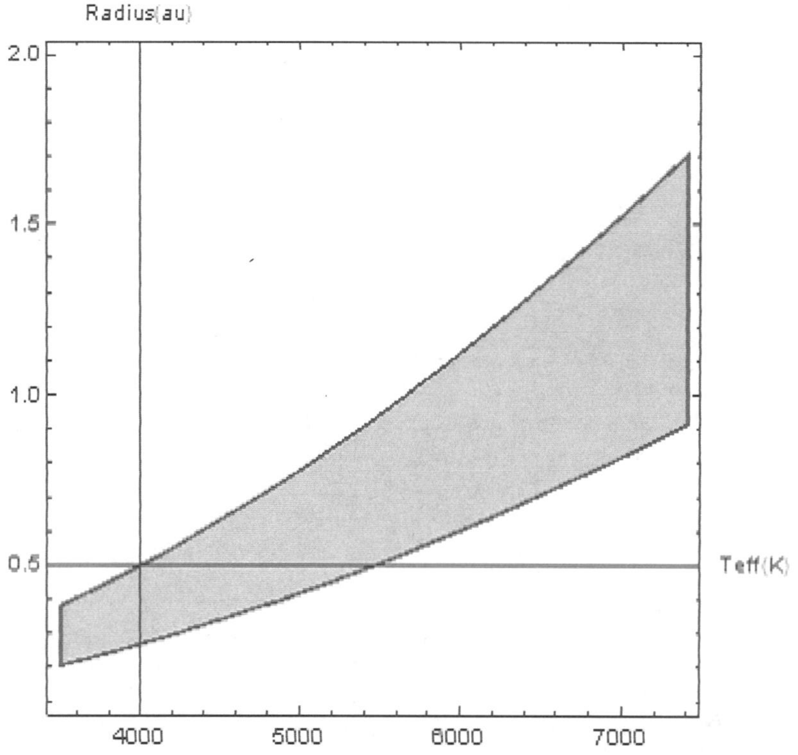

Figure 6. Range of habitable zone around stars with different effective temperature

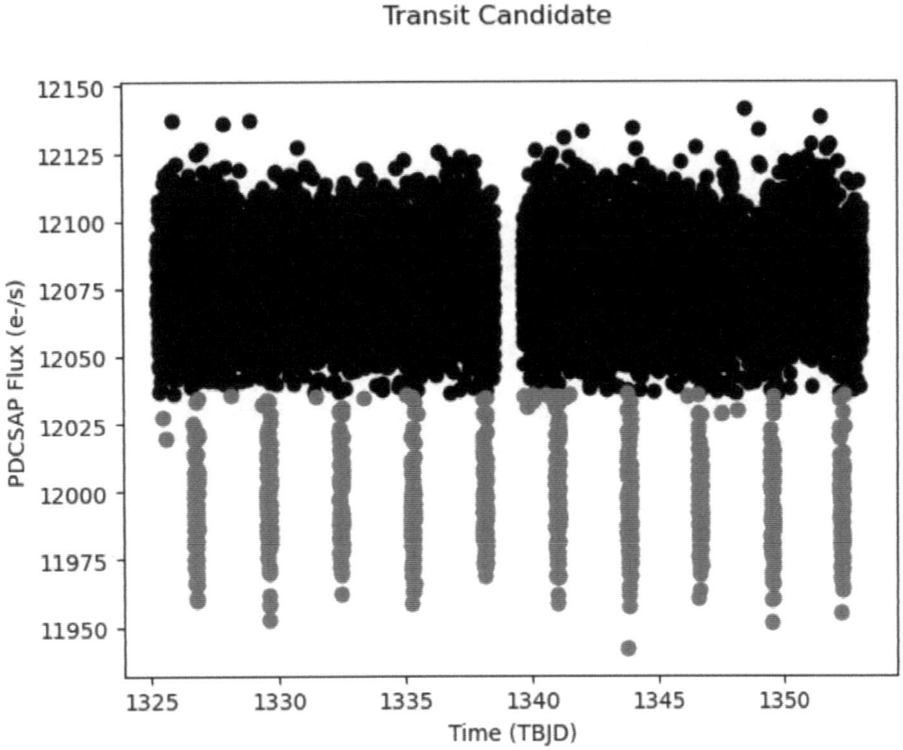

Figure 7. Preprocessed lightcurve of WASP 100b. The outliers were labeled in red, our algorithms identified 10 transit event candidates in this target.

4. Searching for Exoplanet Candidates in TESS archive

4.1. Algorithm for Transit Search

We retrieved data from the TESS catalog to analyze all the 15,888 light curves obtained in the 27.4 days observation on sector 1, centered at 324.6, -33.2. (J2000) We use the TESS light curve of WASP 100b(TIC 38846515), a gas giant orbiting around an F-type star discovered in 2014 and observed in TESS sector 1 by two-minute cadence, to illustrate our method. We first read in the light curve file, including the TBJD time, flux, flux error, and bitmask quality flags. During preprocessing, only data points with reliable measurements and without any quality flags were kept. Next, we want to detect transit events by identifying outliers, or points that are lower than the mean flux level by more than 2.5 times the standard deviation(2.5σ). Figure 7 shows one example of a preprocessed light curve, with the outliers marked in red. Depending on inclination, the duration of a single transit event could range from tens of minutes to hours. Our search would thus focus on finding a consecutive set of outliers in the light curve. All the window of outliers satisfying the following conditions were labeled as transit candidate:

- Consists of consecutive outlier points that is longer than 30 minutes (or 15 points)
- Have a transit depth $\frac{F_{notransit} - Ftransit}{F_{notransit}} = (\frac{R_p}{R_s})^2$ lower than 5% (0.05)

We apply the second cutoff because planets larger than that radius may be massive enough (¿ 13 Mj) to fuse deuterium and thus being brown dwarfs instead of planets. Then, all the transit event candidates were closely examined to determine if it's a planet or a false positive - since that other event such as stellar pulsation or eclipsing binary could also lead to a drop in the star's brightness but were nevertheless distinguishable from transit events.

4.2. Transit Candidates in Result

The algorithm described in section 4.1 yields 160 transit event candidates from 27 unique sources, seven of which were single transit targets. Our results from sector 1 are compared to the Tess Objects of Interest (TOI) catalog [8], a list of exoplanet candidates identified by the TESS Science Office. (TSO) The TOI catalog is produced by a similar algorithm: first, the two data processing pipelines searched for periodic flux decreases in the signal, or Threshold Crossing Events (TCE). Each TCE would be passed to the data validation module where a limb-darkened transit model is fitted and several diagnostic tests were calculated. The TCEs produced by processes other than transits were excluded in the "triage" process, and finally, individual and group vetters will determine the classification for the TCE. Seven unique sources from our catalog were also present in the TOI catalog. Notably, only one single transit candidate events that we discovered are included in the TOI catalog, because the pipelines used in finding TCEs only search for targets with at least two transit signals. Results from section 3.1 and 3.2 has shown that planets with periods above 17 days usually appear as single transit events during the 27.4 days window of one TESS sector observation, and most habitable planets falls within this range.

5. Parameter Fitting

5.1. Modelling the transit lightcurve

We used the GeePea module for gaussian process and Infer module for MCMC during lightcurve fitting. The lightcurve signal from one planet transiting a host star consists of four components:

- Transit signal: the dimming of the host star resulting from the transit event. The transit signal is described by 11 parameters - transit epoch T_0, period P, stellar mass M_{star} (used to derive semi-major axis a), planetary mass M_{planet}, stellar radius R_{star}, ratio of planetary radius to stellar radius p, orbital inclination i, and limb darkening coefficients c1, c2, c3,c4.
- Underlying signal: the signal from the host star, out of transit. It is described by 2 parameters: flux level out of transit F_{oot} and time gradient T_{grad}.
- Systematic noise: noise that results from the variation in the host star's luminosity and the instrumental response. This type of noise usually has higher power density at lower frequencies, and we model it with a gaussian process. Gaussian process is a non-parametric model that can predict the most probable value at different points given the data points we observed, by assuming that the entire data set is drawn from a multivariate Gaussian distribution. The Gaussian process contains three additional parameters.
- Random noise: Poisson noise that arises from the discrete nature of light, or photons. At large numbers, Poisson noise is equivalent to a gaussian noise. This noise can be reduced by increasing the number of transit events observed.

In total, there are 16 parameters for modeling the transit light curve, seven of which could be determined and fixed. M_{star} and R_{star} values were retrieved from the GAIA DR2 catalog, M_{planet} is fixed at 0, and we used the limb-darkening coefficients given in [3]. All the other nine parameters will be fitted in the model (meaning that they are not fixed) and a plausible initial value is estimated from the light curve.

- Transit epoch: The transit search algorithm in section 4.1 gives a rough estimate of the center of each transit event, which is used as the initial value during lightcurve fitting.
- Period: For multiple transit targets, the period can be easily determined; For single transit targets, there's limited about the period. (see sec. 5.2)
- Ratio: The initial value for the ratio between two radii is set to the square root of the transit depth.
- Inclination: Inclination is a free parameter to be fitted in the model.
- Flux level out of transit: initial value of F_{oot} is set as the mean flux level out of transit.
- Time gradient: initial value is 0.

Table 1. Summary of the final candidates

Type	Single	Multiple	Total
New	5	0	5
TOI	2	4	6
Confirmed	0	14	14
Total	7	18	25

5.2. Lightcurve Fitting

We utilized two methods for fitting the above parameters. The first one is based on Nelder–Mead method that finds the optimized set of parameters with maximum likelihood (or lowest χ^2) to fit the light curve. We also used Markov Chain Monte Carlo to estimate the distributions and errors of the fitted parameters, in addition to the

point estimates provided by the first one. At each step in that Markov chain, first a new position is selected by sampling from an arbitrary symmetric proposal distribution, and then this new position is accepted with a probability that depends on the likelihood ratio between the old and new parameters, that is, for new positions with higher likelihoods, the move will be accepted; for new positions with lower likelihoods, the probability of accepting the move is equal to the ratio. This method constructs a Markov chain that has the model parameters' distribution(likelihood) as the equilibrium distribution. Then, recording states from this Markov chain would be equivalent to sampling from the model parameters' distribution.

In the case of a multiple transit target, we have good estimates of all the model parameters except inclination. Therefore, the first simple optimization method could yield a reliable result. In order to reduce the random noise in the data, we combined all the transit signals present in the light curve for multiple transits targets,which then requires high precision of T_0, the center time of each transit event. To improve upon the estimated transit epoch from sec 4.1, a model is fitted on each individual transit event with the transit epoch as a free parameter, and then the phase-folded light curve is used for the final fit.

For single transit events, there are two free parameters in the model without a definite initial value - inclination i and orbital period P. Here, the orbital period is especially important if we want to measure the habitability of that planet. For each single transit target, an MCMC model is fitted with 5 chains, each with 100000 steps and 20000 steps of burn-in. The samples from MCMC also allow us to estimate other statistics, including mean, median and variance.

5.3. Results: Exoplanet candidates

Shown in Table 1 is a summary of our final candidates that entered the parameter fitting process. There are 6 new targets, 14 confirmed planets, and 6 TOI objects. As for transit types, there are 7 single transit targets and 19 multiple transits.

Known TESS planets and candidates: Multiple transits

Although all new exoplanet candidates discovered were single transit, we still tested our maximum likelihood method on the confirmed or cataloged TESS planets with more than one transit event. Three parameters estimated by our model - period, ratio, and inclination - were compared to the known values.

Most of the period values estimated by our model is close to the true value except for TIC 92352620, which is primarily due to the fact that periods were directly estimated from the epochs of the transit events(instead of from lightcurve fitting). TIC 92352620 is a confirmed planet, WASP-94 B b, a hot jupiter orbiting one of the two stars in a binary system with a period of 2.008 days, discovered in [9] by doppler measurements. However, the TESS lightcurve of this target has a period of 3.95 days, and its lightcurve closely resembles another one measured from its cousin - WASP-94 A b. It is possible that the TESS lightcurve for 92352620 is measured from the wrong other member, WASP-94 A, of the twin-star system. The distribution of estimated and true values of ratio and inclination were also shown in the figures in the appendix.

Known TESS planets and candidates: Single transits

We tested our MCMC method on one of the two single transit planets that have been included in the TIC catalog.

- TIC 270341214: For this target, we run the markov chain for 1 million steps for each of the 5 chains. The chains have good convergence, as shown in the posterior distribution plot for each chain in fig x. The period estimate, $33.125^{+2.159}_{-0.748}$, is well-constrained, and the value estimated from the TOI catalog, 29.7536 days, is close to the peak in our MCMC sampled distribution. However, this value does not lie within the 68% percentile region, (which is also true for the ratio) implying that the error range estimated by the algorithm could be too narrow. (see ??)
- TIC 139198430: No available data on the stellar parameters

New Exoplanet candidates

- TIC 211376134: This is a deep (> 2%), V-shaped transit. Initial fitting with five chains, each with 500000 steps, suggests a grazing eclipse by an extremely large body, likely a main-sequence star, with radius ratio $0.427^{+0.030}_{-0.030}$, or around 5 times the Jupiter radius. We also tested the alternative hypothesis of a disc-crossing transit by constraining the planet-star radius ratio to its initial value, around 0.19. The alternative fit is also well-constrained, with a similar period of $15.310^{+0.483}_{-0.487}$ days and a slightly higher inclination of $87.696^{+0.045}_{-0.048}$ days, indicating a more disc-crossing transit. Follow-up radial velocity observations were necessary to rule out the possibility of an eclipsing binary and validate this candidate. The estimated surface temperature is 890.6K, in the case of a disc-crossing transit. (see ?? and ??)
- TIC 214517366: The fit to this deep, V-shaped eclipse suggests an object with radius of 1.6 M_j orbiting a main-sequence K star with a period of $21.283^{+0.500}_{-0.589}$ days. The parameter estimations, especially for radius,were not well-constrained after 500000 steps in each MCMC chain. There's a strong positive correlation between the distributions of inclination and period. The inferred period agrees with the patterns observed in the original light curve. (P¿20 days) The estimated surface temperature is 547K. (see ??)
- TIC 259784884: No available data on the stellar parameters
- TIC 350518984: No available data on the stellar parameters
- TIC 358111274: No available data on the stellar parameters
- TIC 369456296: The initial fit for the target is well-constrained, with a period of 8.71±0.234 days. However, this contradicts with the estimated lower limit(¿15 days) on the period based on the epoch of this single transit event. Previous research shows that the target is an M dwarf masquerading as transiting hot-Jupiter. [14]

References

[1] William Borucki, David Koch, Gibor Basri, Natalie Batalha, Timothy Brown, Douglas Caldwell, Jørgen Christensen-Dalsgaard, William Cochran, Edward Dunham, Thomas N. Gautier, and et al. Finding earth-size planets in the habitable zone: thekepler mission. Proceedings of the International Astronomical Union, 3(S249):17–24, 2007.

[2] William J. Borucki and Audrey L. Summers. The photometric method of detecting other planetary systems.Icarus, 58(1):121–134, 1984.

[3] A Claret. A new non-linear limb-darkening law for lte stellar atmosphere models iii-sloan filters: Calculations for–5.0 log [m/h]+ 1, 2000 k t 50 000 k at several surface gravities. Astronomy & Astrophysics, 428(3):1001– 1005, 2004.

[4] Franc¸ois Fressin, Guillermo Torres, David Charbonneau, Stephen T. Bryson, Jessie Christiansen, Courtney D. Dressing, Jon M. Jenkins, Lucianne M. Walkowicz, and Natalie M. Batalha. The false positive rate ofkeplerand the occurrence of planets. The Astrophysical Journal, 766(2):81, 2013.

[5] Ravi Kumar Kopparapu, Ramses M. Ramirez, James Schottelkotte, James F. Kasting, Shawn

Domagal- Goldman, and Vincent Eymet. Habitable zones around main-sequence stars: Dependence on planetary mass. The Astrophysical Journal, 787(2), 2014.
[6] Kaisey Mandel and Eric Agol. Analytic light curves for planetary transit searches. The Astrophysical Journal, 580(2), 2002.
[7] Michel Mayor and Didier Queloz. A jupiter-mass companion to a solar-type star. Nature, 378(6555):355–359, 1995.
[8] et al. Natalia M. Guerrero. The tess objects of interest catalog from the tess prime mission. The Astrophysical Journal Supplement Series, 254(39), 2021.
[9] M. Neveu-Vanmalle, D. Queloz, D. R. Anderson, C. Charbonnel, A. Collier Cameron, L. Delrez, M. Gillon,C. Hellier, E. Jehin, M. Lendl, and et al. Wasp-94 a and b planets: hot-jupiter cousins in a twin-star system.Astronomy Astrophysics, 572, 2014.
[10] NoblePrize. The nobel prize in physics 2019, 2019.
[11] Tobias Owen. The search for early forms of life in other planetary systems: Future possibilities afforded by spectroscopic techniques. Astrophysics and Space Science Library Strategies for the Search for Life in the Universe, page 177–185, 1980.
[12] M A C Perryman. Extra-solar planets. Reports on Progress in Physics, 63(8):1209–1272, 2000.
[13] Michael Perryman. The Exoplanet Handbook. 2018.
[14] Amaury H. M. J. Triaud, David V. Martin, Damien S´egransan, Barry Smalley, Pierre F. L. Maxted, David R. Anderson, Fran¸cois Bouchy, Andrew Collier Cameron, Francesca Faedi, Yilen G´omez Maqueo Chew, and et al. The eblm project. Astronomy Astrophysics, 608, 2017.

Soliton solution to the time-like curve shortening flow in the pseudo-sphere

Jinwei Chang, Yuyan Wang, Nan Zeng, Yanhua Yu *

Northeastern University, Shenyang, China

*Corresponding author e-mail: yyh_start@126.com

Abstract. A time-like curve on the unit pseudo-sphere can be a solition solution to the time-like curve shortening flow We first introduce the sufficient conditions for it to hold .And by studying the properties of the structure function of non-geodesic curves, I deduce some conditions to be satisfied by the structure function of non-geodesic curves when forming the time-like space plane, and then analyze the influence of the extreme value point of the structure function, and finally discuss the classification according to the time-like space-like situation of the parameter vector.

Keywords: solition solution; time-like curve; curvature flow; pseudo-sphere.

1. Introduction

The Poincaré conjecture is a conjecture proposed by the French mathematician Pangaré and is one of the seven Millennium Prize puzzles offered by the Clay Mathematics Institute. The three-dimensional case was proved by Russian mathematician Grigory Perelman around 2003, and in 2006, the mathematical community finally confirmed that Perelman's proof solved the Poincaré conjecture. The conjecture is a proposition of fundamental importance in topology that will help mankind to better study three-dimensional space, and the results it brings will deepen the understanding of the nature of manifolds.

The analysis of geometric flows, especially the Ricci flow, was an important preliminary step on the road to Perelman's proof of the Poincare conjecture and Thurston's geometrization conjecture in 2004, which counts as one of the most significant breakthroughs in the history of nonlinear partial differential equations and opens up new frontiers in the mathematics and beyond. Hiuri Fellipe Santos Dos Reis And Keti Tenenblat [1] considered soliton solution to the curve shortening flow on the sphere and approached the qualitative behavior of such a curve. They also proved the convergence of the curve to the equator orthogonal to the fixed vector. R_3^1 is affine space endowed with Minkowski metric, such that vector fields or curves R_3^1 can be space-like, time-like and light-like. Depending on the diversity of causal properties of vectors, we must distinguish between the cases where the fixed vectors lie in a space-like cone or a time-like cone, respectively.

This paper is organized as follows. Section 3 is the main part of this paper, and theorem 2.2 descirbes when a regular time-like curve which always lies on the unit pseudo-sphere is a soliton solution of time-like curve shortening flow. Because the fixed vector can either be time-like or space-like, section 3.1 is

Content from this work may be used under the terms of the Creative Commons Attribution 3.0 licence. Any further distribution of this work must maintain attribution to the author(s) and the title of the work, journal citation and DOI.
Published under licence by IOP Publishing Ltd

devoted to space-like fields. Where in section 3.2, we discuss time-like fields. In both subsections, three functions (which can be regarded as structure functions) have been introduced. We end the chapter by mentioning the propeties of such functions.

2. Preliminaries

Let R_1^3 be an affine space endowed with inner product $\langle,\rangle = dx^2 + dy^2 - dz^2$. For any $\mathbf{v} = \{v_1, v_2, v_3\} \in R_1^3$, $\mathbf{v}$ is space-like if $\langle \mathbf{v}, \mathbf{v} \rangle > 0$ or $\mathbf{v} = \mathbf{0}$, time-like if $\langle \mathbf{v}, \mathbf{v} \rangle < 0$, light-like (null) if $\langle \mathbf{v}, \mathbf{v} \rangle = 0$. The length of $\mathbf{v}$ is denoted by $|\mathbf{v}| = \sqrt{|\langle \mathbf{v}, \mathbf{v} \rangle|}$. Similarly, an arbitrary curve $\mathbf{r} = \mathbf{r}(s)$ in R_1^3 is called space-like, time-like, or light-like (null) if all of its velocity vectors $\mathbf{r}'(s)$ are respectively space-like, time-like, or light-like(null). We consider the time-like orientation as follows: A time-like vector $\mathbf{v} = (v_1, v_2, v_3)$ is said to be future pointing (resp. past pointing) if and only if $v_3 > 0$ (resp. $v_3 < 0$).

$H^2 = \{(x, y, z) \in R_1^3 | x^2 + y^2 - z^2 = 1\}$ is called a unit pseudo-sphere in R_1^3.

$H_1^2 = \{(x, y, z) \in R_1^3 | x^2 + y^2 - z^2 = -1\}$ is called a unit anti-pseudo-sphere in R_1^3.

Assume $X: R \to H^2$ be a regular time-like curve in-unit pseudo-sphere with arc length parameter s. We denote the unit tangent vector as $T = X' = \frac{dX}{ds}$, and the normal vector $N = X \times T$. The geodesic curvature denotes as $k_g = \langle T', N \rangle$, then the frenet formulae can be expressed as :

$$\begin{bmatrix} X' \\ T' \\ N' \end{bmatrix} = \begin{bmatrix} 0 & 1 & 0 \\ 1 & 0 & k_g \\ 0 & k_g & 0 \end{bmatrix} \begin{bmatrix} X \\ T \\ N \end{bmatrix} \tag{1}$$

where X, N are both space-like vectors, and T is a time-like vector.

Remark 2.1. The curves with $k_g = 0$are called geodesics. The geodesic on the xoy plane is a circle, where on the yoz and zox plane are both hyperbolic curves.

Let $\mathbf{e}_1 = (1,0,0), \mathbf{e}_2 = (0,1,0), \mathbf{e}_3 = (0,0,-1)$ be the standard orthogonal basis in R_1^3, then $\mathbf{e}_1, \mathbf{e}_2$ are space-like vectors, and e_3 is a time-like vector. The determinant of e_1, e_2, e_3 equals -1, and the cross product are defined by $e_1 \times e_2 = e_3, e_2 \times e_3 = -e_1, e_3 \times e_1 = -e_2$.

3. Qualitative behavior of the soliton solutions

Theorem 3.1 Let $X: J \subset U \to M^2$ be a non-geodesic canonical time-like curve .Then Y is a soliton solution to the time-like curve shortening flow if and only there is a vector $v \in R_1^3 \setminus \{0\}$ such that $\langle v, T \rangle = k_g$, where T is the unit tangent vector and k_g is geodesic curvature of X.

Proof. Assume that $X(s) = \{x_1, x_2, x_3\}$ is parametrized by arc length s. If X is a soliton, then $\bar{X}^t(s) = A(t)X(s)$ is a solution to the TCSF, where we can consider that

$A(t) = A_1(t)A_2(t)A_3(t)$, with

$$A_1(t) = \begin{pmatrix} 1 & 0 & 0 \\ 0 & \cosh\xi_1(t) & \sinh\xi_1(t) \\ 0 & \sinh\xi_1(t) & \cosh\xi_1(t) \end{pmatrix},$$

$$A_2(t) = \begin{pmatrix} \cosh\xi_2(t) & 0 & \sinh\xi_2(t) \\ 0 & 1 & 0 \\ \sinh\xi_2(t) & 0 & \cosh\xi_2(t) \end{pmatrix},$$

$$A_3(t) = \begin{pmatrix} \cos\xi_3(t) & -\sin\xi_3(t) & 0 \\ \sin\xi_3(t) & \cos\xi_3(t) & 0 \\ 0 & 0 & 1 \end{pmatrix},$$

and $\xi_j: J \to R$ satisfy $\xi_j(0) = 0$ for all $j = 1,2,3$.

Taking the derivative of (3.2), at $t = 0$, we have

$$\frac{\partial \overline{X^t}}{\partial t} = \{-x_2\zeta_3'(0) + x_3\zeta_2'(0), x_1\zeta_3'(0) + x_3\zeta_1'(0), x_1\zeta_2'(0) + x_2\zeta_1'(0)\},$$

$$N(s,0) = X(s) \times T(s) = \{x_2'x_3 - x_2x_3', x_1x_3' - x_1'x_3, x_1x_2' - x_1'x_2\}$$

where $x_1^2 + x_2^2 - x_3^2 = 1$ and $x_1x'_1 + x_2x'_2 - x_3x'_3 = 0$.
Such that

$$\langle\frac{\partial \bar{X}^t}{\partial t}, N\rangle = x'_1\xi'_1(0) + x'_2\left(-\xi'_2(0)\right) - x'_3\left(-\xi'_3(0)\right) = \langle T, \mathbf{v}\rangle.$$

Here we denote the vector field $\mathbf{v}$ as $\mathbf{v} = \left(\xi'_1(0), -\xi'_2(0), -\xi'_3(0)\right)$.

Conversely, let X be a time-like curve in $H^2 \subset R_1^3$ such that $\langle \mathbf{v}, \mathbf{T}\rangle = k_g$ for a vector $\mathbf{v} \in R_1^3 \setminus \{0\}$. Up to isometries of H^2 and the property of vector field $\mathbf{v}$, we discuss the following two cases:

Case 1. If $\mathbf{v}$ is a space-like vector, we can think over that $\mathbf{v} = (a, 0,0) = a\mathbf{e}_1$ for some $a > 0$. Here, $\langle T, \mathbf{v}\rangle = ax'_1$.

Define $\bar{X}^t(s) = A(t)X(s)$, and

$$A(t) = \begin{pmatrix} 1 & 0 & 0 \\ 0 & \cosh(at) & \sinh(at) \\ 0 & \sinh(at) & \cosh(at) \end{pmatrix}$$

$$\overline{X}^t(s) = \{x_1, x_2\cosh(at) + x_3\sinh(at), x_2\sinh(at) + x_3\cosh(at)\}, \overline{k}_g = k_g$$

$$\overline{N} = A(t)N(s) = \{N_1, N_2\cosh(at) + N_3\sinh(at), N_2\sinh(at) + N_3\cosh(at)\}$$

where $N = \{x'_2x_3 - x_2x'_3, x_1x'_3 - x'_1x_3, x_1x'_2 - x'_1x_2\}$.
By directly computation, we have

$$\frac{\partial \overline{X^t}}{\partial t} = \{0, ax_2\sinh(at) + ax_3\cosh(at), ax_2\cosh(at) + ax_3\sinh(at)\}$$

$$<\frac{\partial \overline{X^t}}{\partial t}, \overline{N}> = -ax_2N_3 + ax_3N_2 = -ax_2(x_2'x_1 - x_2x_1') + ax_3(x_1x_3' - x_1'x_3) = ax_1' = k_g$$

Case 2. If $\mathbf{v}$ is a time-like vector, we consider that $\mathbf{v} = (0,0,-a) = a\mathbf{e}_3$ for some $a > 0$. In this condition, $\langle T, \mathbf{v}\rangle = ax'_3$.

Define $\bar{X}^t(s) = A(t)X(s)$, and

$$A(t) = \begin{pmatrix} \cos(at) & -\sin(at) & 0 \\ \sin(at) & \cos(at) & 0 \\ 0 & 0 & 1 \end{pmatrix}.$$

At this point, we can go faster in the computation since the strategy is the same as case 1. Thus, a straightforward computation gives us

$$\langle\frac{\partial \bar{X}^t}{\partial t}, \bar{N}\rangle = ax'_3 = k_g.$$

Taking into account the motivations given above, the following questions appear naturally: How dose vector field $\mathbf{v}$ describe the qualitative behavior of the soliton solutions to the TCSF? There are two cases to discuss since $\mathbf{v}$ can either be space-like or time-like.

3.1. v is space-like.

Proposition 3.2. Let $\mathrm{X}: \mathrm{I} \to \mathrm{H}^2$ be a regular time-like curve . Suppose X is not a geodesic curve on the pseudo-sphere. Define the functions

$$\alpha_1(s) = \langle X, \boldsymbol{e}_1\rangle,\ \beta_1(s) = \langle T, \boldsymbol{e}_1\rangle,\ \gamma_1(s) = \langle N, \boldsymbol{e}_1\rangle, \tag{2}$$

Then $\mathrm{k_g} = \mathrm{a}\beta_1(\mathrm{s})$ for all $\mathrm{s} \in \mathrm{I}$, and $\mathrm{a} > 0$ if and only if the function $\alpha_1, \beta_1, \gamma_1$ satisfy the equation

$$\begin{cases} \alpha_1'(s) & = & \beta_1(s) \\ \beta_1'(s) & = & \alpha_1(s) + a\beta_1(s)\gamma_1(s), \\ \gamma_1'(s) & = & a\beta_1^2(s) \end{cases} \quad (3)$$

with initial conditions $\alpha_1^2(0) - \beta_1^2(0) + \gamma_1^2(0) = 1$.

Proof. From definition of functions $\alpha_1(s), \beta_1(s), \gamma_1(s)$, one has

$$\mathbf{e}_1 = \alpha_1(s)X - \beta_1(s)T + \gamma_1(s)N.$$

Comparing with equation (2.1), we obtain

$$\begin{cases} \alpha_1'(s) & = & \beta_1(s) \\ \beta_1'(s) & = & \alpha_1(s) + k_g\gamma_1(s) \\ \gamma_1'(s) & = & k_g\beta_1(s) \end{cases} \quad (4)$$

Suppose that $k_g = a\beta_1(s)$ for all $s \in I$, we have the equation (3.4).

Conversely, suppose $\alpha_1(s), \beta_1(s), \gamma_1(s)$ satisfy equation (3.4), from equation (2.1) we get

$$\gamma_1(s)\left(a\beta_1(s) - k_g\right) = 0,\ \beta_1(s)\left(k_g - a\beta_1(s)\right) = 0.$$

Since $\beta_1(s), \gamma_1(s)$ vanish simultaneously at the point $\mathbf{e}_1$ ad $-\mathbf{e}_1$. It can conclude $k_g = a\beta_1(s)$.

From above, we know the functions $(\alpha_1(s), \beta_1(s), \gamma_1(s))$ can be only determined by a given pseudo-sphere time-like curve X.

Proposition 3.3.

Define $k_g(s) = a\beta_1(s)$. And according to rigid motions, The curve is determined by the initial conditions $\{X(0), T(0), N(0)\}$uniquely ,which can be chosen such that

$$\mathbf{e}_1 = \alpha_1(0)X(0) - \beta_1(0)T(0) + \gamma_1(0)N(0).$$

Straightforward computation shows that (2.1) and (3.4) imply

$$\frac{d(\alpha_1(s) - \beta_1(s)T + \gamma_1(s)N)}{ds} = \mathbf{0}$$

Hence, for all $s \in I$,

$$\alpha_1(s)X - \beta_1(s)T + \gamma_1(s)N = \mathbf{e}_1.$$

Remark 3.4 From the definition of function $\alpha_1(s), \beta_1(s), \gamma_1(s)$, we know $X, N, \mathbf{e}_1$ are space-like vectors, and T is a time-like vector. Hence

$(\alpha_1(s), \beta_1(s), \gamma_1(s))$ lies on H^2.

$\alpha_1(s)$ is the cosine of the angle between X and $\mathbf{e}_1$ if $\{X, \mathbf{e}_1\}$ spans a space-like plane. $\alpha_1(s)$ is the hyperbolic cosine of the angle between X and $\mathbf{e}_1$ if $\{X, \mathbf{e}_1\}$ spans a time-like plane.

$\beta_1(s)$ is the hyperbolic sine of the angle between T and $\mathbf{e}_1$ if T is a future pointing time-like vector, otherwise it is the negative hyperbolic sine of the angle between $-T$ and e_1.

$\gamma_1(s))$ is the cosine of the angle between N and $\mathbf{e}_1$ if $\{N, \mathbf{e}_1\}$ spans a space-like plane. $\gamma_1(s)$ is the hyperbolic cosine of the angle between N and $\mathbf{e}_1$ if $\{N, \mathbf{e}_1\}$ spans a time-like plane.

3.2. v is time-like.

We use the same idea as section 3.1 to study this part. Then, we can get the following common propositions.

Proposition 3.5. Let $X: I \to H^2$ be a regular time-like curve with arc-length s. we now suppose that X is not a geodesic curve on the pseudo-sphere. We define these functions:

$\alpha_2(s) = \langle X, \mathbf{e}_3\rangle$, $\beta_2(s) = \langle T, \mathbf{e}_3\rangle$, $\gamma_2(s) = \langle N, \mathbf{e}_3\rangle$, (3.7)

where T is the unit tangent vector and N is the unit normal vector of X, respectively. Then $k_g = a\beta_2(s)$ for all $s \in I$, and $a > 0$ if and only if the function $\alpha_2(s), \beta_2(s), \gamma_2(s)$ satisfy the equation

$$\begin{cases} \alpha_2'(s) & = & \beta_2(s) \\ \beta_2'(s) & = & \alpha_2(s) + a\beta_2(s)\gamma_2(s), \\ \gamma_2'(s) & = & a\beta_2^2(s) \end{cases} \quad (5)$$

with initial conditions $\alpha_2^2(0) - \beta_2^2(0) + \gamma_2^2(0) = -1$.

Remark 3.6 From the definition of function $\alpha_2(s), \beta_2(s), \gamma_2(s)$, one knows X, N are space-like and $T, \mathbf{e}_3$ are time-like, then

$(\alpha_2(s), \beta_2(s), \gamma_2(s))$ lies on H_1^2.

$\alpha_2(s)$ is the negative hyperbolic sine of the angle between X and $-\mathbf{e}_3$.

β_2 is the negative hyperbolic cosine of the angle between T and $-\mathbf{e}_3$ if T is a future pointing time-like vector otherwise, it is the hyperbolic cosine of the angle between $-T$ and $-\mathbf{e}_3$.

$\gamma_2(s)$ is the negative hyperbolic sine of the angle between N and $-\mathbf{e}_3$.

4. Practical application value and realistic significance

1) For the study of curvilinear shrinkage flow: Using the solution problem of curvilinear shrinkage flow, the explicit expressions of sickle-shaped curves, the density of spherical curves, and the maximum curvature inequality of smooth and simple closed curves on surfaces can be studied.

2) For a pseudosphere: A pseudosphere is a surface of revolution formed by the rotation of a tractor around its asymptote. The full curvature of such a surface is constant and negative at every point. The line lying on this surface is inconsistent with the parallel postulate, so the possibility of constructing this surface provides a relatively compatible proof for non-Euclidean geometry. In 1868, the Italian mathematician Beltrami first proposed that the pseudosphere can be used as a model for realizing hyperbolic geometry, which led to the general recognition of non-Euclidean geometry. In the cooling tower of the thermal power plant, the pseudo-spherical shape is selected, which has the advantage of very uniform heat dissipation.

3) For the soliton solution: In recent years, soliton theory has achieved rapid development, and has been widely used in quantum field theory, condensed polymer physics, fluid mechanics, optics and other physical fields as well as chemistry, biology, communication and other fields, which has stimulated more and more Interest of more and more mathematicians and physicists.In this paper, the soliton solution of curvilinear contraction flow on pseudo-sphere is studied to solve the difficulties in other fields.

4) In this paper, the work of Euclidean space is extended to the indefinite metric space with a negative index, which breaks the conventional method of defining the inner product dot product, and conducts research in the indefinite metric space. Since the curves in this space can be time-like, space-like, or light-like, the research is more complicated, and the inherent properties such as positive definiteness of conventional Euclidean spaces are broken.

5) In the future, we will discuss the multiplicity property of the structure function in depth and further computational proofs. In this paper, we only briefly discuss the case of class-time and class-empty, and we will give rigorous mathematical proofs and derivations in the course of future research

References

[1] H.F.Santos DosReis,K.Tenenblatr,Solition solutions to the curve shortening flow on the sphere, Proc. Amer. Math.Soc., 147(2019), 4955-4967.

[2] J.Urbas,Complete noncompact self-similar solutions of Gauss curvature flows.I.Positive powers, Math.Ann., 311(1998), 251-274.

[3] L'opez,R.,Munteanu,M.I.,Constant angle surfaces in Minkowski space, Bull. Belg. Math. Soc. SimonStevin., 18(2011), 271-286.

[4] M.Fels,P.J.Olver,Moving coframes.II.Regularization and theoretical foundations.Acta Appl. Math., 55(1999), 127-208.

[5] Y.Fu,D.Yang,On constant slope spacelike surfacesin3-dimensional Minkowski space, J. Math. Anal. Appl., 385(2012), 208-220.

FDCS-2022 IOP Publishing
Journal of Physics: Conference Series **2441** (2023) 012032 doi:10.1088/1742-6596/2441/1/012032

The Progress of Investigation on Neutrino Properties

Taosha Sheng [1, *, †], Shitao Qu [1, †] and Chenghao Qian [2, †]

[1] Department of Physics and Astronomy, University of Manchester, Manchester, United Kingdom
[2] Department of Physics, Dalian University of Technology, Dalian Province, China

*Corresponding author's e-mail: taosha.sheng@student.manchester.ac.uk

†These authors contributed equally

Abstract. Generally, particle physics plays an important role in the whole modern physics system. Thereinto, neutrino, as an elementary particle found from beta decay, is of great importance to discover its properties and mechanics. With the help of information retrieve and literature review, this paper aims at helping recollecting sorts of knowledge, e.g., the developing history of the principle, different kinds of matrix expressions, field theories etc. Additionally, some possible future developing directions have been proposed. Moreover, plenty of analytical models and experimental approaches for generating massive neutrinos are demonstrated, e.g., Lagrangian and beta decay. More experiments need to be carried out for the sake of verifying the theoretical predictions. Overall, these results offer a guideline for completing the Standard Model and hopefully the Grand Unified Theory, such as Majorana mass term.

1. Introduction

Neutrino is an elementary fermion (particle with a half spin), which can interact only via the weak interaction and gravity. The neutrino is electrically neutral and was thought to be zero mass since its rest mass is extremely small (much smaller than that of the other known elementary particles excluding massless particles). Since neutrinos have zero charge, they are unable to interact with electromagnetism force. The weak force has a very short range, i.e., it's not significant. Besides, the gravitational interaction is extremely weak due to its tiny mass, and neutrinos do not participate in the strong interaction neither. Thus, neutrinos typically pass through normal matter unimpeded and undetected [1].

This characteristic of neutrino enables it to be widely used in scientific research for many purposes. The low mass and neutral charge of neutrinos means that their interaction with other particles and fields is extremely weak. Therefore, neutrinos can be used to detect environments where other radiation sources cannot penetrate, e.g., the conditions of the solar core, astrophysical sources beyond the Solar System and the observation of supernovae.

The rest of the paper is organized as follows. The section 2 is mainly focused on basic descriptions and brief history neutrino. The section 3 introduces field theory and weak interaction related to neutrino. The section 4 discusses massive neutrino. Finally, a brief summary is given in section 5.

Content from this work may be used under the terms of the Creative Commons Attribution 3.0 licence. Any further distribution of this work must maintain attribution to the author(s) and the title of the work, journal citation and DOI.
Published under licence by IOP Publishing Ltd

2. Basic Descriptions and History of Neutrino

2.1. History of neutrino

In the quantum world, the absorption and emission of energy are discontinuous, which is caused by the transition between different energy levels of the atoms. However, it is strange that the energy spectra of the beta rays released by the substance during the beta decay is continuous. In addition, it is also found that the electrons only take away part of the energy, and some of the energy is missing. On this basis, physicists like Niels Bohr believe that the law of conservation of energy fails in β decay [2].

In 1930, The proposal of the "neutrino" was first put forward by W. Pauli. He believes that in addition to electrons, a new particle with zero rest mass, electrically neutral, and different from photons is emitted in beta decay, taking away another part of the energy. The interaction between this particle and matter is so weak that it is difficult to detect it in experiments. Thus, the law of conservation of energy still holds, the energy distribution ratio between the (anti) neutrinos and the electrons can be varied in different cases. In 1933, the Italian physicist Fermi proposed a quantitative theory of beta decay. According to Fermi's theory, in addition to the known gravitational and electromagnetic forces, there is a third kind of interaction called weak interaction. The beta decay is through the weak interactions, in which a neutron in the nucleus decays into an electron, a proton and an (anti) neutrino. His theory quantitatively described the law of β-ray energy spectrum continuity and β-decay half-life, and the reasons for β-ray energy spectrum continuity have been well explained [3].

In 1956, Reines and Cowan discovered the neutrino through inverse beta decay, which was using antineutrinos emitted by beta decay in nuclear reactors. The antineutrinos undergo anti-beta decay with hydrogen nuclei, forming fast and slow signals in the detector that are related to intensity and time. This is the first time that neutrinos have been detected in an experiment [4]. Madame Wu in 1956 demonstrated that parity conservation was violated in weak interactions following the idea of Lee and Yang [5, 6]. This leads to an awareness of the limitations of the standard model which only involves left-handed neutrino. This is the key issue why neutrino's mass cannot be explained at least in the minimum form of the standard model. It has to be extended, i.e., somehow right-handed neutrino is included.

Another type of neutrinos called Muon-neutrinos were discovered in 1962 by Lederman, M. Schwartz et al [7]. Neutrinos are produced correspondently with its charged leptons. For example, Muon neutrinos are produced in pion decay with Muons. When they performed the experiment, there were only muons but not electrons in the reactor. Therefore, they realized that there must be another type of neutrinos which is Muon neutrinos. Later, the third type of lepton tau was discovered, and it was also expected to have an associated neutrino. Nevertheless, tau neutrinos were announced to be detected until 2000, whose existence had already been inferred by both theoretical consistency and experimental data [8].

2.2. Neutrino oscillation

One of the main focuses of the neutrino research at early stage was the neutrino oscillation, which is a quantum mechanical phenomenon generated by different massive neutrinos. The first idea of neutrino oscillations was considered by B. Pontecorvo in 1957 [9], who developed the mathematical formalism and the modern formulation of neutrino oscillations in the vacuum.

Neutrinos oscillate between different flavours. For example, an electron neutrino produced in a beta decay reaction may interact in a distant detector as a muon or tau-neutrino. This oscillation occurs because the three mass state components of the produced flavour travel at slightly different speeds. Thus, there is a relative phase shift that change the way they combine to produce a varying superposition of three flavours. Each flavour component thereby oscillates as the neutrino travels, with the flavours varying in relative strengths. The relative flavour proportions when the neutrino interacts represent the relative probabilities for that flavour of interaction to produce the corresponding flavour of charged lepton [10, 11].

Journal of Physics: Conference Series **2441** (2023) 012032 doi:10.1088/1742-6596/2441/1/012032

Neutrino oscillations always come along with the idea of mixing. Mixing was introduced at the beginning of the '60 by Maki et al. [12]. The relationship between flavour and mass eigenstates is encoded in the PMNS matrix, which is a unitary mixing matrix containing information on the mismatch of quantum states of neutrinos when they propagate freely and when they take part in weak interactions [13]:

$$U = \begin{bmatrix} c_{12}c_{13} & s_{12}c_{13} & s_{13}e^{-i\delta} \\ -s_{12}c_{23} - c_{12}s_{23}s_{13}e^{i\delta} & c_{12}c_{23} - s_{12}s_{23}s_{13}e^{i\delta} & s_{23}c_{13} \\ s_{12}s_{23} - c_{12}c_{23}s_{13}e^{i\delta} & -c_{12}s_{23} - s_{12}c_{23}s_{13}e^{i\delta} & c_{23}c_{13} \end{bmatrix} \begin{bmatrix} e^{\frac{i\alpha 1}{2}} & 0 & 0 \\ 0 & e^{\frac{i\alpha 2}{2}} & 0 \\ 0 & 0 & 1 \end{bmatrix}, \quad (1)$$

where s_{ij} and c_{ij} are used to denote $\sin\theta_{ij}$ and $\cos\theta_{ij}$ respectively, θ is the mixing angle and there are three of them $(\theta_{12}, \theta_{23}, \theta_{13})$.

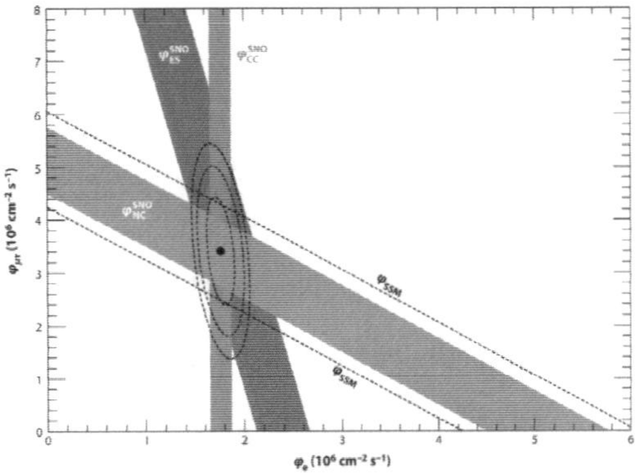

Figure 1. The flux of solar neutrinos of μ or τ flavor versus the flux of electron neutrinos deduced from the three neutrino reactions.

There were two well-known anomalies in the history of neutrinos, one was solar neutrino deficit. First indications of neutrino oscillations came from solar neutrino: R. Davis built the Homestake experiment to detect solar neutrino based on an experimental technique by Pontecorvo. Compared with the predicted solar neutrino fluxes, a significant deficit was first announced. This anomaly received further confirmation from various institutions (SAGE, GALLEX, Super-Kamiokande, SNO, etc.) and was finally interpreted as neutrino oscillations. The three bands intersect at the fit values, indicating that the combined flux results are consistent with neutrino flavour transformation with no distortion in the neutrino energy spectrum as shown in Fig. 1. Contemporarily, it is recognized that solar neutrinos do oscillate into other types of neutrinos like muon and tau neutrinos.

Another anomaly was found in atmospheric neutrinos. The experiments associated with this source of neutrino, including the confirmation of neutrino oscillation, was conducted at the Super-Kamiokande (Super-K) detector. Super-K is a large detector which located 1000 meters under a mountain in Japan, as schematically illustrated in Fig. 2. It contains a cylindrical stainless-steel tank, 39.3m in diameter and 41.4 m in height. The tank is closed during detection and filled with 50 kilotons of ultra-pure water. On each side of the wall, 11129 photo sensors called photo-multiplier-tubes (PMTs) are installed.

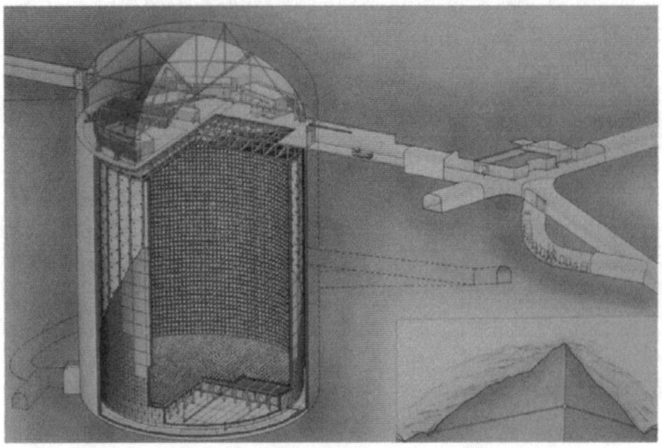

Figure 2. Overview structure of Super-K.

The detection of neutrino in Super-K is through the weak interaction between neutrino and water nuclei. After the interaction, a charged particle is produced and propagating through the water. If the speed of the particle is larger than the speed of light in the water, Cherenkov light will be emitted in a cone shape to the direction of a charged particle, eventually forming a Cherenkov ring on the PMTs sensor as illustrated in Fig. 3.

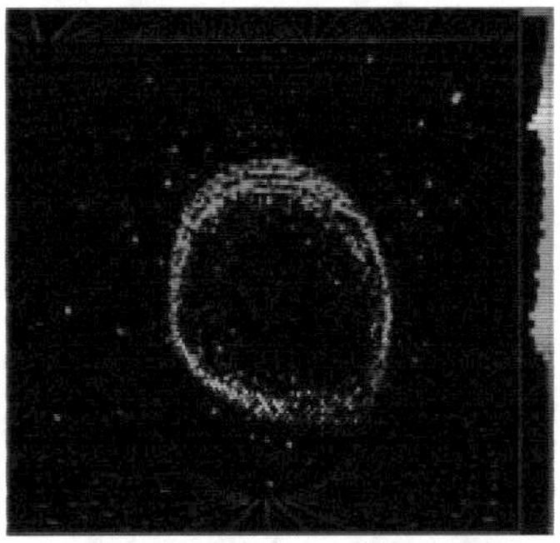

Figure 3. The Cherenkov ring emitted by a muon.

When cosmic rays (e.g., highly accelerated ions) hitting the atmosphere, pions and kaons are produced. Pions can decay into muons and muon neutrinos, and furthermore muons can decay into electrons, muon neutrinos and electron anti-neutrinos. The production of neutrino on the atmosphere is expected to be isotropic, and since neutrinos are hardly to interact with matter, the observed fluxes of up-going and down-going neutrinos in the SK should be equal. This statement had been proved to be true for electron neutrino. However, for the muon neutrino, measured results have shown that the flux of down-going neutrino is much larger up-going neutrino. The phenomenon can be explained by neutrino oscillation. Comparing to the down-going neutrino, the up-going neutrino has to travel longer to reach the detector. During such long distance, some muon neutrinos oscillate into the tau neutrino, decreasing the total flux of muon neutrino. Detailed explanation on experiment in Super-K will be given in the massive neutrino section. In general, the discovery in Super-K, along with solar neutrino deficit are recognized as the result and proof of neutrino oscillations.

3. Field theory and Weak interaction

Before talking in detail about the researches on neutrino, some basic analytical models will be introduced in terms of field theory and weak interaction.

3.1. Field Theory
In field theory, the Lagrangian of a system is different from the formal what we get used to in classical physics. Instead, one needs to rewrite it into the form of the sum of two term: free field and interaction field shown as followed.

$$\mathcal{L} = \mathcal{L}_0 + \mathcal{L}_{int}. \quad (2)$$

For instance, in a charged fermion system with electromagnetism between photons, one can write down the exact expression of free field term within kinetic term and massive term.

$$\mathcal{L}_0 = \bar{\psi}(i\gamma^\mu \partial_\mu - m_0)\psi - \frac{1}{4}F^{\mu\nu}F_{\mu\nu}, \quad (3)$$

where ψ denotes the spinor field and the two F terms denote to electromagnetic field tensor. We can see that the first term of the R.H.S represents free fermions and the second term represents free photons. Based on Lagrangian function, one could easily gain Dirac function and Maxwell function.

Manifestly, the interactions between particles need three or more particles, e.g., the collision of two leads to a new particle. Therefore, it is evident that those two terms with productions only including up to two fields, which means they do not imply any interactions. An expression for the interaction field term is required, for instance:

$$\mathcal{L}_{int} = -eA_\mu \bar{\psi}\gamma^\mu \psi, \quad (4)$$

In which an electromagnetic field A_μ products with two fermion spinor fields ψ and $\bar{\psi}$. With the addition of $\mathcal{L}_{int}$, the Lagrangian function is able to describe the process of positive and negative electrons annihilation or the influence of virtual photons on two charged fermions.

3.2. Weak Interaction
As mentioned in the above sections, the principle of neutrino stemming from β decay. $In\ \beta$ decay, it is known that a neutron decays and produces a proton, an electron and a neutrino according to the conservation of energy, momentum and leptons. In this process, we know that there are four kinds of fermions with spin of 1/2. As a result, the simplest way to derive the Lagrangian of interaction is to product these four terms directly, which origins from Fermi:

$$\mathcal{L}_{int} = \frac{G_F}{\sqrt{2}}\bar{p}\gamma_\mu n \bar{e}\gamma^\mu \nu_e. \quad (5)$$

which is so-called the weak interaction, involving neutrinos.

4. Massive Neutrino
The property of neutrino mass is the most crucial subject in neutrino physics. There are only left-handed components in neutrino fields to satisfy the local symmetry in Standard Model, which means neutrinos' mass-term cannot be derived under the Higgs mechanism. However, experiments about neutrino oscillation demonstrated that neutrinos were massive and mixed, which deviated from the standard model.

4.1. Neutrino oscillation in Super-K
As mentioned above, neutrino oscillation is a quantum mechanical phenomenon, which describes the probability for a certain neutrino to change its flavour spontaneously. As shown in Fig. 4, when a neutrino is moving in space, it can be thought of as superposition of three mass eigenfunctions, which also determines the flavour of this neutrino. Different eigenfunctions oscillate in different frequency, which may result in the change in phase of the superposition wave, hence changing the flavour of the neutrino.

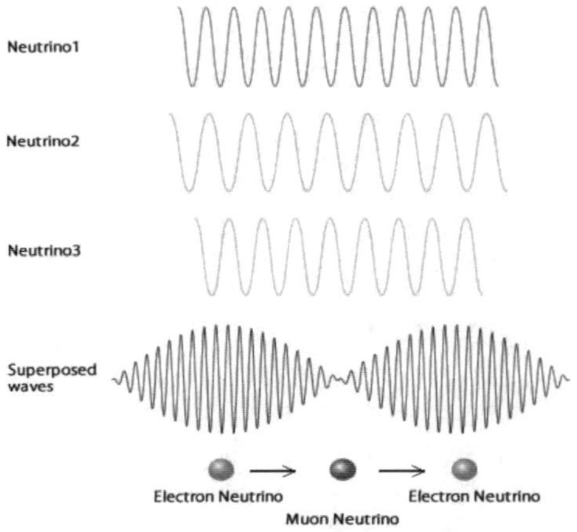

Figure 4. Demonstration of flavour oscillation.

Thus, the observation of neutrino oscillation between two types of neutrinos implies the fact that they have different mass eigenstates, which was the intrinsic behind neutrino oscillation experiments that always measure the squared mass-difference Δm^2 instead of an absolute mass. According to examination different values of Δm^2, neutrinos with different flavors have different non-zero masses.

A typical result of Neutrino oscillation was taken by the Super-K experiment in 1998, in which an analysis of oscillation between atmospheric muon neutrino and tau neutrino was done. As exhibited in Fig. 5, there was inconsistency between the measured muon neutrino flux and expectation value, which meant neutrino oscillation happened when muon neutrino travelled through the Earth. The results showed that Δm^2 between the two oscillating states, under a zenith angle $sin^2 2\theta > 0.82$, were between 5×10^{-4} and $6 \times 10^{-3}\ eV^2$, with 90% confident level.

It should be noted that the magnitudes of neutrino mass were much smaller than these of quarks and leptons. In general, neutrino masses are believed to be a low-energy manifestation of physics beyond the Standard Model. Their masses are suppressed by a new high-energy scale, which may give a clue to the unification of forces.

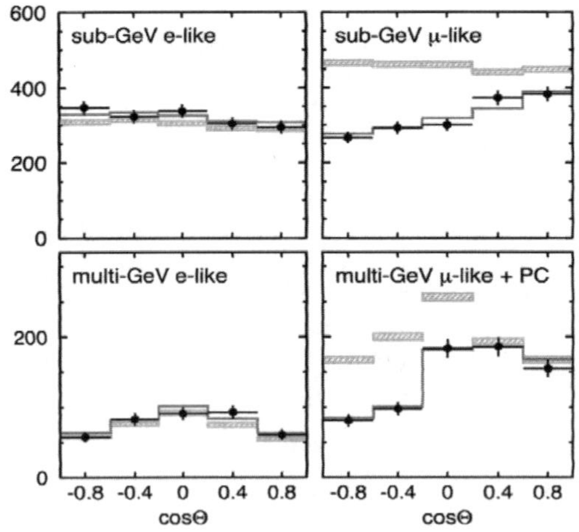

Figure 5. The plots were taken from a similar research of atmospheric neutrino carried out in Super-K detector, which shows the zenith angle distributions of e-like and μ-like events. The red line indicates the results of the best fit and the blue boxes indicates the expectations assuming no neutrino oscillations.

4.2. Dirac Mass term

One way to add a mass term for neutrino is to introduce three right-handed neutrino fields ($\nu_{\alpha R}, \alpha = e, \mu, \tau$) into the Standard Model, deriving the Minimally Extended Standard Model. Then, we can apply the Higgs mechanism to obtain Dirac neutrino mass. Unlike the left-handed component, the right-handed neutrino fields do not participate in any interactions (neither weak, strong, nor electromagnetic), which is why they are called sterile.

By applying the Yukawa interactions to the right-handed sterile neutrino fields, which is in analogy to the derivation of charged fermion masses, one derives the expression of the Dirac mass:

$$M_k = \frac{y_k^{\nu} v}{\sqrt{2}}, (k = 1,2,3)\,, \tag{6}$$

where y_k^{ν} denotes an unknown parameter of the Standard Model, and v denotes the vacuum expectation value of the Higgs field. This expression shares the same form with the masses of the charged leptons.

However, the expression of M_k shows proportionality with v, as the masses of charged leptons and quarks, while it is known that the neutrinos are much lighter than those of quarks and leptons, which means y_k^{ν} must be small. However, the smallness of y_k^{ν} remain unknown under the Higgs-neutrino Yukawa coupling matrix. In fact, under the framework of the Standard Model, lepton and quark masses cannot be predicted and must be obtained from experiments, leading us to believe that Standard Model can be thought as a model that is under the low-energy limit of a complete theory. Thus, the investigation on masses of quarks and leptons may allow us to understand physics beyond the Standard Model.

4.3. Majorana Mass term

Another way of describing the neutrino mass is through the Majorana mass term. To derive this mass term, let's first consider the two-component theory proposed by H. Weil in 1929. For a generic fermion field ψ with left and right-handed component:

$$\psi = \psi_L + \psi_R, \tag{7}$$

its Dirac equation can be split into two different parts:

$$i\gamma^{\mu}\partial_{\mu}\psi_L = m\psi_R, i\gamma^{\mu}\partial_{\mu}\psi_R = m\psi_L. \tag{8}$$

For a massless neutrino we can decoupled the Eq. (7) into Weyl equations:

$$i\gamma^{\mu}\partial_{\mu}\psi_L = 0, i\gamma^{\mu}\partial_{\mu}\psi_R = 0. \tag{9}$$

Hence, one can describe the massless neutrino using a single chiral field (only left-handed) with only two independent components in the Standard Model.

E. Majorana, on the other hand, attempted to construct a massive neutrino using just a single left-handed field. He first tried to express ψ_R in terms of ψ_L by using the charge conjugation matrix C:

$$\psi_R = \xi C\overline{\psi_L}^T, \tag{10}$$

where T indicates the transpose of $\overline{\psi_L}$ and ξ is an arbitrary phase factor. Note that $C\overline{\psi_L}^T$ can be proved to be right-handed, indicating the rationality of this expression.

Thus, the Majorana field can be written as:

$$\psi = \psi_L + \psi_R = \psi_L + C\overline{\psi_L}^T = \psi_L + \psi_L^C, \tag{11}$$

Where the charge conjugate of this field is itself. Therefore, a Majorana particle described by Eq. (10) is its own anti-particle, which can only be a neutral fermion (neutrino). Majorana's statement have changed our understanding on neutrino and anti-neutrino, indicating the fact that we would only have one single particle, the neutrino.

Considering a single neutrino type υ with its right-handed component υ_L^C, and imposing this expression into the Dirac mass term (given by), we obtain:

$$\mathcal{L}_{mass}^{D} = -m\bar{\upsilon}\upsilon, \tag{12}$$

allowing us to derive the expression of Majorana mass term:

$$\mathcal{L}_{mass}^{M} = -\frac{1}{2} m\overline{v_L^C} v_L + H.c. \tag{13}$$

Unlike Dirac mass term, Majorana mass term is a single of the SM gauge group and, as such, it can appear as an essential mass term in Lagrangian. In addition, Majorana mass term differs from the Dirac mass term in that it violates the lepton number conservation. In Dirac field, neutrinos have a lepton number of +1, and antineutrinos have a lepton number of -1, which means during the reaction, the lepton numbers add up to zero and are conserved. Since neutrinos are the same as their antiparticle in the Majorana field, they have the same lepton number. Interactions involving Majorana neutrinos generally violate lepton number conservation by ±2.

4.4. Neutrino-less Double Beta Decay

One way to verify the Majorana neutrinos is through Neutrino-less Double Beta Decay. To understand this process, we first have to introduce the Two-Neutrino Double Beta Decay, in which the nuclear charge changes by two units while the atomic mass is left unchanged:

$$N(A, Z) = N(A, Z + 2) + 2e^- + 2\overline{v_e}, \tag{14}$$

where A denotes the mass number and D denotes the proton number. On the other hand, the Neutrino-less Double Beta Decay:

$$N(A, Z) = N(A, Z + 2) + 2e^-, \tag{15}$$

was proposed by W. H Furry in 1939. Clearly, this decay is forbidden under the Standard Model, as it violates the conservation of the total lepton number by two units. The mechanism of this process is given by Fig. 6.

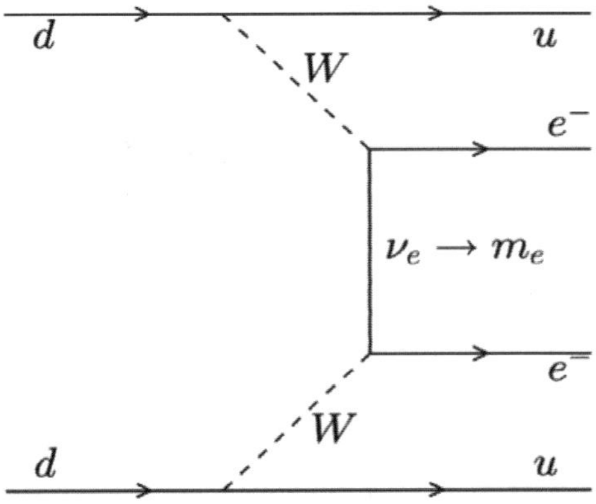

Figure 6. An illustration of the Tree-level quark diagram of Neutrino-less Double Beta Decay.

For this process, a neutrino decays and emits into a right-handed, which has to be absorbed by another neutron as a left-handed. Thus, the only way to fulfil these conditions is that the neutrino and the anti-neutrino are the same. On the other hand, the right-handed emitted by the first decay must evolve a left-handed chiral component as it propagates towards the second, which means the neutrino must have a mass to satisfy the helicity matching. Thus, a Majorana neutrino will fulfil all the conditions of the Neutrino-less Double Beta Decay, which means that proving the existence of Neutrino-less Double Beta Decay implies that neutrinos can be described in Majorana formation.

To sum up, experiments of neutrino oscillation indicate that neutrino indeed contain an absolute mass though really small comparing to other leptons. This result deviates the Standard Model and gives a necessity to derive a more complete theory. There are two ways of generating a neutrino mass term: Dirac and Majorana. Among these, the Majorana mass term suggests that neutrino is its anti-neutrino, which deviates the description from Standard Model. One way to prove that the Majorana mass term can describe neutrinos is by discovering Neutrino-less Double Beta Decay.

5. Conclusion

In summary, we investigate some basic knowledge about neutrino from its history to the recent research based on state-of-art facilities, e.g., massive neutrino. Due to the conservation laws of energy, momentum and lepton, people predict the existence of the new particle called neutrino theoretically. Then, as the development of science, in experiment three kinds of neutrinos have been discovered individually. Moreover, we give out basic knowledge of field theory in this paper and explore more focusing on the massive neutrino according to Yukawa theory and Standard Model. Manifestly, it is just a beginning which is written for introduction, which means more detailed theory and mathematical knowledge need to be given. Additionally, in this area particle physics still need more concrete experiments and evidence to support its theories instead of staying on papers. That is one important direction people have to work hard. Overall, these results pave a path for the aim of experiments need to be carried out and shed light on the possible research directions about future development of the theory.

References

[1] Cowan, C. L., et al. (1956). Detection of the Free Neutrino: a Confirmation." Science 124.3212:103-104.
[2] Pontecorvo, Bruno (1957) Mesonium and antimesonium. Zhur. Eksptl'. i Teoret. Fiz. 33.
[3] Katayama, Yasuhisa, et al. (1962) Possible unified models of elementary particles with two neutrinos. Progress of Theoretical Physics 28.4: 675-689.
[4] Davis Jr, Raymond. Solar neutrinos. ii. experimental. Physical Review Letters 12.11 (1964): 303.
[5] Davis Jr, Raymond, Don S. Harmer, Kenneth C. Hoffman. (1968) Search for neutrinos from the sun. Physical Review Letters 20.21 (1968): 1205.
[6] Jelley Nick, Arthur B. McDonald, RG Hamish Robertson. (2009) The sudbury neutrino observatory. Annual Review of Nuclear and Particle Science 59 (2009): 431-465.
[7] Super-Kamiokande Official Website. [online] Available at: <http://www-sk.icrr.u-tokyo.ac.jp/sk/index-e.html> [Accessed 19 October 2021].
[8] George M. Drake. Advances in Dark Matter Research[M]. Nova Science Publishers, Inc.:2021-09-14.
[9] Nicolai Popov. Neutrino Mass: Past, Present, Future[M]. Nova Science Publishers, Inc.:2021-09-14.
[10] Giunti, C. and Kim, C., 2007. Fundamentals of neutrino physics and astrophysics. Oxford: Oxford University Press.
[11] Ashie, Y., et al. (2005) Measurement of atmospheric neutrino oscillation parameters by Super-Kamiokande I. Physical Review D 71.11: 112005.
[12] Warwick.ac.uk. 2020. [online] Available at: <https://warwick.ac.uk/fac/sci/physics/staff/academic/boyd/warwick_week/neutrino_physics/lec_neutrinomass_writeup.pdf> [Accessed 1 October 2021].
[13] Dell'Acqua, Andrea, et al. (2018) Future Opportunities in Accelerator-based Neutrino Physics. arXiv preprint arXiv:1812.06739.

Black hole growth at early universe: impact of angular momentum of supplied gas

Chenyu liu [1,*], Jingzhou Qiu [2,a]

[1] Shanghai Starriver Bilingual School, Shanghai, China
[2] Huafu International Department, Affiliated High School of South China Normal University, Guangdong, China

*Corresponding author e-mail: Chenyuliu2022@foxmail.com,
[a]qiujz.alice2019@gdhfi.com

Abstract. Over the past decade, more and more high-redshift quasars with $>10^8$ solar-mass black holes (BHs) are observed. These high-mass BHs challenge the BH growth model, and require massive seed BH at redshift 35. In this work, we investigate the growth of high-redshift BHs, with a focus on the impact of angular momentum of surrounding supplied gas. We find that, for the observed 109 solar-mass BH, its seed BH at z=35 should be fairly massive. Applications of our work to the formation channel of seed BH are discussed.

Keywords: black hole, accretion disks, high-redshift galaxies, quasars.

1. Introduction

1.1. Importance of early universe, high-redshift galaxies and quasars

Based on the standard modern Cosmology, our universe began with a dramatic explosion of space, i.e. the so-called Big Bang. As the universe expands, it gradually cools down. We nowadays observe that the universe is fulfilled with the cosmic microwave background (CMB) radiation, which is an almost perfect black body radiation at a temperature 2.7K. These thermal photons are actually emitted at approximately 380k yrs after Big Bang (redshift z~1100), when the universe is cool enough that protons can capture electrons to form hydrogen. The formation of hydrogen is the first and crucial step of structure formation. Indeed, the tiny fluctuations (~10-4 level) of CMB emission indicate the seed of the large-scale structure of our universe.

However, between redshift 1100 (CMB, as observed by COBE and Planck satellites) and redshift ~30, no information is revealed by our universe. Mostly, this is because the universe becomes neutral in this period, and the universe enters to its "dark age". Only after the formation of first stars and black holes, the ionizing photons emitted from these object will re-ionize (or partially at high redshift) the universe – so-called cosmic reionization, the universe can then be observed. Investigations of the universe at z>6 will help to understand how the universe evolve, i.e. when the first stars and galaxies form, when the first black holes (BHs) form. Moreover, high-redshift galaxies and quasars are systems that are far from mature, and they will reveal the early growth and evolution of these systems.

Content from this work may be used under the terms of the Creative Commons Attribution 3.0 licence. Any further distribution of this work must maintain attribution to the author(s) and the title of the work, journal citation and DOI.
Published under licence by IOP Publishing Ltd

Luminous high-redshift quasars are key probes of the history of cosmic reionization. Deep spectroscopy of z>6 quasars indicates that the intergalactic medium is significantly neutral at z>7 (e.g., Banados et al. 2018, Wang et al. 2020, Wang et al. 2021), while highly ionized at z<6 (e.g., Yang et al. 2020). Moreover, the earliest supermassive black holes (SMBHs), the central engines of distant quasars, are very important for our understanding of the formation mechanisms of the first-generation seed BHs (e.g., Inayoshi et al. 2020 for a recent review).

Our work focuses on the BH growth at high redshift. Figure 1 shows the age of our universe at high redshift. Cleary from this plot we can see that the total possible duration for a seed BH at z~=35 to grow to z~=10 is only $3*10^8$ yrs.

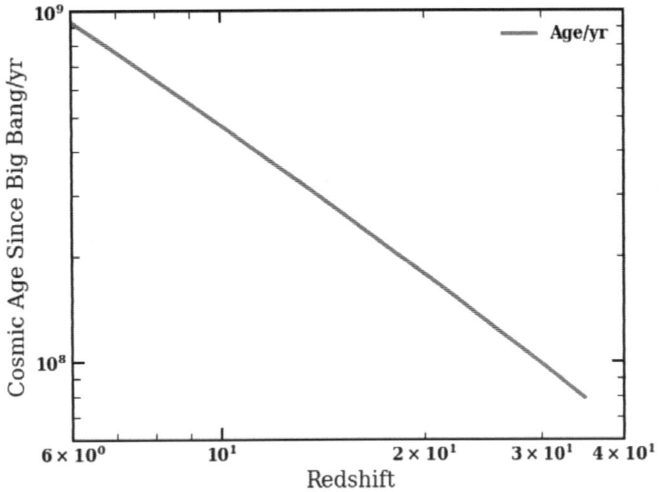

Fig. 1 The age of our universe (since Big Bang) versus redshift. It is clear from this plot that the total evolutionary duration from z~=35 to z~=10 is 3*108 yrs. [2]

1.2. how to observe and search for high redshift quasars

Quasars with redshift greater than 6 are called high-redshift quasars. To date, due to technical reasons, only luminous systems at high redshift can be observed.

There are some technical challenges in finding high-z quasars. Because bright quasars are hard to detect at large redshifts, a large part of the sky needs to be detected. The primary method of identifying quasars is their multicolor wide-band photometry, which can be effectively separated from star locus in the color space, especially by significant Lyman-α break (e.g. Warren et al. 1987). At high redshifts, this requires photometry in the reddest optical bands. For example, the Lyman rupture falls in the center of the public optical u, g, r, i, and z band filters at z = 1.9, 2.9, 4.1, 5.3 and 6.5, respectively. Finally, large amounts of data require efficient automated data processing. [2]

The search of high-z quasars is pioneered by Dr. X. Fan (see e.g., Wang et al. 2021 and references therein for efforts of his group). Table 1 list surveys that discovery high-z quasars. There are in total more than 50 quasars exist at 6.5<z<7. However, because the Lyman-α emission is redshifted to near-infrared, both imaging and spectroscopic observations become challenging, quasars at z>7 is quite limited, which are listed (partially) in Table 2. To date, there are less than 10 quasars with z>7 (e.g., Mortlock et al. 2011, Banados et al. 2018, Wang et al. 2018, Yang et al. 2020, Wang et al. 2021). Sources with z>7.5 are: ULAS J1342+0928 (z=7.54, with BH mass $7.8*10^8$ M*, here M* is the solar mass, Banados et al. 2018), J1007+2115 (z=7.515, with BH mass $1.5*10^9$ M*, Yang et al. 2020), and J0313-1806 (z=7.64, with BH mass $1.6*10^9$ M*, Wang et al. 2021). At high-redshift, the BH mass record goes to SDSS J0100+2802, a z=6.33 source that have a BH mass $1.24*10^{10}$ M* (Wu et al. 2015).

Table 1. List of surveys utilized in the discoveries of high-z quasars at redshift z>6. Adapted from Inayoshi et al. (2020).

Name	Bands	Area (deg^2)	Number of quasi-stellar objects	References
Subaru (including SHELLQs + Subaru SC)	Optical g, r, i, z, y	1,400	78	SHELLQS: Matsuoka et al. 2016, 2018a,b, 2019a
	Optical z_B, z_R	7	2	Subaru SC: Kashikawa et al. 2015
Pan-STARRS1	Optical g, r, i, z, y	31,000	44	Chambers et al. 2016
DELS (including DECaLS, BASS, MzLS)	Optical g, r, z	14,000	27	Dey et al. 2019
DES (including DES SV, Yr1, and DR1)	Optical g, r, i, z, Y	5,000	18	DES Collab. et al. 2005
SDSS	Optical u, g, r, i, z	15,000	26	York et al. 2000
CFHQS (including other CFHTLS)	Optical g, r, i, z	500	15	Willott et al. 2007, 2010b
UKIDSS (including ULAS, UKIDSS-DXS, and UHS)	IR z, Y, J, H, K	7,000^b	64	Lawrence et al. 2007
VISTA (including VHS and VIKING)	IR J, Ks	20,000	62	VHS: McMahon et al. 2013
	IR z, Y, J, H, K	1,500	31	VIKING: Edge et al. 2013; Venemans et al. 2019
VST ATLAS	Optical u, g, r, i, z + IR	4,700	4	Shanks et al. 2015
FIRST + NDWFS + FLAMEX	21 cm + optical + IR	4	1	McGreer et al. 2006
WISE (including unWISE + AllWISE)	mid-IR	All sky	71	Wright et al. 2010
2MASS	IR J, H, Ks	All sky	26	Skrutskie et al. 2006

Table 2. List of z>7 quasars. Adapted from Inayoshi et al. (2020).

Name	Surveys	Redshift	$M_\bullet/M_\odot$^{a}	f_{Edd}	Reference
ULAS J1342+0928	WISE/DELS/ UKIDSS	7.541 [CII]	$7.8^{+3.3}_{-1.9} \times 10^8$	$1.5^{+0.5}_{-0.4}$	Bañados et al. 2018
HSC J1243+0100	SHELLQs	7.07 MgII	$3.3^{+2.0}_{-2.0} \times 10^8$	$0.34^{+0.2}_{-0.2}$	Matsuoka et al. 2019b
ULAS J1120+0641	UKIDSS SDSS	7.085 SiIII/CIII]/MgII	$2.0^{+1.5}_{-0.7} \times 10^9$	$1.2^{+0.6}_{-0.5}$	Mortlock et al. 2011
DELS J0038-1527	DELS/WISE/ Pan-STARRS1	7.021 MgII/OIII	$1.33^{+0.25}_{-0.25} \times 10^9$	$1.25^{+0.19}_{-0.19}$	Wang et al. 2018
DES J0252-0503	DES/VHS/ULAS/ WISE/VIKING	7.021 Lyα/NV	$\sim 1.6 \times 10^9$	Unknown	Yang et al. 2019
HSC J2356+0017	SHELLQs	7.01 Lyα	$\sim 5.5 \times 10^8$	Unknown	Matsuoka et al. 2019a

In this work, we focus on the BH growth. In this case, the BH mass estimation is of fundamental importance. To date, various methods are developed. Among them, some are direct, while the rest rely on various assumptions and calibrations.

1) Direct estimation methods of black hole mass

Mostly, the direct estimation of BH mass relies on the Keplerian motion of surrounding objects, i.e. measuring the gravity of BH. It includes a single measurement of the dynamics of celestial bodies and monitoring of the motion of gas around the central black hole (e.g. aerodynamic method.). Within the gravitational influential radius, the gas behaves as Kepler under the influence of the gravity of the black hole. Detailed monitoring and modeling of these Keplerian motions will reveal the location and mass of central BH. This method is mostly applied to nearby SMBHs.

Journal of Physics: Conference Series **2441** (2023) 012033 doi:10.1088/1742-6596/2441/1/012033

Another way to directly calculate the BH mass is the reverberation mapping method. It assumes that the matter in the broad line region (also close to BH) is keplerian around the central BH. By using the continuous spectrum and the time delay of the broad line region, the radius/location of the broad line region is estimated. Its orbital velocity can also be derived from the Doppler shift of the emission lines. Combing these two measurements, we can then evaluate the mass of the central BH. This method is now widely adopted in literature to evaluate BH mass.

However, we note that both methods cannot be applied to estimate BH mass at high-redshift. Due to observational limitations, we currently rely on indirect methods, as listed below.

2) Indirect estimation methods of black hole mass

First, the empirical correlation between the BH mass and the velocity dispersion of stars in the galaxy, i.e. so-called $M_{BH} - \sigma$ relationship. Since the velocity dispersion is measured from stars far beyond the influence range of BH gravity (about 0.5 pc), it has become an important method to estimate the dynamic mass of BHs. Depending on whether the outflow and host galaxy are "energy driven" or "momentum driven," they can be divided into two categories. In the "energy driven model", we have (theory: Silk & Rees 1998, observation: McConnell & Ma 2013)

$$M_{BH} - \sigma_*^5$$

In the "momentum driven" model, we have (King et al. 2003)

$$M_{BH} - \sigma_*^4$$

Observationally, most systems follow the momentum driven model.

Second, a correlation between the mass of a black hole and the luminosity of the nuclear sphere. This relationship further indicates the co-evolution of SMBH and its host galaxy. The main advantage of this method is that, detailed spectrum is not required, and the luminosity of the nuclear sphere is sufficient (e.g., Graham, & Scott, 2013).

1.3. Seed BH of SMBH: formation pathways and mass range

SMBH can grow from accretion of gas and merger of companioning BHs. Among them it is known that that accreting gas is the main mechanism. For those high-redshift supermassive BHs, it is a challenging task for their seed BHs to grow efficiently to their observed values. As shown in Figure 2, current knowledge of formation pathways of seed BHs is limited (Volentori 2012).

(1) Pop III star collapse. If the primordial gas was dominated by hydrogen, it would have cooled and collapsed to form massive stars. Stellar evolution and collapse can form seed black holes, which can grow into supermassive black holes through super Eddington accretion. Seed BH in this channel is in range of $10^{1\text{-}2}$ M_*.

(2) Direct collapse of primordial galaxies. The direct collapse of a generation of iron-poor primordial galaxies can form seed black holes of intermediate mass. As the temperature gradually drops, the gas in the primordial galactic halo contracts to form stars. Black holes can form from the collapse of supermassive stars. The advantage in this case is that supermassive black holes can form without super Eddington accretion. However, the direct collapse model needs to occur near the galactic halo, so it is not a common mechanism. Seed BH in this channel is in range of $10^{5\text{-}6}$ M_*.

(3) Star mergers in dense galaxies. A third way to produce medium-mass seed black holes is through continuous collisions of stars in dense constellations. Frequent collisions and mergers within stellar groups can produce supermassive stars. The collapse of a supermassive star can form an intermediate seed black hole with BH mass in range of $10^{3\text{-}4}$ M_*.

Journal of Physics: Conference Series **2441** (2023) 012033 doi:10.1088/1742-6596/2441/1/012033

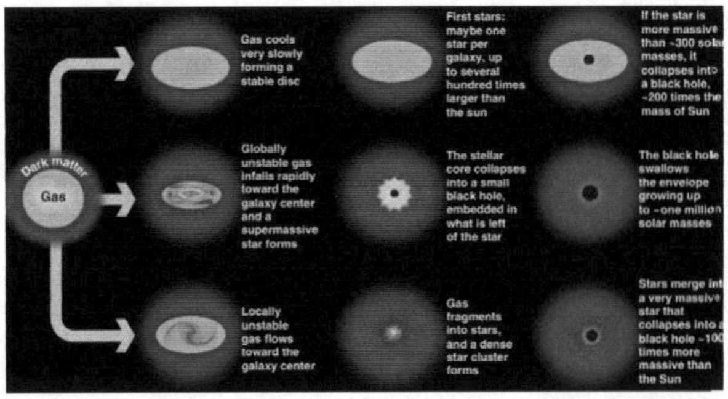

Fig. 2 Illustration showing three pathways to form massive seed BHs that can occur in early galaxies. Adapted from Volentori (2012).

1.4. Salpter Timescale

Salpter timescale is very important for our understanding of BH growth. It considers an ideal situation that the BH growth is dominated by Eddington-limited accretion that has a radiative efficiency of 10%. In this case, a seed BH can grow exponentially, with an e-folding time of the so-called Salpter timescale, $t_{Salpter} \approx 50\ Myr$.Observationally, most low-z SMBHs assemble most of their mass at so-called cosmic noon (z=2-3), where they grow at this radiative efficiency and accrete at near this accretion rate.

For example, a 100 M* Population III seed BH will need to accrete for ~0.8 Gyr to reach 109 M*. This duration is comparable with the age of our universe at z~=6. Moreover, it further requires a duty cycle of near-Eddington growth of 100%, which is very unlikely due to various effects. For examples, the UV radiation and supernova explosion associated with population III stars will also lead strong displacement of the gas reservoir near the post-formed BH. During the BH growth period, the gas near BH will be influenced by BH accretion itself.

1.5. Main factors affecting the growth of black holes

There are several factors that will affect the BH growth, and will be helpful to ease the timescale constraints of 10^{8-9} M* BHs at z~6-7. We should first note that, those SMBHs are observed highly biased, i.e. only bright (thus massive) ones can be observed at high redshift, while the rest fainter ones awaiting for discoveries. In this case, these monsters can be born in highly biased regions of the universe.

Generally there are two ways to create 10^{8-9} M* BHs at z~6-7. One to increase the mass of seed BHs (see above), and the other is to increase the growth rate. Obviously, a high duty cycle of accretion in near-Eddington (or even super-Eddington) phase is required. However, it the accretion can be highly super-Eddington, then the corresponding duty cycle can be somewhat lower. Besides the mass supply of gas, the angular momentum of supplied gas will also have strong effect on the BH growth. The spin of BH will evolve. On the other hand, the radius of BH, i.e. the deepest potential of BH, will then be revised. This will lead to a moderate change in the radiative efficiency of accretion disks. As detailed below, the cold accretion can have a radiative efficiency of 6% for a non-spinning BH, and an efficiency of 43% for a maximal-spinning BH.

2. Theoretical model

In this section, we will provide the basis of our model. In this work, we only consider that the BH is grown through accretion processes, i.e. we assume even at high redshift the merger processes are of minor impact. This is likely true, especially for those bright quasars whose BHs are massive, i.e. greater than 10^8 M*. Black hole accretion is a fundamental physical process in the universe and is the main engine behind active galactic nuclei (AGNs, including quasars), black hole binaries, and gamma-ray bursts.

In this work, we take advantages of progress in accretion theory. For comparison, most work in literature only consider the accretion of SSD (e.g., Shapiro 2005).

2.1. *Black hole evolution model*

In accretion theory, we can define the radiative efficiency as $\epsilon_M \equiv \frac{L}{\dot{M}_0 c^2}$. Here L is the luminosity and $\dot{M}_0$ is the mass accretion rate. The radiative efficiency determines the capability of converting rest-mass (and gravitational) energy into radiation through accretion process. Moreover, there exist a critical luminosity, so-called Eddington luminosity $L_{Edd} \approx 1.3 \times 10^{38} \left(\frac{M}{M_*}\right) erg/s$, above which the radiative force exceeds the gravity and ideally accretion stops. With the Eddington luminosity, we can further define the luminosity in Eddington unit, i.e. $\epsilon_L \equiv \frac{L}{L_{Edd}}$. The BH growth rate must take the loss of accretion mass energy in the form of radiation. After this correction, theoretical we can express the BH growth as (Shapiro 2005):

$$\frac{dM}{dt} = \frac{\varepsilon_L(1\text{-}\varepsilon_M)}{\varepsilon_M}\frac{M}{\tau}$$

where τ is the characteristic accretion timescale $\tau = \frac{Mc^2}{L_{Edd}} \approx 0.45\mu_E{}^{-1} Gyr$ and is independent of M. Note that, if ϵ_L and ϵ_M are constant, the above equation has an analytical expression, i.e.

$$M(t) = M_{seed}\exp[\frac{t}{\tau_0}],$$

where M_{seed} is the mass of seed BH and $\tau_0 = \tau \times \frac{\epsilon_M}{\epsilon_L(1-\epsilon_M)}$. The Saplter timescale is defined when $\epsilon_L = 1$and $\epsilon_M = 0.1$, i.e. $\tau_{Salpeter} \approx 0.05\ Gyr$. In this condition, we can estimate the time for a seed BH to grow its mass up to M to be

$$t = \tau_{Salpeter} \times ln\frac{M}{M_{seed}}.$$

Obviously, it takes much longer for a smaller M_{seed} BH to grow.

Besides the evolution of BH mass, there is another equation that governs the evolution of BH spin. Here the dimensionless BH spin a_* ($a_* = \frac{J}{GM^2c}$ where J is the angular momentum of BH). Considering the angular momentum conservation during accretion process, we can write (Shapiro 2005, Zhang et al. 2020),

$$\frac{da}{dt} = [\tilde{l}(R_{in}) - 2a_*(1-\epsilon_M)]\frac{\epsilon_L}{\epsilon_M\tau}.$$

Here $\tilde{l}(R_{in})$ is the specific angular momentum of accreting gas at R_{in}. With the governing equations of both BH mass and spin, we can then evaluate the growth of BH.

2.2. *Black hole accretion model*

Since the discovery of black hole accretion, great progress has been achieved (e.g., King 1981, Frank et al. 2003 and Yuan & Narayan 2014 for reviews). Currently, there are three distinctive accretion models, depending on the accretion rate.

SSD - When the accretion rate is moderate high, it will enter the Shakura-Sunyaev disk (SSD) regime. This is the first accretion model developed. The optical depth of SSD is extremely high, thus it has a thermal blackbody-like radiation. Besides, SSD is fairly cold, with disk height-to-radius aspect ratio H/R much less than unity. The SSD has been successfully applied to bright AGNs and quasars. It is realized when the mass accretion rate $\dot{M}$ is below the Eddington rate $\dot{M}_{Edd} = \frac{L_{Edd}}{c^2}$.

In SSD, the energy and angular momentum per unit rest mass accreted by a black hole are the energy and angular momentum of a unit mass at the ISCO (the location of R_{in}), immediately prior to its rapid plunge and capture by the hole:

$$\tilde{E}_{ISCO} = \frac{r_{ISCO}{}^2 - 2r_{ISCO} + a_*\sqrt{r_{ISCO}}}{r_{ISCO}(r_{ISCO}{}^2 - 3r_{ISCO} + 2a_*\sqrt{r_{ISCO}})^{\frac{1}{2}}},$$

$$\tilde{I}_{ISCO} = \frac{\sqrt{r_{ISCO}}(r_{ISCO}{}^2 - 2a_*\sqrt{r_{ISCO}} + a_*{}^2)}{r_{ISCO}(r_{ISCO}{}^2 - 3r_{ISCO} + 2a_*\sqrt{r_{ISCO}})^{\frac{1}{2}}}.$$

Here the ISCO radius r_{ISCO} is given by (e.g., Shapiro 2005):

$$r_{ISCO} = \{3 + Z_2 - sign(a_*)[(3 - Z_1)(3 + Z_1 + 2Z_2)]^{\frac{1}{2}}\},$$

where $Z_1 \equiv 1 + (1 - a_*{}^2)^{\frac{1}{3}}[(1 + a_*)^{\frac{1}{3}} + (1 - a_*)^{\frac{1}{3}}]$, and $Z_2 \equiv (3a_*{}^2 + Z_1{}^2)^{\frac{1}{2}}$. The mass accretion efficiency and spin evolution parameters corresponding to the thin disk model are then given by:

$$\epsilon_M = 1 - \tilde{E}_{ISCO}.$$

Note that, ϵ_M varies from 5.7% at $a_* = 0$ and 43% at $a_* \approx 1$, i.e. the difference can be as large as a factor of 7-8. For a very crude estimation, as done in the Salpter timescale calculation, we can adopt $\epsilon_M = 0.1$. Another important (and distinct to the other two models) property of SSD is that, its radiative efficiency is independent of accretion rate.

Slim disk accretion - When the accretion rate is much larger than $\dot{M}_{Edd}$, the accretion can no longer remain cold. In this case, the viscous dissipated accretion energy can heat the accretion flow rather than being radiated away. In other words, a notable fraction of energy will be stored as thermal energy of the gas and will be advected into BH. The important role of advection in accretion flow was first noticed by Ichimaru (1977). This is so-called slim accretion regime, because the aspect ratio H/R~1. Because of high accretion rate, slim disk remains optically thick, thus it also emits thermal emission. In this work, we take the fitting formulae of Mineshige et al. (2000), i.e.

$$\epsilon_M = \frac{2[1+\log\left(\frac{\dot{M}}{50\dot{M}_{Edd}}\right)]}{\dot{M}} \quad \text{when } \dot{M} > 50,$$

$$\epsilon_M = \frac{1}{25} \quad \text{when } 50 > \dot{M} > 3.$$

Interestingly, the radiative efficiency decreases as $\dot{M}$ increases. This is mostly because of photon trapping effect, i.e. because of high optical depth within accretion disk, the emitted photons cannot escape out of the disk surface before they are advected into the BH (Mineshige et al. 2000).

For the impact of BH spin, we assume that slim disk has the same dependence on BH spin to that of SSD.

Hot accretion flow – When the accretion rate is much lower than $\dot{M}_{Edd}$, i.e. below the critical accretion of SSD (taken as $0.5\dot{M}_{Edd}$), then system will enter into the hot accretion flow regime, with H/R~1. Energy advection also plays a crucial role in hot accretion flow, consequently its radiative efficiency is also lower than that of SSD. The emission from hot accretion flow is no longer thermal, because the system is optically thin. Instead, synchrotron, bremsstrahlung and Compton scattering are the dominate radiative mechanisms. Interestingly, opposite to the case of slim accretion, the radiative efficiency of hot accretion flow increases with increasing $\dot{M}$ (Xie & Yuan 2012). This is because of Compton scattering, which strongly depends on optical depth (equivalently, $\dot{M}$). At even higher accretion rate, the cold clumps will form, and they will provide quasi-thermal photons, which can be inverse Compton scattered by hot electrons with the hot accretion flow. The radiative efficiency of this two-phase accretion model is nearly constant, independent of $\dot{M}$.

In this work, we take a simplified version of fitting formulae from Xie & Yuan (2012),

$$\epsilon_M = 8\% \quad \text{when } 1 > \dot{M} > 0.1,$$

$$\epsilon_M = 8\%\left(\frac{\dot{M}}{0.1}\right)^2 \quad \text{when } \dot{M} < 0.1.$$

The impact of BH spin is considered separately. We notice that, in hot accretion flow, the accretion flow extends down to BH horizon. Besides, the innermost regions suffer gravitational redshift for the outcome radiation. With these two reasons, the BH spin will have a weaker impact on the radiative efficiency in hot accretion flow. We here take a simplified treatment, i.e. we take the above formulae to be valid for $a_* = 0$, and further assume that the radiative efficiency has a dependence on BH spin with the form $\epsilon_M \propto \sqrt{1 + a_*/3}$.

2.3. *Chaotic accretion with random orientation in angular momentum*

One important, but not-constrained, property of accreting gas is the direction of their angular momentum. Assuming a fixed orientation during the whole BH growth period is admittedly over-simplified. In this work, we consider, in several cases, a chaotic supply of gas near BH, i.e. the orientation of angular momentum n is random. In this treatment, we can consider a special case that both the accretion rate and/or orientation of supplied gas are different at each epoch. With this treatment, we can investigate the BH growth more realisticly.

When the orientation of supplied gas is considered, we need to compare it to that of BH. The "effective" BH spin will be positive if they are the same, and negative if they are opposite to each other.

3. Numeral Calculations

The key motivation of this project is to investigate the black hole growth at early universe (or equivalently, at high redshift). For this purpose, we relax ourselves from the oversimplified Salpeter assumption, but instead consider more realistic accretion physics. They include the impact of black hole spin evolution, as well as the change in accretion mode. [2] Compared to those in literature, we in this work include three main accretion disks, i.e. slim accretion, SSD, and hot accretion flow.

Another difference to those in literature is that, we here consider the mass accretion rate $f_{Edd} = \frac{\dot{M}_0}{\dot{M}_{Edd}}$. We take this parameter because it directly relates to gas supply. While the $\epsilon_L = \frac{L}{L_{Edd}}$ parameter adopted in Shapiro (2005) includes the effect of the radiative efficiency ϵ_M at given accretion rate, i.e.

$$\epsilon_L = \epsilon_M f_{Edd}$$

3.1. *Fixed BH spin ($a_* = 0$) case*

For this application, we fix the BH spin to $a_* = 0$, and consider only three representative accretion rates, i.e. $f_{Edd} = 0.1$ (hot accretion flow regime), $= 1$ (SSD regime) and $= 10$ (slim accretion regime). Since we consider the non-rotating BH case, the corresponding radiative efficiency is, respectively, 8%, 4% and 4%. For the calculations, we re-calibrate the mass of seed BH, to reach a fixed 10^{10} M* BH at z=6.5. This is similar to the observational data of Wu et al. (2015), where the requirement on BH mass growth is highest. The results are shown in Figure 3.

Obviously, because the BH e-folding timescale $\tau_0 = \frac{\tau}{f_{Edd}(1-\epsilon_M)}$ is inversely proportional to f_{Edd}, systems with highest accretion rate will have lowest τ_0. As shown in Figure 3, only when $f_{Edd} > 1$, the BH will have significant mass growth. The seed BH mass is nearly 109 M* for $f_{Edd} = 1$, while it significantly reduces to 102 M* for $f_{Edd} = 10$.

3.2. *Evolving BH spin case, with fixed orbital orientation of feeding gas*

We then consider the evolution of BH spin. For this application, we still take the three representative accretion rates. We note that, the orbital orientation is fixed in this case. The highest-feeding case ($f_{Edd} = 10$) has the quickest spin evolution, up to $a_* = 1$. For comparison, $f_{Edd} = 0.1$ case only evolve to $a_* = 0.5$. The results are shown in Figure 4.

Because of BH spin evolution (systematically to higher a_*), BH growth rate is systematically lower compare to that with fixed $a_* = 0$ cases. This is most clearly shown by the $f_{Edd} = 10$ curve. We find that, the seed BH mass now is about one order of magnitude higher than previous calculation, i.e. we now require a 103 M* seed BH.

3.3. Chaotic orbital orientation of feeding gas

We then consider that the feeding gas has random orbital orientation. For this, we assume that, the orientation remains unchanged within 10^4 yrs, and changes randomly every 10^4 yrs. Again, for simplifications, we still take three representative mass accretion rates. For this application, we still take the three representative accretion rates. One of the most surprising results is that, because of random orbital orientation, the BH spin remains un-affected, i.e. $a_* = 0$. In other words, the Chaotic orbital orientation cases are very similar to that of fixed $a_* = 0$ cases. The results are shown in Figure 5.
One important application of this result is that, if in future we can measure the spin of the BH in those high-redshift quasars, we may then infer their feeding properties from a_*.

3.4. Fully chaotic gas feeding

We then consider the fully chaotic gas feeding case, i.e. we not only have random orientation of feeding gas, but we also let the feeding rate being random. For this purpose, we let f_{Edd} to uniformly vary between 10^{-2} and 10^1.The result is shown in Figure 6.

One key result is that, in this application, we vary f_{Edd} (accretion rate in Eddington unit) instead of the actual accretion rate. In this case, we will not have over-feeding during early evolutionary period when the BH mass is still low. We find that, despite the fully chaotic luminosity observed (see right panel of Figure 6), the BH growth is admittedly moderate, similar to the case when we fix $f_{Edd} = 1-2$. Again, for this chaotic feeding, the BH remains non-spinning.

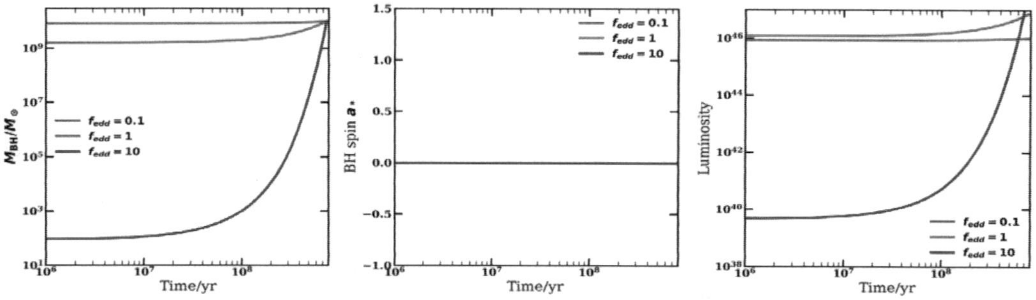

Fig. 3 BH growth for fixed BH spin ($a_* = 0$) case. We consider three representative accretion rates, i.e. $f_{Edd} = 0.1$ (hot accretion flow regime), $= 1$ (SSD regime) and $= 10$ (slim accretion regime).

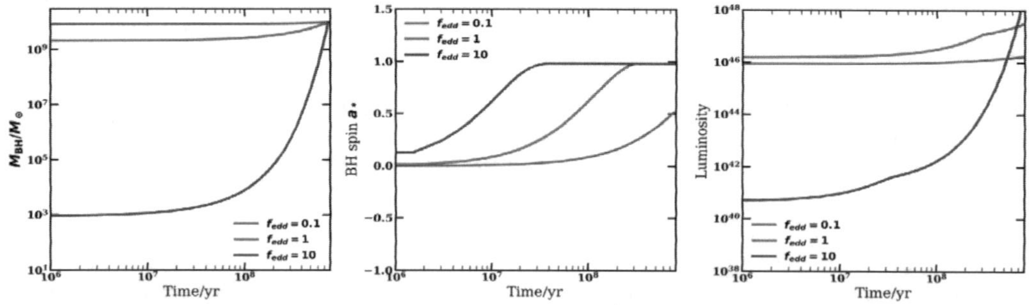

Fig. 4 BH growth. The BH spin now evolves. The feeding gas has the same orbital orientation during the whole evolutionary period. We consider three representative accretion rates, i.e. $f_{Edd} = 0.1$ (hot accretion flow regime), $= 1$ (SSD regime) and $= 10$ (slim accretion regime).

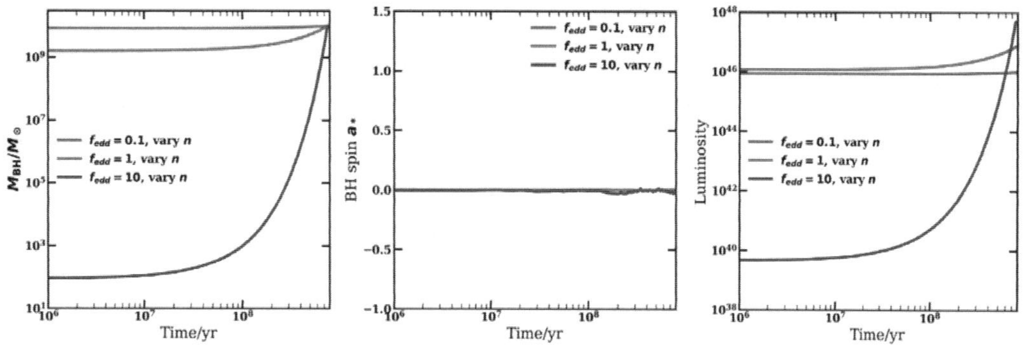

Fig. 5 BH growth. The BH spin now evolves, where we randomly change the orbital orientation of feeding gas. We consider three representative accretion rates, i.e. $f_{Edd} = 0.1$ (hot accretion flow regime), $= 1$ (SSD regime) and $= 10$ (slim accretion regime).

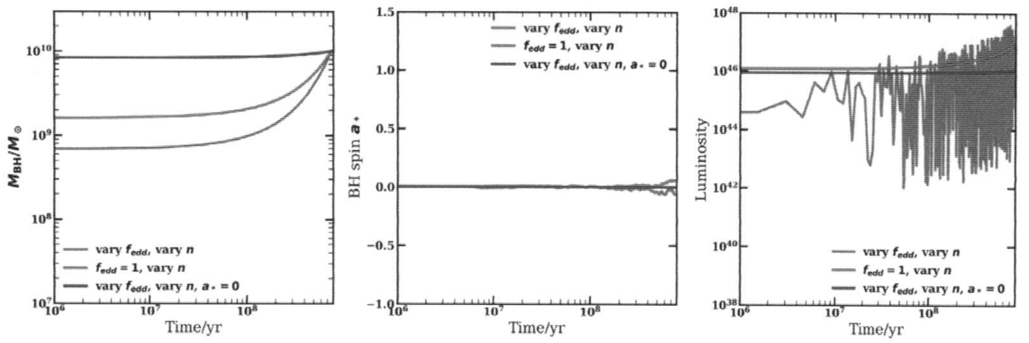

Fig. 6 Chaotic BH growth. For comparison, two other simplified cases are also shown in this plot.

4. Conclusion

Observations of galaxies and quasars at early universe is of crucial importance for our understanding of large-scale structure formation, i.e. between a 10^{-4} level fluctuation at z~1100 from Planck to galaxies and galaxy clusters (and large-scale structures like voids and filaments) since z~3-5. Among them, the co-evolution between BH and its host galaxy at high redshift is also a hot topic. We in this work investigate the BH growth through gas accretion process in early universe. We consider the orientation and mass rate of feeding gas, and find that, if in reality the feed gas has a random orientation, the spin of resultant BH will be nearly non-spinning, this can be tested in future. Besides, we also find that, sufficiently quick feeding is required for those high-redshift quasars, a mass accretion rate in Eddington unit greater than 2-5 will be necessary.

References

[1] Yuan, Feng, Narayan, Ramesh (2014) Hot Accretion Flows Around Black Holes. Annual Review of Astronomy and Astrophysics, 52.
[2] Zhi Cao. The Formation of Supermassive Black Holes at Early Universe. 2021 IOP Conf. Ser.: Earth Environ. Sci., Vol. 658, 012032.
[3] M. Volonteri. The Formation and Evolution of Massive Black Holes[J]. Science, 2012, 337(6094).
[4] Fender R., Belloni T., Stellar mass black holes and ultraluminous x-ray sources. Science 337, 540 (2012).
[5] Madau P., Rees M. J., Massive black holes as population III remnants. Astrophys. J. 551, L27 (2001). 10.1086/319848.

[6] Gültekin K., et al., The M- σ and M -L relations in galactic bulges, and determinations of their intrinsic scatter. Astrophys. J. 698, 198 (2009). 10.1088/0004-637X/698/1/198.
[7] Begelman M. C., Volonteri M., Rees M. J., Formation of supermassive black holes by direct collapse in pre-galactic haloes. Mon. Not. R. Astron. Soc. 370, 289 (2006), arXiv.
[8] Johnson J. L., Whalen D. J., Fryer C. L., Li H., The growth of the stellar seeds of supermassive black holes. Astrophys. J. 750, 66 (2012). 10.1088/0004-637X/750/1/66.

Application of CVOFLS method in multi vortex shear flow field

Liying Cui [1, *], Yingge Yang [2], Cuiping Ren [3]

[1] Xi'an Eurasian University, Xi'an, Shaanxi 710065, China
[2] Xi'an North Electro-Optic Science and Technology Defense Co., Ltd, Xi'an, Shaanxi 710043, China
[3] Xi'an Eurasian University, Xi'an, Shaanxi 710065, China

*Corresponding author e-mail: cuiliying@eurasia.edu

Abstract. Coupled Volume of Fluid and Level Set method inherits the advantages of VOF and Level Set methods, takes VOF function as the main body to simulate the fluid motion interface, and corrects the normal direction of the interface through Level Set function, so as to effectively overcome the shortcomings of the two methods. The numerical simulation example of multi vortex shear flow field shows that this method can ensure better simulation accuracy of moving interface and higher calculation efficiency.

Keywords: motion interface; CVOFLS; Simulation accuracy; computational efficiency.

Introduction many phenomena in real life are closely related to the moving interface, such as water drop falling, gas explosion and so on [1]. Therefore, the study of the numerical simulation of the moving interface is of far-reaching significance. Among Euler methods, Level Set method [2] and VOF method [3] are widely used because of their small memory occupation, easy to solve the problem of fluid moving interface with complex structure and simple implementation. Based on this, some scholars have proposed CLSVOF method, which couples the advantages of the two methods.

Sun Dongliang [4] proposed a new CLSVOF method in 2010. Although this method simplifies the original CLSVOF method, its method theory and the process of reconstructing the interface are relatively complex. Li Qiang [5] created a new CLSVOF method in 2011. This method has its own mass correction formula, and the accuracy of calculating free motion interface is very good, but its calculation efficiency is not high.

The Level Set method can calculate the normal direction and curvature of the interface more accurately in simulating the moving interface. In addition, considering that the VOF method has the advantages of high calculation efficiency and good mass conservation, and its disadvantages of poor normal direction and curvature of the interface; Therefore, if its defect is coupled to VOF method, it can be overcome. In this paper, a new method coupling level set and VOF – CVOFLS method [6] is adopted. Through the simulation of multi vortex shear flow field, it is verified that CVOFLS method has higher efficiency and better interface conservation.

Content from this work may be used under the terms of the Creative Commons Attribution 3.0 licence. Any further distribution of this work must maintain attribution to the author(s) and the title of the work, journal citation and DOI.
Published under licence by IOP Publishing Ltd

1. Transport Equation

CVOFLS method is a new method of coupling VOF and level set. The velocity field of the moving medium is $\boldsymbol{u}(u, v)$, and the expression of the transport equation satisfied by the level set function $\phi(\boldsymbol{x}, t)$ and VOF function $F(\boldsymbol{x}, t)$ are as follows:

$$\frac{\partial \phi}{\partial t} + \mathbf{u} \cdot \nabla \phi = 0 \tag{1}$$

$$\frac{\partial F}{\partial t} + \mathbf{u} \cdot \nabla F = 0 \tag{2}$$

In CVOFLS method, the numerical method of Zhou Wen [6] is used to discretize equations (1) and (2).

During the operation of level set function, the function $\phi(\boldsymbol{x}, t)$ needs to be reinitialized to obtain a new symbolic distance function from $\boldsymbol{x}$ point to interface; This method can ensure the numerical stability of Level Set function and make it have excellent smoothness in the solution domain.

$$\begin{cases} \frac{\partial \phi}{\partial \tau} = sign(\phi_0)(1 - |\nabla \phi|) \\ \phi(\mathbf{x}, 0) = \phi_0(\mathbf{x}) \end{cases} \tag{3}$$

The symbolic distance function is reinitialized by the steady-state solution obtained by discretizing equation (3).

2. Numerical Method

2.1. Discretization of VOF transport equation

For the convenience of calculation, the equation can be written in the conservative form:

$$\frac{\partial F}{\partial t} + \nabla \cdot (\mathbf{u}F) = F\nabla \cdot \mathbf{u} \tag{4}$$

Sussman and Puckett [7] proposed flux splitting algorithm to solve equation (4), and its discrete form is as follows:

$$F_{i,j}^{(n+1)} = \hat{F}_{i,j} - \Delta t \left(\tilde{F}_{i,j} \cdot \frac{u_{i+\frac{1}{2},j}^{(n)} - u_{i-\frac{1}{2},j}^{(n)}}{\Delta x} + \hat{F}_{i,j} \cdot \frac{v_{i,j+\frac{1}{2}}^{(n)} - v_{i,j-\frac{1}{2}}^{(n)}}{\Delta x} \right)$$

3. Numerical Examples

In order to verify the effectiveness of the CVOFLS method proposed in this paper, the simulation effect of cvofls method is tested by simulating the interface motion of multi vortex shear flow field, that is, the velocity field of a circle with a center of (0.5, 0.5) and a radius of 0.15 is investigated.

$$u = sin(4\pi(x + 0.5))\, sin(4\pi(y + 0.5))\, cos(\pi t/T)$$

$$v = cos(4\pi(x + 0.5))\, cos(4\pi(y + 0.5))\, cos(\pi t/T)$$

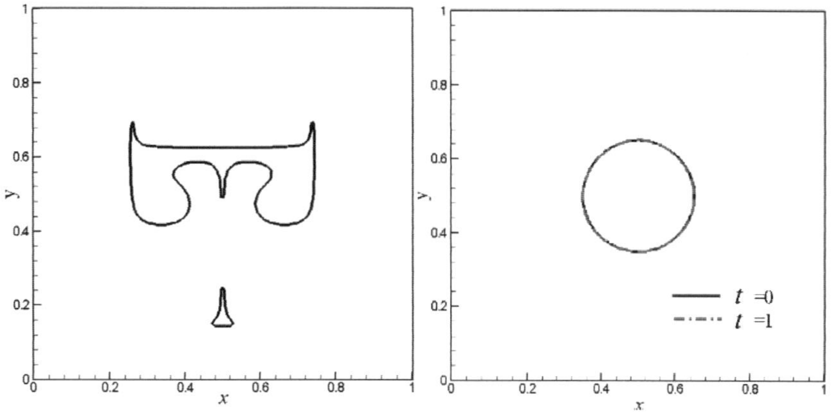

(a) $t = 0.5s$ (b) $t = 1s$

Fig. 1 simulation results of multi vortex shear flow field (T=1s)

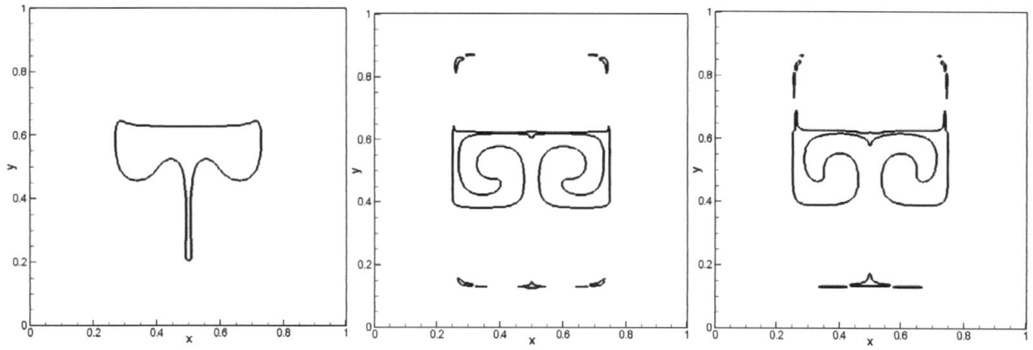

(a) $t = 0.2s$ (b) $t = 1s$ (c) $t = 1.4s$

Fig. 2 simulation results of multi vortex shear flow field (T=2s)

Fig. 1 and Fig. 2 show the numerical results of multi vortex shear flow field simulated by CVOFLS method with periods $T = 1s$ and $T = 2s$ respectively. It can be seen from the figure that the numerical results simulated by CVOFLS method at $T = 1s$ basically coincide with the results at the initial time, indicating that it has good stability and conservation. When $T = 2s$, the numerical results of the simulation are good and have good conservation.

4. Conclusion

The numerical results show that CVOFLS method has a good effect on the numerical simulation of multi vortex shear flow field. The numerical results verify the accuracy of CVOFLS method. At the same time, it shows that this method has higher computational efficiency and better interface simulation accuracy in numerical interface simulation.

Acknowledgments

Xi'an Eurasian University School level Scientific Research Fund Project(2021XJZK03); Xi'an Eurasian University School level Scientific Research Fund Project(2022GCPY01); Shaanxi Provincial Sports Bureau Regular Project (2020289); Shaanxi Province Educational Science "13th Five-Year" Plan 2020 Key Projects (SGH20Z021).

References

[1] Liu R X, Wang Z F. Movement interface tracking and numerical method[M]. Hefei: University of Science and Techn- ology of China Press, 2001.

[2] Osher S.,Sethian J.A. Fronts propagating with curvature dependent speed: algorithms based on Hamilton-Jacobi formulat- ion[J]. Journal of Computational Physics, 1988, 79 (1): 12-49.

[3] Sethian J.A. Fast marching methods[J]. SIAM Review, 1999, 41(2): 199-235.

[4] D.L. Sun, W.Q. Tao. A coupled volume-of-fluid and level set (VOSET) method for computing incompressible two-phase flows[J]. International Journal of Heat and Mass Transfer, 2010, 53(4): 645~655.

[5] Li Qiang, Ouyang J, Yang Bingxin, et al. Numerical simulation of gas-assisted injection molding using CLSVOF method[J]. Applied Mathematical Modelling, 2011, 35(1): 257–275.

[6] Zhou Wen, ou Yangjie, Cui Liying Cvofls method for moving interface tracking [J] Journal of engineering mathematics, 2015,32 (5): 697-708.

[7] M. Sussman, E.G. Puckett. A Coupled Level Set and Volume-of-Fluid method for computing 3D and axisymmetric incompressible two phase flows[J]. Journal of Computational Physics, 2000, 162(2): 301~337.

Morphology of Broad Emission Lines from Binary Supermassive Blackholes

Hongyi Xiong [1, a], Peng Jiang [2, b]

[1] Shenzhen College of International Education, Shenzhen, Guangdong, China
[2] Polar Research Institute of China, Shanghai, China

[a] ryan_xhy@outlook.com, [b] jiangpeng@pric.org.cn

Abstract. Accretion onto a blackhole can indue extreme radiation stimulating emission lines in nearby gases. In the deep gravitational well, these clouds move at speeds up to 10,000 km/s, and so broadens the emission lines via the relativistic Doppler Effect. These broad emission lines on spectra are the most prominent feature of AGNs. The region around central black hole, where such broad emission lines are produced, is known as the broad-line region (BLR). We developed a self-consistent model to describe both the geometry of the BLR and the dynamics of the gas clouds in the BLR. The motion of clouds is described by a series of Keplerian orbits. The spectral shift of emission line can be derived by the radial velocity at each point in Keplerian orbits. Assuming that the BLR is stable, that is, the structure and distribution of gas in BLR doesn't change over a considerable period of time, the accumulated emission line profile can be constructed by simple addition of all Keplerian orbits in the BLR. By comparing the calculated profiles to spectra observed in SDSS (Sloan Digital Sky Survey), we found that our model can satisfactorily match most observed profiles. We attempted to review and summarize potential binary AGNs. We applied the model to the study of binary AGNs and their profiles of emission lines. We considered the velocity offset due to mutual Keplerian motion of black holes in binary and added their respective profiles to produce the final, resulting emission line profile from binary AGNs. Three kinds of emission line profiles are featured in our simulations: twin-peak structures where two strong broad-lines are present; sub-peak structures where a smaller but still visible emission line accompanies the dominant line; single broad emission peak but with significant velocity offset from the systematic redshift that is derived from the narrow emission-lines on the same spectrum. We then analyzed 1,348 AGN spectra with high S/N ratio in SDSS database, and selected 26 as candidate binary supermassive blackhole systems. These candidates are valuable for following-up observation, their binary nature could be confirmed by detecting the profile variation due to orbital motions.

Keywords: Binary systems; Supermassive blackholes; AGN; broad emission lines; emission profile.

Content from this work may be used under the terms of the Creative Commons Attribution 3.0 licence. Any further distribution of this work must maintain attribution to the author(s) and the title of the work, journal citation and DOI.
Published under licence by IOP Publishing Ltd

1. Introduction

1.1. Binary Blackholes

From reverberation mapping techniques, the accretion disk component ranges to approximately the order of light-days, so that angular momentum considerations will affect accretion rates. For gases at distances 10kpc, falling sufficiently close to the blackhole, i.e., ~0.01pc, where friction and viscosity can become important, requires only 10^{-5} of the initial angular momentum per mass (see Peterson, Introduction to AGN). Such loss can be greatly explained by tidal force between interacting binaries, or binary blackholes. Strong gravitation interaction between galaxies at sufficiently close distances can remove angular momentum for accretion to occur. Binary blackholes are important mechanism for accretion and growth of supermassive blackholes. Hence, the study of binary AGN systems can greatly improve our understanding of the nature of such luminous objects, and how they came to become active. Such AGN systems are studied via optical methods, by observing their emission spectra. In particular, Seyfert I galaxies demonstrate "broadline" features, which are important in AGN identification.

The precise physics of the broad-line region (BLR) is unknown. This is because the BLR regions in distance AGNs cannot be spatial resolved. Reverberation techniques identifies a strong response of the BLR from continuum flux, suggesting optically thick clouds and a BLR size of about 10 light-days. Attempts have been made to identify spectra unique to binary sources, most notably by Boroson & Lauer [3]. They identified the source J1536+0441 (See Figure 1) as a binary blackhole at separation ~0.1pc, and velocity separation 3500km/s. Emission spectra demonstrated binary peak systems in all Balmer broadlines, with redshifts $z = 0.3889$ and $z = 0.3727$, respectively, the red and blue system, clear evidence of the superimposing nature of binary blackholes. Different interpretation exists, however. Zhang et al. (2019) interpreted J1536+0441 as a single blackhole due to the lack of the blue system at the He I and the Paschen lines as well as lacking periodical variations [4]. The model suggested that a shocked outflow is responsible for the emergence of double emission lines. Hence, great uncertainty lays in the spectra of binary systems. In addition, double peaked structures have been reportedly been identified in single-blackhole systems, for example NGC 1097 where a broad, double peaked component is seen alongside a superimposed BLR. Storchi-Bergmann et al. (2017) suggested that this is a signature of accretion disk emission, arising from its outer part [5].

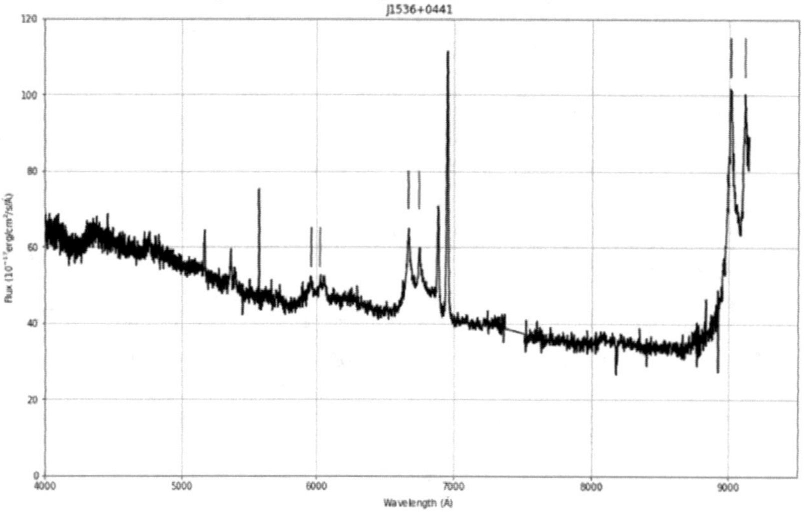

Figure 1. SDSS spectra of J1536+0441.The marked lines indicate the $H\gamma$, $H\beta$, and $H\alpha$ systems. Blue line indicates the lower redshift system $z = 0.3727$, and the red lines indicate the higher redshift system, $z = 0.3889$. Double peaked narrow lines are the $O_{III]}$ $\lambda 5007$ and $O_{III]}$ $\lambda 4959$, at a similar redshift to the red system. Boroson et al. identified the binaries to be at a distance 0.1 pc, and thus would share the NLR region [3].

Some binary systems, too, have been identified with single emission lines that displays no apparent difference to the spectra of single blackholes, but their periodic variation of brightness. Graham et al. (2015) identified the periodic optical variation in the AGN PG1302-102, whose spectrograph (emission lines) displays no apparent difference to that of single AGNs [6]. Skewed lines from their center-of-mass redshifts were also observed in binary candidates, for example J001224 (Eracleous et al. 2012), in which the peak is shifted by $90\ km \cdot s^{-1}$ relative to the narrow line redshift (the O III lines) [7]. This is thought to arise from the weaker nuclear activity of its binary counterpart, erasing the additional set of lines.

Overall, although broad-lines are prominent features to AGN spectra, their profile structure in binary systems remains unclear and unpredictable. It would therefore be of interest to be able to summarize relevant patterns and structures in binary spectra.

2. Model Theory

The line profile can be conveniently expressed in velocity space. The broadline profile is then equivalent to the velocity distribution; areas where a velocity occurs the most contributes the most amount of flux to the profile. In practice, spectrograph resolution cannot be infinitely large, for our interest we take the "resolution" to be ~50 km/s (almost the same resolution of SDSS surveys). That is, we will sample at 50 km/s intervals over the entire orbit for BLR clouds.

For simplicity, our geometric model will assume the following:

a) Mass continuity of the BLR clouds. Although the BLR clouds are assumed to be sparse and non-obscuring, the individual clouds are fast enough in orbit to allow the assumption of mass continuity: At any given point on the orbit the rate of the mass of the clouds entering is equal to the rate at which the clouds are leaving. This assumption is evident from the observed variation of BLR spectra (time taken for the profile to change significantly), which is at the order of ~10 years. With such a long variation timescale, the stability of the orbit is ensured, thus allowing us to find the distribution of clouds, treating it as a continuous element.

BLR clouds can be treated as evenly distributed on the Mean anomaly, represented by $M(t) = \frac{2\pi}{P}(t - t_p)$, where t_p is a time normalization constant. The linear relationship between time and the mean anomaly is sufficient for our uses. Specifically, we assume that the mass distribution is linear with time. For where Δt is large, we expect a larger proportion of BLR clouds and hence contributing more flux to the profile.

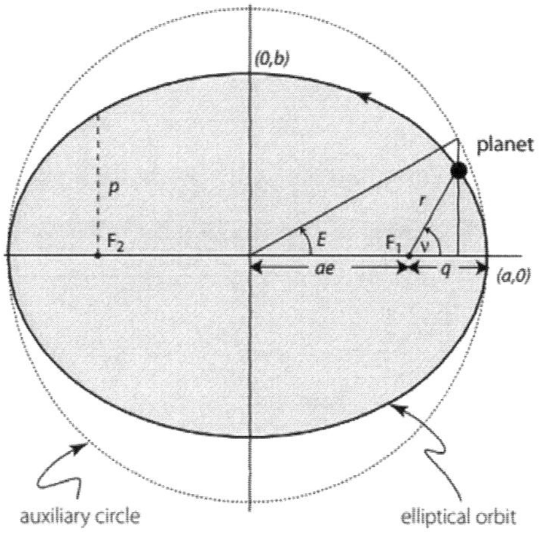

Figure 2. The Eccentric anomaly and the true anomaly. The mean anomaly is defined as the angle the planet would travel (in the same time) on the ellipse's auxiliary circle. Credit: The Exoplanet Handbook, Chapter 2.

The simple relation between time intervals and mass allows us to be able to directly derive the line profile from a radial velocity curve, as the time intervals corresponding to a specific velocity interval is proportional to the mass of BLR clouds at that interval. This will be described in detail in section 2.4.

b) In addition, we assume a simple relationship, in the form of a power law, for the radial emission capabilities. It is clear that the emission capability of an orbit with semi-major axis a must depend on the volume (mass) of gas radiating and the intensity of light reaching the BLR clouds from accretion emission. In addition, photonization power from the central blackhole drops as $\sim r^{-2}$, so we expect an inverse relationship. We can then write the radial weight as:

$$W_R(a) = a^{\gamma}\,,\ \ \gamma \leq 0 \tag{1}$$

Equation 2.5 constrains the possible values of the parameter γ: The weight must decrease as the semi-major axis increases; it is physically improbable for the emission capabilities to increase further out into the BLR region.

c) Emission capabilities must also be taken into account from the angular distribution of clouds. Orbiting bodies at the equatorial plane (of the central blackhole) are expected to have a greater emission capability than those whose orbits are at the face-on configuration. This is mainly due to the rotation of the blackhole; gas particles entering the BLR are likely to be found near the equatorial plane of the blackhole. Similarly, radiation from the accretion disk lie primarily on the equatorial plane. Thus, we expect a greater fraction of contribution from BLR orbits whose inclination are close to that of the blackhole inclination. This can be described using a cosine-power law.

$$W_{\theta}(i) = cos^{\beta}(i - \theta_{inc}) \tag{2}$$

The sum of all contributions from every BLR orbit constitute the broad line profile. Once the flux contribution from all orbits is known, each Kepler orbit is given an appropriate weight depending on the orbit's inclination and the semi-major axis.

2.1. Constructing Broadline profiles

1) Method of construction

The relation between time interval and BLR emitter mass is established in the previous section. To calculate the distribution, or flux contribution, of each velocity interval requires sampling from the radial velocity curve to create a histogram. This is done by taking samples of 50 km/s, corresponding to the resolution of SDSS spectrographs, and noting the time interval the BLR clouds had spent on that velocity interval (see Figure 3). Following the discussion from the previous section, this is equivalent to the relative flux that given velocity contributes, and hence the resulting profile for a single orbit is a histogram of the "relative frequency" of velocity.

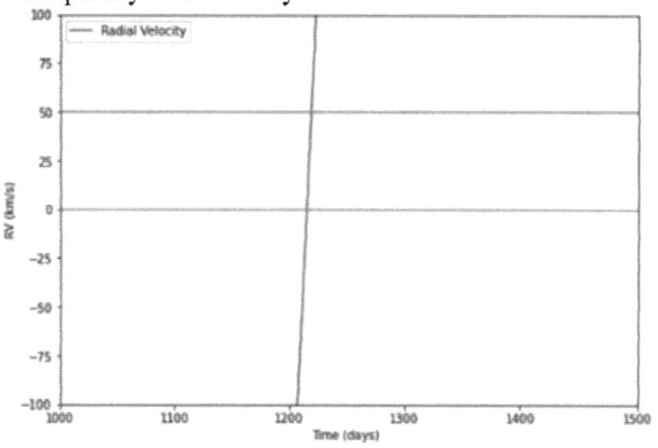

Figure 3. Sampling the radial velocity curve. The blue line indicates the radial velocity curve for one period of orbit. The green and orange lines indicate a velocity interval of 50 km/s. The interval of time corresponding to the section of the curve in the velocity interval is directly associated with its emission capability.

Thus, to construct the model for the full BLR region, we must first create the flux profile for a single orbit. We then must let the geometry parameters, namely the semi-major axis and the inclination, run over the entire allowed region, defined by the opening angle and the inner & outer radius. The final profile is constructed by giving appropriate weight to each of the single-orbit profiles, described by a power law and a cosine-power relation. Finally, we consider effects of the turbulent velocity [9], described by a Lorentz function $f(v) = \frac{1}{\pi}\left(\frac{v_t}{v^2+v_t^2}\right)$, where v_t is the turbulent velocity, usually of the order ~500 km/s, sometimes ranging up to 1000 km/s, and thus having a profound effect on the profile. The effect of turbulent velocity is one of convolution; that is the profile is given by the convolution of the (original) line profile with $f(v)$.

We finally note some differences in BLR modelling between our simplified model and other more sophisticated models, for example those suggested by Pancoast et al. (2014) [9]. Other studies aimed to study and simulate the time-lag of continuum flux and emission line flux response, based on data obtained from reverberation studies. In their model they have considered a more developed distribution of emission abilities by considering the radial distribution as a gamma function, while taking into account the obscuring effects of individual clouds. This model allows statistical inferences based on reverberation data, so that the fitting of time-lag responses may be completed. Although our simplified version of the BLR follows a similar framework in separately considering the geometry of the BLR and the dynamical aspects of the BLR, we have simplified the effects of obscuration and transparencies into the phenomenological parameters (β, γ), summarizing physical effects into a general function. This is done as our work requires fitting of the broad-emission lines, where extensive parameters may prove redundant and computationally impractical for our work.

Table 1. List of model parameters. It must be noted that the eccentricity is described to reflect the overall asymmetry of the BLR region, not of a particular single orbit

Parameter	Value	Description
a_{in}	$10{\sim}20\ ltd$	Inner radius (semi-major axis)
a_{out}	$20{\sim}100\ ltd$	Outer radius (semi-major axis)
θ_{inc}	$0°{\sim}30°$	Viewing angle of BLR
θ_{open}	$0°{\sim}90°$	Opening angle of BLR
e	$0{\sim}1$	Overall eccentricity of orbit
ω	$0°{\sim}180°$	Overall argument of pericenter
γ	$-2{\sim}2$	Radial weight
β	$0{\sim}7$	Angular weight
v_{turb}	$500{\sim}1000\ km/s$	Turbulent velocity
$M_{B,1}$	$10^{8\pm1}M_{sun}$	Blackhole mass
$M_{B,2}$	$10^{8\pm1}M_{sun}$	Blackhole mass
η_1	$0{\sim}0.1$	Mass accretion rate
η_2	$0{\sim}0.1$	Mass accretion rate
r		Separation of binary cores

2) Binary profiles & Double-peaked AGNs

Construction of the binary profile at its core can be taken to be a simple addition of fluxes. Factors that must be taken into consideration are 1. The radial velocity separation of the respective binary cores, calculated from the mass ratios given by $v_1 = \frac{2\pi}{P}\frac{1}{1+r} a \sin i\,(\cos(\omega + v) + e\cos\omega)$ and $v_2 = \frac{2\pi}{P}\frac{r}{1+r} a \sin i\,(\cos(\omega + v) + e\cos\omega)$, where the ratio r is the ratio of blackhole masses: $r = \frac{M_{B,1}}{M_{B,2}}$; and

2. The luminosity ratios which determines the relative contributions from each core to the final profile. An important assumption here is that there is no apparent difference in the self-emission capabilities of the BLR clouds of the two AGNs; their brightness ratio is solely determined by the property of the blackhole itself. Observations and reverberation mappings (Introduction to Galactic Nuclei, 5.5) suggests a strong relation between continuum flux and broadline profile; the continuum flux in turn is determined by blackhole accretion rates, as well as the mass of the central blackhole. Thus, in a simplified model, the "brightness" of a broadline profile can be described as the product of blackhole mass and accretion rate. Thus, the brightness ratio, b, is described as $b = \frac{M_{B,1}\,\eta_1}{M_{B,2}\,\eta_2}$. In particular, if one of the blackhole is not accreting, i.e., no broadline, then $\eta_1 = 0$ and the ratio is 0.

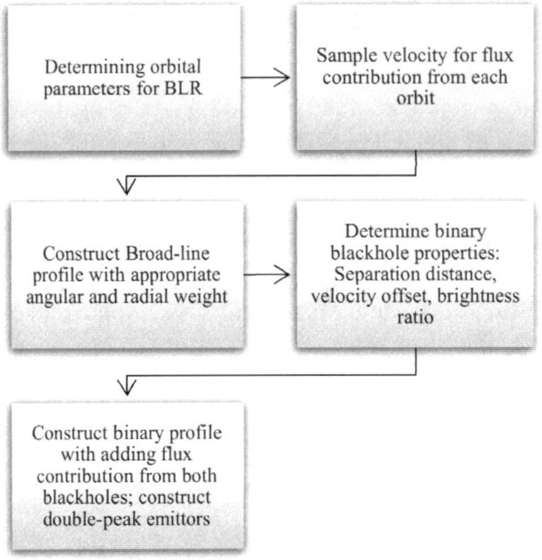

Figure 4. Diagram of modelling process. Selection of orbital parameters will generally be used to determine effects of various parameters on the profile. This constricts the possible values for constructing the true binary profile.

3. Simulations and Observations

This section will focus on inspecting the broadlines generated under different assumptions of the BLR. In Part B we analyze how binary broadlines are formed.

3.1. Single AGN Emission Lines

1) Disk-Like BLR

Disk-like BLR models the whole of the broad-line emission region as a flat disk, that is, the parameter $\theta_{open} = 0$. The disk-like BLR will be entirely on the orbital plane of the central blackhole, its inclination $i = \theta_{inc}$. In our model a disk-like BLR structure requires no angular weight, the addition of flux contributions all come from one plane. Such structures must be highly constrained in order to yield the expected broad and single-peaked BLR profile. Consider, for example, a disk-oriented face-on. The radial velocity in this case is 0 and all of the flux will be concentrated at 0 km/s, removing any broadness (the "broadness" can still exist given the convolution with the turbulent velocity profile). Hence the "disk" must be at least slightly inclined. However, the inclination must also be constrained to satisfy the requirements of the Unification Model, that the inclination must not be too high. The results are plotted in Figure 5.

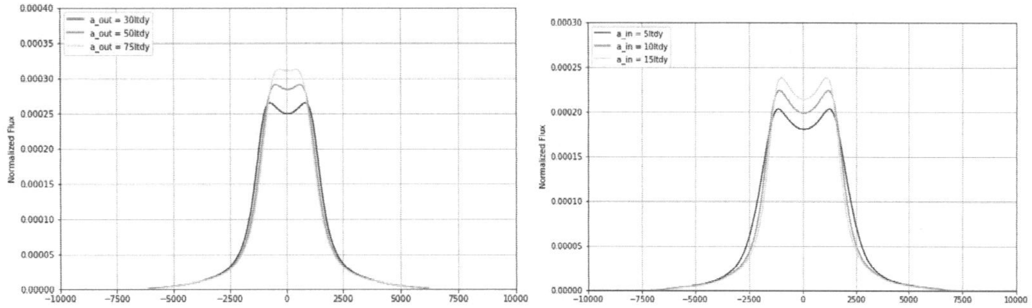

(a) (b)

Figure 5. Results plotted for a disk-like BLR at 15° inclination. The top panel fixes $a_{in} = 20ltdy$ and varies a_{out}, while the bottom panel fixes $a_{out} = 20ltdy$ and varies a_{in}. The other parameters are: $e = 0, \omega = 0, \gamma = 0, M = 10^8 M_{sun}$ and $v_{turb} = 500km/s$.

2) Sphere-Like BLR

Spherical BLR models the line-emitting region as a "sphere", more accurately a shell, whose inner and outer semi-major axis is determined by the model parameters a_{in}, a_{out}. As before, we normalize the emission line curves by setting the area of the space under the profile equal to one: it is the shape rather than the exact value that matters. In our model we first verify that the spherical assumption is a valid model in describing the BLR.

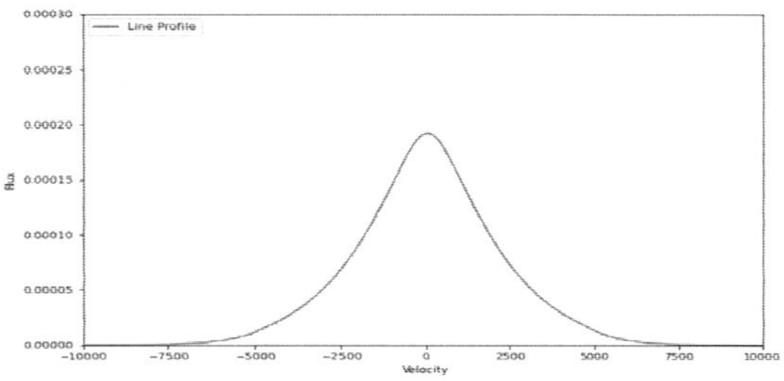

Figure 6. Results for a "spherical" BLR

The profile is generated from two components: One that directly comes addition of Keplerian orbits insider the BLR (the "raw" profile), and one that describes the turbulent velocity. Their respective shapes are plotted in Figure 10. The center part of the "raw" profile in Figure 10 acts as a quasi-delta function and preserves features of the Lorentz profile under convolution. The broad width of the raw profile then adds to the final width of the profile under convolution. The resulting profile is the bottom panel of Figure 10. This feature would be characteristic of a broad line profile. Note that the profiles in Figure 4 contains a "sharp peak", these results from the redshift from velocity space to wavelength. This shift changes the axis accordingly and produces the seemingly "sharp" peak in observed spectra.

The next step is to gauge the appropriate values of the parameters which describe the structure of the BLR, namely $a_{in}, a_{out}, \theta_{open}, e, \omega, \gamma, \beta$ and v_{turb} (see Table. 1). In a spherical model, θ_{open} is set to be 90°. To determine the weight parameters β and γ,We select a typical blackhole mass of $M_B = 10^8 M_S$, and define the line-emitting region to be within 20 light-days to 50 light-days. It is intrinsically clear that the parameters e and ω affect only the shape of a particular BLR orbit and hence determine the shape of the profile. To constrain our variables, we set $e = 0$, that is, a perfectly symmetric BLR, and take ω to be 15°. We set the inclination of the blackhole, θ_{inc}, to be at 15°, a relatively low inclination. Finally, effects from the turbulent velocity must be considered, and we take the value of v_{turb} to be at 500 km/s.

In addition, we note that the gaussian shape characteristic of most BLR profiles are apparent in the model for relatively high angular power. This suggests a higher angular weight should be assigned. The need for a relatively high angular weight calls for an enhanced influence of orbital inclination relative to the blackhole on the final broad-line profile. It is likely that, due to angular momentum considerations, in falling gas towards the BLR will be captured more often along the equatorial plane and thus for a Seyfert I galaxy, a considerable amount of BLR clouds would be located on or near the equatorial plane.

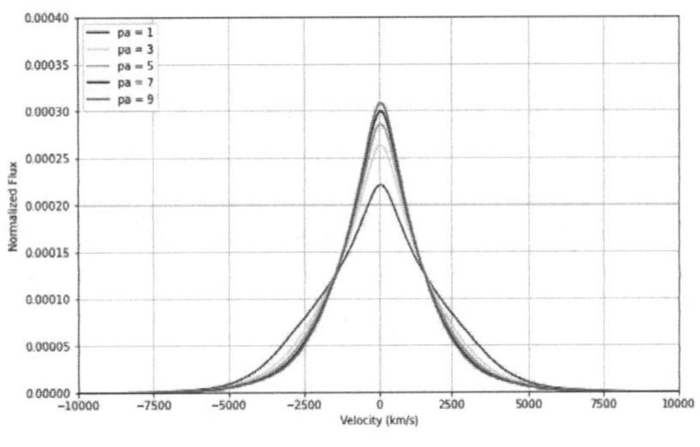

(a)

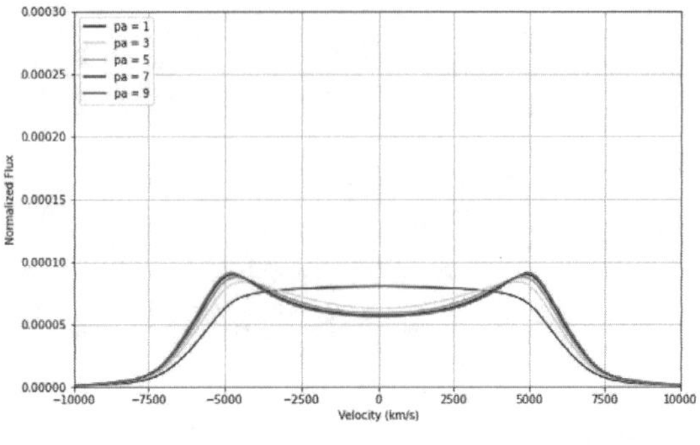

(b)

Figure 7. a) Variation of angular power at inclination 15°, where the broad-line feature is apparent. b) Variation of angular power at a higher inclination of 90°, where the "double peak" structure arises, thought to be the signature of accretion disk emission. "pa" denotes power of angular direction.

Although with increasing angular power the peak became sharper, changes are negligible as power increases higher (see, for example, the blue and magenta line, corresponding to $\beta = 7$ and $\beta = 9$) compared to lower powers. Overall, for further modelling this encourages us to choose higher powers for inclination, as 1. With increasing power, the shape resembles a true profile, and 2. There is safety in overshooting, that is, choosing too high a power will not affect the resulting profile significantly. As expected, with increasing angular weight the profile retains the profile from "disk-like" structures. Similarly, radial weight can be described as a power law with index γ. The resulting flux variation is shown in Figure 8.

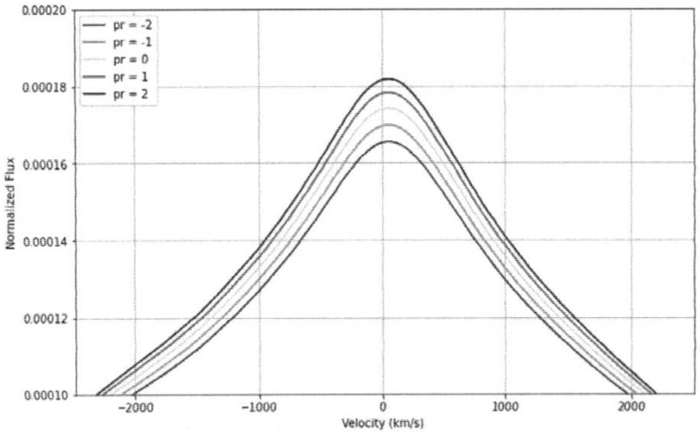

Figure 8. Varying the radial emission capability. Image is blown up to locate small differences in peak height. Slight deviation from symmetric shape results from the "eccentricity", the eccentric factor, to be 0.3.

Figure 8 suggests that although physically important, the effect of radial power is considerably less than that of angular weight. The relatively "parallel" profile lines also suggest only a small change in profile shape due to the change in radial power. This suggests a strong correlation of the profile with angular distribution, but not with radial distribution.

The parameters a_{in}, a_{out} define the span of the broad-line region. With decreasing a_{in} the result would be predicted to be a broadening of the profile, and vice versa. Similarly, decreasing a_{out} results also in a broadening of the profile, as the flux contribution at low radial velocities are lowered, and normalization results in the broadening of the profile. We shall first observe the variation of the parameter a_{out} as it is defined over a larger range (refer to Table. 1), from 20 light-days to 100 light-days. The parameter a_{in} is constrained to a narrow range from 10 to 20 light-days.

To allow variation differences to be distinguishable we plot the profiles at 20 light-days steps from 20 light-days to 100-light-days. The results are plotted in Figure 16. Note that different plots were created for different inclinations (0 and 30°).

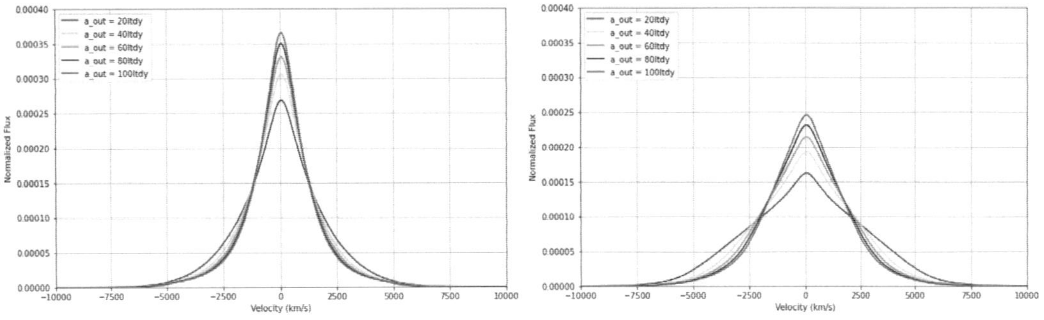

(a) (b)

Figure 9. a) Profiles for 0° inclination. b) profiles for 30° inclination. The broadening of the profile is coupled with the lowering of the profile peak, this is a result of normalization: since more flux is coming from higher radial velocities, the relative flux of the lower radial velocities must be lowered.

Figure 9 Demonstrates the expected results from varying the parameter a_{out}. With increasing a_{out} the profile width is reduced. Notice also that for lower inclinations (Top of Figure 9) the profile's shape is not changed significantly, only its width is reduced. At higher inclinations the variation becomes more extreme. The FWHM (full width at half maximum) is also seen to be larger than most profiles, going well past 5000 km/s. Overall, it can be seen that for higher inclined blackholes, in order to generate the correct profile a higher value of a_{out} must be chosen to make up for the flux lost due to inclination.

It is also apparent that the role of inclination is to play a major role in determining the differences in profiles, with broadlines only appearing in Seyfert I galaxies, thought to be low-inclination AGNs.

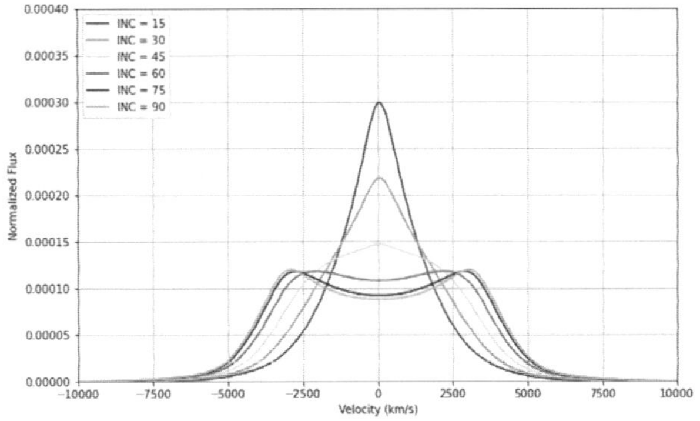

Figure 10. Variation of inclination for a spherical BLR. Note the dramatic change from a gaussian-like profile to that of a double peak structure. Parameters are chosen such that the profile is symmetric (e = 0), $a_{out} = 50\ ltdy$ and $a_{in} = 20 ltdy$.

The effect of inclination to the profile is seen to be dramatic in Figure 10. With increasing inclination, the broad-line feature of the profile slowly reduces to that of a profile from a "disk" structure. This explains the wide variety of broad-line features among single AGNs as those resulting from different inclinations.

3) Ring-Like BLR

In this section we vary the parameter θ_{open} to identify variation features of the profile. For low-inclined orbits, the reduction in the opening angle results in less flux contribution from orbits of the edge-on orientation, so that relatively more flux is coming from the face-on orientation. Such changes to the parameter will result in the profile becoming sharper (more flux at ~0 km/s radial velocity). The BLR is to range from 20 to 50 light-days with an inclination of 15. e and ω are chosen to be 0 so that the profile is symmetric. Angular weight and radial weight are 7 and 0, respectively. Finally, a blackhole mass of $10^8 M_{sun}$ and a turbulent velocity of $v_{turb} = 500km/s$ is selected. The resulting profile is plotted in Figure 11.

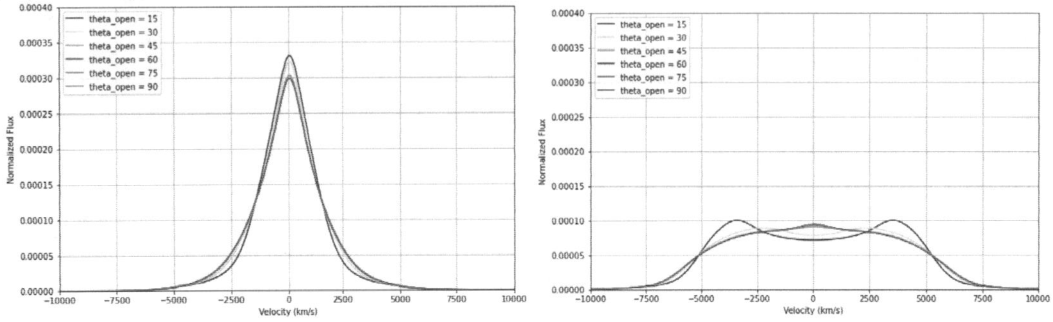

(a) (b)

Figure 11. a) Profiles plotted for a low inclination (15°). b) Profiles plotted at a higher inclination of 45°

3.2. Binary Black Holes Candidates Selected by Their Featured Emission Lines

In this section we explore the structure and profiles of emission lines resulting from binary AGN cores. Binary AGN cores at Keplerian separation (under the influence of gravity) have an additional set of radial velocity calculated as described in Section 2. The period depends on the total mass as well as the cube of the "total" semi-major axis $a_* = a_1 + a_2$. Finally, binary emission lines will evolve and change periodically with period equal to that of the binary AGN orbit. Only through long periods of observation can they be distinguished.

1) Twin Broad Peaks with Velocity Offset

When two AGNs are at large enough radial velocities their respective broadline structures can be easily identified, as shown in Figure 12.

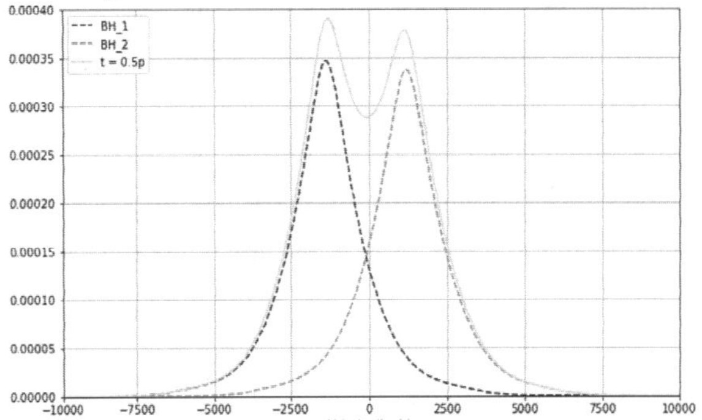

Figure 12. A suggested model for a binary "twin-peak". The AGN masses are $10^{8.3}$ and $10^{8.4}$ solar masses, respectively, and the separation is 100 light-days for higher velocity separation.

The "twin-peaks" are easily isolated from single peaked single AGNs, and hence makes the appearance of "twin-peaks" powerful evidence. There is, however, another explanation. It should however be cautioned that a large velocity separation should be required—and hence smaller separation, which can distort the shape of the BLR. Nevertheless, the period calculated with given orbital parameters ($a_* = 100\ ltdy, M_1 = 10^{8.3} M_{sun}, M_2 = 10^{8.4} M_{sun}$) gives a period of ~100 years.

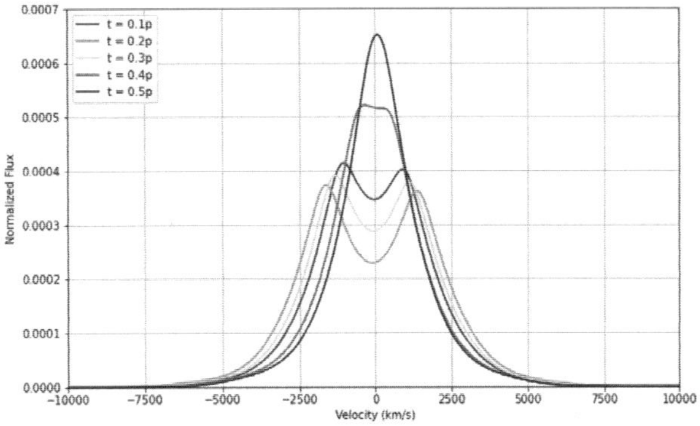

Figure 13. Variation of the profile with time. The lines are plotted with increments of $0.1P$, and plotted until $0.5P$, where the profile begins to repeat the pattern.

The expected result for the periodic variation of the profile is plotted in Figure 13. A key variation feature is the "merging" of the binary peaks, which is important to observe. The observation of the periodic appearance and disappearance of the "twin-peak" feature takes place on a timescale of decades ($0.1P{\sim}10\ yr$). Magnitude change for the "twin-peaked" profile can also serve as an identification of a binary candidate. If a binary candidate was identified based on its twin peaks, an observer can expect to see significant amounts of magnitude variation in times on the order of decades.

These features are highly unstable and are rare. We instead attempt to test our model on one of the most well-known "twin-peaked" binary candidate to date: the AGN(s) J1536+0441 identified by Boroson and Lauer. A velocity separation of ~3500 km/s and AGN masses of $10^{7.3}$ and $10^{8.9}$ solar masses was interpreted [3]. Our goal is then greatly simplified with the orbital parameters of the two AGNs determined, and fitting is done by finding appropriate values for the BLR of both AGNs.

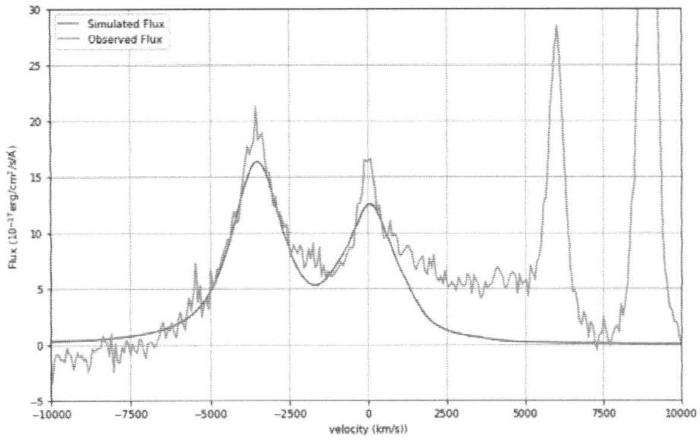

Figure 14. Fitting of the "twin-peak" broad-line J1536+0441.

The objective of fitting the profile is to describe the "twin-peak" structure via our model. In Figure 14 the observed and model flux is plotted (after subtracting continuum emission). The parameters are shown in Table 2.

Table 2. Parameters for the "twin-peaked" binary candidate J1536+0441

Parameter	AGN1	AGN2
a_{in}	20 ltdy	40 ltdy
a_{out}	50 ltdy	50 ltdy
θ_{inc}	0	0
θ_{open}	90	10
e	0	0
ω	130	0
γ	0	0
β	7	7
v_{turb}	500 km/s	300 km/s
M_B	$10^{7.3}M_{sun}$	$10^{8.9}M_{sun}$

The masses of the AGNs in the binary system can be seen to be of a great difference. Peaks with similar masses have radial velocities that are symmetric about v = 0 km/s, for example Figure 13. The profile identified by Boroson et al. suggests that one AGN is essentially stationery at 0 km/s, and the other approximately blue-shifted by 3500 km/s, explaining the large difference in mass ($M_1 v_1 = M_2 v_2$). The peaks from the profile however are seen to be quite similar, and hence parameters for AGN2 must be chosen so its profile is significantly narrowed. The turbulence velocity is hence reduced to 300 km/s and the shape of the BLR chosen to be "disk-like". The model describes the profile quite well, generating the "twin-peak" structure required, roughly following the shape of the two peaks. This structure is repeated in all of the Balmer lines on the SDSS spectrum, and seems to indicate that this can be a binary candidate. However, we do note that such high velocity separation requires the blackholes to be extremely close, and disrupt the shape of the BLR—undermining the model's assumptions. Although Twin-Peaks serve as strong evidence, they must be cautioned against using broadlines as direct evidence.

2) Single Broad Peak with Systematic Redshift

Binary AGN are rarely located on the celestial sphere, and it is even rarer to find close AGNs interacting to produce "twin-peaked" profiles. The generation of "Twin-peaked" profiles requires not only large velocity separations, but also that both blackholes must be active. This is not always the case, considering the short duty cycle for activate blackholes and AGNs can have a high degree of variability in their luminosity. In such a scenario where the velocity separation is not as large, and one of the AGN is weaker than the other, we obtain a "Weak-component" Twin-peaked spectra. In such profiles one broad-line is partially absorbed into the other, generating an altogether different profile than single AGNs and "twin-peaked" structures.

Cases where the magnitude difference is not extreme retain some of the properties of "twin-peaked" structures, for example in Figure 15.

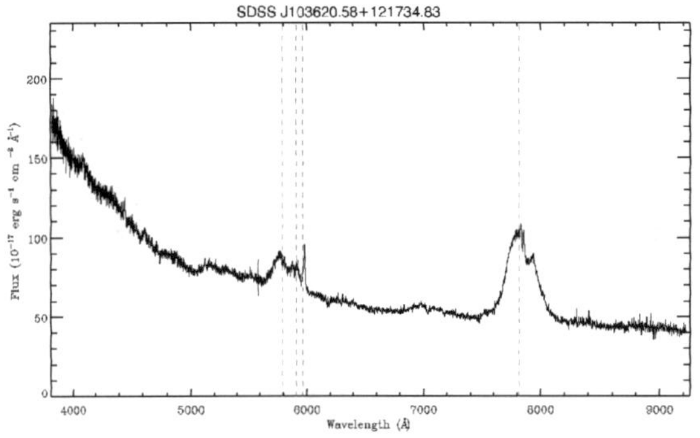

Figure 15. An example of a "Sub-Peak". The difference between single AGN spectra and such spectra is apparent: the H_α region (at $7800\ \dot{A}$), for example, contains an extra, weaker set of emission line.

The H_α region is a fine example of a "weak-component" twin-peak structure. The primary broad-line constitute of most of the profile, centered approximately at the redshift line, suggesting that this stronger component usually is produced by a more massive AGN. The weaker component is at a slightly higher redshift near $8000\dot{A}$. It is not yet strong enough or at higher velocity separations to resemble a "twin-peak" as in Figure 13, and the result is the appearance of a "sub-peak" on the sides of the primary peak.

In Figure 16 we plot a demonstration of how such a "weak-component" spectra may form and evolve with time. The parameters chosen are shown be Table 3.

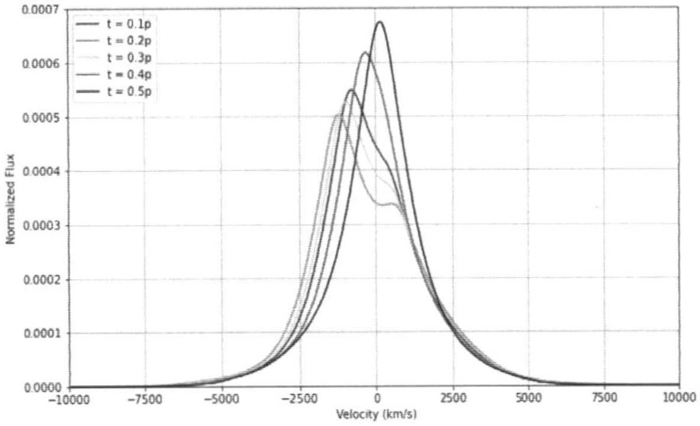

Figure 16. "Weak-component" spectra. Lines are again plotted with 1/10 of the period. Note that there exist two types of spectra: "Weak-component", and a "velocity-offset" profile we shall discuss later.

Table 3. Parameter List for "Sub-Peak"

Parameter	AGN1	AGN2
a_{in}	20 ltdy	10 ltdy
a_{out}	50 ltdy	20 ltdy
θ_{inc}	15	15
θ_{open}	90	90
e	0	0
ω	0	0
γ	0	0
β	7	7
v_{turb}	500 km/s	500 km/s
M_B	$10^{7.8}M_{sun}$	$10^{8.1}M_{sun}$

The orbital parameters are $a_* = 100\ ltdy, e = 0.3, \omega = 40°, i = 60°$. It can be seen that the shape of the profile is different to that of single-AGNs. Our model has reproduced the small "sub-peak" seen in Figure 15, for example the line $t = 0.2P$. This "sub-peak" does not contribute to another emission peak in the profile, but rather deforms the main peak's profile at the slopes. This feature can sometimes be mistaken for other sets of superimposed emission lines, for example strong iron emission in the $H\beta$ region. It would then be required that this feature exists in all Balmer emission lines; in addition, we would require high S/N spectra in order to remove the possibility of noise hiding the "sub-peak". A more significant feature to observe is the deviation of the main peak from the center. Single AGN generates profiles peaked more or less centered at 0 km/s, save the rare cases of a highly eccentric orbit. In the absence of twin-peaked profiles, deviation from the measured emission wavelength can suggest the existence of a binary AGN, as the most probable explanation is the velocity offset by Keplerian motion.

Variations of the "Weak-component" profile is less significant than that of the "twin-peak". This is due to the longer period and hence more difficult to observe, as well as the relatively close broad-lines (on the velocity spectrum), which limits variability in magnitude. The time period of orbital motion for the above parameters is 178 years, and hence any variations would be much longer than that of the "twin-peak" profiles. Despite difficulties in observing its variation, however, its profile stands out quite significantly. In Figure 17 we present another "weak-component" spectra, this time with our model fit.

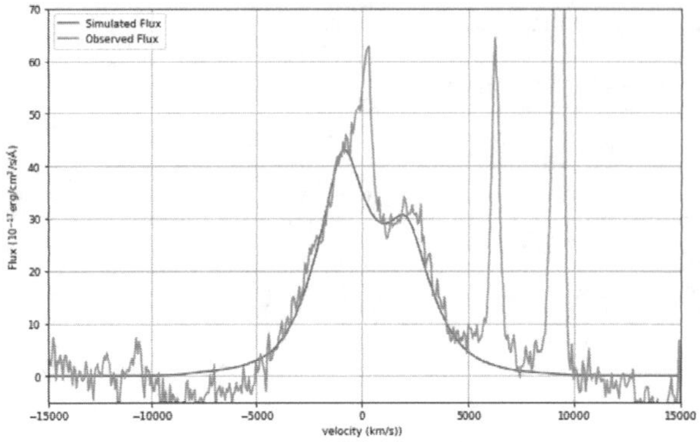

Figure 17. Fitting of the H_β region J153415.41+303435.45. The parameters are chosen.

As seen from the fitted spectra, the profile contains precisely a main peak and a sub-peak, at velocity offset -1000 km/s and 2000 km/s, respectively. It should be noted at such high velocity separation it would be possible to form "twin-peaked" structures; however, from the fitting we can see that one peak, the "sub-peak", comes from an AGN much less luminous than its counterpart. Even so, the high velocity separation suggests that it would be possible to observe some variations in the profile due to Keplerian motion.

It can be seen from Figure 16 that as time evolves, the "sub-peak" is absorbed into the main profile to resemble a typical broad-line. There is, however, another difference. While the profile $t = 0.4P$ contains only one peak, without any "sub-peaks", it can be categorized as a binary candidate based on its velocity offset from the center. As stated before, it would require highly eccentric orbits for single AGNs to achieve this shift, and the best explanation would be that the profile is Doppler shifted (radial velocity). We plot an example of the profile, with $t = 0.6P$ and the same parameter from Figure 16, in Figure 18.

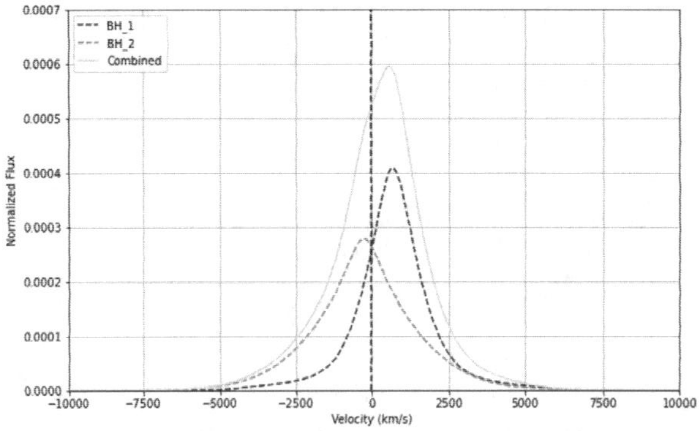

Figure 18. "Velocity-Offset" binary profile.

The most significant feature of this profile is of course the deviation of its peak from the 0 km/s mark. This can be explained by observing the components that makes up the profile (the dashed lines). When the profiles are lined close together, with one weaker than the other, no "sub-pcak" is produced, and the result is a skewed profile. It can be seen that the skewness is shown by the different slopes to either side of the peak: the "blue" side of the profile has a more moderate slope while the "red" side shows a high slope. This asymmetricity, coupled along with a Doppler-shifted peak, can help to identify the presence of a binary candidate.

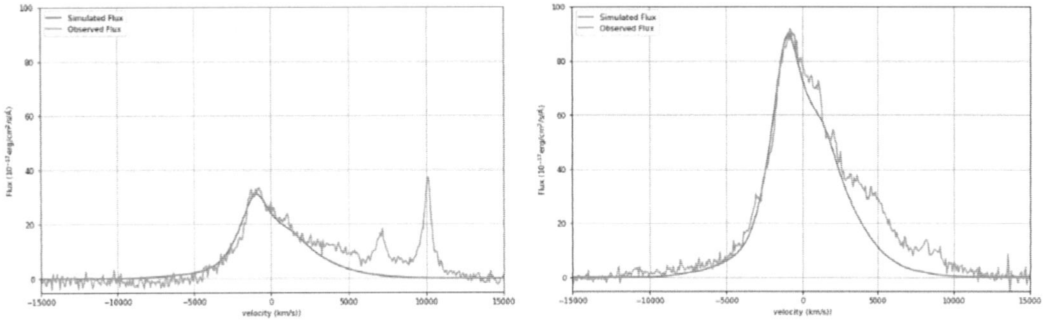

(a) (b)

Figure 19. Fitted spectra for the AGN (system) J001224.02-102226.29. Top: H_β region of the profile. Bottom: H_α region of the profile. The profiles are fitted with the same parameters; only the weights given are different.

In Figure 19 we have fitted the profile for the AGN (system) J001224.02-102226.29, for the strong Balmer lines H_α and H_β. The two profiles are at velocity separation 1100 km/s and -1000 km/s. It can be seen that the modelled profile describes a skewed peak, the "red" slope being gentler than the "blue" slope, which shows a sharp increase in flux. This is because the weaker component is somewhat broader than the main peak, and so the base of the profile is broadened. The system is thought to be a binary system for its peak being shifted by -1000 km/s, in accordance to the results described by our model in Figure 18. The Balmer profiles furthermore follow the same general shape described by our model. In fitting the spectra, we would expect the only parameter difference to be the weights added to each profile. In the case above, H_α emission is relatively stronger in the profile shifted to the right (red-shifted). Its H_β emission is accordingly relatively weaker, suggesting a higher energy level in this AGN.

Table 4. Parameter list for "Velocity Offset"

Parameter	**AGN1**	**AGN2**
a_{in}	10 ltdy	50 ltdy
a_{out}	20 ltdy	60 ltdy
θ_{inc}	15	0
θ_{open}	30	70
e	0	0
ω	0	0
γ	0	0
β	7	7
v_{turb}	700 km/s	500 km/s
M_B	$10^8 M_{sun}$	$10^8 M_{sun}$
Ratio for H_α	1.07	
Ratio for H_β	0.857	

3) Selection of Binary Candidates

Table 5. List of most confident binary candidates. Candidates marked with asterisk are spectra differing from that of single AGNs yet unexplained by our model, which we have included for later verification.

Name	Plate-MJD-FiberID	z	Category
J135829.57+010908.45	0301-51942-0457	0.2439	SP
J172711.81+632241.85	0352-51789-0639	0.2174	VO
J034042.93-073125.56	0462-51909-0133	0.2165	VO
J121018.35+015405.94	0517-52024-0027	0.2159	SP
J143515.65+023221.55	0535-51999-0582	0.3050	VO
J001224.02-102226.29	0651-52141-0072	0.2281	VO
J093653.84+533126.92	0768-52281-0473	0.2281	VO
J103620.58+121734.83	1600-53090-0363	0.1925	SP
J085632.99+595746.94	1785-54439-0481	0.2830	SP
J130534.49+181932.91	2603-54479-0443	0.1180	SP
J133432.35+171147.02	2606-54154-0581	0.2648	VO+SP
J122630.99+252522.18	2659-54498-0353	0.1336	SP
J140336.43+174136.14	2757-54509-0152	0.2223	TP
J140431.42+213415.44	2784-54529-0027	0.0839	TP
J140315.91+225845.34	2784-54529-0535	0.1606	VO
J222435.29-001103.89	4201-55443-0033	0.0579	VO
J001224.01-102226.51	7169-56628-0344	0.2274	VO
J001247.93-084700.59	7169-56628-0665	0.2204	VO
J225307.36+194234.60	7609-56959-0283	0.2820	SP
J225421.65+211816.22	7609-56959-0850	0.1510	SP
J004319.74+005115.41	7856-57260-0132	0.3090	VO+SP
J075403.60+481428.05	8303-57805-0778	0.2741	SP
J170553.87+455115.30	8535-58019-0892	0.2573	VO
J230803.20-013729.91	9153-58022-0174	0.2022	SP
J013521.27+062548.58	11071-58429-0410	0.1492	VO
J094927.67+314109.94	11379-58438-0592	0.3052	VO

Our work above has summarized some properties of binary AGN spectra. In addition, we have seen how the model can describe various features in binary candidates. Following the properties from the previous section, we will now select a list of possible binary candidates for verification (for example, by observing time-variations). We have obtained a dataset of 1348 AGN spectra from the SDSS spectroscopy database, selected for their high S/N ratio. Of these spectra we manually identify features discussed in the previous two sections, including twin-peaked structures (rare) and "weak-component" spectra, as well as emission lines shifted at a velocity offset relative to the center-of-mass redshift. AGNs holding these features are highly probable binary candidates and hence important for future, follow-up observations. In Table 5 we present a list of binary AGN candidates, identified based on the results our model has produced. We have also labelled the feature used to identify said spectra as belonging to that of a binary AGN. In particular, SP refers to spectra with "sub-peaks", with deformation of one of the slopes showing a smaller peak. TP refers to the typical case of a twin-peaked spectra, with both peaks visible and prominent. VO refers to "velocity-offset" profile, profiles showing skewed peaks that deviates from their correct redshift.

Of the 26 profiles we identified as being possible binary profiles, only two are marked as being "twin-peaked": J140336.43+174136.14, and J140431.42+213415.44, and their peak structures are not as easily identifiable as in J1536+0441 (this was not included in the list). A possible explanation for the lack of twin-peaked structures may be that at this stage of proximity, relativistic effects become significant, and orbital decays may alter the structure of the BLR completely so our model can no longer describe it.

Other candidates are much more common, those identified by their "sub-peak" or "velocity-offset" feature. In our model this is much explained by Keplerian orbits and hence those are very reliable candidates. These structures have the property that the radial velocity separation of the profile needed not to be at a great distance apart. This suggests that when undergoing systematic surveys looking for binary galaxy cores, spectra of the TP type, although being convincing, are rare and should not be the focus of such surveys. Systematic searches should be undergone looking for asymmetricities in the broad-line profile as well as velocity-shifts of the order 100~1000km/s from the cent

4. Conclusion

In this paper we have developed a model to simulate the emission-line profile from Seyfert I AGN systems. We have then extended this model to explain and generate profiles for binary AGN systems. The profiles created were able to match and interpret profiles as being a binary system, suggesting the feasibility of our model to identify binary candidates. Discussions were included for the expected time-variation of the profiles, suggesting time-scales for observing changes in binary profiles. We finally included a list of confident binary candidates that may be verified by further observation (i.e., look for changes described in paper) to be true interacting binaries. We concluded our discussions by pointing out unusual cases and their possible explanations.

The model we have developed follows simply manipulated physics that matches observed spectra extremely well. This suggests that, on a whole, AGNs can be explained fully by theories of simplicity waiting to be found. We have considered and assumed a direct relation between time and mass of the BLR clouds, suggesting that we can sample radial velocity curves generated by Python programs. Our model introduced the BLR model as a ring-like structure, its extent defined by the opening angle θ_{open}. The model constructs the profile by summing over all distances from the core as well as all planes withing the opening angle. We then verified certain parameter constrains based on their effects on the final profile. This greatly limits the parameter space, allowing the generation of binary profiles.

Binary AGNs are a key area of study in astronomy, as they provide valuable data on the evolution of galaxy mergers. It is further thought that gravitational interaction between galaxies can initiate the fueling of the accretion process, hence providing us insight into the true physical nature of AGNs. The difficulty of observing a binary AGN remains, as they are found at large distances with no apparent visual differences to common. Despite difficulties and inadequate modelling procedures, our model has still produced satisfying results. Our work has greatly narrowed the range of possible candidates that can be verified by further observation. We hope that the future observations are able to verify or reject our selected list of candidates, so that insights may be provided to improve our model. We are confident that by refining our model in the future, the selection of binary AGNs will become more reliable and efficient.

Acknowledgment

This study originated from readings regarding Quasars and Active Galactic Nuclei (see Peterson, Introduction to Active Galactic Nuclei [11]). Current studies are limited in the fields of Active Galactic Nuclei in binary orbits, while such systems are important in the studying the development of galaxy systems. As the emission-line profile of Active Galactic Nuclei demonstrate different characteristic as to normal galaxy absorption systems, this paper focuses on studying the features of the emission profile of binary AGN.

Particular thanks go to Dr. Jiang Peng of the Polar Research Institute of China. Dr. Jiang Peng acts as the writer's director teacher. Dr. Jiang provided professional background prior to the initiation of the project. These include leading the writer through journals and papers to gain a sound understanding of previous work as well as gaps in studies. Dr. Jiang provided useful insights and improvements to the coded simulation and model. Dr. Jiang had also helped to review the writer's paper, in correcting technical details and format. The writer is sincerely grateful for the support and help from Dr. Jiang Peng.

References

[1] Ghez, A.M. et al., 1998. High Proper - Motion Stars in the Vicinity of Sagittarius A*: Evidence for a Supermassive Black Hole at the Center of Our Galaxy. The Astrophysical Journal, 509(2), pp.678 - 686.
[2] Gillessen, S. et al., 2009. THE ORBIT OF THE STAR S2 AROUND SGR A* FROM VERY LARGE TELESCOPE AND KECK DATA. The Astrophysical Journal, 707(2), pp.L114–L117.
[3] Boroson, T.A. & Lauer, T.R., 2009. A candidate sub-parsec supermassive binary black hole system. Nature, 458(7234), pp.53–55.
[4] Zhang, S. et al., 2019. SDSS J153636.22+044127.0 and Its Analogs: Shocked Outflows, Not Active Binary Black Holes. The Astrophysical Journal, 877(1), p.33.
[5] Storchi-Bergmann, T. et al., 2017. Double-Peaked Profiles: Ubiquitous Signatures of Disks in the Broad Emission Lines of Active Galactic Nuclei. The Astrophysical Journal, 835(2), p.236.
[6] Graham, M.J. et al., 2015. A possible close supermassive black-hole binary in a quasar with optical periodicity. Nature, 518(7537), pp.74–76.
[7] Eracleous, M. et al., 2012. A LARGE SYSTEMATIC SEARCH FOR CLOSE SUPERMASSIVE BINARY AND RAPIDLY RECOILING BLACK HOLES. The Astrophysical Journal Supplement Series, 201(2), p.23.
[8] Moaz, D. (2007) Astrophysics in a Nutshell, Princeton University Press, pp. 94-97,165-170.
[9] Pancoast, A., Brewer, B.J. & Treu, T., 2014. Modelling reverberation mapping data – I. Improved geometric and dynamical models and comparison with cross-correlation results. Monthly Notices of the Royal Astronomical Society, 445(3), pp.3055–3072.
[10] Perryman, M. (2018) The Exoplanet Handbook, Cambridge University Press, 2nd Edition.
[11] Peterson, B. (1997) Introduction to Active Galactic Nuclei, Cambridge University Press.
[12] Wang, X.-W. & Zhou, H.-Y., 2012. DUAL ACTIVE GALACTIC NUCLEI: DEPROJECTING THE BINARY CORES. The Astrophysical Journal, 757(2), p.124.

The wave-particle interactions between chorus waves and electrons in the Earth radiation belts

Yijia Wang

Nanjing University, Nanjing, China

3373469120@nju.edu.cn

Abstract. There exists a large number of high-energy electrons and various plasma waves in the Earth's radiation belts. The chorus waves, as a kind of whistler-mode waves can cause electrons acceleration and loss by wave-particle interactions, thus playing a vital role in the dynamic evolution of the radiation belts. To improve our understandings of such mechanisms and facilitate people to anticipate the dramatic variation of space environment, we utilized the wave and field data from Van Allen Probes to statistically study the acceleration and loss timescale of electrons by chorus waves. The first part in this paper presents the temporal-spatial distribution of the chorus characteristics and background field parameters. The second part exhibits the loss and acceleration timescales of electrons caused by chorus waves. The main results in this paper includes: As geomagnetic activity and L shells increase, the pitch angle and momentum diffusion of electrons driven by chorus waves strengthens. The dramatic acceleration of electrons by chorus waves exists in the region with lower f_{pe}/f_{ce}. Compared with keV electrons, MeV electrons are hard to be scattered by chorus waves, and electrons be precipitated into the atmosphere within several hours during highly disturbed times.

Keywords: chorus wave; wave-particle interaction; electrons diffusion timescales.

1. Introduction

The Earth radiation belts, also known as the Van Allen belt, are firstly discovered by American physicist James A. Van Allen in 1958. Including abundant high-energy particles, the radiation belts are divided into the inner belt and the outer belt by the slot region lying between 2-3Re (Re denotes the Earth radius, Baker et al., 2004). The former with stable structures lies in the range of 2 Re of the earth and the latter composing of many highly variable high-energy particles is located between 3 and 6 Re of the earth (Vamp Ola et al., 1992). During geomagnetic activity, the radiation belts dynamic is dominated by a series of acceleration, transport and loss mechanisms (Horne et al., 2006; Thorne., 2010), such as electrons precipitation and acceleration caused by plasma waves and loss by magnetopause shadow. Influenced by above processes, the flux of relativistic electrons can vary by several orders of magnitudes during few minutes to tens of days (Reeves et al., 1992). More importantly, the highly variable electrons can discharge on the satellite's surface and further disturb its orbiting direction (Baker et al., 2002) or precipitate into the atmosphere, changing the physical and chemical properties of it (Lastovicka.,1996).

Content from this work may be used under the terms of the Creative Commons Attribution 3.0 licence. Any further distribution of this work must maintain attribution to the author(s) and the title of the work, journal citation and DOI.
Published under licence by IOP Publishing Ltd

Thus, understanding the variation mechanism of electron flux is significant for people to predict and respond to the change of space weather.

The Electrons in the radiation belts have three periodic motions. The first one is cyclotron motion around the geomagnetic field line, corresponding to the first adiabatic invariant. The second one denotes the bouncing motion along geomagnetic field line, corresponding to the second adiabatic invariant and the third one refers to the drift motion around the earth, corresponding to the third adiabatic invariant. The three adiabatic invariations conserve when the electromagnetic field varies slightly. However, if electrons encounter electromagnetic waves or other fluctuations with comparative frequencies of it, its adiabatic invariants will be broken and the pitch angle of electrons changes, leading to the precipitation or acceleration of it. (Thorne and Horne., 1998; Shprits., 2009; Thorne., 2010).

One of the significant mechanisms responsible for the variations of electrons flux proves the resonant interaction between the chorus waves and electrons. Chorus waves are right-hand polarized electromagnetic waves with the frequency of few hundreds of hertz to several kHz. They are mainly distributed in the range 23 to 15 MLT (magnetic local time) outside the plasmasphere. Usually, it is divided into two distinct bands, referred to as the upper (0.5-$0.8f_{ce}$, f_{ce} is the electron cyclotron frequency at the equator) and lower band (0.1-$0.5f_{ce}$) chorus, with a distinct gap near $0.5f_{ce}$ (Tsurutani and Smith., 1974). Numerous researches have demonstrated that chorus waves are primarily excited by the cyclotron instability of temperature anisotropic electrons with energy of several to dozens of keV (Hwang et al., 2007; Shprits et al., 2008b). During storms or substorms, the amplitudes of chorus waves is greatly enhanced, which can reach 100pt at highly disturbed times, causing sever electrons scattering (Shprits et al., 2008b, Thorne et al., 2013). The precipitation of low-energy (0.1-30keV) electrons due to chorus waves generates diffuse auroras (Petrinec et al., 1999, Newell et al., 2009, Thorne et al., 2010) and the electrons drifting around earth's magnetosphere lead to the observable pancake distributions (highly anisotropic electron velocity distributions peaked at 90° to the magnetic field were called pancakes from appearance) of it. (Meredith et al., 1999, Thorne et al., 2010). In addition, chorus waves can provide an effective energy transfer mechanism from low-energy electrons (~eV to keV) to relativistic electrons (Horne and Thorne., 2003, Thorne., 2010), resulting in the the peak of electron phase space density (PSD) in radiation belts core (L=4-5) during magnetic storms (Horne et al., 2005b; Tsurutani et al., 2006; Thorne et al., 2014).

To measure the variability of electron flux in the radiation belt, some scientists have conducted systematic researches on the electrons loss and acceleration timescales caused by chorus waves based on the quasi-linear theory (assuming the average wave amplitude is small and the wave spectrum is wide enough to consider the wave-particle resonance interaction is a stochastic process determined by relatively slow processes). (Kennel and Petschek,.1966). Shprits (2006) showed that electrons will decay exponentially with the same rates determined by the electrons near the loss cone despite of pitch angles, if the pitch angle diffusion coefficients of electrons with the pitch angle of less than 30° doesn't vary over one order of magnitude. Based on this study, Summers (2007) and Orlova (2014) adopted the pitch angle diffusion rates at the loss cone to measure the electrons loss timescales. Some established simple mathematical models to estimate the timescales (Albert and Shprits., 2009; Aryan et al., 2019; Artemiyev et al., 2013). The acceleration timescales of electrons can be roughly estimated by the energy diffusion rates at the equatorial pitch angle between 60 ° and 80 °. Moreover, Agapitov (2020) used the ratio of energy diffusion to pitch angle diffusion coefficients to characterize the efficiency of electron acceleration. If the ratio of the energy and pitch-angle diffusion coefficients is greater than 1, the electrons will accelerate and if not the electrons will be dominated by loss mechanisms, and the electron flux will decrease. If the two are close, the electron flux remains basically unchanged (Summers et al., 2004a). These studys exhibit that chorus waves can accelerate the electrons to relativistic energy in nearly one day (Horne et al., 2005) as well as to precipitate it in a few hours and days.

In order to realize the statistic distribution of chorus waves and the acceleration and loss mechanisms of electrons driven by chorus waves in the radiation belts under different conditions, a statistical distribution model of chorus waves and plasma density in three different geomagnetic conditions is established utilizing the data from the Van Allen Probes. Based on quasi-linear theory, we used the

PADIE model (Glauert and Horne,.2005) to calculate pitch angle and momentum diffusion coefficients of electrons with different energies and further acquired the electrons diffusion timescales. Comparing the acceleration and loss of electrons caused by chorus waves under various conditions and studying their temporal-spatial variation can lay a foundation for people to predict and respond to the changes of space weather.

2. Data Analysis

The data used in this paper comes from Van Allen Probes launched by NASA in 2012. With perigee of about 600km, apogee of about 30500km and flight period of 9h, it provides an unprecedented opportunity for particular studies of the radiation belts since its high resolution and extensive spatial coverage of path. The Electric and Magnetic Field Instrument Suite and Integrated Science (EMFISIS) Waves instrument onboard the Van Allen Probes can provide the wave and field data used to analyze the wave properties and differentiate the wave emissions. EMFISIS includes the three-axis fluxgate magnetometer and the three-axis magnetic field search coil magnetometer, of which the latter includes WFR (wave form receiver) and HFR (high frequency receiver). WFR records the complete spectral matrices in the frequency range of 10 Hz – 12 kHz, which are transformed from scientific coordinates into the field-aligned coordinates to determine the wave normal angle, planarity and polarization by the Singular Value Decomposition (SVD) technique to further distinguish different wave modes. HFR records the single spectral information in the higher frequency range of 10 – 400 kHz, which can be utilized to compute the plasma density from the upper hybrid resonance frequency.

The criteria used to filtrate the chorus waves are listed as follows: (1) Regions with density < 124×(3/L)4 cm^{-3} are located outside the plasmasphere. (2)The wave frequency is in the range of 0.05-$0.5f_{ce}$, (3)The wave planarity is larger than 0.5. (4) The wave ellipticity is larger than 0.7.

Then we use the PADIE (Pitch Angle and Energy Diffusion of Ions and Electrons) model to calculate the pitch angle and momentum diffusion coefficients of electrons. Based on the cold plasma assumption, the PADIE model employs the full electromagnetic dispersion relation in order that the interactions between various linear electromagnetic wave modes and electrons can be addressed. Moreover, the PADIE model can measure the effects of oblique waves on electrons and the research isn't limited by the assumption of filed-aligned wave propagation. The distribution of wave amplitude and normal angles φ are assumed to have a Gaussian distribution and the main parameters are listed as follows:

Table 1. λ is the magnetic latitude, ω_m and $\delta\omega$ is the frequency of maximum wave power and bandwidth, ω_{lc} and ω_{uc} are lower and upper bounds to the wave spectrum.

	λ	ω_m	ω_{lc}	ω_{uc}	$\delta\omega$
Nightside MLT(k)>=18||MLT(k)<=6	20	0.25	0.05	0.5	0.175
Dayside 6<=MLT(k)<=18	35	0.2	0.05	0.4	0.2

Table 2. $X = \tan\varphi$, X_{min} is the minimum and X_{max} is the maximum, X_w is the angular width and X_m is the peak value.

X_{min}	X_{max}	X_m	X_w
0	1	0	$\frac{\sqrt{3}}{3}$

Finally, we calculate the electrons loss timescale using the formula

$$\tau = \int_{\alpha_L}^{\frac{\pi}{2}} \frac{\cos\alpha}{2D_{\alpha\alpha}\sin\alpha} d\alpha$$

and the acceleration timescale by

$$\tau_e = \frac{1}{D_{pp}}$$

Note that the $D_{\alpha\alpha}$ and D_{pp} are the bounce and drift averaged diffusion coefficients.

3. Statistical results

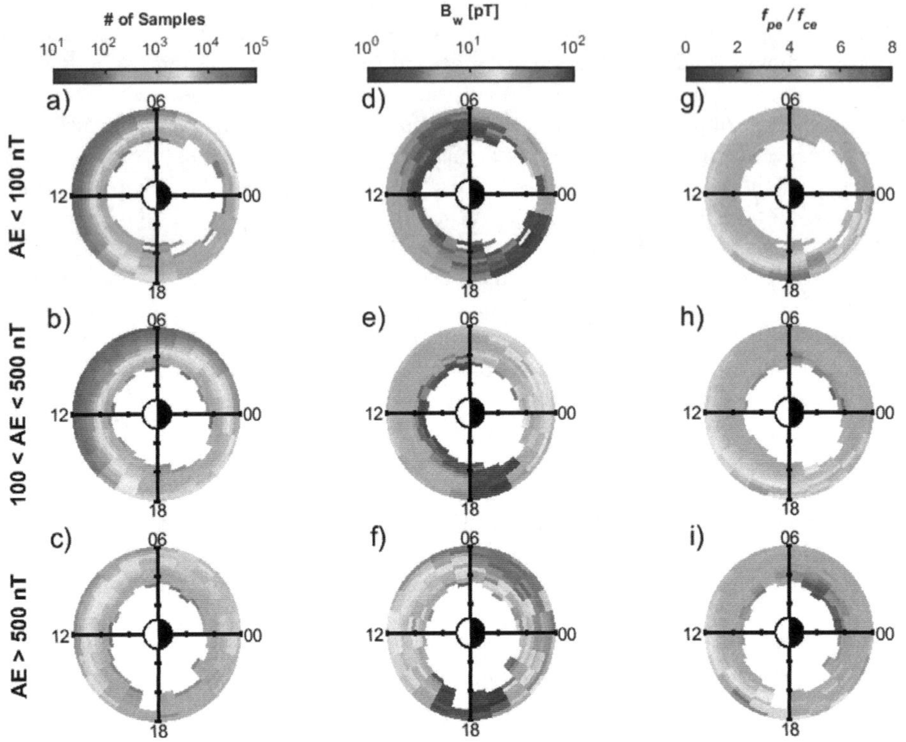

Fig. 1 (a-c) Statistical distribution of chorus wave samples detected by Van Allen Probes as a function of L and MLT in three ranges of geomagnetic activity. (d-f) The wave amplitudes distribution of chorus waves in three ranges of geomagnetic activity. (g-i) The ratio of the plasma frequency to the equatorial electron cyclotron frequency (f_{pe}/f_{ce}) outside the plasmasphere in three ranges of geomagnetic activity.

Figure 1 shows the statistical distribution of chorus waves and background field parameters in quiet(top), moderately disturbed(middle) and highly disturbed(bottom) times. The spatial distribution of chorus samples detected by Van Allen Probes as a function of L and MLT is shown in figure1 (a-c). Note that the number of chorus samples in figure1 (a-c) is more than 100, justifying the reliability of the statistical results. Figure1 (d-f) present the wave amplitudes distribution of chorus waves and figure1 (g-i) show the distribution of f_{pe}/f_{ce} outside the plasmasphere in three ranges of geomagnetic activity. With the increase of geomagnetic activity, the chorus wave amplitudes significant enhanced globally, even getting 100pt during highly disturbed times (AE > 500nT) (Figures d-f), while f_{pe}/f_{ce} outside the plasmasphere decreases especially in the range of 0 < L < 12 (Figure1 g-i). In figure1 (d-f), the chorus waves is mainly located between 16:00 and 23:00 MLT during disturbed times (AE > 100nT) with the largest amplitude seen from the midnight to the morning sector. In addition, the radial wave amplitude enhances as L increases.

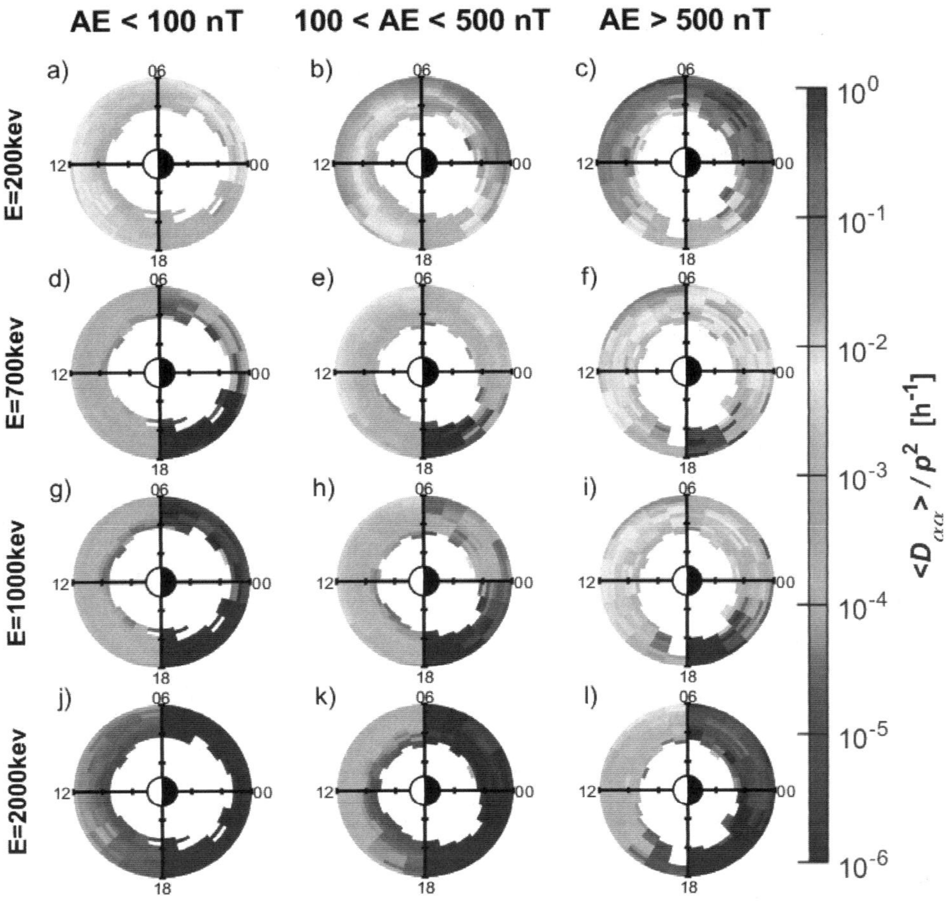

Fig. 2 The two-dimensional plots of pitch angle diffusion rates of electrons with different energies at a pitch angle of 10 ° as a function of L and MLT. From left to right, the results are presented for quiet (AE <100nT), moderate (100 < AE <500nT), and active (AE >500nT) conditions. From top to bottom, the electron energies are 200keV, 700kev, 1000kev and 2000 keV respectively.

To further study the temporal-spatial distribution of electrons diffusion driven by chorus waves, the pitch angle and momentum diffusion coefficients are calculated using the statistical data in Figure 1. Figure 2 shows the two-dimensional panels of electrons pitch angle diffusion coefficients in different energies at equatorial pitch angle of 10 ° as a function of L and MLT. From left to right, the results are presented for quiet (AE <100nT), moderately disturbed (100 < AE <500nT), and highly disturbed (AE >500nT) times. From top to bottom, the electron energies are 200, 700, 1000 and 2000 keV respectively. In figure 2, the pitch angle diffusion coefficients of electrons with different energies argument as AE increases and such variation proves more intense for keV electrons. With the increase of electrons energy, the pitch angle diffusion coefficients lets up, demonstrating that the keV electrons are easier to be precipitated into atmosphere by chorus waves compared to MeV electrons. Moreover, for hundreds of keV electrons, the pitch angle diffusion coefficientss are of the same order on the dayside and nightside, while for MeV electrons, the pitch angle diffusion on the dayside is greater than on the nightside, showing dramatic pitch angle scattering of MeV electrons driven by chorus waves on the dayside.

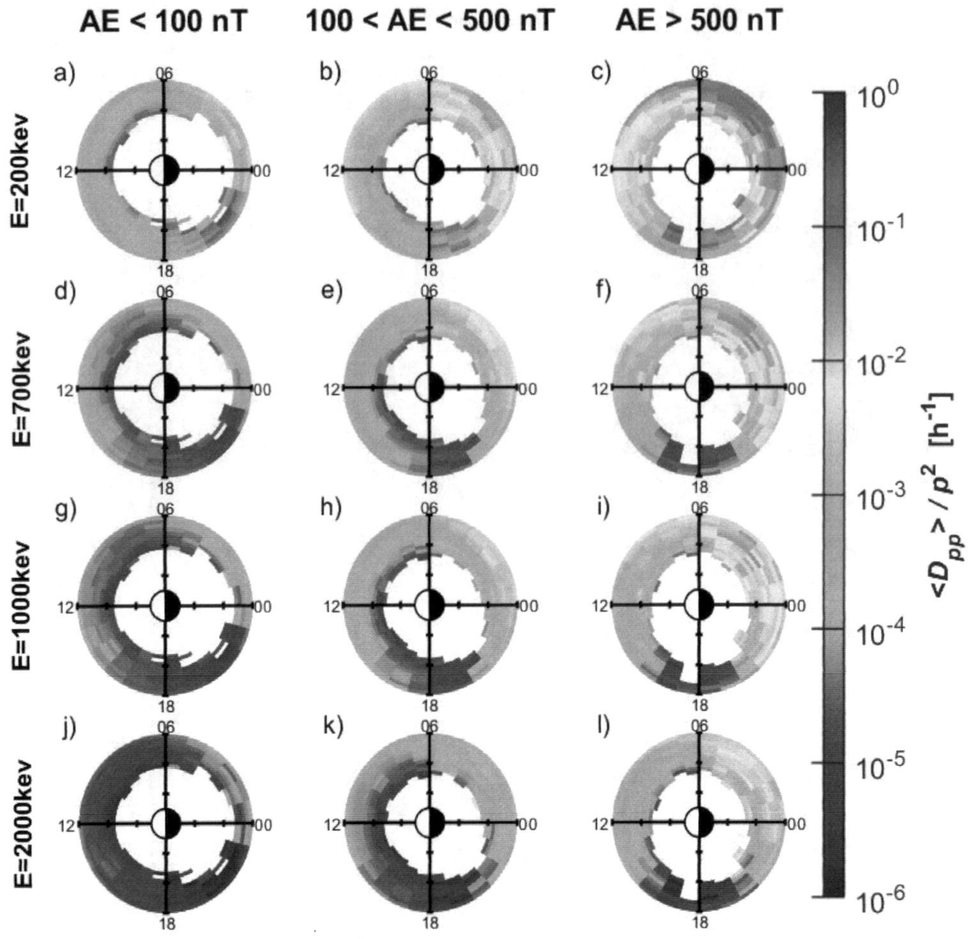

Fig. 3 The two-dimensional plots of momentum diffusion rates of electrons with different energies at a pitch angle of 70 ° as a function of L and MLT. From left to right, the results are presented for quiet (AE <100nT), moderate (100 < AE <500nT), and active (AE >500nT) conditions. From top to bottom, the electron energies are 200keV, 700kev, 1000kev and 2000 keV respectively.

Figure 3 shows the two-dimensional panels of the electrons momentum diffusion coefficients at the equatorial pitch angle of 70 ° as a function of L and MLT under different circumstances. From left to right, the results are presented for quiet (AE <100nT), moderately disturbed (100 < AE <500nT), and highly disturbed (AE >500nT) times. From top to bottom, the electron energies are 200, 700, 1000and 2000 keV respectively. In figure 3, the momentum diffusion coefficients argument with the enhancement of geomagnetic activity by two orders of magnitude, while they decrease as electron energy increases. For instance, during the highly disturbed times on the dawn side, the momentum diffusion coefficients of 200keV electrons are about$10^{-1}h^{-1}$, and that of 2MeV electrons are about $10^{-3}h^{-1}$, varying by two orders of magnitude. Meanwhile, the peak of the electrons momentum diffusion coefficients exists on the dawn side, corresponding to the distribution of the chorus waves amplitude. In general, the electrons pitch angle and momentum diffusion coefficients increase with wave amplitude, and the peak of the momentum diffusion coefficients In line with the minimum regions of f_{pe}/f_{ce}. Accordingly, electrons acceleration caused by the chorus waves is located in the area with lower f_{pe}/f_{ce}.

Comparing figure 2 with figure 1, the momentum diffusion coefficients of electrons with both keV and MeV energies are always greater on the dayside than on the nightside, while the distribution of pitch angle diffusion coefficients are inconsistent in different MLT. Therefore, to focus on the disparities of the pitch angle and momentum diffusion coefficients of electrons with different energy on the dayside and nightside, the MLT-averaged pitch angle and momentum diffusion coefficients of 237kev and 1.3MeV electrons on the dayside and nightside are selected.

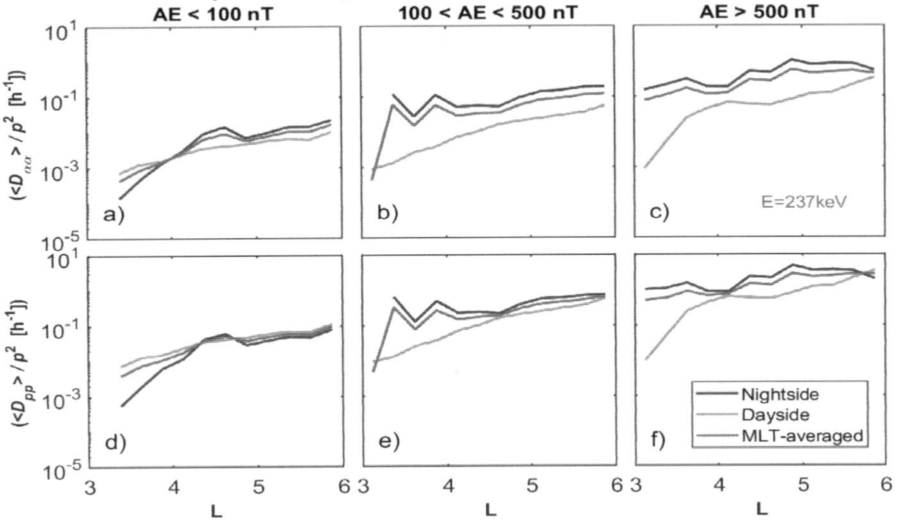

Fig. 4 The variability of 237keV electrons' diffusion rates as a function of L during quiet conditions(left), moderately disturbed conditions(middle) and active conditions(right). The blue and green curves represent averaged diffusion rates on the nightside and dayside respectively, while the red curve represents MLT-averaged diffusion rates.

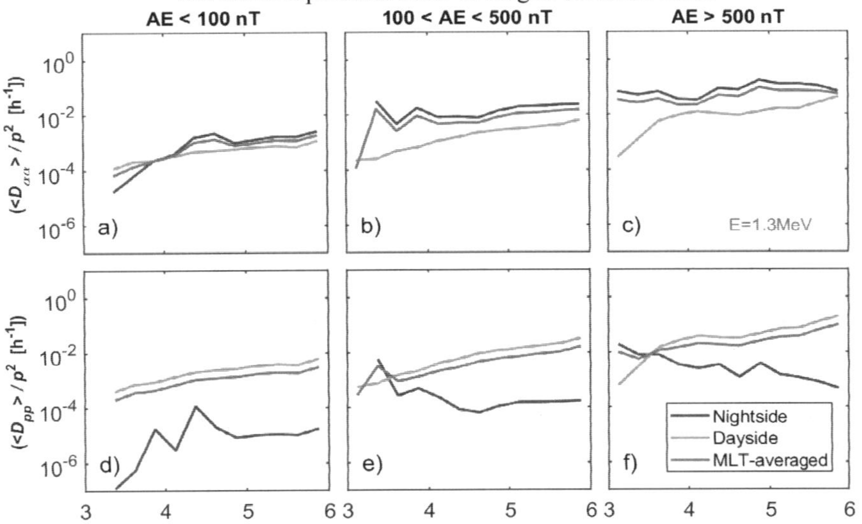

Fig. 5 The variability of 1300keV electrons' diffusion rates as a function of L during quiet conditions (left), moderately disturbed conditions (middle) and active conditions (right). The blue and green curves represent averaged diffusion rates on the nightside and dayside respectively, while the red curve represents MLT-averaged diffusion rates.

Figure 4 and 5 show the variability of 237keV and 1300keV electrons diffusion coefficients as the variation of L during quiet (left), moderately disturbed (middle) and highly disturbed times (right). The blue and green curves represent the MLT-average diffusion coefficients on the nightside and dayside respectively, and the red curve represents whole MLT-averaged diffusion coefficients. Overall, the pitch angle and momentum diffusion coefficients increase with the geomagnetic activity and L. the pitch angle and momentum diffusion coefficients on the nightside are greater than those on the dayside during disturbed times. During quiet conditions, there exist a critical point at the position of L=4-5. On the left of the point, the dayside averaged pitch angle and momentum diffusion coefficients are larger while on its right the opposite is true. As shown in Figure 5 for 1.3MeV electrons, the night side averaged pitch angle diffusion coefficients are always greater, while the momentum diffusion coefficients are smaller on the nightside than those on the dayside under different geomagnetic conditions. At the same time, it is notable that the nightside averaged momentum diffusion coefficients of 1.3MeV electrons remains unchanged with L during quiet times, but decreases with L during disturbed times.

4. The electrons loss and acceleration timescales caused by the chorus waves

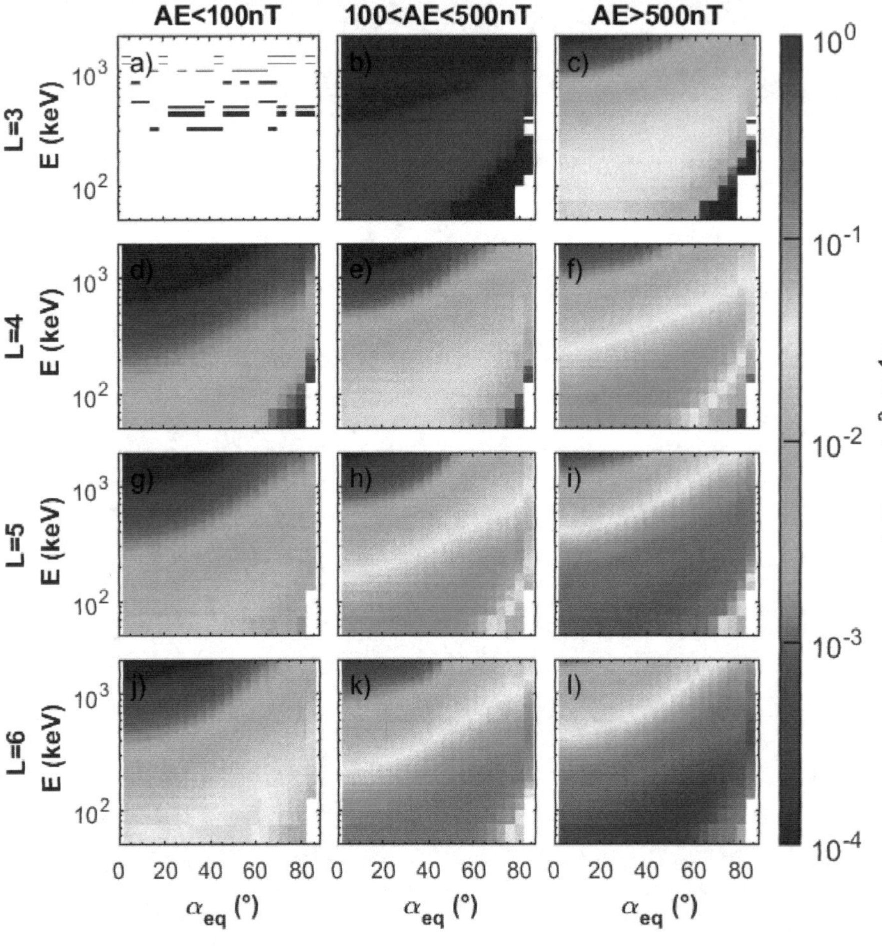

Fig. 6 MLT-averaged pitch angle diffusion rates of electrons with different energies and pitch angles due to chorus waves between L=3 and L=6 (from top to bottom increase by 1L) during quiet conditions (left), moderately disturbed conditions (middle) and active conditions (right).

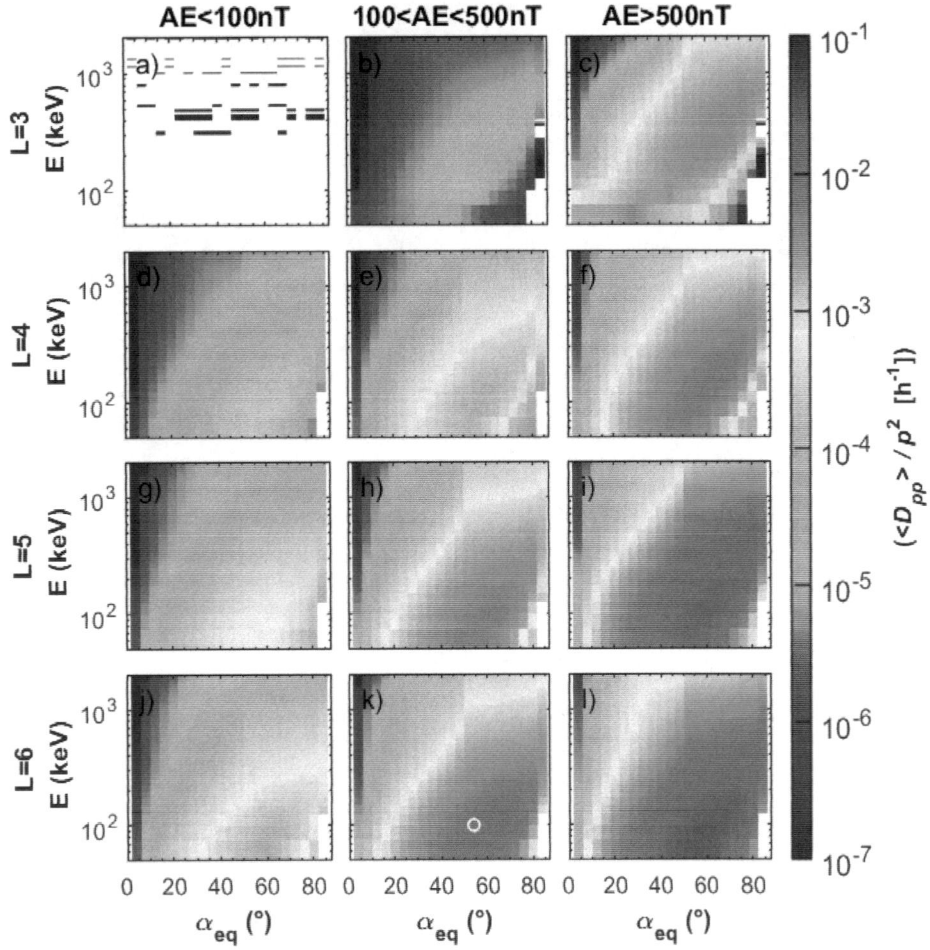

Fig. 7 MLT-averaged momentum diffusion rates of electrons with different energies and pitch angles due to chorus waves between L=3 and L=6 (from top to bottom increase by 1L) during quiet conditions (left), moderately disturbed conditions (middle) and active conditions (right).

To obtain the electrons diffusion timescales driven by chorus waves, the two-dimensional distribution panels of pitch angle and momentum diffusion coefficients as a function of the electrons energy and pitch angle under different conditions are presented firstly. Figure 6 and 7 respectively demonstrate the MLT-averaged pitch angle and momentum diffusion coefficients of electrons in different energies and pitch angles due to chorus waves between L=3 and L=6 (from top to bottom increased by 1L) during quiet (left), moderately disturbed conditions (middle) and highly disturbed times(right). To ensure the reliability of statistics, samples less than 100 are excluded, thus the data in Figure 6 (a) and Figure 7 (a) are missing. In Figure 6, from left to right, the pitch angle diffusion coefficients of electrons grows with AE and alters by about two orders of magnitude from quiet to highly disturbed times. From top to bottom, the pitch angle diffusion coefficients increases with L and changes by one to two orders of magnitude from L=3 to L=6. In different geomagnetic activities and space regions, the variability of the pitch angle diffusion coefficients in the framework of the electrons energy and pitch angle exhibit coherent. As it is shown in figure6 (i), the peak of the pitch angle diffusion coefficients moves to the position with larger pitch angles as the electrons energy diminishes, indicating that the higher-energy electrons are tend to be scattered into the atmosphere by the chorus waves relative to the lower-energy electrons. Similarly, the momentum diffusion coefficients also increase with the intensity of geomagnetic activity and L (Figure 7), whose variation tendencies are equal to the pitch angle diffusion coefficients.

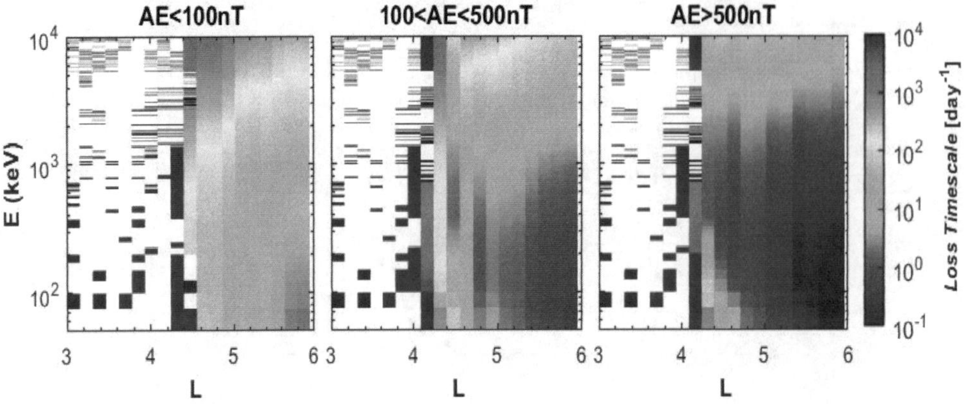

Fig. 8 Electron loss timescale as a function of energy and L due to chorus wave during quiet conditions (left), moderately disturbed conditions (middle) and active conditions (right).

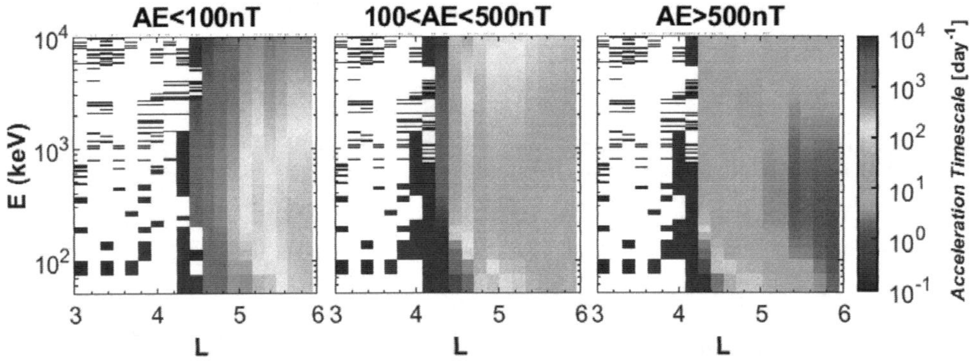

Fig. 9 Electron loss timescale as a function of energy and L due to chorus wave during quiet conditions (left), moderately disturbed conditions (middle) and active conditions (right).

Figure 8 and Figure 9 respectively present the two-dimensional panels of the electrons loss and acceleration timescales as the function of the electrons energy and L caused by the chorus waves during diverse conditions. From it, the electrons loss and acceleration timescales both decrease with the raise of AE and L. During quiet times, the electrons loss timescale can reach up to 104, while during highly disturbed times, the keV electrons can be scattered into the atmosphere within a few hours by the chorus waves. Compared with the electrons loss timescale, the electrons acceleration timescale is longer. Additionally, the electron loss and acceleration timescales grows as the electrons energy increases, corresponding to the Figure 6 and 7.

5. Conclusions

To study the interactions between the chorus waves and electrons in the radiation belts, this paper establish the statistics of the chorus waves amplitude and the background field parameters in the framework of MLT and L during different geomagnetic conditions utilizing the data from the Van Allen Probes. Based on the quasi-linear theory, the electrons pitch angle and energy diffusion coefficients are calculated and the acceleration and loss timescales of electrons with different energies are obtained as well. The principal conclusions of this study are described below:

1. The chorus waves are mainly distributed on the dayside and the dawn side. With the reinforce of substorms, its amplitudes enhance significantly, even reaching 100pt during highly disturbed times and the f_{pe}/f_{ce} outside the plasmasphere dips.

2. The electrons pitch angle and momentum diffusion coefficients caused by the chorus waves get enhanced with the AE and L, while decrease with the electrons energy. And the places with intense momentum diffusion of electrons is consistent with the maximum of the chorus waves amplitude and the minimum of the f_{pe}/f_{ce}.

3. The pitch angle diffusion coefficients are of the same order on the dayside and nightside for keV electrons but more dramatic on the dayside for MeV electrons.

4. Compared with the lower-energy electrons, the higher-energy electrons are tend to be scattered into the atmosphere by the chorus waves facilely.

5. The electrons loss and acceleration timescales both raise with the augment of AE and L and the electrons can be precipitated into the atmosphere by the chorus waves within a few hours in highly disturbed times.

References

[1] Agapitov, O., Mourenas, D., Artemyev, A., Hospodarsky, G., & Bonnell, J. W. (2019). Time scales for electron quasi - linear diffusion by lower - band chorus waves: The effects of ωpe/Ωce dependence on geomagnetic activity. Geophysical Research Letters, 46, 6178–6187. https://doi.org/10.1029/ 2019GL083446.

[2] Aryan, H., Agapitov, O. V., Artemyev, A., Mourenas, D., Balikhin, M. A., & Boynton, R., et al. (2020). Outer radiation belt electron lifetime model based on combined Van Allen Probes and Cluster VLF measurements. Journal of Geophysical Research: Space Physics, 125, e2020JA028018. https:// doi.org/10.1029/2020JA028018.

[3] Albert,J.M., Shprits, Y. Y.,(2009), Estimates of lifetimes against pitch angle diffusion,Journal of Atmospheric and Solar-Terrestrial Physics 71(2009)1647-1652.

[4] Artemyev, A., D. Mourenas, O. Agapitov, and V. Krasnoselskikh (2013), Parametric validations of analytical lifetime estimates for radiation belt electron diffusion by whistler waves, Ann. Geophys., 31, 599–624, doi : 10.5194 / angeo-31-599-2013.

[5] Baker.D.N, How to Cope with Space Weather, (2002) DOI: 10.1126/science.1074956, Science 297 (5586), 1486-1487.

[6] Baker, D. N., S. G. Kanekal, X. Li, S. P. Monk, J. Goldstein, and J. L. Burch (2004), An extreme distortion of the Van Allen belt arising from the Halloween solar storm in 2003, Nature, 432, 878–880.

[7] Glauert, S. A., and Horne, R. B. (2005), Calculation of pitch angle and energy diffusion coefficients with the PADIE code, J. Geophys. Res., 110, A04206, doi:10.1029/2004JA010851.

[8] Hwang, J. A., D.-Y. Lee, L. R. Lyons, A. J. Smith, S. Zou, K. W. Min, K.-H. Kim, Y.-J. Moon, and Y. D. Park (2007), Statistical significance of association between whistler-mode chorus enhancements and enhanced convection periods during high-speed streams, J. Geophys. Res., 112, A09213, doi:10.1029/2007JA012388.

[9] Horne, R. B., and R. M. Thorne (1998), Potential waves for relativistic electron scattering and stochastic acceleration during magnetic storms, Geophys. Res. Lett., 25, 3011–3014, doi:10.1029/98GL01002.

[10] Horne, R. B., and R. M. Thorne (2003), Relativistic electron acceleration and precipitation during resonant interactions with whistler-mode chorus, Geophys. Res. Lett., 30(10), 1527, doi:10.1029/2003GL016973.

[11] Horne, R. B., et al. (2005b), Wave acceleration of electrons in the Van Allen radiation belts, Nature, 437, 227–230, doi:10.1038/nature03939.

[12] Kennel, C. F., and H. E. Petschek (1966), Limit on stably trapped particle fluxes, J. Geophys. Res., 71, 1.
[13] Lastovicka, J. (1996), Effects of geomagnetic storms in the lower ionosphere, middle atmosphere and troposphere, J. Atmos. Terr. Phys., 58, 831.
[14] Li, W., et al. (2014), Radiation belt electron acceleration by chorus waves during the 17 March 2013 storm, J. Geophys. Res. Space Physics, 119, 4681– 4693, doi:10.1002/2014JA019945.
[15] Meredith, N. P. et al. "Pancake" electron distributions in the outer radiation belts. J. Geophys. Res. 104, 12431–12444 (1999).
[16] Newell, P. T., Sotirelis, T. & Wing, S. Diffuse, monoenergetic, and broadband aurora: the global precipitation budget. J. Geophys. Res. 114, A09207 (2009).
[17] Orlova, K., and Y. Shprits (2014), Model of lifetimes of the outer radiation belt electrons in a realistic magnetic field using realistic chorus wave parameters, J. Geophys. Res. Space Physics, 119, 770–780, doi:10.1002/2013JA019596.
[18] Petrinec, S. M., Chenette, D. L., Mobilia, J., Rinaldi, M. A. & Imhof, W. L. Statistical X-ray auroral emissions-PIXIE observations. Geophys. Res. Lett. 26, 1565–1568 (1999).
[19] Reeves, G. D., Cayton, T. E., Gary, S. P., and Belian, R. D. (1992), The great solar energetic particle events of 1989 observed from geosynchronous orbit, J. Geophys. Res., 97(A5), 6219–6226, doi:10.1029/91JA03102.
[20] Shprits, Y. Y., Thorne, R. M., Horne, R. B., Glauert, S. A., Cartwright, M., Russell, C. T., Baker, D. N., and Kanekal, S. G. (2006), Acceleration mechanism responsible for the formation of the new radiation belt during the 2003 Halloween solar storm, Geophys. Res. Lett., 33, L05104, doi:10.1029/2005GL024256.
[21] Shprits, Y. Y., et al. (2008b), Review of modeling of loses and sources of relativistic electrons in the outer radiation belt. II: Local acceleration and loss, J. Atmos. Sol. Terr. Phys., 70, 1694–1713, doi:10.1016/j.jastp.2008.06.014.
[22] Shprits, Y. Y. (2009), Potential waves for pitch-angle scattering of near equatorially mirroring electrons due to violation of the second adiabatic invariant, Geophys. Res. Lett., 36, L12106, doi:10.1029/2009GL038322.
[23] Summers, D., C. Ma, and T. Mukai (2004a), Competition between acceleration and loss mechanisms of relativistic electrons during geomagnetic storms, J. Geophys. Res., 109, A04221, doi:10.1029/2004JA010437.
[24] Summers, D., B. Ni, and N. P. Meredith (2007), Timescales for radiation belt electron acceleration and loss due to resonant wave-particle interactions: 2. Evaluation for VLF chorus, ELF hiss, and electromagnetic ion cyclotron waves, J. Geophys. Res., 112,A04207, doi:10.1029/2006JA011993.
[25] Tsurutani, B. T., and E. J. Smith (1974), Postmidnight chorus: A substorm phenomenon, J. Geophys. Res., 79, 118–127, doi:10.1029/JA079i001p00118.
[26] Tsurutani, B. T., et al. (2006), Corotating solar wind streams and recurrent geomagnetic activity: A review, J. Geophys. Res., 111, A07S01, doi: 10.1029 / 2005JA011273.
[27] Thorne, R. M. (2010), Radiation belt dynamics: The importance of wave-particle interactions, Geophys. Res. Lett., 37, L22107, doi:10.1029/2010GL044990.
[28] Thorne, R. M., et al. (2010), Scattering by chorus waves as the dominant cause of diffuse auroral precipitation, Nature, 467, 943–946, doi:10.1038/nature09467.
[29] Thorne, R., Li, W., Ni, B. et al. (2013), Rapid local acceleration of relativistic radiation belt electrons by magnetospheric chorus. Nature 504, 411–414. https://doi.org/10.1038/nature12889.
[30] Vamp ola, A. L., J. V. Osborne, and B. M. Johnson (1992), CRRES magnetic electron spectrometer AFGL-701-5A (MEA), J. Spacecraft Rockets, 29, 592–595, doi:10.2514/3.25504.

A Numerical Investigation of Tidal Disruption Rate Enhancement in Post-Starburst galaxies

Marco Tinyeung Xue [1], Fuguo Xie [2,*]

[1] Shenzhen Hong Kong Pui Kiu College Longhua Xinyi School, Shenzhen, China
[2] Shanghai Astronomical Observatory, Chinese Academy of Sciences, Shanghai, China

*Corresponding author e-mail: fgxie@shao.ac.cn

Abstract. Tidal disruption is an uncommon but important phenomenon that occurs when a star's gravity is overcome by that of the SMBH which it is orbiting. In this process, researchers may gleam various properties of the host galaxy's nucleic region, such as the accretion process, debris fallback, behavior of active galactic nuclei and quasars, etc. In this work we present a numerical analysis of TDE rates using the loss cone theory, a leading theoretical picture in this field, based on changes to several pivotal variables, such as SMBH mass, stellar mass, velocity dispersion etc. We then extend this to fit the profiles of E+A post starburst galaxies and the occasion of binary black hole systems.

Keywords: galaxies, black holes, tidal disruption, stellar dynamics, active galactic nuclei.

1. Introduction

1.1. Introduction & Overview

An infrequent yet important astrophysical phenomenon is tidal disruption, as seen in Fig. 1. The majority of galaxies host at least one supermassive black hole (SMBH) in their nuclei (e.g., Kormendy & Ho 2013 for a related review), which can perturb the relative motion of nearby stars and celestial objects (Hills 1975, Lidskii & Ozernoi 1979). Typically the orbits of these bodies are governed by the gravitational potential of the galaxy, and they are regular. But if a star in orbit around the SMBH is brought within close enough proximity due to various phenomena such as n-body relaxation (see Sec. 2.4 below), then it may be tidally disrupted (i.e. the gravity of central black hole overcomes that of the star' and causes the collapse of the latter). During this disruption, a fraction of the star to be captured by the black hole, leaving the remainder to be flung out into space. The process of the black hole capturing the fraction of the star emits a burst of intense radiation, which is one of modern science's current best indicators for a MBH's existence at the heart of a galaxy. The fallback rate, or the rate at which tidal debris returns to the presence of the black hole, can be as short as months. Therefore TDEs are a useful way to study formation of accretion disks and jets. For TDEs to occur, the tidal disruption radius must be larger than the event horizon. This radius is defined as

Content from this work may be used under the terms of the Creative Commons Attribution 3.0 licence. Any further distribution of this work must maintain attribution to the author(s) and the title of the work, journal citation and DOI.
Published under licence by IOP Publishing Ltd

$$R_T = R_* \left(\frac{M_{BH}}{M_*}\right)^{1/3} \quad (1)$$

Here R_* and M_* are the radius and the mass of the star respectively (Rees 1988, Evans & Kochanek 1989). For SMBHs with mass $M_{BH} > 10^8\ M_\odot$ (here $M_\odot$ is the mass of Sun), the tidal radius will be smaller than the size of the SMBH, i.e. TDE will not happen for a main-sequence star around massive ($M_{BH} > 10^8\ M_\odot$) black holes. For a non-spinning BH, its size (event horizon) is the Schwartzchild radius $R_s = \frac{2\,GM_{BH}}{c^2} = 3 \times 10^8\ km \frac{M_{BH}}{10^8 M_\odot}$ ($\cong 2\ AU$).

The TDE investigation is crucial for several reasons. Firstly, they provide a probe through which to observe the inertia of a star. In most cases the inertia of a star is hard or almost impossible to observe/detect. But, during the disruption process, the gas from the inner zones of a star will be expelled and can then be identified through emission/absorption line analysis. Indeed, spectroscopic and timescale observations of TDEs have been used to determine properties of the disrupted star (see e.g., Gezari 2021 for a related discussion). This provides valuable insight into the properties of the now disrupted star. Secondly, the fallback process, disk accretion process, and other associated processes, provides pivotal information on black hole accretion physics, including the launching of a jet and the formation of winds. Usually, the variability of accretion onto SMBH is low, with an atypical evolutionary timescale of greater than 10^{2-4} yrs, but with spontaneous events such as TDEs, the evolutionary timescale is reduced to a more convenient months – yr timescale.

1.2. Basic Theoretical Picture

Due to various processes, a star's orbit may shift in a way such that it reaches or approaches periapse at a point at or within the SMBH's tidal disruption radius (Rees 1988, Alexander 2012). At this point, the SMBH's gravity overcomes that of the star, and the star is disrupted (Rees 1988). A portion of the star is captured and accreted by the SMBH while the remainder is ejected outward into space in a family of highly eccentric orbits, each defined by different binding energies for each little "clump" of star matter. This is shown in the left panel of Fig. 1. The details of this process is now updated, thanks to the advance in numerical simulations that solve both high-contrast and large dynamical range issues (see, e.g., Lodato et al. 2020 for a review), which is shown in the right panel of Fig. 1. The key differences from the original idea is that the star is actually distrupted before it passes the periapse. The collapse and accretion of the star releases high amounts of radiation (typically in the visible, UV, X-ray and gamma ray wavebands) which can be readily observed, as well as energy in the form of gravitational waves (Gezari 2021).

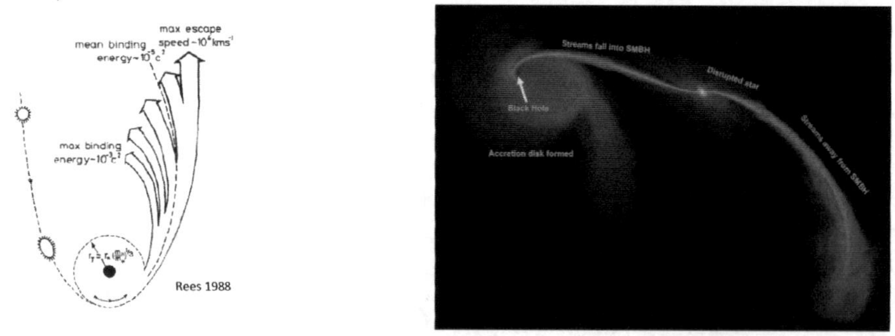

Fig. 1 (Left) the original idea of tidal disruption event as proposed in Rees (1988). Approaching from the left, it elongates as it closes in on the SMBH, and then is disrupted upon passing it. A portion of the stellar debris is ejected outward in the arrows on the right while the portion accreted is not shown. Figure adapted from Rees (1988). (Right) snapshot of a star disrupted by a SMBH. Here the locations of both SMBH and the center of the star are labeled in the plot. Obviously, the star is tidally disrupted to form a long, elongated stream, with approximately half gas being attracted towards the SMBH, forming an accretion disk. Figure adapted from Kohler & Stephen 2015 (https://news.ucsc.edu/2015/09/tidal-disruption.html).

As expected, galactic properties such as its type/classification, nuclear stellar density and orbital distribution can greatly affect the outcome TDE rate. Numerical approximations for stars diffusing into the MBH's loss cone specific angular momentum phase space generally range from $10^{-4} \sim 10^{-5}$ per year per galaxy (Wang & Merritt 2004, Stone & Metzger 2016, etc.), but can be in the range of 10^-3 per year per galaxy in galaxies with steep nuclear density profiles, as well as nucleated dwarf galaxies and post-starburst (E+A) galaxies. These ranges can be magnified if the host galaxy has recently undergone a galaxy merger event (within ~1Gyr), which would cause a sizeable perturbation in the local star density, and in certain situations, such merger events could also form a binary MBH system, which could either increase or decrease the TDE rate by attracting or ejecting stars at a higher frequency respectively before stabilizing (Milosavljevic & Merritt 2003, Lezhnin & Vasiliev 2016). The spin distribution of the MBH can also greatly affect the corresponding TDE rate, as due to the spin of the MBH decreasing the size of its Schwarzchild radius, the maximum mass of a SMBH the tidal disruption radius of which exceeds the Schwarzchild radius decreases to 7×10^{-5} star masses.

Corresponding observations made regarding TDEs involve the peak luminosity and blackbody temperatures of various individual events (Gezari 2021). An interesting thing worth noting is that in recent years there has been a significant increase in the number of detected TDEs in various (e.g., infrared, optical, UV and X-ray) wavebands with the most being the optical waveband. When the peak luminosity of the 56 TDEs examined in the literature are plotted against their blackbody temperatures, it can be concluded that TDEs observed in the optical waveband have a significantly lower average blackbody temperature than TDEs observed in the X-ray waveband. Interestingly, neither of these averages coincide with the expected thermal emission from a circular debris disk (Gezari 2021). Another observable property of TDEs is the fallback rate of the portion of the disrupted star that is ejected outward. The fallback rate of the stellar debris is defined as the orbital period of the most bound debris and scales with the square root of the mass of the black hole, which implies that observing debris fallback should be used to calculate black hole mass. The fallback debris stream also has a uniform energy distribution, and the rate at which mass falls back to the pericenter follows a power law (Rees 1988, Guillochon & Ramirez-Ruiz 2013).

1.3. Limitations and Other Factors

It is important to note here that there is a significant difference between posited and observed TDE rates in the majority of galaxies types that indicates a lower average TDE rate (Stone & Metzger 2016), yet there is a considerably higher TDE rate in "E+A", or post starburst galaxies (Arcavi et al. 2014) due to the high concentration and asymmetry of the distribution of young stars in the nucleus region of E+A galaxies (French 2017), as shown in Fig. 2. Thus, an overview of TDE rate observational constraints is of high relevance to this research.

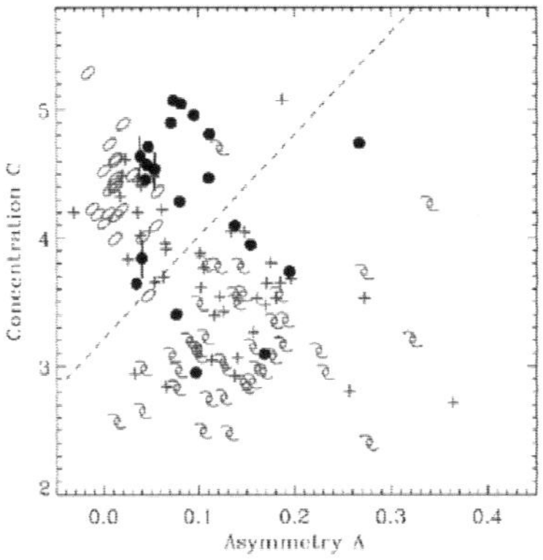

Fig. 2 A plot of young star population asymmetry against concentration. Each point represents a TDE candidate. Evidently the majority of them occur in galaxies with high concentrations of young stars (i.e. post starburst galaxies). Figure adapted from French (2017).

Over the course of decades, researchers have produced an array of estimated TDE rates based on small samples of soft X-ray emission TDEs that span a range of roughly 1.5 orders of magnitude. This disagreement can be attributed to many factors. One such common factor is the unfortunate consequences of small-scale statistical analysis. Another influential factor is the fact that the majority of TDE surveys are poor in temporal resolution, which forces researchers to make critical assumptions regarding the X-ray luminosity functions and characteristic decay times of TDEs. Using these assumptions to infer a volumetric TDE rate, which can then be used to infer a per-galaxy TDE rate, inevitably leads to more assumptions about which galaxies can even host TDEs in the first place (Hung et al. 2018).

This issue is partially alleviated by TDEs discovered via UV and optical thermal emissions by the GALEX satellite (Gezari et al. 2006, van Velzen et al. 2011), the TDE rates derived from which were consistent with those posited by Wang & Merritt 2004. The icing on the cake, so to speak, is that with optically observed TDEs, there is little need to assume the temporal evolution of light curves due to a higher clarity in time domain surveys. The volumetric TDE rate estimated through this method is approximately 1 order of magnitude smaller than that of the previous method (van Velzen & Farrar 2014), and higher TDE rates can be found using empirical luminosity functions, such as in van Velzen 2018.

There are also a variety of phenomena that influence the radiation emission bandwidth of TDEs. Three in particular are primarily responsible for shifting emission toward the UV and optical ends of the spectrum: 1) trapping of radiation in outflowing material by electron scattering and the following losses of radioactive energy, 2) absorption of accretion disk photons by other materials at a distance, and 3) Compton scattering (Roth et al., 2020). These phenomena, all associated with an early formation of an accretion disk, occur to varying degrees in all TDEs. Another scenario involves a very slow circularization of disrupted debris (on a timescale of several months), which prompts emission of soft X-ray, UV and optical emissions. These provide an explanation for the different bandwidths in which TDEs are discovered.

Interestingly, there appears to be an unprecedented amount of TDEs occurring in quiescent galaxies with no ongoing star formation. While the majority of these events have to do with A stars, which indicate that the galaxy underwent a recent major star formation episode, there is also a strong correlation between this enhanced rate and the strength of Balmer absorption features. It is found that in the most extreme Balmer-strong galaxies, the ratio of the absolute rate density of TDEs in post starburst galaxies to the average volumetric rate $R = \frac{n_{PS}}{n}$ can reach $R = 190^{+191}_{-100}$, while it attains a value of $R = 33^{+7}_{-11}$at the other end of the spectrum (French et al. 2016). Later analysis by Graur et al. 2018 provided less extreme ranges for the same levels of Balmer strength, which may be considered more precise due to a larger sample size. In order to hash these values out more accurately, and also resolve overarching problems faced in this field as illustrated above, a larger number of observations of higher quality is required.

2. Loss Cone Theory and TDE Rate

In this section, we will provide basic equations to estimate the TDE rate, following the convention of this community (e.g., Wang & Merritt 2004; Stone & Metzger 2016) to introduce two basic physical quantities $\psi(r) = -\Phi(r)$ (here $\Phi(r)$ is the gravitational potential, which is defined that $\Phi(r) < 0$.) and $\epsilon(r) = -E(r)$ (here the total energy per unit mass of individual stars is $E(r)$). For convenience, below we will call $\psi(r)$ and $\epsilon(r)$ respectively, the gravitational potential and total energy. Obviously, for the problem interested to us, both $\psi(r)$ and $\epsilon(r)$ will be non-negative.

2.1. Galaxy Properties

Host galaxies characteristically include one or a system of SMBHs and a surrounding cluster of stars in their nuclei, which result in a bulge and occasionally a bar shape for some galaxy structures. Beyond the nuclei are disks consisting of stars and other celestial matter, all through which large masses of gas permeates.

We need only to concern ourselves with the region near the nucleus of host galaxies. Galaxies can have vastly different structures on a macro scale (i.e., elliptical, spiral, etc.), but they are mostly similar at a relatively micro scale near the nucleus, and their nuclei can similarly be approximated by a bulge. Additionally we may omit the influence of gasses and dark matter in the system, and assume the gravitational potential $\psi(r)$ of the star-SMBH system can be expressed as $\psi(r) = \psi_*(r) + \psi_{BH}(r)$. Here $\psi_*(r)$ and $\psi_{BH}(r) = \frac{GM_{BH}}{r}$ are respectively, the gravitational potential of the star(s) and SMBH, where the potential of stars varies among different galaxies types. With knowledge of the system's potential, we can express the number density of stars in phase space (also named the distribution function) $f(\epsilon)$. Taking Eddington's formula, it can be expressed as

$$f(\epsilon) = \frac{1}{\sqrt{8}\pi m_*}\frac{d}{d\epsilon}\int_0^{\epsilon}\frac{d\rho}{d\psi}\frac{d\psi}{\sqrt{(\epsilon-\psi)}} \quad (2)$$

Where m_* is the stellar mass and ϵ is the stellar energy defined above (Wang & Merritt 2004, Eq. 4).

It is known that host galaxies feed SMBHs with large amounts of gas, fractions of which are accreted and hence convert into a huge amount of energy and momentum, in the form of radiation, wind and jet, that will have significant impact on the gas dynamics of the galaxy (Gezari 2021). Through this cyclic feedback process of active galactic nuclei (AGN, i.e. accretion onto SMBH), the galaxy and its SMBH are coupled together (Kormendy & Ho 2013, Heckman & Best 2014). Now, two relations are established (e.g., Kormendy & Ho 2013 for a review), i.e. $M_{BH} \approx 10^{-3}\, M_{bulge}$, where M_{bulge} is the mass of (pseudo-)bulge, and $M_{BH} \propto \sigma^4$, where σ is the velocity dispersion of the stars in the bulge region. The uncertainty of derived black hole mass, is similar in these two relations, i.e. M_{BH} is accurate within a factor of 2 (i.e., 0.3 dex).

It is known that a large fraction of spiral galaxies have bar structures in their central regions. Such a non-axismetric structure will introduce an additional torque on the gas and stellar dynamics, where for the purposes of this research will induce a change in $\psi_*(r)$, necessitating another variable input $\psi_*(r, \phi)$, where ϕ is the azimuthal angle.

Another significant factor that needs to be considered is whether the host galaxy is a post starburst galaxy (E+A) or not (Wang & Merritt 2004, Stone & Metzger 2016). These E+A galaxies have recently undergone major galaxy-galaxy merger events, which could lead to the formation of a binary-SMBH system. Recently, based on a novel so-called Gaia multipeak method, Mannucci et al. (2022) find a large population of binary-SMBH systems, with separations much closer than 1kpc. This observation confirms the theoretical expectation that galaxies may harbor multi SMBHs. This increases the TDE rate by many orders of magnitude up to $\mathrm{N} \approx 0.1\,\mathrm{yr}^{-1}$(Ivanov et al. 2005, Chen et al. 2009, 2011). This enhancement, however, is ultimately temporary over a timescale of $\sim 10^5 - 10^6$yrs, rendering its impact on the total TDE rate resulting from hardening SMBH binaries barely even noticeable, at a mere $\sim 3\%$.

2.2. Loss Cone Theory

Among the various models for SMBH feeding, the loss cone model presents an especially clear and coherent visualization and quantification. Essentially, it describes the specific angular momentum and energy requirements for a star to fall into an orbit in which it reaches periapse within the SMBH's tidal disruption radius. Here we focus on the simpler steady-state (time-independent) loss-cone theory in a spherical geometry. More general discussions can be found in literature (Stone et al. 2020). One key and reasonable assumption here is that stars in the bulge region at an equal distance to the SMBH have equal specific kinetic energies but the directions of their motion do not necessarily need to agree.

We first note that in Newtonian gravity, a star on a highly eccentric elliptical orbit will have a specific orbital angular momentum $J = \sqrt{2GM_{BH}R_p}$, where R_p denotes the pericenter of the orbit. In other words, stars whose $J \leq J_t = \sqrt{2GM_{BH}R_t}$ will be disrupted when passing within R_t of the SMBH, thus creating a specific angular momentum phase space "cone" in which a star is doomed to be tidally disrupted, which we will henceforth dub as the loss cone. Meanwhile, the maximal J of stars with total energy ϵ will be that of circular motions, i.e. $J_{circ}(\epsilon)$ $(e = 0)$. The basis of this picture is illustrated in Figure XXX. It is known that, as the star moves closer to the SMBH, its total energy ϵ increases as well as the J/J_{circ} range by which the loss cone is defined. In an ideal world, assuming that angular momentum is conserved for every stellar orbit, then the number of stars inside the loss cone would drop to zero within one orbital period. However, in the real world, there are many processes that have impacts on the orbits of stars in the bulge region that must be accounted for, such as two-body relaxation, resonant relaxation, torques induced by aspherical stellar potential, etc, therefore there is always an averaged nonzero number of stars within the loss cone. Conventionally we define a new variable $\mathcal{R} = \left(\frac{J}{J_{circ}(\epsilon)}\right)^2$, which ranges from [0, 1]. Obviously, for motions in a Keplerian potential, $\mathcal{R} = 1 - e^2$ (Stone et al. 2020).

2.3. Accounting for Two-Body Relaxation and Angular Momentum Diffusion

Changes to the stellar distribution function over time may be expressed as cumulative interactions between systems of two bodies. For our purposes, we may define the relaxation time as $T_{rel} = \frac{\sigma^3}{G^2 n < m_*^2 > ln\Lambda}$, where $ln\Lambda \sim 10$ is the Coulomb logarithm (Chandrasekhar 1942, Spitzer & Hart 1971). Defining the orbital period as P_{orb}, we focus primarily on cases where $T_{rel} < P_{orb}$, where we observe that for stars nearing the boundary of the loss cone $J \sim J_{lc}$, the relaxation in specific angular momentum occurs much more rapidly than that in energy, thus we may conclude that in energy space, stellar diffusion eventually leads to a quasi-steady state.

Focusing then on angular momentum diffusion, we adopt the one dimensional orbit-averaged Fokker-Planck equations for the evolution of the stellar distribution function $N(\mathcal{R}, t)$ in the space of dimensionless angular momentum $\mathcal{R}$ and fixed total energy ϵ (Stone et al. 2020):

$$\frac{\partial}{\partial t} N(\mathcal{R}, t) = \frac{\partial F}{\partial \mathcal{R}}, \quad F = D(\epsilon, \mathcal{R}) \frac{\partial N}{\partial R}, \quad D(\epsilon, \mathcal{R}) \approx R * D(\epsilon) \tag{3}$$

Where F is the flux of stars through the integral space in the direction of the inner boundary $\mathcal{R}_t$. The dimensionless orbit-averaged diffusion coefficient can be approximated as $D(\epsilon) \sim 1/T_{rel}$, which also depends on other factors of the stellar distribution function (Wang & Merritt 2004):

$$D(\epsilon) = \frac{64\,\pi^2 r^2 G^2 m_*^2 ln\Lambda}{3\,J_{circ}^2}(3I_{1/2} - I_{3/2} + 2\,I_0), \tag{4}$$

$$I_0 = \int_0^{\epsilon} f(\epsilon')d\epsilon', \quad I_{n/2} = \{2[\psi(r) - \epsilon]\}^{-\frac{n}{2}} \int_{\epsilon}^{\psi(r)} \{2[\psi(r) - \epsilon']\}^{\frac{n}{2}} f(\epsilon')d\epsilon' \tag{5}$$

With the above expressions, we solve the Fokker-Planck equation. The result can be expressed as (Cohn & Kulsrud 1978):

$$F(\epsilon)d\epsilon = 4\pi^2\, J_{circ}^2\, P_{circ}(\epsilon)\, D(\epsilon)\, \frac{f(\epsilon)}{ln\mathcal{R}_0^{-1}}\, d\epsilon. \tag{6}$$

Here $\mathcal{R}_0(\epsilon)$ is the value of R at which $f(\epsilon) < 0$ because of the removal of stars that scatter into the loss cone (note that $\mathcal{R}_0(\epsilon)$ is different from $\mathcal{R}_{lc}(\epsilon) = J_{lc}^2/J_{circ}^2(\epsilon)$). Cohn & Kulsrud (1978) also provides a good numerical approximation,

$\mathcal{R}_0(\epsilon) = \mathcal{R}_{lc}(\epsilon)\, exp(-q)$ if $q(\epsilon) > 1$; and $\mathcal{R}_0(\epsilon) = \mathcal{R}_{lc}(\epsilon)\, exp\left(-0.186q - 0.824\sqrt{q}\right)$ if $q(\epsilon) < 1$

Where $q(\epsilon) = \frac{P_{circ}(\epsilon)D(\epsilon)}{\mathcal{R}_{lc}(\epsilon)}$. The function of $q(\epsilon)$ can also be interpreted as the ratio of the orbital period ($P_{circ}(\epsilon)$) to the timescale for the diffusional refilling of the loss cone ($t_{LC.refill} \cong \frac{\mathcal{R}_{lc}(\epsilon)}{D(\epsilon)}$). Hence the TDE rate is therefore simply the integral of $F(\epsilon)$, i.e.

$$\dot{N} = \int F(\epsilon)d\epsilon. \tag{7}$$

2.4. *Rapid Relaxation Mechanisms*

TDE rates depend upon the supply channel of stars and their influx into the loss cone, and thus, two-body relaxation is unavoidable during this process. However, other factors that greatly contribute to this influx cannot be neglected.

One such factor is the presence of massive perturbers, in whose presence the relaxation rate may be approximated as $\gamma \sim n_p {M^2}_p/\sigma^3$, where n_p is the number density of stars with mass M_p and σ is the velocity dispersion. The relaxation rate's dependence on ${M^2}_p$ implies that even a small population of massive perturbers in the region can cause a sizeable impact on the TDE rate. This is indeed congruent with observations in many galactic nuclei, where giant molecular clouds, stellar clusters (and possible intermediate-mass black holes) lower the estimated 2-body relaxation time by orders of magnitude relative to that due to stars alone.

The other factor is resonant relaxation, which is a rapid process of angular momentum relaxation that takes place in near-spherical symmetrical potential. This characteristically takes place in near-circular symmetrical orbits where orbital evolution is limited. The RR timescale can be approximated as $t_{RR} \sim (QP)^2/N_* t_{coh}$, where N_* is the number of stars, t_{coh} is the time it takes for perturbations to break orbital symmetry, P is the orbital timescale, and $Q = \frac{M_\cdot}{M_*}$ is the ratio between the SMBH mass and star mass. This process is also detected in n-body numerical simulations and there is evidence that it also operates in the global cluster (ref. 9 Alexander 2012). However, it seems that its effect on TDE rate is less than a factor of 2.

Another factor is the influence of a non-spherical system. Highly symmetric stellar potential is associated with conserved quantities, which bound orbital evolution and can decrease the flux of stars into the LC to the rate induced by two-body relaxation. The converse of this discovery is also true in that asymmetry leads to more orbital evolution and a higher rate of flux. This is most noticeable in triaxial systems, which can support a large fraction of chaotic orbits, but this depends on whether chaotic orbits can exist in the presence of the stabilizing influence of the SMBH.

2.5. *Stellar Kinematics*

Though it is difficult to extrapolate galactic nuclei properties to encompass the larger galactic system, there are many models with which to correlate nucleic property with galaxy properties on kpc scales. The most relevant of these models to this research is the $M - \sigma$ relation, which relates the mass of the SMBH at the galaxy's nucleus with the stellar velocity dispersion of the surrounding region (Kormendy & Richstone 1995). The $M - \sigma$ relation is also an indirect evidence of the feedback of nuclei AGN on its host galaxy.

Although there is a high observational dependence of TDE hosts on galaxies with high stellar mass densities, there seems to be no such dependence on galaxy central stellar distribution. (French 2020) From this it can be intuitively derived that TDE hosts are more concentrated on galactic scales (Law-Smith et al. 2017, Graur et al. 2018). In an asymmetry-concentration phase space, where elliptical galaxies are more concentrated and spiral galaxies are more asymmetrical, it was also discovered that merger events in the recent history of galaxies with a lack of high asymmetry cannot be ruled out due to this property alone (French 2017).

3. Computation & Analysis

In the following section we review the impact of various galaxy and nuclear properties on TDE rate, the observable trends and their interpretations, and an extension of existing models to fit E+A and post-starburst galaxies with binary black hole systems.

3.1. *Impact of Galaxy Properties on Various Tidal Disruption Characteristics*

Curiously, for galaxies with fixed properties, the TDE rate $\dot{N}$ appears to increase as M_{BH} decreases. In our work, we present two ways to understand this phenomenon through energy.

As mentioned in Sec. 2.1, We separate the galaxy potential into two parts, one relates to the nuclei SMBH, and the other relates to the stellar (and gas, dark matter) of the galaxy, i.e.

$$\psi(r) = \psi_*(\mathrm{r}) + \frac{\mathrm{GM_{BH}}}{\mathrm{r}}. \tag{8}$$

Here we provide more information on the properties of the galaxy. Galaxies with low luminosity, otherwise classified as faint galaxies, have the highest consumption rates. They also tend to follow a power-law nuclear density profile $\rho \sim r^{-\gamma}$, where γ is a dimensionless quantity.

There exist two simple models. First, for the $\gamma = 2$ density profile, the potential of the stars (and also gas and dark matter) can be collectively described as

$$\psi_*(r) = -2\sigma^2 \ln\frac{r}{r_h} \tag{9}$$

Where σ is the one dimensional stellar velocity dispersion which is independent of orbital radius for $r \geq r_h$, $r_h = \frac{GM_{BH}}{\sigma^2}$.

For a general power-law distribution case where $\gamma \neq 2$ (obviously we should always have $\gamma > 0$.), the galaxy density distribution and the potential can be expressed as,

$$\rho(r) = \rho_0\left(\frac{r}{r_h}\right)^{-\gamma}, where\ \rho_0 = M_{*r_0}\frac{3-\gamma}{4\pi r_0^3} \tag{10}$$

$$\psi_*(r) = \frac{M_{*r_0}}{(2-\gamma)r_0}\left(\frac{r}{r_h}\right)^{2-\gamma} \tag{11}$$

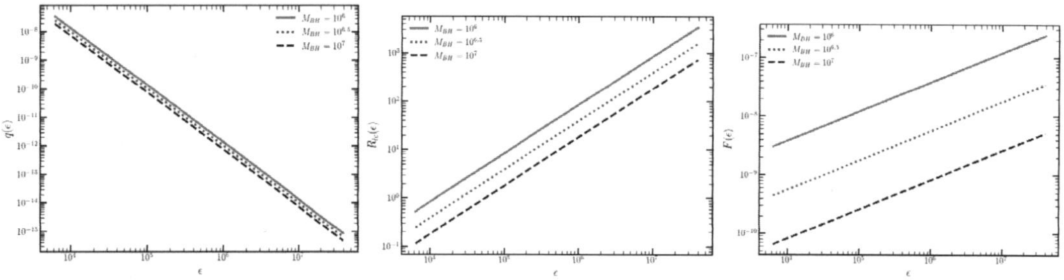

Fig. 3 Impact of stellar energy on TDEs, where we considered three characteristic representatives, i.e. the orbital-to-refilling timescale ratio $q(\epsilon)$ (top panel), the (square) of the loss cone-to-circular angular momentum ratio $\mathcal{R}_{lc}(\epsilon)$ (middle panel), and the TDE rate at given stellar energy $F(\epsilon)$ (bottom panel). In this figure, the basic parameters of the host galaxy are set by $\sigma = 200\ km/s$, and the stars are having mass $M_* = 1\ M_\odot$. In the three plots, we also consider three choices of black hole mass, i.e. $\frac{M_{BH}}{M_\odot} = 10^6$ (red solid), $= 10^{6.5}$ (blue dotted), $= 10^7$ (black dashed). From the left plot, we find that, as the (absolute value of) stellar energy increases (i.e., increase in ϵ), the $q(\epsilon)$ decreases sharply, i.e. $q(\epsilon) \propto \epsilon^{\sim -2}$. Moreover, the increase in black hole mass also lead to a decrease in $q(\epsilon)$. The $\mathcal{R}_{lc}(\epsilon)$ plot follows a very similar trend to the $F(\epsilon)$ plot, and obviously, while the loss cone angular momentum threshold increases with stellar energy, a lower black hole mass leads to a higher angular momentum threshold. For the $F(\epsilon)$ plot, it is found that while the TDE rate changes at the same rate for each black hole mass, the lower the black hole mass, the higher the $F(\epsilon)$ value and hence the lower the TDE rate.

With the above host galaxy property, we can evaluate the TDE rate. Before that, we first evaluate the impact of stellar energy ϵ on TDE happening efficiency. Here we take $q(\epsilon)$, $\mathcal{R}_{lc}(\epsilon)$ and $F(\epsilon)$ as representatives (see Sec. 2.3, and Cohn & Kulsrud 1978). For this investigation, the properties of the host galaxy are set through $\sigma = 200\ km/s$ and $M_* = 1M_\odot$. We consider three choices of black hole mass, i.e $M_{BH}/M_\odot = 10^6, 10^{6.5}, 10^7$, which are shown in Figure 3 by respectively, red solid, blue dotted and black dashed curves.

We note that, $q(\epsilon)$ defines the ratio between the orbital period of a star and the timescale of refilling the loss cone, i.e. larger values of $q(\epsilon)$ means more efficient loss cone refilling thus more frequent

TDE case. As shown in the left panel of Fig. 3, for our chosen parameters, we always have $q(\epsilon) \ll 1$, i.e. only a tiny fraction of stars will have their angular momentum be diffused into their corresponding loss cone value and will eventually be captured by the black hole. This is consistent with our expectation, since TDE is not a frequent phenomenon for individual galaxies. Moreover, as the energy of star increases (increase in ϵ; actually a decrease of total energy, since we adopt the opposite sign, i.e. $\epsilon(r) = -E(r)$), we observe a sharp decrease in $q(\epsilon)$. This also indicates that, most of the disrupted stars are from small energy regions. A simple fitting to this result reads that $q(\epsilon) \propto \epsilon^{\sim -2}$. On the other hand, we find that, there is a strong decrease in $q(\epsilon)$ when $M_{BH}/M_{\odot} > 10^7$; but it becomes less sensitive when $M_{BH}/M_{\odot} \geq 10^8$.

The middle panel of Fig. 3 shows $\mathcal{R}_{lc}(\epsilon)$, which defines the square of the angular momentum ratio between the loss cone threshold and that of an ideal circular orbit: $\mathcal{R}_{lc}(\epsilon) = \left(\frac{J_{lc}}{J_{circ}(\epsilon)}\right)^2$. As the stellar energy ϵ increases (for most general cases, the stars with larger ϵ locate at smaller radius, i.e. more close to the supermassive black hole), $\mathcal{R}_{lc}(\epsilon)$ increases accordingly. In other words, the loss cone will become wider. This can be easily understood through the tidal disruption radius, where $J_{lc} \propto \sqrt{R_t}$ and $R_t \propto M_{bh}$, i.e. it reflects the foundational conclusion that TDEs happen more frequently in systems with smaller black hole mass.

Before providing the total TDE rate, we in the bottom panel shows the TDE rate at a given energy ϵ, i.e. $F(\epsilon)$. Clearly, we always have $F(\epsilon) \ll 1$, meaning that the TDE rate is very low, i.e. statistically TDE on average is a rare phenomenon. Besides, we also make the observation that as the total energy of the stars ϵ increases (i.e. systematically locate more close to BH), the TDE rate increases, which fits the expectation outlined earlier in this section. Again, we observe that $F(\epsilon)$ also decreases with increasing SMBH mass.

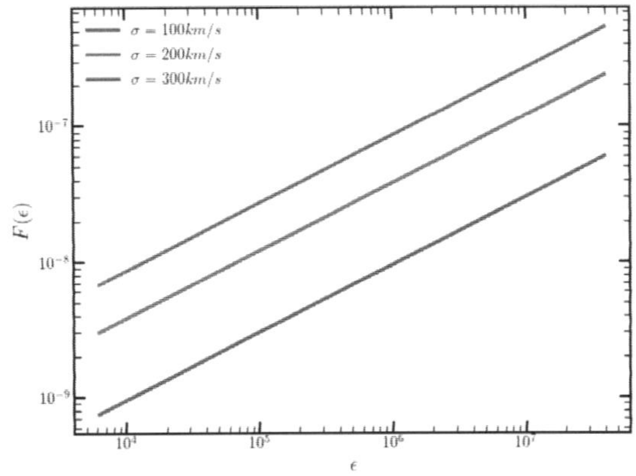

Fig. 4 Graphs of TDE rates for three different choices of σ. It follows a similar trend as Fig. 3, in that as stellar energy increases, the TDE rate increases as well.

All above investigations focus on the energy (or equivalently location) of stars and black hole mass. Obviously the TDE rate also depends on the property of the whole galaxy, i.e. not only the distribution but also its normalization (i.e. total stellar mass). This is mostly set by the one dimensional velocity dispersion σ. For an ideal virialized system as the case of a galaxy, the potential of the galaxy as well as the energy of individual stars directly depend on the stellar velocity dispersion, as outlined previously in Sec. 3.1. We in Fig. 4 explore the impact of total stellar mass, where we considered three choices of σ, i.e. $\sigma = 100\ km/s$ (purple), $200\ km/s$ (red), and $300\ km/s$ (green). The TDE rate at given stellar energy ϵ increases in accordance with ϵ. The physical reason for this dependence has been given in Fig. 3. If the galaxy becomes more massive (i.e., for given stellar mass, there are more abundant stars and hence there is an increase of the number density of stars at given stellar energy ϵ), we should observe an enhancement in TDE rate. This is clearly shown in Fig. 4.

Finally, we notice that, in this plot, we only show results with given ϵ, it may be more interesting to explore the impact on ϵ/σ^2 (i.e., stars with the same "normalized" stella energy ϵ/σ^2 in general locate at the same radius in different galaxies). But clearly from Fig. 4, we may find that, the TDE rate will have an much higher enhancement at given ϵ/σ^2 compared to that at given ϵ. Again, this is due the increase of stellar number density.

3.2. Impact of Galaxy Properties on TDE Rate

Now we numerically investigate the impact of SMBH and stellar mass on the TDE rate. As reviewed in the introduction, (Sec. 1.1), the threshold $M_{BH} = 10^8\ M_\odot$ is the boundary beyond which an SMBH can no longer tidally disrupt nearby stars due to its tidal disruption radius being within its event horizon. For this investigation, we explore a range of SMBH masses $M_{BH} \in [10^6, 1.5 \times 10^7]$. Other model parameters are fixed to $\sigma = 200\ km/s$ and $M_* = 1M_\odot$.

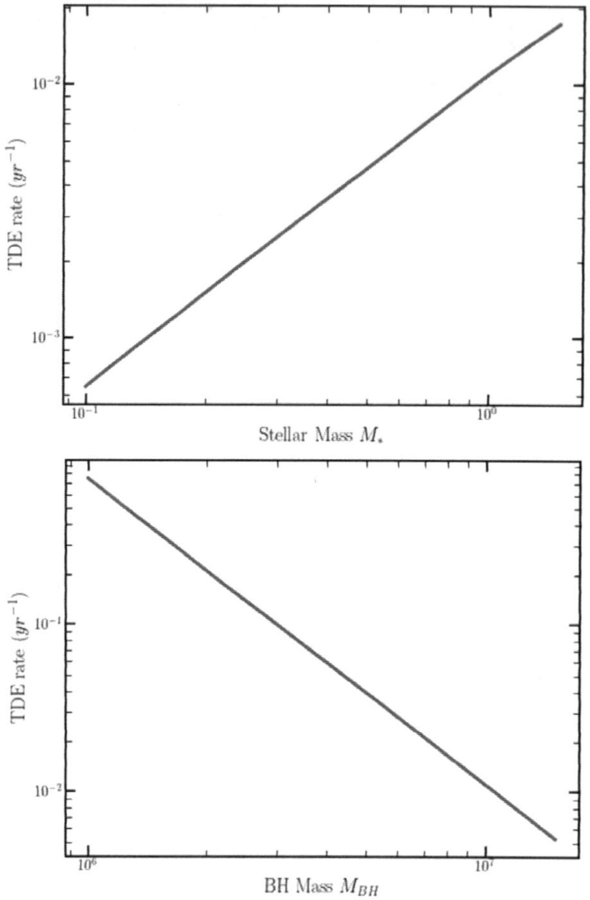

Fig. 5 (Top) TDE rate as a function of average stellar mass in the nuclei regions of a galaxy, where we explore $M_* \in [10^{-1}, 1.5]$. (Bottom) TDE rate as a function of BH mass. We explore a narrow range, with $M_{BH} \in [10^6, 1.5 \times 10^7]$. Evidently, it follows an exponential decreasing model.

The first variable in this model is the "average stellar mass" in the nuclear region that is at risk of being tidally disrupted. The results of taking a realistic stellar mass range $M_* \in [10^{-1}, 1.5]$ is shown in Fig. 5 (top). Other model parameters are fixed to $M_{BH} = 10^7$ and $\sigma = 200\ km/s$. The impact of M_* is shown in two aspects. First, for given σ (total stellar mass), the number of stars decreases (linearly) with increasing average stellar mass M_*. Second, the increase in stellar mass also leads a weaker increase in tidal radius ($R_t \propto R_* M_*^{-1/3}$ and note that, based on stellar physics, we approximately have $R_* \propto M_*^{0.5-0.8}$). Combing these two effects, we may expect that, for given black hole mass and total stellar mass of the galaxy, the increase in average stellar mass M_* will lead to an increase in TDE rate. This is indeed observed from our numerical calculations.

The bottom panel of Fig. 5 are for the TDE rate with average stellar mass $M_* = 1M_\odot$. We find that the TDE rate decreases at a fast but decreasing rate throughout. This can be considered a statistical representation of how considering just black hole mass, only small black holes can produce high TDE rates, and increments to a black hole's mass exponentially decreases its ability to tidally disrupt nearby stars, which remains consistent with predictions outlined in previous sections. The graph has an average of $\sim 8.57 \times 10^{-2}$ year^{-1}galaxy^{-1}, which coincides with estimated loss cone diffusion rates of galaxy with steep nuclear density profiles as we take into account the fact that extremely high TDE rates caused by abnormally small SMBHs are statistical outliers (Wang & Merritt 2004, Stone & Metzger 2016, etc.). This confirms that TDE rate is very sensitive to black hole mass, especially as SMBH masses hover about the $10^8 M_\odot$ threshold.

3.3. Application: Impact of Binary Black Holes in Post-Starburst (E+A) Galaxies

In this binary SMBH model, we determined two primary parameters, those being the relative masses of the two black holes and the distance between their centers. For this it is convenient to define a new variable $q_M \in [\frac{1}{3}, \frac{1}{2}]$ which represents the ratio of the mass of the smaller BH to the bigger BH. This allows us to make the following distinctions:

$$M_{BHP} = \frac{M_{BH}}{1+q_M}, M_{BHS} = q_M \frac{M_{BH}}{1+q_M} \tag{12}$$

Where M_{BHP}, M_{BHS} are the masses of the primary (larger) and secondary (smaller) black holes respectively based on the parameters q_M and M_{BH}. For this model we will keep the stellar mass constant at one solar mass.

For this application, we consider a new variable, i.e. the separation between two black holes a_{sep}. One notable difference between single BH system to that of binary BH system is that, the two black holes will orbit each other. Consequently, their exact location (thus specific angular momentum to be accreted by BH) will vary with their orbital locations. Taking the center of mass of the binary black hole system as the epicenter, the maximal allowed angular momentum of stars to disrupted can enlarge from j_{lc} of the single BH system to a much larger value

$$j_{lcMAX} = j_{lc}\sqrt{\frac{r_{tMAX}}{R_t}} \tag{13}$$

Where $r_{tMAX} = \frac{a_{sep}}{1+q_M}$. Considering the dynamical motions of the two black holes, we in this work adopt a mean value of loss cone angular momentum as

$$j_{lcMEAN} = \sqrt{j_{lcMAX} \times j_{lc}}. \tag{14}$$

Correspondingly, the effective tidal radius will be $r_{tMEAN} = r_t(\frac{j_{lcMEAN}}{j_{lc}})^2$.

These values are particularly helpful in our simplified model. These also allow us to make two crucial predictions about binary systems: 1) the effective tidal disruption radius of a binary system is enhanced, and 2) so is the loss cone angular momentum threshold. These properties, therefore, lead to a greatly enhanced TDE rate.

For this application, we fix the galaxy (stellar) properties through $\sigma = 200\ km/s$ and $M_* = 1M_\odot$. We also fix the total mass of the two black holes to be $M_{BH} = 10^7 M_\odot$ and their mass ratio as $q_M = 1/3$. The only variable here is the separation of two black holes a_{sep}. The results are shown in Figure 6.

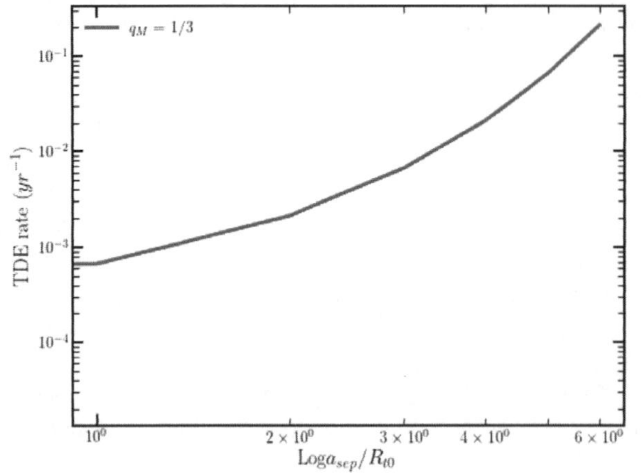

Fig. 6 A graph of TDE rate against the common log of the ratio between the separation of the black holes and the tidal disruption radius given by the central parameter M_{BH}. Evidently TDE rate is highly sensitive to the separation between binary black holes, and this sensitivity increases as said separation increases. For reference, we also plot in this figure the TDE rate at $a_{sep} = r_t$, which marks the TDE rate of a quasi-single black hole case.

Evidently, TDE rate is highly sensitive to the separation between binary black holes. This is both intuitively and theoretically sound since as described above, greater separation between binary black holes means a larger effective tidal disruption radius and a higher loss cone angular momentum threshold. For reference, we also show in this figure the TDE rate at $a_{sep} = r_t$, the point at which the graph is flat. This can be considered as a quasi-single black hole case, where small changes to the separation for $a_{sep} \leq r_t$ are inconsequential . In reality, binary black holes with separations $a_{sep} < 10^4 R_s$ will suffer strong gravitational wave emission and will merge within 10^2 yrs. It is therefore reasonable to infer that most TDEs observed to date, if they are of binary black hole system, have $a_{sep} \gg 10^4 R_s$. In this case, their event rate is highly enhanced.

4. Conclusions & Summary

Tidal disruption of stars (and planets or gas clouds) is of interest to wider astronomy for various reasons. It not only provides the information of the internal composition of individual stars, but also reveals the properties of both stars of nuclei regions and supermassive black hole. Moreover, this is the only way to investigate the supermassive black hole in quiescent/normal galaxies, where the black hole will remain in a quiescent state for 10^{3-5} yrs (see Sec 1.1). This work focused on the impact of binary BH systems on TDE rates. We have reviewed the standard theoretical background for TDEs, the conditions necessary for them to happen, how various seemingly unrelated properties such as those of the host galaxy and electromagnetic emission allow for predictions and observations, why it is important for the wider scientific community to investigate this phenomenon, and what recent projects could aid in wider TDE research. Among various investigations of the TDE phenomena, TDE rate can reveal the properties of both the nuclei (e.g., single BH or binary BHs, as investigated by this work) and the host galaxy (central region), which is of high value to astronomers. With expected fruitful observational data in near future (see Sec. 1.1), such investigation will become much advanced on statistical basis.

In this work, we follow mostly Wang & Merritt (2004) to numerically compute and analyze the TDE rate based on the loss cone theory, a leading theoretical picture in this field. We confirm how TDE rate varies based on several different parameters: TDE rate decreases as black hole mass increases, and increases as the one-dimensional velocity dispersion and kinetic energy of the system decreases.

We then extended the aforementioned model to include E+A post-starburst galaxies and TDEs under the influence of binary black holes. The post-starburst galaxies suffer a major merger within the past

1Gyr. On the other hand, theoretically, because the merger timescale of two black holes is sufficiently long (around and beyond 1 Gyr), we may expect that such system may harbor two supermassive black holes at its center. We emphasize one notable difference of binary black holes to single black hole, i.e. because of the relative orbital motions of two black holes, the space occupation of two black holes will be much larger than that of a single black hole. With this in mind, we estimate the "effective" tidal disruption radius of binary black holes to be tightly related to its orbital separation. With this modification, we find that, the TDE rate is far greater than that of a single black hole. This theoretical enhancement can partially answer why a significant fraction of observed TDEs are in the post-starburst galaxies.

References

[1] Hills, J. Possible power source of Seyfert galaxies and QSOs. Nature 254, 295–298 (1975). https://doi.org/10.1038/254295a0

[2] Lidskii, V.V.; Ozernoi, L.M. Tidal triggering of stellar flares by a massive black hole. Soviet Astronomy Letters, vol. 5, Jan.-Feb. 1979, P. 16-19(1979). Translation Pisma v Astronomicheskii Zhurnal, vol. 5, Jan. 1979, p. 28-33. https://ui.adsabs.harvard.edu/abs/1979PAZh....5...28L/abstract

[3] Rees, M. Tidal disruption of stars by black holes of 106–108 solar masses in nearby galaxies. Nature 333, 523–528 (1988). https://doi.org/10.1038/333523a0

[4] Evans, C. R.; Kochanek, C. S. The Tidal Disruption of a Star by a Massive Black Hole. Astrophysical Journal Letters v. 346, p.L13 (1989). https://ui.adsabs.harvard.edu/abs/1989ApJ...346L..13E/abstract

[5] Gezari, S. Tidal Disruption Events. Annu. Rev. Astron. Astrophys. 2021. 59:21-58. https://arxiv.org/abs/2104.14580

[6] Khabibullin I, Sazanov S, Sunyaev R. 2014. MNRAS 437:327-37.

[7] Alexander, T. Stellar dynamics and tidal disruption events in galactic nuclei. EPJ Web of Conferences 39, 05001 (2012) DOI: 10.1051/epjconf/20123905001

[8] Wang, J.; Merritt, D. Revised Rates of Stellar Disruption in Galactic Nuclei. The Astrophysical Journal, 600:149-161, 2004.

[9] Stone, N. C.; Metzger, B. D. Rates of stellar disruption as probes of the supermassive black hole mass function. MNRAS 455, 859-883 (2016)

[10] Milosavljevic, M. D. Long-Term Evolution of Black Hole Binaries. The Astrophysical Journal, 596, 860-878 (2003). https://ui.adsabs.harvard.edu/abs/2003ApJ...596..860M/abstract

[11] Lezhnin, K.; Vasiliev, E.. Tidal Disruption Rates in Non-Spherical Galactic Nuclei Formed by Galaxy Mergers. arXiv:1609.00009v1 [astro-ph.GA] 31 Aug 2016

[12] Guillochon, J.; Ramirez-Ruiz, E.. HYDRODYNAMICAL SIMULATIONS TO DETERMINE THE FEEDING RATE OF BLACK HOLES BY THE TIDAL DISRUPTION OF STARS: THE IMPORTANCE OF THE IMPACT PARAMETER AND STELLAR STRUCTURE. The Astrophysical Journal, 767, 25 (2013). doi:10.1088/0004-637X/767/1/25

[13] Arcavi, I. et al. A CONTINUUM OF H- TO He-RICH TIDAL DISRUPTION CANDIDATES WITH A PREFERENCE FOR E+A GALAXIES. The Astrophysical Journal, 793, 38 (2014). doi:10.1088/0004-637X/793/1/38

[14] French, K.D., Arcavi, I., Zabludoff, A. The Post-starburst Evolution of Tidal Disruption Event Host Galaxies. Astrophys. J. 835, 176 (2017). doi:10.3847/1538-4357/835/2/176

[15] French K.D. et al. The Host Galaxies of Tidal Disruption Events. Space Sci Rev (2020) 216:32 https://doi.org/10.1007/s11214-020-00657-y

[16] Hung T. et al. Sifting for Sapphires: Systematic Selection of Tidal Disruption Events in iPTF. 2018 ApJS 238 15

[17] Gezari, S. et al. Ultraviolet Detection of the Tidal Disruption of a Star by a Supermassive Black Hole. The Astrophysical Journal, 653: L25-L28, 2006 December 10.

[18] Van Velzen, S. et al. "Optical discovery of probable stellar tidal disruption flares." The

Astrophysical Journal 741.2 (2011): 73.
[19] Roth, N. et al. Radiative Emission Mechanisms. Space Sci Rev (2020): 216-114. https://doi.org/10.1007/s11214-020-00735-1
[20] French K. D. et al. Tidal Disruption Events Prefer Unusual Host Galaxies. ApJL 2016. https://arxiv.org/abs/1601.04705v1
[21] Kormendy, J.; Ho, L. C. Coevolution (Or Not) of Supermassive Black Holes and Host Galaxies. Annual Review of Astronomy and Astrophysics, 51, 511 (2013) https://doi.org/10.48550/arXiv.1304.7762
[22] Heckman, T., Best, P.. The Co-Evolution of Galaxies and Supermassive Black Holes: Insights from Surveys of the Contemporary Universe. Annual Review of Astronomy and Astrophysics (ARAA), vol. 52. https://doi.org/10.48550/arXiv.1403.4620
[23] Ivanov, P. B., Polnarev, A. G. and Saha, P. "The tidal disruption rate in dense galactic cusps containing a supermassive binary black hole." Monthly Notices of the Royal Astronomical Society 358.4 (2005): 1361-1378.
[24] Chen, X., et al. "Enhanced tidal disruption rates from massive black hole binaries." The Astrophysical Journal 697.2 (2009): L149.
[25] Chen, X., et al. "Tidal stellar disruptions by massive black hole pairs. II. Decaying binaries." The Astrophysical Journal 729.1 (2011): 13.
[26] Schönberg, M., and Chandrasekhar S.. "On the evolution of the main-sequence stars." Astrophysical journal 96 (1942): 161-172.
[27] Spitzer, L., and Michael H. H. "Random gravitational encounters and the evolution of spherical systems. II. Models." The Astrophysical Journal 166 (1971): 483.
[28] Cohn, H., and Russell M. K. "The stellar distribution around a black hole-Numerical integration of the Fokker-Planck equation." The Astrophysical Journal 226 (1978): 1087-1108.
[29] Kormendy, J., and Douglas R. "Inward Bound---The Search For Supermassive Black Holes In Galactic Nuclei." Annual Review of Astronomy and Astrophysics 33 (1995): 581.
[30] Law-Smith, J., et al. "Tidal disruption event host galaxies in the context of the local galaxy population." The Astrophysical Journal 850.1 (2017): 22.
[31] Graur, Or, et al. "A dependence of the tidal disruption event rate on global stellar surface mass density and stellar velocity dispersion." The Astrophysical Journal 853.1 (2018): 39.
[32] Toscani, M., Rossi, E. M., & Lodato, G. (2020). The gravitational wave background signal from tidal disruption events. Monthly Notices of the Royal Astronomical Society, 498(1), 507-516.

Research on Atmospheric Turbulence Distortion Measurement Based on Compressed Sensing Technology

Kaichen Huang*, Hengliang Deng, Yanlong Lin

Jiangnan University, Wuxi, China

*Corresponding author e-mail: blakehuang2001@gmail.com

Abstract. The Hartmann wavefront sensor can measure the phase change of atmospheric turbulence in real time and realize the measurement of the distorted wavefront signal of atmospheric turbulence. However, existing Hartmann wavefront sensor microlens arrays are generally small, which limits their spatial resolution capabilities. By introducing compressed sensing technology into wavefront detection, this paper breaks through the limitations of the traditional Nyquist sampling theorem and achieves higher resolution spatial wavefront detection with a smaller microlens array.

Keywords: sensor, atmospheric turbulence, compressed sensing, Nyquist sampling theorem.

1. Introduction

The diameter of telescopes in astronomical observation is getting larger and larger. In order to ensure the observation efficiency of telescopes, corresponding adaptive optics systems are equipped to detect and correct the distortion caused by atmospheric turbulence in real time. The Shack Hartmann wavefront detector (hereafter referred to as Hartmann) is one of the core devices of the adaptive optics system. Its detection ability directly determines the resolution of the target imaging [1-4]. In order to improve its spatial resolution, we introduce compression perception technology. Compressed sensing technology was proposed by Donoho et al. It uses an observation matrix that is "uncorrelated" with the sparse basis and has a much lower dimension than the signal, and projects the high-dimensional signal on a low-dimensional space to obtain a sparse measurement of the signal. value to reconstruct the signal, breaking the limitation of Nyquist theorem, that is, reconstructing the high-dimensional original signal from the low-level measurement. In the field of signal processing, compressed sensing can sample data for sparse or compressible signals in a way that is much lower than the standard of Shannon-Nyquist sampling theorem, and it can still achieve accurate reconstruction of sparse or compressible signals, which makes it have Outstanding advantages and potential application prospects [5,6].

In 2014, James et al. studied the principle and algorithm of compressional wavefront detection. The simulation method he adopted is shown in Figure 1. He proposed a fast wavefront detection algorithm SPARZER for sparse representation. The partial derivatives of the light wavefront were sparsely sampled using Hartmann by randomly sampling the raw data down to 5% [7]. Reconstructed by a sparse representation algorithm utilizing Zernike polynomial basis. Experiments on real and simulated data demonstrate the accuracy of the proposed technique compared to traditional sampling and reconstruction

Content from this work may be used under the terms of the Creative Commons Attribution 3.0 licence. Any further distribution of this work must maintain attribution to the author(s) and the title of the work, journal citation and DOI.
Published under licence by IOP Publishing Ltd

methods. Compressed wavefront sensing technology has the potential to increase wavefront acquisition speed and reduce the cost of equipment.

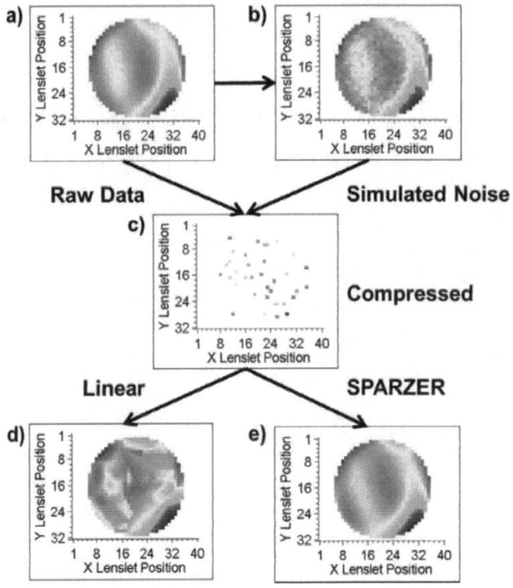

Figure 1. Compression wavefront detection simulation flow chart

If only the traditional Hartmann is used, there will be a problem of low spatial sampling rate. In this paper, the method of compressive sensing wavefront sensing is used to solve the defects of the traditional Hartmann [8]. The use of compressed wavefront detectors can greatly improve the efficiency of information collection and the spatial sampling rate, achieve sampling and complete compression at the same time, extract as much information as possible from as little data as possible, and avoid massive data storage and excessively demanding hardware equipment.

2. Research methods

2.1. Wavefront Compressed Sensing Measurement Principle

Definition $\phi(x, y)$ is the phase distribution of the atmospheric turbulent wavefront, and its slope signals in the x and y directions are ϕ_x and ϕ_y respectively [9]. The slope is sparsely transformed, that is, it is put into the transform domain and expressed as:

$$\begin{cases} \phi_x = \Psi \theta_x \\ \phi_y = \Psi \theta_y \end{cases} \tag{1}$$

In the formula: $\theta_x \in R^{M\times N}$, $\theta_y \in R^{M\times N}$ are sparse matrices, that is, the number of non-zero elements of the matrix satisfies $K_x \ll N^2$, $K_y \ll N^2$. Design an observation matrix ϕ of size $M \times N$ that satisfies the RIP criterion (where M is much smaller than N) to sparsely measure the wavefront slope, and obtain the slope measurement f_x f_y,

$$\begin{cases} f_x = \Phi \phi_x \\ f_y = \Phi \phi_y \end{cases} \tag{2}$$

Where,

$$\begin{cases} f_x \in R^{M\times N} \\ f_y \in R^{M\times N} \end{cases} \tag{3}$$

Then,

$$\begin{cases} f_x = \phi\psi\theta_x = A\theta_x \\ f_y = \phi\psi\theta_y = A\theta_y \end{cases} \tag{4}$$

Therefore, it can be concluded that,

$$A = \phi\psi \in R^{M\times N} \tag{5}$$

SHWFS is based on the microlens array to measure the slope values ϕ_x and ϕ_y of the wavefront respectively. On this basis, we add compressed sensing to the wavefront, and obtain the sparse random measurement values f_x and f_y through the observation matrix Ψ. The wavefront slope is reconstructed from random measurements, and the phase distribution of the atmospheric turbulent wavefront is reconstructed by this algorithm $\phi(x, y)$.

The advantage of the method of optimizing wavefront detection through compressed sensing is that it can effectively reduce the number of wavefront detection values and increase the measurement speed. At the same time, with the reduction of the amount of data, the hardware and data storage requirements of the detector are also significantly reduced. Compared with the conventional measurement process, the compressed sensing method adds a reconstruction process from the sparse measurement values f_x, f_y to the wavefront slope value $\phi(x, y)$. Due to the real-time change and high precision requirements of the adaptive optics system, the reconstruction algorithm adopted must meet these two conditions, that is, compared with the traditional SHWFS, the two requirements are qualitatively improved. Only in this way can the compressive sensing method have practical significance in the wavefront measurement of atmospheric turbulence[10].

2.2. B-SLO reconstruction algorithm

The reconstruction algorithm is discussed below by taking the reconstruction process of the wavefront slope ϕ_x as an example. The core problem of solving the reconstruction of the sparse measurements f_x of the wavefront slope ϕ_x can be expressed as,

$$\min_{\theta_x} \|\theta_x\|_0 \quad \text{s.t.} \quad f_x = A\theta_x \tag{6}$$

Define function $l(\theta_x)$ as,

$$l(\theta_x) = \begin{cases} 1 & \theta_x = 1 \\ 0 & \theta_x = 0 \end{cases} \tag{7}$$

Therefore, the L0 norm can be equivalent to,

$$|\theta_x\| = L(\theta_x) = \sum_{x=1}^{N} l(\theta_x) \tag{8}$$

Because the norm L0 solution is an NP-hard problem and cannot be solved directly, many reconstruction algorithms have emerged. According to the requirement of real-time measurement of atmospheric turbulence wavefront, through the analysis of multiple reconstruction algorithms, the SL0 algorithm does not need to know the sparsity of the signal in advance.

1) Regional parallelization of atmospheric turbulence signals

Atmospheric turbulent wavefront slope signals belong to a two-dimensional distribution. For two-dimensional signals, most of the existing reconstruction algorithms convert them into one-dimensional signals and perform reconstruction through serial calculation, which not only prolongs the reconstruction algorithm time, but also destroys the wavefront slope signal column and column. Contact, reduce the reconstruction accuracy of the wavefront slope, and affect the recovery of the wavefront phase. In order to solve these problems, this paper considers the reconstruction of the wavefront slope signal in different regions. The following takes the slope ϕ_x in the x direction as an example. Since SHWFS measures the wavefront slope value ϕ_x, ϕ_y through an N×N microlens array regularly arranged, the N×N regularly arranged microlens array is divided into n A×A size areas, of which $n = N^2 / B^2$, and then n B×B size ϕ_{x_i} are obtained. Each column is connected at the end to obtain n $B^2 \times 1$ size column matrix y_{f_i}. The n column vectors y_{f_i} are merged together to obtain the processed wavefront slope signal y_f, and finally a sparse random measurement value Y_f of y_f is obtained by random sampling ϕ_f.

Although the method of converting two-dimensional signals into one-dimensional signals is also adopted in this paper, it does not completely destroy the relationship between signal columns and columns. Pre-phase reconstruction accuracy.

So formula (6) is transformed into formula (9),

$$\min_{\theta_f} |\theta_f| \quad \text{s.}t. \quad Y_f = A_f \theta_f \tag{9}$$

Where, $A_f = \Phi_f \Psi$.

For each block y_f, equation (6) is transformed into equation (10),

$$\min_{\theta_{f_i}} |\theta_{f_i}| \quad \text{s.}t. \quad Y_{f_i} = A_{f_i} \theta_{f_i} (i = 1,2,3\ldots,n) \tag{10}$$

For each area, they are independent of each other, so the obtained wavefront slope ϕ_x is also independent of each other, that is, the columns of the processed slope signal y_f are independent of each other, and can be used for y_f Each column is reconstructed in parallel using the SL0 algorithm and then merged into the processed slope reconstruction value $\hat{y}_f$, and then the final wavefront slope reconstruction signal is obtained through inverse transformation. This not only improves the reconstruction speed of the signal, but also protects the slope signal column and column to a certain extent. The correlation between them improves the reconstruction accuracy.

2) Principle of bslo algorithm

The basis of the B-slo algorithm is the slo algorithm. In 2009, Mohimani et al. proposed the problem of using a smooth function to replace the non-convex minimum norm, and the extreme value problem of the minimum norm was solved by solving the optimal problem of the smooth 0-norm under certain conditions. To obtain it, a smooth 0-norm algorithm is proposed, that is, the SL0 algorithm. The SL0 algorithm uses a smooth Gaussian function with variable parameters to approximate the replacement norm, which is a more general algorithm UW for sparse solution of underdetermined linear problems. Compared with the BP algorithm, the SL0 algorithm can obtain 2~3 times faster speed than the BP algorithm, and at the same time obtain the same recovery quality as the BP algorithm.[11]

The L_0 norm of a signal can be defined as the number of sparse signals $\theta = \{\theta_1, \theta_2, \ldots, \theta_n\}$, θ that are not 0 in the support set, if there is a function as follows,

$$l(\theta) = \begin{cases} 1 & \theta \neq 0 \\ 0 & \theta = 0 \end{cases} \tag{11}$$

Then the norm of can be expressed as formula (12),

$$\|\theta\| = \sum_{i=0}^{n} v(\theta_i) \tag{12}$$

Therefore, whether $v(\theta)$ is continuous in formula (12) determines whether the norm is continuous or not. If you replace $v(\theta)$ with a continuous function, you can get a L_0 smooth estimation function.

At present, there are many functions that can be used for estimation. In the SL0 algorithm, the 0-mean Gaussian function is used as the estimation function, because this function has good differentiability. Define the function:

$$f_\sigma(\theta) = \exp(-\frac{s^2}{2\sigma^2}) \tag{13}$$

Thus,

$$\lim_{\sigma \to 0} f_\sigma(\theta) = \begin{cases} 1 & s \neq 0 \\ 0 & s = 0 \end{cases} \tag{14}$$

$$\lim_{\sigma \to 0} f_\sigma(\theta) = 1 - \nu(\theta) \tag{15}$$

$$F_\sigma(\theta) = \lim_{\sigma \to 0} \sum_{i=1}^{n} f_\sigma(\theta_i) \tag{16}$$

$$F_\sigma(\theta) = n\text{-}\sum_{i=0}^{n} \nu(\theta_i) = n - \|\theta\|_0 \tag{17}$$

$$\|\theta\|_0 = n - F_\sigma(\theta) \tag{18}$$

The value of σ controls the approximation degree of the L_0-norm estimation and the smoothness of the estimation function. The smaller the value of σ, the more accurate the estimation of the L_0-norm is; the larger the value of σ, the smoother the estimation function[12]. It can be seen from equation (17) that when σ is small enough, the minimum L_0 norm problem is the solution to the maximum $F_\sigma(\theta)$ problem. For smaller σ, $F_\sigma(\theta)$ discontinuities will directly cause F to contain multiple local extreme points. Therefore, when σ is small, the existing optimization algorithm is easy to fall into the local extreme point, so it is difficult to directly obtain the global maximum value of $F_\sigma(\theta)$ through the existing algorithm. But as σ gets bigger, $F_\sigma(\theta)$ gets smoother. Therefore, when σ is large enough, the solution of the L_0-norm optimization problem is equivalent to the solution of the L_2-norm optimization, and $F_\sigma(\theta)$ will not fall into a local extreme point at this time.

In order to avoid the algorithm falling into the local extreme point in the optimization process, in the SL0 algorithm, iterative optimization is carried out by gradually reducing σ, and the initial value of σ selects a larger value relative to the processing signal. For each given σ, the current extreme point is obtained by the steepest descent method, and each σ attenuates a very small value ∂, so it can be guaranteed that the starting point of the maximum value of each σ is close to the global extreme point. By gradually decaying σ to find the global optimal solution, the algorithm will not fall into the local extreme point. The idea of this algorithm is somewhat similar to the simulated annealing algorithm, but in SL0, the step size of each drop is small, and the definition of the drop sequence is simple.

The B-SL0 algorithm is an optimization of the SLO algorithm. First, the sparse measurement of the wavefront slope y_f processed by the observation matrix Φ_f is used to obtain the observation data

Y_f, and then the estimated value $\hat{\theta}_f$ of the sparse matrix θ_f is reconstructed by parallel operation combined with the SL0 algorithm. The inverse matrix transformation results in an estimate $\hat{y}_f$ of the subregionally processed signal y_f . Since what is obtained is the estimated value of the wavefront slope processed by each sub-region, $\hat{y}_f$ needs to be inversely transformed to obtain the estimated value $\hat{\phi}_x$ of the wavefront slope signal ϕ_x. Similarly, the estimated value $\hat{\phi}_y$ of ϕ_y can be obtained, and finally, use $\hat{\phi}_x$, $\hat{\phi}_y$ to recover the estimated value $\hat{\phi}(x, y)$ of the wavefront phase distribution $\phi(x, y)$.

3. Research results

Based on the SL0 algorithm, Li Can of Taiyuan University of Technology measured the wavefront slope signal in sub-regions, and combined with parallel operations, through theoretical analysis and simulation experiments, a sub-regional parallel algorithm capable of fast and high-precision signal reconstruction was realized. B-SL0. Experimental results show that B-SL0 significantly outperforms other existing reconstruction algorithms in both computational time and accuracy.

On this basis, we write a program to simulate and verify the compression wavefront detection. We use the power spectrum to generate a 256x256 phase spectrum, use the sub-regional measurement method, and set the size of each region to 16x16, 32x32, and 64x64 for simulation verification(Figure 2, Figure 3 and Figure 4), and finally achieve the expected calculation time and recovery accuracy.

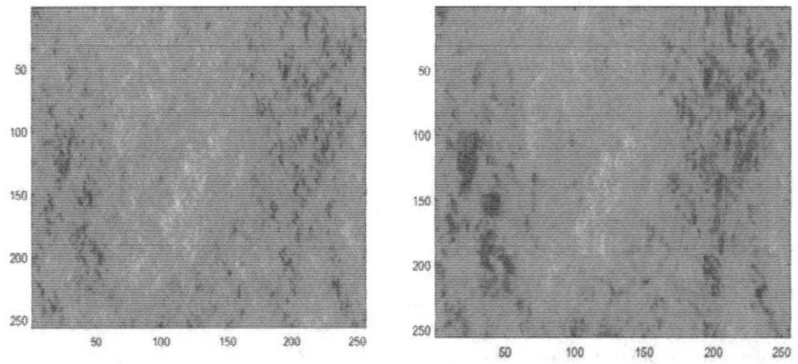

Figure 2. Comparison of before (left) and after (right) restoration in the case of 16x16 blocks

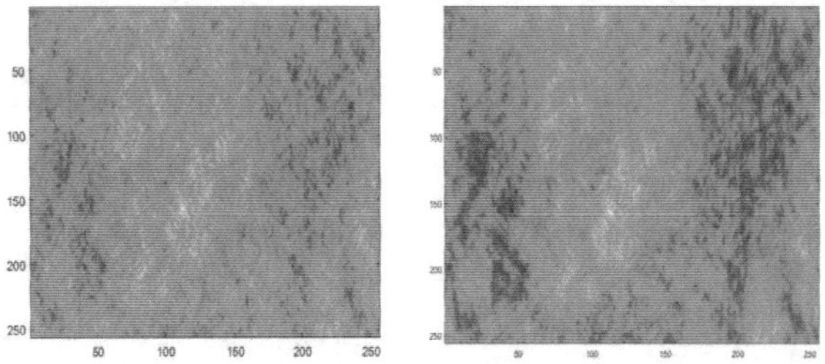

Figure 3. Comparison of before (left) and after (right) restoration in the case of 32x32 blocks

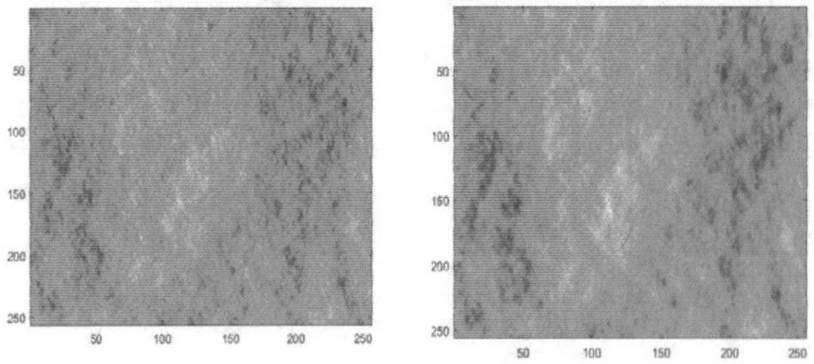

Figure 4. Comparison of before (left) and after (right) restoration in the case of 64x64 blocks

In addition, we also tried to use the basis pursuit noise reduction (BPDN) method, using the Zernike matrix as the sparse matrix, using its characteristics to make the coefficients sparse naturally, and artificially adjusting the size of the noise to compare the results obtained before and after restoration. Large relative error, but the recovered wavefront array is almost the same as the original wavefront, and we can continue to optimize the algorithm (Figure 5).

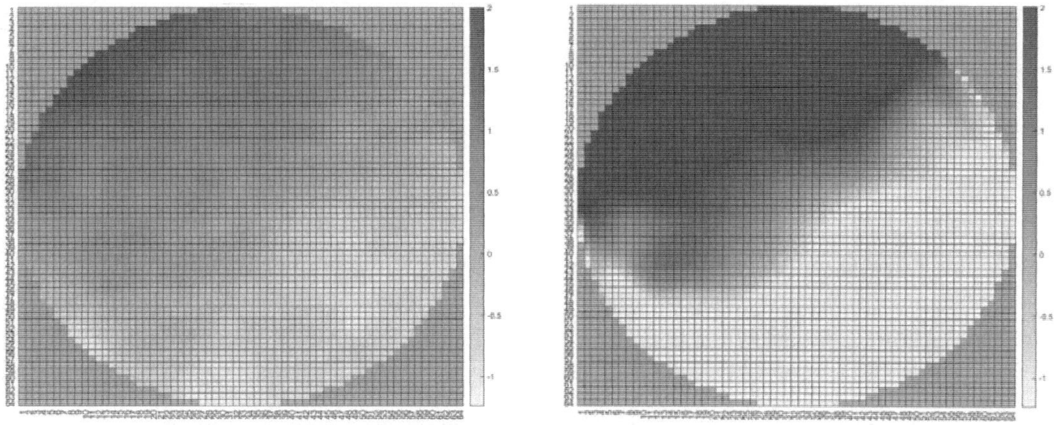

Figure 5. Comparison of input wavefront (left) and recovered wavefront (right)

In order to further confirm the correctness of the algorithm, the following two methods are used for test confirmation.

1) Customize the input coefficient of wavefront CS restoration, use the slope obtained by the coefficient to obtain the restoration coefficient and compare it with the input coefficient

Set the self-determined coefficient column vector to be composed of 0 or 1, and the probability of 0 is p. By changing the size of p, the effect of the coefficient recovery algorithm is detected, that is, the influence of the sparsity of the input signal on the recovery effect is explored. The figure below shows the results for p=0.1 and 0.9, comparing the coefficients before and after recovery (Figure 6).

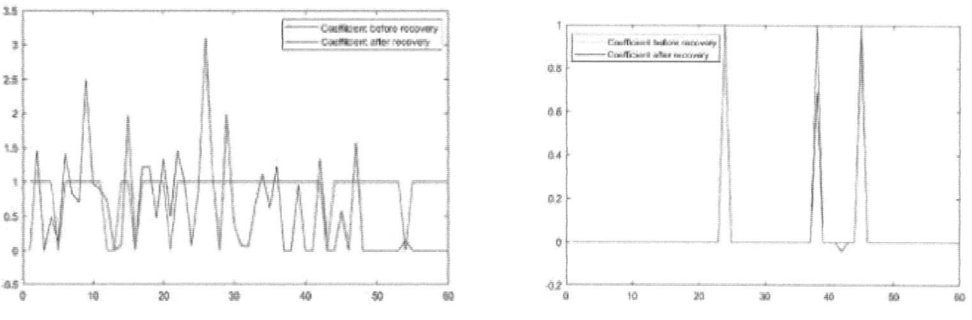

Figure 6. Comparison of coefficients before and after recovery when p=0.1 (left) and p=0.9 (right)

2) In order to further test the reliability of the simulation algorithm, we customize the input coefficient as a random coefficient, and the elements in it are randomly generated between -1 and 1. At the same time, in order to correlate the influence of sparsity, we set a threshold for the custom vector, so that the signals below the threshold are all zero to approximate the sparse signal. Figure 7 below shows the result when the threshold is 0.1.

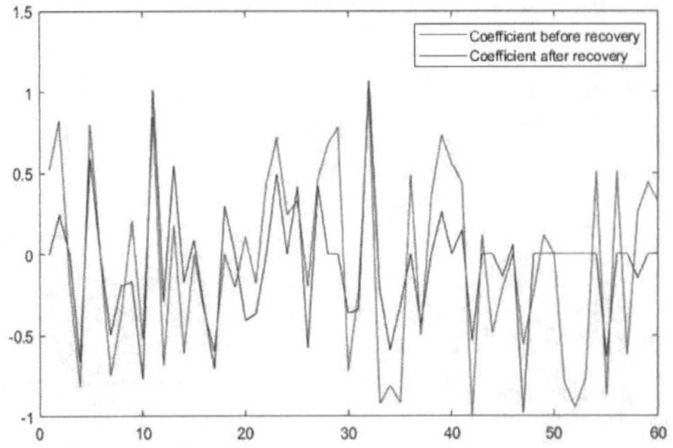

Figure 7. Comparison before and after when the threshold is 0.1

Since the coefficient and the slope array have a certain matrix relationship, we use the slope obtained by the custom coefficient as the input of the CS algorithm to restore the restored coefficient, and then reconstruct the restored wavefront. Finally, the original wavefront and the Input wavefronts for comparison(Figure 8). Furthermore, we change the size of the threshold and observe the change of the recovery accuracy.

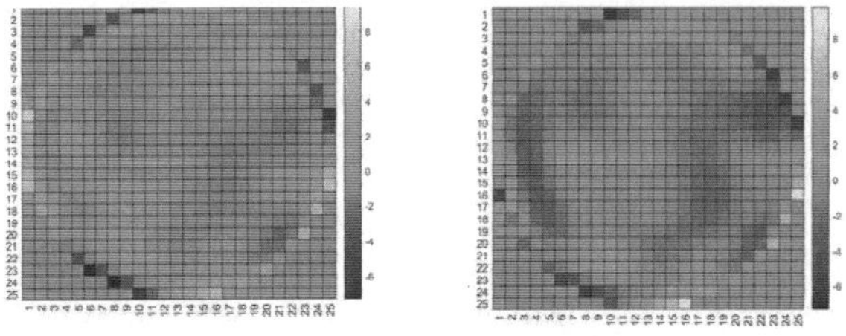

Figure 8. Comparison of recovered wavefront (left) and input wavefront (right) after CS algorithm input

3) Using oomao to generate the initial wavefront and slope, use the input slope to get the recovery coefficient, and use the zernike polynomial relationship to restore the recovery wavefront and compare(Figure 9). The following figure shows one of the results. The recovery algorithm needs to be further optimized. In the future, it is necessary to explore the influence of the number of zernike items on the recovery results.

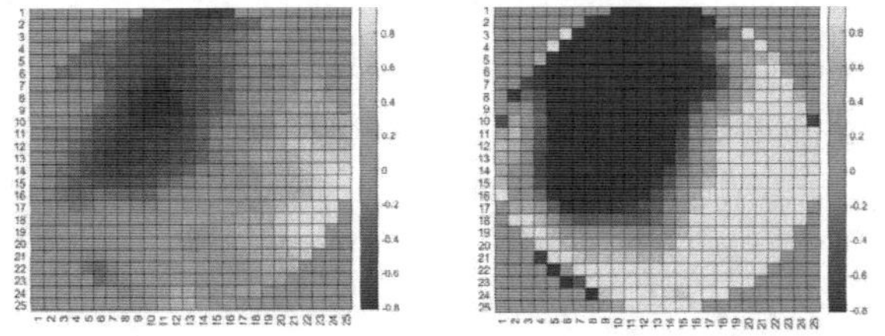

Figure 9. Recovery wavefront (right) versus input wavefront (left) using zernike polynomials

4. Conclusion

The number of microlenses in the traditional Hartmann microlens array is 20*20, and only a few Hartmanns can achieve 50*50, so the actual Hartmann space sampling rate is relatively low, which is limited in many aspects. Considering the characteristics of compressional wavefront sensing that can accurately reconstruct the signal under the condition of undersampling, this paper innovatively combines compressional sensing with Hartmann, so that Hartmann can obtain higher performance with a smaller number of microlens arrays. Wavefront detection with spatial resolution. As a result, the spatial resolution of the wavefront signal is no longer limited by the number of microlens arrays, breaking the hardware limitation of the detection resolution of the wavefront sensor.

References

[1] D. C. Yuan, H. Y. Zhao, X. Tao, S. B. Li, X. L. Zhu, and C. P. Zhang, "Aspheric surface measurement using capacitive sensors," Sensors, 2017, 17(6): 13551–1355-16.
[2] C. R. Forest, C. R. Canizares, D. R. Neal, M. McGuirk, and M. L. Schattenburg, "Metrology of thin transparent optics using Shack-Hartmann wavefront sensing," Optical Engineering, 2004, 43: 1–12.
[3] D. L. Donoho, "Compressed sensing," IEEE Transactions on Information Theory, 2006, 52(4): 1289–1306.
[4] R. J. Noll, "Zernike polynomials and atmospheric turbulence," Journal of the Optical Society of America, 1976, 66(3): 207–211.
[5] J. Polans, R. P. McNabb, J. A. Izatt, and S. Farsiu, "Compressed wavefront sensing," Optics Letters, 2014, 39(5): 1189–1192.
[6] Mohammad Rostami, Oleg Michailovich, Zhou Wang, Deblurring using derivative compressed sensing for optical imaging application, IEEE transactions on image processing, 2012, 21(7): 3139-3147.
[7] Eddy Mun Tick CHOW1, Ningqun GUO, Edwin CHONG, and Xin WANG, Surface Measurement Using Compressed Wave-front Sensing.
[8] Polans Ryan P. McNabb, Joseph Izatt, and Sina Farsiu, Compressed Wavefront Sensing, Gregory A. Howland, Compressive wave front sensing with weak values.
[9] Li Can. Research on sparse reconstruction of distorted wavefront signal of atmospheric turbulence[D]. Taiyuan University of Technology, 2019.
[10] Li Can, Cai Dongmei, Jia Peng, et al. Research on the reconstruction of atmospheric turbulent wavefront compressive sensing measurement [J]. Optoelectronic Engineering, 2018, 45(04): 41-49.
[11] Beck. A, Teboulle. A fast iterative shrinkage-thresholding algorithm for linear inverse problems [J]. SIAM Journal on Imaging Sciences, 2009, 2(1): 183-202.
[12] Mohimani. G. H, Babaie-Zadeh M, Jxitten.C. Fast Sparse Representation based on Smoothed L0 Norm [M] // Independent Component Analysis and Signal Separation. Springer Berlin Heidelerg 2007: 396.

Method of using theodolite's observation data of fixed star to evaluate INS's attitude accuracy based on time-delay compensation

Xianglong Kong *, Chenghao Liu, Xingcun Wu, Kecheng Liu, Junhao Liang

China Satellite Maritime Tracking and Control Department, Jiangyin 214431, China

*Corresponding author e-mail: Kongxianglong2022@foxmail.com

Abstract. In order to evaluate the attitude accuracy of the TT&C's SINS under dynamic condition, a new method of using theodolite's observation data of fixed star to evaluate SINS's attitude accuracy based on time-delay compensation is proposed. The detailed design is given, and the influence of the time delay is analyzed based on real data experiments. The data analysis and experimental results shows that the new method can improve the computation accuracy, and it is valuable for engineering application of evaluating SINS's attitude accuracy under dynamic conditions.

Key words: strap-down inertial navigation system; attitude error; theodolite.

1. Introduction

Photoelectric measuring equipment is a comprehensive optical measuring instrument integrating optics, mechanics, electronics, servo control, information and computer multidisciplinary technology, and its development and application in the fields of national defense, aviation and aerospace are becoming more and more extensive [1, 2]. At present, the photoelectric theodolite of the survey ship has multiple visible and infrared optical sensors. The optical axes of the optical systems of these sensors should be strictly parallel to maintain the same orientation. The parallelism of the optical axis of the optical system is one of the important indicators of the multi-optical axis system [3].

The calibration of the optical axis parallelism of the measuring ship theodolite equipment is generally carried out in the dock. The entire calibration process is time-consuming and labor-intensive, and the degree of automation is not high. In addition, due to environmental factors such as temperature and hull vibration during the voyage, the parallelism between the optical axes of the optical systems may still change [4]. Therefore, in order to ensure the measurement accuracy and consistency, it is necessary to dynamically detect the optical axis of each optical system before the task, and provide necessary data support for the subsequent analysis of the measurement accuracy.

Commonly used optical axis parallelism calibration methods include collimator method, aiming star method and projection target method. However, the collimator method is difficult to process and assemble, and it is not convenient to use at sea. The star aiming method is restricted by sea conditions. It must be carried out at night with no wind, no swells, and small ship sway. It also requires manual visual staring, and the calibration efficiency is low and the speed is slow. The projection target method also has the problems that the target is difficult to erect and complicated to operate.

Content from this work may be used under the terms of the Creative Commons Attribution 3.0 licence. Any further distribution of this work must maintain attribution to the author(s) and the title of the work, journal citation and DOI.
Published under licence by IOP Publishing Ltd

The purpose of the present invention is to overcome the above-mentioned shortcomings and propose a method for dynamically calibrating the parallelism of the optical axis of a ship-borne multi-optical system based on inertial navigation information. This method uses the existing inertial navigation information of the survey ship to isolate the optical theodolite from the ship. At the same time, using the principle that the position difference of the visual axis can be ignored at the infinite distance of the target, the star that can be locked by the visible light and the infrared system is selected as the target, and the optical axis parallelism calibration is achieved by analyzing the difference between the visible light and the infrared system miss distance during the stable tracking process.

2. Self-stabilization of shipborne theodolite based on inertial navigation information

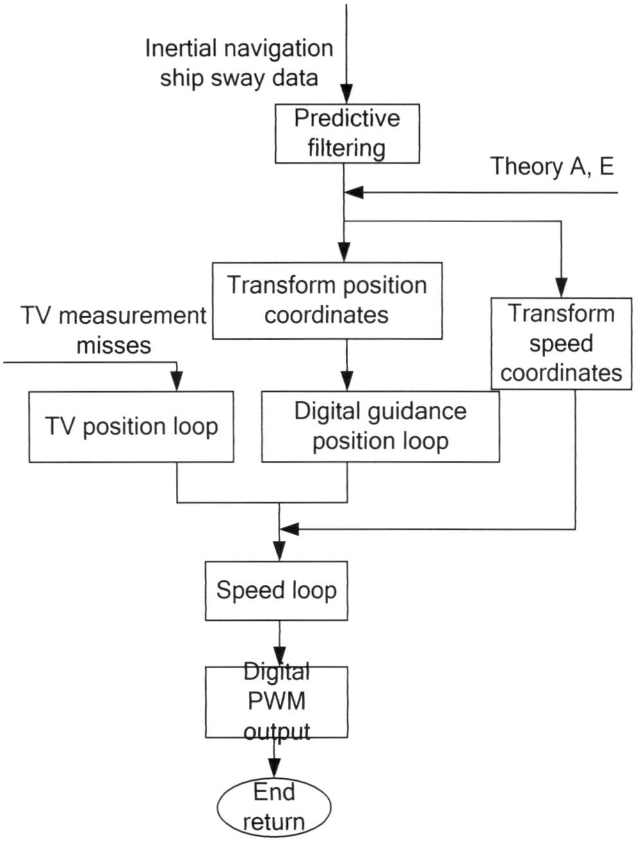

Figure 1. Ship sway self-stabilization algorithm implementation process

The theodolite servo system consists of two independent and coordinated systems of azimuth and height. It is a position follow-up system, which is composed of a position loop and a speed loop. In order to isolate the influence of ship sway, the control scheme of the boresight self-stabilization method is that the position loop input adopts two-dimensional composite control of digital TV and the speed loop introduces the feed forward control of ship sway. The computer guides the star catching, completes the self-stabilization function, and corrects the star error caused by the change of the ship's position and attitude through the TV, so that the effect of high-precision star measurement can be obtained even in bad sea conditions. Taking tracking stars as an example, the calculation process of the feedforward signal guidance value is shown in Figure 1. The detailed calculation process is as follows.

2.1. Calculation of theoretical position of stars

The original star database only provides parameters such as asterisk, magnitude, magnitude, horizontal right ascension, horizontal declination, right ascension, declination, radial velocity, parallax, spectrum, etc., and the observation epoch should be based on these parameters Solving the theoretical position of stars. The main process includes stellar self-correction, precession correction, nutation correction, aberration correction and oblique aberration correction, etc., and finally the theoretical azimuth and elevation angle are obtained. Furthermore, the theoretical position of stars in the INS horizon system can be calculated as follows:

$$\begin{cases} X_g = D\cos E\cos A \\ Y_g = D\sin E \\ Z_g = D\cos E\sin A \end{cases} \tag{1}$$

In the formula, D is the distance from the star to the origin of the coordinate.

2.2. Ship sway data prediction filtering

Measure the hull attitude angle data given by the ship's inertial navigation system. In the process of transmitting the ship sway data measured by the inertial navigation to the theodolite system, there will be a certain delay, and the inertial navigation does not give the ship sway speed, so it is necessary to predict and filter the ship sway data. The least squares method is used here to achieve, the specific details will not be introduced.

2.3. Calculation of ship sway and angular velocity of ship sway under deck tether

The theoretical position of the star in the deck coordinate system is calculated as follows:

$$\begin{bmatrix} X_c \\ Y_c \\ Z_c \end{bmatrix} = \begin{bmatrix} \cos\varphi_c & \sin\varphi_c & 0 \\ -\sin\varphi_c & \cos\varphi_c & 0 \\ 0 & 0 & 1 \end{bmatrix} \begin{bmatrix} \cos K_c & 0 & \sin K_c \\ 0 & 1 & 0 \\ -\sin K_c & 0 & \cos K_c \end{bmatrix} \begin{bmatrix} 1 & 0 & 0 \\ 0 & \cos\theta_c & \sin\theta_c \\ 0 & -\sin\theta_c & \cos\theta_c \end{bmatrix} \begin{bmatrix} X_g \\ Y_g \\ Z_g \end{bmatrix} \tag{2}$$

According to the definition of the theodolite measurement coordinate system, the azimuth angle A_c and the pitch angle E_c of the star below the deck system can be calculated as follows:

$$\begin{cases} A_c = \tan^{-1}(Zc/Xc) \\ E_c = \sin^{-1}(Yc/D) \end{cases} \tag{3}$$

Combining formula (3), formula (4), and formula (5) can be obtained:

$$\begin{cases} Ac = \tan^{-1}\{[\cos E[\cos\theta_c \sin(A-K_c) + \sin\theta_c \sin\varphi_c \cos(A-K_c)] - \sin E\sin\theta_c\cos\varphi_c]/ \\ [\cos E\cos\varphi_c\cos(A-K_c) + \sin E\sin\varphi_c]\} \\ Ec = \sin^{-1}\{\cos E[\sin\theta_c \sin(A-H) - \cos\theta_c\sin\varphi_c\cos(A-K_c)] + \sin E\cos\theta_c\cos\varphi_c\} \end{cases} \tag{4}$$

According to the rotation sequence and the principle of angular velocity vector superposition, the angular velocity vector of the ship's swaying off the deck can be calculated as follows from the angular velocity vector superimposition corresponding to three rotations:

$$\begin{cases} \omega_{xc} = \omega_\theta - \omega_K \sin\varphi_c \\ \omega_{yc} = \omega_\varphi \sin\theta_c - \omega_K \cos\varphi_c \cos\theta_c \\ \omega_{zc} = \omega_\varphi \cos\theta_c + \omega_K \cos\varphi_c \sin\theta_c \end{cases} \tag{5}$$

In the formula, $(\omega_\theta, \omega_\varphi, \omega_K)$ is the ship's yaw angular velocity under the inertial navigation horizon system.

Further, the angular velocity of the theodolite's boresight perturbation can be calculated as follows:

$$\begin{cases} Ecv = -(\omega_\varphi \cos\theta_c + \omega_K \cos\varphi_c \sin\theta_c)\cos Ac + (\omega_\theta - \omega_K \sin\varphi_c)\sin Ac \\ Acv = \omega_\varphi \sin\theta_c - \omega_K \cos\varphi_c \cos\theta_c - \tan Ec[(\omega_\theta - \omega_K \sin\varphi_c)\cos Ac \\ +\omega_\varphi \cos\theta_c + \omega_K \cos\varphi_c \sin\theta_c)\sin Ac] \end{cases} \tag{6}$$

3. Visible light and infrared public star database establishment

In order to facilitate the detection of the parallelism of the optical axis of the visible light and infrared systems, it is necessary to establish a common star catalog of the two. Due to the limited number of known infrared standard star catalogs in China, a public star catalog is established based on the existing FK5 catalog. In the daily follow-up process, gradually accumulate the stars that can be observed by infrared in the existing star catalog, and establish the star catalog according to the original format to facilitate parallelism detection.

4. Optical axis non-parallelism detection and compensation

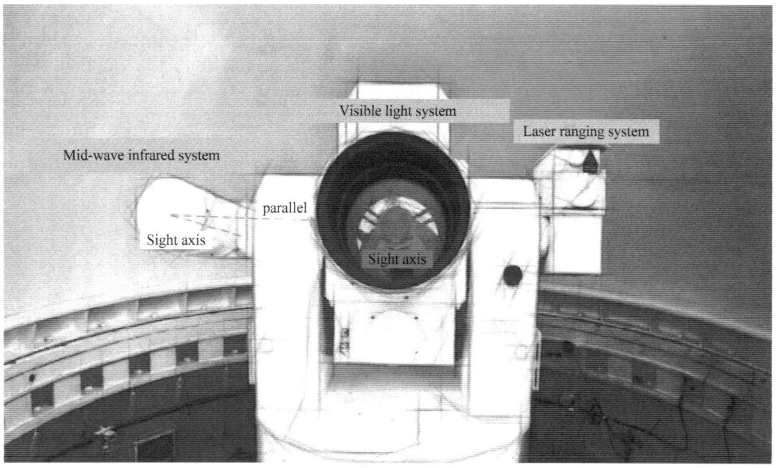

Figure 2. Relative position of optical instruments

In theory, the visual axis of the mid-wave infrared system of the photoelectric theodolite should be parallel to the visual axis of the main visible light mirror, as shown in Figure 2. When the target is far away, the direction deviation caused by the horizontal deviation of the visual axis can be ignored. When there is an angle between the two visual axes, that is, when the two visual axes are not parallel, the position of the target image in the visual field of the visible light system and the mid-wave infrared system is inconsistent. The deviation of the visual axis between the two in the horizontal and vertical directions can be obtained by calculating the difference between the amount of missed targets for visible light and the amount of missed targets for mid-wave infrared. The miss-target calculation is realized by the centroid detection method. After the image is preprocessed by denoising, threshold segmentation and centroid extraction are performed, and the centroid coordinates can be calculated as follows:

$$x = \frac{\sum_{i=1}^{M}\sum_{j=1}^{N} G_{ij} \times j}{\sum_{j=1}^{N}\sum_{i=1}^{M} G_{ij}}, \quad y = \frac{\sum_{j=1}^{N}\sum_{i=1}^{M} G_{ij} \times i}{\sum_{j=1}^{N}\sum_{i=1}^{M} G_{ij}} \tag{7}$$

Furthermore, the amount of missed targets for visible light and mid-wave infrared can be calculated by the centroid coordinate calculation. Finally, the deviation of the parallelism of the two sight axes is obtained. After obtaining the deviation of the sight axis, it can be compensated in the post-data processing. If the deviation is too large, adjust the mechanical angle of the mid-wave infrared boresight.

5. Experimental results and discussion

The algorithm was tested three times, and the test results are shown in Table 1.

Table 1. Visible light and mid-wave infrared parallelism calibration results

	1	2	3
X	-10.4″	-21″	-16″
Y	-2.8″	-8″	-8″

6. Conclusion

This paper studies a calibration method for the parallelism of visible light and mid-wave infrared optical axes based on star observations. The ship-sway isolation algorithm is given, the public star catalog is established, and the method of detecting and compensating the non-parallelism of the optical axis is given. Experimental results show that the algorithm quickly and accurately calibrate the non-parallelism of the optical axis, ensuring the accuracy and consistency of the equipment measurement.

References

[1] Shilong Jian. Introduction to the marine measurement and control technology of aerospace survey ship. National Defense Industry Press, 2009.

[2] Liang Pan, Wenhua Zhao, Xinming Liu, Lijian Zhao. Theodolite measuring satellite to evaluate the attitude angle error method of inertial navigation system. Journal of Chinese Inertial Technology, 2011, 19(3): 369-373. DOI: 10.13695/j.cnki.12-1222/o3.2011.03.028.

[3] Li You. Research on Autonomous Deformation Measurement Rotation Demodulation Technology Based on Single-axis Rotating Inertial Navigation. Master's Thesis, National University of Defense Technology, 2014.

[4] Xia Zhang, Di Chang, Weimin Qi, Zhiming Zhan. A Study on Different Functionalities and Performances among Different Activation Functions across Different ANNs for Image Classification. Journal of Physics: Conference Series, Vol. 1732, No. 1, p. 012026. DOI: 10.1088/1742-6596/1732/1/012026.

Effect of collision of particles on migration behaviour in hydrocyclone

Yang Gao [1, a], Xiaohan Pei [1], Gang Cao [1], Jiaqing Yu [1], Xiaojie Zhao [1], Feng Li [2]

[1] Research Institute of Petroleum Exploration and Development, CNPC, China
[2] School of Mechanical Science and Engineering, Northeast Petroleum University, China

[a] gaoyang573@petrochina.com.cn

Abstract. The CFD-DEM coupling calculation method is used to study and analyse the influence of different particle sizes on the particle motion behaviour when they collide in the hydrocyclone. The results show that when the particles move in the hydrocyclone, they hardly collide with the inner wall of the hydrocyclone, and the collision position of the particles is mainly concentrated in the 140mm-80mm range of the hydrocyclone. When the particles collide, the particles are suddenly subjected to the contact reaction caused by the collision, and the trajectory will be adjusted from following the fluid streamline to circling at a certain longitudinal position, and the particle size increases with the increase of the particle size.

1. Introduction

As an important separation equipment in mineral processing industry, solid-liquid separation hydrocyclone has the advantages of small size, convenience, and high efficiency [1]. However, according to different particle parameters, it is often necessary to redesign the parameters and operation parameters of the hydrocyclone [2]. With the gradual maturity of CFD-DEM method, more and more scholars use this method to analyze the particle group motion in the hydrocyclone [3]. The analysis results show that the random collision of particles is obvious because the trajectory of particles in the hydrocyclone is unstable in the early stage of flow [4]. Liu Yu [5] studies the distribution of solid-liquid two-phase flow in turbulence in the boundary layer and the motion characteristics of particles. It is considered that Saffman lift is necessary in modeling, and at a certain density ratio, the following of particles to the fluid will become worse with the increase of particle size. But the applicable particle size range is in the micron level. Li Chun yang [6] aimed at the cyclone separator used in agriculture, the motion behavior of particles was studied without considering the interaction force between particles, but because the velocity of gas phase was not large, it is concluded that the velocity has little effect on the trajectory of particles [7].

2. CFD-DEM coupling theory

2.1. CFD-DEM coupling calculation method

A brief introduction of Coupling principle. During the CFD-DEM coupling simulation, CFD and DEM are calculated alternately. First, the CFD simulation is carried out, which is responsible for calculating

Content from this work may be used under the terms of the Creative Commons Attribution 3.0 licence. Any further distribution of this work must maintain attribution to the author(s) and the title of the work, journal citation and DOI.
Published under licence by IOP Publishing Ltd

the flow field. Simulate ahead of time during simulation, and then pass any required data to DEM to allow it to simulate to the same point in time.

DEM integrates fluid drag and torque into a single particle level particle simulation. When DEM performs a simulation step, an external force acts on the particles in addition to any gravity or collision force. The various stages of the DEM simulation cycle and the points at which it interacts with the CFD solver are shown in Figure 1.

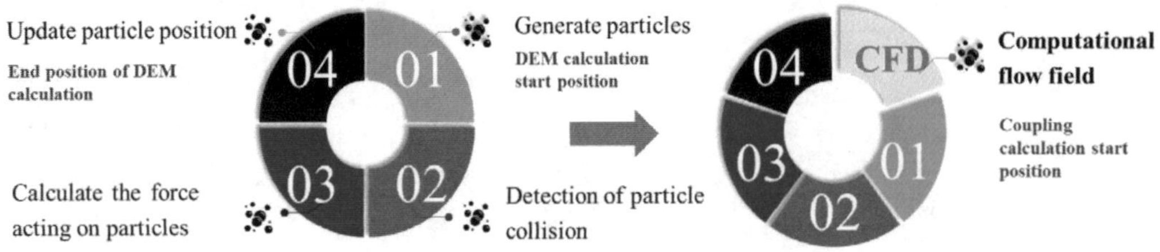

Fig. 1 comparison of DEM simulation cycle and CFD-DEM simulation cycle

When the calculation of the DEM solver starts, the particles are first generated and detected whether there is a collision. After the collision is detected, the force on the collided particles is calculated, and the motion trajectory is calculated according to the force balance equation to update the position of the next time node of the particles, to cycle. After the coupling of the CFD solver, the first step of the simulation calculation is changed to the CFD solver to calculate the flow field.

Scale relationship of coupled simulation grid. In the CFD-DEM model, the effect of particle motion on fluid flow is calculated by the volume fraction of each phase and the momentum exchange through resistance. While the micro-scale of particles ensures the calculation accuracy of particle motion, the mesoscopic fluid phase greatly improves the computational efficiency.

2.2. Flow field calculation method

Reynolds stress equation model (RSM) is used to calculate the flow field, which is the most complete classical turbulence model. In these models, the eddy viscosity hypothesis is avoided, and the components of Reynolds stress tensor are calculated directly. These models are formulated using exact Reynolds stress transfer equations. They explain the directional effect of Reynolds stress and the complex interaction in turbulence.

2.3. Calculation method of particle drag force

During the CFD-DEM coupling, the generation of particles and the prediction of trajectories are controlled and calculated by the DEM solver. In the DEM solver, the fluid resistance and the related drag force in the CFD solver need to be involved in the formula calculation as additional field forces, but the main body of the formula is provided by the CFD solver. The research object of this paper mainly invokes the relevant formula of drag force in DPM model in CFD to calculate the drag force of particles.

2.3.1. Equation of motion of particles. CFD predicts the trajectory of a discrete phase particle (or droplet or bubble) by integrating the force balance on the particle, which is written in a Lagrangian reference frame. This force balance equates the particle inertia with the forces acting on the particle, and can be written (for the x direction in Cartesian coordinates) as

$$\frac{du_p}{dt} = F_D(u - u_p) + \frac{g_x(\rho_p - \rho)}{\rho_p} + Fx \tag{1}$$

where F_x is an additional acceleration (force/unit particle mass) term, $F_D(u - u_p)$ is the drag force per unit particle mass and

$$F_D = \frac{18\mu}{\rho_p d_p^2} \cdot \frac{C_D \operatorname{Re}}{24} \tag{2}$$

Here, u is the fluid phase velocity, u_p is the particle velocity, μ is the molecular viscosity of the fluid, ρ is the fluid density, ρ_p is the density of the particle, and d_p is the particle diameter. Re is the relative Reynolds number, which is defined as

$$\operatorname{Re} \equiv \frac{\rho d_p |u_p - u|}{\mu} \tag{3}$$

Equation 1 incorporates additional forces (F_x) in the particle force balance that can be important under special circumstances. The first of these is the "virtual mass" force, the force required to accelerate the fluid surrounding the particle. This force can be written as

$$Fx = \frac{1}{2} \frac{\rho}{\rho_p} \frac{d}{dt}(u - u_p) \tag{4}$$

and is important when $\rho > \rho_p$. An additional force arises due to the pressure gradient in the fluid:

$$Fx = \left(\frac{\rho}{\rho_p}\right) u_{p_i} \frac{\partial u}{\partial x_i} \tag{5}$$

2.3.2. Random tracking of particles. When the flow is turbulent, CFD will predict the trajectories of particles using the mean fluid phase velocity, $\bar{u}$, in the trajectory equations (Equation 1). Optionally, you can include the instantaneous value of the fluctuating gas flow velocity,

$$u = \bar{u} + u' \tag{6}$$

to predict the dispersion of the particles due to turbulence.

3. Establishment of numerical simulation model

3.1. Geometric models and boundary conditions

The three main parameters of the hydrocyclone used in this paper are the main diameter D_1=100mm, the double tangential inlet diameter D_2=25mm and the bottom flow pipe diameter D_3=22mm. As shown in Figure 2, because of the density difference between the solid and liquid phases, the denser solid particles move downward along the hydrocyclone wall to escape from the bottom flow mouth.

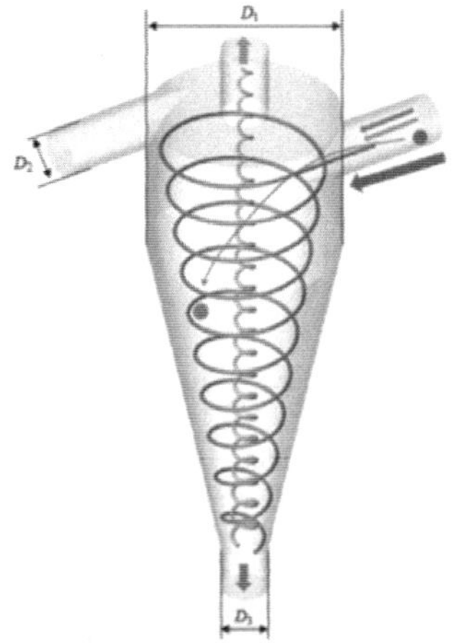

Fig. 2 Modelling and principle

To determine the grid cell size, the grid independence test is carried out for five different numbers of grids: 143631, 251318, 303746, 450276 and 506492. The grid division is not similar to the general CFD simulation. Taking the tangential velocity value in the flow field as the index, the steady-state calculation of the flow field is carried out. As shown in Figure 3, the Line-1 straight-line output feature point of coordinate Z=40mm at the entrance of the Y plane is selected in the flow field for comparison.

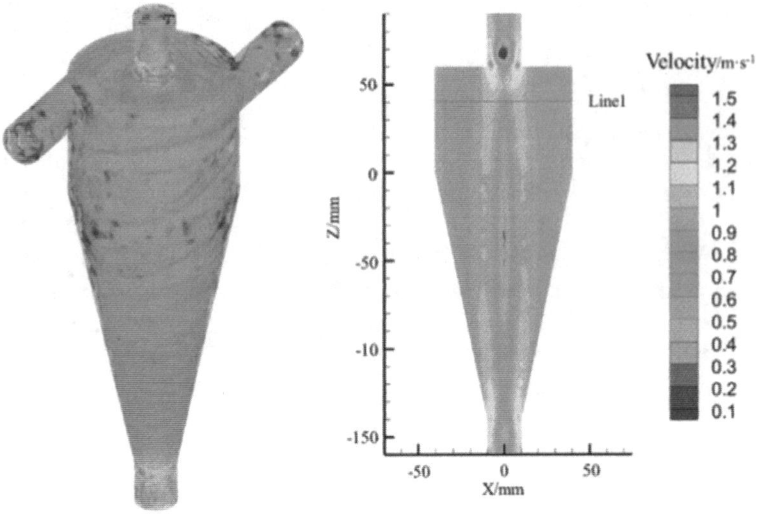

Fig. 3 velocity cloud picture and eigenvalue position

The results show that the number of grids increases from 143631 to 303746, and the change is obvious. When the number of grids increases to 506492, the data change has converged compared with 450276 grids, so the grid with 506492 grids is used in the subsequent calculation.

The quartz sand with Poisson's ratio 0.3 and density (ρ p) 2300kg/m3 is selected. The particle sizes are 1mm, 1.5mm, 2mm, 2.5mm and 3mm.

4. Establishment of numerical simulation model

To convert the vectors in X, Y and Z directions into centripetal, tangential and gravity vectors in the process of motion, it is necessary to convert the simulation results into vectors. In the simulation software, the coordinates and vectors are represented by the components of X, Y and Z directions, as shown in Figure 4.

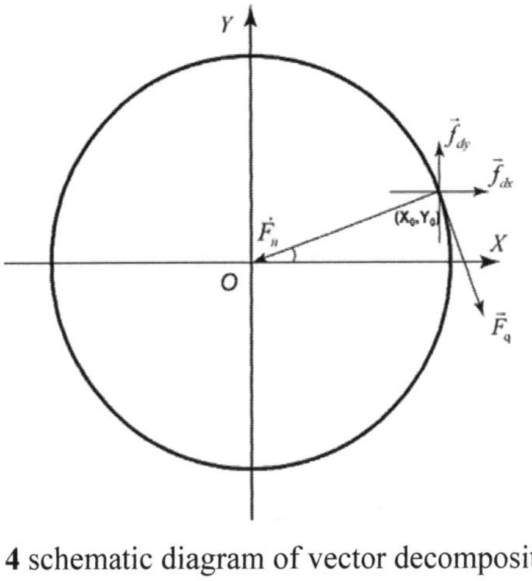

Fig. 4 schematic diagram of vector decomposition

In order to explore the force of the particles moving in the hydrocyclone, the force of the particles in the process of movement was analysed. The particles are mainly subjected to three types: coupling force, gravity and contact reaction force. After the force analysis, the contact reaction data are extracted by combining total-force data with coupling-force, and the statistical table of particle collision in Table 1 can be obtained.

Table 1. Particle collision statistics

Particle size	Number of collisions
1mm	17
1.5mm	13
2mm	21
2.5mm	25
3mm	25

The maximum value, mean value and variance of particle force are summarized in Table 2.

Table 2. Statistical table of centripetal components of contact reaction

Particle size	Contact the reaction centripetal component/N		
	Max	Average	Variance
1mm	0.001533689	0.000455945	2.3635×10^{-7}
1.5mm	0.017669589	0.007325146	2.38644×10^{-5}
2mm	0.025485756	0.008377587	5.40588×10^{-5}
2.5mm	0.080259614	0.017833648	0.000378371
3mm	0.113812743	0.025344542	0.001121971

To explore the change of tangential velocity of particles with different particle sizes moving in the hydrocyclone, the tangential velocity of particles is compared with the direction of gravity, as shown in Figure 5. When the particle size ≤ 2mm, the tangential velocity fluctuates, but the fluctuation is not strong, and between different particle sizes. When the particle size > 2mm, the tangential velocity begins to change.

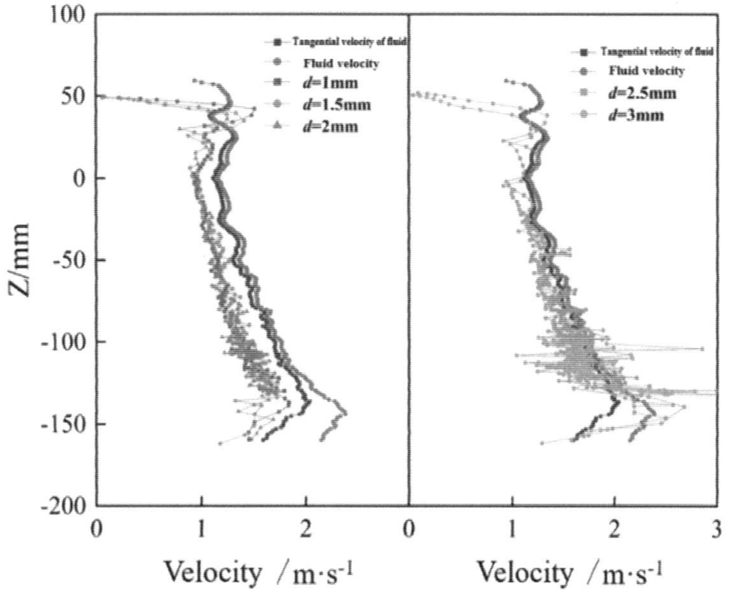

Fig. 5 comparison of tangential velocities of five kinds of particles

The simulation results of 1mm, 1.5mm, 2mm, 2.5mm and 3mm are shown in Figure 6. When the particle is > 1mm, the difference can be found obviously when the particle trajectory is compared with the streamline of the flow field.

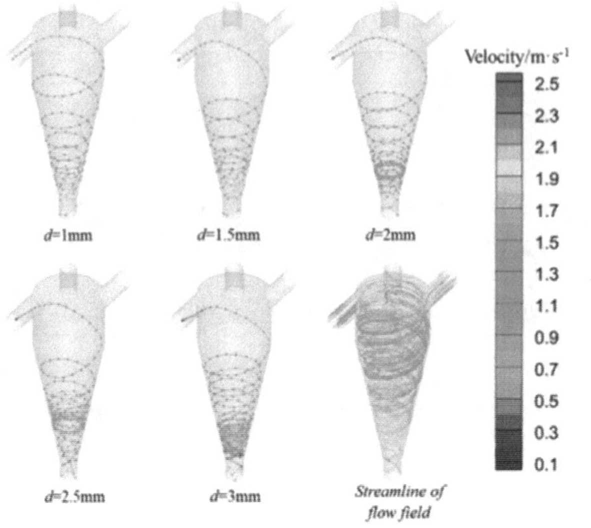

Fig. 6 comparison of trace and streamline of five kinds of particle sizes

The contact reaction force of the particles in different positions, as shown in Figure 7, compared with the particle trace, it can be seen that the particles mainly collide in the middle and lower part of the hydrocyclone, within the range of-140mm~-80mm, which is consistent with the range of the particles' spiral movement. when the particles collide, the particles suddenly suffer the contact reaction caused by the collision.

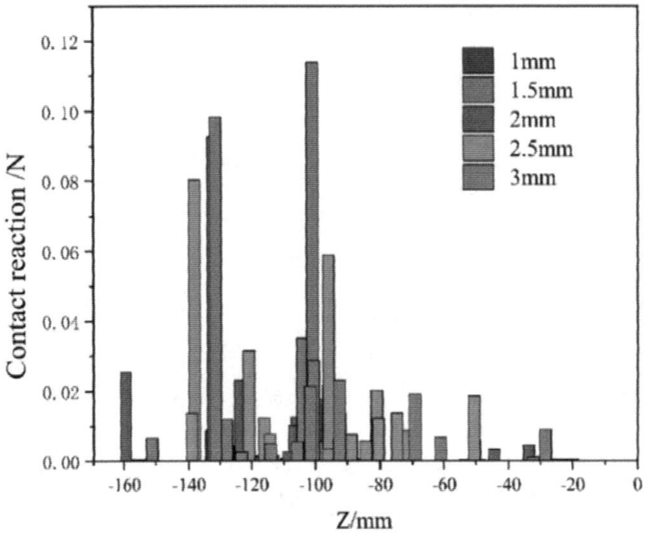

Fig. 7 contact reaction of particles of different sizes at different longitudinal positions under the same working condition

5. Conclusion

When the particles move in the hydrocyclone, they hardly collide with the inner wall of the hydrocyclone, and the collision position of the particles is mainly concentrated in the-140mm-80mm range of the hydrocyclone. When the particles collide, the particles are suddenly subjected to the contact reaction caused by the collision, which overcomes the following of the fluid, and the trajectory will be adjusted from following the fluid streamline to circling in a certain longitudinal position, and with the increase of the particle size. The range of hovering motion will become wider.

References

[1] Gao B. Application of Heavy Media Cyclone in Coal Preparation Process [J]. Shanxi Chemical Industry, 2021, 41(03): 136-137+153.
[2] WANG Zunce, GAO Ye, LV Fengxia, et al. Effect of Liquid-liquid Hydrocyclone's Inlet Structure on Pressure Characteristic [J]. FLUID MACHINERY, 2003, 31(02): 16-19+26.
[3] Yang S L. CFD-DEM coupling study on the two-phase flow mechanism of dense gas and solid in fluidized bed [D]. Hangzhou: Zhejiang University, 2014.
[4] Liu H B, Zhang J, Xiao H N, et al. Motion analysis of solid phase particles in the formation of swirl field[J]. Chemical Industry and Engineering Progress, 2019, 38(03): 1236-1243.
[5] Liu Y. Analysis of motion characteristics of solid particles in the boundary layer of two-phase turbulence [D]. Xi'an: Xi'an Shiyou University, 2018.
[6] Li C Y. Simulation of the Influence of Offset Velocity on the Trajectory of Forage Particles in Cyclone Separator [J]. Heilongjiang Agricultural Sciences, 2011(09): 136-138.
[7] ZHOU H, HU Z, ZHANG Q, et al. Numerical study on gas-solid flow characteristics of ultra-light particles in a cyclone separator [J]. Powder Technology, 2019, 344: 784-796.
[8] Zhang Z P. Study on agglomeration characteristics of particle turbulence at microscale [D]. Dalian: Dalian University of Technology, 2021.

Longitudinal stability analysis of underwater towed body based on CFD

Xianlong Liu*, Cheng Liu, Cheng Chen, Haixing Kui

CSSC 750 Test Range, Kunming, China

*Corresponding author e-mail: liuxianlong@yeah.net

Abstract. The angle of the underwater towed body is very important when it is towed passively. The steady angle is calculated by cubic spline interpolation and CFD. Generally, in order to obtain a stable pitch angle, it is necessary to adjust the angle of the tail rudder plate. With this method, stable angle can be obtained quickly and effectively, which can be used as reference for corresponding scheme design.

Keywords: CFD; Finite difference method; Cubic spline interpolation.

1. Introduction

Underwater towing system is widely used in underwater environmental dynamic monitoring system. It has special applications in underwater environmental monitoring of oceans, lakes and inland rivers, underwater engineering structure inspection and marine resources investigation. The towing system is usually composed of tugboat, towing cable, towing body and attitude control mechanism. The towing body can carry corresponding loads according to different needs, such as temperature, salinity, pressure, sonar, optical camera and other acoustic and optical physical detection sensors.

The depth and attitude control of the towed body can be achieved by changing the speed of the towed ship, retracting and releasing cables, changing the angle of attack of the submerged hydrofoil of the towed body, and controlling the speed of the propeller. With the increasing requirements of countries around the world for rapid on-site measurement of underwater environment, the market demand for such devices is increasing. The working nature of the underwater environment monitoring sensor carried in the towed body requires that the towed body must be stable in attitude, and have flexible and rapid depth and attitude adjustment and control capabilities.

Compared with underwater robots, underwater towing systems have the following characteristics:

1). The towed body of the underwater towing system has simple structure and high maintainability. Many towed vehicles are only composed of an underwater vehicle, with various sensors, sonar or advanced cameras placed inside. In this way, the towing body is not easy to be damaged during operation, and it is easy to repair and replace in case of failure.

2). The underwater towing system has wide working range, long time and high efficiency. In the case of knowing nothing about the marine landform, the underwater towing system will generally be selected to detect the landform, hydrology and other conditions. Because the power of the underwater towing system is the tug, the whole system can work for a long time.

3). The underwater towing system is easy to control. Underwater robots usually input the task program in advance when designing, and can only work according to the program after entering the

Content from this work may be used under the terms of the Creative Commons Attribution 3.0 licence. Any further distribution of this work must maintain attribution to the author(s) and the title of the work, journal citation and DOI.
Published under licence by IOP Publishing Ltd

water; The motion of the underwater towed body can be controlled artificially, and the speed, steering, lifting and sinking of the towed body can be controlled according to the actual underwater situation.

2. The parameters of the underwater towing body

The shape of the underwater towing body is shown in Figure 1. The equipment includes: battery cabin, transmitting transducer, receiving hydrophone, electronic cabin, attitude sensor and depth sensor.

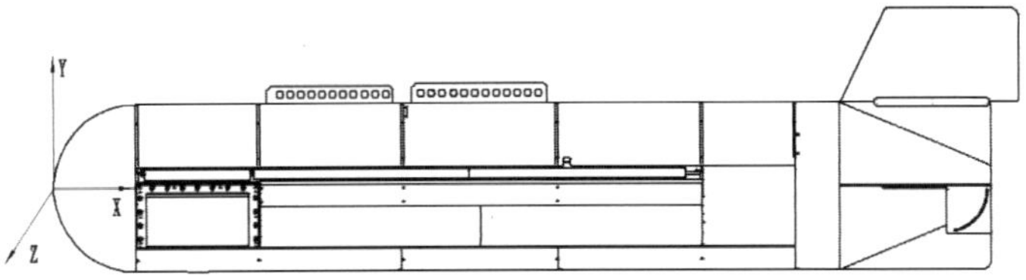

Figure 1. The Shape of underwater towing body.

The diameter of the underwater towing body is 535 mm, the length is 3186 mm, and the total weight is 350 kg. The weight in the water is 228 kg. Before the corresponding experiment, adjust the towed body to a horizontal state in the water. The location coordinates of the drag point can be seen in Table 1.

Table 1. The points of towed.

No.	1	2	3	4	5	6	7	8
X/mm	744	779	814	849	884	919	954	989
Y/mm	298	298	298	298	298	298	298	298
Z/mm	0	0	0	0	0	0	0	0
No.	**9**	**10**	**11**	**12**	**13**	**14**	**15**	**16**
X/mm	1024	1059	1094	1209	1244	1279	1314	1349
Y/mm	298	298	298	298	298	298	298	298
Z/mm	0	0	0	0	0	0	0	0
No.	**17**	**18**	**19**	**20**	**21**	**22**	**23**	/
X/mm	1384	1419	1454	1489	1524	1559	1594	/
Y/mm	298	298	298	298	298	298	298	/
Z/mm	0	0	0	0	0	0	0	/

The stress analysis of underwater towing is shown in Figure 2. It is subject to the pull force F, gravity g, buoyancy V, resistance R and lift L of the streamer. The coordinate origin is established at the leftmost point of the towed body, and the Cartesian coordinate system is adopted. Generally, the motion stability of the towing body is divided into three types: lateral stability (refers to the stability of motion around the x-axis), longitudinal stability (refers to the stability of motion around the Z-axis) and directional stability (refers to the stability of motion around the y-axis).

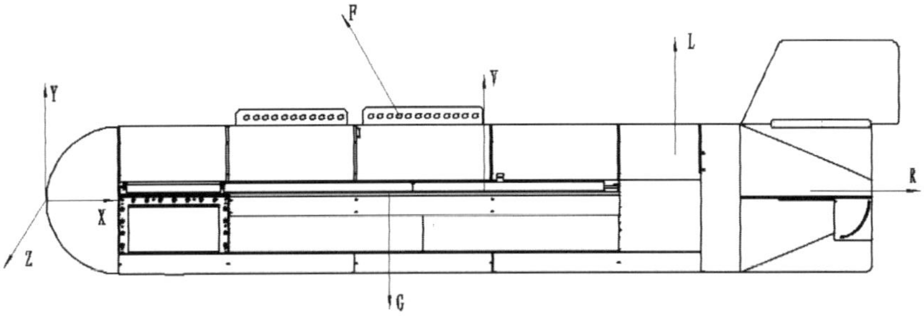

Figure 2. Force analysis diagram of towing body.

Lateral stability mainly affects the change of rolling angle. In order to ensure the stability in this direction, it is mainly to ensure that the center of gravity and the center of buoyancy are on the xoy plane, and the center of gravity is lower than the center of buoyancy (in the y-axis direction), so as to ensure a better recovery torque. Therefore, it is required that the equipment arranged in the towing body should be symmetrical to the left and right as far as possible, and the adjustment and test of the heavy floating center should be carried out after the installation of the equipment. The directional stability is mainly guaranteed by the vertical rudder plate.

The following focuses on the analysis of longitudinal stability. The stress analysis of Figure 2 shows:

$$F_x - R = 0 \tag{1}$$

$$V + L + F_y - G = 0 \tag{2}$$

Take a moment for the tension point, and get:

$$M_\alpha = M_V + M_L + M_R - M_G \tag{3}$$

The simulation adopts workbench software, and its coordinate system needs to be processed in the result module. The coordinate system of CFX adopts Cartesian coordinate system, but it is special to move the origin position to establish a new coordinate system. To understand it simply, the first point is the coordinate origin, the vector direction between the second point and the first point is the z-axis direction, and the third point determines the xoz plane and the y-axis direction.

In order to ensure that when towing at a certain towing point, the towing body can maintain a horizontal state at a certain speed, it is necessary to adjust the tail rudder angle.

3. Principle of curve interpolation

Curve interpolation has exponential Lagrange interpolation, function interpolation, Cubic spline interpolation, B spline curve, NURBS curve and so on [1].

3.1. Cubic spline curve interpolation

There, $a = x_0 < x_1 < \cdots < x_n = b$, if function $y(x)$, there are two condition:

(a) $y(x)$ is Cubic Polynomial on each interval $[x_{i-1}, x_i], i = 1, 2, \cdots, n$;

(b) $y(x)$ is twice continuously differentiable function on interval $[a,b]$.

Then, $y(x)$ is Cubic spline on interval $[a,b]$. $x_i\,(i = 0, 1, \cdots, n)$ is the node. The interpolation points are shown in Table 2.

Table 2. Interpolation points.

x	x_0	x_1	$\cdots$	x_n
y	y_0	y_1	$\cdots$	y_n

Cubic spline interpolation function is like this:

$$s(x) = \frac{h_k + 2(x - x_k)}{h_k^3}(x - x_{k+1})^2 y_k + \frac{h_k - 2(x - x_{k+1})}{h_k^3}(x - x_k)^2 y_{k+1} + \frac{(x - x_k)(x - x_{k+1})^2}{h_k^2} m_k + \frac{(x - x_{k+1})(x - x_k)^2}{h_k^2} m_{k+1} \tag{4}$$

In equality (1),

$$h_k = x_{k+1} - x_k (k = 0,1,\cdots,n-1) \tag{5}$$

$$\lambda_k m_{k-1} + 2m_k + \mu_k m_{k+1} = g_k (k = 1,2,\cdots,n-1) \tag{6}$$

There,

$$\lambda_k = \frac{h_k}{h_k + h_{k-1}}, \mu_k = \frac{h_{k-1}}{h_k + h_{k-1}}, g_k = 3(\mu_k \frac{y_{k+1} - y_k}{h_k} + \lambda_k \frac{y_k - y_{k-1}}{h_{k-1}})(k = 1,2,\cdots,n-1)$$

4. Basic theory of fluid mechanics

4.1. Control equation

The continuity equation and momentum equation for incompressible viscous fluid is [2]:

$$\frac{\partial \overline{u_i}}{\partial x_i} = 0 \tag{7}$$

$$\rho \frac{\partial \overline{u_i}}{\partial t} + \rho \partial \overline{u_j} \frac{\partial \overline{u_i}}{\partial x_i} = \rho \overline{F_i} - \frac{\partial \overline{p}}{\partial x_i} + \frac{\partial}{\partial x_j}(\mu \frac{\partial \overline{u_i}}{\partial x_j} - \rho \overline{u_i' u_j'}) \tag{8}$$

There, ρ is density, μ is Viscosity Coefficient, $\overline{p}$ is average pressure, $\overline{F_i}$ is external force, $\overline{u_i}$ is mean velocity, u_i' is fluctuating velocity, $-\rho \overline{u_i' u_j'}$ is reynolds stress.

4.2. SST $k-\omega$

The $k-\omega$ based SST model accounts for the transport of the turbulent shear stress.

The BSL model combines the advantages of the Wilcox and the $k-\varepsilon$ model. The proper transport behavior can be obtained by a limiter to the formulation of the eddy-viscosity:

$$v_t = \frac{a_1 k}{\max(a_1 \omega, SF_2)} \tag{9}$$

$$\text{Where } v_t = \frac{\mu_t}{\rho} \tag{10}$$

F_2 is a blending function similar to F, S is an invariant measure of the strain rate.

5. Calculation results

Drag point actually select point No. 17, and adjust the angle of the tail rudder plate to make the whole drag body horizontal. In actual calculation, the actual rudder angle of the tail when the towed body is horizontal is unknown. Therefore, assuming the attack angle of the tail rudder plate, the corresponding M_α.

Using cubic spline interpolation, we can get the angles of attack when the torque is zero. $\beta = 0.99$ when v=5kn. $\beta = 0.70$ when v=4kn (Table 3).

Table 3. Torque and resistance at different speeds and angles of attack.

v/(kn)	β/°	M_{α}	Fy/N	Fz/N
5	-12	234.90	94.06	136.21
5	-9	178.06	89.93	98.78
5	-6	122.85	87.14	62.44
5	-3	68.51	85.51	26.41
5	0	12.25	85.00	-11.34
5	3	-20.21	56.17	-7.33
5	6	-97.58	87.18	-85.50
5	9	-150.90	90.14	-121.99
5	12	-205.89	94.43	-159.52
4	-12	150.22	61.94	86.79
4	-9	113.99	59.30	62.92
4	-6	78.77	57.53	39.74
4	-3	44.11	56.48	16.75
4	0	8.23	56.17	-7.33
4	3	-27.29	56.52	-31.25
4	6	-61.81	57.56	-54.62
4	9	-95.83	59.46	-77.92
4	12	-130.91	62.17	-101.85

The pressure distribution cloud image and velocity streamline image of the towing body are shown in Figure 3 and Figure 4.

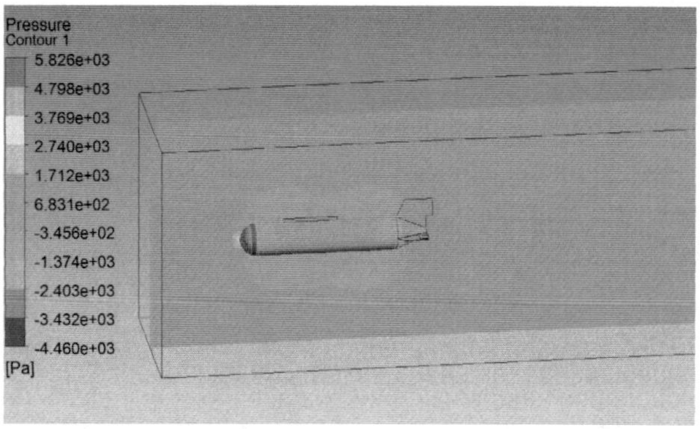

Figure 3. The pressure distribution cloud image of towing body.

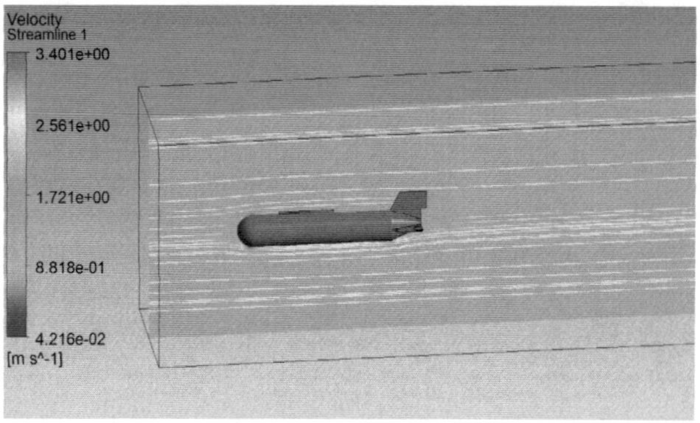

Figure 4. The velocity streamline image of the towing body.

6. Conclusion

By using CFD and cubic spline interpolation, the stable pitching angle of the towed body can be easily obtained. This method can also be used to solve the stable roll angle and pitch angle. It can quickly and effectively solve the practical problems in the project, guide the design of the trailer in advance, and facilitate the development of related work.

References

[1] Xianlong Liu, Miao Wang. Calculation the total resistance and squat of underwater towed vehicle based on CFD [J]. Journal of Physics: Conference Series, Volume 1064, 2018 the 2nd International Conference on Fluid Mechanics and Industrial Applications (FMIA 2018).

[2] Xianlong Liu, Jinwang Luo, Cheng Liu. Analysis of water jet propeller hydrodynamic performance based on CFD [J]. Journal of Physics: Conference Series, Volume 1600, 4th International Conference on Fluid Mechanics and Industrial Applications (FMIA 2020).

[3] Yang X, Wu J, Xu S. Dynamic analysis of underwater towed system under undulating motion mode of towed vehicle [J]. Applied Ocean Research, 2022, 121:103083.

[4] X Xiang, C Yu, Q Zhang. On intelligent risk analysis and critical decision of under-water robotic vehicle. Ocean Engineering, 140: 453-4658, 2017.

[5] Htun T Z, Suzuki H, D García-Vallejo, et al. On Singularity-free Multi-body Modeling of Rigid-flexible Coupled Dynamics of an Underwater Tethered System. 2021.

[6] Ma S, Wang Z, Li S, et al. Similarity Research on the Action of a Centrifuge Scaled Near-Field Underwater Explosion on a Concrete Gravity Dam [J]. International Journal of Structural Stability and Dynamics, 2022, 22(08).

[7] C Cheng, V Pallayil. Noise cancellation for an autonomous underwater vehicle - towed thin line array through recursive adaptive filtering. Journal of the Acoustical Society of America, 141(5):3925-3925, 2017.

[8] Zhang T, Wang R, Wang Y, et al. Design and Locomotion Control of a Dactylopteridae-Inspired Biomimetic Underwater Vehicle With Hybrid Propulsion[J]. IEEE Transactions on Automation Science and Engineering, 2021, PP(99):1-13.

[9] Montes-Herrera J C, Cimoli E, Cummings V, et al. Underwater Hyperspectral Imaging (UHI): A Review of Systems and Applications for Proximal Seafloor Ecosystem Studies[J]. Remote Sensing, 2021, 13(17):3451.

Frequency Analysis of Submerged Cylindrical Shell near the Rigid Wall with Hydrostatic Pressure

Peng Wang [1], **Qingyong Niu** [2,*], **Hao Zhang** [1]

[1] Wuhan Second Ship Design and Research Institute, Wuhan, 430205, China
[2] College of Science, Wuhan University of Science and Technology, Wuhan 430065, China

*Corresponding author Email: niuqingyong@wust.edu.cn

Abstract. Fluid boundary is quite common to be noted in engineering, for example, water tank, or the ocean. The tank wall could be regarded as rigid wall. Besides, for structures submerged in water, the hydrostatic pressure actually existes, and its effect is important especially for big pressure level. In this work, the rigid wall and hysrostatic pressure are both accounted, and natural frequency is predicted for submerged cylindrical shell near rigid wall with high hydrostatic pressure. Results show that rigid wall effect will reduce modal frequency. And rigid wall effect will be more and more small with increase of distance between shell and rigid wall. It's finally be ignored while the gap is large enough. Hysrostatic pressure will also reduce modal frequency, and its influence is obvious for high pressure level.

Key words: Cylindrical shell; Rigid wall; Hydrostatic pressure.

1. Introduction

Fluid domain is usually bounded, e.g., the ocean is bounded by free surface, seabed and land around it. Besides, hydrostatic pressure is also very important for structural vibration due to the gravity effect. To ensure good dynamic performance of cylindrical structures, it's very important to make clear the effect of fluid boundary and hydrostatic pressure on structural vibration.

For cylindrical shell in bounded space, Chang [1], Skidan [2] and Jouaillec [3] have done some works, but most of them focus on the shell with infinite length. For shell vibration under hydrostatic pressure, Liu [4-5] studies the effect of hydrostatic pressure on dispersion characteristics and input power flow for infinite long shell.

Here, the vibration of a finite submerged cylindrical shell near the rigid wall under hydrostatic pressure is studied.

2. Theoretical Model

Fig.1 shows a cylindrical shell immersed near the rigid wall. The shell length is of L, mean radius is R and thickness is h. The shell material, E is of elastic modulus, μ is the Poisson's ratio, and ρ_s is density. The density of fluid is ρ_f. Shell is of distance H away from the rigid wall.

Content from this work may be used under the terms of the Creative Commons Attribution 3.0 licence. Any further distribution of this work must maintain attribution to the author(s) and the title of the work, journal citation and DOI.
Published under licence by IOP Publishing Ltd

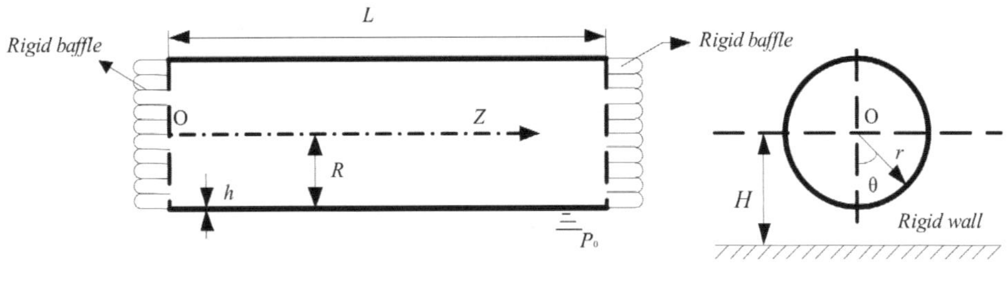

Fig. 1 Submerged cylindrical shell near rigid wall under hydrostatic pressure

Equation for cylindrical shell should be using the Flügge shell theory [6]

$$[\mathbf{L}]\begin{bmatrix} u \\ v \\ w \end{bmatrix} = \frac{(1-\mu^2)R^2}{Eh}\begin{bmatrix} 0 \\ 0 \\ -p|_{r=R} \end{bmatrix} \tag{1}$$

where, $[\mathbf{L}]$ is the differential operator matrix, p is the sound pressure. u, v, w are the shell displacements in the axial, circumferential and radial directions, respectively.

In acoustic media, sound pressure should satisfy the following equation:

$$\begin{cases} \nabla^2 p - \dfrac{1}{c_f^2}\dfrac{\partial^2 p}{\partial t^2} = 0, & \text{in fluid domain;} \\ \dfrac{\partial p}{\partial \bar{n}} = 0, & \text{on rigid wall.} \end{cases} \tag{2}$$

where, cf is the sound speed in fluid media.

In fluid, rigid wall will reflect acoustic wave. So sound pressure in fluid domain should contain two parts, while the scattering effect is neglected

$$p(r,\theta,z) = p^r + p^i \tag{3}$$

where, pr, pi denote the sound pressure of radiated and reflected acoustic wave. The relation between pr and pi could be connected using the Graf addition theory for cylindrical functions [7].

Displacements of shell could be set

$$\begin{cases} u = \sum_{m=0}^{+\infty}\sum_{n=0}^{+\infty} U_{mn}\cos(n\theta) sin(k_m z) \\ v = \sum_{m=0}^{+\infty}\sum_{n=0}^{+\infty} V_{mn}\sin(n\theta)\cos(k_m z) \\ w = \sum_{m=0}^{+\infty}\sum_{n=0}^{+\infty} W_{mn}\cos(n\theta)\cos(k_m z) \end{cases} \tag{4}$$

where, U_{mn}, V_{mn}, W_{mn} are separately the displacement amplitudes in axial, circumferential and radial directions. m, n are axial and circumferential mode order. For the shell with both ends simply-supported, km=mπ/L. The time harmonic item exp (iωt) in variable is omitted for brevity, ω is the circular frequency.

Similarly, sound pressure could be set as

$$\begin{cases} p^r = \sum_{m=0}^{\infty}\sum_{n=0}^{\infty} p_{mn}^r H_n^{(2)}(k_r r)\cos(n\theta)\cos(k_m z) \\ p^i = \sum_{m=0}^{\infty}\sum_{n=0}^{\infty} p_{mn}^i H_n^{(2)}(k_r r^{'})\cos(n\theta^{'})\cos(k_m z) \end{cases} \tag{5}$$

where, p_{mn}^r, p_{mn}^i are the amplitudes of mode (m, n). $H_n^{(2)}()$ is the nth order Hankle function of the second kind. $k_r^2 = k_f^2 - k_m^2$, $k_f = \omega / c_f$.

To ensure that the shell remains in contact with the fluid media, the following relation should be satisfied at the fluid-structure interface

$$\left.\frac{\partial p^r}{\partial r}\right|_{r=R} = -i\rho_f \omega \frac{\partial w}{\partial t} \tag{6}$$

Substitution of Eqs.(4), (5) into Eq.(1) may yield the coupling equation

$$[\mathbf{T}]\begin{Bmatrix} U_{mn} \\ V_{mn} \\ W_{mn} \end{Bmatrix} = \begin{Bmatrix} 0 \\ 0 \\ 0 \end{Bmatrix} \tag{7}$$

where, $[\mathbf{T}]$ is the coefficient matrix for mode (m, n),

$$T_{11} = \Omega^2 - (1+F_1)\lambda^2 - n^2[F_2 + (1+K)(1-\mu)/2],\ T_{12} = -i\lambda n(1+\mu)/2,\ T_{21} = -T_{12},$$
$$T_{13} = -i[(\mu - F_2)\lambda + K\lambda^3 - n^2 K\lambda(1-\mu)/2],\ T_{23} = n[1 + F_2 + K\lambda^2(3-\mu)/2],$$
$$T_{22} = [F_1 + (1+3K)(1-\mu)/2]\lambda^2 + (1+F_2)n^2 - \Omega^2,\ T_{31} = TT_{13},\ T_{32} = -T_{23},$$

$T_{33} = 1 + K - (F_2 - 2K)n^2 + F_1\lambda^2 + K(\lambda^2 + n^2)^2 - \Omega^2 + FL$, $\Omega = \omega R\sqrt{\rho_s(1-\mu^2)/E}$, $F_1 = -\frac{R^2(1-\mu^2)}{2Eh}p_0$, $F_2 = -\frac{R^2(1-\mu^2)}{Eh}p_0$, the fluid loading term,

$$FL = R^2(1-\mu^2)\left(\frac{\rho_f\omega^2 H_n^{(2)}(k_r R)}{k_r H_n^{(2)'}(k_r R)} + \sum_{a=0}^{\infty}\frac{(-1)^a \alpha\rho_f\omega^2 \Delta J_n(k_r R)}{k_r H_a^{(2)'}(k_r R)}\right)\Big/ Eh,$$

$\Delta = (-1)^{a+n}H_{a-n}^{(2)}(2k_r H) + (-1)^a H_{a+n}^{(2)}(2k_r H)$. The effect of rigid wall is shown in second item of expression for FL.

3. Numerical analysis

3.1. Validation

Parameters for numerical model is, L=1.284m, R=0.18m, h=0.003m, ρs=7850kg/m³, μ=0.3, E = 2.1E11Pa, ρf=1024kg/m³, cf=1460m/s.

The FEM is applied to verify the accuracy of present method, and numerical calculation is conducted with ANSYS software. Comparative analysis on coupled modal frequencies from present method and FEM are listed in Table 2. The hydrostatic pressure is set to be zero. Coupled modal frequencies from present method match well with those from the FEM, so the present method has good accuracy to predict coupled modal frequencies of submerged cylindrical shell near rigid wall under hydrostatic pressure.

Table 1. Validation of present method

Mode	*H*=0.36m		
	present	*FEM*	*difference*
(1,2)	95.9	97.5	-1.6%
(1,3)	108.7	110.4	-1.5%
(1,4)	201.7	205.8	-2.0%
(2,3)	216.3	218.1	-0.8%

3.2. Rigid wall effect
The modal frequency distribution of mode (1,1) are plotted against the distance ratio H/R in Fig.4, while hydrostatic pressure is, p0=0.4MPa. The corresponding modal frequencies of a cylindrical shell ideally immersed in infinite fluid domain are also given for comparative analysis.

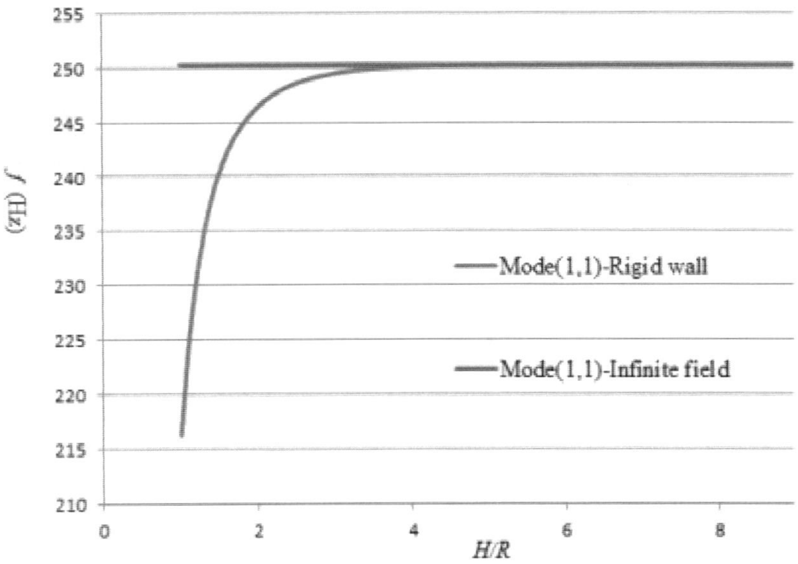

Fig. 2 Modal frequency distribution of an immersed cylindrical shell with rigid wall nearby versus immersion depth for mode (1,1)

When the distance between shell and rigid wall is small, modal frequency in case of "rigid wall" is obvious smaller than that in case of "Infinite field". When the distance between shell and rigid wall increase, modal frequency in case of "rigid wall" will increase and approach that in case of "Infinite field". Fninally, the difference between modal frequencies in both case could be ignored if the distance H is large enough.

3.3. Hydrostatic pressure effect
To study the effect of hydrostatic pressure on natural vibration of submerged shell near the rigid wall, Distribution of natural frequency of mode (1, 2) versus increase of hydrostatic pressure is plotted in Fig.3. In this case, H=0.3m.

It's noted from Fig.3 that modal frequency will dicrease with the increase of hydrostatic pressure. This phenomenon is mainly due to the resuction of structural flexural stiffncss in the presence of hydrostatic pressure. Therefore, in the design of cylindrical shell type structure for application in the oceans, the effects of both sea bottom and hysdrostatic pressure should be taken into consideration.

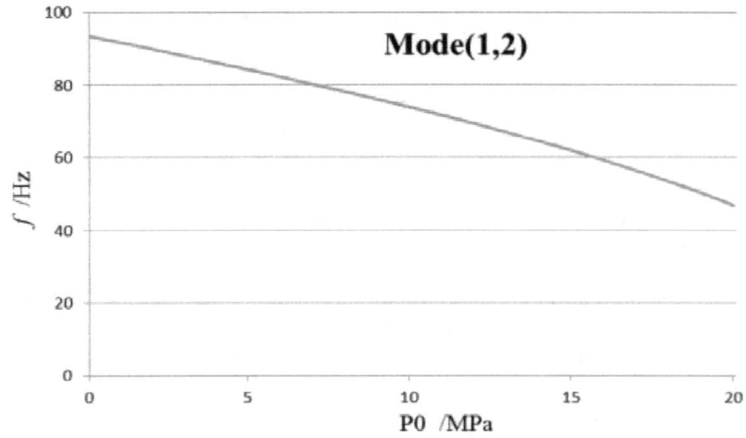

Fig. 3 Distribution of natural frequency of mode (1, 2) versus increase of hydrostatic pressure

4. Conclusions

In this work, one approach is put forward to predict the natural frequencies of submerged cylindrical shell near the rigid wall under hydrostatic pressure. The rigid wall effect will result in reduction of coupled modal frequencies, and, it will redecrease and finally be ignored while the shell is far enough from rigid wall. When the distance between shell and rigid wall is constant, the increase of external hydrostatic pressure effect will also led to reduction of coupled modal frequencies.

Acknowledgements

We sincerely thank the financial support from the Hubei Provincial Natural Science Foundation (No. ZRMS2020002231) and the Young Top-notch Talent Cultivation Program of Hubei Province.

Reference

[1] K.Y. Chang, F.L. DiMaggio. Vibrations of cylindrical shells in a semi-infinite acoustic medium[J]. The Journal of the Acoustical Society of America, 1971, 49(3B): 759-767.

[2] O. Skidan, J.M. Klosner, M.L. Baron. Sound radiation from a cylinder immersed in an acoustic fluid bounded by an elastic half-space[J]. The Journal of the Acoustical Society of America, 1974, 56(2): 427-439.

[3] F. Jouaillec, G. Jacquart. The acoustic radiation from a semi-submerged elastic cylinder. The Journal of the Acoustical Society of America 2005, 84(S1): S58-S58.

[4] Liu Z Z, Li T Y, Zhu X, et al. The effect of hydrostatic pressure fields on the dispersion characteristics of fluid shell coupled system. Journal of Marine Science and Application, 2010, 9(2): 129-136.

[5] Liu Z Z., Li T Y, Zhu X, et al. Effect of hydrostatic pressure on input power flow in submerged ring-stiffened cylindrical shells. Journal of Ship Mechanics, 2011, 15(3): 301-312.

[6] S. Fluegge. Stresses in shells[M]. Berlin: Springer, 1973, 2nd ed.

[7] G.N. Watson. A treatise on the theory of Bessel functions. Cambridge university press, 1995.

Simulation study on active cavitation suppression for a typical hydrofoil

Yun Wang[1], Liang Dong[1], Jianwei Wu[1], Zhijian Li[2] and Wei Wang[2,*]

[1]Wuhan Second Ship Design and Research Institute, Wuhan, China
[2]School of Energy and Power Engineering, Dalian University of Technology, Dalian, China

*Corresponding author e-mail: wangw@dlut.edu.cn

Abstract. An active cavitation suppression structure with jet holes on its surface is proposed for a typical hydrofoil (NACA66 MOD). Based on the homogeneous equilibrium model and the Zwart-Gerber-Belamri cavitation model, the cavitation characteristics around the hydrofoil before and after jet holes are provided on hydrofoil surface at a typical operating condition are investigated through simulation. Simulation results show that the jet holes on hydrofoil surface are capable of regulating and controlling the cavitation flow field, and to a certain extent, the jet holes on hydrofoil suction surface inhibit the development and shedding of cavitation. Four groups of different jet flow rates are set up, and it is found that with the increase of jet flow rate, the cavitation suppression effect gradually increases, and the lift-to-drag ratio gradually decreases. The research shows that the jet can achieve a cavitation suppression effect of 69.35% under the premise of ensuring that the range of hydrodynamic performance does not exceed 5%. By coupling the distribution of cavitation shape, re-entrant jet and pressure field of a hydrofoil with fluid flowing around and studying the periodic change of flow field, the mechanism of the jet to suppress cavitation is analyzed.

1. Introduction

Cavitation refers to the process of formation, development, shedding and collapse of vapor or gas cavities within liquid or on the interface between liquid and solid when local pressure within liquid drops below the saturated vapor pressure. In essence, it is the phase transition between liquid and vapor [1]. Cavity rupture leads to shock waves and micro jet, promoting cavitation development, and causing vibration, shock and noise of varying degrees to hydraulic machinery. For underwater vehicles, due to their high movement speed and changing underwater depth, conditions for cavitation formation and development are subject to constant change, and unsteady cavitation fluid behavior appears on their surface, resulting in changes in hydrodynamic performance and great fluctuation of underwater loads and underwater motion behavior. Also, they may cause disastrous vibration and dynamic instability. Therefore, delay of cavitation formation and suppression of cavitation shedding have been a focus of, as well as a challenge to the research on hydraulic machinery. Researchers performed a large number of theoretical and experimental studies on the mechanism of cavitation development and the mechanism in which cavitation leads to unsteady state, in an attempt to propose new cavitation suppression techniques [2-4].

Content from this work may be used under the terms of the Creative Commons Attribution 3.0 licence. Any further distribution of this work must maintain attribution to the author(s) and the title of the work, journal citation and DOI.
Published under licence by IOP Publishing Ltd

By shape, cavitation can be classified as bubble cavitation, sheet cavitation, spot cavitation and cloud cavitation. The generation mechanism of cloud cavitation reveals that re-entrant jet is a major contributor to cavitation destabilizing and shedding. Thus, re-entrant jet control holds the key to controlling cloud cavitation.

Active control of re-entrant jet refers to controlling the cavitation flow in the entire flow field by introducing external energy into internal flow field to enable interaction between external and internal flow, with the help of man-made interference, such as injecting a polymer or water, or a non-condensable gas into flow field, providing air jet on the suction surface of hydrofoil, etc. In this paper, an active cavitation suppression structure with jet holes on its surface for a typical hydrofoil (NACA66 MOD) is proposed. Based on the homogeneous equilibrium model and the Zwart-Gerber-Belamri cavitation model, the cavitation characteristics of the hydrofoil before and after jet holes are provided on its surface at a typical operating condition is investigated through simulation and calculation, the effect of jet holes on cavitation suppression is evaluated, and the suppression mechanism is analyzed.

2. Mathematical model

Cavitation involves a wide range of complex flow processes, including phase transition between vapor and liquid, transport, shedding, compression, and collapse, etc. Based on the Navier-Stokes equations, numerical simulation research on hydrofoil cavitation is carried out by employing the computational fluid dynamics software Ansys Fluent. The homogeneous equilibrium model (HEM) is used to handle the interface between the gas and liquid phases of cavitation. The Zwart-Gerber-Belamri (Z-G-B) cavitation model is employed to describe the mass transport between gas and liquid phases. The RNG k - ε turbulence model is used to deal with the turbulence in flow field.

Energy, mass and momentum control equations can be expressed in the following uniform form.

$$\frac{\mathrm{d}}{\mathrm{d}t}\int_{V_S} \rho\phi dV + \int_{L_S} \rho\phi\mathbf{u}\cdot\mathrm{d}\mathbf{A} = \int_{L_S} \Gamma\nabla\phi\cdot\mathrm{d}\mathbf{A} + \int_{V_S} S_\phi \mathrm{d}V \quad (1)$$

where, ϕ is a common variable, such as velocity or pressure; ρ is density; V_S represents the control volume and its boundary conditions; $\mathbf{A}$ is a standard vector of the boundary surface of control volume; $\mathbf{u}$ is velocity; Γ is diffusion coefficient; and S_ϕ is a source term with respect to a variable ϕ.

3. Study Object and computational Settings

The study is performed on the typical NACA66 MOD hydrofoil, as shown in Figure 1a, with wing length of 0.07m, span length of 0.07m, and the maximum thickness ratio of 12%. Figure 1b shows the remodeled NACA66 (MOD) hydrofoil, which has the same basic parameters as the original NACA66 (MOD) hydrofoil: chord length c=0.07 m; span length a=0.07 m; the maximum thickness ratio 12%. The difference is that the remodeled hydrofoil has jet holes provided on its suction surface, from which a certain amount of water can be injected into flow field. The center of these holes are x=0.19c away from hydrofoil front edge. 24 holes were provided in a single role, with diameter of d=1.4 mm, and center-to-center distance between adjacent holes of Δx=2.5 mm. The jet direction is the normal direction of the center of the jet hole on hydrofoil surface. Grid division is as shown in Figure 2, with 7.6 million grids generated by employing a C structured grid division approach.

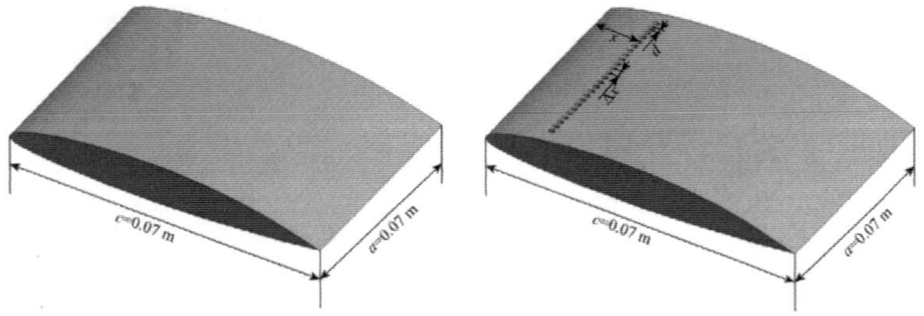

a. The NACA66 MOD hydrofoil. b. The remodeled hydrofoil.

Figure 1. The outline structure diagram of hydrofoils.

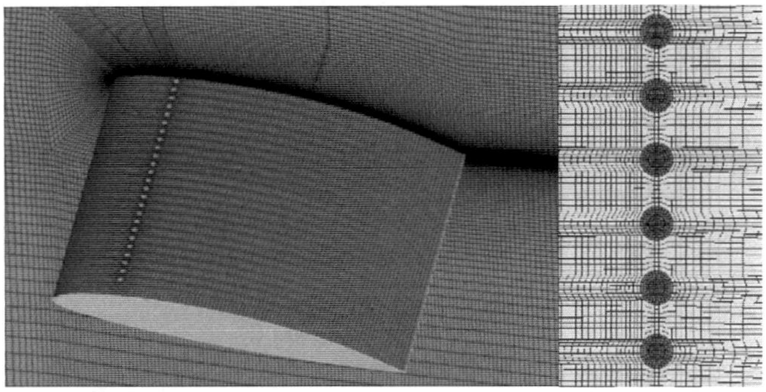

Figure 2. The grid division of the jet hydrofoil.

The numerical computation domain of the cavitation flow field around hydrofoils is set as shown in Figure 3. The hydrofoil's attack angle is $\alpha=8^{o}$, the entrance and the exit are 4c and 11c away from hydrofoil front edge, respectively. Boundary conditions are set as velocity inlet and pressure outlet conditions, with no slip on wall surface.

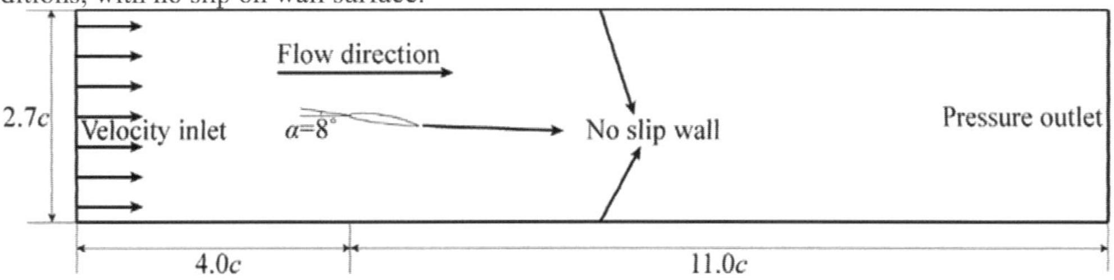

Figure 3. The schematic diagram of the computational domain.

The momentum equation and the continuity equation are discretized with the finite volume approach, and are solved with the help of a pressure-velocity coupling algorithm in SIMPLE format, with the first-order upwind scheme adopted for the calculation of convective term and PRESTO scheme for the pressure term, and the first-order implicit format used for discretizing unsteady terms. The time step is set at 0.1ms.

In the above numerical analysis of hydrofoil cavitation flow, the main flow velocity $U_\infty=7.832$ m/s, and the corresponding Reynolds number $R_e=c*U_\infty/\upsilon=5.1\times10^5$, where, c is the chord length of hydrofoil. Density and dynamic viscosity are $\rho_l=998.7$ kg/m^3 and $\mu_l=1079.85\times10^{-6}$ Pa•s, respectively, for liquid phase. $\rho_v=0.01449$ kg/m^3 and $\mu_v=9.65\times10^{-6}$ Pa•s, respectively, for vapor phase. Corresponding temperature is 17°C, and the saturated vapor pressure at this temperature is 1940 Pa.

4. Study on active cavitation suppression

The development of cloud cavitation is accompanied by a strong unsteady process, and the complex interaction between re-entrant jet and cavities plays an important role in the unsteady development of cavitation. Because cavitation develops in the presence of a low pressure region, the re-entrant jet forms under the action of adverse pressure gradient, and marches towards hydrofoil front edge, causing an impact to cavities. Based on the design principle of improving the pressure distribution in low-pressure region, stopping re-entrant jet from traveling from cavitation region's tail part to hydrofoil front edge, and weakening re-entrant jet's intensity, equipping jet holes on hydrofoil surface is proposed, so that water flow sprays into the hydrofoil's main flow field along given channels at a certain angle and flow velocity, to blend with the main fluid. This means controlling cavitation development and shedding by injecting flow momentum into flow field.

In order to clearly show the change of the cavitation volume, the dimensionless cavitation volume is obtained by comparing the volume of the cavitation with the volume of the hydrofoil, as shown in the following equations 2 and 3. In order to study the influence of the active jet on the cavitation flow characteristics rule, compare the cavitation volume of the jet foil with the cavitation volume of the original foil, and the cavitation suppression effect index is expressed by η, as shown in the following formula 4.

$$V_{nd} = V_{cav} / V_{hydrofoil} \tag{2}$$

$$V_{\mathrm{cav}} = \int_V \alpha_{\mathrm{vapor}} \mathrm{d}V \tag{3}$$

$$\eta = 1 - \frac{V_{\mathrm{remodel}}}{V_{\mathrm{original}}} \tag{4}$$

Where, V is the volume of computation domain; αvapor is the vapor-phase volume fraction on hydrofoil surface, V_{original} is the cavity volume of the original hydrofoil, and V_{remodel} is the cavity volume of the remodeled hydrofoil.

4.1. *Influence of jet flow rate on cavitation suppression effect*

In order to study the influence of the active suppress method on the cavitation around the hydrofoil, four groups of jet flow rates are set up, from small to large, 350 L/h, 400 L/h, 450 L/h, and 500 L/h. Figure 4 shows the variation law of the cavity volume above the suction surface of the original hydrofoil and the jet hydrofoils based on four different jet flow rates. It can be seen that the cavity volume on the surface of the hydrofoil is significantly reduced compared with the original hydrofoil after using the active jet. With the increase of the jet flow rate, the cavitation period T also gradually lengthens, which reflects the effectiveness of the jet foil in suppressing cavitation. When the jet flow rate increases, the cavitation suppression effect is more obvious. When the jet flow rate increases to 500 L/h, the cavitation period T is shortened, but the cavity volume is still effectively reduced.

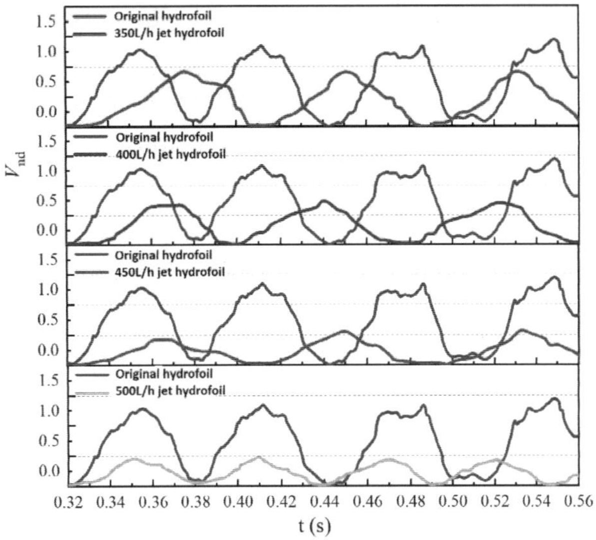

Figure 4. Variation of hydrofoil cavitation volume with time

4.2. Influence of jet flow on hydrodynamic performance

When receiving the paper, we assume that the corresponding authors grant us the copyright to use the paper for the book or journal in question. Should authors use tables or figures from other Publications, they must ask the corresponding publishers to grant them the right to publish this material in their paper. Table 1 shows the time-averaged statistical results of the hydrofoil cavitation flow field with or without jets to suppress cavitation. It can be seen that using the active suppress technology of the jet, the cavity volume decreases significantly compared to the original hydrofoil. In general, the frequency of cavity shedding is reduced. The first three groups of jets increase the development period of holes, which indicates that the frequency of cavity shedding decreases and also increases the stability of cavitation flow. When the jet flow rate is at a larger level of 500L/h, the cavity volume is reduced the most, but the period is almost the same as that of the original hydrofoil. In addition, compared with the 69.35% inhibition effect of the 450L/h configuration, the increase in the inhibition effect of the fourth group of 500L/h is no longer obvious. In order to comprehensively evaluate the effect of active control, it is also necessary to analyze the hydrodynamic performance of the hydrofoil.

Table 1. Cavitation performance index.

	V_{nd}	η	T/ms	f/Hz
original hydrofoil	0.7199	-	$T_0 = 58.3$	16.09
350 L/h	0.4194	41.75%	$T_1 = 80.2$	12.46
400 L/h	0.3203	55.52%	$T_2 = 85.7$	11.67
450 L/h	0.2206	69.35%	$T_3 = 105.5$	9.47
500 L/h	0.2141	70.27%	$T_4 = 56.3$	17.76

4.3. Jet interference cavitation morphology structure

From the analysis in the previous two sections, it can be seen that the arrangement of jet holes on the surface of the hydrofoil, on the basis of ensuring the dynamic performance of the hydrofoil, the optimal cavitation suppression effect is 69.35%, which comes from the third group of jet rate of 450 L/h. Therefore, this working condition is taken as a typical jet control scheme, compared with the original hydrofoil flow field, and the morphological and structural characteristics of the jet cavitation cluster are studied. Figure 5 is a three-dimensional cloud diagram of the coupled pressure field of the isosurface when σ=0.83 and the vapor phase volume fraction αv=0.1 in a typical cycle.

It can be seen from Figure 5 that for the original hydrofoil, the first half cycle is the growth stage of the cavity, and with the development of the cavity, the low-pressure area almost covers the entire suction surface. Afterwards, the sheet-like cavities attached to the suction surface begin to rupture at the position where the velocity stagnates and evolve into cloud cavitation. At the last moment, the cloud-like cavities and the attached flake-like cavities are completely separated, fall off, and form free cavitation groups. Similar to the original hydrofoil, the cavitation flow of the diffractive hydrofoil also exhibits the characteristics of periodic variation. The cavities develop backward along the surface of the hydrofoil, and the cavities on both sides continue to develop backward. Then, under the action of the back jet from the trailing edge of the hydrofoil to the head, the cavities in the middle of the hydrofoil retract, and the cavities in the side jets retreat. Under the action of the cavity falls off to both sides of the hydrofoil, showing a typical U-shaped cavity structure. On the whole, the cavity shape on the surface of the jet foil exists in the form of sheet cavitation, which lasts until the next development cycle. During the entire process of the evolution of the cavity shape of the jet foil, the volume of the cavity obviously reduces, there is no large-scale cavitation that occurs with the original hydrofoil.

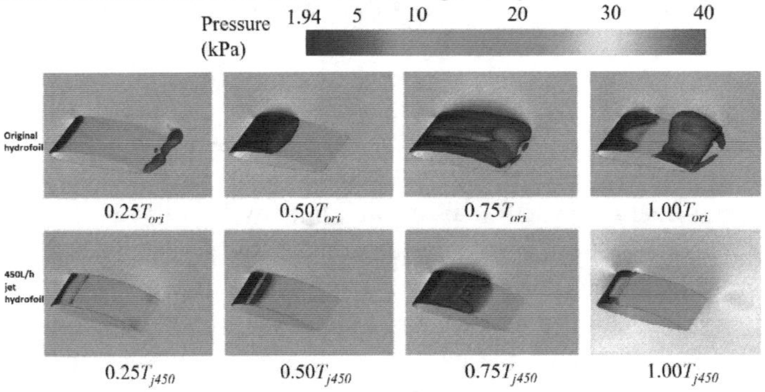

Figure 5. The evolution process of cavities in diffractive jet foils in a typical period.

5. Mechanism Analysis of Jet Suppression of Cavitation

Based on the previous research and analysis, this section takes the working condition of 450 L/h hydrofoil as the important research object. By studying the characteristics of the cavitation flow field under this working condition, the temporal and spatial evolution of the return jet, the velocity distribution of the boundary layer and the pressure pulsation, the mechanism of jet suppression of cavitation is revealed.

First of all, Fig. 6 shows the analysis results of the evolution process of the unsteady cavitation shape around the hydrofoils. The left column of Figure 6 is the instantaneous shape change of the cavity at the mid-section position of the original hydrofoil in one cycle, and the right column of Figure 6 is the instantaneous shape change of the cavity at the mid-section position of the 450 L/h jet hydrofoil in one cycle. In order to study the interaction between the retro-jet region and the cavitation region, in addition to the instantaneous morphological changes of the cavity, the figure also shows the instantaneous pressure distribution and the instantaneous morphological changes of the retro-jet region, respectively.

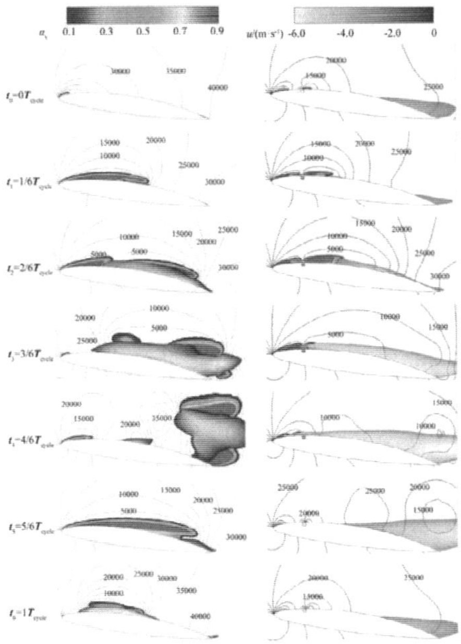

a. Original hydrofoil b. 450L/h jet hydrofoil

Figure 6. The evolution process of unsteady cavitation shape around hydrofoil.

According to Figure 6, numerical analysis well indicates the quasi-periodic process of cavities from growth, retraction, development to collapse, when cloud cavitation took place. As shown in Figure 6a, the original hydrofoil's front edge and tail edge have a high-pressure region each. Due to the velocity stagnation of flow field at hydrofoil front edge, the high-pressure region at this edge is stable. At this time, the pressure distribution and changes over suction surface play a critical role in the development and shedding of cavitation. At t_0 when cavitation was born, high pressure gradient at hydrofoil front edge enabled the cavitation region to move towards hydrofoil tail edge, leading to reduced pressure gradient at the front edge. As the cavitation grew, the low-pressure region where cavitation occurred extended rapidly, causing an area with high pressure gradient at the cavitation's tail part at t_1. High pressure gradient enabled the liquid at cavitation's tail part to return. In other words, the adverse pressure gradient on suction surface facilitated the presence of re-entrant jet. At the t_1-t_2 stage in which cavitation developed, re-entrant jet and cavities grew rapidly, and the re-entrant jet caused the primary separation of attached cavities from hydrofoil suction surface. At the t_2-t_3 stage in which cavitation development continued, the re-entrant jet began to move from a position with the highest pressure gradient at t_2, and it "pinched off" attached cavities to form cloud cavitation when flowing quickly to hydrofoil front edge, leading to primary cavity shedding. In the process of re-entrant jet's continuous movement, a high and positive pressure gradient area appeared at its head because of its interaction with main flow. The positive pressure gradient caused secondary cavitation development and re-entrant jet retraction at t_3-t_5. Attached cavities underwent rapid secondary growth at t_5 due to the presence of a low-pressure region at hydrofoil tail edge caused by primary cavity shedding. Likewise, expansion of the low-pressure region gave rise to an area with high negative pressure gradient at hydrofoil tail edge again, causing re-entrant jet. At t_5-t_6, the attached cavities subject to secondary growth shed, and high pressure caused by cavity collapse enabled the attached cavities on the original hydrofoil to retract, so that new attached cavitation formed at hydrofoil front edge, marking the beginning of a new cavitation period.

The development of cavitation on the surface of jet hydrofoil is shown in Figure 6b. As on the original hydrofoil, cavities on the jet hydrofoil were born at t_0. Cavities at hydrofoil front edge began to grow from a short length, and the active jet provided by hydrofoil surface divided the low-pressure region over hydrofoil suction surface into two parts, which was absent on the original hydrofoil. As a result of the momentum provided by jet in flow field, cavity development in an area in front of jet holes was prevented, and it stayed at the front edge. The development of attached cavities behind jet holes were extremely limited. Due to its reduced velocity, re-entrant jet's intensity decayed, and to some extent, it was blocked by jet when reaching jet holes at t_3, as shown in Figure 7. Because the local pressure at jet holes were slightly higher than that at hydrofoil front edge, there was weak re-entrant jet moving towards front edge. At t_4, the re-entrant jet behind jet crossed over the jet and moved to hydrofoil front edge, and the low-pressure region shrank because of the blending of jet, re-entrant jet and main flow, leading to reduced cavity volume. At t_5-t_6, active jet improved the pressure distribution in the low-pressure region, and prevented the transition of water from liquid phase to vapor phase, which extended the cavitation development period by 70.7%, and thus enhanced the stability of flow field, compared to the original hydrofoil. There was no cloud cavitation occurrence and shedding in the whole process.

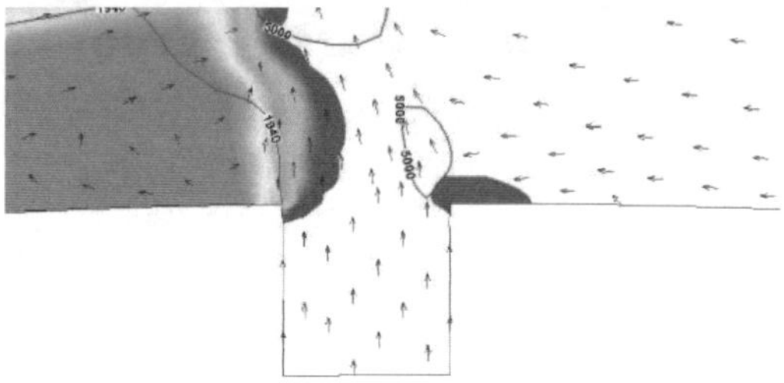

Figure 7. The cavitation flow characteristics near jet holes.

In order to better illustrate the distribution of the retro-jet, the temporal and spatial distribution cloud map of the retro-jet in the mid-section of the hydrofoil is extracted, as shown in Figure 8, where the abscissa represents the chord direction of the hydrofoil, and the ordinate represents the development time of cavitation. The initial retro-jet of the original hydrofoil is located at the tail of this transiently attached cavity. After that, the existence area of the back jet continues to expand, and the cavitation area is retracted by advancing toward the leading edge of the hydrofoil. The retro-jet intensity on the surface of the jet hydrofoil is significantly reduced. At the 0.19c position of the jet hole, the forward trend of the flow is blocked and the intensity is weakened.

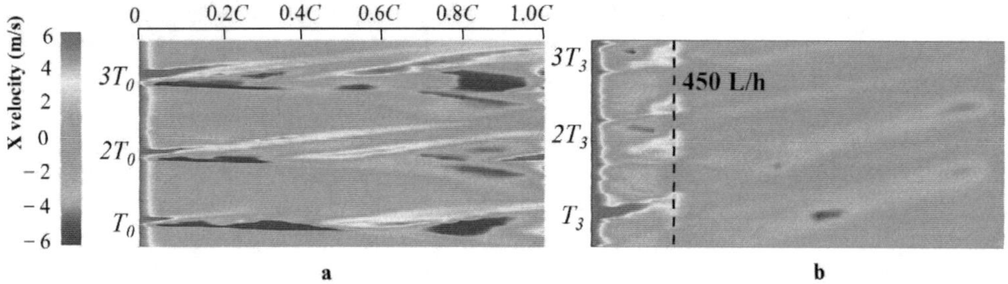

Figure 8. The velocity distribution of the return jet on the suction surface of the original hydrofoil.

The velocity distribution in the boundary layer at different positions of the mid-section of the selected hydrofoil is shown in Figure 9. The abscissa represents the ratio of the local velocity to the mainstream velocity, the ordinate represents the ratio of the distance from the suction surface to the chord length, and the green line represents the jet foil distribution of the velocity boundary layer. It can be seen that at the front, middle and rear positions on the suction surface of the hydrofoil, the velocity boundary layer of the jet hydrofoil is obviously thinned, the velocity gradient in the boundary layer increases, and the resistance to reverse pressure is enhanced. At the position of x/c=0.1, due to the proximity of the jet hole, the development of the retro-jet on the surface of the hydrofoil is blocked. From the point of view of x/c=0.4 and x/c=0.7, the strength of the retro-jet is weakened. In general, the change of the velocity boundary layer of the jet hydrofoil is smaller than that of the original hydrofoil. Therefore, the jet improves the stability of the entire flow field because of this.

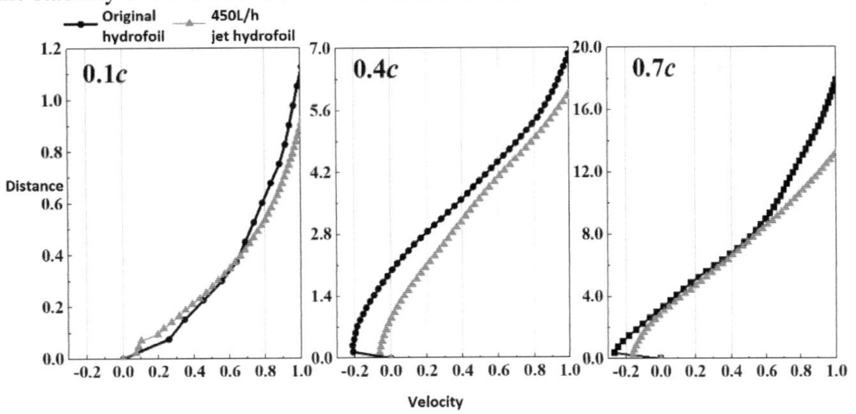

Figure 9. Time-averaged boundary layer velocities around the original hydrofoil and the 450 L/h hydrofoil.

Figure 10 quantitatively analyzes the changes of transient pressure monitored at each characteristic position on the hydrofoil surface (x=0.30C, x=0.45C, x=0.60C, and x=0.75C from the leading edge of the hydrofoil, respectively). It can be clearly seen from this figure that, using the active jet suppress method, on the one hand, the pressure peaks at each characteristic position are reduced to varying degrees, indicating that the impact generated in the process of cavity shedding and collapsing is reduced, and the impact on the surface of the hydrofoil is reduced. The damage caused is also greatly reduced. On the other hand, compared with the original hydrofoil, the duration of the low-pressure zone at each characteristic location during the whole cycle is also greatly reduced, which is consistent with the duration of the cavitation on the suction surface of the hydrofoil.

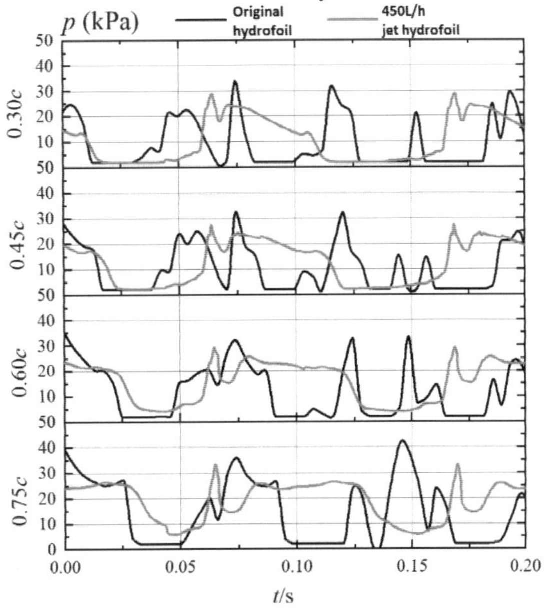

Figure 10. Time domain diagram of instantaneous pressure on suction surface of original hydrofoil and 450L/h jet hydrofoil.

6. Conclusion

Based on the active suppress technology, it is proposed to arrange jet holes on the surface of the hydrofoil, so that the water flow is sprayed into the main flow field of the hydrofoil at the angle of the vertical suction surface and a certain flow rate along the set flow channel, and mixed with the main fluid, that is, through the flow direction. The development and shedding of cavitation can be regulated by injecting fluid momentum into the field. The research results show that under the same cloud cavitation conditions, the active jet method can better suppress the occurrence, development and shedding of cavitation compared with the original hydrofoil. The specific conclusions are as follows.

(1) When the jet flow rate is 350L/h – 450 L/h, the cavitation suppression effect can reach 41.75% - 69.35%, and the hydrodynamic performance of the hydrofoil is reduced within 5%. Therefore, the control of the cavitation flow can be achieved by changing the jet flow rate, and the hydrodynamic performance is not changed too much.

(2) Excessive jet flow rate (500 L/h) can also achieve a good 70.27% cavitation suppression effect, but the lift-to-drag ratio drops by more than 10%. Such a jet scheme will destroy the hydrodynamic performance of the hydrofoil.

(3) The jet eases the pressure pulsation on the suction surface and has the effect of reducing drag. The jet thins the boundary layer, which makes the overall flow around the hydrofoil relatively stable, the jet strongly blocks the development of the return jet, thus avoiding sheet cavitation development, and subsequent cloud shedding is also inhibited.

References

[1] Sensen Pan, Xiaoxing Peng, Physical Mechannism of Cavitation, Beijing, 2012.

[2] Wei Wu, Ying Xiong, Wanjinag Qi, Cavitation Control of a 2-D Hydrofoil Under Section Reshaping, Chinese Journal of Ship Research. 3 (2012) 36-40.

[3] Wei Wang, Tao Tang, Shengpeng Lu, Jianxiong Jiao, Qingdian Zhang, Xiaofang Wang, Investigation of cavitation suppression by arranging pits on hydrofoil suction side Transactions of the Chinese Society of Agricultural Engineering, 2 (2019) 40-47.

[4] Wei Wu, Ying Xiong, A Reshaping Method for Anti-Cavitating Hydrofoil Design, Chinese Journal of Shanghai Jiao Tong University. 6 (2013) 878-883.

Prediction on the axial water thrust characteristics of pump turbine runner under partial load

Jianqiang Guo[1], Huili Bi[2], Guohu Luo[1], Zhengwei Wang[2, *]

[1] Anhui Jinzhai Pumped Storage Power Co., Ltd., Anhui 237333, China;
[2] Department of Energy and Power Engineering, Tsinghua University, Beijing 100084, China

*Correspondence: wzw@mail.tsinghua.edu.cn

Abstract. For the high head pump turbine, the unit instability during its operation mainly dues to the hydraulic vibration inside the machinery, which will cause the mechanical fatigue failure under certain circumstances. In this paper, the flow pattern and the runner axial water thrust in turbine model under partial load were analyzed by numerical simulation. The internal flow field instability and vortex load characteristics of the runner were analyzed to obtain the reasons for the different axial water thrust characteristics of the runner under two different loads. The results showed that due to the reduction of the guide vane opening, the flow rate is smaller at 50% load, and the free flow vortex structure on the high-pressure side of the runner blade inlet strengthens the impact of the water flow on the runner. Moreover, when the pressure difference between the runner blades decreases and the axial water thrust of the blades decreases, the total axial water thrust of the runner decreases and fluctuates frequently with time.

Keywords: Pump turbine; Runner; Axial water thrust; Numerical simulation.

1. Introduction

As an important part of new energy development, pumped storage power station has the characteristics of cleanness, economy, environmental protection, long life cycle and large capacity. Pump turbine is widely used in pumped storage power station. The unit adopts the same runner, pumping in forward rotation is the working condition of pump mode, and power generation in reverse is the working condition of turbine mode. Compared with simple pump units or turbine units, the instability of the pump turbine runner is particularly prominent, and the force characteristics of the runner are complex, which can easily cause accidents such as structural vibration, unit lifting and noise [1-2]. Therefore, the analysis of the internal flow characteristics of the pump turbine under partial load is of great significance.

Compared with the empirical formula of runner axial water thrust [3-4], with the development of computational fluid dynamics, numerical simulation has become a necessary means to study the force characteristics of pump turbine. The calculation results show that the operating parameters of the unit, such as speed, load and head, will affect the runner axial water thrust characteristics [5-6].

Liu Jintao et al [7] solved Reynolds stress based on turbulent kinetic energy, and compared the effects of pre-opened guide vane and synchronous guide vane on the internal flow and pressure fluctuation in the vaneless region. The study by Zhu B et al. [8] showed that under partial load conditions, a positive

Content from this work may be used under the terms of the Creative Commons Attribution 3.0 licence. Any further distribution of this work must maintain attribution to the author(s) and the title of the work, journal citation and DOI.
Published under licence by IOP Publishing Ltd

slope is likely to cause larger pressure pulsations in the leafless area. The negative pitch blade effectively controls the flow separation on the front of the blade, changes the position where the separation vortex appears, and effectively reduces the pressure pulsation. With the strongest working condition of the vortex band, the damage caused by the vortex band is one order of magnitude higher than the optimal working condition point [9-10]. Liu Demin [11] used PIV test technology to observe and study the internal flow field of the model pump turbine, and found that the separated vortex has a strong influence on its pressure distribution.

At present, in order to meet the energy demand, the pump turbine is developing towards the characteristics of large capacity, high head, large size and frequent startup and shutdown. The number of pump turbine operations under partial load conditions has increased. At this time, the relationship between the internal flow characteristics of the pump-turbine and its force characteristics needs to be further studied.

2. Numerical calculation

2.1. Mathematical Model

Considering the influence of water hammer wave, the numerical calculation of one-dimensional pipeline is as follows:

$$V\frac{\partial H}{\partial x}+\frac{\partial H}{\partial t}-V\sin\alpha+\frac{a^2}{g}\frac{\partial V}{\partial x}=0 \tag{1}$$

$$g\frac{\partial H}{\partial X}+V\frac{\partial V}{\partial x}+\frac{\partial V}{\partial t}+\frac{fV|V|}{2D}=0 \tag{2}$$

H is the piezometric head, V is the average velocity at pipe, g is the gravitational acceleration, f is the Darcy-Weisbach friction factor, α is the pipe line slop, D is the pipe diameter, t is the time, and a is the speed of pressure pulse.

The flow medium inside the pump turbine can be approximated as an incompressible Newtonian fluid. The flow is a three-dimensional unsteady turbulent flow, which follows the laws of mass conservation and momentum conservation:

$$\frac{\partial\rho}{\partial t}+\frac{\partial}{\partial x_j}\left(\rho u_j\right)=0 \tag{3}$$

$$\frac{\partial}{\partial t}(\rho u_i)+\frac{\partial}{\partial x_j}\left(\rho u_i u_j\right)=-\frac{\partial p}{\partial x_i}+\frac{\partial}{\partial x_j}\left(\mu\frac{\partial u_i}{\partial t}-\rho\overline{u_i' u_j'}\right)+S_M \tag{4}$$

where ρ is water density, p is the pressure, u is the flow velocity, μ is the dynamic viscosity, S_M is the external momentum source term, $-\overline{\rho u_i' u_j'}$ is the Reynolds stresses. SST *k-ω* model is selected to solve the Reynolds stresses [12].

2.2. Calculation Model

The research object of this paper is a pump turbine. In order to provide accurate boundary conditions for three-dimensional transient process calculation of pump turbine, one-dimensional modeling of unit pipeline system is carried out in this paper, as shown in Fig. 1. The three-dimensional fluid domain computational model is shown in Fig. 2.

This paper is based on the Ansys CFX platform to calculate the internal flow of the unit under the working conditions of the turbine. The boundary condition is one-dimensional calculation. The total pressure inlet is set in the spiral case and the static pressure outlet is set in the draft tube. The medium is water, and the turbulence model is SST k-ω model. The time step for unsteady calculation is 1/360 of the rotation period. The total calculation time is 2.7 s.

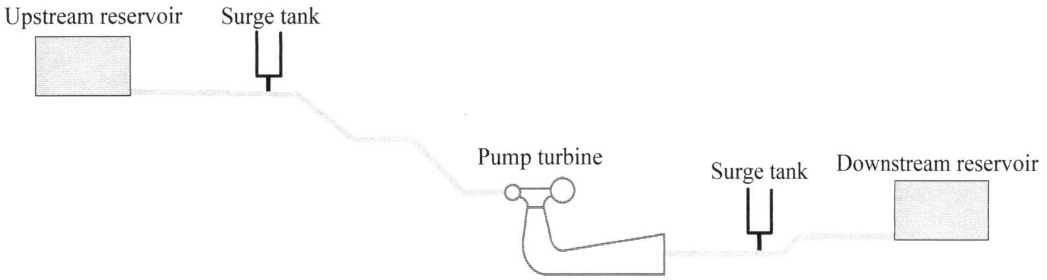

Fig. 1 Water conveyance system of Pumped Storage Power Station.

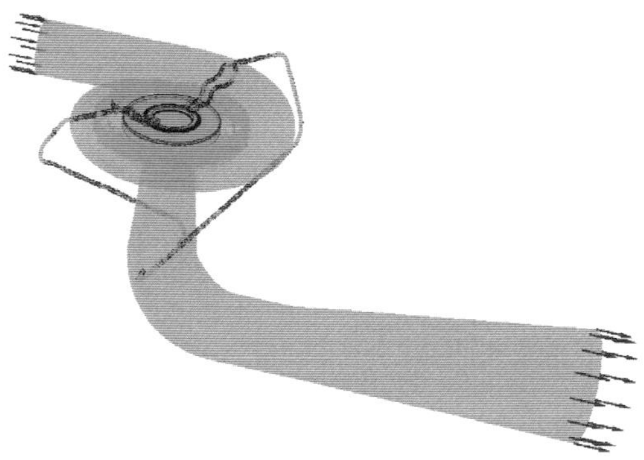

Fig. 2 Three-dimensional computing domain.

3. Results and discussion

Through the analysis of the internal flow characteristics of the pump turbine under different loads, in order to obtain the flow mechanism in the overflow channel of the pump turbine. Fig. 3 (a) and Fig. 3 (b) show the flow structure in the channel of hydraulic turbine unit under 50% load and 100% load operation conditions respectively.

The flow separation vortex structure on the high-pressure side of the runner blade inlet intensifies the impact of water flow on the runner. Under 100% load, the opening of the guide vane is increases, the flow rate is large, and the overall flow field stability of the unit channel will be increased.

(a) 50% load (b) 100% load

Fig.3 The flow pattern of the pump turbine unit.

Journal of Physics: Conference Series **2441** (2023) 012044 doi:10.1088/1742-6596/2441/1/012044

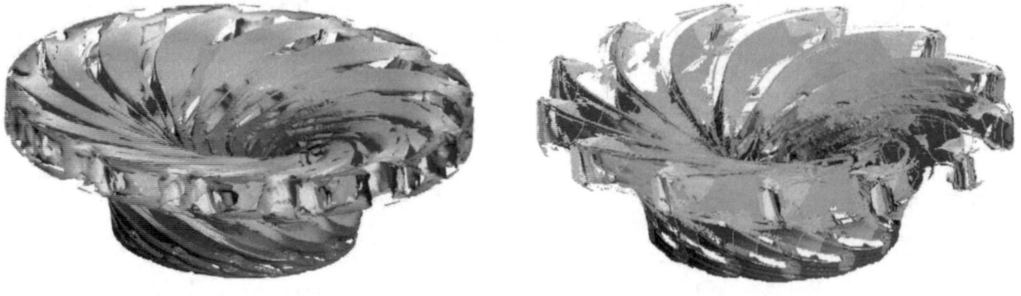

(a) 50% load (b) 100% load

Fig. 4 Vortex distribution in runner

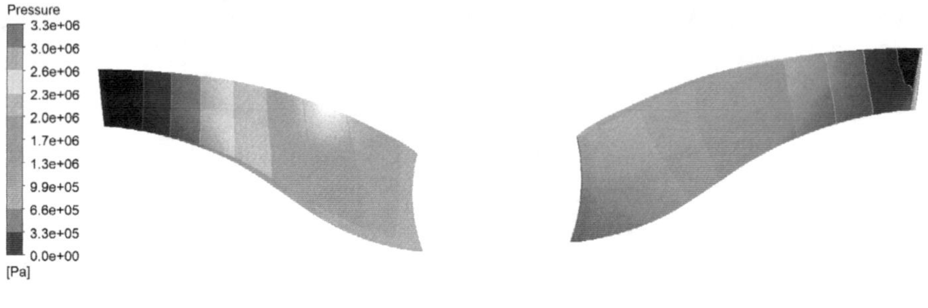

(a) 50% load

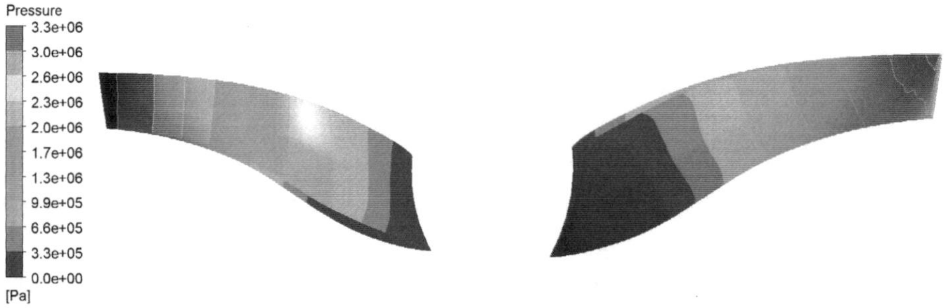

(b) 100% load

Fig. 5 Runner blade pressure distribution.

The complex flow in the runner affects the work efficiency and the runner axial water thrust. Fig. 4 shows the vortex structure distribution in the runner under 50% load and 100% load. Because the pressure load affects the stress distribution on the runner wall, the visual display of the pressure on the vortex is adopted. Under the two loads, the runner is full of complex vortices, and the pressure load on the vortex surface will directly affect the runner surface axial water thrust. Under 100% load, the scale and number of vortices in the runner channel are significantly weakened, and the disturbance intensity of vortices to the flow field in the runner is reduced.

Fig. 5 shows the runner surface pressure distribution under 50% load and 100% load. It can be seen from the figure that with the decrease of load, a high-pressure area is generated at the blade inlet, and the pressure difference between pressure surface and suction surface decreases.

The complex internal flow in the runner directly affects the force characteristics of the unit, especially the runner blade axial water thrust distribution. In this paper, the relative axial water thrust is used to analyze the axial water thrust under 50% load and 100% load.

The axial water thrust coefficient (F^*_z) is written as follows:

$$F^*_z=F_z/m_t g$$

Where, m_t is the weight of rotating parts (including runner and shafting) of pump turbine unit. In this paper, it is specified that the upward F^*_z is positive and the downward F^*_z is negative.

Fig. 6 (a) and Fig. 5 (b) show the F^*_z of the blade surface under 50% load and 100% load respectively. The absolute value of axial water thrust under 100% load is significantly higher than that under 50% load. This is because under partial load, a high pressure zone is generated at the inlet of runner blade, and the pressure difference between blade pressure surface and suction surface is reduced. The axial water thrust of the blade is reduced at 50% load. Moreover, the F^*_z of the runner changes periodically with time under the two loads. Compared with the time-domain diagram of axial water thrust, the axial water thrust of the blade fluctuates frequently under partial load.

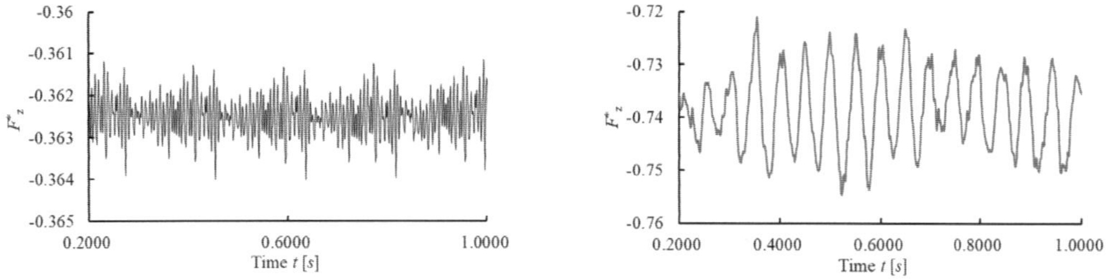

(a) 50% load (b) 100% load

Fig. 6 Axial water thrust of the blades

Fig. 7 (a) and Fig. 6 (b) are the time-varying curves of local F^*_z on the runner surface under 50% load and 100% load respectively. Under the two working conditions, the local F^*_z on the runner surface decreases gradually with time, and fluctuates in varying degrees. Compared with 100% load, under partial load, the unstable flow in the runner leads to the frequent fluctuation of the F^*_z curve of the runner with time.

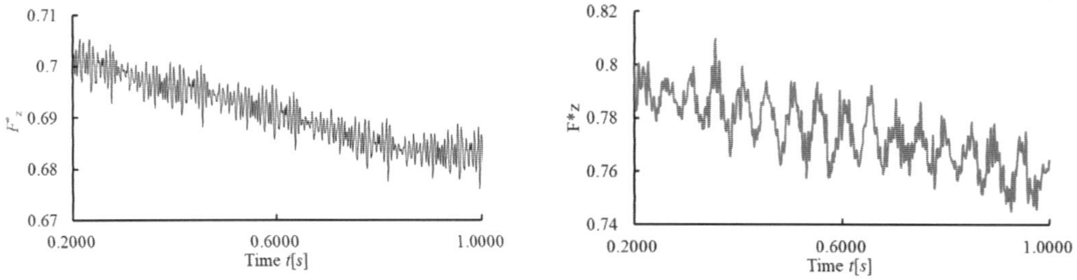

(a) 50% load (b) 100% load

Fig. 7 Axial water thrust of the runner

4. Conclusion

In this paper, the flow pattern and runner force of pump turbine model under 50% load and 100 load are analyzed by numerical simulation. Under 50% load condition, the opening of guide vane is low, the flow is small, the flow separation vortex structure appears on the high-pressure side of runner blade inlet, and the pressure difference between pressure surface and suction surface decreases. Under 100% load, the opening of the movable guide vane is large and the flow is large, and the overall flow field stability of the unit channel will be relatively increased. Compared with 100% load, under partial load, the axial water thrust F^*_z of blade decreases, the total runner axial water thrust decreases, and the variation curve of axial water thrust fluctuates frequently with time.

Acknowledgements
This work was supported by National Natural Science Foundation of China (51876099).

References

[1] ZUO Z G, LIU S H. Research on flow instability of China pump turbine. Engineering, 2017, 3(4): 162-178.
[2] Ji X Y, Li X B, Su W T, et al. On the hydraulic axial thrust of Francis hydro-turbine. Journal of Mechanical Science & Technology, 2016, 30(5): 2029-2035.
[3] Ran T, Xiao R, Liu W. Investigation of the flow characteristics in a main nuclear power plant pump with eccentric impeller. Nuclear Engineering & Design, 2018, 327: 70-81.
[4] Shi W, Wang C, Li Z. Calculation Formula Optimization and Effect of Ring Clearance on Axial Force of Multistage Pump. Mathematical Problems in Engineering, 2013: 206-226.
[5] Mao Z, Tao R, Chen F, Wang Z. Investigation of the Starting-Up Axial Hydraulic Force and Structure Characteristics of Pump Turbine in Pump Mode. Journal of Marine Science and Engineering, 2021, 9(2): 158.
[6] Wang W, Wang X, Wang Z, Ni M, Yang C. Analysis of Internal Flow Characteristics of a Startup Pump Turbine at the Lowest Head under No-Load Conditions. Journal of Marine Science and Engineering, 2021, 9(12): 1360.
[7] Liu S, Sun Y, Zuo Z, et al. Numerical simulation and analysis of pressure pulsation transfer characteristics of prototype water pump turbine [C]// The 19th China Hydropower Equipment Academic Symposium. Dalian, 2013. (in Chinese)
[8] Zhu B, Tan L, Wang X, et al. Investigation on flow characteristics of pump-turbine runners with large blade lean. Journal of Fluids Engineering, 2018, 140 (3): 031101.
[9] Bouajila S, Colombel T D, Lowys P, et al. Hydraulic phenomena frequency signature of Francis Turbines operating in part load conditions. IOP Conference Series: Earth and Environmental Science, 2016, 49: 082001.
[10] Valentín D, Presas A, Egusquiza E, et al. Power swing generated in francis turbines by part load and overload instabilities. Energies, 2017, 10 (12): 2124.
[11] LIU D, ZHAO Y, XU W. Experimental research on flow field of a high head model pump turbine based on PIV test. Chinese Journal of Mechanical Engineering, 2020, 40(2020): 1-9.
[12] Menter F.R. Two-equation eddy-viscosity turbulence models for engineering applications. AIAA J. 1994, 32, 1598-1605.

Simulation study on passive cavitation suppression for a typical hydrofoil

Jinlan Zhang[1], Yun Wang[1], Liang Dong[1], Jianwei Wu[1], Wei Wang[2], Zhijian Li[2,*]

[1] Wuhan Second Ship Design and Research Institute, Wuhan, China
[2] School of Energy and Power Engineering, Dalian University of Technology, Dalian, China

*Corresponding author e-mail: litiken@mail.dlut.edu.cn

Abstract. A passive cavitation suppression structure with flow barriers on its surface was proposed for a typical hydrofoil (NACA66 MOD). Based on the homogeneous equilibrium model and the Zwart-Gerber-Belamri cavitation model, the cavitation characteristics around the hydrofoil before and after flow barriers on hydrofoil surface at a typical operating condition are simulation investigated. Results show that the flow barriers provide on hydrofoil surface are capable of regulating and controlling the cavitation flow field, suppressing 47.1 percent of cavitation flow. By coupling the distribution of cavitation shape, re-entrant jet and pressure field of the hydrofoil with fluid flowing around, and analysing the periodic change of flow field, it is found that to a certain extent, the flow barrier provided on hydrofoil suction surface inhibited the development and shedding of cavitation.

1. Introduction

Cavitation refers to the process of formation, development, shedding and collapse of vapor or gas cavities within liquid or on the interface between liquid and solid when local pressure within liquid drops below the saturated vapor pressure. It is a unique liquid flow phenomenon in the field of flow dynamics [1]. Cavity rupture results in high local pressure, exacerbating the cavitation on an object surface and thus inflicting fatigue failure on the object. For the propellers of surface ships and underwater vehicles, cavitation occurred at blade tip and blade surface during their rotation caused cavitation on them, leading to ship body vibration and strong impulse noise and thus affecting ship's safety and tranquility. It may also influence the ability of an object moving fast to control its stability. Therefore, delay of cavitation formation and suppression of cavitation shedding have been a focus of, as well as a challenge to the research on hydraulic machinery. Researchers both at home and abroad performed a large number of scientific studies on the mechanism of cavitation development and the mechanism in which cavitation leads to unsteady state, in an attempt to propose new cavitation suppression techniques [2, 3, 4].

Cloud cavitation is a typical form of cavitation. Studies on the generation mechanism of cloud cavitation reveal that re-entrant jet is a major contributor to the destabilizing and shedding of cloud cavitation. Therefore, to a certain extent, it is possible to control cloud cavitation through re-entrant jet control. Passive control of re-entrant jet is achieved by changing the wall surface properties of an object, without providing energy into flow field, such as choosing a wall surface material with special properties;

Content from this work may be used under the terms of the Creative Commons Attribution 3.0 licence. Any further distribution of this work must maintain attribution to the author(s) and the title of the work, journal citation and DOI.
Published under licence by IOP Publishing Ltd

equipping regular or irregular barriers on object surface; adopting through-hole design on object surface, etc. In this paper, we propose a passive cavitation suppression structure with a flow barrier arranged on its surface for a typical hydrofoil (NACA66 MOD). Based on the homogeneous equilibrium model and the Zwart-Gerber-Belamri cavitation model, the cavitation characteristics around a hydrofoil before and after a flow barrier on hydrofoil surface at a typical operating condition are simulation investigated, the effect of the flow barriers on cavitation suppression is evaluated, and their suppression mechanism is analyzed.

2. Mathematical model

As an extremely complex flow phenomenon, cavitation involves phase transition between vapor and liquid, transport, shedding of turbulence vortexes of various sizes, compression of cavitation, collapse of cavitation bubbles, etc. Based on the Navier-Stokes equations, numerical simulation research on passive cavitation suppression of hydrofoil is carried out by employing the computational fluid dynamics software Ansys Fluent. The homogeneous equilibrium model is used to handle the interface between the gas and liquid phases of cavitation. The Zwart-Gerber-Belamri cavitation model is employed to describe the mass transport between gas and liquid phases. The RNG k - ε turbulence model is used to deal with the turbulence in flow field.

Energy, mass and momentum control equations used for the calculation on fluid can be expressed in the form of Navier-Stokes equations, as shown in equation (1).

$$\frac{\mathrm{d}}{\mathrm{d}t}\int_{V_S} \rho\phi dV + \int_{L_S} \rho\phi\mathbf{u}\cdot \mathrm{d}\mathbf{A} = \int_{L_S} \Gamma\nabla\phi\cdot \mathrm{d}\mathbf{A} + \int_{V_S} S_\phi \mathrm{d}V \tag{1}$$

In these equations, ϕ is a common variable, such as velocity or pressure; ρ is density; V_S represents the control volume and its boundary conditions; $\mathbf{A}$ is a standard vector of the boundary surface of control volume; $\mathbf{u}$ is velocity; Γ is diffusion coefficient; and S_ϕ is a source term with respect to a variable ϕ.

3. Analysis of cavitation flow of original hydrofoil

The research is performed on the typical NACA66 MOD hydrofoil, as shown in Figure 1, with wing length of 0.07m, span length of 0.07m, and the maximum thickness ratio of 12%.

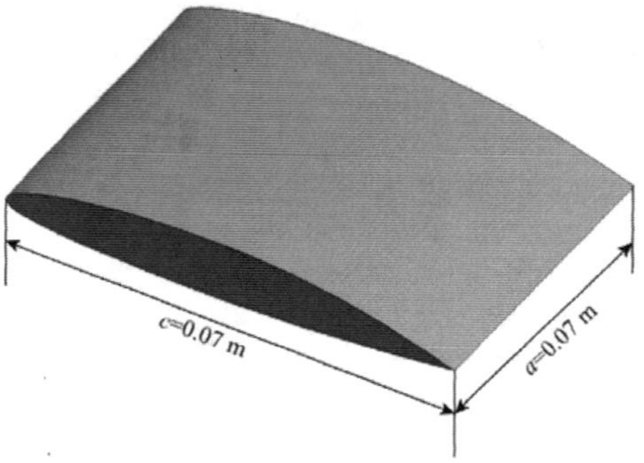

Figure 1. The typical NACA66 MOD hydrofoil.

The computation domain of the cavitation flow field around original hydrofoil is set as shown in Figure 2. The hydrofoil's attack angle is α=8°; the entrance and the exit are 4c and 11c away from hydrofoil front edge, respectively. Boundary conditions are set as velocity inlet and pressure outlet conditions, with no slip on wall surface.

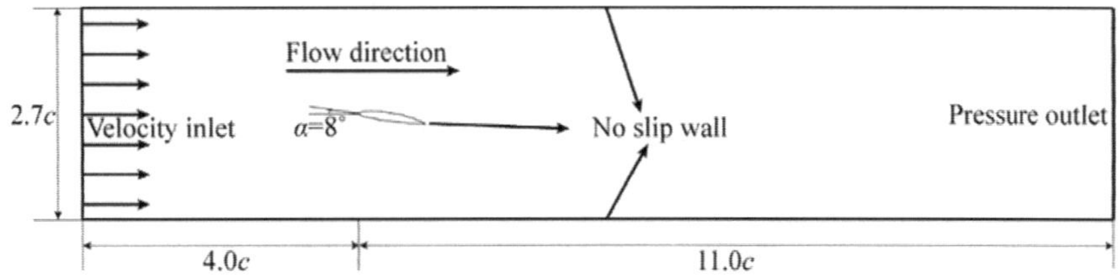

Figure 2. The computation domain.

We discretize the momentum equation and the continuity equation with a finite volume approach, and solve them with the help of a pressure-velocity coupling algorithm in SIMPLE format, with the first-order upwind scheme adopted for the calculation of convective term and PRESTO scheme for the pressure term, and the first-order implicit format used for discretizing unsteady terms. In calculation, time step is set at 0.1ms.

In the above numerical analysis of hydrofoil cavitation flow, the attack angle of the hydrofoil is $\alpha=8°$, main flow velocity $U_\infty=7.832$ m/s, and corresponding Reynolds number $Re=c*U_\infty/\upsilon=5.1\times10^5$. Where, c is the chord length of hydrofoil. Density and dynamic viscosity are $\rho_l=998.7$ kg/m3 and $u_l=1079.85\times10^{-6}$ Pa•s, respectively, for liquid phase; $\rho_v=0.01449$ kg/m^3 and $\mu_v=9.65\times10^{-6}$ Pa•s, respectively, for vapor phase. Corresponding temperature Is 17°C, and the saturated vapor pressure at this temperature is 1940 Pa.

Figure 3 shows over a typical period, the evolution process of cavities around the original hydrofoil and the pressure distribution in flow field at cavitation number $\sigma=0.83$ and vapor phase volume fraction $\alpha v=0.1$.

For the original hydrofoil, t_1-t_2 represents a period over which cavities grew, and they stay at 50%c-90%c for a long time after rapidly growing to 80%c. From Figure 3b, we knew that as cavities grew, a low pressure region almost covered the entire suction surface at t_2. Because of the existence of a high pressure region at hydrofoil tail edge, re-entrant jet formed on hydrofoil suction surface at t_2. Taking shape at the tail parts of cavities, the re-entrant jet flowed in the direction counter to main flow, and is of the same order of magnitude with the main flow. These two streams of flow met at 19%c from hydrofoil front edge, resulting in a local high-pressure region as shown in Figure 3c, where sheet cavities attached to the suction surface started to break at a position where speed stagnated, and evolved into cloud cavitation. After cloud cavities were completely separated from attached sheet cavities at t_4, secondary cavity growth and collapse happened at t_4-t_6 because of the existence of the low pressure region at tail edge.

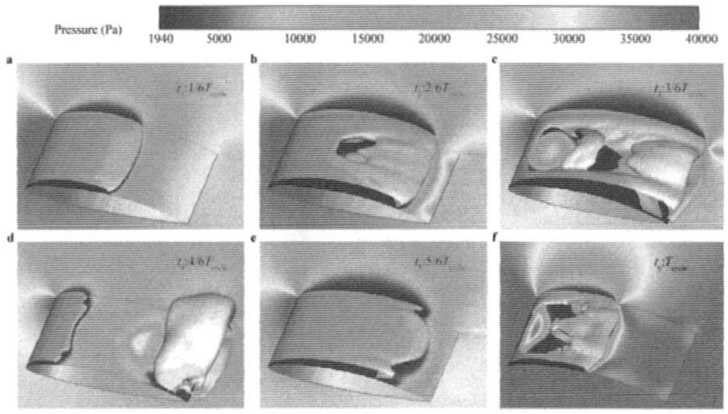

Figure 3. The evolution process of cavities around the original hydrofoil.

From the analysis performed with the help of Figure 4, it is found that cavity shedding and collapse are attributable to the re-entrant jet caused by the high pressure at hydrofoil tail edge. In Figure 4b and 4c, the re-entrant jet stopped at 19%c and 75%c, respectively. At corresponding attack angle, the flow field near the highest point of hydrofoil suction surface had the highest velocity, where flow separation and loss often occurred. In subsequent studies, we will block the re-entrant jet by providing flow barriers at 19%c and 75%c, to achieve the goal of suppressing cavitation.

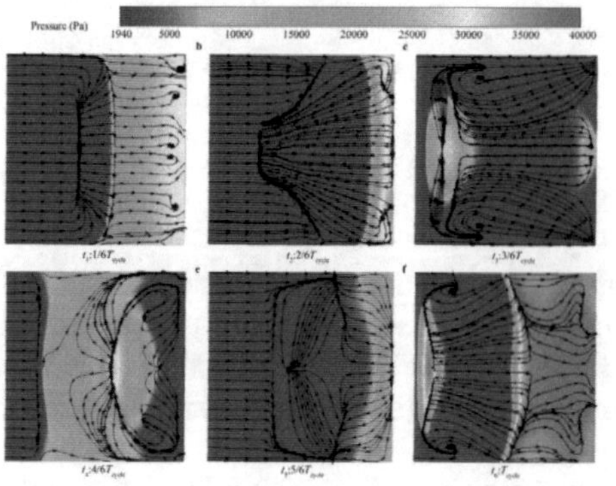

Figure 4. The evolution process of hydrofoil suction surface streamline of the original hydrofoil.

4. Study on passive cavitation suppression

Passive cavitation control is an approach in which the flow properties of near-wall fluid are controlled by changing the surface structure or material properties of an object. Compared with active control, passive control features simple structure and easy operation, without resorting to external power. By their action mechanism, passive control approaches are classified into two types. One of them acts on boundary layers, especially on the flow on the boundary layer which lies at the front edge of attached cavitation, including changing roughness, using a vortex generator, etc. The other works on re-entrant jet, including providing flow barriers on surface to control cavitation generation, development and shedding by stopping re-entrant jet development.

4.1. Design of flow barriers on hydrofoil surface

Figure 5 shows flow barriers provided on the surface of the original hydrofoil. Their height h0 and thickness l0 are designed to be 2%c, with their length sufficient to cover the hydrofoil in span direction. Symbols P19-2-2 and P75-2-2 ARE used to represent the situation that they are 19%c and 75%c away from hydrofoil front edge, respectively; where, P indicates a passive control technique.

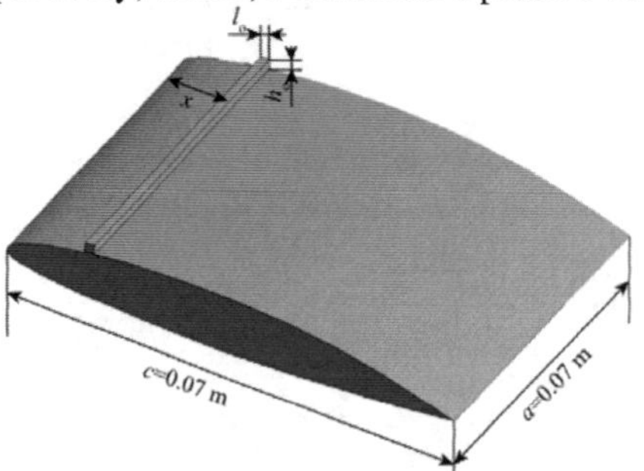

Figure 5. The flow barrier provided on the surface of the original hydrofoil.

When comparing the hydrofoil with flow barriers with the original hydrofoil, η is used to express cavitation suppression performance. η is defined by equations (2) and (3).

$$\eta = 1 - \frac{V_{\text{remodel}}}{V_{\text{original}}} \tag{2}$$

$$V_{\text{cav}} = \int_V \alpha_{\text{vapor}} \mathrm{d}V \tag{3}$$

Where, V is the volume of computation domain; α_{vapor} is the vapor phase volume fraction on hydrofoil surface; V_{original} is the cavity volume of the original hydrofoil; and V_{remodel} is the cavity volume of the remodeled hydrofoil.

4.2. Effects of flow barriers on the structure of cavitation flow field around hydrofoil

Figure 6 gives the evolution process of cavities in the cavitation flow field around P75-2-2 hydrofoil over a typical period. By comparing the hydrofoil with flow barriers with the original hydrofoil, we found that its flow barriers limited the growth speed of cavities. Comparison between Figure 17 and Figure 18 suggests that the flow barriers improved the pressure distribution in low-pressure region. Flow barriers' upstream pressure increased, and downstream pressure decreased when they were provided on hydrofoil suction surface. A shrank low-pressure region on hydrofoil suction surface resulted in slowed-down cavity growth and weakened cavity development. Flow barriers prevented the low-pressure region from expanding on suction surface, and at the same time, affected the development of re-entrant jet.

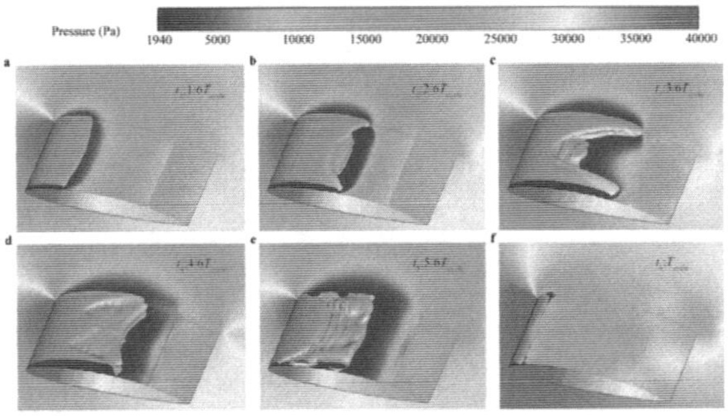

Figure 6. The evolution process of cavities in the cavitation flow field around P75-2-2 hydrofoil.

As a result of the flow barriers equipped on hydrofoil suction surface as shown in Figure 7, re-entrant jet was blocked when moving upstream, so that it can't reach cavity front edge. Because few side re-entrant jet existed over the period in which partial cavities vibrated, and re-entrant jet flowed in the direction perpendicular to flow barriers, these barriers achieved good obstruction performance. Cavities shed and collapsed in the form of sheet, so that large-scale overall cloud cavity shedding would not occur.

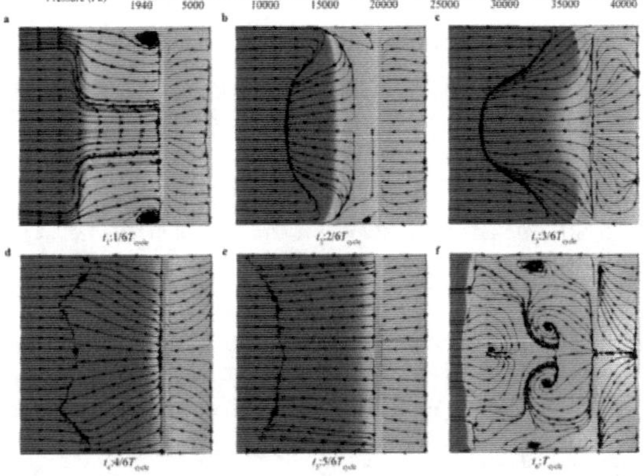

Figure 7. The evolution process of hydrofoil suction surface streamline of the P75-2-2 hydrofoil.

4.3. Effects of flow barriers on the structure of cavitation flow field around hydrofoil

Figure 8 is the spatial-temporal diagram of lift and drag change around different hydrofoils. In passive control, the position of flow barriers have a remarkable effect on cavitation flow. When they are arranged at 19%c, the stability of flow field deteriorated, and cavity shedding frequency f increased. In addition, the hydrodynamic performance of hydrofoil is greatly affected, although cavity volume decreased by 34.4%. In contrast, better cavitation suppression performance is obtained, with suppression efficiency η reaching 47.1%, and cavity shedding frequency decreases by 34.7% when they are arranged at 75%c, as shown in Table 1. Though the flow barriers reduce the lift coefficient and the drag coefficient of hydrofoil, the ratio of lift to drag increase slightly. For the hydrofoil with flow barriers, its time-averaged ratio of lift to drag rose by 13.2%, indicating the arrangement of flow barriers at 0.75c on hydrofoil suction surface improved the hydrodynamic performance of hydrofoil.

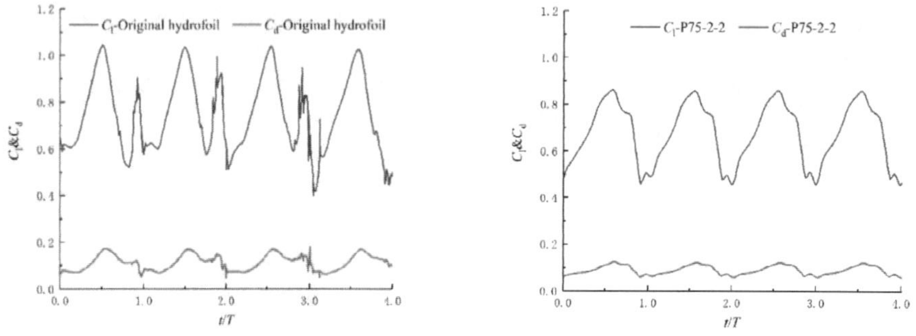

Original hydrofoil P75-2-2

Figure 8. The lift coefficient and the drag coefficient of hydrofoils.

Table 1. The influence on cavitation and hydrodynamic performance of hydrofoils.

	$V_{cav}/10^{-6}$ m^3	η	f	C_l	C_d	C_l/C_d
Original hydrofoil	6.9194	-	16.09	0.7320	0.1100	6.6545
P75-2-2	3.6638	0.471	10.51	0.6764	0.0898	7.5323
P19-2-2	4.5409	0.344	20.07	0.4945	0.1703	2.9034

5. Study on passive cavitation suppression

To analyze the mechanism of cavitation suppression by flow barriers, we selected the central section in the span direction of hydrofoil, coupled the distribution of cavitation shape, re-entrant jet and pressure field of a hydrofoil with fluid flowing around, and analyzed the periodic change of its flow field, as shown in Figure 9.

In Figure 9b, the hydrofoil with flow barriers had a much lower pressure gradient at its front edge than the original hydrofoil, and thus cavities grew at a low speed. Over the period of t_0-t_2, cavities grew, and the adverse pressure gradient in the middle of hydrofoil increased at t_2, leading to the presence of re-entrant jet. As revealed in Figure 9b, despite the persistent existence of re-entrant jet at the tail edge behind flow barriers due to turbulence, flow barriers stopped it from marching towards hydrofoil front edge, and thus it can't exert a strong impact on cavitation vibration or shedding in the early stage of cavitation formation and development. Over the period of t_2-t_3, as a result of cavity retraction and re-entrant jet growth caused by the adverse pressure gradient at cavity tail edge at t_2, the re-entrant jet expanded upstream while the sheet cavitation attached to hydrofoil continued to grow. However, at this moment, the re-entrant jet extended at a much lower speed than the growth speed of cavities, so that it can't reach cavity front edge to "pinch off" cavities, and was separated from attached cavities at their tail edges to form small shed cavities. The reduction of low-pressure region resulted in decreased pressure gradient in the middle of hydrofoil at t_3, and secondary cavitation growth at t_3-t_4. When cavity tail edge approached the flow barriers, adverse pressure gradient at the front part of these barriers intensified the re-entrant jet, which "pinched off" cavity tail edge and led to secondary cavity shedding. Then, the high-pressure region moved quickly to hydrofoil front edge, resulting in the situation that attached cavities collapsed and a new cavitation evolution period began.

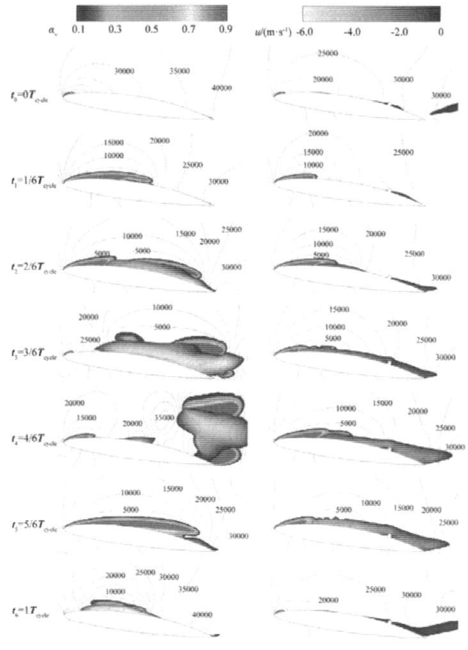

Figure 9. The cavitation forms around the hydrofoils.

From above analysis, to some extent, cavitation development and shedding can be suppressed by providing jet or flow barriers on the suction surface of hydrofoil. To understand the cavitation suppression mechanism, the evolution of cavitation shape, boundary layer speed, re-entrant jet, and the pressure fluctuation on suction surface are analyzed.

By analyzing the spatial-temporal distribution of re-entrant jet and boundary layer speed of the original hydrofoil and P75-2-2 hydrofoil, we found that for the original hydrofoil, the re-entrant jet on suction surface developed fast, and arrived at a position which was close to hydrofoil front edge, leading to heavy shedding of cavities attached to the suction surface. However, the development of re-entrant jet was greatly affected after flow barriers were provided at 75%c from front edge, as shown in Figure 10.

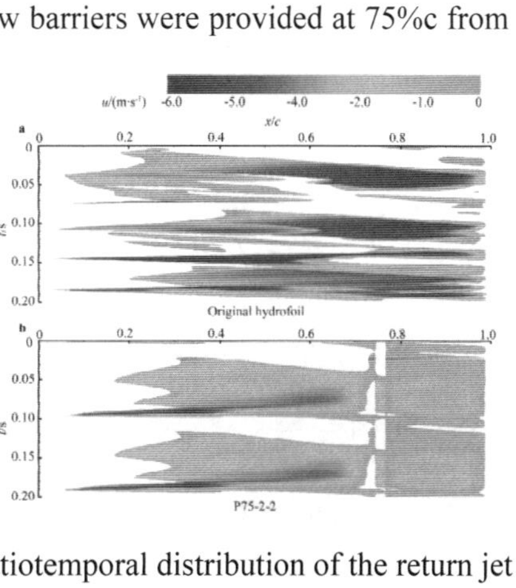

Figure 10. The spatiotemporal distribution of the return jet of the hydrofoils.

Over a period of cavitation development, the re-entrant jet grew in two stages. At the first stage, there was no re-entrant jet over flow barriers, and at this time, they split re-entrant jet into upstream and downstream parts. Since the re-entrant jet had not crossed flow barriers, its intensity was much lower than that of the suction surface of the original hydrofoil, and thus cavity development progressed slowly. At the second stage, the re-entrant jet crossed flow barriers in an extremely short period of time, and a re-entrant jet with great intensity appeared at hydrofoil front edge, leading to the shedding of cavities at front edge. Analysis of the spatial-temporal distribution of re-entrant jet shows that compared to the original hydrofoil, the P75-2-2 hydrofoil had a lower re-entrant jet intensity, and its maximum re-entrant jet rate fell from 11.3 m/s to 6.5 m/s, with a decrease of 42.5%.

Figure 11 shows the time-averaged boundary layer and time-averaged vapor-phase volume fraction distribution on the central sections of the original hydrofoil and P75-2-2 hydrofoil, with the region enveloped by velocity contour 7.832 m/s representing the time-averaged boundary layer, and the region enveloped by velocity contour 0 m/s indicating the time-averaged re-entrant jet area. In the figure, an area with time-averaged vapor-phase volume fraction αv greater than 0.1 is used to represent the time-averaged cavitation region. The P75-2-2 hydrofoil had a much smaller cavitation region and a thinner boundary layer than the original hydrofoil. The re-entrant jet region expanded slightly after flow barriers were provided at tail edge. However, because the expanded part mainly sat behind flow barriers, it didn't leave a strong impact on the separation of boundary layer. To compare the effects of flow barriers on the separation of flow boundary layer around hydrofoil, approximate calculation of the separation point of boundary layer from hydrofoil suction surface was carried out by employing the criterion of two-dimensional steady boundary layer separation, namely Prandtl separation criterion. Point with (du/dy) y=0=0 on suction surface served as boundary layer separation point. Numerical calculation statistics suggest that for both hydrofoils, time-averaged boundary separation occurred at 9.6%c from hydrofoil front edge, which means the expansion of re-entrant jet region would not produce adverse effect on the boundary layer separation of flow around a hydrofoil.

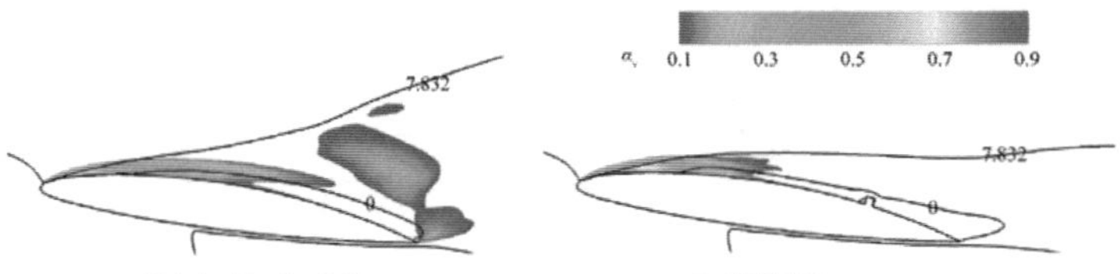

Figure 11. The time-averaged boundary layers of the hydrofoils.

To further analyze the flow behavior within boundary layer, boundary layer velocities at different chord lengths were extracted and compared, and results were provided in Figure 12. The x-axis represents the ratio of local velocity to incoming flow velocity, and the y-axis indicates the ratio in percentage of the distance between local position and hydrofoil suction surface to hydrofoil chord length. As shown in Figure 12, the thickness of boundary layer decreased greatly due to the presence of flow barriers. In other words, the velocity gradient near wall increased, which means enhanced momentum of liquid and its capability to resist adverse pressure gradient. The P75-2-2 hydrofoil had a slightly greater re-entrant jet velocity than the original hydrofoil at x/c=0.3 and x/c=0.9 alone, which, however, didn't result in boundary layer enlargement caused by cavity shedding, and thus didn't lead to overall deterioration of flow field. At x/c=0.9, a low-pressure region formed at the back edge of flow barriers because of the turbulence they caused, resulting in re-entrant jet here. As a result, the re-entrant jet thickness of P75-2-2 hydrofoil at this position increased by 34.3%, compared with that of the original hydrofoil. At x/c=0.7 and x/c=0.8, flow barriers rapidly weakened the re-entrant jet of P75-2-2 hydrofoil. Compared to the original hydrofoil, P75-2-2 hydrofoil's re-entrant jet thickness dropped by 23.6% at x/c=0.7, with a decrease of 67.5% in the maximum re-entrant jet velocity.

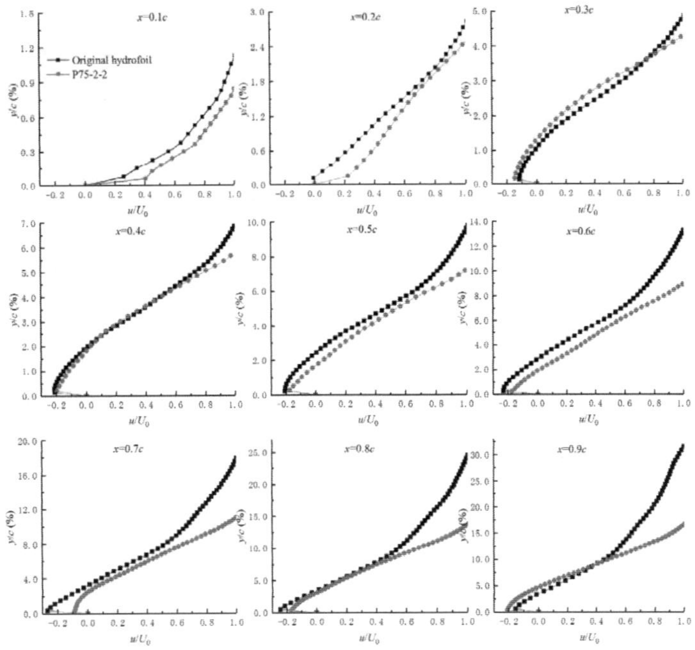

Figure 12. The time-averaged boundary layer velocities of the hydrofoils.

In order to compare the pressure change on suction surfaces of the original hydrofoil and P75-2-2 hydrofoil, the pressure changes at 0.19c, 0.30c, 0.45c, 0.60c, 0.75c and 0.90c of central section in different hydrofoil span directions were monitored, and transient pressure changes were presented in Figure 13.

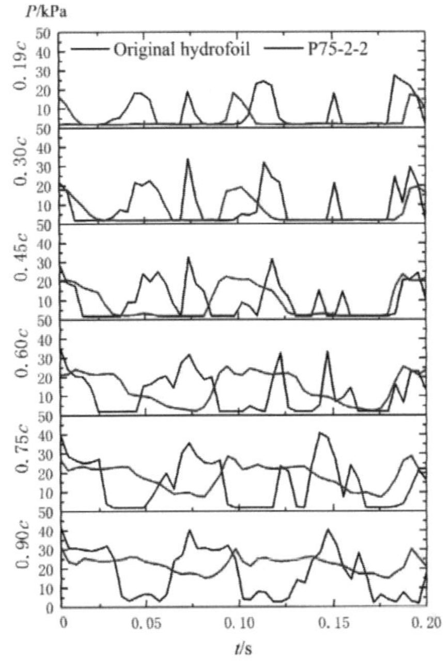

Figure 13. The pressure change curves of the hydrofoil suction surface.

At 0.19c from hydrofoil front edge, the low-pressure region on the suction surface of original hydrofoil existed for a short period of time, which was attributable to a short cavitation development period, frequent cavity shedding, and easy arrival of re-entrant jet at hydrofoil front edge. From the pressure-time domain diagram of the hydrofoil with flow barriers, we can see that compared with the original hydrofoil, its time needed for high-pressure pulsation to arrive at front edge and the frequency of such arrival declined significantly. As shown in Figure 13, on the original hydrofoil there remained a low-pressure region at 0.75c which stayed for a long time, and attached cavities were able to grow steadily to this position. For the P75-2-2 hydrofoil, the continuous low-pressure region just existed before 0.45c from hydrofoil front edge, which means flow barriers effectively reduced the growth length of attached cavities. Behind 0.45c from hydrofoil frond edge, stable high pressure lingered on the suction surface of P75-2-2 hydrofoil for a significantly longer period than on the original hydrofoil, indicating that flow barriers can control the flow on suction surface, and hence, changed the pressure distribution on the surface and suppressed cavity growth. In addition, in spite of its long high pressure duration on suction surface, the P75-2-2 hydrofoil had lower maximum pressure and pressure pulsation frequency than the original hydrofoil, with its pressure pulsation amplitude dropping by 24.6%~42.6%, compared with the original hydrofoil. The reduction in pressure pulsation amplitude and peak led to reduced adverse pressure gradient and weakened re-entrant jet intensity on hydrofoil suction surface, which enabled cavity shedding to be inhibited.

6. Conclusion

Based on passive control techniques, controlling cavitation flow by providing flow barriers on the surface of hydrofoil is proposed. Research results reveal that compared to the original hydrofoil, passive jet achieved better cavitation suppression performance at same cloud cavitation conditions. Flow barriers enable a cavitation suppression performance of 47.1%, with an increase of 13.2% in hydrodynamic performance when they were arranged at 75%c from hydrofoil front edge. But, despite a suppression performance of 34.4%, they caused a decrease of 56.4% in hydrodynamic performance when being provided at 19%c from hydrofoil front edge, which prevents hydrofoil from operating normally.

References

[1] Sensen Pan, Xiaoxing Peng, Physical Mechannism of Cavitation, Beijing, 2012.
[2] Wei Wu, Ying Xiong, Wanjinag Qi, Cavitation Control of a 2-D Hydrofoil Under Section Reshaping, Chinese Journal of Ship Research. 3 (2012) 36-40.
[3] Wei Wang, Tao Tang, Shengpeng Lu, Jianxiong Jiao, Qingdian Zhang, Xiaofang Wang, Investigation of cavitation suppression by arranging pits on hydrofoil suction side Transactions of the Chinese Society of Agricultural Engineering, 2 (2019) 40-47.
[4] Wei Wu, Ying Xiong, A Reshaping Method for Anti-Cavitating Hydrofoil Design, Chinese Journal of Shanghai Jiao Tong University. 6 (2013) 878-883.

An investigation of flow characteristics between different mediums

Zihan Wang [1, a, †], **Yulong Yi** [2, b, †] **and Siyuan Yin** [3, *, †]

[1] The Department of Physics, University of Toronto, Toronto, Ontario, M5S 1A7, Canada
[2] Department of Engineering, Rensselaer Polytechnic Institute, New York, New York, 12180, United States
[3] Pittsburgh Institute, Sichuan University, Chengdu, Sichuan, 610000, China

[a] rayzihan.wang@mail.utoronto.ca, [b] yiy4@rpi.edu,
*Corresponding author's e-mail: 2018141522009@stu.scu.edu.cn

† These authors contributed equally.

Abstract. In this research topic, two-dimensional tunnels made of different mediums (including the contrast of gas-solid condition and solid-solid condition and the contrast of solid-solid condition with or without a bump) are designed to compare how they influence the velocity and the pressure of the fluid when the fluid is flowing through them. The Navier-Stokes momentum equation is used to guide the project and set up the equations of motion for these models. At the same time, Newton's law of viscosity is applied to set the boundary conditions. Numerical methods involving finite differences are used to help discretize those equations and simulate the models in computers. The result shows that the velocity and the pressure of the fluid are influenced significantly by different tunnels settings. A nozzle with or without a bump can also significantly affect the velocity and pressure of the fluid. As the materials of tunnels change, the dependence of these flow characteristics in terms of x and y-direction will change, and the shape of the tunnel will also influence these traits.

1. Introduction

Generally, nozzle flow is a well-studied topic. As is well known, nozzle flow means a length of tube that causes a fall of pressure head in a liquid flowing through it from which the rate can be calculated. In addition, predecessors have discussed aspects such as how the shape of the nozzle affects the shock wave of a transonic gas [1]; the acoustics of a flow affected by a chevron nozzle [2]; how to diameter of the nozzle can affect the fuel combustion, and emission, etc. [3]. Hence, to discover something new, the project to conduct simulations to compare the fluid characteristics between nozzle flow and fluid in a normal tunnel is attempted. Meanwhile, the mediums on each side of the tunnel are changed to enrich comparisons.

The characteristics of Nozzle flow, like velocity or viscosity, can be well demonstrated in the diesel injector. The team of S Som used FLUENT V6.2 software to model the diesel injector and simulated the mixing model. Then, they verified the results by comparing measurements of two-phase nozzle flows

Content from this work may be used under the terms of the Creative Commons Attribution 3.0 licence. Any further distribution of this work must maintain attribution to the author(s) and the title of the work, journal citation and DOI.
Published under licence by IOP Publishing Ltd

at different pressures, and particularly their injection rates. The experimental results show that the practical operation of the diesel engine is closely related to the cavitation mode of the diesel injector. In addition, they also mentioned a new cavitation criterion [4].

In addition, critical nozzles have been widely used in many kinds of industries in recent years. As we all know, the critical nozzle means the critical velocity nozzle. It is a convergent nozzle installed between two flanges. It has convergent end points so that it is generally used to control the flow of downstream fluid. When the diameter of the critical nozzle differs, and the mass flow rate that can be measured is also different. The research of the team of JK Kim aimed to work on the change of mass flow rate of the critical nozzle by changing the effective cross-sectional area of the nozzle [5]. Their research uses a large number of two-dimensional compressible Navier-Stokes equations to measure the required experimental data, which is very helpful. After learning their use of the Navier-Stokes equations, this topic is gradually completed and refined.

To sum up, based on the previous research, this paper will analyse the change of fluid velocity and pressure under the guidance of the Navier-Stokes momentum equation. In this project, three scenarios are set to compare how different circumstances will influence these two features of the fluid. Meanwhile, Newton's law of viscosity in the isotropic case is used to set boundary conditions. Furthermore, to study and investigate the change of those two traits when analytic solutions are difficult to get, a numerical method that involves finite difference approximation is used to generate the simulation in a computer [6]. Details of these settings, the boundary conditions and the numerical methods will be illustrated in later sections.

2. Method

2.1. Principle and Physical Models

The first thing to conduct the research is to make the continuum assumption. It is clear that those fluids are made of small particles. Suppose the project wants to ignore the discrete nature of these materials. In that case, it is necessary to assume that under the control volume ΔV, the density of the fluid is not fluctuating, which makes the fluid continuous. Therefore, there is no need to involve statistical mechanics, and all equations listed below are valid. Fortunately, in the macroscopic world, the total number of particles in ΔV will not vary significantly, and the assumption is valid in this project [7, 8].

In this project, the change of velocity and pressure of the fluid is the major subject that is being discovered. Three two-dimensional scenarios are set to compare how different circumstances will influence these two features of the fluid.

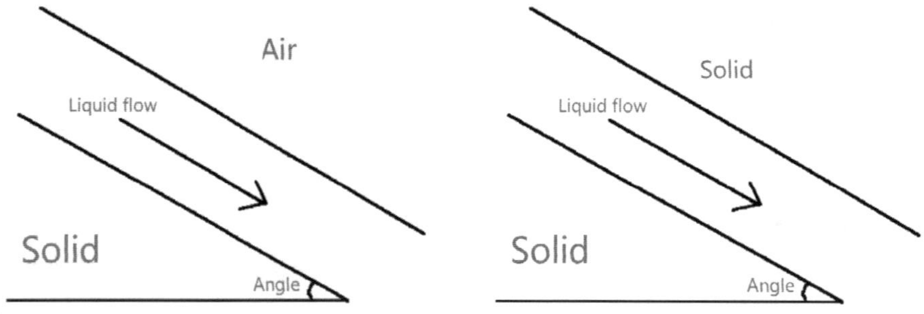

a) b)

Figure 1. (a) Schematic of the first scenario with fluid between different mediums. (b) Schematic of the second scenario when the fluid flows between solids.

In the first two settings, investigations are conducted to compare the change of flow characteristics when the fluid is flowing between different mediums. In the first one, it flows between the solid and the air, and in the second circumstance, both sides of the fluid interact with solids, and it looks like a tunnel. In both conditions, two interfaces between fluid and the mediums are parallel. The flow characteristics will be different as the way fluids interact with different mediums differ. All factors are excluded to make the simulation under control except the gravitational force, which means that the fluid is steady and laminar [9]. Since gravity is the only factor that can accelerate the flow, and to calculate all possible effects of it, the angle β between solids and the horizontal line should be considered. The restriction of the angle is that: $0 < \beta < \frac{\pi}{2}$.

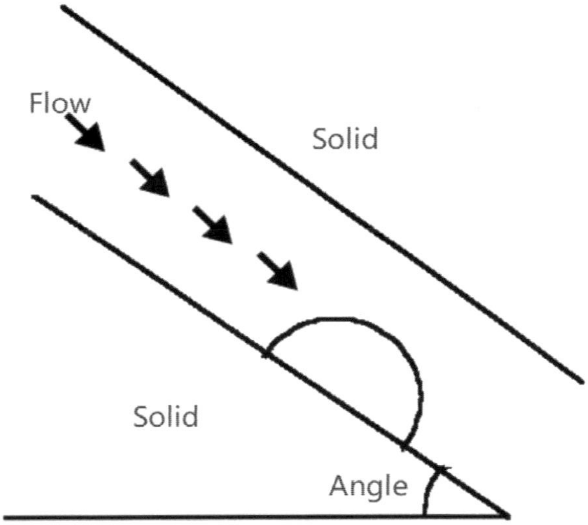

Figure 2. Schematic of the third setting, which makes it a comparison set to the second one, and it is an analogy of the nozzle flow.

The third circumstance is set to compare with the second one. As the fluid is flowing in a similar way between two solids, its difference compared to the second setting is that a "bump" is introduced near the end of the tunnel. This bump is used to simulate the narrowing of the tunnel that occurred in the normal model of a nozzle flow, and it will significantly change the velocity and the pressure of the fluid.

Clearly, the density of air will fluctuate in terms of pressure and altitude, and the density of the liquid will vary when the type of it is changed. So, it is assumed that the temperature is between 20 to 25 degrees Celsius, and therefore, the density of air will not fluctuate significantly. Also, it is assumed that the type of the liquid is water, which makes the density 1000 kg/m^3. Also, viscosity is related to temperature, so the viscosity of water at 20 degrees Celsius can be applied here.

2.2. Computation and Numerical Method

In fluid dynamics, one of the most dominant equations is the Navier-Stokes momentum equation. It is widely used to describe the motion of liquid and gas, and it also takes an important role in fields like aerodynamics, atmospheric physics and automobile engineering. Thus, it will play an important part in this topic and guide the research process. The general form of it in the Lagrangian frame in index notation is that [10, 11]:

$$\rho \frac{Du_j}{Dt} = -\frac{\partial p}{\partial x_j} + \rho g_j + \mu \frac{\partial^2 u_j}{\partial x_i{}^2} + (\mu_v + \frac{1}{3}\mu) \frac{\partial}{\partial x_j} \frac{\partial u_m}{\partial x_m} \quad (1)$$

Where $\frac{D}{Dt}$ is the material derivative; μ is the fluid's viscosity, and μ_v is the coefficient of bulk viscosity, and it is related to μ.

Also, in this model, the assumption that the fluid is incompressible is made before conducting the investigation, which means that the density of the flow is independent of pressure, and therefore, $\nabla \cdot u = \frac{\partial u_m}{\partial x_m} = 0$. The assumption is made because the speed of the flow in this model is low, and it is known that when the flow speed is below 300 miles per hour, the flow can be approximated as incompressible [11]. Hence, it changes the general form of the N-S momentum equation to [12]:

$$\rho \frac{Du}{Dt} = -\nabla p + \rho g + \mu \nabla^2 u \tag{2}$$

where the bold u and g are vector functions. This equation will lead to differential equations that describe the motions of fluids.

Meanwhile, Newton's law of viscosity in the isotropic case is used to set boundary conditions [13]:

$$\tau_{ij} = -p\delta_{ij} + \mu\left(\frac{\partial u_i}{\partial x_j} + \frac{\partial u_j}{\partial u_i}\right) \tag{3}$$

where the index notation is used here to state the different components of the velocity vector; δ_{ij} is the Kronecker delta, and τ is the viscous stress tensor of the fluid. The use of this equation will be explained later when setting the boundary conditions.

Before introducing the simulation, it is necessary to give differential equations that describe the motion of the fluids. To make the equation and simulation neat, the coordinate system is changed to adapt to the problem. The x-axis is parallel to those two interfaces, and the y-direction is perpendicular to the. The fluid depth is set to be h, and since it is parallel to the y-axis, its value varies from 0 to h along the axis. Thus, the gravitational force in these two directions become $g \sin \beta$ and $-g \cos \beta$.

The equations of motions that describe the models should be deducted from the Navier-Stokes equation, and boundary conditions must be found. Since the fluid is steady and in the Lagrangian frame, the assumption $\frac{Du}{Dt} = 0$ could be made for simplicity. This will lead the two-dimensional Navier-Stokes equation to two parts:

x-direction

$$0 = -\frac{\partial p}{\partial x} + \rho g \sin \beta + \mu \nabla^2 u \tag{4}$$

y-direction

$$0 = -\frac{\partial p}{\partial y} + \rho g \sin \beta + \mu \nabla^2 u \tag{5}$$

to further simplify these two equations, boundary conditions must be set.

At the fluid-solid interface, because there is no emission from it, it is easy to get:

$$\mathrm{u} \bullet \mathrm{n} = \mathrm{b} \bullet \mathrm{n} \tag{6}$$

for n is the normal of the surface; u and b are the velocity of the fluid and the solid, respectively.

Therefore, since the solid is not moving, b = 0, which makes the fluid's velocity 0 at this interface. Then, it is obvious that $U_y = 0$ throughout the pipe and $Ux = 0$ at $y = 0$.

At the fluid-air interface, because normally, the fluid's density and viscosity are much greater than these two values of the air, the free interface will not easily be deformed. Thus, manipulation of equation (3) will get:

$$\frac{\partial u_{x@fluid}}{\partial y} = \frac{\partial u_{x@air}}{\partial y} \approx 0 \tag{7}$$

and meanwhile, the boundary condition for pressure is that $P = P_{atm}$ since it is a free surface.

Now, with equation (7) and the boundary condition $Ux = 0$ at $y = 0$ and $U_y = 0$, Navier-Stokes equation in those two directions become:

x-direction

$$\frac{\partial^2 u_x}{\partial y^2} = \frac{1}{\mu}(\frac{\partial p}{\partial x} - \rho g \sin\beta) \tag{8}$$

y-direction

$$\frac{\partial p}{\partial y} = pg \sin\beta \tag{9}$$

these two equations will be used repeatedly in the second and the third setting.

In the second setting, the boundary conditions for the bottom interface are the same. For the top one, since it is also a fluid-solid interface, it is not reasonable to assume that $P = P_{atm}$ and similarly get equation (7). Therefore, it will resemble the boundary conditions for the bottom interface, which gives that $Ux = 0$ at $y = h$. At the same time, because the fluid is laminar, naturally at the central layer of the fluid, $\frac{\partial u_x}{\partial y} = 0$, where $y = \frac{h}{2}$. These boundary conditions will be used again in the third setting, but a new restriction is added since the end of the tunnel in the third setting is narrowed in the beginning and widened again. It is reasonable to consider a periodic change of the fluid velocity occurs at that point, which gives a boundary condition $dU_x = \sin(\pi x)$.

Once the equations of motion and boundary conditions are set, the numerical method solves them numerically in computers. The finite difference method and second-order central difference scheme are immediately recalled to deal with differential equations like equations (8) and (9) [14]:

$$\frac{\partial^2 U_x}{\partial y^2} = \frac{U_{i,j+1} - 2U_{i,j} + U_{i,j-1}}{\Delta y^2} \tag{10}$$

$$\frac{\partial p}{\partial y} = \frac{P_{i,j+1} - P_{i,j-1}}{\Delta y} \tag{11}$$

when substituting these two equations into equations (8) and (9), discretized formulas are generated, and they can work in computers with boundary conditions set above.

However, this simple discretization method is restricted because it is difficult to solve numerically $P_{i,j+1}$ and $U_{i,j+1}$ simultaneously [15]. To simplify the problem, a possible way is to simulate equation (9) first due to its simpler PDEs. Once this equation is solved and, therefore, more information about the pressure is known, equation (8) will be further restricted, and the computer can simulate it.

3. Results and Discussion

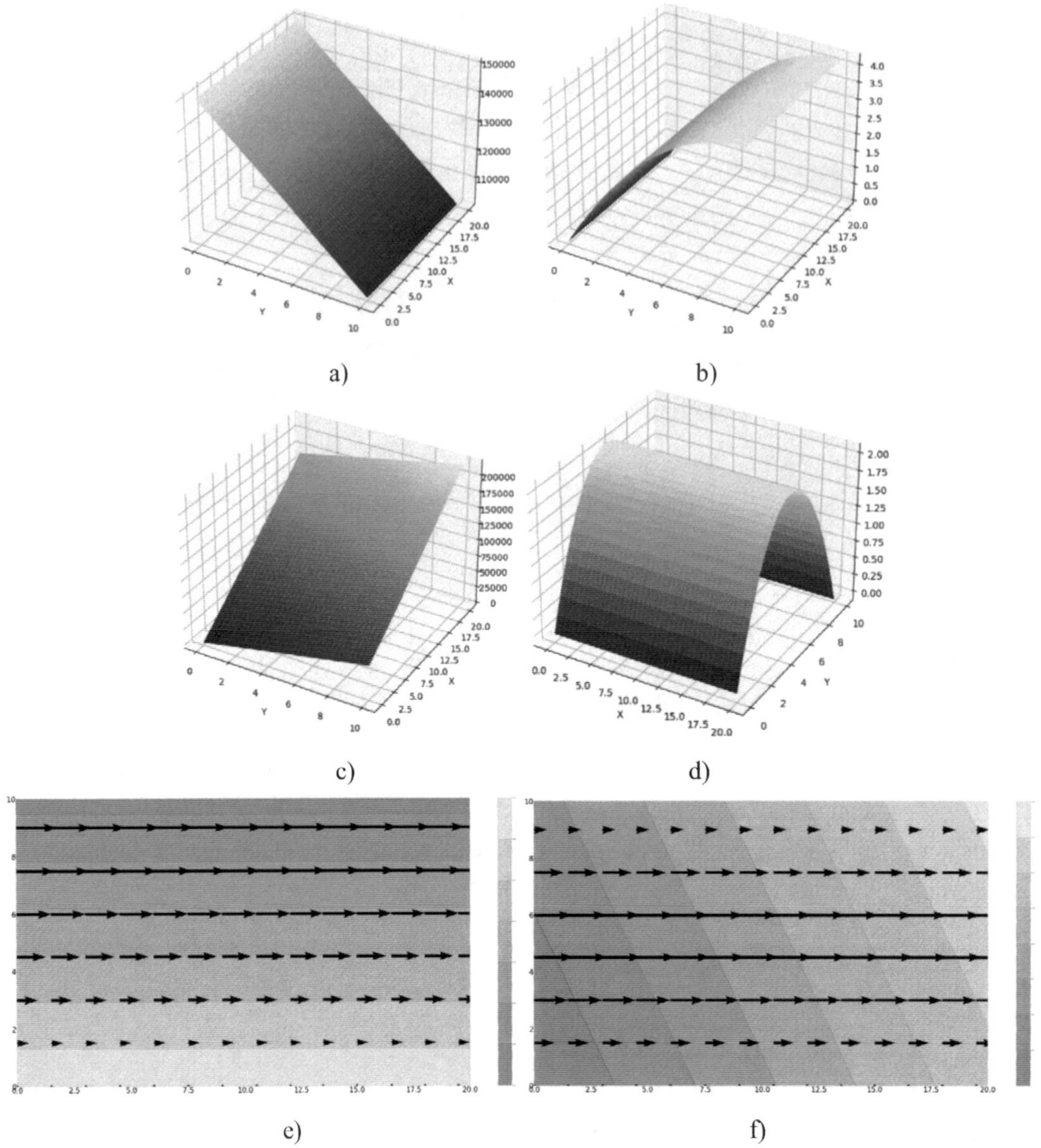

Figure 3. Two characteristics of the fluid in the first two sets, where $\beta = \frac{\pi}{3}$, and in the numerical setting, the lengths in x and y-directions are set to be 20 units and 10 units, respectively. (a)The change of pressure in the first setting. (b)The change of velocity in the first setting. (c)The change of pressure in the second setting. (d)The change of velocity in the second setting. (e)contour plot of the first, where the background colors represent the pressure gradients and the arrows represent the velocity fields. (f)It shows the pressure gradients and the velocity fields in the second setting.

In the first two sets, changes of these two characteristics are compared. In the first scenario, the pressure is only dependent on y-direction, and it is obvious that the pressure is at its smallest value at the liquid-air interface, which is close to instincts based on hydrostatic equilibrium [16]. Also, it is shown that the pressure varies linearly, which verifies the observation from the first-order PDE equation

(9). For velocity, due to the boundary condition set at the liquid-solid interface, the x-component of this vector is also a constant. Its shape is a half parabola, which means that the velocity increases as it approaches the surface. From the contour Figure 3 (e), it is clearly shown that this is a laminar flow.

These two traits have different appearances in the second set. Due to the boundary condition changed by the upper liquid-solid interfaces, the pressure is no longer affected by the atmospheric pressure. Therefore, the pressure looks like a slanted plane in the coordinate system since it also depends on x-direction under this circumstance. Meanwhile, the velocity reaches its greatest value at the middle of the fluid layer. Then it decelerates as it touches another fluid-solid interface, which makes its shape of a complete parabola. The contour clearly shows that the pressure gradient varies in both directions, and the velocity field has the highest density in the middle.

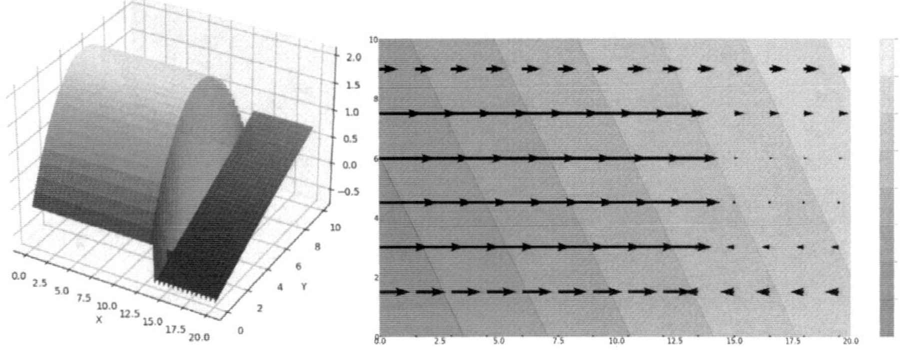

a) b)

Figure 4. (a) The velocity change when the "bump" occurs in the third setting. (b) The contour plot of the third setting, and it is obvious to see the difference between it and the contour of the second set.

The third situation compares with the second set, where all boundary conditions are the same except a bump that occurs at the last third of the tunnel. Due to the emergence of this bump, the velocity is changed significantly. When the flow hits the bump, a backflow happens at the bottom, and at the same time, the middle part of the flow becomes static, and the top layer is not affected a lot. This could happen due to a sudden collision between the flow and the bump, which changes the characteristics of the flow. Nevertheless, it somehow resembles the velocity field of a normal nozzle flow [17].

4. Conclusion

Based on previous research, we analyze the contrast of gas-solid condition and solid-solid condition and the contrast of solid-solid by using the Navier-Stokes momentum equation Condition with or without a bump, numerical methods are used to study the velocity and other characteristics of the nozzle fluid in different cases, and the main conclusions are as follows:

(1) By comparing the solid-solid condition with the solid-gas condition, we conclude that the pressure is minimal at the liquid-gas interface, which is close to the instinct based on hydrostatic equilibrium. The pressure in the Nozzle changes linearly, which verifies the observation result of the first-order PDE equation. The shape of the velocity at the liquid-solid interface is a semi-parabola, meaning that the velocity increases as it approaches the surface. In this case, the flow of the nozzle is laminar.

(2) The second comparison is whether the Nozzle has a bump. Since pressure is no longer affected by atmospheric pressure, pressure, in this case, depends not only on the y direction but also on the X direction. The velocity reaches its maximum in the middle of the fluid layer and then slows down when it touches another fluid-solid interface, forming a complete parabolic shape. The velocity field is the densest in the middle.

References

[1] Kluwick, A. (1993) Transonic nozzle flow of dense gases. Journal of Fluid Mechanics., 247: 661–688.

[2] Callender, B., Gutmark, E., Martens, S. (2008) Near-Field investigation of chevron nozzle mechanisms. AIAA Journal., 46: 556–563.

[3] Bergstrand, P., Denbratt, I. (2001) Diesel combustion with reduced nozzle orifice diameter. International spring fuels and lubricants meeting., 78: 678–683.

[4] Som, S., Aggarwal, S. K., El-hannouny, E. M., Longman, D.E. (2010) Investigation of nozzle flow and cavitation characteristics in a diesel injector. Journal of Engineering for Gas Turbines and Power., 132: 357–365.

[5] Kim, J.H., Kim, H.D., Park, K.A. (2006) Computational/experimental study of a variable critical nozzle flow. Flow Measurement and Instrumentation., 17: 81–86.

[6] Cavanagh, P.R., Rodgers, M.M., Liboshi, A. (1987) Pressure Distribution under Symptom–Free Feet during Barefoot Standing. Foot & Ankle International., 7: 262–278.

[7] Goyal, K., Goyal, M., Samuel, A.J. (2020) Core stabilization exercises impact on balance and foot pressure distribution in overweight children: A randomized controlled trial protocol. Revista Pesquisa em Fisioterapia., 10: 724–736.

[8] Sun, X.Y., Yuan, Z.H., Yuan, Y.C. (2020) Numerical simulation of wind pressure distribution outside cooling tower and coefficient analysis of the regression curve. Journal of Physics Conference Series., 1676: 012234.

[9] Picoche, J.C., Guillot, M., Marchand, A. (1089) Automatic device for precise magnetic measurements in high continuous magnetic fields. Physica B Physics of Condensed Matter., 155: 407–410.

[10] Dahmen, U. (2004) Encyclopedia of Physical Science and Technology Volume 16 || Phase Transformations, Crystallographic Aspects., 2004: 821–853.

[11] Bilei, P.V. (1984) Fundamentals of aerodynamics of timber drying chambers with perforated partitions. Izvestiia Vysshikh Uchebnykh Zavedenii Lesnoi Zhurnal., 1984: 856–867.

[12] Adrian, R.J. (1991) Particle-Imaging Techniques For Experimental Fluid-Mechanics. Annual Review of Fluid Mechanics., 23: 261–304.

[13] Reddy, J.N. (2013) An Introduction to Continuum Mechanics, Second Edition: Frontmatter. Principles of Engineering Mechanics., 210: 215–246.

[14] Ohchi, M., Furukawa, T., Tanaka, S.I. (2010) Development of Education Support System for Numerical Electromagnetic Analysis Based on Server-Client Model using Java. Ieej Transactions on Fundamentals & Materials., 163: 8–17.

[15] Bristeau, M.O., Glowinski, R., Periaux, J. (1987) Numerical methods for the navier-stokes equations. Applications to the simulation of compressible and incompressible viscous flows. Computer Physics Reports., 6: 73–187.

[16] Scott, D.E. (2015) The Nebulous, Essential Dimensions in Effective University Teaching: The Ethic of Care and Relational Acumen. Am J Case Rep., 1: 21–24.

[17] Jang, I., Lee. (2012) A Study on University Students'Essential Skills and University Admission Methods. Journal of Agricultural Education and Human Resource Development., 44: 73–96.

Numerical simulation of steam flooding in offshore heavy oil reservoir after water flooding

Qingquan Zhang, Xiaofei Sun *

School of Petroleum Engineering, China University of Petroleum (East China), Qingdao, China

*Corresponding author e-mail: 1209475702@foxmail.com

Abstract. Steam flooding is an effective method to enhance oil recovery in heavy oil reservoirs. However, the influence factors of steam flooding will affect the development of heavy oil reservoir after the traditional water flooding. So the dynamic prediction of steam flooding needs to be performed before the development scheme is implemented. In this study, the 3D reservoir model was established by the reservoir numerical simulation software CMG; Combined with operating parameters of oilfield, the influence of steam temperature, steam injection timing and steam quality on the development effect of L7 block in offshore heavy oil reservoir was studied by using 3D reservoir model. The results show that compared with traditional water flooding, steam flooding has better development effect, and the cumulative oil production can reach $100.673 \times 10^4 m^3$; With the increase of steam temperature and steam quality, the cumulative oil production, incremental oil production and heat efficiency have an increasing trend; This result also suggests that the development method of higher steam temperature and steam quality conforms to the principle of economic development; For steam injection timing, the earlier the steam injection is more conducive to better development results.

1. Introduction

In recent years, with the depletion of land petroleum and the rising demand for energy, the development of offshore heavy oil resources has attracted more attention [1]. There are many heavy oil resources in Bohai Bay of China. However, because heavy oil has high viscosity and low fluidity, the traditional water flooding leads to low oil production. It severely limits the improvement of oil recovery, and it is necessary to find a suitable replacement technology. Steam flooding is an effective and mature development method in the late stage of heavy oil reservoir development [2]. Steam flooding (SF) is to inject steam into the reservoir to achieve effective recovery of heavy oil by increasing reservoir temperature, reducing viscosity and improving mobility of heavy oil [3].

However, there are few studies on SF after the traditional water flooding in offshore heavy oil reservoirs. The influence of factors in the process of SF on the development of offshore heavy oil is still unclear. Therefore, CMG-SRATS software is applied in this study to solve this problem. The main influencing factors were selected according to the reservoir characteristics of L7 Block, including steam temperature, injection timing and steam quality. The research on influencing factors of SF can provide basic parameters and basis for subsequent steam flooding in offshore heavy oil reservoirs.

Content from this work may be used under the terms of the Creative Commons Attribution 3.0 licence. Any further distribution of this work must maintain attribution to the author(s) and the title of the work, journal citation and DOI.
Published under licence by IOP Publishing Ltd

2. Establishment of numerical simulation model

Reservoir numerical simulation has the characteristics of low cost and high precision [4]. At present, it has become a powerful tool for oilfield development scheme design, production dynamic analysis, remaining oil distribution research and development plan adjustment, so as to provide data support and guidance for future reservoir development prediction [5].

In order to study the oil displacement effect of SF, the whole reservoir model was established, so as to lay the foundation for the subsequent study of SF.

2.1. *Establishment of geological model*

Based on the actual field data of geology, seismic, drilling, production and other tests, a relatively fine geological model was established. The model has 26 × 17 × 35 (15470) grids, as shown in Figure 1.

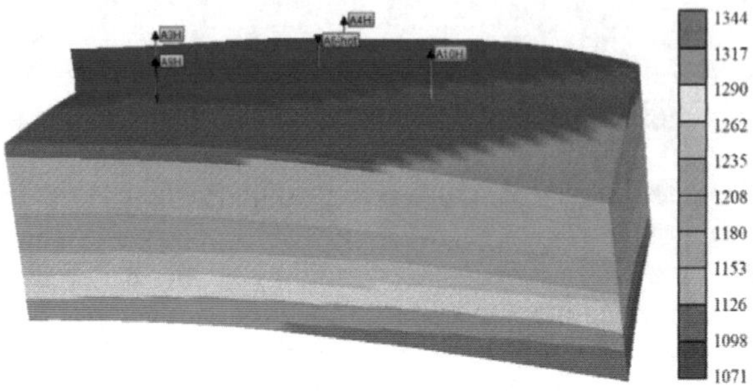

Figure 1. Geological Model

2.2. *Establishment of the fluid model*

To fully simulate the process and the mechanism of SF, two components are added, including the heavy oil and H_2O. In addition, since the heavy oil has different fluidity at different temperature, the viscosity-temperature curve is also added to the model (Figure 2).

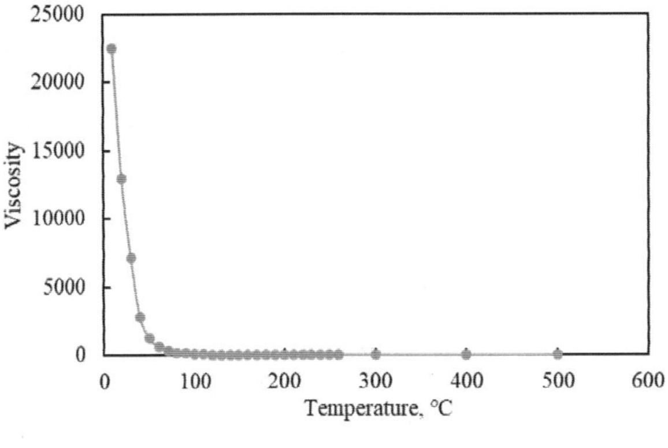

Figure 2. The viscosity-temperature curve of the heavy oil

2.3. *Establishment of production dynamic model*

Based on the production parameters of the oilfield, the production dynamic model of the L7 block is established. The established 3D model has five wells, including one injector(A6hot) and four production wells. For the injection well, the constraints were constant water injection rate; For production wells, the constraint conditions were constant liquid rate. The early development mode of the reservoir was water flooding, and the production time was 6 years. Then steam flooding replaced water flooding for 15 years.

3. Research and discussion

Combined with operating parameters of oilfield, the influence of steam temperature, steam injection timing and steam quality on the development effect of L7 block in offshore heavy oil reservoir was studied by using 3D reservoir model. This study has certain application value for guiding the specific production of oilfield.

3.1. Demonstration on development effect of steam flooding

Table 1 shows the development scheme design and simulation results of traditional water flooding and steam flooding. The cumulative oil production (COP) under steam flooding is $100.673 \times 10^4 m^3$ (Table 1), which is significantly higher than the traditional water flooding. This result supports the feasibility of SF, and also provides a guarantee for subsequent research on different influencing factors.

Table 1. Scheme design and simulation results under different development modes

Development method	Injection timing (water cut)	Injection Temperature °C	Steam quality	Cumulative oil production $10^4 m^3$
Water flooding	-	-	-	81.92
Steam flooding	90%	340	0.5	100.673

3.2. Effect of steam temperature

The fluidity of heavy oil is very sensitive to temperature, so it is necessary to explore the influence of steam temperature on development effect. Figure 3 shows the effect of different steam temperature (300 °C~ 380 °C) on heavy oil reservoir production.

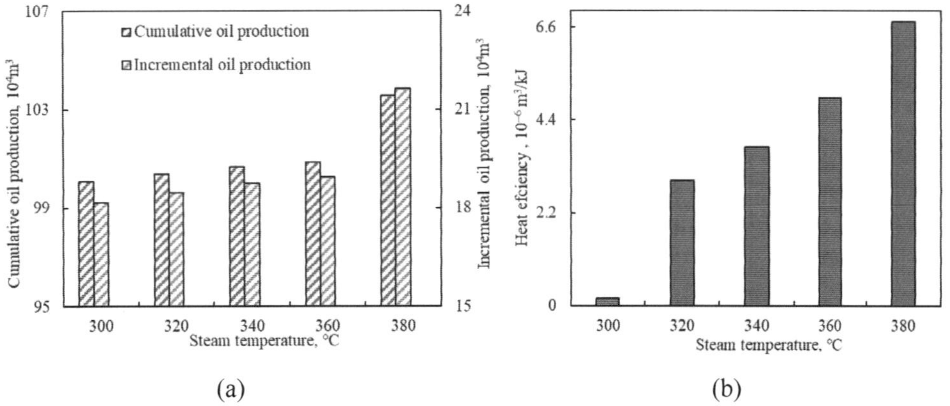

Figure 3. Simulation results: (a) COP and incremental oil production; (b) heat efficiency

The COP, incremental oil production (compared with water flooding) and heat efficiency have an increasing trend with the increase of steam temperature (Figure 3). This is due to that the higher steam temperature can better improve the mobility ratio and increase the oil discharge area. In addition, higher steam temperature can increase heavy oil production under the action of rock elastic expansion. Higher heat efficiency also shows that higher temperature steam can fully heat the reservoir to avoid more heat loss. The development method of high steam temperature in L7 Block conforms to the principle of economic development.

3.3. Effect of steam quality

Steam quality has an obvious effect on steam enthalpy, which further affects the heating effect of steam on reservoir. Figure 4 shows the influence of different steam quality (0.3, 0.4, 0.5, 0.6, 0.7) on the development effect.

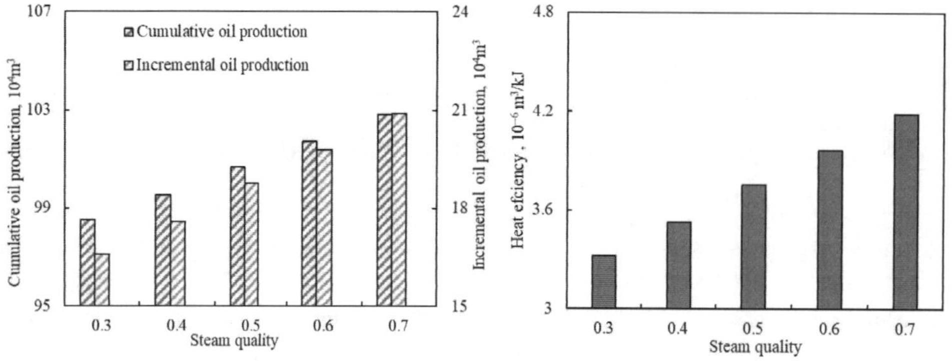

(a) (b)

Figure 4. Simulation results: (a) COP and incremental oil production; (b) heat efficiency

As shown in Figure 4, when the steam dryness increased from 0.3 to 0.7, the development effect of steam flooding is gradually better. It is mainly manifested in the gradual increase of the COP, incremental oil production and heat efficiency. This is due to that the steam has high latent heat of vaporization and large specific volume, and wet steam with higher quality has higher enthalpy. Once the steam is injected into the reservoir, the heating volume becomes larger and the heat transfer effect is better, which greatly improves the fluidity of heavy oil and increases the cumulative oil production. Therefore, when the oilfield is ready to inject steam, the steam with high quality should be injected as much as possible.

3.4. Effect of injection timing

Reasonable steam injection timing can effectively improve the economic benefits of heavy oil reservoir. The water cut in different periods was used as the time point for steam injection, and the production effects of steam flooding were studied when the water cut was 50 %, 60 %, 70 %, 80 % and 90 %, respectively. The results are shown in Figure 5.

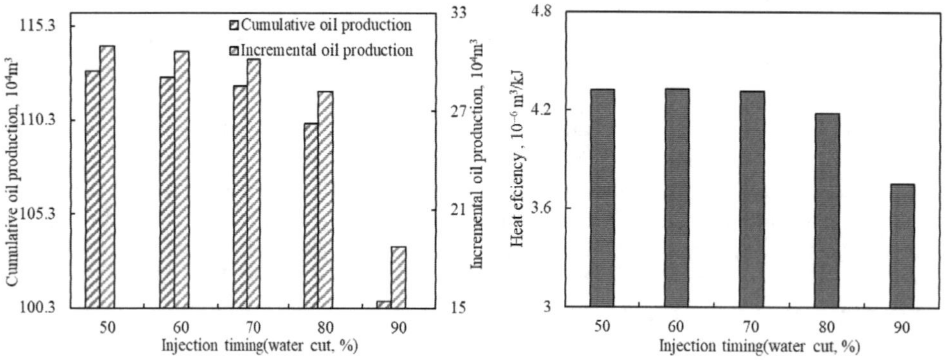

(a) (b)

Figure 5. Simulation results: (a) COP and incremental oil production; (b) heat efficiency

The earlier steam injection is more beneficial to improving COP, incremental oil production and heat efficiency (Figure 5). This is due to that the early injection of steam can give full play to the role of steam. When steam is injected at high water cut, steam channeling occurs in the later injected steam due to more crossflow channel in the reservoir caused by early water flooding. Therefore, the heat efficiency of steam and the cumulative oil production are reduced.

4. Conclusion

In this paper, the development effect of the influencing factors of steam flooding is studied by CMG, and the main conclusions are as follows:

(1) Compared with traditional water flooding, steam flooding has better development effect. This result supports the feasibility of steam flooding in offshore heavy oil reservoirs, and it also provides a guarantee for subsequent research on different influencing factors.

(2) Higher steam temperature and steam quality are beneficial to obtain higher COP and incremental oil production. The heat efficiency increases with the increase of steam quality and temperature, which indicates that the development mode of higher steam quality and steam temperature conforms to the principle of economic development. The earlier the steam injection is more conducive to better development results. And it can also ensure the efficient utilization of steam.

References

[1] Zhao, Y.. Laboratory experiment and field application of high pressure and high quality steam flooding. J Pet Sci Eng. 2020, 189(4): 107016.

[2] Yanbin C, Dongqing L, Zhang Z, et al. Steam channeling control in the steam flooding of super heavy oil reservoirs, Shengli Oilfield. Petroleum Exploration and Development, China University of Petroleum (East China), 2012, 39(6): 785-790.

[3] Sun, X.F., Cai, J.M., Li, X.Y., et al. Experimental investigation of a novel method for heavy oil recovery using supercritical multithermal fluid flooding. Appl Therm Eng, 2020, 185(1): 116330.

[4] Pang Z, Liu H, Zhu L. A laboratory study of enhancing heavy oil recovery with steam flooding by adding nitrogen foams. Journal of Petroleum Science and Engineering, China University of Petroleum (Beijing), 2015, 128: 184-193.

[5] Zheng Q, Liu H Q, Li F, et al. Quantitative description of steam channels after steam flooding. Science China Technological Sciences, China University of Petroleum (Beijing) ,2013, 56(5): 1164-1168.

CFD analysis and prediction of centrifugal pump cavitation performance based on three-dimensional two-phase flow between impeller and worm gear

Shi Ma [1], **Yiou Liu** [2,*]

[1] Huludao Military Representative Office, Huludao, Liaoning, China
[2] Wuhan Second Ship Design & Research Institute, Wuhan 430064, Hubei, China

*Corresponding author e-mail: best_20_22@163.com

Abstract. Cavitation performance is an important performance indicator of centrifugal pumps. The occurrence of cavitation will trigger a complex gas-liquid two-phase flow, which will reduce the efficiency of a low ratio centrifugal pump and cause machine vibration, as well as damage to the impeller, seriously affecting the working performance and preventing its normal operation. In this paper, CFD numerical simulation is used to analyze the cavitation characteristics of a centrifugal pump, obtain its internal flow law, calculate the cavitation margin of the device and the head generated by the pump under each inlet condition, and thus predict the critical cavitation margin of the pump at this flow point.

Keywords: two-phase flow, CFD analysis, centrifugal pump, cavitation performance.

1. Introduction

When cavitation occurs, there will be complicated gas-liquid two-phase flow in the impeller, which makes the efficiency and head of the low specific speed centrifugal pump decrease and causes the vibration of the machine. At the same time, the impeller will be damaged, seriously affect the performance and hinder its normal operation. In this paper, two-phase flow model is established to simulate the cavitation performance of centrifugal pump and reveal the law of its internal flow. The medium and its temperature are important parameters in cavitation simulation. Under the design condition, the inlet pressure of the pump can be adjusted by adjusting the outlet static pressure under the design flow to adjust the cavitation degree. Therefore, the cavitation flow phenomenon in the pump under the stable condition of various inlet pressures is calculated. Through the simulation of cavitation flow, the cavitation allowance of the device and the pump head are calculated, and the critical cavitation allowance of the pump at this flow point is predicted.

2. CFD Analysis

Volume of Fluid (VOF) method is used in this paper to model and simulate the system, and the cavitation characteristics of centrifugal pump are analyzed and predicted.

Content from this work may be used under the terms of the Creative Commons Attribution 3.0 licence. Any further distribution of this work must maintain attribution to the author(s) and the title of the work, journal citation and DOI.
Published under licence by IOP Publishing Ltd

2.1. Control equations

Two-phase flow has the characteristics of complex shape and constantly changing spatial distribution, so it is not easy to accurately simulate two-phase flow. The VOF method constructs and tracks two-phase flow according to the function f. When f equal 1, it means that all the fluid in the unit is occupied by the specified phase. When f equal 0, it means the fluid element is a fluid element with no specified phase. When f is from 0 to1, it means the element is an interface element. The function f(x, y, T) is defined as follows.

$$f(x,y,t)=\begin{cases}1\\0\end{cases} \tag{1}$$

The conservation form of the transport equation is expressed as follows.

$$\frac{\partial f}{\partial t}+\frac{\partial u_f}{\partial x}+\frac{\partial v_f}{\partial y}=0 \tag{2}$$

2.2. Model building

The centrifugal pump grid model is based on its three-dimensional geometric model, and the hydraulic model is obtained by Bourg subtraction. In order to avoid outlet backflow, a 5.1m pipe is added at the outlet end. Also, in order to make the flow into the impeller more stable, the inlet end of the 0.6m pipe. The model after meshing is shown in figure 1 below.

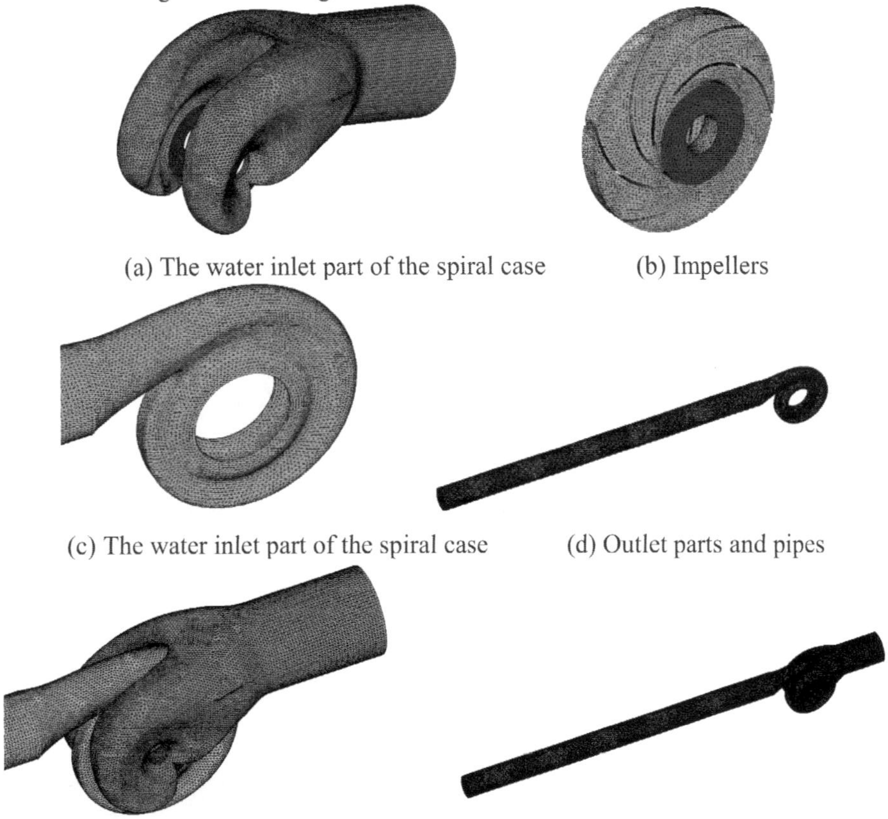

(a) The water inlet part of the spiral case (b) Impellers

(c) The water inlet part of the spiral case (d) Outlet parts and pipes

(e) Local grid model (f) Integrated assembly grid model

Fig. 1 Grid Model

In steady state, the impeller spins at a constant angular speed. Create a rotating Cartesian coordinate system that rotates in sync with the impeller and coincides with the axis. The Cartesian coordinate system is fixed to the impeller and rotates at a constant angular speed around the z axis.

2.3. Boundary Condition

① Inlet conditions

The selection of velocity inlet is very important. The axial velocity at the inlet should follow the assumptions of mass conservation and no rotation. The relative velocity distribution at the inlet section of the impeller must consider the relative motion of the impeller and fluid flow. Assume uniform pressure distribution in the inlet section. Turbulent kinetic energy at the inlet k_{in} and turbulent kinetic energy dissipation rate at the inlet ε_{in} can be calculated according to the following formula.

$$k_{in} = 0.005 u_{in}^2 \tag{3}$$

$$\varepsilon_{in} = \frac{C_\mu^{3/4} k_{in}^{3/2}}{l} \tag{4}$$

$$l = 0.07 D_{inlet} \tag{5}$$

where u_{in} is inlet velocity and D_{inlet} is the diameter of inlet.

② Outlet conditions

The outlet boundary condition is the pressure outlet. After the static pressure at the outlet is given, the definition of the outlet boundary condition can be completed.

③ Wall conditions

Due to the low Reynolds number near the wall, the development of turbulence is not sufficient. Therefore, the condition of no slip is met on the fixed wall, that is, the relative velocity is 0. $\partial p/\partial n = 0$. In the area close to the solid wall, the wall forces the flow to produce a large velocity gradient. It is assumed that the distance of the point P to the wall is y_p. The wall functions are as follows.

$$\frac{u_P}{u_\tau} = \frac{1}{\kappa}\ln\left(E y_p^+\right) \tag{6}$$

$$k_p = \frac{u_\tau^2}{\sqrt{C_u}} \tag{7}$$

$$\varepsilon_p = \frac{u_\tau^3}{\kappa y_p} \tag{8}$$

$$y_p^+ = \frac{\rho u_\tau y_p}{\mu} = \frac{\rho C_\mu^{1/4} k_p^{1/4} y_p}{\mu} \tag{9}$$

The wall friction factor $u_\tau = \sqrt{\tau_w/\rho}$. τ_w is the wall Shear stress. The constant number E and K is 9.011 and 0.419.

Tab. 1 Working conditions

working condition	flow velocity (m/s)	Q (m^3/h)
Case1	0.349311261	200
Case2	0.698622521	400
Case3	1.047933782	600
Case4	1.397245042	800
Case5	1.746556303	1000
Case6	2.095867563	1200
Case7	2.445178824	1400
Case8	2.794490084	1600
Case9	3.143801345	1800
Case10	3.493112606	2000
Case11	3.842423866	2200
Case12	4.191735127	2400

2.4. Analysis results
Figure 2 shows the CFD analysis results under different operating conditions.

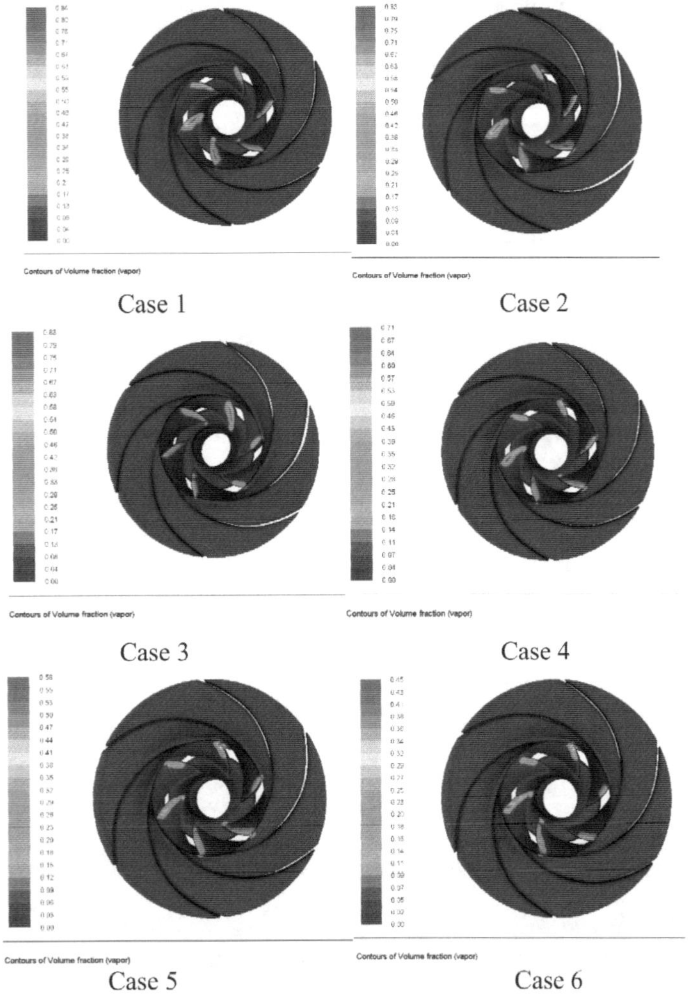

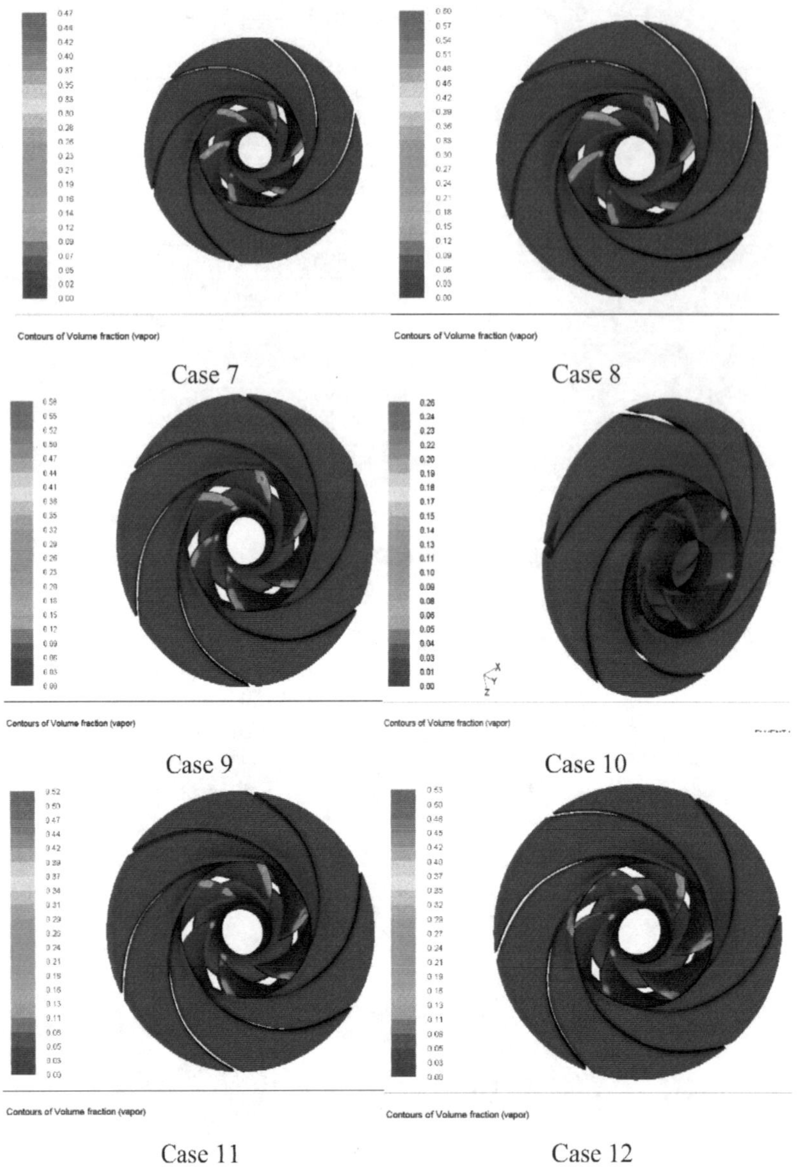

Case 7 Case 8

Case 9 Case 10

Case 11 Case 12

Fig. 2 Volume fraction of gas phase

3. Prediction of cavitation characteristics

3.1. Cavitation allowance

During the process of decreasing the outflow pressure, the steam structure appears in the low hydrostatic pressure region. The lower the pressure, the faster the cavitation zone grows, which will have an impact on the pump head and efficiency. According to the simulation results, the static pressure value of the inlet and outlet of the pump can be calculated, and the cavitation allowance of the device can be obtained.

3.2. Classical forecasting methods

At present, the most commonly used method for cavitation prediction is Thoma-Moody cavitation coefficient method. This coefficient is defined as follows.

$$\sigma = \frac{(NPSH)_r}{H} \tag{10}$$

It is shown that there is a relationship between the cavitation allowance of pump and pump head H, in fact, there is no relationship between the two. According to the similarity theory, the relationship between Thoma cavitation coefficient and pump specific rotation is as follows.

$$\sigma = kn_s^{4/3} \tag{11}$$

where k is constant number. For single-suction pump and double-suction pump using different values. This difference is with the single-suction pump and double-suction pump import design different mutual confirmation.

The calculation formula of specific revolution is introduced into the upper formula to obtain the prediction formula of cavitation allowance of pump.

$$(NPSH)_r = k\left(3.65n\sqrt{Q}\right)4/3 \tag{12}$$

As can be seen from the top, the pump cavitation allowance with the head independent, only with the pump speed n and pump flow Q. For single suction pump, $k=2.16\times10^{-4}$. For double suction pump, $k=1.37\times10^{-4}$.

3.3. Simulation results

The simulated inlet pressure is put in, and the result is shown in table 2, and the curve is drawn as shown in figure 3.

Tab. 2 The value of (NPSH)a and (NPSH)r under different working conditions

Q (m^3/h)	Velocity (m/s)	inlet pressure (Pa)	$(NPSH)_a$(m)	$(NPSH)_r$(m)
200	0.349311261	98861.76	10.0282	1.7556
400	0.698622521	97876.77	9.9071	2.7869
600	1.047933782	101017.8	10.2025	3.6519
800	1.397245042	101072.5	10.1647	4.4239
1000	1.746556303	101029.2	10.1042	5.1335
1200	2.095867563	101276.46	10.0614	5.797
1400	2.445178824	99172.97	9.7619	6.4244
1600	2.794490084	97034.21	9.4462	7.0225
1800	3.143801345	93267.48	8.9489	7.5962
2000	3.493112606	94251.15	8.9329	8.1489
2200	3.842423866	93200.39	8.6931	8.6835
2400	4.191735127	92180.78	8.444	9.2021

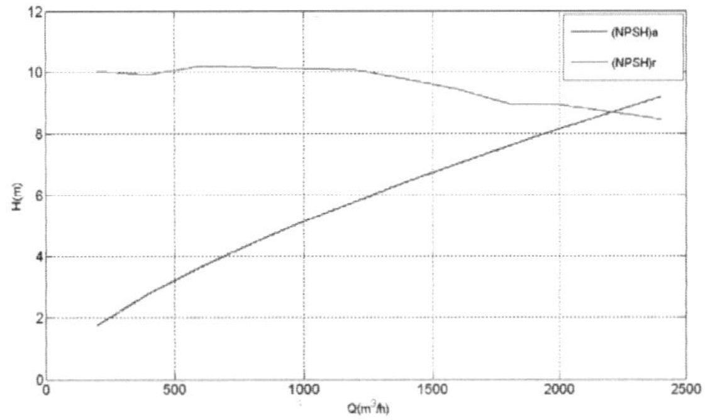

Fig. 3 Q-NPSH curse

4. Conclusions

In this paper, the cavitation performance of centrifugal pump is analyzed by CFD numerical simulation method, and the cavitation performance is predicted based on the simulation data. As can be seen from the Q-NPSH curve, (NPSH)a decreases as the flow rate increases, and (NPSH)r increases as the flow rate increases. With the increase of the flow rate, before the intersection of the two curves, the pump will not produce cavitation phenomenon. When the flow rate increases to a certain extent, the two are equal, then cavitation begins. With the increase of flow rate, cavitation phenomenon will become more serious. Therefore, the pump flow should be selected below $2000m^3/h$.

References

[1] LIN Qunqing, XUAN Yimin, HAN Yuge. Study on Flow and Heat Transfer Coupling Mechanism in Two-phase Flow of High-temperature Underwater Exhaust [J]. Journal of Engineering Thermophysics, 2020, 41(10): 198–204.

[2] WANG Meng, SUN Nan, WANG Yi, et al. Experimental Study and Numerical Simulation of Gas-Liquid Two Phase Flow Velocity Field Distribution in Aeration Tank[J]. Journal of Hydraulic Engineering, 2016, 47(10): 1322–1331.

[3] Nguyen T T, Tsuzuki N, Murakawa H, et al. Measurement of the Condensation Rate of Vapor Bubbles Rising Upward in Subcooled Water by Using Two Ultrasonic Frequencies[J]. International Journal of Heat and Mass Transfer, 2016,99: 159–169.

[4] WANG Shidong, WANG Zhenghui, NI Mingjiu. Ultrasonic Doppler Experimental Research and Numerical Simulation of Single Bubble Rising in Static Liquid Metal [J].Journal of University of Chinese Academy of Sciences,2013, 30(5): 598–602.

[5] WIKLUND J, STADING M, TRAGARDH C, Monitoring Liquid Displacement of Model and Industrial Fluids in Pipes by In-Line Ultrasonic Rheometry [J]. Journal of Food Engineering, 2010, 99(3): 330–337.

[6] NGUYEN T T, KIKURA H, MURAKAWA H, et al. Measurement of Bubbly Two-phase Flow in Vertical Pipe Using Multiwave Ultrasonic Pulsed Dopller Method and Wire Mesh Tomography [J]. Energy Procedia, 2015, 71(7): 337–351.

Water Hammer Protection Measures and Calculation Analysis of Reservoir Water Pipelines

Sen Xu [1, *], Xiaolong Zhu [2], Yuwu Tan [1] and Chuping Chen [1]

[1] Guangdong Polytechnic of Water Resources and Electric Engineering, Guangzhou, China
[2] Xi'an Water Supply Co., Ltd., Xi'an, China

*Corresponding author e-mail: xusen@gdsdxy.edu.cn

Abstract. In fluid mechanics, the large pressure fluctuation caused by the rapid change of fluid flow in a closed pipeline is called water hammer. It has huge destructive power for pressurized water pipelines, which often lead to pipeline leakage or even burst. In this paper, through the analysis of the dynamic characteristics of the gas in the pressurized water pipeline, the necessity of setting the air valve is drawn. Then based on the Bentley HAMMER model, combined with the example of the reservoir water pipeline engineering, the water hammer analysis was carried out. Compared with the maximum pressure head before and after adding the air valve, it is reduced from 126.89 m to 62.11 m, a decrease of about 50%. At the same time, the negative pressure in the pipeline under transient conditions is also eliminated. The results show that the air valve used in the pressure water pipeline has a good effect on the prevention of water hammer.

key word: Reservoir water pipeline, Water Hammer, Protection Measures, Calculation Analysis.

1. Introduction

The design of reservoir water delivery projects in mountainous areas is affected by the characteristics of large undulating terrain, crossing rivers, valleys and important traffic roads. In the case of accidental pump stop, sudden valve closing, etc., due to the existence of water hammer phenomenon, pipe bursts often occur in the convex part of the pipeline. In the light of the damage of the water pipeline, the normal water supply of the reservoir will be affected; in the worst case, the water transmission system will be paralyzed, and even the safety of nearby life and property will be threatened. Based on the analysis of the dynamic characteristics of the gas in the pipeline, this study establishes a water hammer analysis model of reservoir water pipelines through Bentley HAMMER, simulates the instantaneous pressure of the pipeline, and compares the changes in pipeline pressure before and after the installation of water hammer protection facilities. The development of the work is of great significance to the stable and safe operation of the pressure water pipeline.

Content from this work may be used under the terms of the Creative Commons Attribution 3.0 licence. Any further distribution of this work must maintain attribution to the author(s) and the title of the work, journal citation and DOI.
Published under licence by IOP Publishing Ltd

2. Analysis of Dynamic Characteristics of Gas in Pressure Water Pipeline

2.1. Characteristics of the gas in the pipeline

The gas carried in the pipeline system of the pressure water pipeline project mainly comes from: accidental emergency repair and emptying; too fast flow adjustment; insufficient submerged depth of the water inlet, resulting in inhalation; release of dissolved gas in the water body; gas in the pipeline before the pipeline is not filled. Due to the different water supply conditions, the water flow pattern of the water supply pipeline is complex, especially when the pipeline is filled with water and drained. The flow state of water flow is roughly divided into laminar flow, wavy flow, slug flow, air mass flow, foam flow, annular flow, and the like. The schematic diagrams of various gas-liquid two-phase flow states are shown in Figure 1. Moreover, various flow states will be converted to each other according to different factors, including the velocity, temperature, viscosity of the gas-liquid two phases, and the filling velocity of the water flow. Under different working conditions of the water supply pipeline, the gas in the water flow roughly presents as free air bubbles, small air pockets and retained air masses. Under normal circumstances, the gas in the pipeline exists in the following places: at the change of the pipe diameter and the intersection of the pipeline, when the water flow speed is small, the gas will gather into an airbag at the interface; at the top of the pipeline slope, the gas will gather at the top when it ascends; At the pipe, when the thrust of the water flow is greater than the buoyancy of the bubble, the gas accumulates on the pipe wall and gradually forms an air pocket.

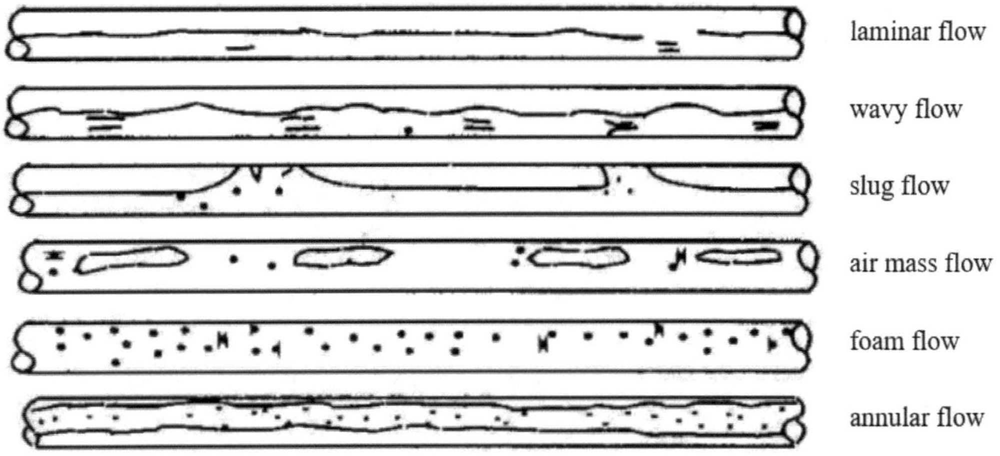

Figure 1. Schematic diagram of various gas-liquid two-phase flow states.

Therefore, during the water filling process of the pipeline, it is not the position where the water body replaces the gas in an ideal state, and all the gas is discharged out of the pipeline, but there will be many discontinuous airbags attached to the wall of the pipeline, and sometimes with the flow of water in the pipeline. move in. And these elastic airbags will have a great impact on the safety of water supply.

2.2. Necessity of Exhaust Valve Setting

The exhaust valve has positive functions such as negative pressure protection and running exhaust. When setting and designing the exhaust valve, factors such as installation position, caliber size, and spacing must be comprehensively considered. Based on engineering experience, the exhaust valve is generally arranged at the elevated point of the water supply pipeline and the position of the long straight pipeline. There is no unified calculation method for the diameter and spacing of the exhaust valve, so it can only be determined based on engineering practice and experience. If the air valve is not set properly, it will cause waste of equipment and insufficient air intake and exhaust capacity of the pipeline, which will affect the operation safety of the water supply project.

Generally, the pipelines of pressure water conveyance projects have the characteristics of large diameter, long distance, multiple fluctuations and reverse siphon, and the water flow is often between the two states of gas and water. Since the airbag in the pipeline occupies a certain cross section of the water, the local resistance coefficient becomes larger, which will reduce the flow area of the pipeline, resulting in poor water supply and even blocking the water flow. Moreover, the elasticity of the air bubbles is the key factor that causes the pipeline to vibrate and cause water hammer. In the lighter case, the socket-intubation pipe interface becomes loose, which reduces the overall pressure of the pipeline and increases the water leakage of the pipeline. suffered severe damage. In addition, in the process of water delivery, the existence of air bubbles will increase the energy consumption of the water pump by about 1/10 and accelerate the corrosion of metal pipes.

3. Application analysis of water hammer prevention measures based on Bentley HAMMER model——taking air valve as an example

The project is part of the reservoir's pressurized water delivery project. The water from the river is transported to the replenishment point through the pump station set up in this project. The design flow rate of the pump station is 1.0m3/s, and three pumps (two-use and one-standby) are installed. The design flow of a single pump is 0.50 m3/s, the rated power of the pump is 355 kW, and the pump head is 45.00 m. The water source located at an elevation of 72.50 m is transported to the reservoir at an elevation of 103.96 m by a water pump, and the water transmission pipeline adopts DN800 ordinary steel pipe.

The project uses Bentley HAMMER model to establish a water pipeline model to analyze and simulate the steady state and transient state of the water pipeline. Through the analysis of the transient conditions of water hammer when the pump is stopped, corresponding preventive measures for water hammer are taken to alleviate or eliminate the water hammer of the pipeline.

3.1. Steady-state model analysis

For this reservoir water pipeline project, Bentley HAMMER was used to calculate and analyze the pipeline water delivery status of the pump station under steady-state operation. After the two working pumps are turned on at the same time and continue to run for a period of time, the pumps and pipelines are running under rated conditions. At this time, the water flow in the pipeline is a constant flow, the starting point of the pipeline pressure is 45.00 m, and the free water head at the end of the pipeline is 7.80 m. Due to the consumption of pipeline resistance, the water pressure gradually decreases from the back of the pump to the end of the pipeline. At this time, the maximum pressure envelope and the minimum pressure envelope of the pipeline are coincident and both decrease uniformly. It can be seen that the pressure on the pipeline at this time is consistent with the design pressure, and there is no problem of fluctuation in the pressure in the pipeline. The calculation results are shown in Figure 2.

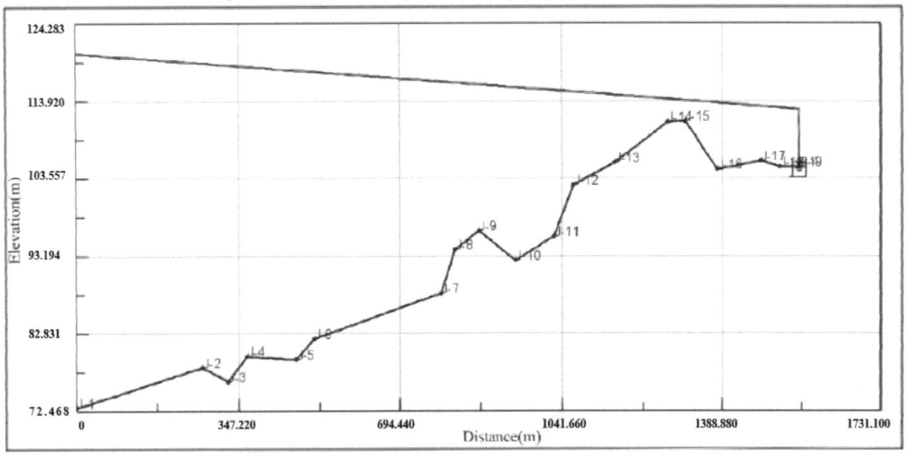

Figure 2. Steady-State Analysis Chart.

3.2. Transient Model Analysis

On the basis of the steady state of the pipeline model, the operating conditions under the condition of instantaneous pump stop are set based on the engineering example. The two pumps operate normally according to the design conditions until a sudden power failure occurs at a certain moment, and the two pumps lose power at the same time. No water hammer prevention facilities were installed, and the rest of the fittings installed on the pipeline were operating normally according to the design conditions.

Through Bentley HAMMER's analysis of pump stop water hammer, the calculation results are shown in Figure 3. The analysis shows that the maximum pressure envelope of the pipeline fluctuates greatly, and the change is more severe. The maximum pressure head of different pipe sections is also quite different. It should be pointed out that the maximum pressure of the pipeline under transient conditions is 78.48 m higher than that under steady-state conditions, reaching 123.48 m. The pressure generated by the water hammer when the pump is stopped is equivalent to 2.74 times the design water pressure. It can be seen that without water hammer protection measures, the pressure resistance of the pipeline itself is very high, and there is a huge risk of pipe bursting during operation.

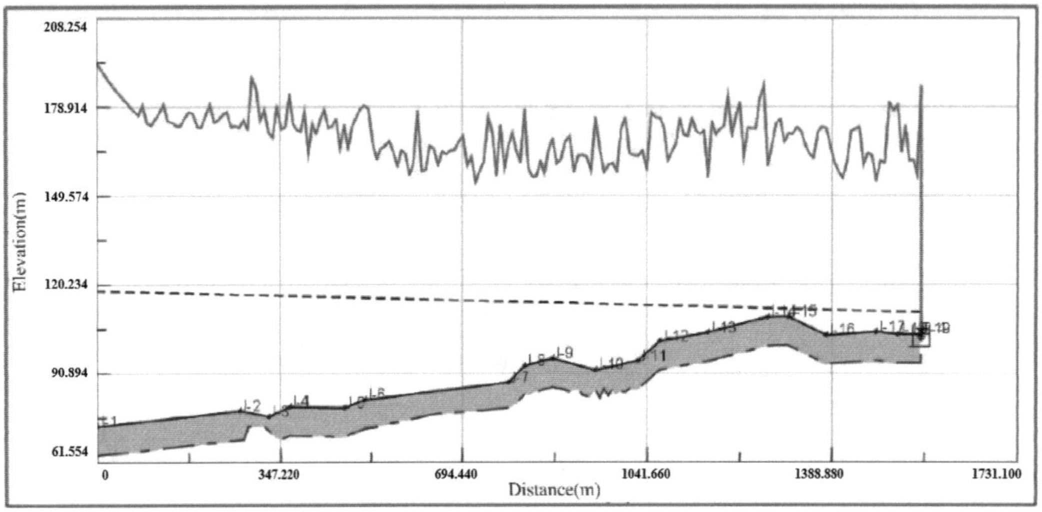

Figure 3. Transient Analysis Chart.

At the same time, it can be seen from the minimum pressure envelope in the model that there will be a certain degree of negative pressure along the pipeline, and these pipeline sections will produce vaporization in actual operation. The existence of the negative pressure section will cause the pipeline to generate bridging water hammer, which in turn will generate larger water hammer pressure fluctuations and increase the risk of damage to the pipeline.

3.3. Pre-water hammer protection measures

Since the main research of this paper is the water transmission pipeline of the reservoir,, the water hammer protection measures widely used in practical projects are selected for case analysis. The air valve has been widely used in water conservancy pipelines because of its simple equipment, convenient installation and disassembly, and easy later management. Therefore, this paper chooses the air valve as the water hammer protection measure for this case analysis, and carries out the analysis and calculation of water hammer protection.

According to the elevation of the water transmission line and the working principle of the air valve, in engineering practice, it is often chosen to set the air valve at the local convex point of the pipeline. In this project, the nodes J-8, J-9, J-14, and J-15 are selected to set air valves respectively, and the air valve labels are set to AV-8, AV-9, AV-14, and AV-15. When the water hammer occurs, in order to quickly fill the vacuum in the pipeline, prevent the liquid from vaporizing, and at the same time, when the pipeline pressure rises, the gas in the pipeline can be released slowly. After calculation, the diameter of the air inlet of the air valve used in this project is 250 mm, and the diameter of the air outlet is 20 mm. The calculation results are shown in Figure 4.

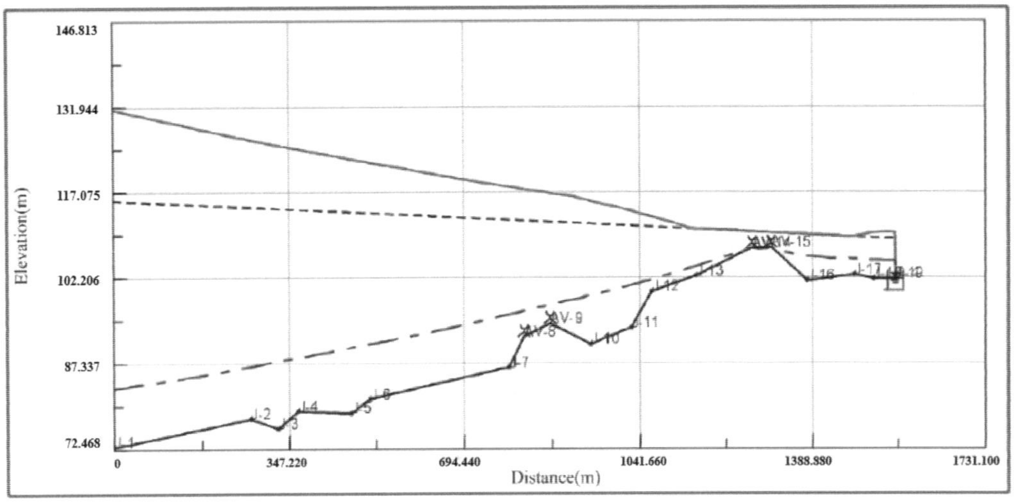

Figure 4. Analysis diagram of pre-water hammer protection measures

After the pipeline is set with the air valve, the transient state of the pipeline is also analyzed and simulated based on the Bentley HAMMER software. According to the calculation results of transient analysis, after adding an air valve to the pipeline, the maximum pressure envelope of the pipeline changes from a violent fluctuation to a smooth slope. The maximum pressure head is also reduced from 123.48m before adding protective measures to 59.4 m after adding protective measures. In addition, the minimum pressure envelope of the pipeline is located above the center line of the pipeline, which basically eliminates the negative pressure under transient conditions and avoids the vaporization in the pipeline due to excessive vacuum.

4. Conclusion

Water hammer analysis is an important work in the design of pressure water pipelines. Based on the analysis of the dynamic characteristics of the gas in the pressure water pipeline, this paper illustrates the necessity of the air valve in the design of the pressure pipeline. Secondly, based on the Bentley HAMMER model, the water hammer protection measures taken by a reservoir water pipeline project are analyzed. The results show that setting appropriate water hammer protection measures on the pressure water pipeline can play a significant role in restraining water hammer damage. It is hoped that through the analysis of water hammer protection and engineering cases, it will serve as a reference for colleagues in the design of water pipelines, and also provide a reference for further research on water hammer.

Acknowledgments

This work was financially supported by the Guangdong University Students' Science and Technology Innovation Cultivation Special Fund Funding Project.

References

[1] Huade Cao, Magdi Mohareb, Ioan Nistor. Finite element for the dynamic analysis of pipes subjected to water hammer[J]. Journal of Fluids and Structures, 2020, 93(C).

[2] Arefi Mohammad Hossein, Ghaeini Hessaroeyeh Mahnaz, Memarzadeh Rasoul. Numerical modeling of water hammer in long water transmission pipeline [J]. Applied Water Science, 2021, 11(8).

[3] AlKhayat Rasha Hayder, Kadhim Ameer A., AlBaghdadi Maher A. R. Sadiq, AlWaily Muhannad. Flow parameters effect on water hammer stability in hydraulic system by using state-space method [J]. Open Engineering, 2022, 12(1).

[4] Hossein Mahdizadeh. Numerical modelling of one- and two-dimensional water hammer problems using a modified wave propagation algorithm and turbulence model [J]. Journal of Hydraulic Research, 2019, 57(3).

[5] Jerzy Marcinkiewicz, Adam Adamowski, Mariusz Lewandowski. Experimental evaluation of ability of Relap5, Drako ® , Flowmaster2™ and program using unsteady wall friction model to calculate water hammer loadings on pipelines [J]. Nuclear Engineering and Design, 2007, 238(8).

[6] Lyu Jiawen, Zhang Jian, Wang Xingtao, Xu Tingyu. A combined water hammer protective method for optimizing the volume of the air vessel in water supply systems [J]. AQUA — Water Infrastructure, Ecosystems and Society, 2021, 70(8).

[7] Kubrak Michał, Malesińska Agnieszka, Kodura Apoloniusz, Urbanowicz Kamil, Bury Paweł, Stosiak Michał. Water Hammer Control Using Additional Branched HDPE Pipe [J]. Energies, 2021, 14(23).

[8] Kandil M., Kamal A.M., El Sayed T.A.. Study the effect of pipematerials properties on the water hammer considering the fluid-structure interaction, frictionless model [J]. International Journal of Pressure Vessels and Piping, 2021(prepublish).

[9] Hibiki Takashi, Rassame Somboon, Liu Wenbing,Wang Lutao,Zhao Quanbin,Chong Daotong,Yan Junije. Modeling and simulation of onset of condensation-induced water hammer[J]. Progress in Nuclear Energy,2020,130.

[10] Mounir Trabelsi, Ali Triki. Exploring the performances of the dual technique-based water hammer redesign strategy in water supply systems [J]. Journal of Water Supply: Research and Technology—AQUA, 2020, 69(1).

[11] Danciu Daniela, Răsvan Vladimir. Stability by the First Approximation of a Water Hammer Model [J]. IFAC PapersOnLine, 2020, 53(2).

Analysis of flow resistance characteristics of solenoid valve based on numerical simulation

Zhengjiang Wu [1, *], Wei Xia [1], Yuexia Fan [2], Chenhui Li [3], Baoping Zeng [1]

[1] Wuhan Second Ship Design & Research Institute, Hubei, Wuhan, 430064, China
[2] Wuhan Maritime Communication Research Institute, Hubei, Wuhan, 430079, China
[3] School of Automotive Engineering, Wuhan University of Technology, Wuhan 430070, China

*Corresponding author e-mail: hjy15927442154@163.com

Abstract. Flow resistance characteristic is an important parameter in the design of solenoid valve. It not only affects the determination of structural parameters of valve, but also plays a decisive role in the selection of control method of valve. In this paper, the flow resistance characteristic curve of the solenoid valve is obtained based on a 3D steady-state CFD simulation.

Keywords: CFD, solenoid valve, Flow resistance characteristic.

1. Introduction
Solenoid valve is a very important component and widely used in industrial field. Its physical characteristic parameters are very important when using the solenoid valve. One of the commonly used technology to study the physical characteristics of a solenoid valve is CFD simulation. In this paper, a 3D CFD model of the valve is build by a commercial software. Then, a steady simulation is implemented to obtain the flow resistance of it. The physical prototype testing result meets the simulation well.

2. CFD basic

2.1. Governing equation
Euler's method is usually used to describe the fluid principle, which are described by the control equations. [1].

Based on the standard k- ε turbulence assumption, the governing equations of the flow can be described as following [2]:

$$\frac{\partial \rho}{\partial t}+\frac{\partial(\rho u_i)}{\partial x_j}=0 \tag{1}$$

The momentum conservation equation can be deduced [2]:

$$\frac{\partial}{\partial t}(\rho u_i)+\frac{\partial}{\partial x_j}(\rho u_i u_j)=\rho f_i-\frac{\partial}{\partial x_i}+\frac{\partial \tau_{ji}}{\partial x_j} \tag{2}$$

Content from this work may be used under the terms of the Creative Commons Attribution 3.0 licence. Any further distribution of this work must maintain attribution to the author(s) and the title of the work, journal citation and DOI.
Published under licence by IOP Publishing Ltd

The equation [3] is the energy conservation equation:

$$\frac{\partial}{\partial t}(\rho E)+\frac{\partial}{\partial x_j}(\rho E u_j)=\rho f_j u_j-\frac{\partial}{\partial t}(\rho u_j)+\frac{\partial}{\partial x_j}(\tau_{ji}u_i)+\frac{\partial}{\partial x_j}(k\frac{\partial T}{\partial x_j})+S_E \tag{3}$$

In the above three governing equations:

$$\tau_{ji}=(\mu+\mu_t)(\frac{\partial u_i}{\partial x_j}+\frac{\partial u_j}{\partial x_i})-\frac{2}{3}(\mu+\mu_t)\frac{\partial u_j}{\partial x_i}\delta_{ij} \tag{4}$$

$$E=e+\frac{1}{2}u_i u_j \tag{5}$$

In which, δ_{ij} is the Kronic tensor and S_E means the source term.

2.2. Turbulence Model

This paper intends to use the Realizable k-epsilon turbulence model [4]. Using mathematical constraints to improve the performance of the model, it can be used to predict medium intensity swirls. The Realizable k-epsilon turbulence model can be described as following:

$$\frac{\partial(\rho k)}{\partial t}+\frac{\partial(\rho k u_i)}{\partial x_i}=\frac{\partial}{\partial x_j}\left[\left(\mu+\frac{\mu_t}{\sigma_k}\right)\frac{\partial k}{\partial x_j}\right]+G_k-\rho\varepsilon \tag{6}$$

$$\frac{\partial(\rho\varepsilon)}{\partial t}+\frac{\partial(\rho\varepsilon u_i)}{\partial x_i}=\frac{\partial}{\partial x_j}\left[\left(\mu+\frac{\mu_t}{\sigma_\varepsilon}\right)\frac{\partial \varepsilon}{\partial x_j}\right]+\rho C_1 E\varepsilon-\rho C_2\frac{\varepsilon^2}{k+\sqrt{v\varepsilon}} \tag{7}$$

$$G_k=\mu_t\left(\frac{\partial u_i}{\partial x_j}+\frac{\partial u_j}{\partial x_i}\right)\frac{\partial u_i}{\partial x_i} \tag{8}$$

In which, μ denotes the dynamic viscosity and μ_t means the turbulent viscosity,

$$\mu_t=\rho C_\mu\frac{k^2}{\varepsilon}, C_\mu=\frac{1}{A_0+A_s U^* k/\varepsilon} \tag{9}$$

$$A_0=4.0 \tag{10}$$

$$A_s=\sqrt{6}\cos\phi \tag{11}$$

$$\phi=\frac{1}{3}\cos^{-1}\left(\sqrt{6}W\right) \tag{12}$$

$$W=\frac{E_{ij}E_{jk}E_{k_j}}{(E_{ij}E_{ij})^{1/2}} \tag{13}$$

$$U^*=\sqrt{E_{ij}E_{ij}} \tag{14}$$

C2 is an empirical parameter, which needs to be obtained by best fitting the experimental results and calculation examples of typical flows.

2.3. Near wall treatment

The existence of wall has a significant effect on turbulent flow. On the outside of the area near the wall, due to the large gradient of the average velocity, the generation of turbulent kinetic energy makes the turbulence increase rapidly. The near wall is modeling by formula (15)~(16)

$$u^+ = \frac{u}{u_\tau} = \frac{u}{\sqrt{\tau_w / \rho}} \tag{15}$$

$$y^+ = \frac{y\rho u_\tau}{\mu} = \frac{y}{v}\sqrt{\tau_w / \rho} \tag{16}$$

In general, there are two methods to simulating the near wall phenomena. The first one does not address the viscous effects of the inner region. This method is called "wall function method". The use of wall function avoids modifying the turbulence model to consider the existence of wall. The second method is to modify the turbulence model so that the viscous influence area can be resolved to the wall through the grid, including the viscous bottom layer. This method is called the "near wall model" method

Equations (2-16) and (2-17) respectively define the u+ and the y+, but only the time mean, and no turbulence parameters. In order to reflect the influence of turbulent pulsation, u+ and y+ The definition of is expanded:

$$u^+ = \frac{u(C_\mu^{1/4} k^{1/2})}{\tau_w / \rho};\ y^+ = \frac{u(C_\mu^{1/4} k^{1/2})}{v} \tag{17}$$

The wall function method first needs to determine the equivalent viscosity coefficient u^+ of the flow velocity on the wall. The determination of k_p and ε_p of the first inner node P on the near wall, k_p is given by the equation of k, and the boundary conditions are:

$$(\frac{\partial k}{\partial n})_W = 0 \tag{18}$$

N denotes the normal of the wall.

Calculated according to the equation of k, the boundary conditions are:

$$\varepsilon_p = \frac{C_\mu^{3/4} k_p^{3/2}}{k y_p} \tag{19}$$

3. CFD modelling of the internal flow of a solenoid valve

3.1. Gemetry

As shown in Figure 1., firstly, the 3D geometric model of the valve was build in CATIA. And the internal flow field is modeled by bool method. Figure 2. shows the final 3d geometry model of the valve's flow domain.

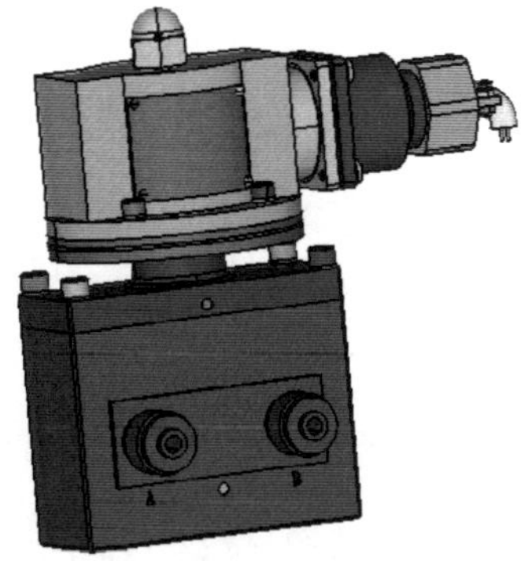

Figure 1. geometry demonstration

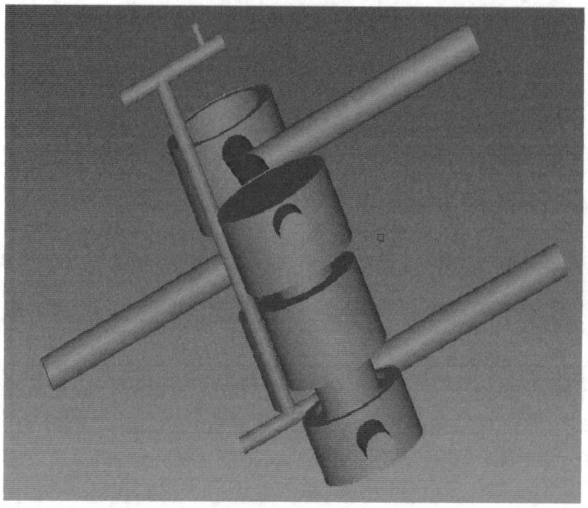

Figure 2. flow domain of the valve

3.2. meshing

The idea of grid is derived from the idea of discretization. The discretization discretizes the continuous solution domain into a number of limited sub regions, respectively solving the physical variables of each sub region. Each sub region is adjacent to each other continuously and harmoniously, so as to achieve the coordination and continuity of the entire variable field. The discrete grid is just a "token" of physical quantities. The grid is tangible, but the object to be discretized can be tangible (all kinds of solids) or intangible (heat conduction, gas). The key is the mathematical and physical formula hidden behind the grid. Therefore, to put it simply, the visible grid discretization is the form, while the invisible physical quantity discretization is the essential core. The paper uses structured grids to discretize the whole flow domain of the solenoid valve. Finally, more than 500,000 tetrahedral structured meshes is obtained. The result of mesh discretization is shown in the figures 3 and 4.

Figure 3. 3D meshes of the solenoid valve

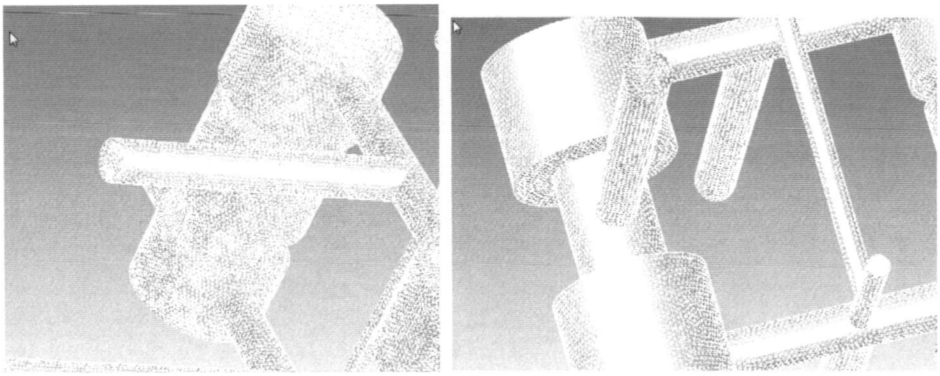

Figure 4. other views of the meshing

3.3. Simulation results

The simulation results is got based on such boundary conditions: the inlet velocity=0~0.09kg/s; The outlet meets the continuous outlet conditions; The rest boundary situations are fixed walls.

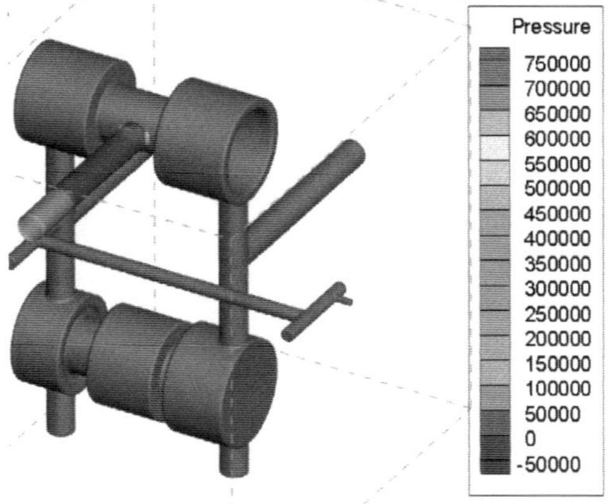

Figure 5. The Pressure nephogram of the flow field in the solenoid valve (port A)

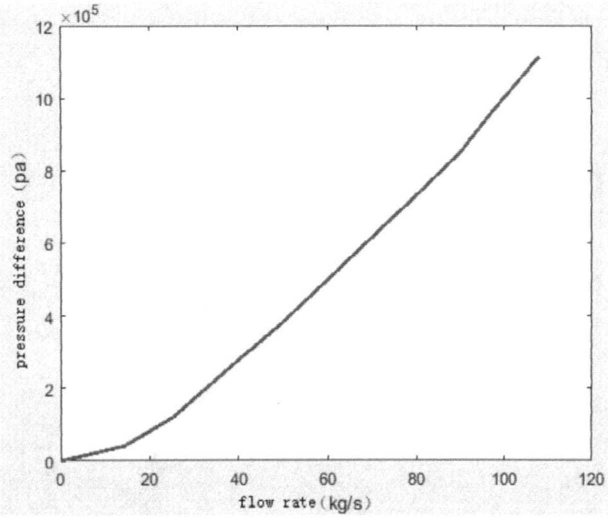

Figure 6. Solenoid valve flow resistance characteristic curve (port A)

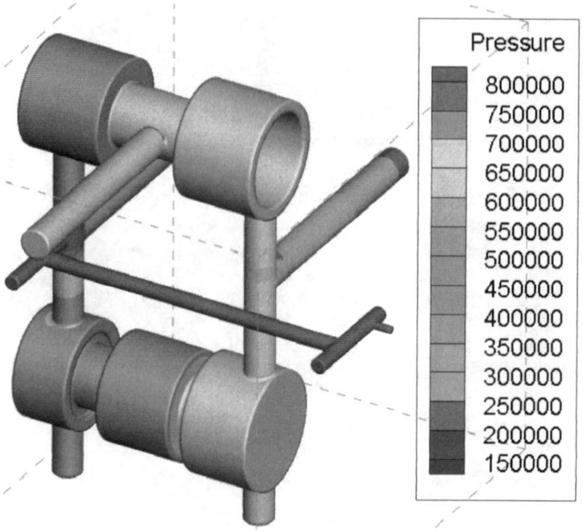

Figure 7. The Pressure nephogram of the flow field in the solenoid valve (port O)

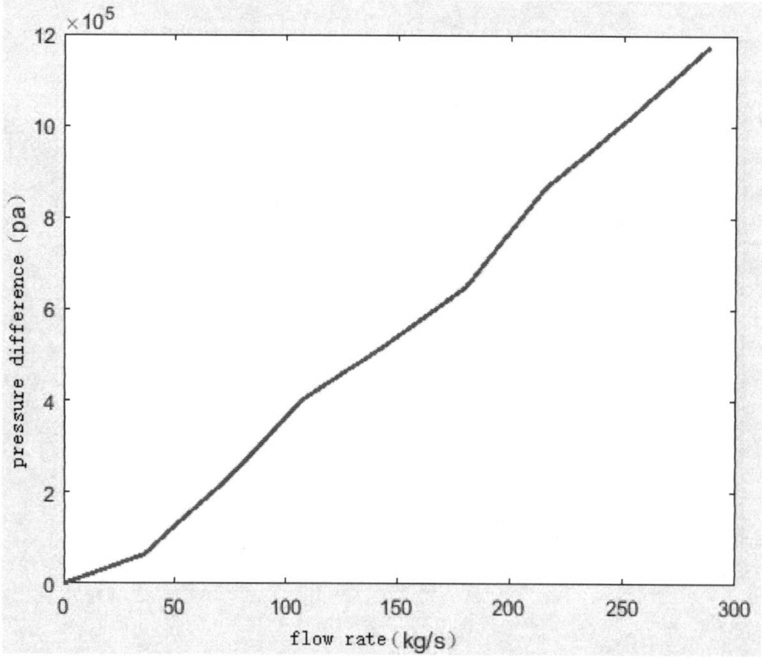

Figure 8. Solenoid valve flow resistance characteristic curve (O port)

The flow resistance of the solenoid valve is close to the quadratic function distribution, which is parabolic, and the quadratic term coefficients are about 43.25 (A port) and 2.9 (O port).

4. Conclusion

The 3D steady-state CFD simulation of an solenoid valve was established, and the pressure difference under different flow rates was calculated, based which the resistance properties of the valve is obtained.

References

[1] Chen Chunji. Discussion on the causes and solutions of cavitation phenomenon in liquefied petroleum gas pump[J]. Chemical Enterprise Management, 2018, (26): 37-38.

[2] PARK S, YANG J, RHEE S H. Parametric study on ship's exhaust-gas behavior using computational fluid dynamics [J]. ENGINEERING APPLICATIONS OF COMPUTATIONAL FLUID MECHANICS, 2017, 11(1): 159-171.

[3] Zhang Bo,Fu Weibiao,Wu Xianghui.Effect of Hydrogen on Fuel Consumption of Diesel Engine[J]. Journal of Combustion Science and Technology, 2006, (03): 238-242.

[4] MURAI Y. Frictional drag reduction by bubble injection [J]. EXPERIMENTS IN FLUIDS, 2014, 55(7).

[5] KANG C, ZHANG W, ZOU Z W, et al. Effects of initial bubble size on geometric and motion characteristics of bubble released in water [J]. JOURNAL OF CENTRAL SOUTH UNIVERSITY, 2018, 25(12): 3021-3032.

[6] XIAO H, GENG S J, CHEN A Q, et al. Bubble formation in continuous liquid phase under industrial jetting conditions [J]. CHEMICAL ENGINEERING SCIENCE, 2019, 200: 214-224.

[7] LIU H, WU X F, XIE M Z, et al. Numerical simulation of predict bubble generation diameter [Z]. 3RD INTERNATIONAL SYMPOSIUM ON MODERN MINING & SAFETY TECHNOLOGY PROCEEDINGS. 2008: 169-173.

[8] DENG B, CHIN R J, TANG Y, et al. NewApproach to Predict the Motion Characteristics of Single Bubbles in Still Water [J]. APPLIED SCIENCES-BASEL, 2019, 9(19).

Research in Velocity Profile and Pressure Distribution in 2D Conduit Flow with Stagger Arranged liquid separator blocks

Feiyang Ou *

Department of Chemical and Biomolecular Engineering, University of California, Los Angeles, California, 90024, United States

*Corresponding author's e-mail: zjchen@buct.edu.cn

Abstract. This work uses ANSYS Fluent commercial software to develop a pipe liquid separator model with inside staggered arranged diamond-rectangular shaped blocks for liquid separating. This model serves to liquid separating process in large diameter pipes. All computations are conducted under laminar flow with 200 iterations, and the residues are constantly monitored to ensure accuracy. Two different block sizes and two different inlet velocities are tested for comparison and analysis. Streamline plots and pressure contours are plotted and analyzed for separating performance and pressure drop. The model works well on separating for any combination of inlet velocity and block size. Velocity does not influence performance a lot, but larger blocks have better performance. The pressure drop is significant in these models. Increasing velocity or block size would greatly increase pressure drop, which means the model may work at low velocity for a pressure-sensitive environment. The fitted math relationship between parameters is not fully analyzed and could be developed in future work.

1. Introduction

Fluid dynamics has a long history for researchers, but the cost of in-field fluids experiments, especially for large scale experiments, was very expensive and time-consuming. Under this situation, the numerical simulation of fluid dynamics, mostly known as Computational Fluid Dynamics (CFD in the rest of the paper), has developed for several years. Since the establishment of Navior-Stokes (N-S in rest of paper) equations, proposed by C. L. Navior and G. G. Stokes in the 19th century and widely analysed in the late 20th century, numerous researches in CFD with different solid configuration and boundary conditions aided by the discretized form of N-S equations and computers have been proposed. Liquid splitting and mixing are important research areas in CFD simulations, especially the splitting and mixing of liquid in conduit flows. It is essential in the chemical industry to effectively split and mixing liquid in pipes to achieve well mixing and relatively turbulent flow. To deal with the conduit flow liquid splitting and mixing problems, multiple solid structures inside pipes are analysed, such as the trapezoid structure proposed by I.A. Ghani (2017) and hump structures at walls proposed by D. Sun (2019) and staggered arranged long hexagonal structure. [1-3] All these works focus on the small pipe diameter and splitting flows by just a few separator blocks in the middle or on the wall for well mixing or well heat transfer with the outer environment. There are fewer papers research in liquid splitting in large pipes with substantial regular arranged separating blocks.

Content from this work may be used under the terms of the Creative Commons Attribution 3.0 licence. Any further distribution of this work must maintain attribution to the author(s) and the title of the work, journal citation and DOI.
Published under licence by IOP Publishing Ltd

The wall turbulent structures in previous works like wall hump on narrow pipes completely have no value for large diameter pipes liquid separation. The coverage and disturbance of a single small block in the middle of pipe or structures attached to walls is not enough. A single large separator located at the middle in large diameter flow may also be able to do liquid separation. Still, the quick decrease of effective diameter would cause large head loss, generating great pressure on flowing regions at two sides and developing into obstruction or damage to the separator. Because of the speciality of large diameter pipe liquid separation, multi-block, multi-layer structure in the middle could be a good solution for liquid separation in large diameter pipes. [4, 5] This work proposes a new staggered arranged diamond-rectangle shape solid splitter structure in large diameter pipes. The diameter of the pipe and region of separating blocks are fixed for all tests in this article. The fluid is assumed isothermal and incompressible. [2, 3]

In addition, only steady-state conditions are analysed. Although it is possible to develop the numerical calculation scheme by hand in a scientific programming language like Python, it is very time-consuming and hard to debug. ANSYS Fluent, the commercial software developed by ANSYS Inc. with the help of other components of the whole ANSYS software, is used to conduct the modelling, meshing and simulation. Two different block sizes and two different inlet fluid velocities are tested in the model to fully analyse the performance and profiles under the most possible conditions and could provide a fitting line of simulation data. There are four lines of staggered arranged square blocks for splitting and create turbulence from uniform inlet flows. The Reynolds number is controlled under 2500 during the whole simulation process to ensure the whole process is in the laminar flow region to avoid turbulence problems, which is very complex to analysed with numerical simulation and beyond the scope of this paper. The velocity profile is analysed by plotting with ANSYS CFD-Post software by streamline plots for the degree of turbulence and direction. The pressure profile is also analysed for both the influence of separating blocks on the total pressure distribution and the influence of block sizes on overall pressure drop, which is an important issue to be considered in the real industry process by plotting the contour plots also in CFD-Post software.

2. Method

The finite element method is used to numerically simulate the problem. The total testing area is created in ANSYS SpaceCaim software with 50mm wide and 94mm long with 9x6 mm separation blocks in the diamond-rectangular hybrid shape shown as below:

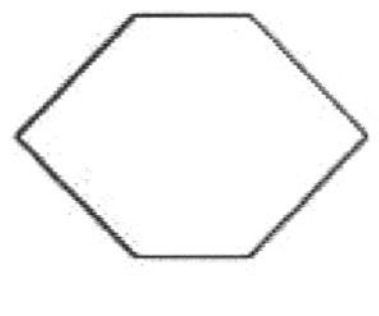

Figure 1. The shape of a single separator block

The upper trapezoid part and lower trapezoid part are fully symmetrical. Each one in these three-line segments has the same length in the horizontal direction. All bevel edges are 45 degrees with the either vertical or horizontal direction. This bi-direction sharp-angle design could effectively guide the flow to two directions but not completely obstruct the flow, reducing pressure drop since the effective local flow diameter decrease is gradually but not immediately. Both 9mm x 6mm version block and 6mm x 4mm version block is tested for velocity and pressure profile for testing separating performance. In the model creation process, the areas of the blocks are deleted as holes in the 2D model, which has the same effect as solid blocks, so the meshing process would not create grids inside the blocks.

The next step is using ANSYS Meshing to create a mesh grid for the Fluent solver to do the simulation. There are two types of meshing, structured and unstructured. Structured grids are hexahedron in 3D and quadrilateral shape in the 2D situation. However, better accuracy brings longer calculation times and a bad ability to fit in complex shapes. [6] The unstructured grids have some kinds of randomness and involve grids like tetrahedral in 3D or triangle in the 2D situation, so it could fit complex shapes well, but with worse accuracy. In this work, the shape is not so complex, and the whole structure is in 2D. Besides, for a grid size of 0.0002 meters, or 0.2 millimeters, which is sufficiently small, there are just 108401 points and 106736 grids. As the total calculation workload for the software is not large, and the computer has 16 cores and GTX 1080 GPU, the whole area is covered with structured grids and conducted by automatic meshing in ANSYS Meshing software.

After creating the mesh for the model, the mesh is sent to ANSYS Fluent solver. The central physical governing equations for the Fluent are the Navior-Stokes equation and continuity equation. Since the simulation doesn't include any mass diffusion or heat transfer, the N-S equation for mass transfer and energy are not set up. [7, 8] The assumptions are made that the fluid is incompressible, and the pipe is placed fully horizontal, which means ignoring gravity influence on the flow. After these simplification assumptions, the physical N-S equation is in the following form [9, 10]:

$$\frac{\partial u}{\partial t}+u\frac{\partial u}{\partial x}+v\frac{\partial u}{\partial y}=-\frac{1}{\rho}\frac{\partial p}{\partial x}+\upsilon\frac{\partial^2 u}{\partial x^2}+\upsilon\frac{\partial^2 u}{\partial y^2} \tag{1}$$

$$\frac{\partial v}{\partial t}+u\frac{\partial v}{\partial x}+v\frac{\partial v}{\partial y}=-\frac{1}{\rho}\frac{\partial p}{\partial y}+\upsilon\frac{\partial^2 v}{\partial x^2}+\upsilon\frac{\partial^2 v}{\partial y^2} \tag{2}$$

u(cm/s) is the horizontal velocity, v(cm/s) is vertical velocity, t(s) is time, x is horizontal direction unit, y is vertical direction unit, ρ(g/cm^3) is density, p(g/cm·s^2) is pressure, υ(g/cm·s) is viscosity of fluid.

Since the fluid is incompressible, the continuity equation equals zero,

$$\frac{\partial u}{\partial x}+\frac{\partial v}{\partial y}=0 \tag{3}$$

The pressure governing equation is derived from the Navior-Stokes equation and the continuity equation, known as the Poisson equation, which is shown below as:

$$\frac{\partial^2 p}{\partial x^2}+\frac{\partial^2 p}{\partial y^2}=-\rho\left(\frac{\partial u}{\partial x}\frac{\partial u}{\partial x}+2\frac{\partial u}{\partial y}\frac{\partial v}{\partial x}+\frac{\partial v}{\partial y}\frac{\partial v}{\partial y}\right) \tag{4}$$

At boundary conditions, the pressure governing equations, which is also derived from Navior-Stokes Equations, is shown below:

$$\frac{\partial p}{\partial x}=\rho\upsilon\frac{\partial^2 u}{\partial x^2} \tag{5}$$

$$\frac{\partial p}{\partial y}=\rho\upsilon\frac{\partial^2 v}{\partial y^2} \tag{6}$$

In addition, because of no-slip boundary conditions, both horizontal and vertical velocities around the blocks and on the walls are all zero. The calculation method is the internal built Fluent Second-Order Upwind method. Since Second-Order calculation could improve accuracy and because of the small amount of calculation workload, the time complexity of the Second-Order method is not considered.

To determine convergence, The minimum 10-6 residue values for velocity and pressure are enforced in Fluent software, and 200 iterations are set to guarantee convergence to a steady-state. The iteration-independent experiment is conducted with 200, 300 and 400 iterations, and the results are identical for them, so 200 iterations are acceptable to reach a steady-state and save computing times. The inlet velocity is assumed to be uniform distribution, which means all points in the grid at the inlet side are always equal to inlet velocity, which is adjustable as a parameter. All points in the grid at the outlet side are calculated by Fluent and not an enforced boundary condition.

In this simulation, the length of the pipe is set to a fixed 94 mm long, and the width of 2d pipe, or diameter in 3D structure, is 50 mm. The resolution of the grid is 250×470 with empty blocks in it with grid size 0.2 mm. The dt, which is the time increment, is controlled by the automatic built-in method in Fluent with a time ratio of 1, except inlet velocity = 0.03, no block case, the time ratio is 0.1 since 1 time ratio cause divergence and negative pressure near outlet region. The fluid is set as liquid water at any time. The inlet velocity is 0.05 m/s and 0.03 m/s in the simulation, and the material is always liquid water, which has a density of 998.2 kg/m3 and viscosity of 0.001 Pa•s. So, the Reynolds number is:

$$Re = \frac{\rho v D}{\mu} = \frac{998.2 \times 0.05 \times 0.05}{0.001} = 2495.5 \tag{7}$$

$$Re = \frac{\rho v D}{\mu} = \frac{998.2 \times 0.05 \times 0.03}{0.001} = 1497.3 \tag{8}$$

It is obvious that the flows are all in the laminar region, so the turbulence intensity is set to 1% (ANSYS, Inc. (2018), "ANSYS Fluent User's Guide, Release 19.0") and the hydraulic diameter at the inlet is the same as the original diameter since it is a fully open inlet.

3. Results

The streamline plot of velocity profile and colored contour plot of pressure distribution is shown below. For comparison, the flow regime without blocks is also demonstrated. The legends are always shown close to the plot for reference. Plots are generated by ANSYS CFD-Post software with free-edge sampling in streamlined plots from the inlet and the pressure contour.

Inlet Velocity = 0.05 Streamline Plot:

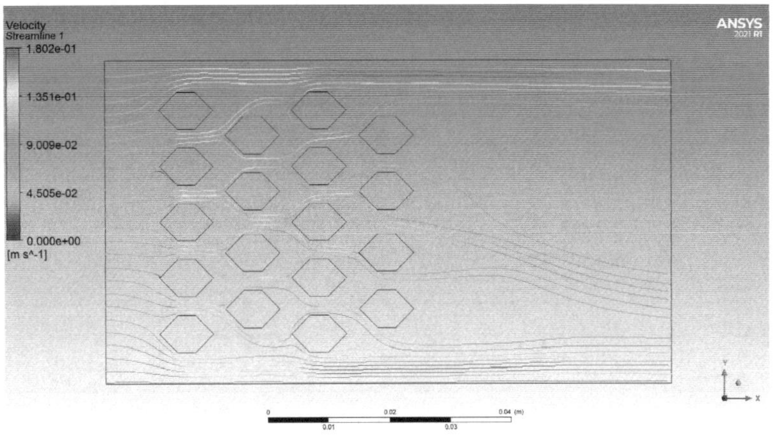

Figure 2. 9 mm x 6 mm Blocks Streamline Plot with 0.05 m/s uniform inlet velocity at Steady State.

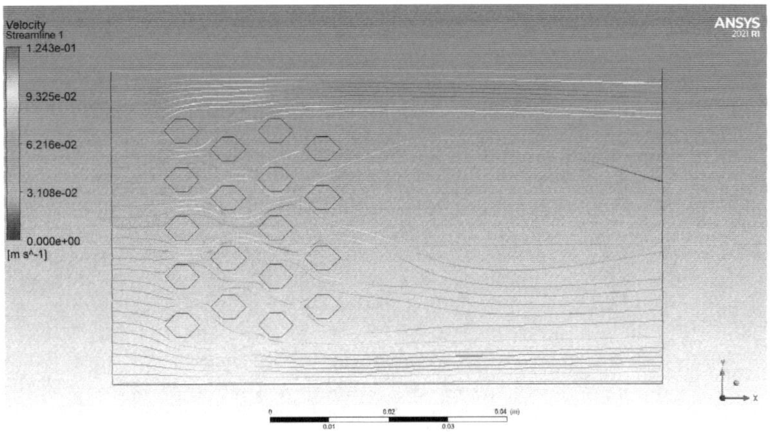

Figure 3. 6 mm x 4 mm Blocks Streamline Plot with 0.05 m/s uniform inlet velocity at Steady State.

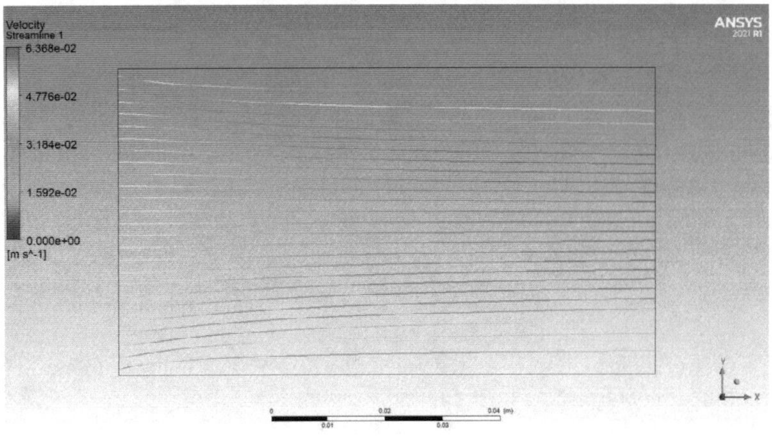

Figure 4. Zero Blocks(comparison) Streamline Plot with 0.05 m/s uniform inlet velocity at Steady State.

Inlet Velocity = 0.03 Streamline Plot:

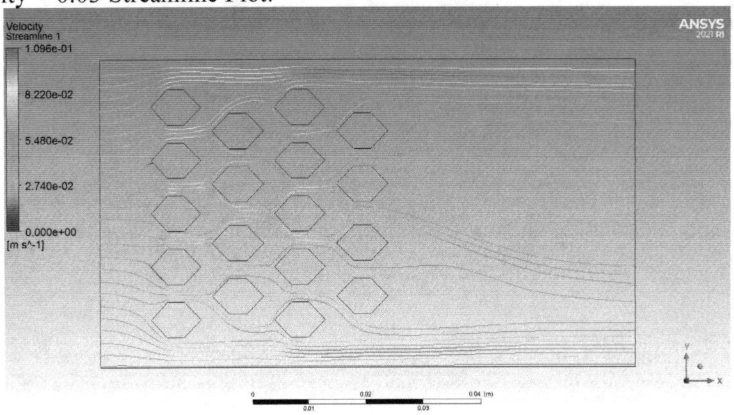

Figure 5. 9 mm x 6 mm Blocks Streamline Plot with 0.03 m/s uniform inlet velocity at Steady State.

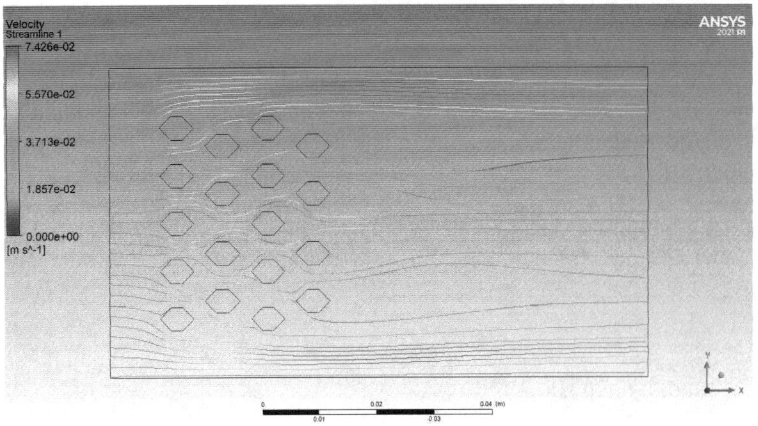

Figure 6. 6mm x 4mm Blocks Streamline Plot with 0.03 m/s uniform inlet velocity at Steady State.

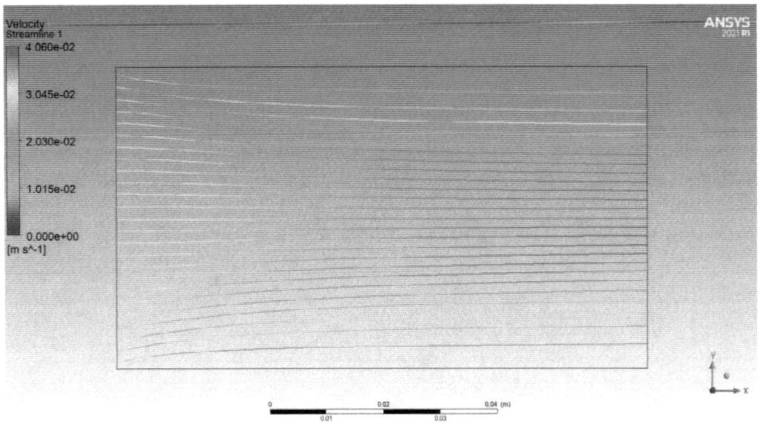

Figure 7. Zero Blocks(comparison) Streamline Plot with 0.03 m/s uniform inlet velocity at Steady State.

Inlet Velocity = 0.05 m/s Pressure contour:

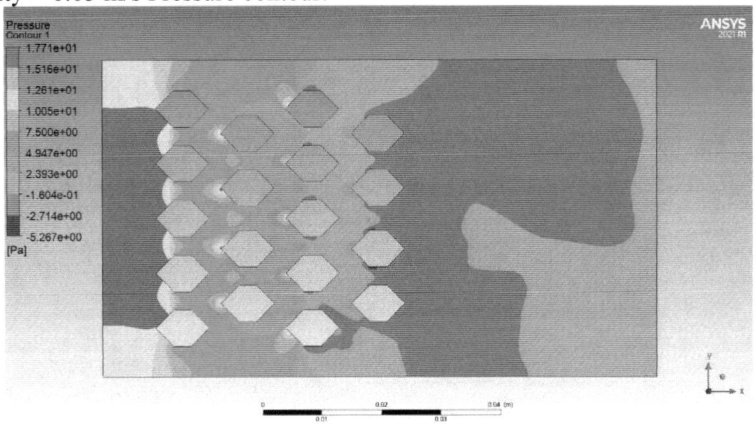

Figure 8. 9 mm x 6 mm Blocks pressure contour with 0.05 m/s uniform inlet velocity at Steady State.

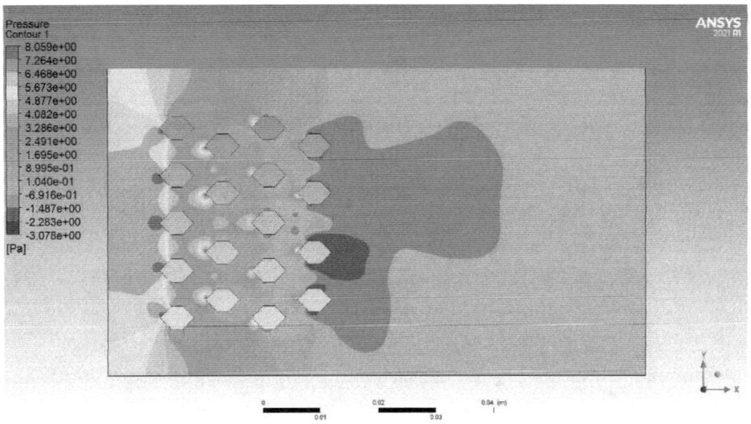

Figure 9. 6 mm x 4 mm Blocks pressure contour with 0.05 m/s uniform inlet velocity at Steady State.

Figure 10. Zero Blocks(comparison) pressure contour with 0.05 m/s uniform inlet velocity at Steady State.

Inlet Velocity = 0.03 m/s Pressure Contour:

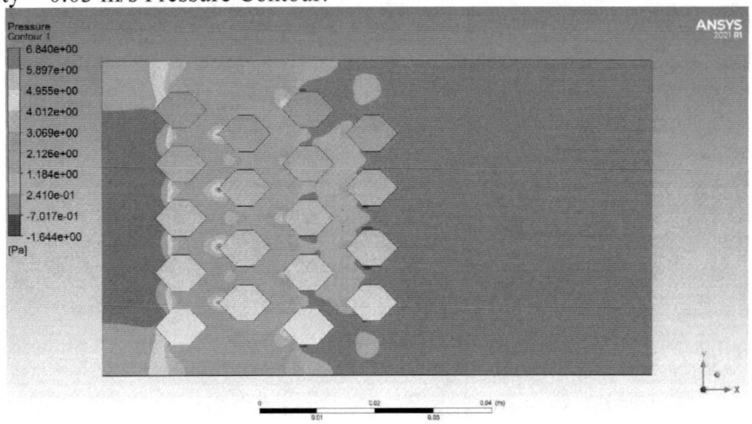

Figure 11. 9 mm x 6 mm Blocks pressure contour with 0.03 m/s uniform inlet velocity at Steady State.

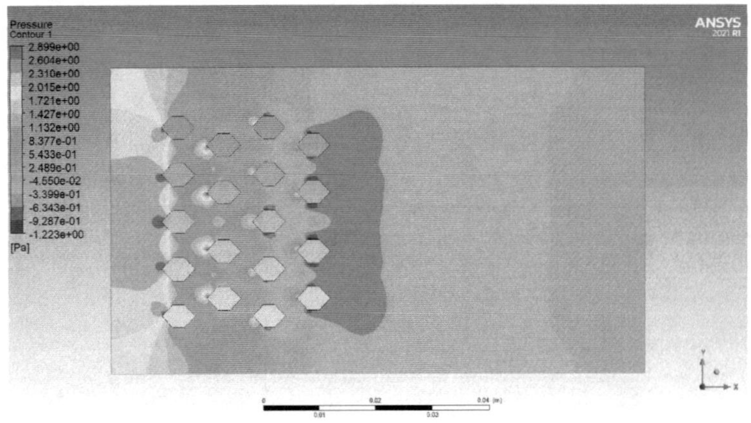

Figure 12. 6 mm x 4 mm Blocks pressure contour with 0.05 m/s uniform inlet velocity at Steady State.

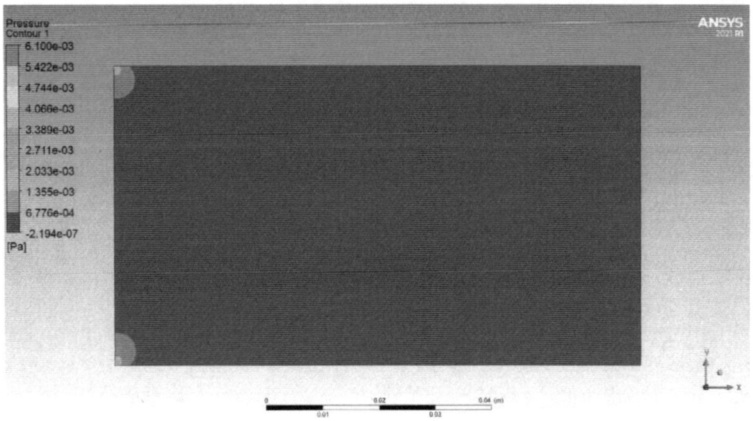

Figure 13. Zero Blocks(comparison) pressure contour with 0.03 m/s uniform inlet velocity at Steady State.

4. Discussion

From the pressure contour plots and streamline plots (Figure 3-5, Figure 8-10), the flows get turbulence, and the fluid separation is successful at all sides of the block. It is very hard to quantify how much turbulence or extent of fluid separating in each structure without the simulation of concentration diffusion or other third-party elements as an indicator, but conduct a qualified analysis for the effects from plots is indeed feasible and intuitionistic. One assessment standard is judging the flow velocity close to the wall and reach the end of the model region from the streamline and see if most of the flows from the inlet in the streamline plot are separated into two walls. From Figure 2, which is 0.05 m/s inlet velocity for 9mm x 6mm blocks, and Figure 3, which is the 6mm x 4mm blocks, it is apparent that the flow velocity at regions close to walls is much larger than the inlet velocity. For the 9mm x 6mm blocks model, the highest velocity achieved is 0.1243 m/s, and the velocity reaches the end of the model, which is 94mm location, still has over 0.095 m/s velocities. In comparison, flow velocity close to the wall of the model without blocks is only about 0.016 m/s. Even the largest velocity at the middle of flow is just 0.0637 m/s, which is substantially smaller than the one with separating blocks. In the stream plot without blocks, the streams always stay on a nearly straight-line flow path. However, in the stream plots with blocks, the streamlines change directions frequently and always flow to the direction of sidewalls. Besides, under the same inlet velocity, the flow velocity close to the wall when reaching the end of the model also differs from different block sizes. It is apparent that the larger blocks model streamlines converge less to the middle of the pipe. The direction of middle streams are also deviating more to the wall directions but not the original horizontal one. As a result, it is responsible for saying that larger blocks have better liquid separating ability than the small ones. Nevertheless, under the same model with different inlet velocity, the separating effects differs little. For the 9mm x 6mm large blocks model, the ratio of final wall velocity to inlet velocity at 0.03 m/s inlet condition is 3.2. This value is 3.153 for 0.05 m/s inlet velocity case, whose difference is negligible. In addition, the extent of convergence to the middle and the extent of streamlines deviation to walls are also similar for the two cases. The same phenomenon also happens on the 6mm x 4mm small blocks model. The ratios of 0.05 m/s case and 0.03 m/s case are 2.18 and 2.166, respectively, and the streamline behaviors at the middle of the pipe are also similar. Consequently, the model performance is not sensitive to inlet velocity, at least under the laminar flow region, which indicates that the model could be applied to different inlet velocity working conditions without much performance change.

The overall pressure drops for the large 9mm x 6mm model at 0.05 m/s inlet velocity and at 0.03 m/s inlet velocity is 16.6 pascal and 6.43 pascal, respectively. For 6mm x 4mm one, the pressure drop value is 6.93 and 2.46 pascal, respectively for 0.05 m/s and 0.03 m/s. The calculation method assumes the outlet pressure is zero and picks the largest inlet pressure value as the standard base. The result shows that pressure drop increases with velocity, and the pressure drop increases significantly, which is about 2.6 times for pressure drop under 1.67 times of velocity for 9mm x 6mm large blocks. For the small 6mm x 4mm blocks, this number increases to 2.8 times. This phenomenon implies that the increase in pressure drop is not proportional to velocity increase but increase significantly with velocity. In addition, the ratio of pressure drop of large blocks and small blocks are around 2.5. However, the effective diameter of larger blocks is 20 mm, and 30 mm for small blocks. Because friction loss equation [2, 3]:

$$S = \frac{32uv}{gD^2} \tag{9}$$

The friction loss is second-order inverse proportional to diameter, and 1.52 = 2.25, which is close to 2.5. A result that the pressure is sensitive to both the inlet velocity and block sizes is obvious from these data. Because of that, for industry works that have strict pressure loss requirements, the velocity and block size are restricted. Since the separation process would not be highly influenced by velocity under the same block size, this block could work well on slow inlet velocity for pressure drop sensitive works. Still, it may generate problems on high velocity one. The math relationship between flow velocity and pressure is not analyzed in this work. Still, since 1.672 = 2.79, which is close to the pressure drop ratio, it is possible that the second-order relationship also occurs between these two parameters.

5. Conclusion

This work used ANSYS Fluent software to conduct research on a proposed staggered arranged multi-blocks liquid separating structure used in large diameter pipes. The structures were arranged in four layers, with 5-4-5-4 staggered arrangements in the middle of the pipe. The block shape was a diamond-rectangular shape with a shape corner placed horizontally. Two sizes blocks are tested. One is 9 mm total length and 6mm total width, and the other one is 6 mm total length and 4 mm total width. Besides, both 0.05 m/s and 0.03 m/s inlet velocity were chosen to conduct the test.

These velocities are chosen to control the flow in the laminar region under the diameter of the whole reactor structure. These combinations could analyze the relationship between performance and block sizes, and inlet velocity. The modeling process was conducted in SpaceClaim software with a total area of 50 mm width and 94 mm length, and the areas of the internal blocks were deleted as empty holes. The meshing process was conducted in ANSYS Meshing software with a grid size of 0.0002 m, or 0.2 mm, so the whole mesh grid is 250 x 470 units with center blocks holes. The method in Fluent for calculation is second order upwind for pressure and velocity. Since only a steady state solution is necessary, so steady state calculation option was chosen, and the time ratio is set 1 except for the zero block comparison condition, whose time ratio was set to 0.1. The inlet flow turbulence intensity is set to 1% since the flow is always laminar. [6, 7] The time control was set to Fluent built-in automatic method. 200 iterations were performed, and the residue plot was lively monitored to guarantee convergence of simulation results.

The streamline plots and pressure contour plots were plotted with CFD-Post software. From the plots, under all conditions with different inlet velocities and different block sizes, the model could work well on liquid separating because of significantly higher-end velocity close to the wall than the comparison no-block condition. Besides, from the streamline plots, most inlet flows were separated into two walls areas. The middle flows are also in the wall direction. Also, the streamline plots found that the velocity has little influence for separating performance, which means this model could fit different inlet velocities in the laminar region. Nevertheless, larger block sizes could do the separating works better than the small ones. Besides, the pressure drop is significantly larger for larger velocities and larger block sizes, which possibly followed the second order relationship, so pressure drop is sensitive to velocity and block size. This phenomenon implied that if the process is sensitive to pressure drop or has strict pressure drop requirements in the industry, the model in this paper could only work well at a relatively slower velocity. The exact math relationship between velocity and pressure drop was not analyzed in this paper and could be researched in the future.

References

[1] Khadem, M.H., Shams, M., Hossainpour, S. (2009) Numerical simulation of roughness effects on flow and heat transfer in microchannels at slip flow regime. International Communications in Heat & Mass Transfer., 36: 69–77.

[2] Wu, J., Chen, Y., Shi, M. (2009) One dimensional numerical simulation for steady annular condensation flow in rectangular microchannels. Heat & Mass Transfer., 13: 13–29.

[3] Swift, G., Migliori, A., Wheatley, J. (1985) Construction of and measurements with an extremely compact cross-flow heat exchanger. J Heat Transfer Eng. 6 (2): 39–47.

[4] Yang, D.Y., Liu, Y. (2009) Numerical simulation of electroosmotic flow in hydrophobic microchannels. Science in China., 2009: 2460-2465.

[5] Guo, Z. (2000) Characteristics of microscale fluid flow and heat transfer I MEMS. In: Proc. Heat Transfer and Transport Phenomena in Microscale. Banff. Canada. pp.24–31.

[6] Vontas, K., Andr, E., Aki, M., Georgoulas, A. (2021) The effect of surface wettability on flow boiling characteristics within microchannels. International Journal of Heat and Mass Transfer., 172

[7] Li, Z.X., Wang, W., Guo, Z.Y. (2003) Effects of Axial Heat Conduction in the Wall on Convection in Microtubes. ASME 2003 1st International Conference on Microchannels and Minichannels. pp.327–333.

[8] Shaukat, R., Anwar, Z., Imran, S. (2021) Numerical Study of Heat Transfer Characteristics of mPCM Slurry During Freezing. Arabian Journal for Science and Engineering., 46: 7977–7988.

[9] Hetsroni, G., Mosyak, A., Pogrebnyak, E., Yarin, L.P. (2005) Fluid flow in micro-channels. Int. J. Heat Mass Tran., 48: 1982–1998.

[10] Harirchian, T., Garimella, S.V. (2009) Effects of channel dimension, heat flux, and mass flux on flow boiling regimes in microchannels. Int. J. Multiphas. Flow., 35: 349–362.

Coupled double-distribution-function lattice Boltzmann model with streaming-collision process

Zhuojie He [1, a], Ruifang Liang [2]

[1] Department of Micro/Nano Electronics, Shanghai Jiao Tong University, Shanghai, People's Republic of China
[2] Huaibei Normal University, Huaibei, People's Republic of China

[a] h729973599@sjtu.edu.cn

Abstract. A coupled double distribution function lattice Boltzmann model with streaming-collision process is presented. In this model, continuity and momentum equations are recovered by a density distribution function and energy equation is recovered by a total energy distribution function, and the specific-heat ratio and Prandtl number are flexible. At the same time, the streaming-collision process is used as update rule. Couette flow is simulated as a numerical test. The numerical results agreed relatively excellent with the analytical solutions.

Keywords: streaming-collision process; lattice Boltzmann method; compressible flow.

1. Introduction

The lattice Boltzmann method (LBM) has become a remarkably attractive method in kinetic computational fluid dynamics approach since its appearance in 1988 [1-2]. Unlike traditional computational fluid dynamics (CFD) based on discretization of macroscopic equations, the LBM is based on discretization of mesoscopic kinetic equations. The traditional and most used LBM achieved great success mainly in isothermal and incompressible problem. Comparatively, compressible and thermal problems need more complex models.

The current thermal compressible LBM can be generally classified into three main categories [3]: the hybrid method, the multi-speed method, and the double distribution function (DDF) method. Among these methods, the hybrid method decouples flow and temperature simulation, and the temperature is solved by traditional CFD method like finite difference method, then the simplicity of LBM is naturally destroyed [4]. The multi-speed approach [5] introduces additional particle speeds and other one or two order velocity terms in the most used Taylor series expansion of the equilibrium distribution functions to recover the accurate macroscopic energy equation. Despite the visible complexity because of the huge number of velocity, the multi-speed approach also suffers from numerically instability, narrow-range temperature and fixed Prandtl number.

DDF approach was first devised by He et al. [6], this kind of model introduced another distribution function which is for fluid's internal energy. There were two distribution functions in total: one for the fluid's density and the other one for the fluid's internal energy. This model is more stable and has an attractive adjustability of the Prandtl number, and thus get much attention. Later, Guo et al. [3] proposed a new DDF model which changed the second distribution function: now it is for the total energy. This

Content from this work may be used under the terms of the Creative Commons Attribution 3.0 licence. Any further distribution of this work must maintain attribution to the author(s) and the title of the work, journal citation and DOI.
Published under licence by IOP Publishing Ltd

model eliminated the complicated gradient terms in He's model, but it was a decoupling model, which means that the flow field was not effected by temperature. Li et al. [7] then solved this problem by proposed a coupled DDF model. This model was then developed to Multiple-Relaxation-Time (MRT) scheme and complex geometry conditions. However, the Li' model and the following models are all finite difference lattice boltzmann method, which means that these models are different from the traditional streaming-collision process on standard lattice, and so destroyed the simplicity and potential of high-performance on parallel architectures of the traditional LBM.

In this study, we change Li's model to streaming-collision process on standard method, which is a simpler and still accurate model. There are two sections apart from this part. In section 2, the coupled total energy double distribution function model with streaming-collision process is described. In section 3, Couette flow is simulated as a numerical test. Finally, we give the conclusion in section 4.

2. Coupled Double Distribution Function Lattice Boltzmann Model

2.1. Compressible coupled double distribution function model

In this compressible coupled double-distribution-function model, the basic equations from the mesoscopic kinetic Boltzmann equation are in the following form: (with the BGK collision term)

$$\frac{\partial f_\alpha}{\partial t} + (\vec{e_\alpha} \bullet \nabla) f_\alpha = -\frac{1}{\tau_f}(f_\alpha - f_\alpha^{eq}) \tag{1}$$

$$\begin{aligned}\frac{\partial h_\alpha}{\partial t} + (\vec{e_\alpha} \bullet \nabla) h_\alpha &= -\frac{1}{\tau_h}(h_\alpha - h_\alpha^{eq}) \\ &\quad + \frac{1}{\tau_{hf}}(\vec{e_\alpha} \bullet \vec{u} - u^2/2)(f_\alpha - f_\alpha^{eq})\end{aligned} \tag{2}$$

where α=0,1,2,...,N represents the direction of discrete velocity, N is the number of discrete velocity directions, $f_\alpha, h_\alpha, f_\alpha^{eq}$ and h_α^{eq} respectively represent distribution function of the density, the total energy and their equilibrium counterparts, $\mathbf{e}_\alpha$ represents the discrete particle velocity in α direction, τ_f and τ_h are relaxation times for density and total energy respectively, and $\tau_{hf} = \tau_h\tau_f/(\tau_f - \tau_h)$, $\mathbf{u}$ is the macroscopic velocity of the fluid.

The macroscopic quantities are defined as:

$$\rho = \sum_\alpha f_\alpha \tag{3}$$

$$\rho \vec{u} = \sum_\alpha f_\alpha \vec{e}_\alpha \tag{4}$$

$$\rho E = \sum_\alpha h_\alpha \tag{5}$$

$$T = \frac{2}{bR}(E - \frac{u^2}{2}) \tag{6}$$

$$p = \rho RT \tag{7}$$

where ρ is the density, E is the total energy, T is the temperature, p is the pressure, R is the specific gas constant, b is an introduced constant and is related to the specific-heat ratio γ by γ=(b+2)/b, so the specific heat ratio at constant volume is c_v=bR/2 and the specific heat ratio at constant pressure is c_p=(b+2)R/2.

The equilibrium distribution functions need to satisfy some velocity moment conditions:

$$\sum_\alpha f_\alpha^{eq} = \rho \tag{8}$$

$$\sum_\alpha f_\alpha^{eq} e_{\alpha i} = \rho u_i \tag{9}$$

$$\sum_\alpha f_\alpha^{eq} e_{\alpha i} e_{\alpha j} = \rho u_i u_j + p\delta_{ij} \tag{10}$$

$$\sum_\alpha f_\alpha^{eq} e_{\alpha i} e_{\alpha j} e_{\alpha k} = \rho u_i u_j u_k + p(u_k\delta_{ij} + u_j\delta_{ik} + u_i\delta_{jk}) \tag{11}$$

$$\sum_\alpha f_\alpha^{eq} e_\alpha^2 = \rho u^2 + Dp \tag{12}$$

$$\sum_\alpha f_\alpha^{eq} e_\alpha^2 e_{\alpha i} = [\rho u^2 + (D+2)p]u_i \tag{13}$$

$$\sum_\alpha h_\alpha^{eq} = \rho E \tag{14}$$

$$\sum_\alpha h_\alpha^{eq} e_{\alpha i} = (\rho E + p)u_i \tag{15}$$

$$\sum_\alpha h_\alpha^{eq} e_{\alpha i} e_{\alpha j} = (\rho E + 2p)u_i u_j + p(E + RT)\delta_{ij} \tag{16}$$

where D is the dimension of the space, and the subscripts i, j, and k indicate the x, y, or z component.

Through the traditional Chapman-Enskog expansion, we can finally arrive at the macroscopic equations which are continuity, momentum and energy equation:

$$\frac{\partial \rho}{\partial t} + \nabla \bullet (\rho \vec{u}) = 0 \tag{17}$$

$$\frac{\partial (\rho \vec{u})}{\partial t} + \nabla \bullet (\rho \vec{u}\vec{u}) = -\nabla p + \nabla \bullet \Pi \tag{18}$$

$$\frac{\partial}{\partial t}(\rho E) + \nabla \bullet [(\rho E + p)\vec{u}] = \nabla \bullet (\lambda \nabla T) + \nabla \bullet (\vec{u} \bullet \Pi) \tag{19}$$

$$\Pi = \mu[\nabla\vec{u} + (\nabla\vec{u})^T - \frac{2}{D}(\nabla \bullet \vec{u})\vec{I}] \tag{20}$$

where $\mu=\tau_f p$ is the dynamic viscosity, I is the unit tensor, $\lambda=\tau_h c_p p$ is the thermal conductivity.

2.2. *The discrete velocity model*

To ensure up to 6th-rank lattice tensor isotropic, the 2-dimensional 12-velocity (D2Q12) square lattice is chosen. This kind of lattice is defines as: (see Fig. 1)

$$\vec{e}_\alpha = \begin{cases} c\{\cos[(\alpha-1)\pi/2], \sin[(\alpha-1)\pi/2]\}, \\ (\alpha = 1,2,3,4) \\ \sqrt{2}c\{\cos[(2\alpha-1)\pi/2], \sin[(2\alpha-1)\pi/2]\}, \\ (\alpha = 5,6,7,8) \\ 2c\{\cos[(\alpha-9)\pi/2], \sin[(\alpha-9)\pi/2]\}, \\ (\alpha = 9,10,11,12) \end{cases} \tag{21}$$

where c is the characteristic speed of the lattice fluid and $c=(RT_c)^{1/2}$, Tc is the characteristic temperature.

The specific form of equilibrium distribution functions is based on the discrete Maxwellian equilibrium distribution function:

$$f_\alpha^{eq} = \frac{\rho}{(2\pi RT)^{D/2}} \exp(-\frac{(\vec{e}_\alpha - \vec{u})^2}{2RT}) \tag{22}$$

The equilibrium distribution functions of density and total energy are:

$$\begin{aligned} f_{1,2,3,4}^{eq} = \rho[&\frac{1}{12}(5 - \frac{4p}{\rho c^2}) + (2 - \frac{3p}{\rho c^2})(\frac{\vec{e}_\alpha \bullet \vec{u}}{3c^2}) \\ &- \frac{u^2}{6c^2} + \frac{(\vec{e}_\alpha \bullet \vec{u})^2}{6c^4} - \frac{(\vec{e}_\alpha \bullet \vec{u})u^2}{2c^4} + \frac{(\vec{e}_\alpha \bullet \vec{u})^3}{3c^6}] \end{aligned} \tag{23}$$

$$\begin{aligned} f_{5,6,7,8}^{eq} = \rho[&\frac{1}{8}(\frac{2p}{\rho c^2} - 1) + \frac{p}{\rho c^2}\frac{(\vec{e}_\alpha \bullet \vec{u})}{4c^2} - \frac{u^2}{8c^2} \\ &+ \frac{(\vec{e}_\alpha \bullet \vec{u})^2}{8c^4} - \frac{(\vec{e}_\alpha \bullet \vec{u})u^2}{8c^4} + \frac{(\vec{e}_\alpha \bullet \vec{u})^3}{8c^6}] \end{aligned} \tag{24}$$

$$f^{eq}_{9,12,11,12} = \rho[\frac{1}{24}(\frac{2p}{\rho c^2}-1)+(\frac{3p}{\rho c^2}-1)\frac{(\vec{e}_\alpha \bullet \vec{u})}{24c^2} + \frac{u^2}{24c^2}+\frac{(\vec{e}_\alpha \bullet \vec{u})^2}{48c^4}+\frac{(\vec{e}_\alpha \bullet \vec{u})^3}{96c^6}] \tag{25}$$

$$h^{eq}_{1,2,3,4} = \frac{5\rho E}{12} - \frac{p}{3c^2}(E+RT)+\frac{1}{6}\{(\rho E+p)\frac{(\vec{e}_\alpha \bullet \vec{u})}{c^2} + (\rho E+2p)[\frac{(\vec{e}_\alpha \bullet \vec{u})^2}{c^4}-\frac{u^2}{c^2}]\} \tag{26}$$

$$h^{eq}_{5,6,7,8} = -\frac{\rho E}{8} + \frac{p}{4c^2}(E+RT)+\frac{1}{8}\{(\rho E+p)\frac{(\vec{e}_\alpha \bullet \vec{u})}{c^2} + (\rho E+2p)[\frac{(\vec{e}_\alpha \bullet \vec{u})^2}{c^4}-\frac{u^2}{c^2}]\} \tag{27}$$

$$h^{eq}_{9,10,11,12} = -\frac{\rho E}{24} + \frac{p}{12c^2}(E+RT)+\frac{1}{48}\{(\rho E+p)\frac{(\vec{e}_\alpha \bullet \vec{u})}{c^2} + (\rho E+2p)[\frac{(\vec{e}_\alpha \bullet \vec{u})^2}{c^4}+2\frac{u^2}{c^2}]\} \tag{28}$$

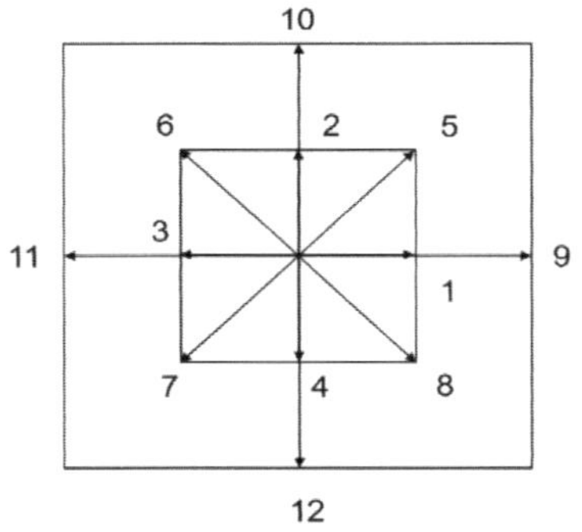

Fig. 1 D2Q12 discrete velocity model

2.3. *Time and space discretization*

Many people abandoned the standard LBM and chose to use finite difference method to discrete time and space [7], that was because they need a model suitable for non-uniform grid. When we just need standard lattice, the traditional discrete method and update rule is enough. So this study choose to use the standard LBM model, which means that spacial discretization and velocity discretization are coupled, and every particle with a discrete velocity will jump to the nearest point on the lattice.

To construct a standard lattice Boltzmann model, (1) and (2) need to be discretized. First, integrate (1) and (2) along the characteristic line:

$$f_\alpha(\vec{x}+\vec{e}_\alpha t, t+\delta t) - f_\alpha(\vec{x},t) = \int_0^{\delta t} \Omega_f(\vec{x}+\vec{e}_\alpha t', t+t')dt' \tag{29}$$

$$h_\alpha(\vec{x}+\vec{e}_\alpha \delta t, t+\delta t) - h_\alpha(\vec{x},t) = \int_0^{\delta t} \Omega_h(\vec{x}+\vec{e}_\alpha t', t+t')dt' \tag{30}$$

Where δt is the time step and

$$\Omega_f = (f^{eq}_\alpha - f_\alpha)/\tau_f \tag{31}$$

$$\Omega_h = (h^{eq}_\alpha - h_\alpha)/\tau_h + (\vec{e}_\alpha \bullet \vec{u} - u^2/2)(f - f^{eq})/\tau_{hf} \tag{32}$$

The trapezoidal rule can guarantee second-order accuracy and gives such time discrete scheme:

$$f_\alpha(\vec{x}+\vec{e}_\alpha\delta t,t+\delta t)-f_\alpha(\vec{x},t)=\frac{\delta t}{2}[\Omega_f(\vec{x}+\vec{e}_\alpha\delta t,t+\delta t)+\Omega_f(\vec{x},t)] \tag{33}$$

$$h_\alpha(\vec{x}_\alpha+\vec{e}_\alpha\delta t,t+\delta t)-h_\alpha(\vec{x},t)=\frac{\delta t}{2}[\Omega_h(\vec{x}_\alpha+\vec{e}_\alpha\delta t,t+\delta t)-\Omega_h(\vec{x},t)] \tag{34}$$

Then, introduce two new distribution functions:

$$\bar{f}_\alpha = f_\alpha - \frac{\delta t}{2}\Omega_f \tag{35}$$

$$\bar{h}_\alpha = h_\alpha - \frac{\delta t}{2}\Omega_h \tag{36}$$

Now, the equation (33) and equation (34) can be rewritten as:

$$\bar{f}_\alpha(\vec{x}+\vec{e}_\alpha\delta t,t+\delta t)-\bar{f}_\alpha(\vec{x},t)=-\omega_f[\bar{f}_\alpha(\vec{x},t)-f_\alpha^{eq}(\vec{x},t)] \tag{37}$$

$$\bar{h}_\alpha(\vec{x}+\vec{e}_\alpha\delta t,t+\delta t)-\bar{h}_\alpha(\vec{x},t)=-\omega_h[\bar{h}_\alpha(\vec{x},t)-h_\alpha^{eq}(\vec{x},t)]+(\omega_h-\omega_f)(\vec{e}_\alpha\bullet\vec{u}-u^2/2)(\bar{f}_\alpha-f_\alpha^{eq}) \tag{38}$$

where $\omega_f=2\delta t/(2\tau_f+\delta t)$, $\omega_h=2\delta t/(2\tau_h+\delta t)$. According to Li [7], the right hind side in (38) can be simplified and (38) can be rewritten as :

$$\bar{h}_\alpha(\vec{x}+\vec{e}_\alpha\delta t,t+\delta t)-\bar{h}_\alpha(\vec{x},t)=-\omega_h[\bar{h}_\alpha(\vec{x},t)-h_\alpha^{eq}(\vec{x},t)]+(\omega_h-\omega_f)(\vec{e}_\alpha\bullet\vec{u})(\bar{f}_\alpha-f_\alpha^{eq}) \tag{39}$$

And the macroscopic quantities can be computed by:

$$\rho = \sum_\alpha \bar{f}_\alpha \tag{40}$$

$$\rho\vec{u} = \sum_\alpha \vec{e}_\alpha \bar{f}_\alpha \tag{41}$$

$$\rho E = \sum_\alpha \bar{h}_\alpha \tag{42}$$

So $\bar{f}_\alpha$ and $\bar{h}_\alpha$ follow the same rules as f_α and h_α, at the same time guarantee second-order accuracy. Above is what the full model looks like.

3. Numerical Tests

Couette flow is a classical heat-transfer problem and is simulated by this model as a test. The distance between two planks is H, the top plank with a constant temperature T2 moves at speed U, while the bottom plank with a constant temperature T1 is stationary. The temperature profiles in an analytical solution is as follows:

$$T=\frac{y}{H}(T_2-T_1)+\frac{\Pr Ec}{2}\frac{y}{H}(1-\frac{y}{H})(T_2-T_1)+T_1 \tag{43}$$

where y is the distance from the bottom plank. Prandtl number is $\Pr=\mu c_p/\lambda$, Eckert number is $Ec=u^2/[c_p(T_2-T_1)]$. In this model, we can get $\Pr=\tau_f/\tau_h$, $Ec=2u^2/[(b+2)R(T_2-T_1)]$, and (43) can be rewritten as:

$$T = T_1 + \frac{y}{H}(T_2 - T_1) + \frac{U^2 \tau_f}{(b+2)R\tau_h}\frac{y}{H}(1 - \frac{y}{H}) \tag{44}$$

The simulation takes $N_x \times N_y = 40 \times 40$, the characteristic temperature is chosen to be $T_c = 5/3T_1$. The periodic boundary condition is applied to the left and right boundary, and the non-equilibrium extrapolation method is applied to the top and bottom planks. In Fig. 2 the results for U=0.1, $\gamma=7/5$ (namely b=5), Pr=1 (namely $\tau_f=\tau_h=0.9$), $T_2=1.0005T_1$, $1.001T_1$, $1.005T_1$, are shown. In Fig. 3 the results for U=0.1, $\gamma=5/3$, 7/5, 9/7 (namely b=3, 5, 7), Pr=1(namely $\tau_f=\tau_h=0.9$), $T_2=1.001T_1$, are shown. In Fig. 4 the results for U=0.1, $\gamma=7/5$ (namely b=5), Pr=1, 2, 3(namely $\tau_f=0.9$, 1.8, 2.7, $\tau_h=0.9$), $T_2=1.001T_1$, are shown. It can be seen that simulation recovers the analytical solution with good accuracy.

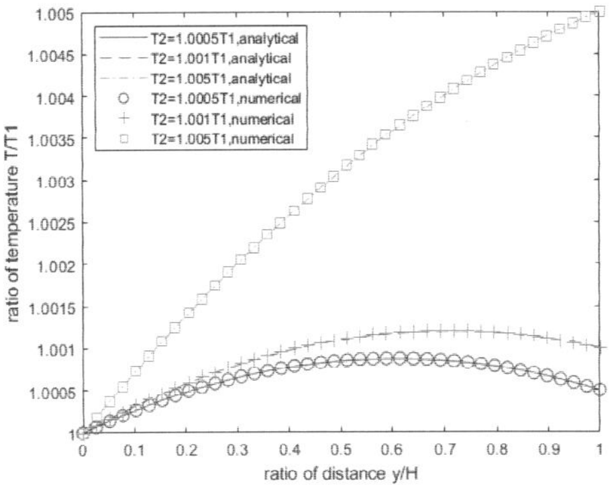

Fig. 2 Dimensionless temperature profiles for different T2

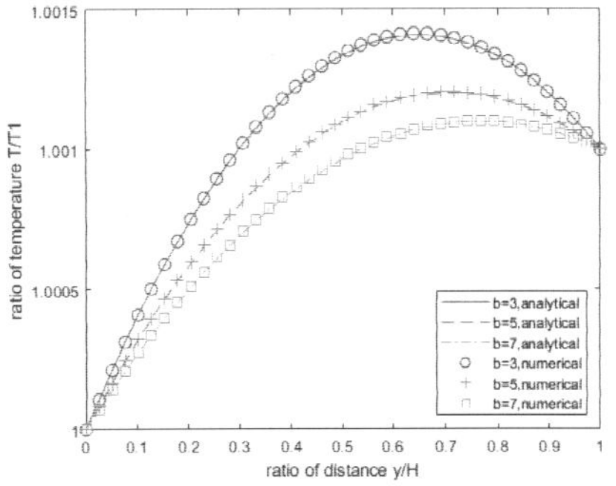

Fig. 3 Dimensionless temperature profiles for different b

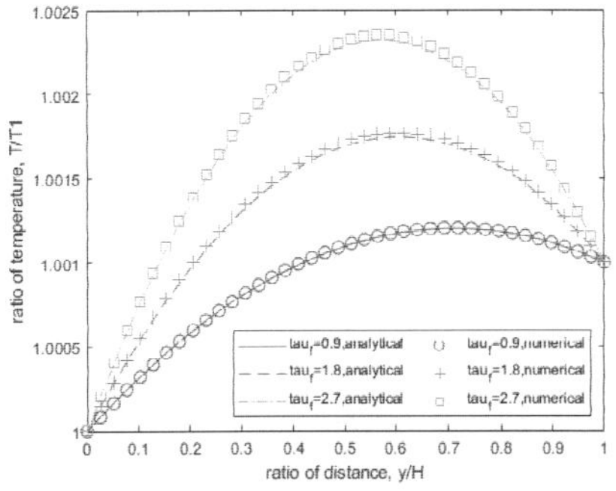

Fig. 4 Dimensionless temperature profiles for different tauf

4. Conclusion

In this paper, a coupled double distribution function lattice Boltzmann model with streaming-collision process is developed. The density distribution function together with total energy distribution function recovered continuity, momentum and energy equations, and they are coupled through gas' equation of state. The specific-heat ratio and Prandtl number are adjustable through two introduced constants. Different from existing coupled DDF lattice Boltzmann model, the traditional streaming-collision process is used in the new model to get linearity in streaming and locality in collision, and thus parallelism is retained. The simulation results show good agreement between this model and analytical solutions.

References

[1] Guy R. McNamara, Gianluigi Zanetti, Use of the Boltzmann equation to simulate lattice gas automata, Physical review letters, 1988, 61(20): 2332-2335.
[2] Xu Aiguo, Chen Jie, Song Jiahui, Chen Dawei, Chen Zhihua, Progress of discrete Boltzmann study on multiphase complex flows, Acta Aerodynamica Sinica, 2021, 39(3): 138-169.
[3] Guo Zhaoli, Zheng Chuguang, Shi Baochang), TS. Zhao, Thermal lattice Boltzmann equation for low Mach number flows: Decoupling model, Physical Review E, 2007, 75: 036704.
[4] P. Lallemand, LS. Luo, Theory of the lattice Boltzmann method: Acoustic and thermal properties in two and three dimensions, Physical Review E, 2003, 68: 036706.
[5] M. Watari, M. Tsutahara, Possibility of constructing a multispeed Bhatnagar-Gross-Krook thermal model of the lattice Boltzmann method, Physical Review E, 2004, 70: 016703.
[6] X. He, S. Chen, GD. Doolen, A novel thermal model for the lattice Boltzmann method in incompressible limit, Journal of Computational Physics, 1998, 146(1): 282-300.
[7] Q. Li, Y.L. He, Y. Wang, W.Q. Tao, Coupled double-distribution-function lattice Boltzmann method for the compressible Navier-Stokes equations, Physical Review E, 2007, 76: 056705.

Dynamic Load Analysis of an Centrifugal Pump Axle using FSI Simulation

Biaohua Cai *, Zheng Zhang, Yinzhong Chen, Zhou Tang

Wuhan Second Ship Design & Research Institute, Wuhan 430064, Hubei, China

* Corresponding author e-mail: jxs204204204@163.com

Abstract. The mechanical characteristics of the impeller are important indicators that determine the working characteristics of the pump. The pump is subjected to periodic hydraulic loads during the working process. This load will cause the dynamic deformation of the impeller and the impeller shaft, thereby further affecting the distribution of the fluid. Liquid-solid coupling is used. Method, the rated operating point of centrifugal pump is analyzed, and it is proved that the dynamic stress of the impeller and the impeller shaft meets the requirements, which lays a foundation for the formulation of the control strategy of the impeller.

Keywords: CFD, impeller, liquid-solid coupling.

1. Introduction

During the working process of centrifugal pump, because of the high-speed rotation, the flow field structure in the pump keeps changing. Therefore, the impeller caused by the load will also change and produce fluid exciting force, causing the centrifugal pump impeller shaft vibration. Due to the interaction of liquid and solid, the process is a highly coupled nonlinear process, thus Liquid-solid coupling method must be used to solve the problem. Based on this, he transient CFD simulation of three-dimensional unsteady flow field in the flow passage of centrifugal pump and the transient structural dynamic coupling analysis of impeller and impeller shaft are adopted in this paper. The solution is shown in figure 1.

2. CFD modeling of three-dimensional unsteady fluid flow

2.1. FSI simulation scheme of the centrifugal pump

This article employe the joint simulation of the commercial engineering software FLUENT and ABAQUS to carry out the dynamic load caculation of a centrifugal pump. The solving scheme is shown in the figure below.

Content from this work may be used under the terms of the Creative Commons Attribution 3.0 licence. Any further distribution of this work must maintain attribution to the author(s) and the title of the work, journal citation and DOI.
Published under licence by IOP Publishing Ltd

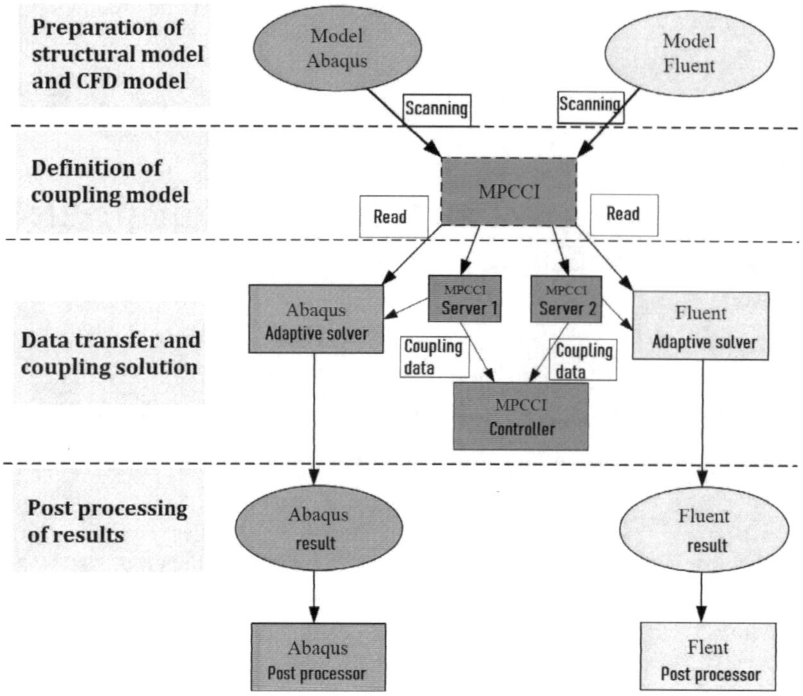

Fig. 1 The solution process of liquid-solid coupling problem

3. Establishment of multi-flexible body dynamic model of impeller shaft

3.1. Grid Model

The centrifugal pump impeller shaft model is built in UG software, as shown in the figure 2. On this basis, the grid is divided.

3.2. Boundary Condition

Because the liquid-solid coupling model mainly considers the dynamic response of the impeller and the impeller shaft under the fluid periodic load, the force on the impeller is reasonably simplified here. The axial load in the process of impeller assembly is not considered and neglects the bearing flexibility.

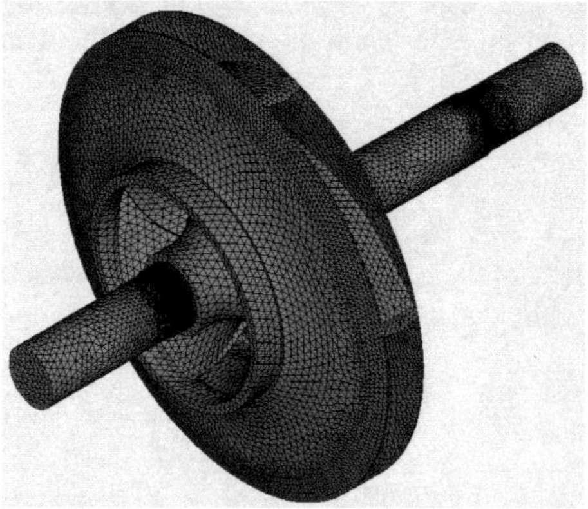

Fig. 2 Finite element mesh model of impeller shaft

4. Results of three-dimensional unsteady flow analysis

T=0.0015s

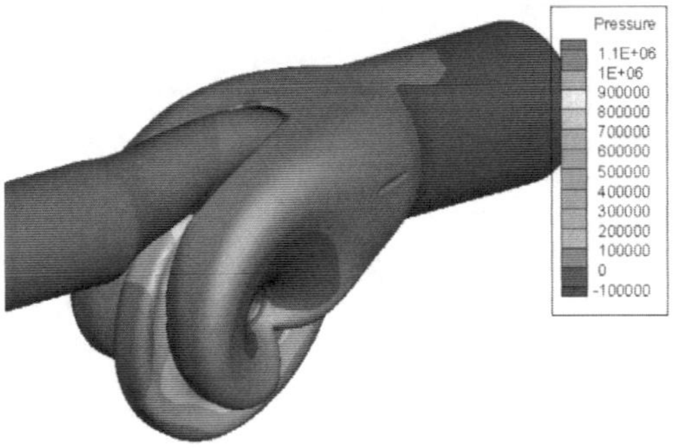

Fig. 3 The pressure distribution of the whole flow between the impeller and the volute

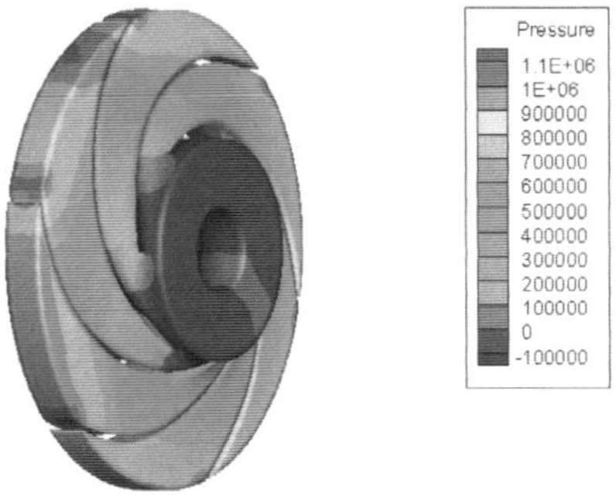

Fig. 4 The pressure distribution of the flow in the impeller

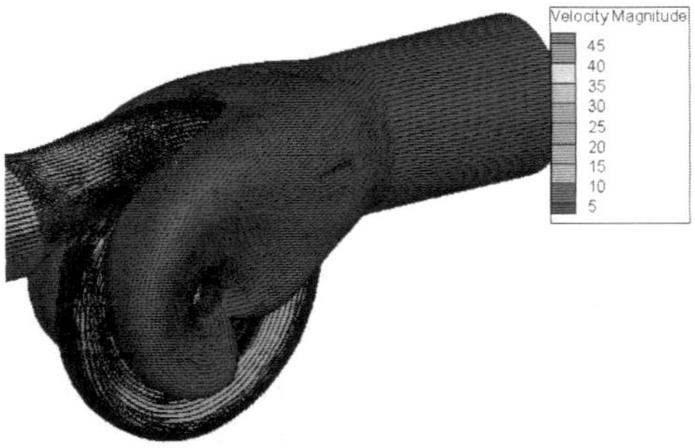

Fig. 5 The velocity distribution of the overall flow between the impeller and the volute

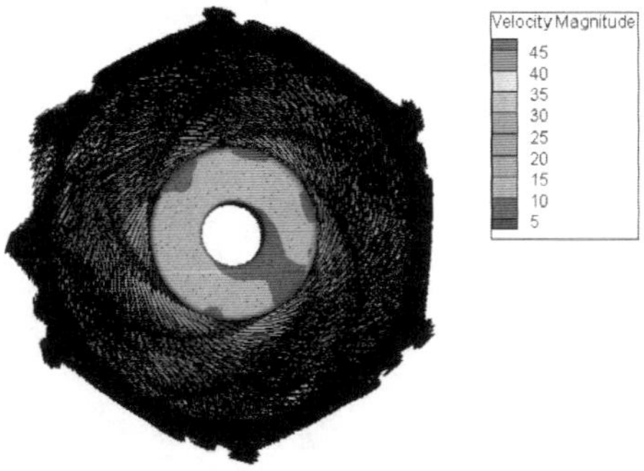

Fig. 6 The velocity distribution of the flow in an impeller

5. Dynamic load analysis of impeller shaft

In this paper, the FSI method is used to simulate a centrifugal pump at its rated operating point. The results of the simulation analysis are listed below.

5.1. Force analysis of impeller shaft

We can know from the above modeling and analysis process that the coupling pressure load, gravity and inertia force produced by the circular motion of the bearing are balanced by the support and reaction force of the bearing and the input torque of the steam turbine. Through the analysis of liquid-solid coupling, the dynamic values of the bearing and the turbine input torque can be obtained. Furthermore, the calculated dynamic force on the impeller shaft is of great significance to the performance analysis of the centrifugal pump mechanical system. The figure below shows the force curves of bearing 1 and bearing 2.

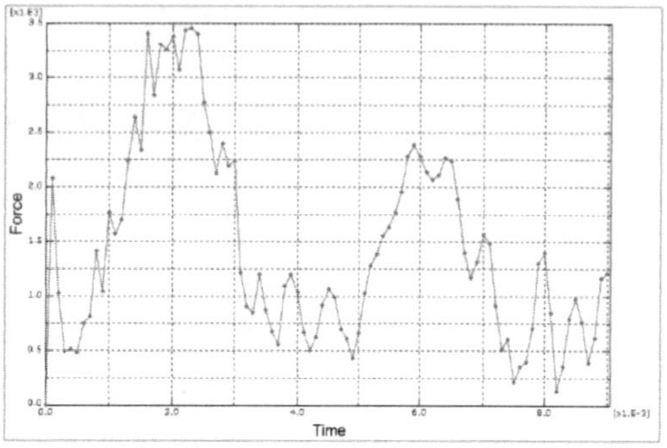

Fig. 7 Bearing 1 radial force

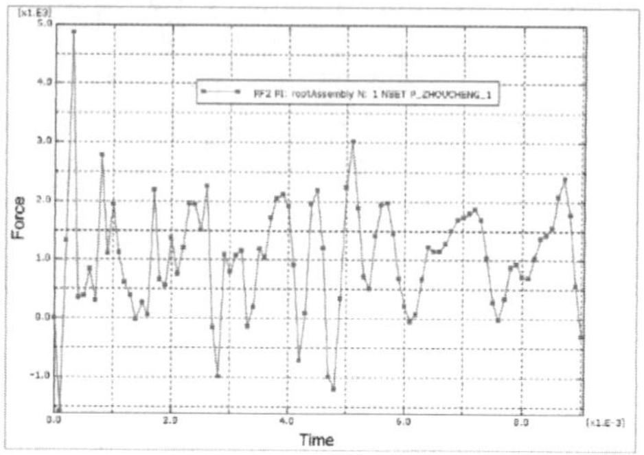

Fig. 8 Bearing 1 axial force

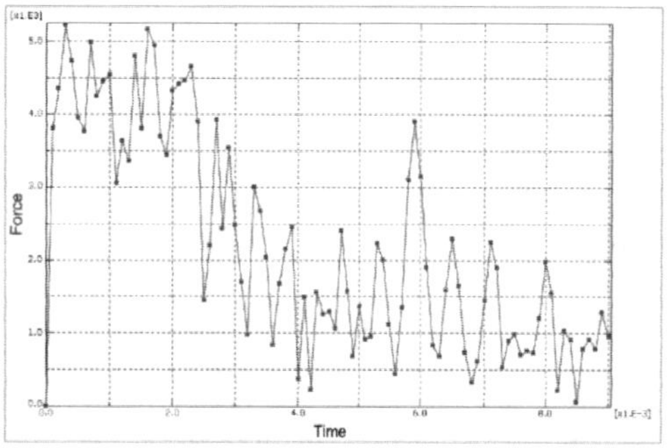

Fig. 9 Bearing 2 radial force

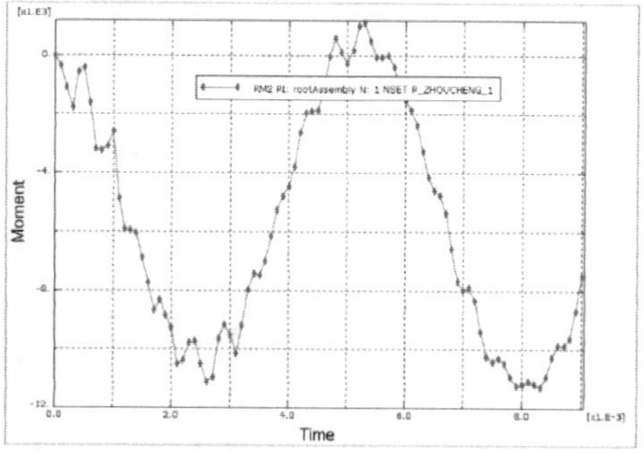

Fig. 10 Turbine input torque curve My

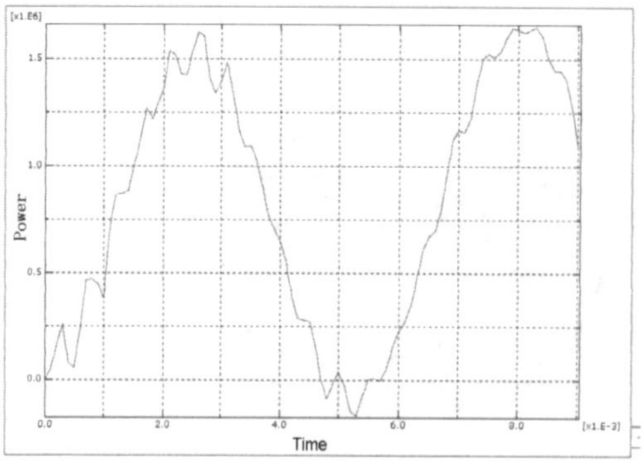

Fig. 11 Steam turbine input power

5.2. Dynamic stress analysis of impeller shaft and impeller
Figure 12 and Figure 13 are the dynamic stress diagram of the impeller shaft at different times.

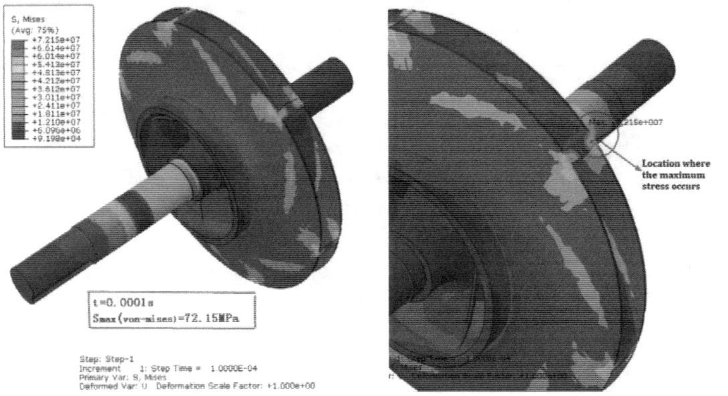

Fig. 12 Dynamic stress distribution of impeller and impeller shaft when t=0.0001s

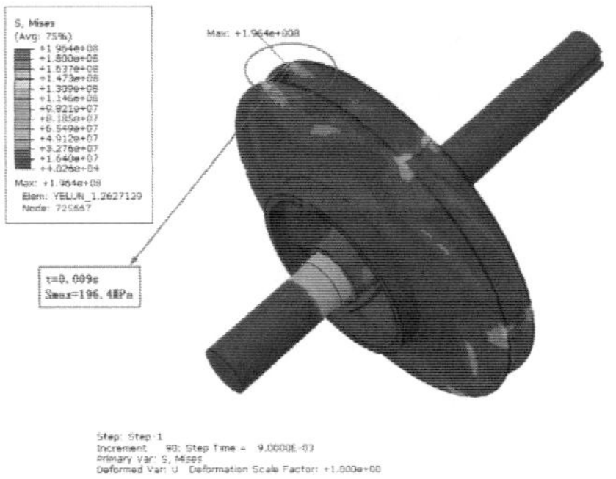

Fig. 13 Dynamic stress distribution of impeller and impeller shaft when t=0.009s

In the time of 0.009s, the turning angle of the impeller shaft is reached 75.6o>60o. This impeller has six blades which are uniformly distributed in circumferential direction, and its motion law has periodicity. The simulation time is longer than one cycle, and the simulation result can reflect the whole working state of the impeller shaft.

5.3. Dynamic deformation analysis of impeller shaft and impeller
The following diagram shows the results of the dynamic deformation analysis of the impeller and the impeller shaft.

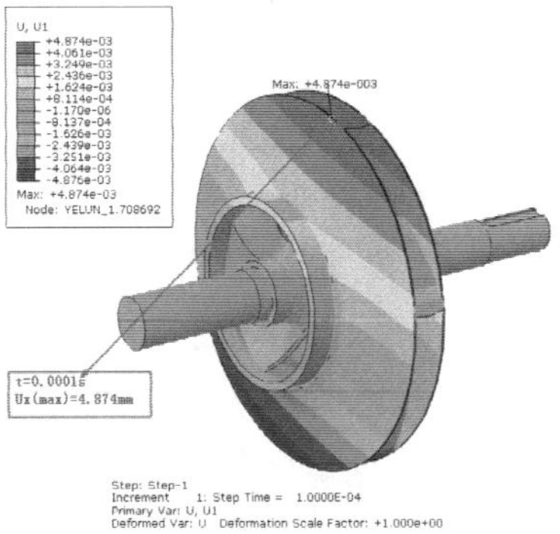

Fig. 14 Dynamic deformation distribution diagram of impeller and impeller shaft when t=0.0001s

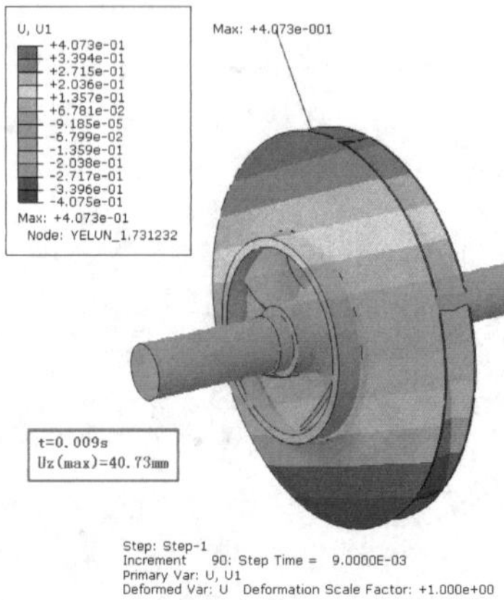

Fig. 15 Dynamic deformation distribution diagram of impeller and impeller shaft when t=0.009s

As can be seen from the graph above, the displacement distribution of the dynamic deformation graph at 0.009 is very similar to that at 0.0001s. But because the absolute value of the deformation is based on the global coordinate system, therefore, the actual deformation value of the impeller and the impeller shaft in its own local coordinate system should also subtract the displacement caused by the rigid rotation of the impeller and the impeller shaft. The calculated deformation is about 5.0 mm, and there is a certain fluctuation.

6. Conclusions

The mechanical characteristics of impeller is an important index to determine the working characteristics of pump. The pump is subject to periodic hydraulic load in the working process, which will lead to the dynamic deformation of the impeller and the impeller shaft, thus further affecting the distribution of fluid. Therefore, this process is a typical coupling process, which must be analyzed by FSI method. In this paper, such a implementation is used to analyze the centrifugal pump. The simulation results said that the dynamic stresses of the impeller meet the requirement. The dynamic deformation of impeller and shaft fluctuates about 5.0 mm, which also meets the design requirements. By means of the liquid-solid coupling method, the dynamic supporting force curve of the bearing and the dynamic input torque curve of the steam turbine are obtained.

References

[1] Chen Chunji. Discussion on the causes and solutions of cavitation phenomenon in liquefied petroleum gas pump [J]. Chemical Enterprise Management, 2018, (26): 37-38.

[2] PARK S, YANG J, RHEE S H. Parametric study on ship's exhaust-gas behavior using computational fluid dynamics [J]. ENGINEERING APPLICATIONS OF COMPUTATIONAL FLUID MECHANICS, 2017, 11(1): 159-171.

[3] Zhang Bo,Fu Weibiao,Wu Xianghui.Effect of Hydrogen on Fuel Consumption of Diesel Engine [J]. Journal of Combustion Science and Technology, 2006, (03): 238-242.

[4] MURAI Y. Frictional drag reduction by bubble injection [J]. EXPERIMENTS IN FLUIDS, 2014, 55(7).

[5] KANG C, ZHANG W, ZOU Z W, et al. Effects of initial bubble size on geometric and motion characteristics of bubble released in water [J]. JOURNAL OF CENTRAL SOUTH UNIVERSITY, 2018, 25(12): 3021-3032.

[6] XIAO H, GENG S J, CHEN A Q, et al. Bubble formation in continuous liquid phase under industrial jetting conditions [J]. CHEMICAL ENGINEERING SCIENCE, 2019, 200: 214-224.

Research on the Arrangement Process of Marine Submersible Buoy System based on Computational Fluid Dynamics

Xiaohan Chen *

School of Mechanical Engineering, Nanjing University of Science and Technology, Nanjing 210094, China

*Corresponding author e-mail: chenxiaohan@njust.edu.cn

Abstract. This paper establishes the computational fluid dynamics models of marine submersible buoy system. The models use incompressible Navier-Stokes equations, Renormalization-Group (RNG) turbulence model and the fractional areas/volume obstacle representation method. Using the lumped mass mothed to simulate the anchor line and power supply cable. The result show that the great tension forces increase occurs at the moment when the mooring lines was straightened and after a marine submersible buoy system sinks to the sea floor.

Keywords: Submersible Buoy System, computational fluid dynamics, lumped mass method.

1. Introduction

With the development and utilization of marine resources, the exploration of the seabed environment has become more in-depth. Marine submersible buoy system is an important equipment for underwater environment observation. The submarine buoy system can obtain the hydrological data of the seabed, and has the characteristics of low cost, concealment and good mobility.

This paper researched the submarine buoy system connected by anchor line and power supply cable. In the submarine buoy system, the anchor block is connected with the battery compartment through the anchor line. The battery compartment is connected with the submersible buoy through the power supply cable. In the whole system, the anchor block is made of reinforced concrete, and its density is $3.5g/cm^3$, which is greater than the density of water. While the density of submersible buoy and battery compartment is less than that of water.

During the arrangement process of the submersible buoy system, the submersible buoy and battery compartment are first put on the sea, and then the anchor block is thrown into the water. The anchor block will sink quickly after being thrown into the water. When the distance between the anchor block and the battery compartment is greater than the length of the anchor line, the anchor block will pull the battery compartment into the water together. Subsequently, the submersible was also pulled into the water. To make sure that the anchor line and power supply cablc not broken during the arrangement process, this paper researched the mechanical behavior of the anchor line and power supply cable.

Content from this work may be used under the terms of the Creative Commons Attribution 3.0 licence. Any further distribution of this work must maintain attribution to the author(s) and the title of the work, journal citation and DOI.
Published under licence by IOP Publishing Ltd

2. The numerical method

2.1. Fluid model method

Based on fractional areas/volume obstacle representation method [1], the three dimensional incompressible Navier-Stokes equations in cartesian coordinate system are established as follows.

$$\begin{cases}\frac{\partial u}{\partial t}+\frac{1}{V_F}\left(uA_x\frac{\partial u}{\partial x}+vA_y\frac{\partial u}{\partial y}+wA_z\frac{\partial u}{\partial z}\right)=-\frac{1}{\rho}\frac{\partial p}{\partial x}+G_x+f_x\\ \frac{\partial v}{\partial t}+\frac{1}{V_F}\left(uA_x\frac{\partial v}{\partial x}+vA_y\frac{\partial v}{\partial y}+wA_z\frac{\partial v}{\partial z}\right)=-\frac{1}{\rho}\frac{\partial p}{\partial y}+G_y+f_y\\ \frac{\partial w}{\partial t}+\frac{1}{V_F}\left(uA_x\frac{\partial w}{\partial x}+vA_y\frac{\partial w}{\partial y}+wA_z\frac{\partial w}{\partial z}\right)=-\frac{1}{\rho}\frac{\partial p}{\partial z}+G_z+f_z\end{cases} \tag{1}$$

$$\begin{cases}f_x=-\frac{1}{\rho V_F}\left[\frac{\partial}{\partial x}(A_x\tau_{xx})+\frac{\partial}{\partial y}(A_y\tau_{xy})+\frac{\partial}{\partial z}(A_z\tau_{xz})\right]\\ f_y=-\frac{1}{\rho V_F}\left[\frac{\partial}{\partial x}(A_x\tau_{xy})+\frac{\partial}{\partial y}(A_y\tau_{yy})+\frac{\partial}{\partial z}(A_z\tau_{yz})\right]\\ f_z=-\frac{1}{\rho V_F}\left[\frac{\partial}{\partial x}(A_x\tau_{xz})+\frac{\partial}{\partial y}(A_y\tau_{yz})+\frac{\partial}{\partial z}(A_z\tau_{zz})\right]\end{cases} \tag{2}$$

$$\begin{cases}\tau_{xx}=-2\mu\left[\frac{\partial u}{\partial x}-\frac{1}{3}\left(\frac{\partial u}{\partial x}+\frac{\partial v}{\partial y}+\frac{\partial w}{\partial z}\right)\right]\\ \tau_{xy}=-\mu\left(\frac{\partial v}{\partial x}+\frac{\partial u}{\partial y}\right)\\ \tau_{yy}=-2\mu\left[\frac{\partial v}{\partial y}-\frac{1}{3}\left(\frac{\partial u}{\partial x}+\frac{\partial v}{\partial y}+\frac{\partial w}{\partial z}\right)\right]\\ \tau_{xy}=-\mu\left(\frac{\partial u}{\partial z}+\frac{\partial w}{\partial x}\right)\\ \tau_{zz}=-2\mu\left[\frac{\partial w}{\partial z}-\frac{1}{3}\left(\frac{\partial u}{\partial x}+\frac{\partial v}{\partial y}+\frac{\partial w}{\partial z}\right)\right]\\ \tau_{yz}=-\mu\left(\frac{\partial v}{\partial z}+\frac{\partial w}{\partial y}\right)\end{cases} \tag{3}$$

where ρ is the fluid density, u, v and w are the fractional quantities of the surface product of the fluid velocity above x, y and z respectively. A_x, A_y and A_z are the fluid in the x, y, and z directions respectively, V_F is the volume fraction of the fluid; P is fluid pressure, G_x, G_y and G_z are gravitational acceleration, f_x, f_y and f_z are the accelerations of viscous forces.

The RNG K - ε turbulence model [2,3] is used in the turbulence model. Compared with the standard K-ε model, it is more suitable to describe the flow in the region with strong shear. The equation k of turbulent kinetic energy is

$$\frac{\partial}{\partial t}(\rho k)+\frac{\partial}{\partial x_i}(\rho k u_i)=\frac{\partial}{\partial x_j}\left[\sigma_k(\mu+\mu_t)\frac{\partial k}{\partial x_j}\right]+G_k-\rho\varepsilon \tag{4}$$

where u_i and u_j are velocity components in the direction of x_i and x_j, μ is fluid viscosity, μ_t is dynamic viscosity, G_k is the generation term of turbulent kinetic energy k caused by average velocity gradient, constant coefficient σ k = 1. 39.

Turbulent dissipation rate ε The equation is

$$\frac{\partial}{\partial t}(\rho k)+\frac{\partial}{\partial x_i}(\rho k u_i)=\frac{\partial}{\partial x_j}\left[\sigma_k(\mu+\mu_t)\frac{\partial k}{\partial x_j}\right]+G_k-\rho\varepsilon+C_{1\varepsilon}^*\frac{\varepsilon}{k}G_k-C_{2\varepsilon}\rho\frac{\varepsilon^2}{k} \tag{5}$$

where, equation coefficient $C_{1\varepsilon}^*$, Constant coefficient $C_{2\varepsilon}$ and σ_ε are given by

$$\begin{cases}C_{1\varepsilon}^*=C_{1\varepsilon}-\frac{\eta(1-\eta/\eta_0)}{1+\beta\eta^3},\eta=\left(2E_{ij}\cdot E_{ij}\right)^{1/2}\frac{k}{\varepsilon}\\ E_{ij}=\frac{1}{2}\left(\frac{\partial u_i}{\partial x_j}+\frac{\partial u_j}{\partial x_i}\right),\sigma_\varepsilon=1.39\\ C_{1\varepsilon}=1.42,C_{2\varepsilon}=1.68\\ \eta_0=4.377,\beta=0.012\end{cases} \tag{6}$$

2.2. Anchor line and power supply cable model
Power supply cable and anchor line are simulated using lumped mass method. This method divides the rope into several segments for calculation. It is assumed that each section of the rope is a cylinder with uniform diameter and uniform material distribution. Each segment has its own length, diameter, mass density, and physical properties. The mooring motion is calculated by the centralized method. A mooring line is evenly divided into several discrete segments represented by mass particles in the center of segment.

The kinetic equation of the piecewise motion is

$$m_p \frac{dV_p}{dt} = G + B + T_1 + T_2 + D_n + D_t \tag{7}$$

where, G is gravity, B is buoyancy, $m_p$is segment mass, V_p is velocity at the center of segment mass, T_1 and T_2 are mass at both ends of segment respectively, D_n and D_t are normal and tangential resistance of segment respectively.

$$\begin{cases} D_n = -C_{D,n}\rho_f A_n v_{r,n}|v_r| \\ D_t = -C_{D,t}\rho_f A_t v_{r,t}|v_r| \end{cases} \tag{8}$$

where, $C_{D,n}$ and $C_{D,t}$ are tangential and normal drag coefficients of segments respectively, v_r is the relative velocity between the segment and the surrounding fluid, $v_{r,n}$ and $v_{r,t}$ are the tangential and normal components of v_r in segment respectively, ρ_f is the fluid density. A_n is the cross-sectional area of the segment, A_t is the longitudinal area of the segment.

3. Calculation model
The computing domain size is 51m×3m×6m. The anchor block falls to the water surface from a height of three meters, so the anchor block has an initial downward velocity of -7.67m/s. The initial velocity of the submersible buoy and the battery compartment is zero. The highest point of the anchor block, battery compartment and submersible buoy is on the same plane. To simplify the calculation, only the z-direction and x-direction degrees of freedom of the three objects are open. The length of anchor rope and supply cable is 12m. The underwater net weight of anchor block was calculated at 60kg, 80kg and 100kg respectively.

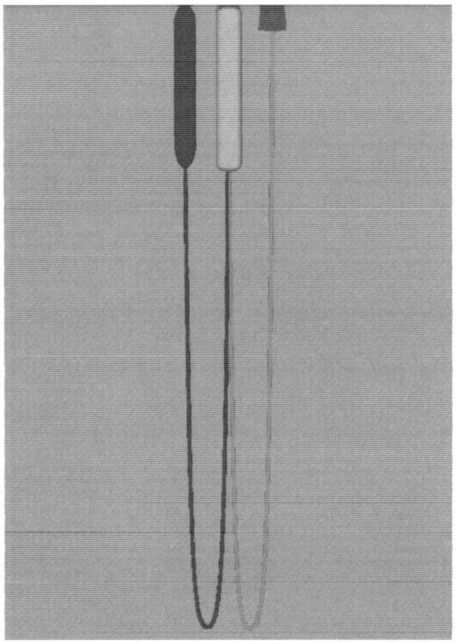

Figure 1. Calculation model of submersible buoy system

4. Calculation results and analysis

4.1. System speed variation

After the anchor block enters the water, the speed decreases rapidly to a stable value, and then pulls the battery compartment and the submersible buoy into the water. The velocity of the battery compartment and the submersible buoy showed a tendency of vibration change after entering water. The amplitude of vibration changes gradually decreases to a steady state. After the anchor block falls to the sea floor, the velocity of the submersible buoy and the battery compartment shows a trend of oscillation decreasing.

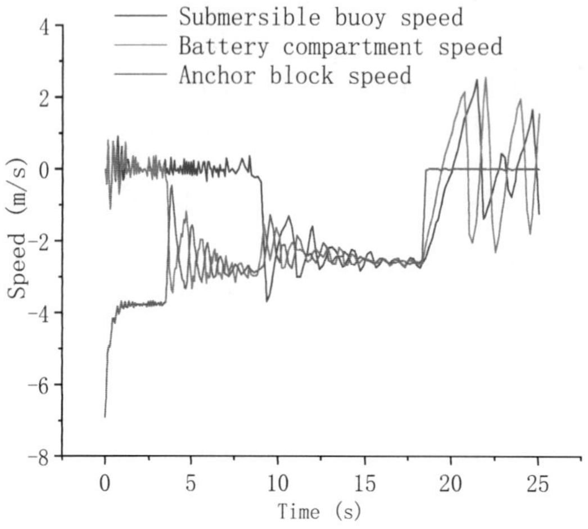

Figure 2. Speed variation of submersible buoy system with 60kg underwater net weight of anchor block

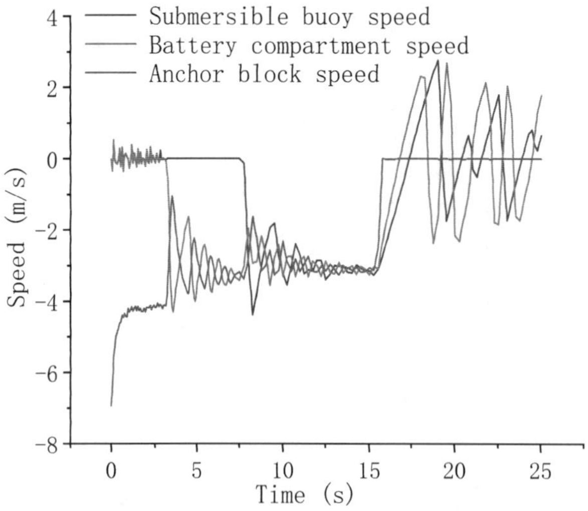

Figure 3. Speed variation of submersible buoy system with 80kg underwater net weight of anchor block

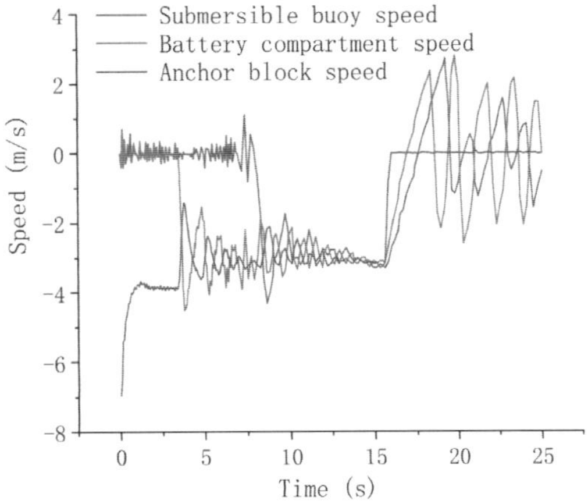

Figure 4. Speed variation of submersible buoy system with 100kg underwater net weight of anchor block

4.2. Stress changes of anchor rope and supply cable

When the anchor line and power supply cable are straightened, a greater tension occurs. Subsequently, the tensile force shows a trend of oscillation change and decrease. When the anchor block lands on the sea floor, the battery compartment and submersible buoy continue to move downward for some distance due to inertia before rising. In the process of rising, there is a great tension in the anchor line and power supply cable.

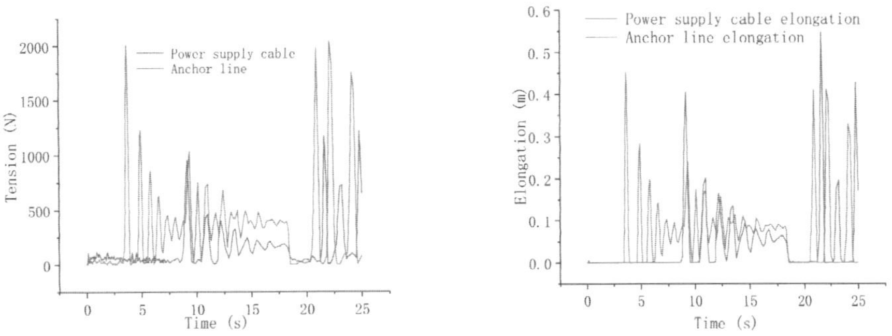

Figure 5. Tension and elongation of anchor line and power supply cable of submersible buoy system with 60kg underwater net weight of anchor block

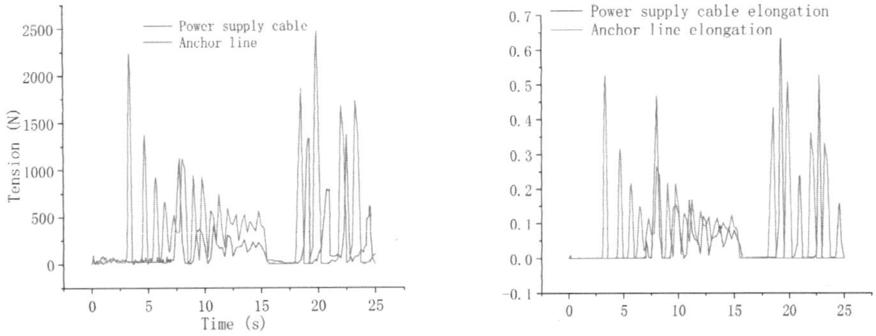

Figure 6. Tension and elongation of anchor line and power supply cable of submersible buoy system with 80kg underwater net weight of anchor block

Journal of Physics: Conference Series **2441** (2023) 012054 doi:10.1088/1742-6596/2441/1/012054

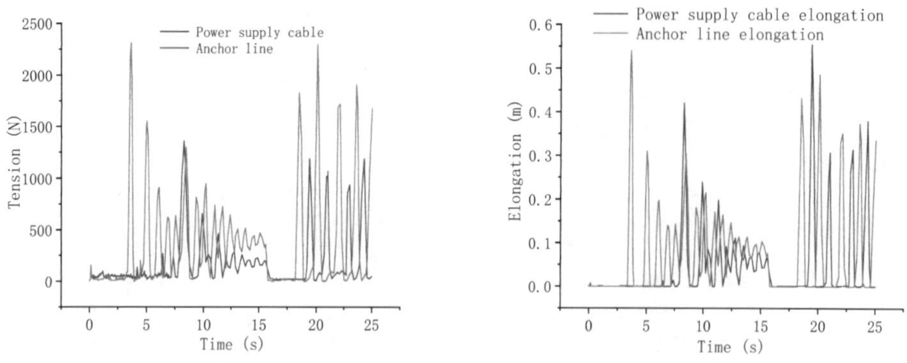

Figure 7. Tension and elongation of anchor line and power supply cable of submersible buoy system with 100kg underwater net weight of anchor block

5. Conclusion

When the anchor line and power supply cable are straightened, there will be a large tension.

The greater the weight of the anchor block, the greater the tension on the anchor line and the power supply cable when straightening.

After the anchor block sinks to the sea floor, the positive buoyancy of the battery compartment and the buoy will cause a large tension on the anchor line and power supply cable. The pulling force is roughly the same as when it is straightened.

Reference

[1] Hirt C W, Sicilian J M. A porosity technique for the definition of obstacles in rectangular cell meshes. 1985.

[2] Yakhot V, Orzag S A . "Renormalization group analysis of turbulence-i: basic theory" [J]. Journal of Scientific Computing, 1986:1-51.

[3] Yakhot V, Smith L M. The renormalization group, the -expansion and derivation of turbulence models [J]. Journal of Scientific Computing, 1992, 7(1):35-61.

High-Resolution Hemodynamic Analysis of a Wearable Oxygenator Using High-Performance Computing

Wei Chen, Rongliang Chen*

Shenzhen Institutes of Advanced Technology Chinese Academy of Sciences, Shenzhen, China

*Corresponding author e-mail: rl.chen@siat.ac.cn

Abstract. In this paper, we study the hydrodynamics of a wearable oxygenator using computational fluid dynamics on a supercomputer. In this computation, the fiber bundle of the oxygenator is modeled by a porous medium model, and the incompressible Navier-Stokes equations with a K-ε turbulence model are used to model the flow in the channels. The blood flow velocity, pressure, and shear stress are carefully studied and the hemolysis index of the oxygenator NIH is calculated by using the fast three-dimensional numerical hemolysis approximation method, which is in the range 0.006 ~ 0.094 g/100L. The results show that the pressure loss of the wearable oxygenator mainly occurs in the fiber bundle area. The inlet, outlet, and channels have relatively high shear stress, which may damage the red blood cells. When the flow rate ranges from 2.0 L/min to 5.0 L/min, the estimated hemolysis meets the applicable range allowed by human physiology. In addition, the parallel performance is studied on a supercomputer, which shows that, for the simulation with over 9 million mesh cells, it scales up to 720 processors and the parallel efficiency is over 60%.

Keywords: oxygenator; hemolysis index; computational fluid dynamics; OpenFOAM; parallel computing.

1. Introduction

Extracorporeal Membrane Oxygenation (ECMO) is a medical equipment which provides a long-term gas exchange for patients with severe but potentially reversible respiratory and heart failure [1]. There are several forms of ECMO; the two most common are the veno-venous (VV) and veno-arterial (VA) ECMO. The core component of the VV ECMO is the artificial lung, also known as an oxygenator, which is used to provide oxygenation function for patients with respiratory failure.

During the last few decades, in order to meet the respiratory support needs of patients with lung diseases, researchers have developed a variety of oxygenators with different structures and materials [2]. The development of oxygenators can be divided into four generations according to their structures and materials: bubble oxygenator before the 1950s [3,4], plate oxygenator from the 1950s to 1980s [5–7], membrane oxygenator after the 1980s [8,9], and the latest cardiopulmonary bypass circuit [10,11].

In the past, the design of oxygenators is mainly based on

the experiment which is highly material and time cost, and it is difficult to directly measure the hydrodynamics in the oxygenator due to the opacity of the device. In contrast, computational fluid

Content from this work may be used under the terms of the Creative Commons Attribution 3.0 licence. Any further distribution of this work must maintain attribution to the author(s) and the title of the work, journal citation and DOI.
Published under licence by IOP Publishing Ltd

dynamics (CFD) provides a fast, flexible and accurate way for the design and optimization of oxygenators. Zhang et al. designed a novel pump oxygenator made of microporous hollow fiber membranes and optimize the whole geometry and flow path using CFD [12], and then verified its performance by vitro test [2]. Kaesler et al. established a micro-scale CFD model [13] to simulate oxygen transfer in blood at the fiber level. Taskin et al. conducted CFD simulations of a wearable artificial pump lung, and the simulation results show that it has good oxygen transfer performance and biocompatibility [14]. Madhani et al. designed an oxygenator with circulating blood flow and a portable oxygenator [15–17], and carefully analyzed its hydrodynamic properties and shear stress using CFD.

In this study, a wearable oxygenator is designed based on Affinity NT Oxygenation System (Medtronic Inc., Minneapolis, MN, U.S.A.) and its hydrodynamics are studied using the open-source computational fluid dynamics library OpenFOAM, where the oxygenator's internal hemodynamic simulation is carried out, and the velocity, pressure, and shear stress of the blood flow are carefully analyzed. The hemolysis index NIH of the oxygenator is calculated and analyzed using the fast hemolysis numerical prediction model. In addition, the parallel performance of the simulation method is studied on a supercomputer. Results of this study have certain guiding significance for the follow-up designs and simulations of the wearable oxygenator.

2. Materials and methods

In this section, we first introduce the concept of wearable oxygenator and show the detailed geometry parameters of the model that used in this study, and then the specific method and process of the oxygenator hemodynamic simulation are given.

2.1. Concept of wearable oxygenator

The external outline and internal structure of the wearable oxygenator (hereinafter referred to as Oxyg-I) are shown in Fig. 1 (a) and Fig. 1 (b), respectively.

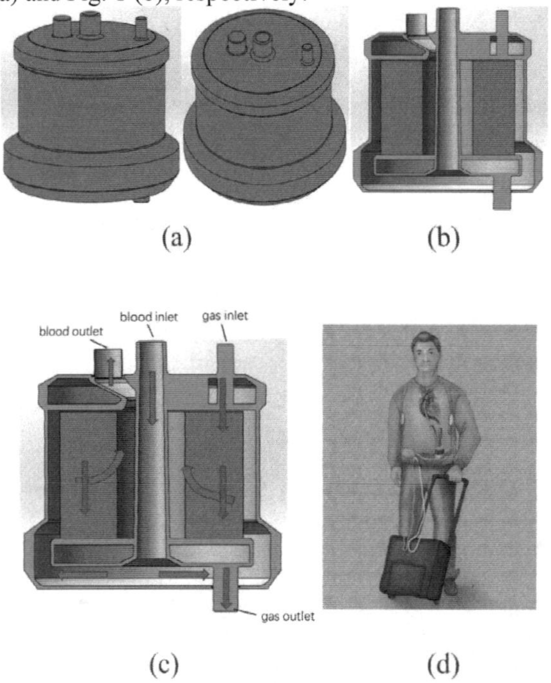

Figure 1. Geometric characteristics, Conceptualization, and usage of Oxyg-I: (a) Top and front views of the wearable oxygenator, (b) Internal view of the wearable oxygenator, (c) Conceptualization of the flow path and components for the wearable oxygenator, (d) Schematic of the patient implanted with the wearable oxygenator

Oxyg-I is composed of the inlet and outlet of blood and oxygen as well as chambers, The chambers are filled with fiber bundles. Oxyg-I is the key component of ECMO, which

exchanges oxygen and carbon dioxide in the blood of a patient with severe respiratory. The blood enters from the blood inlet and then goes upward through the fiber bundle, exchanging gas with the sweep oxygen from top to bottom at the membrane hole of the fiber wall. Finally, the oxygenated blood goes out of the oxygenator from the blood outlet. The whole process is shown in Fig. 1 (c), and as shown in Fig. 1 (d) [18], the Oxyg-I and the controller can be wearable to enable the patient to move around.

2.2. Design parameters

The prime volume of the oxygenator is about 260 ml, with a prime device length of 118 mm. The fiber material of the wearable oxygenator is Celgard x30-240 (3M Medical Membranes, Wuppertal, Germany), which is one of the most widely used microporous Polypropylene materials for commercial oxygenators [19]. Celgard x30-240 has an outer diameter of 300 μm and an inner diameter of 240 μm, and the total surface area of the fiber tube is 1.05 m2. The detailed parameters of the model are shown in Table I [2].

Table 1. Specifications of the wearable oxygenator

Model name	**Oxyg-I**
Recommended Blood flow rate	1~5 L/min
Diameter of arterial outlet port	9.525 mm
Diameter of venous inlet port	9.525 mm
Diameter of gas inlet/outlet port	6.35 mm
Fiber bundle thickness	22.225 mm
Inner diameter of the fiber bundle	27 mm
Outer diameter of the fiber bundle	73 mm
Effective fiber length	50.8 mm
Device length	118 mm

2.3. Computational fluid dynamics

In this study, the blood flow is regarded as an incompressible Newtonian fluid, and the transient Navier-Stokes equations are used to model the motion of blood in the Oxyg-I model,

$$\nabla\cdot\vec{v}=0 \tag{1}$$

$$\frac{\partial\vec{v}}{\partial t}+\rho\nabla\cdot(\vec{v}\vec{v})=-\nabla p+\nabla\cdot\bar{\tau}+\rho\vec{g}+S \tag{2}$$

where ρ is the density of the blood, $\vec{v}$ is the velocity vector of blood, $\bar{\tau}$ is the stress tensor, p is the pressure, $\vec{g}$ is the gravitational force, and S is the source term. The standard k-ε turbulence model is applied to model the turbulence in the blood flow [20].

1) Porous media model

The porous media model is one of the main CFD models used to calculate the internal flow field of the oxygenator, that is, assuming that the fiber bundle in the oxygenator is porous media, the momentum loss of the fiber bundle to fluid is estimated through permeability and porosity. This method has been proved to be able to simulate the performance of oxygenators effectively [21–23].

According to the Darcy-Forchheimer law, the momentum source term S superimposed in the Navier-Stokes equations is used to describe the momentum loss of fluid in the porous media region, and the source term at the ith direction (i=1,2,3) can be written as:

$$S_{r\_i}=-\left(\frac{\mu}{C_1}v_i+C_2\frac{1}{2}\rho V_{mag}v_i\right). \tag{3}$$

The first term on the right represents the viscosity loss and the second term represents the inertia loss. The permeability (C_1) and inertial loss coefficient (C_2) were estimated by the following equations:

$$C_1 = \frac{d_0^{\,2}}{150}\frac{\varepsilon^2}{(1-\varepsilon)^2} \tag{4}$$

$$C_2 = \frac{3.5}{d_0}\frac{(1-\varepsilon)}{\varepsilon^2} \tag{5}$$

where d_0 is the diameter of the hollow fibers, and ε is the porosity of the fiber bundle. And for equation (3), when the flow Q is less than 6.0 L/min, the inertia loss can be ignored [24].

2) Fast three-dimensional numerical hemolysis approximation method

For the prediction of the hemolysis of the oxygenator, a numerical method called Fast Three-dimensional Numerical Hemolysis Approximation, introduced by André Garon [25], is used to calculate the hemolysis index (NIH) defined in the American Society for Testing and Materials (ASTM) in vitro test standard. This method directly calculates the hemolysis index from the velocity field and shear stress field, thus avoiding the problems caused by the previous calculation of blood damage along several particle paths. After obtaining a stable flow field, the linear average hemolysis index inside the oxygenator $\overline{D_I}$ is defined as:

$$\overline{D_I} = \frac{1}{Q}\int \sigma \, dV, \tag{6}$$

where Q is the volume flow, and σ is the hemolytic destruction rate per unit time, which is calculated as:

$$\sigma = (3.62\times10^{-7})^{\frac{1}{0.785}}\tau^{\frac{2.416}{0.785}}. \tag{7}$$

The coefficients in this equation were provided by Giersepen et al. through in-vitro experiments. And the hemolysis value is obtained by the index conversion:

$$D = \overline{D_I}^{\,0.785}. \tag{8}$$

Finally, the hemolysis value is converted to the final standard hemolysis parameter value (NIH) as follows:

$$NIH = Hb\times D\times 100, \tag{9}$$

where Hb represents total hemoglobin concentration, with a value of 140 g/L.

2.4. Geometry, mesh, and settings

The commercial 3D CAD software package SolidWorks was used to create the three-dimensional model of the oxygenator. The computational domain consists of the inlet channel, outlet channel, and porous media domain, and the unstructured grid was generated by using the OpenFOAM-snappyHexMesh. Nonuniform meshes are used in the simulation, that is, the mesh near the boundary is relatively fine in order to capture the detailed geometry and resolve the vortex near the boundary. In order to see the impact of the mesh on the simulation results, six meshes with different mesh sizes are generated, which have 1.9 million, 3.36 million, 4.7 million, 8.74 million, 9.08 million, and 11.88 million mesh cells, respectively. The details of the six meshes are shown in Table II, where "#Cells" refers to the total number of mesh cells, "Mesh Size" refers to the average mesh size, and "Boundary Layers" refers to the number of layers been refined near the boundary.

For this simulation, the open-source CFD tool OpenFOAM was used to complete the whole calculation process. The inlet adopted the fixed value boundary condition, with a corresponding flow rate of 2L to 5L, and the operating pressure was the default atmospheric pressure (101325 pa). The standard K-ε model has been selected, and the pressure-velocity coupling is carried out by using the pimpleFoam solver, which combines the Pressure Implicit with Splitting the Operators (PISO) and Semi-Implicit Method for Pressure-Linked Equation (SIMPLE) algorithms. A detailed explanation of the PISO and SIMPLE algorithms can be found in [26].

For the discretization of the velocity $\vec{v}$, the Gauss-linear upwind scheme is used, and Gauss-linear scheme is used for the turbulent kinetic energy k as well as turbulent energy dissipation rate ε.

The blood flow is regarded as an incompressible Newtonian fluid with a constant density of 1055 kg/m^3 and viscosity 0.0035 kg/(m·s). The walls adopt a no-slip boundary condition. The hollow fiber membrane was treated as an isotropic porous medium with a porosity of 0.45.

All of our numerical experiments are run on a supercomputer in the Chinese Academy of Sciences, which has a 32-cores Hygon C86-7185 2.5 GHz CPU and 128 GB RAM per computing node. The operating system of the cluster is Red Hat 4.8.5-36, and the OpenFOAM version 8.0 is used for the simulation. The stopping condition is when all of the residuals of velocity, pressure, and turbulence variables are less than 1E-5, and the simulation run until the flow is fully developed.

The pressure drops with respect to the blood flow rates for different mesh sizes are shown in Fig. 2. From the figure, we see that the difference of the pressure drop between two neighboring meshes decreases with the mesh refinement, for example, the pressure drops of Mesh2.1 are basically consistent with that of Mesh3, with an error of less than 0.1%, which means that mesh refinement is able to improve the simulation accuracy but the improvement decreases with the mesh refine.

Considering the calculation time and accuracy of calculation results, Mesh2.1 has been selected as the default mesh for the rest of the simulation. Fig. 3 shows the details of the Mesh2.1 and we can see that the mesh near the wall is relatively finer than the inner domain.

Table 2. Specifications of different global and local mesh sizes

Mesh	#Cells	Mesh Size	Boundary Layers
1	1903949	5.73E-4 m	5
1.1	4700780	4.24E-4 m	6
1.2	8741078	3.45E-4 m	8~9
2	3365301	4.74E-4 m	5~6
2.1	9086710	3.4E-4 m	9
3	11885543	3.11E-4 m	11~12

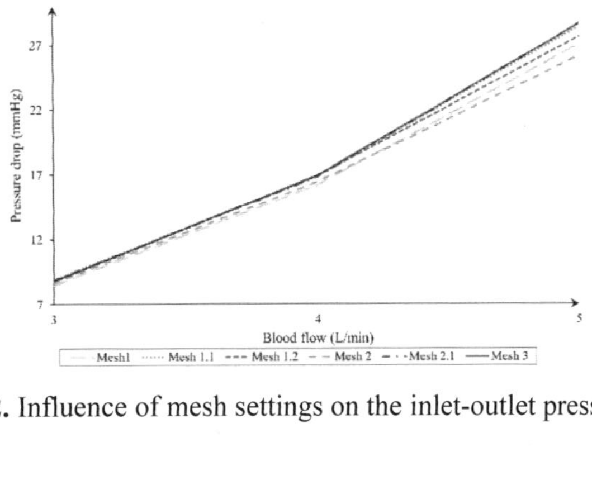

Figure 2. Influence of mesh settings on the inlet-outlet pressure drop

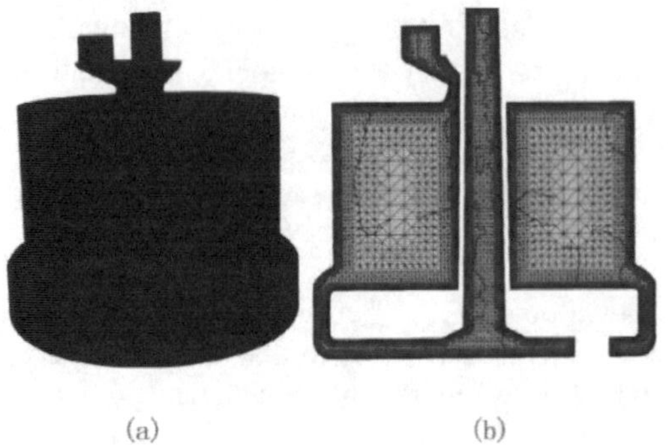

Figure 3. The computational mesh used in the simulation: (a) the surface mesh and (b) the inner mesh (vertical slice)

3. Results and analysis

This section shows the simulation results, which include: the analysis of the flow field, the shear stress of the blood in Oxyg-I, and the predictions of blood damage based on these results.

3.1. Analysis of the flow field

In order to study the internal fluid motion characteristics of Oxyg-I, velocity distribution, velocity vectors, and streamlines are visualized and analyzed using ParaView. Fig. 4 shows the velocity magnitude distribution on cross-sections for the case that the inlet flow is 5.0 L/min.

From Fig. 4 we can see that blood goes into the oxygenator

at a constant speed of 1.16 m/s (5.0 L/min) and following the inlet channel, it goes through the fiber bundle from outside to inside under the action of fiber bundle resistance. In this process, due to the dispersion of flow, the speed is relatively small, less than 0.5 m/s. Finally, the blood is collected in the arc-shaped outlet channel and goes out of the oxygenator from the outlet. The whole process can be seen in Fig. 4 (a). And from Fig. 4 (b) we can see that the distribution of the blood flow velocity at the inlet is almost uniform but the flow at the outlet is turbulent. The main reason is that the flow is disturbed in the channels and bundles.

The streamline is shown in Fig. 5. From the figure, we can see that the blood flow is scattered around at the bottom of the inlet channel and the velocity decreases following the flow direction. We can see how the blood is collected in the outlet channel and the flow is rotated in the outlet channel. Because of the symmetrical design and radial flow, there is no significant stagnancy or vortex in the flow. When the blood goes through the porous medium area located in the central area of the oxygenator, the flow presents a state of longitudinal divergence at a lower flow rate, which can be more clearly observed in the cross-section view; see Fig. 5 (b).

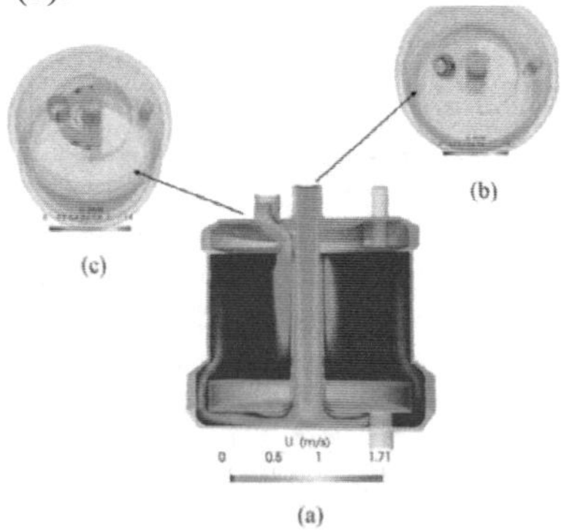

Figure 4. Velocity magnitude distribution of the blood flow in Oxyg-I: (a) vertical cross-section, (b) inlet and outlet sections, (c) the cross-section in the outlet channel

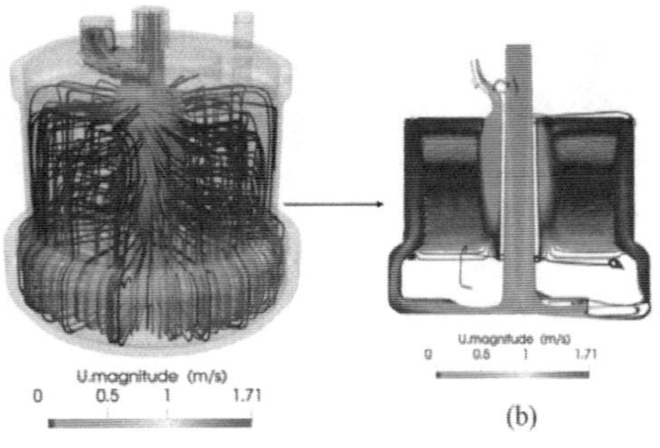

(a) (b)

Figure 5. Streamline distribution of the blood flow in Oxyg-I: (a) the global view and (b) a cross-section view

From Fig. 6 (a) we can see that because of the inclined angle of the inlet channel (about 2o), the blood flow velocity gradually decreases from top to bottom in the inlet channel, accompanied by the increase of the channel area. There are a few small vortices that appear at the outlet channel.

Fig. 7 shows the velocity distribution of blood flow and the velocity vector distribution at the inlet and outlet under different blood flow rates (2 ~ 4 L/min). From the figure, we can see that under different blood flow rates, the velocity distribution in the oxygenator is roughly the same. Under the condition of low flow rate (2 L/min), the blood flow at the outlet is slightly different from that under the condition of high flow rate (4 L/min, 5 L/min). The velocity at the outlet is more evenly distributed when the flow rate is 3.0 L/min or 4.0 L/min, while it is more concentrated on the outside of the channel when the flow rate is 2.0 L/min.

Fig. 8 shows the total pressure distribution of the Oxyg-I under a blood flow rate of 5.0 L/min. The total pressure drop between the inlet and outlet (ΔP) is about 30 mmHg. Due to the high viscous resistance of the fiber bundle, most of the pressure loss is located in the fiber bundle, about 20 mmHg. The influence of the inlet pipe and the intermediate pipe on the pressure drop can be ignored. Due to the 3 ~ 4 times change of the flow area when the blood enters the outlet channel, a negative pressure area (purple parts of Fig. 8 (b)) is generated in the outlet channel, which basically coincides with the distribution of small vortices mentioned in Fig. 8.

Fig. 9 shows the comparison of the pressure loss predicted by the computational model in this study and the experiment measured value of the Affinity NT oxygenator from J. Zhang et al., where they used bovine blood and tested the pressure drop of the affinity oxygenator under constant temperature and pressure. We can see from Fig. 9 that with the increase of the blood flow, the pressure drops of the Oxyg-I calculated in this study are basically consistent with the experimental test results, while slightly lower than it, about 25%. This research is aimed at the structural optimization of the wearable oxygenator, so Oxyg-I has a more compact internal structure and a more concentrated flow channel. That can also explain why it has a better performance in pressure drop than that of the previously developed oxygenators.

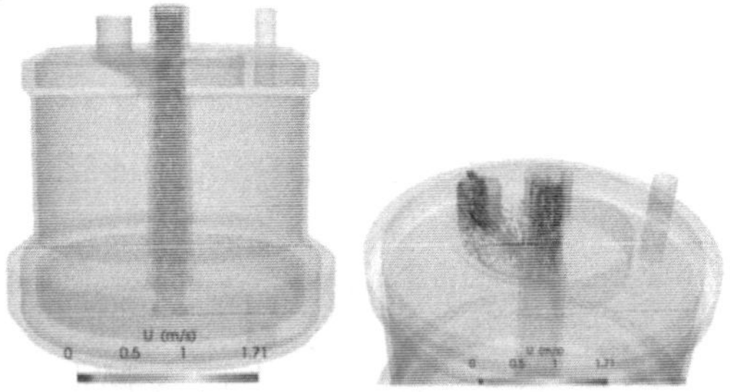

(a) (b)

Figure 6. Velocity vector distribution of the flow field in Oxyg-I: (a) inlet channel, (b) outlet channel

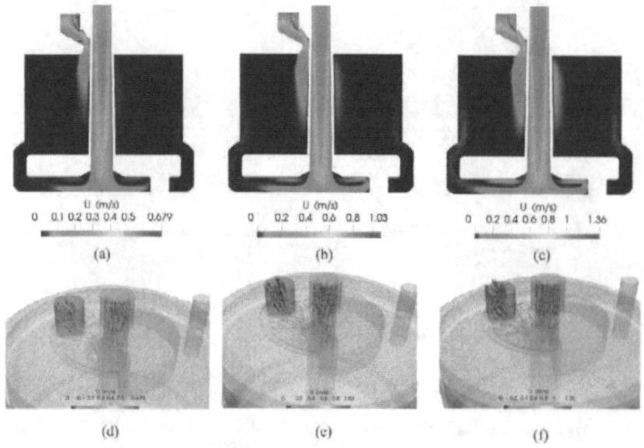

Figure 7. Velocity magnitude distribution of the blood flow with different inflow rate in Oxyg-I: (a) 2.0 L/min, (b) 3.0 L/min, (c) 4.0 L/min, and velocity vector distribution under: (d) 2.0 L/min, (e) 3.0 L/min, (f) 4.0 L/min

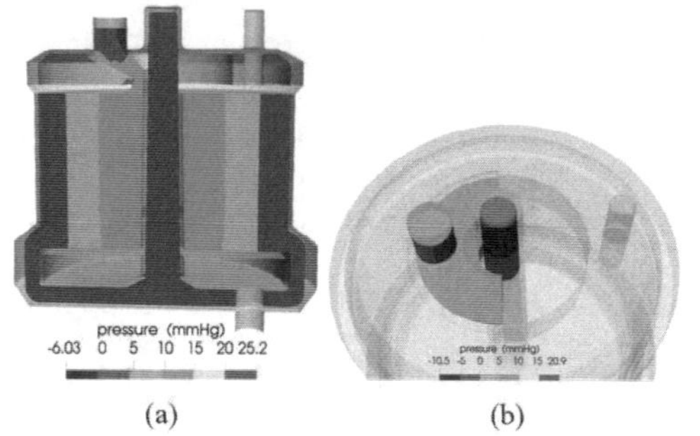

Figure 8. Pressure distribution in Oxyg-I: (a) a slice view, (b) parts of inlet and outlet

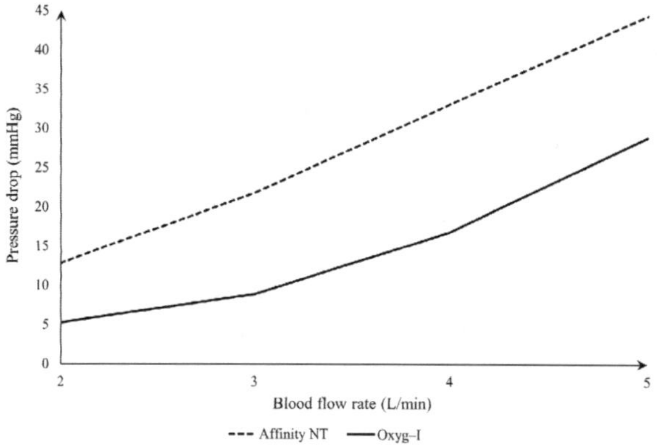

Figure 9. Comparison of the computationally predicted and experimentally measured pressure drops between the inlet and outlet with different flow rates for the Affinity NT oxygenator and Oxyg-I.

3.2. Blood damage prediction

In this study, the standard hemolysis index (NIH) of Oxyg-I was calculated in the way described in part II, which was used to analyze the degree of blood damage in the oxygenator. As shown in Fig. 10, the maximum shear stress at the flow rate of 5.0 L/min is about 134 Pa, which is located on the inlet channel wall.

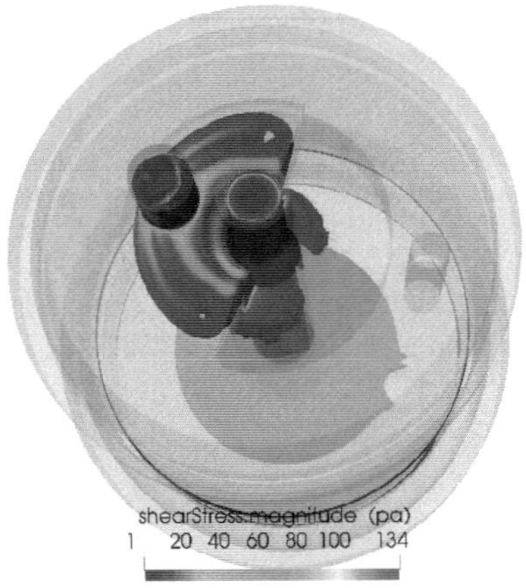

Figure 10. The shear stress distribution

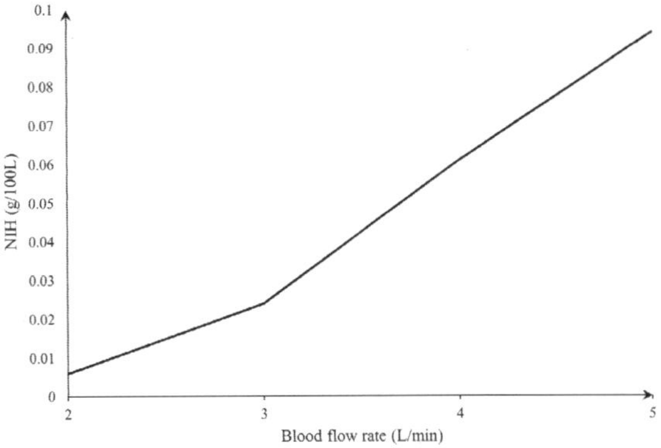

Figure 11. The relationship of the standard hemolysis index NIH and flow rate

It can be seen from Fig. 10 the high shear stress region is mainly located in the inlet and outlet area (about 100~130 pa), According to the theory mentioned in part II, the blood damage is positively correlated with the power of shear stress, which means that the possible blood cell damage mainly occurs at the inlet, outlet, and channels.

In this research, after the blood flow reached a steady state, we used the Fast Three-dimensional Numerical Hemolysis Approximation method mentioned in part II to estimate the hemolysis value of the Oxyg-I at a different flow rate (2.0 ~ 5.0L). The result of the estimation is shown in Fig. 11.

As we can see from Fig. 11, the standard hemolysis index NIH ranges from 0.009 to 0.094 (g/100L) with the blood flow rate from 2.0 L to 5.0 L. Note that the maximum hemolysis index allowed by human physiology is 0.1 g/100L, so the hemolysis performance of Oxyg-I meets the requirements.

3.3. Efficiency analysis of parallel computing

The simulation in this study is very time-consuming and in the design procedure, we usually need to do tens or hundreds of such simulations. In order to reduce the simulation time, we usually use the supercomputer to do the simulation in parallel. In parallel computing, the parallel performance of the algorithm is very important. In this subsection, we study the parallel performance of the simulation algorithms that have been used in this study. Note that we only count the total simulation time of the first 30 timesteps in this parallel performance study. The simulation time with different processors is shown in Fig. 12 and the speedup of the simulation algorithm is shown in Fig. 13. As we can see from Fig. 12, with the increase of the number of processors, the total simulation time is reduced quickly.

When the number of processors reaches 720, the simulation time is reduced to 22.41s, which saves over 80% time compared with the case using 90 processors.

As we can see from Fig. 13 the speedup is almost linear when the number of processors is less than 360 and it drops when the number of processors increases to 720. From the parallel performance results, we can make a conclusion that the simulation algorithm used in this study is robust and parallel scalable with hundreds of processors.

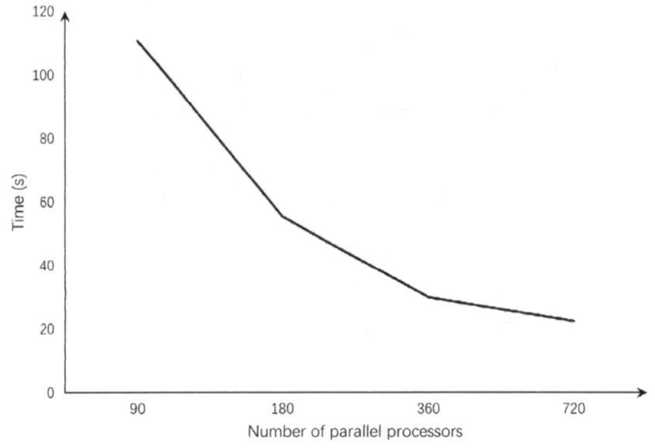

Figure 12. The total simulation time (in seconds) of the first 30 time steps with a different number of processors. The mesh has 9086710 cells.

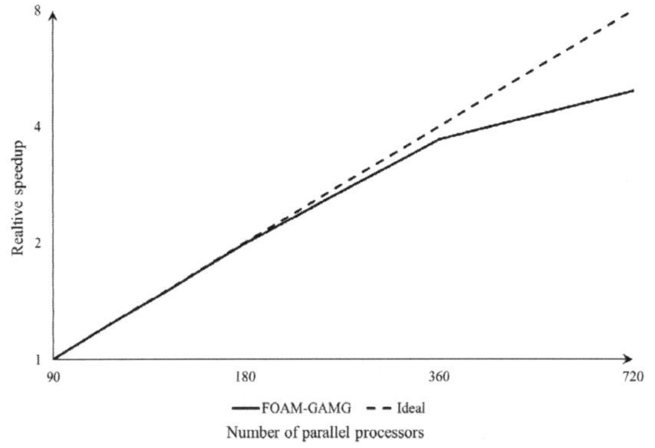

Figure 13. The speedup of the simulation algorithm

4. Conclusion

In this paper, the hemodynamics of a wearable membrane oxygenator is carefully studied using a K-ε Turbulence model and a hemolysis numerical prediction model on a supercomputer. From the numerical results, we can draw the following conclusions.

The large blood flow velocity in the oxygenator is located at the inlet and outlet, and most of the pressure loss is located at the fiber bundle area, accounting for about 60% of the total pressure drop.

The high shear stress is mainly located at the inlet and outlet of the oxygenator, especially the intersection of outlet and channel, where the shear stress reaches the maximum (>= 130 Pa). From the shear stress distribution, one can optimize the shape of the oxygenator to reduce the damage of the blood cell.

When the blood flow rate is 2.0 ~ 5.0 L/min, the hemolysis value calculated by the hemolysis numerical prediction model ranges from 0.006 g/100L to 0.094 g/100L, which meets the allowable range of human physiology.

The simulation algorithm scales up to 720 processors for solving a problem with over 9 million mesh cells and the parallel efficiency is over 60%, which shows the potential to do higher-resolution simulations with a larger number of processors.

Acknowledgments
This work was partially supported by the Shenzhen Project (No. RCYX20200714114735074) and NSFC (No. 12071461).

References

[1] J. Philip, Wolfson, "The development and use of extracorporeal membrane oxygenation in neonates," The Annals of Thoracic Surgery, Elsevier. vol. 76, 2003, pp. S2224–S2229.
[2] J. Zhang, M. E. Taskin, A. Koert, et al., "Computational design and in vitro characterization of an integrated Maglev pump-oxygenator," Artificial Organs, Wiley Online Library. vol. 33, 2009, pp. 805–817.
[3] DT. and B. McArdle, Pearson, "Haemocompatibility of membrane and bubble oxygenators," Perfusion, Sage Publications Sage CA: Thousand Oaks. CA, vol. 4, 1989, pp. 9–24.
[4] Jr. Clark, C. Leland and Gollan, Frank and Gupta, B. Vishwa, "The oxygenation of blood by gas dispersion," Science, vol. 111, 1950, pp. 85–87.
[5] Iwahashi, Hidehiko and Yuri, Koichi and Nos{\'e}, Yukihiko, "Development of the oxygenator: past, present, and future," Journal of Artificial Organs, vol. 7, 2004, pp. 111–120.
[6] E. B. Kay, H. A. Zimmerman, R. M. Berne, et al., "Certain clinical aspects of the use of a pump oxygenator," Journal of the American Medical Association, vol. 162, 1956, pp. 639–641.
[7] F. S. Cross, R. M. Berne, Y. Hirose, et al., "Evaluation of a rotating disc type reservoir-oxygenator," Proceedings of the Society for Experimental Biology and Medicine, vol. 93, 1956, pp. 210–214.
[8] P. A. M. Segers, J. F. Heida, De Vries I, et al., "Clinical evaluation of nine hollow-fibre membrane oxygenators," Perfusion, Sage Publications Sage CA: Thousand Oaks. CA, vol. 16, 2001, pp. 95–106.
[9] S. R. Wickramasinghe, C. M. Kahr, B. Han, "Mass transfer in blood oxygenators using blood analogue fluids," Biotechnology Progress, vol. 18, 2001, pp. 867–873.
[10] A. Rungatscher, M. Tessari, C. Stranieri, et al, "Oxygenator is the main responsible for leukocyte activation in experimental model of extracorporeal circulation: a cautionary tale," Mediators of Inflammation, 2015.
[11] E. Tatsumi, "Artificial lungs: current state and trends of clinical use and research and development," Journal of Artificial Organs, vol. 10, 2015, pp. 1–5.
[12] T. Zhang, G. Cheng, A. Koert, et al., "Functional and biocompatibility performances of an integrated Maglev pump-oxygenator," Journal of Artificial Organs, vol. 33, 2009, pp. 36–45.
[13] A. Kaesler, M. Rosen, Schmitz–T. Rode, et al., "Computational modeling of oxygen transfer in artificial lungs," Journal of Artificial Organs, vol. 42, 2018, pp. 786–799.
[14] M. E. Taskin, T. Zhang, C. Wu, et al., "CFD Assisted Design of a Wearable Artificial Pump Lung Device," ASME International Mechanical Engineering Congress and Exposition, vol. 48630, pp. 187–193.
[15] S. P. Madhani, May A G, B. J. Frankowski, et al., "Blood recirculation enhances oxygenation efficiency of artificial lungs," ASAIO Journal, vol. 66, pp. 565–570.
[16] S. P. Madhani, B. J. Frankowski, S. H. Ye, et al., "In-Vivo 5-day Animal Studies of a Compact, Wearable Pumping Artificial Lung," ASAIO Journal, vol. 65, pp. 94.
[17] S. P. Madhani, B. J. Frankowski, G. W. Burgreen, et al., "In vitro and in vivo evaluation of a novel integrated wearable artificial lung," The Journal of Heart and Lung Transplantation, vol. 36, 2017, pp. 806–811.
[18] M. E. Taskin, T. Zhang, K. H. Fraser, et al., "Design optimization of a wearable artificial pump-lung device with computational modeling," Journal of Medical Devices, vol. 6, 2012, pp. 031009.
[19] G. He, T. Zhang, J. Zhang, et al., "Model-based design and optimization of blood oxygenators," Journal of Medical Devices, vol. 6, 2020, pp. 041001.
[20] M. Jahangiri, M. Saghafian, M. R. Sadeghi, "Numerical study of turbulent pulsatile blood flow

through stenosed artery using fluid-solid interaction," Computational and Mathematical Methods in Medicine, vol. 2015, 2015, pp. 525613.

[21] C. Cameron, Jones, McDonough, et al., "Improved computational fluid dynamic simulations of blood flow in membrane oxygenators from X-ray imaging," Annals of Biomedical Engineering, vol. 41, 2013, pp. 2088–2098.

[22] Khanafer, Khalil and Cook, Keith and Marafie, Alia, "The role of porous media in modeling fluid flow within hollow fiber membranes of the total artificial lung," Journal of Porous Media, vol. 15, 2012, pp. 113-122.

[23] G. Florez, Jonathan and others, "Numerical study of bio-fluids and mass transfer processes through membranes," Universitat Rovira i Virgili, 2018.

[24] Consolo, Filippo, Fiore et al., "A numerical performance assessment of a commercial cardiopulmonary by-pass blood heat exchanger," Medical Engineering and Physics, vol. 37, pp. 584 – 592, 2015.

[25] A. Garon, and Farinas, M.-I, "Fast three-dimensional numerical hemolysis approximation," Artificial Organs, vol. 28, 2004, pp. 1016-1025.

[26] H. Jasak, "Error analysis and estimation for the finite volume method with applications to fluid flows," Imperial College London (University of London, 1996.

Layout of measuring points for flue gas flow velocity in coal-fired power plants based on CFD technology

Junhao Wu, Chunling Liu*

Liaoning University of Technology, Jinzhou, China

*Corresponding author e-mail: 554353350@126.com

Abstract. In view of the existence of 90° bends in the flue after desulfurization in coal-fired power plants and the complicated distribution of flue gas flow velocity, the CFD simulation technology is proposed to simulate and analyze the flow characteristics of the flue flow field, and to find suitable flow velocity measurement points. On this basis, a set of on-line monitoring system for flue gas flow velocity and pollutants is designed by using LabVIEW.

Keywords: flue gas velocity, CFD, LabVIEW, SQL Server.

1. Introduction

From the perspective of power generation structure, the national thermal power generation in the first half of 2022 will be 2,727.68 billion kWh, accounting for 68.8% of the total power generation. Thermal power has become the most important part of my country's coal consumption and gas pollutants SO2, NOx and greenhouse gases major source of CO2 emissions. The "Environmental Protection Tax Law of the People's Republic of China", which has been implemented since January 1, 2018, clearly requires thermal power plants to install automatic pollutant monitoring equipment, and calculate taxes and fees according to the automatic monitoring data of pollutants. In addition, enterprise-related environmental protection equipment such as desulfurization, denitrification and other devices also need to be managed and controlled based on the data monitored by CEMS. Accurate measurement of various pollutant emissions from coal-fired power plants will not only help the country's air pollution prevention and control work, but also help improve the economic benefits of coal-fired power plants themselves.

Among them, the measurement of flue gas flow velocity directly affects the calculation of various pollutant emissions, which is of great significance to the country and enterprises. In HJ 75-2017 [1], there are clear requirements for the installation position of the flow velocity measurement device. The measurement point position should avoid the flue bend and the part where the section changes sharply. However, due to the limitations of the thermal power plant site space and installation conditions, the length of the straight pipe section cannot meet the specification requirements for the arrangement of flue gas flow measurement points [2], In addition, the flue gas has the characteristics of high temperature, high humidity, and many particles and impurities, the flow is uneven, complex and changeable, and it is difficult to ensure that the location of the flow velocity measurement is in the sampling area that can represent the real flow velocity of the flue gas. Due to the large diameter of the flue, complex flow, and high turbulence in the flue flow measurement, in most cases, the selection of flowmeters and sampling

Content from this work may be used under the terms of the Creative Commons Attribution 3.0 licence. Any further distribution of this work must maintain attribution to the author(s) and the title of the work, journal citation and DOI.
Published under licence by IOP Publishing Ltd

points is lacking [3]. In this paper, CFD simulation technology is used to numerically analyze the flue flow field to find suitable flow velocity measurement points.

2. Model building and simulation

CFD is the abbreviation of Computational Fluid Dynamics. It is an emerging cross-cutting technology that integrates fluid mechanics, mathematics and computer science. It starts from computational methods and uses computer and numerical methods to quickly solve various conservation control partial differential equations in fluid flow to obtain an approximate solution to the fluid governing equations. Compared with traditional experimental-based theoretical fluid mechanics, CFD technology has the advantages of strong adaptability, not limited by physical models and experimental models, and low cost of use. It is widely used in aerospace design, automotive design, chemical industry and other fields. Coal-fired power plant flues are often large in size, and the fluid conditions in the flue are difficult to reproduce under laboratory conditions, therefore, the computational fluid dynamics method is used to analyze the fluid flow characteristics of the flue [3].

As shown in Fig.1, CFD simulation calculation generally consists of three parts: problem definition, pre-processing and solution process, and post-processing process.

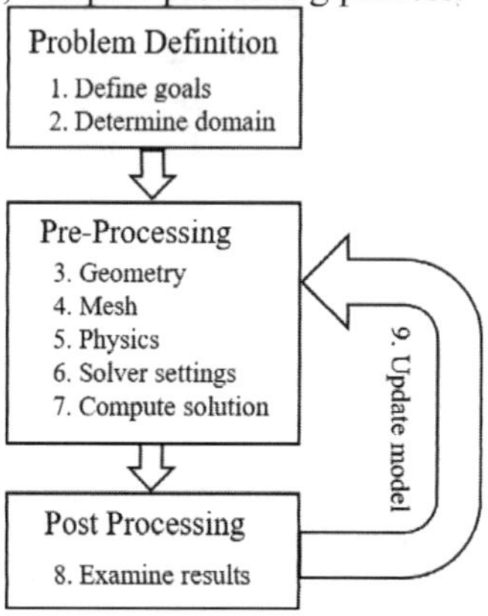

Figure 1. CFD simulation process

2.1. 3D model

The flue gas emitted by the coal-fired power plant is discharged from chimney through the tail flue after desulfurization and other environmental protection treatments, and part of the monitoring equipment of the flue gas emission continuous monitoring system is installed in the flue. Due to the short length of flue after desulfurization in most coal-fired power plants, the flue gas is not fully developed in the flue, and the flue often has a 90°bend [4], which further chaotic the internal flow field of the flue, making it difficult to meet the requirements of CEMS-related sampling point layout. In this paper, referring to the flue size of a 600MW unit after desulfurization, a model as shown in Fig.2 is established in SolidWorks, in which the flue cross-section is a rectangle of 4m×4m. After desulfurization, the length of the flue before the 90° bend is 4m, and the length of the flue after the 90° bend is 8m.

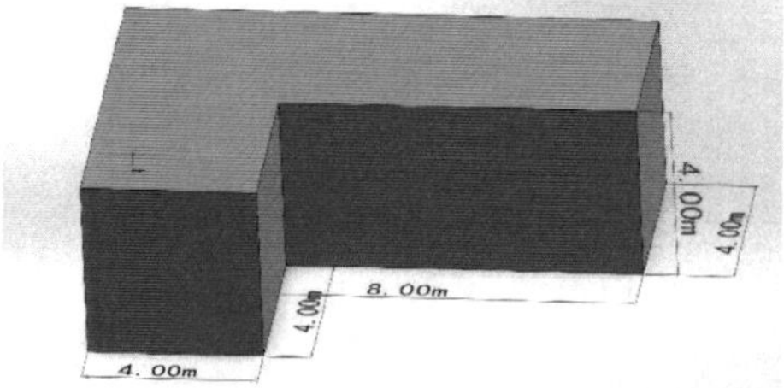

Figure 2. 3D flue model

2.2. Mesh
Meshing is an important step in CFD simulation. The quality of the mesh will directly affect the calculation accuracy of the simulation. When the mesh quality is poor, it may even cause the solution to fail to converge. There are two main types of grids: structured grids and unstructured grids. The structured grid has the advantages of fast grid generation, good grid generation quality, and easier access to the actual model. At the same time, the structured grid has the disadvantage that it has a narrow range of use and is only suitable for models with regular shapes. Unstructured grids use certain criteria for optimization judgment in the generation process, so that high-quality grids can be generated. However, compared with structured grids, for the same physical space, the grid filling efficiency is not high. Under the flow field calculation conditions, the number of generated grids is much larger than that of structured grids, and more computing resources are required. Due to the regular shape of the flue model, and considering the grid quality and computing resources, this paper chooses to divide it into structured grids. Define the relevant grid parameters to complete the structural grid division of the flue model shown in Fig.3.

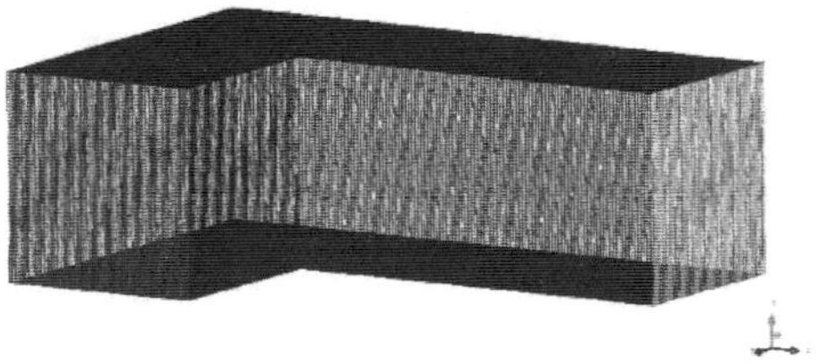

Figure 3. Structured meshing

2.3. Boundary Condition Settings
The temperature of flue gas discharged from coal-fired power plants after desulfurization treatment is generally 50°C, and the average flue gas flow velocity is about 10m/s. According to relevant information, the density of flue gas at 50°C is about 1.16kg/m^3, and the dynamic viscosity is about 1.72×10^{-5}Pa·s, the Reynolds number of flue gas in the flue can be calculated according to (1).

$$Re = \frac{F_g}{F_m} = \frac{\rho v d}{\mu} \tag{1}$$

In (1), Re is the Reynolds number, Fg is the inertial force, Fm is the viscous force, v is the average velocity of the fluid, μ is the dynamic viscosity, and d is the characteristic length. In (1), circular pipes often take the diameter as d, non-circular pipes often take its hydraulic diameter as the characteristic length d for calculation, and the hydraulic diameter is expressed as:

$$D_H = \frac{4A}{X} \tag{2}$$

In (2), A is the cross-sectional area of the rectangular pipe through which the fluid flows, and X is the perimeter of the cross-section of the rectangular pipe through which the fluid flows. For the tail flue of a coal-fired power plant, the flue gas Reynolds number is generally greater than 106 [6], which is a typical turbulent flow.

The solver chooses the pressure-based solver, the SIMPLE algorithm, which solves with momentum and pressure as basic variables. The operating environment is selected under the standard atmospheric pressure, that is, 101325Pa, and the temperature is set to 50 °C, that is, 323.15K. The inlet boundary is defined as the velocity inlet, and the flow field distribution under different conditions can be obtained by defining different inlet velocities. First, the velocity is defined as the average flue gas velocity of 10m/s. The standard k-ε model was chosen for the turbulence model. Select the turbulence intensity and hydraulic diameter to set the turbulence parameters, where the hydraulic diameter can be calculated as 4m from (2), and the turbulent intensity can be calculated from the following formula:

$$I = 0.16(Re)^{-\frac{1}{8}} \tag{3}$$

Calculate the flue gas Reynolds number according to (1), and bring it into (3) to obtain the turbulent flow intensity I, which is 3.35%. The outlet boundary condition does not need to be set, it is obtained from the simulation calculation after setting the inlet condition. Set the wall as a standard no-slip static wall.

2.4. Analysis of simulation results

Intercept the velocity contour of the horizontal section at 2m from the bottom of the flue and the vertical section at 1m from the exit of the flue. It can be seen from the figure that the flue gas with an average velocity of 10m/s enters from the entrance of the flue, and the flue gas flows steadily. After the 90-degree bend, the flue gas flow begins to become chaotic, and the flue gas flow velocity gradually increases from the inside of the flue to the outside, showing three states of high flow velocity area, medium flow velocity area, and low flow velocity area. If the location of the flow measurement point is not selected properly, it is very easy to cause the deviation between the measurement result and the actual value to be too large.

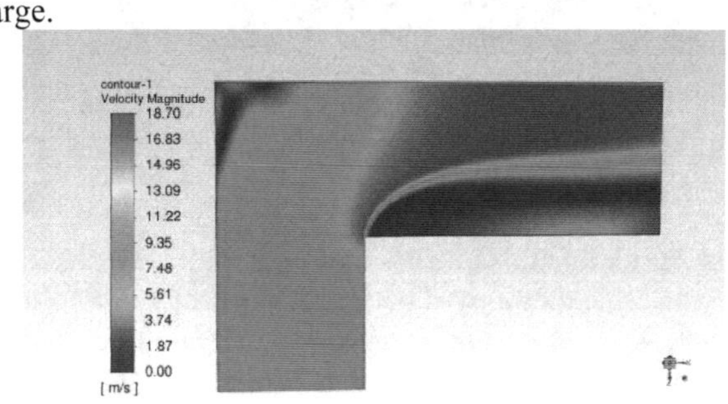

Figure 4. Horizontal section velocity contour

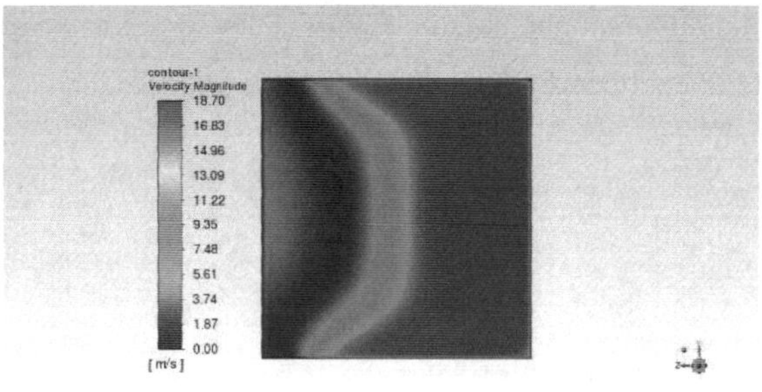

Figure 5. Vertical section velocity contour

2.5. Grid independence verification

From the point of view of CFD simulation principle, the finer the grid is divided, the more the number of grids, the higher the accuracy of the solution results. However, in practical engineering applications, the increase in the number of grids will increase the time cost of calculation and require more computing resources, and when the number of grids reaches a certain level, increasing the number of grids does not significantly improve the calculation accuracy. At this time, it can be considered that a grid-independent solution has been obtained. In the subsequent calculation process, the number of grids at this time is used as the benchmark for dividing the grid. The four grid division plans are shown in Table 1.

Table 1. Different meshing plans

	Maximum grid size(mm)	Total grids
Plan A	80	545000
Plan B	65	1022504
Plan C	55	1651990
Plan D	50	2163200

The same boundary conditions are set in the four different plans. The vertical section height is 2m at a distance of 1m from the flue outlet, and 7 points are selected at 0.5m intervals. The velocities at 7 points under four different meshing plans under the 10m/s inlet condition were extracted respectively, and draw the line graph.

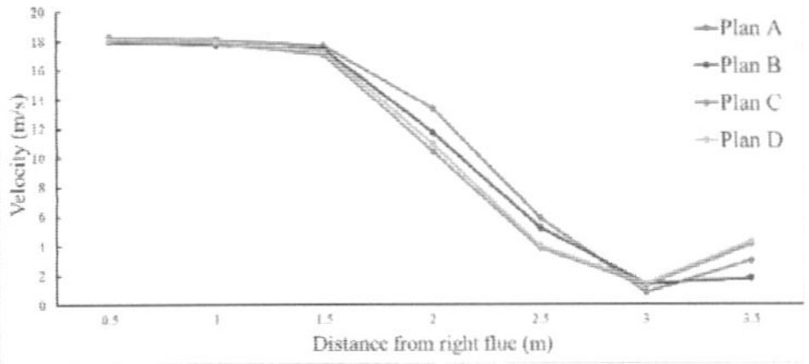

Figure 6. Line chart of velocity under different plans

It can be seen from the figure that with the increase of the number of grids, there are obvious differences in the calculation results in the medium and low flow velocity regions, but when the number of grids is further increased, the change of the calculation results becomes very small, it can be seen that the curves of plan C and plan D are basically the same, and the maximum change is about 5%. At this time, it can be considered that the grid number of plan C has achieved grid independence, and the grid number of plan C is used as the benchmark in subsequent calculations.

3. Velocity measuring point selection

It can be seen from the fig.7 that the flow velocity of the flue gas begins to change sharply after passing through the bend, and the first half area is mainly composed of a low-speed area and a high-speed area. In the middle of the flue, there is a relatively violent backflow phenomenon of the flue gas, and the flow velocity measurement point should be avoided here. At the tail of the flue, the backflow is weakened and the area of the medium-velocity zone reaches the maximum, and the flow velocity measurement point should be arranged at the tail of the flue.

Figure 7. Sectional streamline diagram at height 2m

The velocity contours of vertical sections at 1m and 0.5m away from the flue outlet were intercepted. It can be seen from the figure that the two velocity contours are basically the same. The flow field distribution in this area is relatively stable and the area of the medium-velocity area is large. Define the midpoint of the section at 1m from the flue outlet as point A, and define the midpoint of the section at

0.5m from the flue outlet as point B, and read the speeds of points A and B as 10.399m/s and 9.826m/s respectively, the average velocity of the two points is 10.1125 m/s, which is only 1% error from the inlet velocity of 10 m/s.

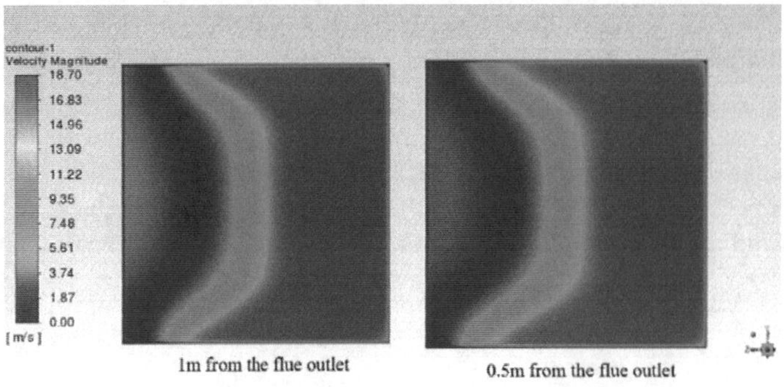

Figure 8. 10m/s velocity contour

In order to further verify the universality of the selected flow velocity measurement points, plan C is selected for mesh division. Under the condition that other boundary conditions remain unchanged, the inlet flow velocity is changed to obtain the inlet flow velocity of 5 m/s and 15 m/s, the cross-sectional velocity contour at 1m and 0.5m away from the flue outlet respectively, it can be seen from the fig.9 that the flow velocity distributions are basically the same under different inlet flow velocities. The velocities of points A and B at 5 m/s are read as 5.200 m/s and 5.027 m/s, the average velocity is 5.1135 m/s, and the error with the inlet velocity of 5 m/s is 2.27%. The velocities of points A and B at 15 m/s are read as 15.044 m/s and 14.698 m/s, the average velocity is 14.871 m/s, and the error with the inlet velocity of 15 m/s is 0.86%. The results show that points A and B are universal, and with the increase of flue gas flow rate, the error will decrease.

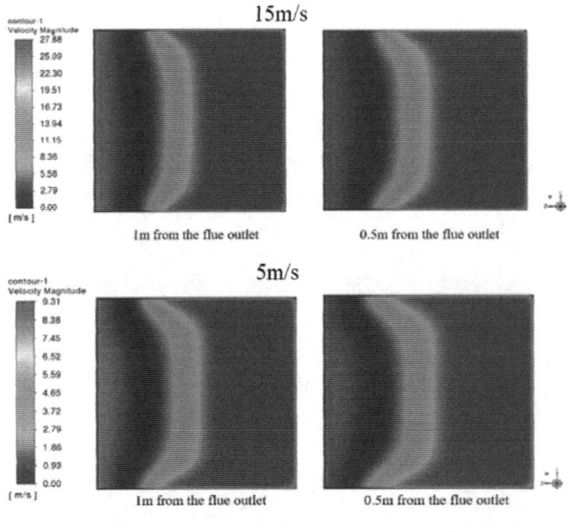

Figure 9. 5m/s and 15 m/s velocity contour

4. Flow velocity and pollutant monitoring system

In this paper, based on the selection of flue gas flow velocity measurement points, a set of flue gas flow velocity and pollutant monitoring system is developed and designed using LabVIEW.

Figure 10. login interface

The system can only be used after user login. In order to prevent some important parameters from being modified arbitrarily, the system has dual operation rights of ordinary users and administrators.

The system can display various monitoring data in real time, and can complete the conversion of working conditions and standard states, wet base concentration and dry base concentration of relevant data according to HJ-75 requirements, and calculate the corresponding daily, monthly and annual emissions. The data of the system is stored in the SQL Server database, and the corresponding day, month and year Excel reports can be exported.

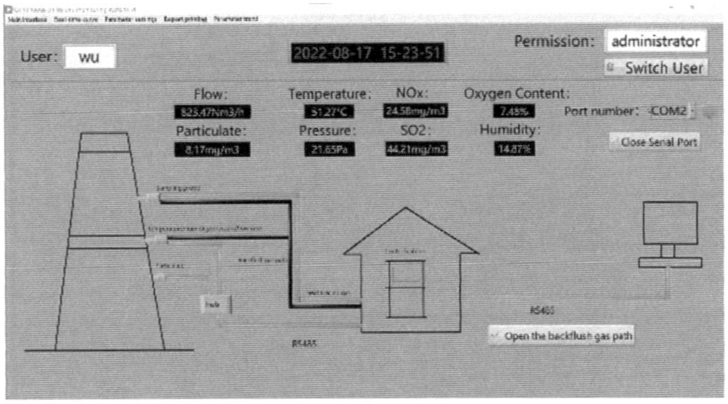

Figure 11. Upper computer interface

5. Conclusion

In this paper, in view of the 90° bend in the flue after desulfurization in the coal-fired power plant, and the complex flow velocity distribution of the flue gas in the flue, the CFD simulation technology is used to simulate and analyze the flow characteristics of the flue flow field, and two suitable flow velocity measurement points are found. Based on this, a set of online monitoring system for flue gas flow rate and pollutants is designed by LabVIEW. The system can display various monitoring data in real time and complete corresponding numerical calculations. The system runs stably and has great practical application value.

References

[1] HJ 75-2017. Specifications for continuous emissions monitoring of SO2, NOx, and particulate matter in the flue gas emitted from stationary sources [S].2017

[2] Li Hu, Wu Xinping. Application of matrix flow measurement device in desulfurization flue gas [J]. Electric Power Technology and Environmental Protection, 2016,32(05):37-38.

[3] Z Liu, D Guan, W Wei, et al. Reduced Carbon Emission Estimates from Fossil Fuel Combustion and Cement Production in China [J]. Nature, 2015, 524 (7565): 335-8.
[4] Chen Hongwei, Cheng Kai, Jia Jiandong, Zhu Dongqi. Optimal design of gas-solid two-phase flow in flue with 90° square cross-section [J]. Electric Power Science and Engineering, 2017, 33(11): 54-60.
[5] Qian Conghao, Feng Xuan, Zhu Xiaoliang. Study on CFD simulation of flue gas flow velocity measurement by multi-line method [J]. Power Generation Equipment, 2019, 33(06):375-380.
[6] Dong Honglin, Zhu Xiaoliang. Research on the arrangement of flue gas flow velocity measuring points based on self-modeling theory [J]. Power Generation Equipment, 2021, 35(05): 319-324.

Design and Fluid Analysis of a Novel Power Supply Unit for Unmaned Marine Equipment

Changlei Wang, Zhongyue Lu, Zirong Luo *, Mangkuan Wang

National University of Defense Technology, Changsha, China

*Corresponding author e-mail: luozirong@nudt.edu.cn

Abstract. The problem of insufficient power supply for unmanned marine equipment at sea limits its development and utilization. In this paper, we designed a power supply unit capable of absorbing wave energy and converting it into electricity, which helps to solve the problem of limited power supply ability. Its main structure is demonstrated by theoretical design and modeling, and its operating principle is presented. Its performance (power coefficient and torque coefficient) was numerically simulated via Computational Fluid Dynamics tools, and the effect of blade groups' solidity on its operating performance was investigated. The study found that the blades of the blade groups can absorb more energy by arranging along the outer ring, and the energy conversion efficiency of the blade groups is related to their load.

Keywords: wave energy, wave energy converter, power, numerical simulation, computational fluid dynamics.

1. Introduction

The oceans cover 70% of the earth's surface and are rich in mineral resources and ocean energy[1], such as oil, natural gas, wave energy, tidal energy, etc... The vastness of the ocean and the complexity and variability environment make it difficult for humans to explore, so many countries develop unmanned marine equipment to detect and study the ocean. At this stage, a lot of unmanned marine equipment has been widely used, such as various navigation buoys, unmanned marine robots, meteorological and hydrological monitoring equipment, etc... They can promote the development of the national maritime economy, safeguard national maritime rights, and assist in the realization of maritime defense tasks. However, most of the existing unmanned marine equipment is moored offshore or powered by cables or batteries, and there is no mature solution for powering unmanned equipment in the deep and distant sea[2]. The battery-powered supply method can hardly support the operation of unmanned marine equipment for several months, and the cable-powered supply limits its mobility, so it is an inevitable choice to develop an autonomous power supply system for unmanned marine equipment[3].

Unmanned marine equipment is surrounded by the ocean, so it is economical and easy to choose offshore clean energy to power it according to local conditions[4]. Wave energy is a clean energy source that is formed by wind blowing on the ocean surface[5]. It has the following advantages over other forms of ocean energy: highly energetic, widely distributed, and easy to collect (in the form of mechanical energy). Absorbing and converting wave energy into electrical energy for unmanned marine equipment will greatly improve its cruising range and autonomous survivability. It can also power unmanned

Content from this work may be used under the terms of the Creative Commons Attribution 3.0 licence. Any further distribution of this work must maintain attribution to the author(s) and the title of the work, journal citation and DOI.
Published under licence by IOP Publishing Ltd

marine equipment in the deep and distant sea[6]. However, the focus of wave absorption and conversion technology research at this stage is on large Wave Energy Conversions (WEC), so the existing wave technology cannot be directly used for energy supply modules for small and medium-sized unmanned marine equipment.

This paper designs an autonomous Power Supply Unit (PSU) that absorbs wave energy and generates electricity to power the unmanned marine equipment or to charge its battery. This article mainly describes its design, modeling, working principle, and investigates the effect of its parameters on the working performance via numerical simulation.

The other chapters of this paper are briefly described below: Chapter II describes its design, modeling, and working principle; Chapter III investigates the effects of its parameters on the working performance through Computational Fluid Dynamics (CFD) numerical simulations; Chapter IV summarizes the entire text.

2. Design and working principle

2.1. Design and structure

The detailed structure of the novel PSU (14) is shown in 0. It has two counter-rotating blade groups (12,13), arranged up and down, with their blades (5) facing in opposite directions. The middle of the two blade groups contains four stabilizer wings (6), also outside the two generators (7). The PSU is connected to the unmanned marine equipment (1) by a drag rope (2) and the counterweight (11) is attached to the bottom via a drag rope (10). The blades (5) are fixed between the flywheel (9) and the blade assembly (4) and they form the blade groups. The blade groups are connected to the generator through the spindle (3), and the stabilizer wings are connected to the outside of the generator through the fixed stabilizers (8). As shown in Figure 1b, this PSU can be made in small size, which helps to integrate within the unmanned marine equipment or move with it.

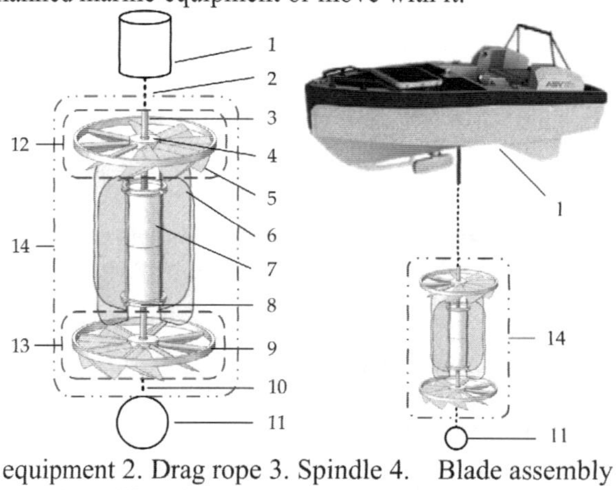

1. Unmanned marine equipment 2. Drag rope 3. Spindle 4. Blade assembly 5. Blade 6. Stabilizer wing 7. Generator 8. Fixed stabilizer 9. Flywheel 10. Drop rope 11. Counterweight 12. Up-blade-group 13. Low-blade-group 14. Power supply unit

Figure 1. Power supply unit detail structural

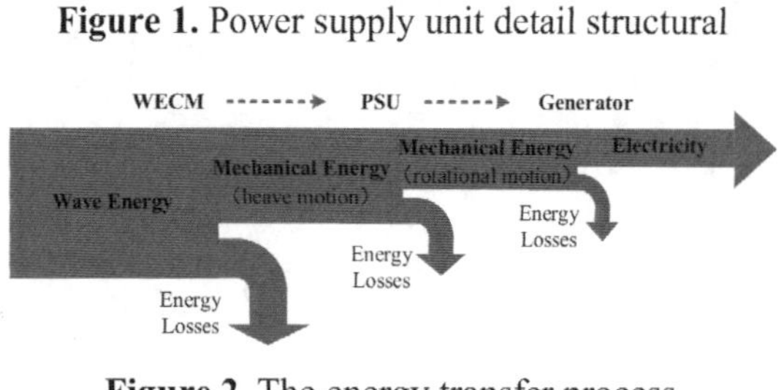

Figure 2. The energy transfer process

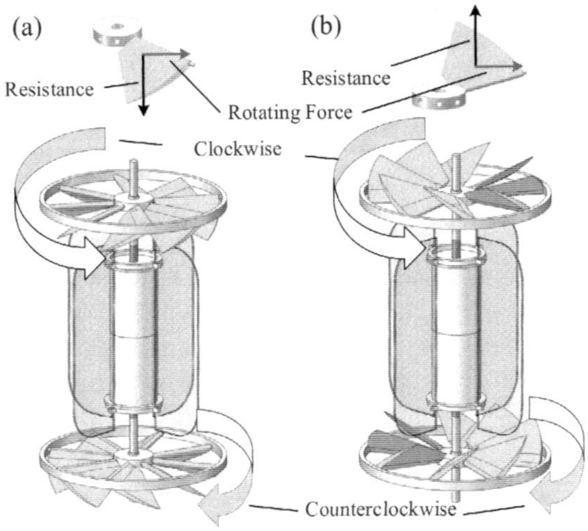

Figure 3. Working princile

2.2. Working principle

Similar to the WEC[7], the unmanned marine equipment can be considered as a Wave Energy Capture Module (WECM), the PSU can be considered as an energy conversion module (also referred to as PTO in WECs), and the counterweight and ropes as auxiliary modules. Obviously, the PSU is the core of the device and its operating performance will be mainly studied.

The WECM absorbs wave energy and converts it into heave motion (mechanical energy) for the PSU's input energy, then the PSU converts its heave motion into rotational motion to generate electricity[8, 9], as shown in 0. The PSU converts wave energy into electrical energy through its unique mechanical structure and transmission mechanism. The blade group and the generator are connected by a speed increaser, when the blade group rotates under the impact of water flow, it can drive the generator to generate electricity[10], where the speed increaser can convert the blade group' high-torque and low-speed rotational motion to the generator spindle's high-speed and low-torque rotational motion. The two generators are fixed back-to-back and connected to their own blade group via a speed increasers. As auxiliary module, the stabilizer wings and counterweight mainly assist in the normal operation of the PSU and are not involved in its transmission.

The PSU is a small device operating on the principle of adaptive counter-rotation, which mainly absorbs and converts its pendular motion into electricity. 0 shows a sketch of the conversion mechanism: 1. As shown in 0(a), when the unmanned marine equipment ascends, the PSS also ascends under the drag of the rope, and the two blade groups' upper surfaces are impacted by the water and the blades swing downward adaptively. Due to the limiting facility, the blades stop deflecting and remain tilted when the maximum deflection angle is reached. Under the continuous impact of the water, the tilted blade produces a horizontal force and pushes it to move. Circumferentially arranged blades generate circumferential torque to drive the blade group to rotate. Because the blade arrangement of the two blade groups is opposite, the upper blade group rotates counterclockwise and the lower blade group rotates clockwise. 2. As shown in 0(b), when the unmanned marine equipment descends, the PSS also descends under the drag of its weight and counterweight. The lower surfaces of the two blade groups are impacted by water and the blades adaptively oscillate upwards. Similarly, the blades remain tilted after deflecting upward to the maximum deflection angle due to the limiting facility. The tilted blades generate a horizontal force and push them to move under the continuous impact of the water. The horizontal forces generated by the tilted blades are in the same direction, whether the PSS is ascending or descending. So, the blade groups rotate in a single direction, allowing the generators to rotate in one direction too, which helps to provide continuous power generation.

3. Numerical simulations and results

3.1. Numerical simulation settings

CFD analysis results can be experimentally validated to predict device performance, reducing costs and accelerating research progress. ANSYS Fluent is a numerical simulation software for solving fluid flow and related physical quantities. It has many turbulence models, advanced numerical simulation methods, and good post-processing capabilities that help to calculate and simulate the motion of PSS. In this paper, Fluent 2021 is used to calculate the Navier-Stokes equation and simulate the motion process.

As a key component of PSS, the blade group has a direct impact on its performance. The blade group's solidity is one of its important parameters, and the blade group's energy absorption efficiency is also influenced by the arrangement method under the same solidity condition. To investigate the effect of solidity and arrangement on the energy absorption efficiency, we propose two kinds of blade arrangement: a. Average along the fan, b. Outermost along the radial direction, the schematic diagram is shown in 0.

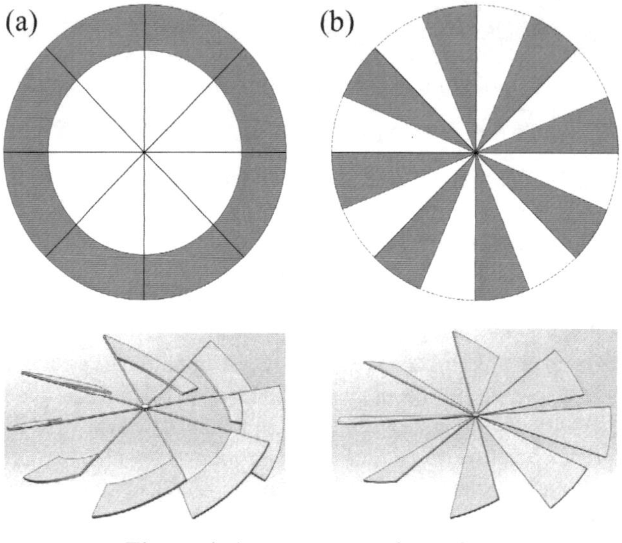

Figure 4. Arrangement schematic

Table 1. The main configuration of numerical analysis

Content	Numerical configuration
Simulation method	Multiple references (MRF)
Models	k-epsilon RNG
Materials	Water-liquid
Inlet condition	Velocity inlet (1.0m/s)
Outlet condition	Press outlet
Wall condition	Stationary wall
Methods	Coupled

To analyze and predict the performance characteristics of the PSS, ANSYS FLUENT, an efficient and accurate commercial CFD software, was chosen as the analysis tool in this chapter. Its software obtains the parameter distribution of the flow field and the performance characteristics of the device by solving the Navier-Stokes equations. To facilitate the analysis and study of the performance characteristics of the PSS, we only take the upper blade set as the study object. Its numerical simulation configuration is shown in 0.

The torque T_{max} and the dynamic power P values of the blade groups are highly dependent on the impinging water's flow rate. For this reason, we propose two coefficients C_T and C_P to describe the blade group operating performance: The torque coefficients C_T was introduced to describe the ability of getting torque for the blade group; The power coefficients C_P was introduced to describe the operating performance under preload conditions. They are defined as follows:

$$C_T = \frac{T_{max}}{0.5\rho v^2 A_d} = \frac{T_{max}}{0.5\rho v^2 \pi R^2} \tag{1}$$

$$C_P = \frac{P}{0.5\rho v^3 A_d} = \frac{M\Omega}{0.5\rho v^3 \pi R^2} \tag{2}$$

Where, T_{max} is the maximum torque that blade groups can be obtained under the impact of steady flow field ($N \bullet m$), M is the dynamic torque of blade group under preload conditions ($N \bullet m$), Ω is the angular velocity of blade groups (rad/s), ρ is the density of sea water (kg/m^3), v is the flow rate (v/s), R is the radius of blade group (m).

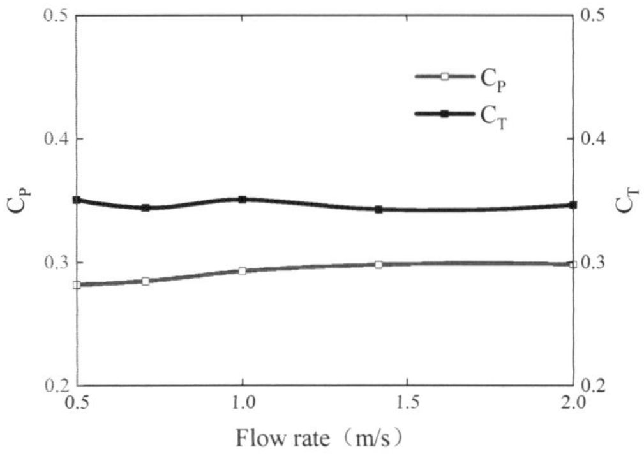

Figure 5. Effect of flow rate

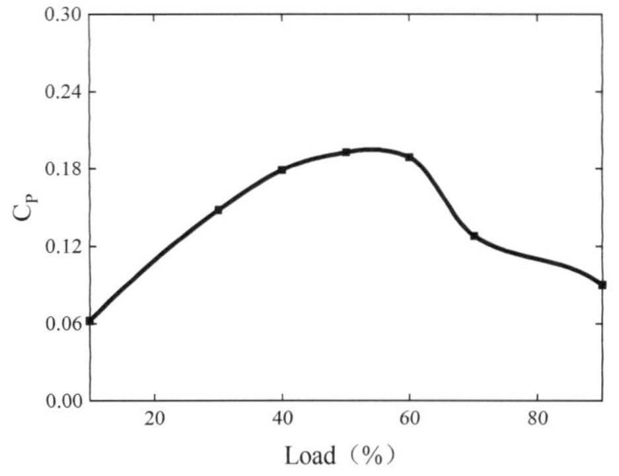

Figure 6. Effect of blade load

3.2. Numerical simulation results

Although the high solidity blade groups have a better performance in acquiring torque, the difference between the two types of arrangements (0) of the high solidity blade groups is not noticeable. Similarly, the performance parameters between the low solidity blade groups are not significant for PSS, so the blade groups with medium-high solidity are selected for numerical simulation.

1) Effect of flow rate

0 shows the effect of flow field velocity on power coefficient C_P and torque coefficient C_T. With the increases of flow field velocity, the power P and the maximum torque T_{max} of the blade groups also gradually increase, but the power coefficient C_P and torque coefficient C_T are basically unchanged, which indicates that the energy conversion performance of the blade groups are not affected by the water velocity.

2) Effect of blade load

The effect of the load ratio on the power coefficient C_P is shown in 0. When the load ratio is lower than 50%, the power coefficient C_P gradually increases with increasing load; when the load ratio is higher than 50%, the power coefficient C_P gradually decreases with increasing load ratio. Generally speaking, the power coefficient C_P is desirable at 40%-60% load conditions, and the PSS should design the load circuit within this load range to obtain the maximum energy conversion efficiency.

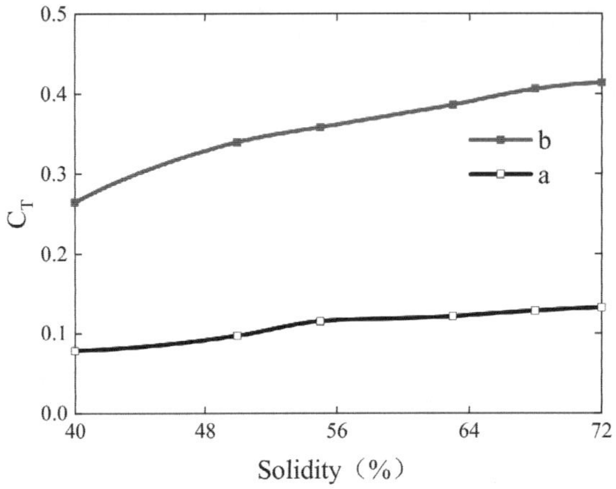

Figure 7. Effect of blade solidity on C_T

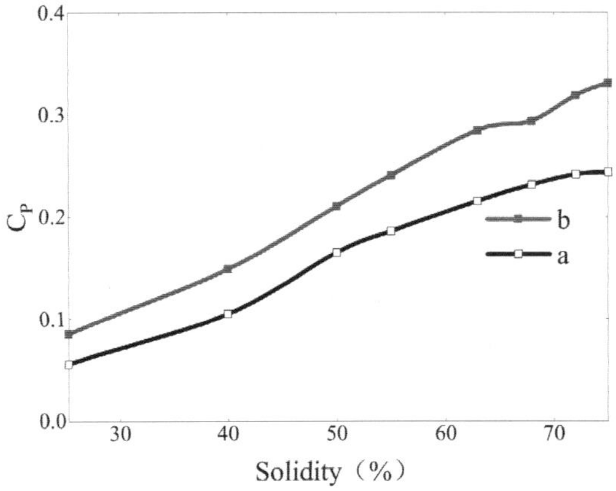

Figure 8. Effect of blade solidity on C_P

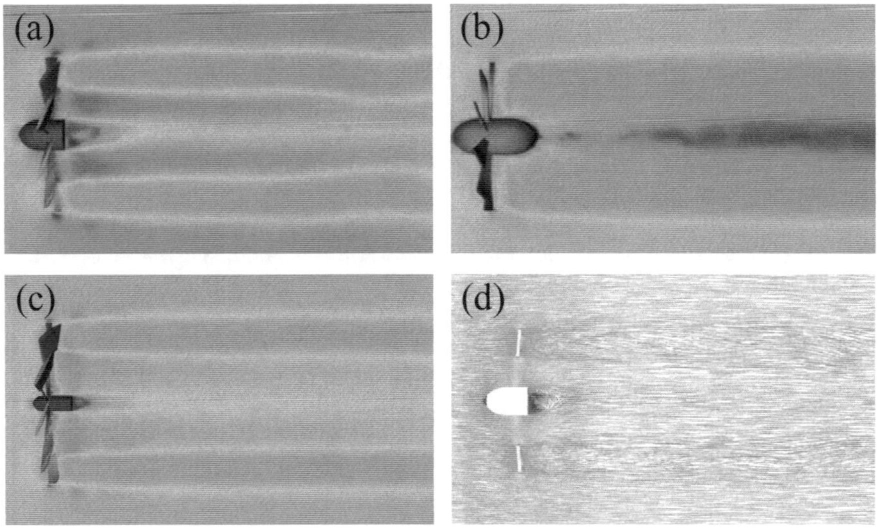

Figure 9. Velocity numerical simulation results

3) Effect of blade solidity on C_T

0 shows the effect of different arrangements of the two blade groups on their torque coefficients C_T. The figure shows that the torque coefficients C_T of both arrangements increase with the increase of blade group solidity, but the torque coefficients C_T of the b-type arrangement are larger and increase more rapidly than that of the a-type.

4) Effect of blade solidity on C_P

The effect of different arrangements of the two blade groups on their power coefficients C_P is shown in 0. Similar to 0, the power coefficients C_P of both blade groups' arrangement increase with the increase of blade group solidity, and the power coefficients C_P of type b are still greater than that of type a, and the increase rate appears to be the same.

5) Analysis of velocity numerical simulation results

The wake field of the blade groups also has a strong influence on its operating performance. 0a shows the wake field of the a-type blade group, and 0b. shows b-type. It can be seen that the wake field region (including turbulence and return flow) of a-type blade group is smaller than that of b-type. Moreover, the smaller the diameter of the spindle inside the blade group, the smaller the influence on the wake field (as shown in 0a and 0c), and if the spindle is smaller than 0.1 of the blade group diameters the influence is almost negligible. 0d shows the patelline of a-type blade group, whose trajectory is similar to the wake field.

4. Conclusion

In this paper, we design a PSU that can absorb wave energy and supply power to unmanned devices in the deep and distant sea. Through theoretical design, modeling and numerical simulation, the effect of blade group solidity on its operating performance was investigated. The main findings are as follows:

1) Flow field velocity has no effect on torque coefficient C_T and power coefficient C_P;

2) The power coefficient C_P of the blade groups is affected by its load, and it can obtain a large value when working in 40%-60% (the maximum torque that blade groups can be obtained) load conditions.

3) The power coefficient C_P and torque coefficients C_T of the blade group will positively increase with the solidity, but the b-type blade group has bigger torque coefficients C_T, which means higher starting performance, and it also has higher power coefficient C_P, which can absorb more energy.

Acknowledgments
This work was financially supported by National Natural Science Foundation of China, grant number 52,075,537; and Guangdong Province Key Field R&D Program, grant number 2020B1111030002.

References

[1] E. Al Shami, R. Zhang, X. Wang, Point Absorber Wave Energy Harvesters: A Review of Recent Developments, Energies 12 (2019).
[2] Z. Liu, R. Zhang, H. Xiao, X. Wang, Survey of the mechanisms of power take-off (PTO) devices of wave energy converters, ACTA MECHANICA SINICA 36(3) (2020) 644-658.
[3] R. Ahamed, K. McKee, I. Howard, Advancements of wave energy converters based on power take off (PTO) systems: A review, OCEAN ENGINEERING 204 (2020).
[4] I. López, J. Andreu, S. Ceballos, I. Martínez de Alegría, I. Kortabarria, Review of wave energy technologies and the necessary power-equipment, Renewable and Sustainable Energy Reviews 27 (2013) 413-434.
[5] E. Al Shami, Z. Wang, X. Wang, Non-linear dynamic simulations of two-body wave energy converters via identification of viscous drag coefficients of different shapes of the submerged body based on numerical wave tank CFD simulation, Renewable Energy 179 (2021) 983-997.
[6] C. Sun, J. Shang, Z. Luo, Y. Zhu, Z. Lu, G. Wu, D. Cong, Performance characteristics of a Novel point Absorber-type WEC based on counter-rotating self-adaptable movement mechanism, Energy Sources, Part A: Recovery, Utilization, and Environmental Effects 43(7) (2021) 783-799.
[7] X. Li, D. Martin, C. Liang, C. Chen, R.G. Parker, L. Zuo, Characterization and verification of a two-body wave energy converter with a novel power take-off, Renewable Energy 163 (2021) 910-920.
[8] G. Wu, Z. Lu, Z. Luo, J. Shang, C. Sun, Y. Zhu, Experimental Analysis of a Novel Adaptively Counter-Rotating Wave Energy Converter for Powering Drifters, Journal of Marine Science and Engineering 7(6) (2019).
[9] S. Barbarelli, M. Amelio, T. Castiglione, G. Florio, N.M. Scornaienchi, Design and analysis of a new wave energy converter based on a point absorber and a hydraulic system harvesting energy from waves near the shore in calm seas, International Journal of Energy Research 45(1) (2021) 661-690.
[10] A. Shehata, Q. Xiao, K. Saqr, S. Day, Wells turbine for wave energy conversion: A review, International Journal of Energy Research 41 (2017) 6-38.

Underway Conductivity-Temperature-Depth profiler system cable force analysis and vortex vibration phenomenon research

Libin Du [a], Chen Liu [b], Jie Liu * and Ding Gao [c]

School of Shandong University of Science and Technology University, Qingdao, China

*Corresponding author e-mail: liujie0232@sdust.edu.cn, [a]dulibinhit@163.com, [b]lcljhlc@163.com, [c]gaoding888@126.com

Abstract. In order to analyze the vibration response of the traveling temperature salt depth system cable at a certain velocity, its stress and fracture stress are analyzed. Based on the ANSYS Workbench platform, a numerical simulation model of two-dimensional underwater cable vortex induced vibration is built using the fluid structure coupling calculation method. Six groups of different speeds, including 1kn, 3kn, 6kn, 9kn, 12kn and 15kn, are selected for the ship's sailing speed, and the vortex induced vibration response of the cable is numerically simulated and analyzed. The simulation results of lift, drag coefficient, eddy viscosity and velocity change provide a certain research basis for studying the influence of the vibration response of the cable of the traveling temperature salt depth system on the underwater detection system.

1. Introduction

The characteristic of the Underway Conductivity Temperature Depth Profiler (UCTD) is that the measuring probe can be towed at the stern of the ship during the course of navigation to conduct the navigation measurement. The system is mainly composed of mother ship, measuring probe towed underwater and cable for connection, as shown in Figure 1. In the UCTD measurement system, the cable, as the main connector, plays an important role. On the one hand, the cable is the power for the underwater probe to move forward with the mothership; on the other hand, the stability and safety of the cable also affect the acquisition of measurement signals. With the need of ocean exploration, the application of thermohaline depth profile measurement system is more and more extensive, especially when it is used in the military field, higher requirements will be put forward for such equipment to ensure its high security and stability.

Content from this work may be used under the terms of the Creative Commons Attribution 3.0 licence. Any further distribution of this work must maintain attribution to the author(s) and the title of the work, journal citation and DOI.
Published under licence by IOP Publishing Ltd

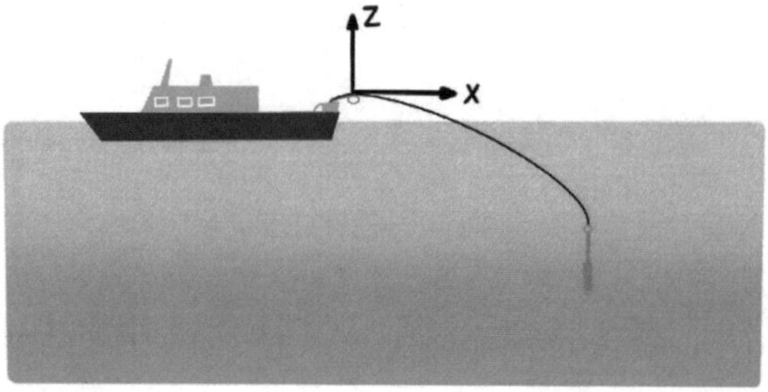

Figure 1. Composition of UCTD system towing system

As the underwater working environment is relatively complex, it is a key technical problem to ensure that the UCTD measurement system has good anti-interference capability and stability under complex and changeable sea conditions. There are many factors that affect the UCTD measurement. Because the UCTD working depth reaches 3000 m underwater and the cable length is long, the resistance between the cable and the water will increase, which will affect the stability and motion attitude of the probe, At the same time, the response interference caused by the coupling of UCTD underwater probe and its towed cable will affect the stability of the system.

From the system dynamics study, the dynamic characteristics of the cable are an important part of the whole system. This flexible cable system with a large aspect ratio can be regarded as a structure with the upper end attached to the moving hull and the lower end free and subject to the tension of the hull and the UCTD probe. The cable can be approximated as a flexible cylinder with a large aspect ratio, and in the study of its dynamics, it is necessary to consider not only the tension of the cable structure itself and the reliability problems caused by fatigue damage, but also the influence caused by the eddy vibration phenomenon that is easily generated by such structures. Vortex vibration phenomenon refers to the flow of downstream water flowing through the cable and perpendicular to the longitudinal flow of the cable to generate periodic excitation force, which is due to the cable surrounding the winding vortex off caused by. [1]

2. Selection of cable

At present, the commonly used cable types on the market include aramid, polyester, carbon fiber and polyethylene fiber materials. The UCTD system uses the ultra-high molecular weight polyethylene fiber material known as the "king of fibers". The performance indexes of this fiber are at a high level, which is significantly superior to the commonly used fiber materials in the market, such as aramid fiber and carbon fiber. UHMWPE fiber has excellent weather resistance, corrosion resistance, abrasion resistance and bending resistance due to its unique physical and chemical structure, which is single and inert. Because of its superior performance, this material has been widely used in various fields of production and life, among which the application in cable is also extremely extensive. Now it is widely used in towing rope, fishing rope, heavy load rope, sailing rope, scientific research ship, oil tanker, lighthouse and other marine operation platforms. Compared with steel cable and aramid fiber, the cable made of ultra-high molecular weight polyethylene fiber has a fracture length of eight times that of steel cable and twice that of aramid fiber under its own weight. At the same time, because it is not easy to be corroded, rusted and degraded by ultraviolet light, it can effectively ensure the strength stability of the cable, avoid fracture, and reduce the replacement problem.

2.1. Characteristics of UHMWPE fiber cable

Ultra high molecular weight polyethylene fiber, also known as high-strength and high modulus polyethylene fiber, is a new type of high-performance fiber, which is made of thermosetting polymer with an average molecular weight of more than 2 million polymerized by ethylene and butadiene monomers under the action of catalyst through spinning and high power drafting. The properties of UHMWPE fiber materials are as follows: [2]

High abrasion resistance: the abrasion resistance is 4-7 times higher than that of steel and 10 times higher than that of high-density polyethylene.

High impact resistance: It has extremely strong impact resistance and impact energy absorption characteristics (including strong external impact force, internal pressure fluctuations). Its impact strength is 10 times that of nylon, 20 times that of polyvinyl chloride; 4 times that of high-density polyethylene

Special flexibility: The fiber has good bending performance, and will not break to form knitting loops and knots.

High corrosion resistance: it can be immersed in organic solvents such as acid, alkali, seawater, kerosene, toluene, etc. for a long time.

High weather resistance: it will not age for decades, and will still maintain more than 70% of the mechanical properties at - 70 °C.

Self lubrication: the surface of the cylinder is extremely smooth, and it is difficult for microorganisms in the water to climb.

Light weight: the density is about 0.97g/cm^3, which can float with water.

2.2. *Fracture strength*

As for the cables made of fibers, their tensile fracture properties are different due to different materials and thicknesses. We call the strength of the fiber cables converted into the specified thickness when they break strength, and compare the fracture properties of different fibers based on the strength. There are three common expressions of fracture strength.

The first way is the fracture stress, which refers to the maximum stress that the fiber can withstand per unit cross-sectional area, and is calculated by the formula

$$\sigma = \frac{\mathrm{P}}{\mathrm{A}} \tag{1}$$

where σ is the fracture stress of the fiber, MPa; P is the strength of the fiber, N; A is the cross-sectional area of the fiber, mm^2

The second way is the fracture strength, which refers to the maximum stress per ter, or per denier of fiber, and is calculated by the formula

$$\begin{cases} P_{tex} = \frac{P}{T_t} \\ P_{den} = \frac{P}{D} \end{cases} \tag{2}$$

Where, P_{tex} is the linear density system strength, N/tex; P_{den} is the fiber strength, N/dan; P is the strength of fiber, N; T_t is the linear density of the fiber, tex; D is the fiber size, dan.

The third is the breaking length, that is, the elongation of the fiber when it is pulled off. In the actual production process, the breaking length of the fiber can be expressed by strength conversion, and the formula is as follows:

$$L_p = \frac{\mathrm{P}}{\mathrm{g}T_t} \times 1000 \tag{3}$$

Where L_p is the fracture length of the fiber, km; P is the strength of the fiber, N; g is the acceleration of gravity, taking $\mathrm{g} = 9.8\,m/s^2$; T_t is the linear density of the fiber, tex.

UCTD system in the work process, because the probe in the lifting process will be subject to gravity and seawater buoyancy and wave force, will directly affect the UCTD cable tension size, so the strength of the selected cable has an important role in the safety and stability of the UCTD probe drop process.

According to the fracture stress formula 1 of the cable, the strength of the cable can be calculated.

$$\mathrm{P} = \sigma \mathrm{A} \tag{4}$$

The critical breaking strength of the cable is:

$$P_m = \frac{\sigma A}{s} \tag{5}$$

In the formula, s is the safety factor, which is obtained from the experience of taking 6-10 as the safety factor for general ships and 5-8 as the safety factor for hoisting. The fracture stress of UHMWPE fiber rope used in UCTD is 2.6~3.8 GPa, and the diameter of the rope is 2mm.

According to Equation 4, the critical breaking force of the cable is

$$P_m = \frac{\sigma A}{s} = \frac{2.6 \times 1000 \times \pi}{8} \tag{6}$$

Where, taking s=8, $\sigma = 2.6GPa$, $A = \pi \times 1mm^2$, P_m is about 1021N.

The mass of the UCTD probe is known to be about 7.5kg, and the tension caused to the cable is not enough to break it.

3. Vibration characteristic analysis

In the process of underwater movement, UCTD is affected by incoming flow, and its vibration characteristics will affect the attitude of the UCTD probe, thus affecting the acquisition of seawater data by the sensor. The vibration characteristics of the cable are analyzed below.

In the process of underwater movement, the factors that cause the cable to produce vortex vibration phenomenon is mainly caused by the current. When the seawater flows through the cable, it will be divided by the cable to produce a diversion at the back of the rope, which will produce the vortex street phenomenon when the Reynolds number is low, and will produce the vortex region when the Reynolds number is high. No matter in which case, the vortex generated by the water behind the cable on both sides due to the asymmetry, will produce alternating pressure field, because the pressure direction is perpendicular to the cable, so the cable in the cross-current direction will produce vibration, that is, "vortex vibration".

Vortex induced vibration does not always occur around the cable. The following two conditions must be met to cause vortex induced vibration:

(1) Reynolds number based on feature size Re>120;

(2) The cable is in the locked area.

That is, when the vortex shedding frequency of the cable is close to the natural frequency of the sea water at a certain seawater inflow speed, the structure will lock and produce resonance. In order to facilitate the research of the cable, considering that the cable is in a tensioned state in the dragging state, it can be simplified as a cylinder for research. The dynamic equation of the cylinder when vortex induced vibration occurs is

$$m\ddot{y} + c\dot{y} + ky = F_y \tag{7}$$

Where m: cable mass; y: cross-current displacement of the cable; c: cable damping coefficient; k: cable stiffness coefficient; F_y: the force of seawater in the cross-current direction, its calculation formula is $F_y = \frac{1}{2} C_L \rho D V^2$; Where ρ: seawater density; D: cable diameter; V: current velocity; C_L: lift coefficient.

The Reynolds number Re and the vortex off frequency fs are calculated as

$$R_e = \frac{\mathrm{DV}}{\mathrm{v}} S_t \tag{8}$$

$$f_s = \frac{\mathrm{V}}{\mathrm{D}}\left(1 - \frac{20}{R_e}\right) \tag{9}$$

where D: cable diameter; V: current velocity; v: seawater viscosity; S_t: Strouhal number, which is a dimensionless similarity number characterizing the effect of periodicity in a non-stationary flow.

According to the actual work of UCTD, take the cable diameter D = 2mm; water kinematic viscosity $v = 1.006 \times 10^{-6}\, m^2/s$, get the Reynolds number of the cable at different speeds, according to Equation (9) can be derived from the eddy off frequency fs of the cable at different speeds.

4. Mathematical model verification

4.1. Control equation

There are many turbulence models that can be used for fluid dynamics simulation analysis, and the most common model is k- ε Model, RNG k- ε Model, k- ω Model, SST k- ω Model, BSL k- ω Model, zero equation turbulence model, Reynolds Stress model, etc. In this paper, k, which is widely used- ε Model is also one of the most mature computing models. Since the working environment of UCTD is ocean, which is a turbulent flow field and belongs to the situation of high Reynolds number, k suitable for high Reynolds number simulation is adopted- ε The model is the most appropriate. For k- ε The basic control equation of the model is as follows

Continuity equation:

$$\frac{\partial \rho}{\partial t} + \nabla \cdot (\rho V) = 0 \tag{10}$$

Momentum (vector) equation:

$$\frac{\partial \rho V}{\partial t} + \nabla \cdot (\rho V \times V) - \nabla \cdot (\mu_{eff} V) = \nabla \cdot P' + \nabla \cdot (\mu_{eff} V)^T \tag{11}$$

where μ_{eff} is the effective viscosity coefficient, and $\mu_{eff} = \mu + \mu_t$, μ is the dynamic viscosity coefficient, μ_t is the turbulent viscosity coefficient, where $\mu_t = C_\mu \rho k^2/\varepsilon$, C_μ is a constant, generally take $C_\mu = 0.99$, P' prime is the correction pressure, $P' = P + \frac{2}{3}\rho k$, k and ε values can be derived from both sides of the equation.

Turbulent kinetic energy k equation:

$$\frac{\partial \rho k}{\partial t} + \nabla \cdot (\rho V k) - \nabla \cdot \left(\frac{\mu_{eff}}{\sigma_k} \nabla \varepsilon\right) = G + \rho \varepsilon \tag{12}$$

Turbulent flow energy dissipation rate varepsilon equation:

$$\frac{\partial \rho k}{\partial t} + \nabla \cdot (\rho V k) - \nabla \cdot \left(\frac{\mu_{eff}}{\sigma_\varepsilon} \nabla \varepsilon\right) = \frac{\varepsilon}{k}(C_{\varepsilon 1} G - C_{\varepsilon 2} \rho \varepsilon) \tag{13}$$

Where $C_{\varepsilon 1}, C_{\varepsilon 2}, \sigma_k, \sigma_\varepsilon$ is a constant, $C_{\varepsilon 1}, C_{\varepsilon 2}, \sigma_k, \sigma_\varepsilon$, the value of shear generated item G is obtained according to the following formula:

$$G = \mu_t \nabla V \cdot (\nabla V + V^T) - \frac{2}{3} \nabla \cdot V(\mu_t \nabla \cdot V + \rho k) \tag{14}$$

In Fluent software, select k in the mixture multi stream model- ε Turbulence module performs steady calculation, and the calculation iteration step is set as 1000 steps. Second order upwind difference scheme is adopted to calculate momentum and turbulent kinetic energy, and SIMPLE algorithm is adopted for pressure velocity coupling to improve calculation accuracy.

4.2. Grid division and parameter selection

The two-dimensional calculation domain is constructed as shown in Figure 2. The diameter of the cylinder is D=2mm, and the total size of the watershed is 20D × 50D. In order to reduce the influence of the boundary of the computational domain on the flow process near the cylinder wall, the distance from the cylinder wall to the inlet and upper and lower walls is set as 10D, and the distance from the cylinder wall to the outlet is set as 25D. Velocity inlet is set at the inlet boundary of the basin, and the uniform incoming velocity is 1kn, 3kn, 6kn, 9kn, 12kn, 15kn, and pressure outlet is set at the outlet, and symmetry wall is set at the top and bottom.

The calculation area is divided into quadrilateral structured grids. At the same time, in order to ensure uniform grid transition, the grid uses a fixed growth rate to control grid changes, so that the grid densification area and the overall grid can achieve a linear transition at the junction, avoiding grid mutation, so as to improve the calculation convergence. The grid division results are shown in Figure 3.

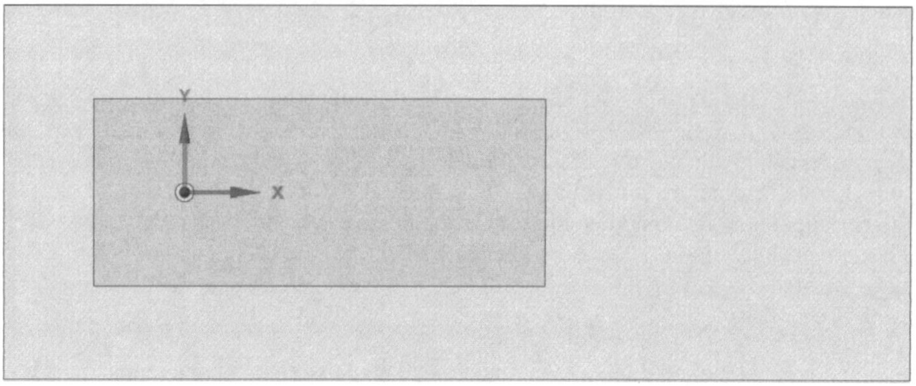

Figure 2. Schematic diagram of the two-dimensional computational domain

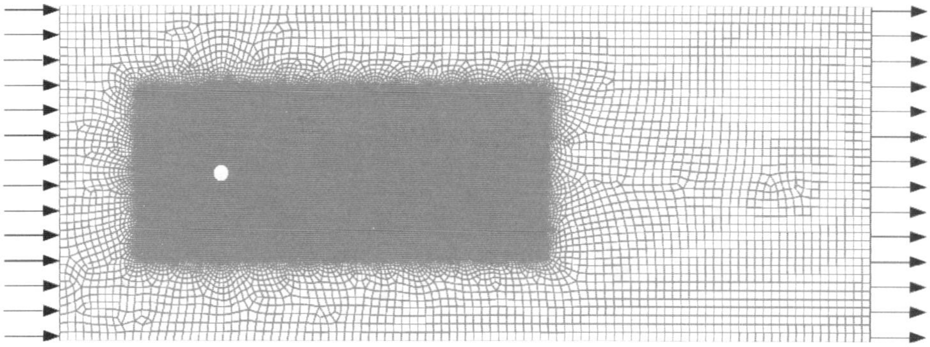

Figure 3. Schematic diagram of grid division

5. Numerical calculation results and analysis

Based on the two-dimensional numerical simulation calculation framework, the numerical simulation calculation of vortex induced vibration of the streamer under different velocity conditions is completed, and the fluid motion simulation results under various conditions, such as lift coefficient, drag coefficient, downstream velocity change, downstream displacement eddy viscosity coefficient change, are extracted. The change curve of the lift coefficient is shown in Figure 4. The lift represents the component of the resultant force in the direction perpendicular to the incoming flow. It can be seen from the figure that the force on the cable in the direction perpendicular to the incoming flow changes obviously. With the increase of the speed, the lift coefficient increases continuously and the change amplitude also increases, indicating that the force on the cable in this direction is significant and the vibration phenomenon is obvious. The change curve of resistance coefficient is shown in Figure 5. It can be seen that the resistance coefficient remains stable at different speeds. It can be seen that the force on the cable in the downstream direction is stable and the vibration is weak. The visual diagram of eddy viscosity around the cable is shown in Figure 6, which shows the change of eddy viscosity around the cable. Figure 7 shows the change curve of downstream displacement eddy viscosity coefficient. With the increase of speed, the eddy viscosity increases, indicating that the effect of turbulence around the cable is more significant. Figure 8 shows the displacement velocity change curve of downstream flow. It can be seen that the current will fluctuate in a small range around the cable after impacting the cable, and then tend to be gentle. The visual diagram of the impact of this flow velocity is shown in Figure 9.

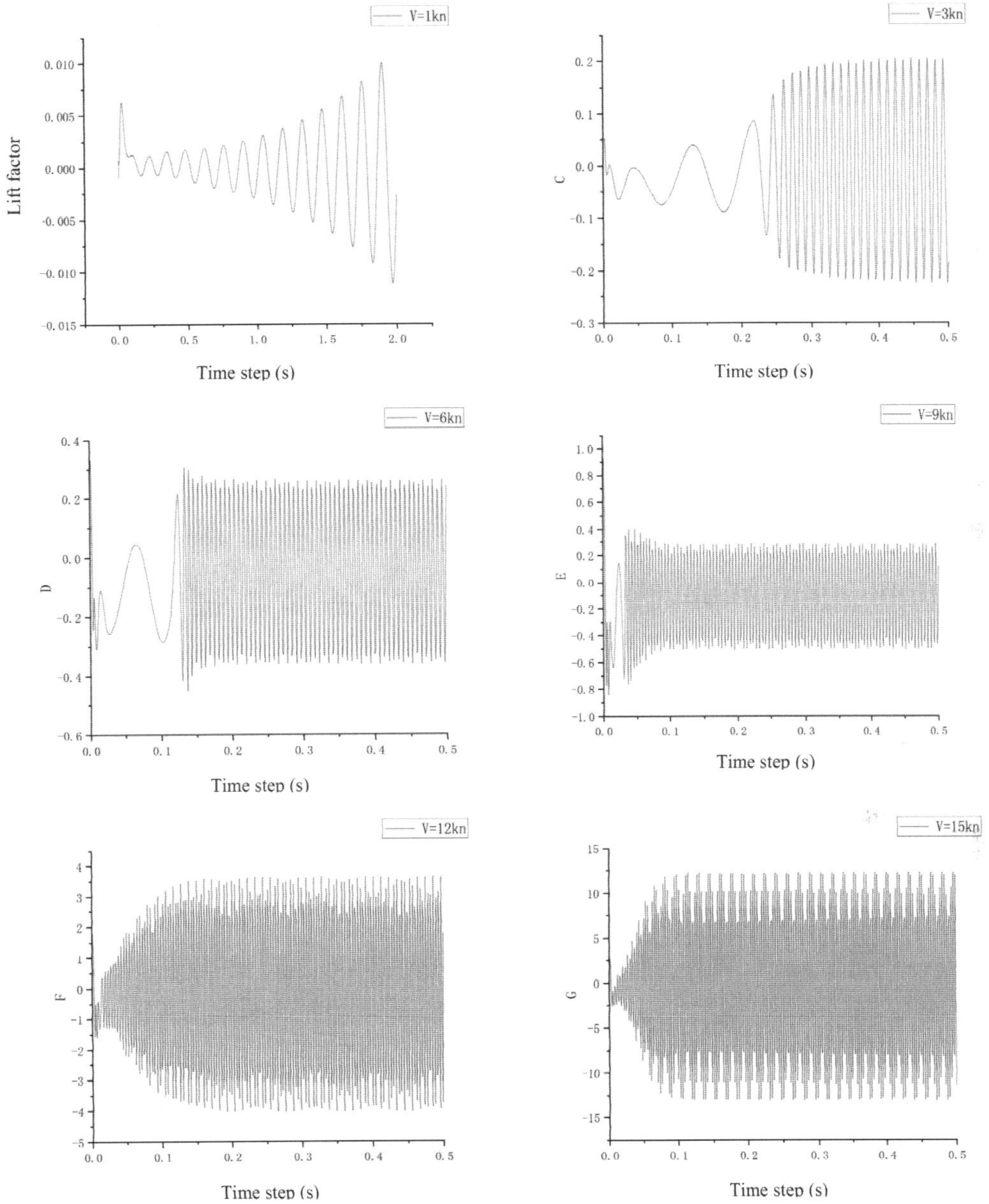

Figure 4. Variation Curve of Lift Coefficient

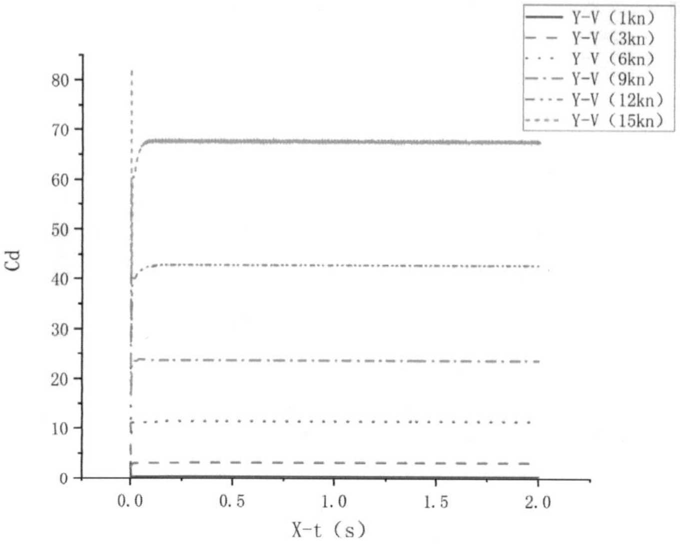

Figure 5. Resistance coefficient variation curve

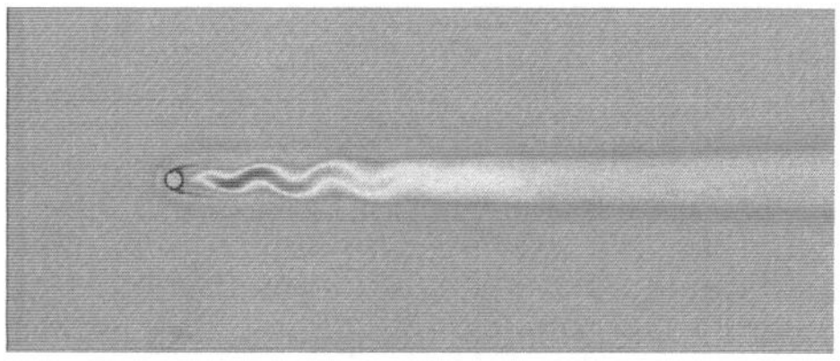

V=1kn

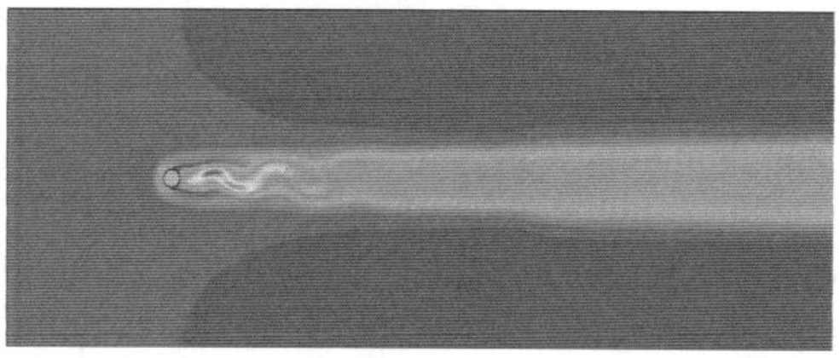

V=3kn

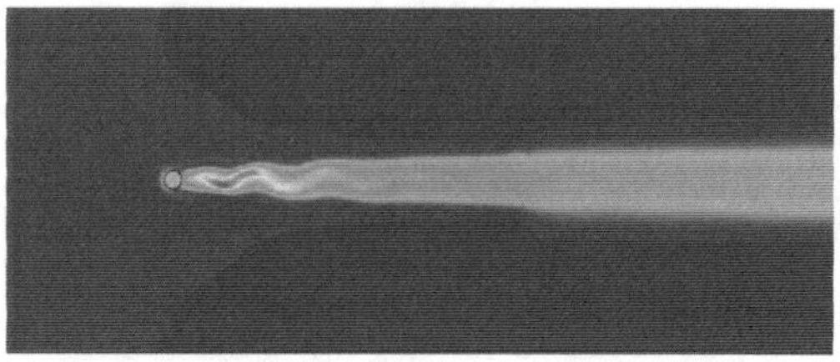

V=6kn

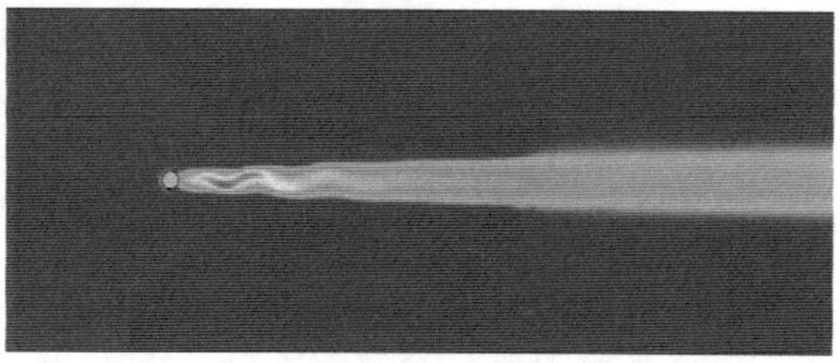

V=9kn

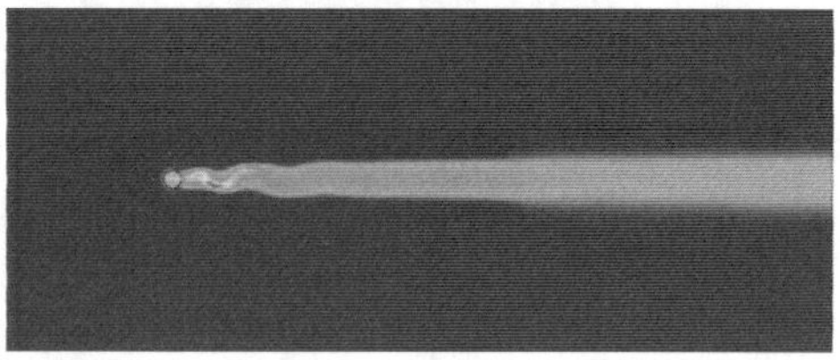

V=12kn

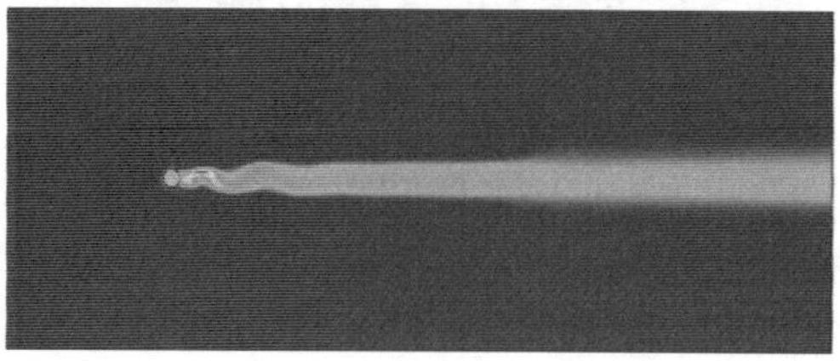

V=15kn

Figure 6. Visual diagram of eddy viscosity

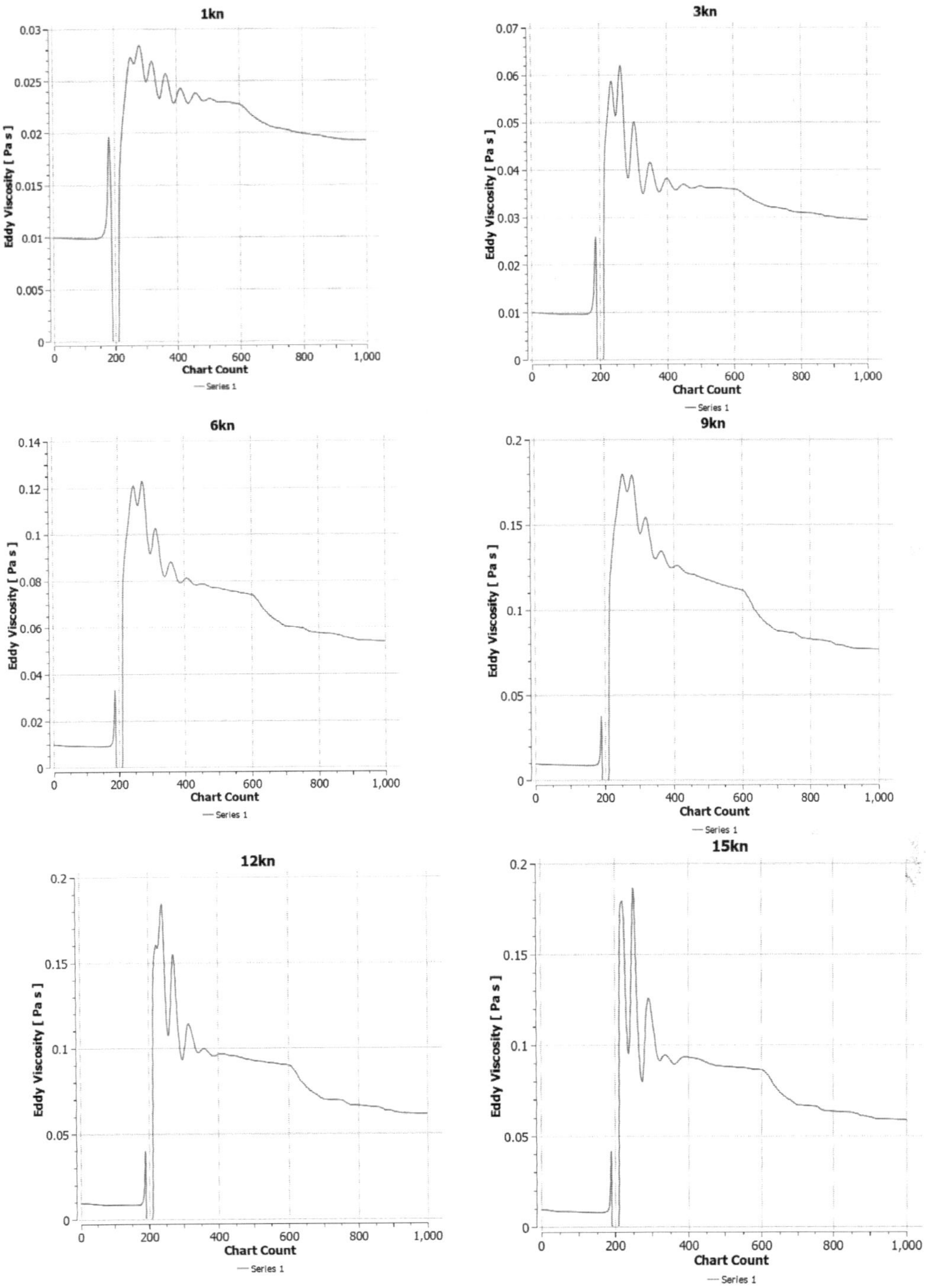

Figure 7. Change curve of downstream displacement eddy viscosity coefficient

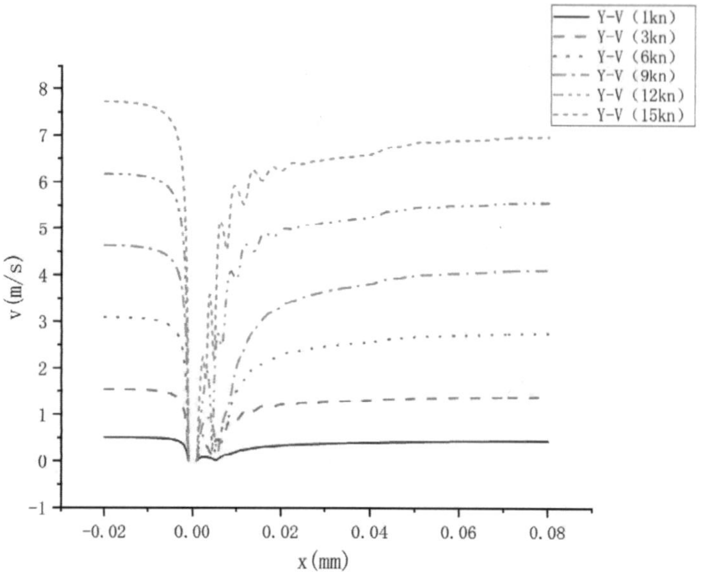

Figure 8. Displacement velocity variation curve of downstream flow

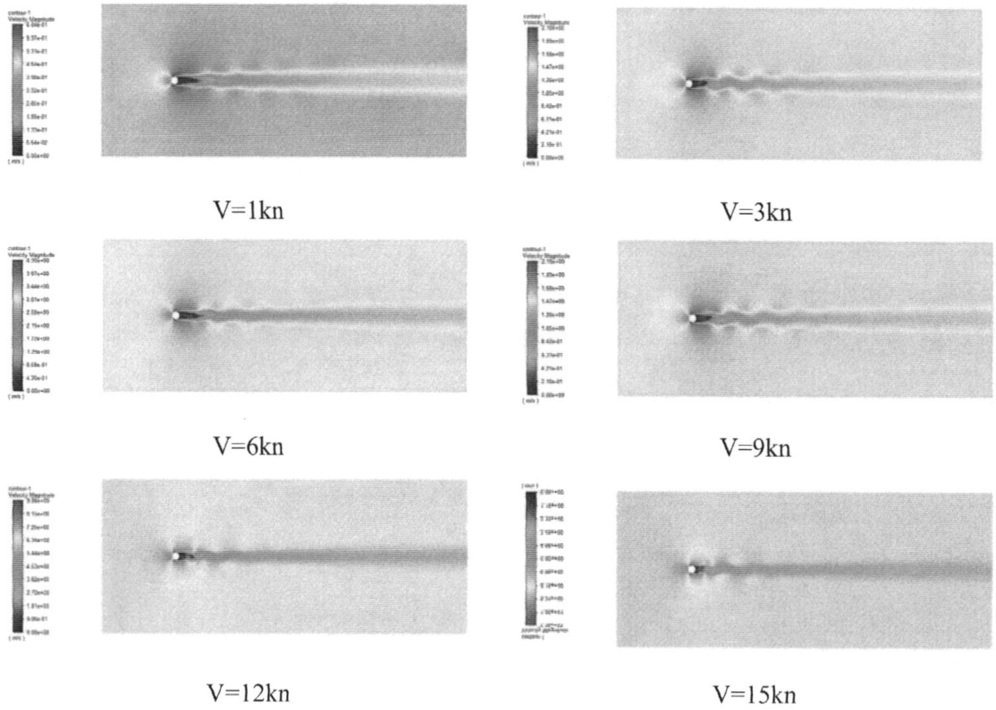

Figure 9. Water Flow Velocity Variation Diagram

6. Conclusion

In order to meet the application requirements of UCTD in high-speed and deep towing, this paper has carried out a comprehensive analysis, calculation and verification of the design of the towing system.

Through the mechanical analysis of the towing system, based on the analysis model, the motion of the cable under six groups of speed is analyzed, and the lift and viscosity coefficients of the cable under various application conditions are obtained, which meets the requirements of cable towing when the speed is 0-12kn in practical applications.

(1) The UCTD cable is made of ultra-high molecular weight polyethylene fiber, which has good creep resistance and fatigue resistance characteristics, can ensure multiple reuse of the cable, and has good cost performance and utilization efficiency.

(2) The UCTD cable will have vortex induced vibration during underwater movement, but the vibration is weak and will not affect the use process.

(3) With the increase of ship speed, the impact of current on the cable increases and the vortex induced vibration increases.

(4) Based on the ANSYS WorkBench platform, the two-dimensional streamer vortex induced vibration numerical simulation analysis framework can effectively capture the vortex induced vibration response at a certain flow rate, providing guidance for studying the impact of streamer vibration response on the underwater detection system.

Acknowledgments
This work was financially supported by the Shandong Provincial Natural Science Foundation (No. ZR2020QD071); the National Natural Science Foundation (No.42106072).

References
[1] Gang Zhao, Li Zhao,Xiongjun Xie. Ultra High Molecular Weight Polyethylene Fiber Material Technology and Market Development Prospec [J]. Fiber Composites, 2011 (1): 50-56
[2] Li Jianli, Zhang Xinyuan,Jia Zhekun. Property and Production Situation of UMWPE Fiber [J]. Knitting Industries, 2016 (6): 21-25.
[3] Franke J, Frank W. Large eddy simulation of the flow past a circular cylinder at ReD= 3900[J]. Journal of wind engineering and industrial aerodynamics, 2002, 90(10): 1191-1206.
[4] Pan Z Y, Cui W C, Zhang X C. An overview on VIV of slender marine structures[J]. Journal of Ship Mechanics, 2005, 9(6): 135-154.
[5] Wu X, Ge F, Hong Y. A review of recent studies on vortex-induced vibrations of long slender cylinders[J]. Journal of Fluids and structures, 2012, 28: 292-308.
[6] Gabbai R D, Benaroya H. An overview of modeling and experiments of vortex-induced vibration of circular cylinders[J]. Journal of sound and vibration, 2005, 282(3-5): 575-616.
[7] Holmes S, Constantinides Y. CFD modeling of long risers with buoyancy modules and complex shapes[C]//International Conference on Offshore Mechanics and Arctic Engineering. American Society of Mechanical Engineers, 2014, 45400: V002T08A083.
[8] Huang Weiping, Liu Juan, Tang Shizhen. Nonlinear model of vortex induced vibration of flexible cylinder in consideration of fluid-structure interaction[J]. Journal of Vibration and Shock, 2012, 31(9): 140-143.
[9] Zhao Jing. Numerical simulation on vortex-induced vibrations with fluid-structure interaction for risers by CFD[D]. Qingd- ao: Ocean University of China, 2012.
[10] Chao Jing. Numerical simulation on vortex-induced vibrations with fluid-structure interaction for risers by CFD[D]. Qingd- ao: Ocean University of China, 2012.

Research on wind direction measurement of wind turbine based on fluid simulation

Na Ren*, Guangwei Zhu, Shaonan Fu, Xiaocong He, Jianbin Hu and Yangguang Zhu

MingYang Smart Energy Group Co., Inc., Zhongshan, Guangdong 510428, China

*Corresponding author e-mail: renna@mywind.com

Abstract. For horizontal-axis wind turbines, wind turbines typically alignment nacelle to the wind using yaw system, realizing max energy capture. If the wind turbine's nacelle has a large error to the wind, the captured wind energy loss will be large, and it will also cause an increase in the load of the unit, which will pose a major risk to the safety of the wind turbine. Affecting the wind measurement error in addition to the performance of the sensor itself, the installation position of the wind measurement equipment also accounts for an important factor, this paper on the basis of the calculation of fluid dynamics simulation results, through the lateral and longitudinal comparison of the simulation results, pointed out the best installation location of the wind direction sensor, optimize the loss caused by wind measurement error of wind turbine in design, and provide guidance for the installation of the wind direction sensor on site.

Keywords: Wind turbine, Wind direction sensor, Installation position.

1. Introduction

With the proposal of the "3060" goal of carbon peaking and carbon neutralization, wind power generation has ushered in rapid development, especially offshore wind power. At present, the development of offshore wind power takes large capacity and large impeller as the main development direction. With the increase of impeller diameter, the diameter of blade bottom is also increasing. For the wind turbine with upwind direction and wind measuring device installed at the tail of engine room, there is a large error in wind measurement of the unit due to the influence of blade shielding and wake. A large wind direction error will lead to a large yaw to wind error, which will directly reduce the energy capture of the unit.

However, there is little research on the installation location of wind direction equipment in the wind power industry. Wang Xiaoyu [1] and others used the method of combining horizontal lidar test and numerical simulation to study the accuracy of mechanical wind direction meter in measuring wind direction. In his patent, Martin [3] used the historical data of his target fan for pre-learning and used the wind direction meter of other similar fans for error correction.

Content from this work may be used under the terms of the Creative Commons Attribution 3.0 licence. Any further distribution of this work must maintain attribution to the author(s) and the title of the work, journal citation and DOI.
Published under licence by IOP Publishing Ltd

2. Simulation model establishment

(1) Test point selection

Set the highest point just above the center of the hub is the origin of the coordinates, the horizontal direction is the X-axis direction, and the vertical direction is the Z-direction, as shown in Figure 1.

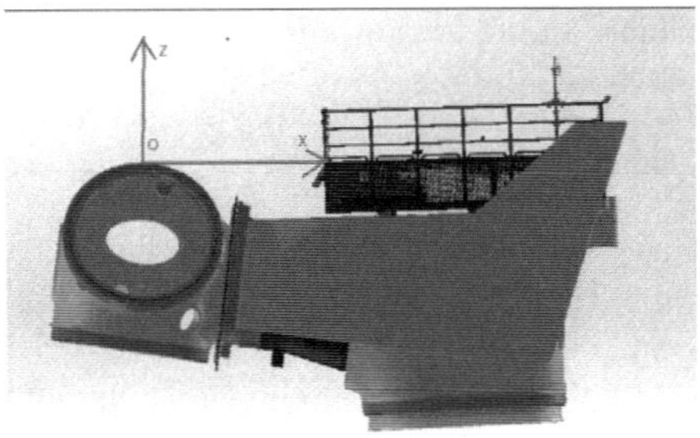

Figure 1. Test point axis

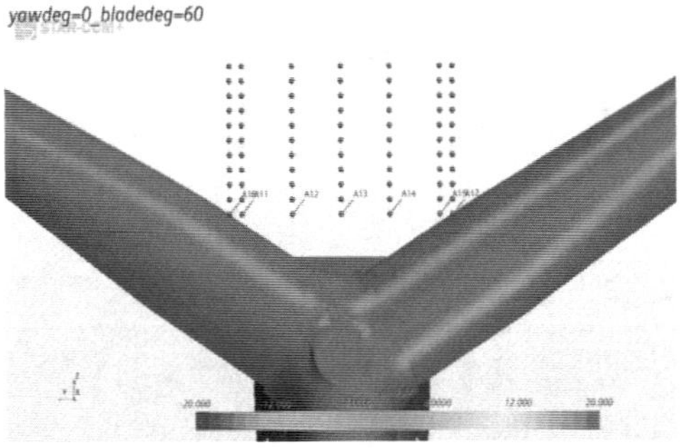

Figure 2. Simulation test point

Table 1. Simulation point coordinates

Coordinate number	Yaw 0°		
	x(m)	y(m)	z(m)
A11	12.497	3.94	2.279
A12	12.497	1.94	2.279
A13	12.497	0	2.279
A14	12.497	-1.94	2.279
A15	12.497	3.94	2.279
A16	12.497	4.44	2.279
A17	12.497	-4.44	2.279
A21	12.497	3.94	2.779
A22	12.497	1.94	2.779
A23	12.497	0	2.779
A24	12.497	-1.94	2.779
A25	12.497	3.94	2.779
A26	12.497	4.44	2.779
A27	12.497	-4.44	2.779

By analogy, the longitudinal first floor is A1, the second floor is A2, and the eleventh floor is a11. The difference between each floor is 0.5m. The horizontal arrangement points are 6, 1, 2, 3, 4, 5 and 7. For example, the fourth measurement point on the third floor is named A34

(2) Simulation condition determination

This paper takes a company's MySE8-195 wind turbine as an example for simulation. When the nacelle is facing the wind, the yaw error is 0°, the counterclockwise direction is negative yaw error, the clockwise direction is positive yaw error, and the blade orientation The angle is 0° vertically upward for a blade. The simulation environment conditions are typhoon environment, the wind speed is 50m/s, the simulation conditions include blade azimuth angles of 0°, 15°, 30°, 45°, and 60°, respectively, and yaw errors of 0°, 8°, and 15°, respectively. There are a total of 15 simulation conditions, each of which contains the wind measurement error data of 77 measuring points. When the yaw error is negative, the simulation data when the yaw error is positive can be used for symmetry processing.

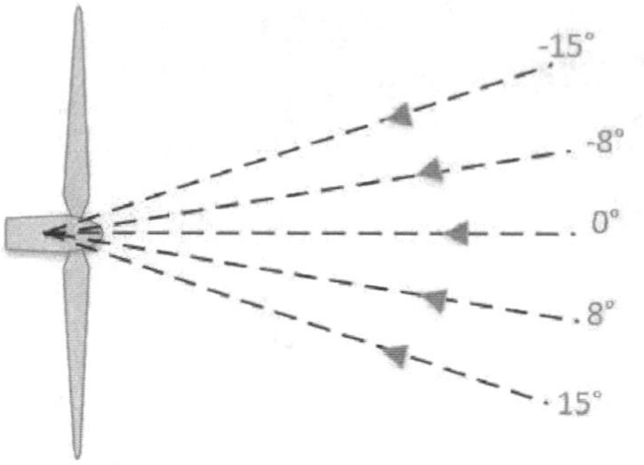

Figure 3. Yaw angle definition

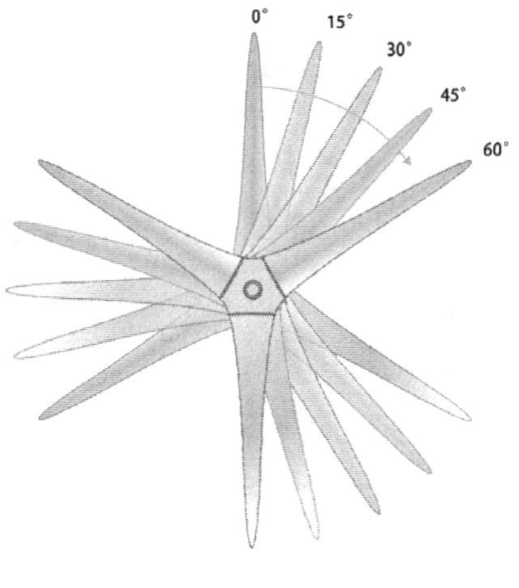

Figure 4. Blade azimuth

3. Simulation data analysis

By observing the wind measurement errors under various working conditions and heights of the simulated test points, it is found that the wind measurement errors at A1 and A2 heights are relatively large, so the measurement point heights are selected at least from A3, and the first point 2 is randomly selected from the A3 height. Check the wind measurement error under each working condition. Specify X°_Y°: X represents the yaw error angle, and Y represents the blade azimuth angle.

(1) Simulation test point 2

According to the wind measurement errors of all working conditions at the A3 height at point 2 as shown in Figure 5, it is found that the mean absolute value of the wind measurement errors exceeds 8°

under the working condition 15_60 (yaw error 15°, blade azimuth angle 60°). To ensure the accuracy of the measurement, it is necessary to increase the height of the measuring point.

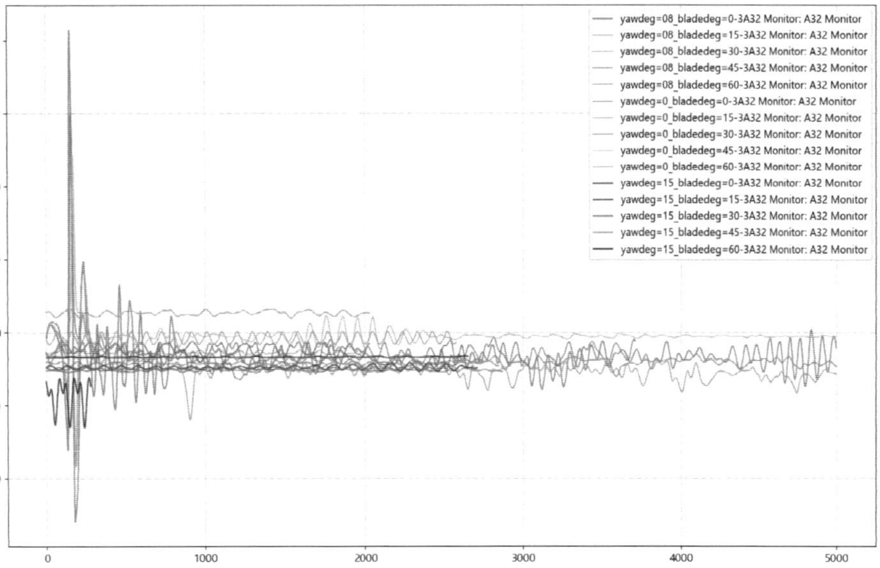

Figure 5. Wind direction error of all working conditions at point 2 at A3 height

Check the wind measurement errors of test point 2 at various heights under working conditions 15_60. As shown in Figure 6, when the height is A4 and above, the average measurement error is within 8 degrees.

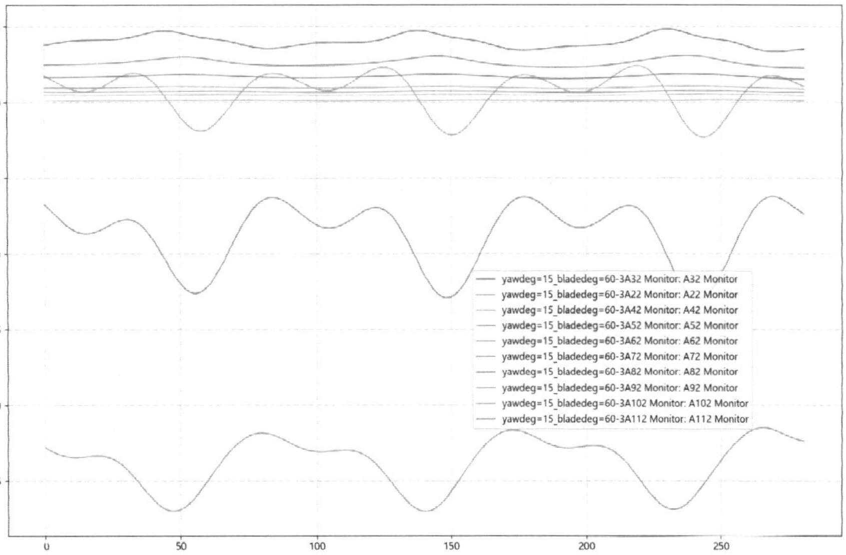

Figure 6. Wind measurement error at various heights at point 2 at 15_60

According to Figure 7, check the wind measurement errors of all the working conditions of test point 2 at A4 height. After stabilization, the average error value is within 8 degrees. If test point 2 is selected, at least A4 height is required.

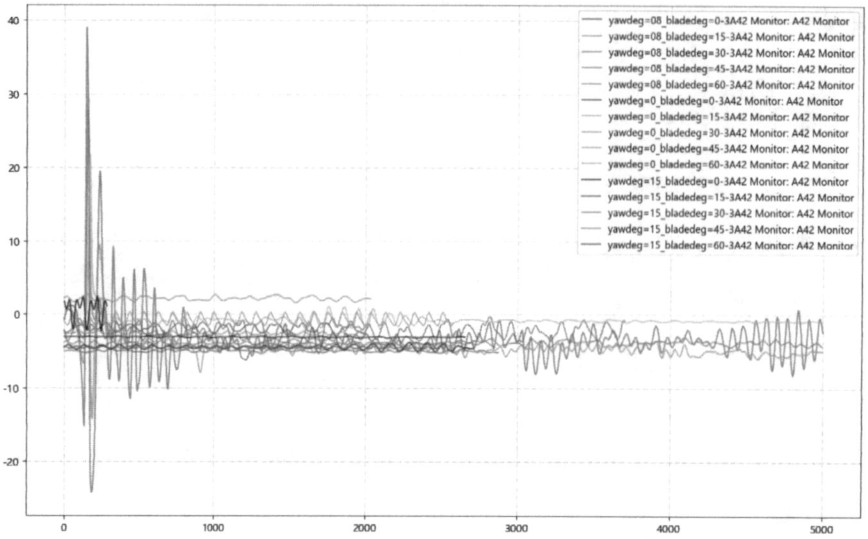

Figure 7. Wind measurement error of point 2 at A4 height under all conditions

By analogy, the other 6 measuring points are analyzed one by one, the height is at A4 height and above, the working condition data is listed in 15 simulated working conditions, and the measuring point selection criteria should meet the wind measurement error within 8° as much as possible. By comparison, it is found that for measuring point 4, the height should be A5; for measuring point 3, the height should be A5; for measuring point 6, the height should be A5; for measuring point 1, the height should be Select A5; for measuring points 5 and 7, the height must reach the height of A10 and A11.Based on the analysis of the wind measurement error of each working condition, each height and each point, it is found that A5 is more suitable for the height, and the measurement data is more stable when the measurement points 2, 3, and 4 are selected. Therefore, A52, A53, and A54 are finally determined. The points are selected measuring points, and the coordinates are A52(12.497, 1.94, 4.279), A53(12.497, 0, 4.279), A54(12.497, -1.94, 4.279).

4. Test verification

According to the selected A52, A53, A54 measuring points, three anemometers are installed on site, and a cabin-type lidar anemometer is installed for verification.

Table 2. Field Validation Test Results

condition	A52	A53	A54	Lidar	Lidar -A52	Lidar -A53	Lidar -A54
Blade azimuth 0 degrees	17.005	12.435	17.2	14.332	-2.673	1.897	-2.868
	13.93	12.46	17.03	13.898	-0.032	1.438	-3.132
	19.807	12.957	17.527	15.45	-4.357	2.493	-2.077
Blade azimuth 0 degrees	19.544	14.974	19.544	17.689	-1.855	2.715	-1.855
	16.371	14.901	19.471	17.377	1.006	2.476	-2.094
	22.347	15.497	20.067	18.869	-3.478	3.372	-1.198
Blade azimuth 30 degrees	15.422	15.422	10.852	13.924	-1.498	-1.498	3.072
	12.278	15.378	10.808	13.355	1.077	-2.023	2.547
	17.899	15.619	11.049	15.213	-2.686	-0.406	4.164
Blade azimuth 30 degrees	14.712	14.712	10.142	10.96	-3.752	-3.752	0.818
	11.808	14.908	10.338	9.924	-1.884	-4.984	-0.414
	17.186	14.906	10.336	12.302	-4.884	-2.604	1.966
Blade azimuth 30 degrees	22.643	22.643	18.073	22.056	-0.587	-0.587	3.983
	19.366	22.466	17.896	21.901	2.535	-0.565	4.005
	25.471	23.191	18.621	20.843	-4.628	-2.348	2.222

After on-site tests on representative working conditions, the results show that the errors of A52, A53, and A54 points relative to the actual incoming wind direction (lidar wind direction) are all within 8°, which is acceptable in practical applications. It is verified that these three points are reasonable inferences for the location of the wind measuring instrument.

5. Summary

Based on the analysis of computational fluid dynamics simulation results, this paper analyzes the measurement error of the preset simulation measuring points, and compares it with the actual measurement to verify the accuracy of the simulation results. The main conclusions are as follows:

1) According to the simulation data of the height of the anemometer A1-A11, in general, the higher the installation position of the anemometer, the smaller the wind direction error. However, due to the limitation of the actual anemometer installation process on site, when the installation position of the anemometer is too high, the rigidity of the mast is not enough and is not conducive to maintenance. In general, it is more appropriate to install the anemometer at the height of A5;

2) After the installation height is determined, comprehensive analysis finds that the wind direction data obtained by installing the wind vane is relatively stable by selecting the three points in the middle, namely 2, 3, and 4;

3) The same research method can be applied to other wind turbines.

References

[1] Wang Xiaoyu, Ding Tongguang, Bai Nianzong, Gao XIn, Xu Bingkun, Feng Xinjian. Measurement error analysis and correction method of wind turbine wind direction meter [J]. Distributed energy 2019, 4(06).

[2] Xu Fengyou, Wind turbine system and wind energy data error correction method and yaw control method. Henan province, Xuji Group Co., Ltd, 2020-02-21.

[3] Ma Ding An alternative learning method for measuring error curve of anemometer, 2015(12).

[4] Liu Zhijing. A wind speed correction method for wind turbine nacelle, 2016.

Analysis of fluid excited noise characteristics of bulkhead opening of an Underwater Unmanned Vehicle

Zhengjiang Wu *, Chao Xiang, Junhua Hu, Baoping Zeng

Wuhan Second Ship Design & Research Institute, Wuhan 430064, China

* Corresponding author e-mail: hjy15927442154@163.com

Abstract. Acoustic stealth performance is an important performance index of Underwater Unmanned Aerial Vehicles. Due to equipment installation and other reasons, there will be a sea port on the bulkhead of the aircraft. Because the sea port destroys the flow field on the surface of the aircraft and produces noise, which affects the underwater communication of the aircraft. In this paper, the numerical simulation method is used to analyze the flow induced noise characteristics of the bulkhead of an underwater unmanned vehicle, and the main noise frequency characteristics at the bulkhead of the estuary are obtained, which provides data reference for the structural design of the estuary and the layout of the underwater communication system.

Keywords: Underwater Unmanned Vehicle, computational fluid dynamics, fluid excited noise.

1. Introduction

Underwater unmanned aerial vehicle underwater navigation, the free incoming flow through the opening, the fluid at the guide edge of the separation of the shear layer. The flow velocity of the cavity above the shear layer is low, and the flow velocity below it is close to the free flow velocity. Due to the existence of a large velocity gradient, the instability of the shear layer produces an amplifying effect on the disturbance, so that the shear layer oscillates up and down, increasing the amplitude, and when it reaches the vicinity of the following edge, the shear layer strikes the following edge, so that the following edge produces pressure pulse. Under certain conditions it is possible to produce monotonic tone or broadband noise, it is also possible to excite the cavity around the structure to produce vibration and noise. In other words, the shear oscillation at the opening resonates with the cavity acoustic mode and the structural acoustic vibration coupling mode, which may be the main reason for the abnormal flow excitation noise. According to the above qualitative analysis, the research of this paper is carried out in order to find out the mechanism of abnormal flow excitation noise generation and to quantitatively analyze the flow excitation noise source.

This paper studies the mechanism of opening shear oscillation and its coupling resonance and acoustic radiation with the cavity sound field. Structural finite element analysis, fluid dynamics simulation, acoustic boundary element simulation and other methods are used to establish relevant analytical models to analyze the structure, flow field and acoustic field, and a preliminary numerical prediction method for the analysis of hydrodynamic noise characteristics of flow-excited open-hole cavity structures is established. The main contents include: (1) cavity fluid modeling and acoustic modal

Content from this work may be used under the terms of the Creative Commons Attribution 3.0 licence. Any further distribution of this work must maintain attribution to the author(s) and the title of the work, journal citation and DOI.
Published under licence by IOP Publishing Ltd

analysis; (2) chamber structure modeling and structural and flow field acoustic vibration coupling analysis; (3) flow-excited noise source characteristics simulation analysis.

Through the above analysis, the causes of abnormal noise generation are identified, targeted measures are taken for improvement, and the improvement plan is simulated and verified.

2. Characterization of flow-excited noise sources in open oral chambers

In this paper, computational fluid dynamics and acoustic boundary element methods are used to model and simulate the system and analyze its flow-excited noise characteristics.

2.1. Analysis Method

The steps for the analysis of flow-excited noise characteristics using computational fluid dynamics and acoustic boundary element methods are as follows.

(1) Establish a detailed three-dimensional geometric model of the system (including: hull shape, installation compartments, and valve system) using software including CATIA.

(2) Establish the fluid domain geometry model by Boolean operations through the 3D geometry model of the system and hull compartments and shape.

(3) Import the fluid domain geometry model into Hypermesh to establish the CFD mesh model.

(4) Import the established CFD mesh model into Fluent software to analyze the internal flow field of the hydraulic system and obtain the results of pressure pulsation at the wall boundary layer.

(5) Import the pressure pulsation results calculated by Fluent into Virtual. Lab and load them on the acoustic grid boundary as a sound source to establish the boundary element model of flow excitation noise.

(6) Start the Sysnoise solver to solve for the noise characteristics of the hydraulic system.

(7) View the results and prepare the analysis report. The main steps can be summarized in the following Figure 1.

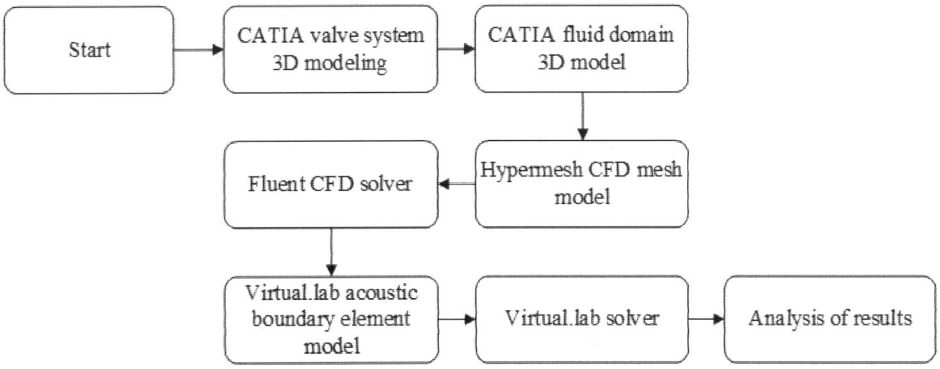

Figure 1. Flow noise analysis flow chart

2.2. FLUENT-based modeling and analysis of non-constant flow fields

(1) CFD mesh modeling

The CFD mesh model of the fluid system is given in Figure 2. to Figure 4. In order to improve the model accuracy, the mesh near the bar spacer is refined, as shown in Figure 3., and its mesh edge length is about 6 mm, which leads to a larger scale of the overall model. Considering the symmetry of the structure, the simulation model is reduced to 1/2 of the overall model by setting symmetric boundary conditions, which reduces the calculation scale and improves the solution accuracy at the same time. Figure 4. shows the schematic diagram of the boundary conditions of the CFD model. Among them, the velocity inlet is selected as u=6.6878m/s, which corresponds to a speed of 13 Kn. The back pressure of the pressure outlet is set to 101325 Pa, which is an atmospheric pressure. In order to calculate the pressure pulsation of the flow field, the non-constant LES Large Eddy Simulation (LES) method is used, with a calculation step of 0.0002s and a total simulation time of 2.22s.

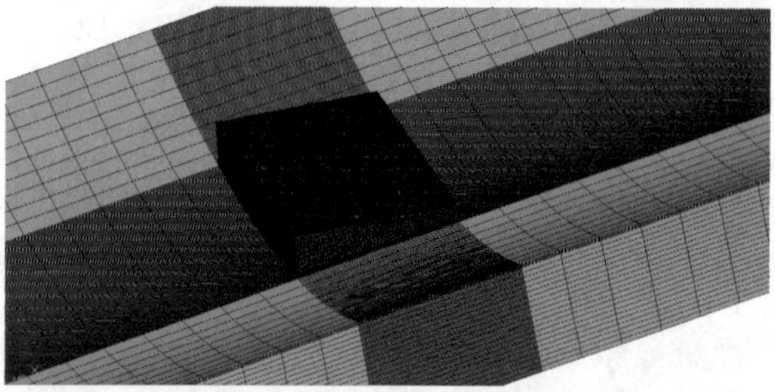

Figure 2. Mesh models of fluid systems

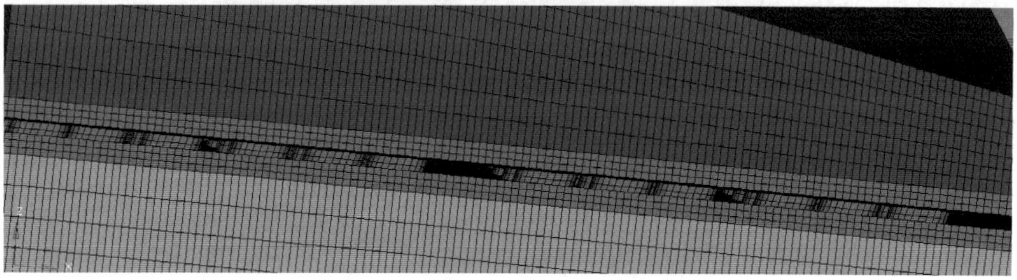

Figure 3. Refined fluid mesh model near the open spacer

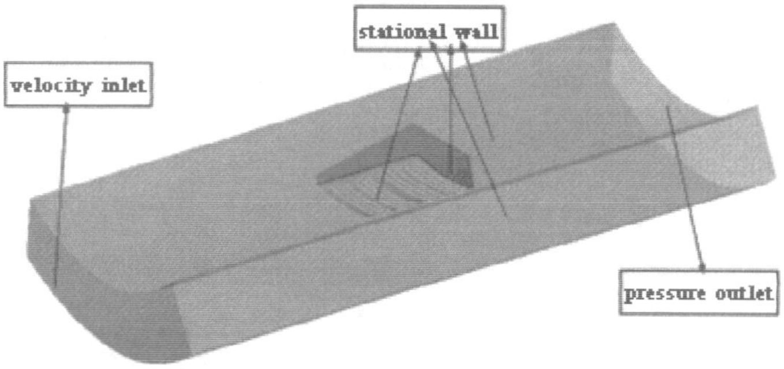

Figure 4. Boundary conditions of the fluid model

(2) Flow field analysis results

The fluid simulation results at different moments are given in Figures 5 to 10. Comparing the pressure clouds at different moments, we can see that there is a large pressure fluctuation at the open bulkhead. This is due to the fact that when the fluid flows through the bulkhead, its upper part is located inside the cavity with low flow velocity; its lower part is located outside the hull with flow velocity close to the free flow velocity, so there is a large velocity gradient at this very small size, which makes the fluid separate at the leading edge of the bulkhead and produces a shear layer. As the shear layer instability further increases the disturbance, so that the shear layer up and down swing, increasing the amplitude, reaching the rear of the bulkhead near the edge, the shear layer hit with the edge, so that the edge of the larger pressure pulsation, outward radiation flow noise. When the mode of the bulkhead or chamber acoustic cavity is comparable to the frequency of the shear layer flow excitation noise, it may cause resonance, the flow noise will be amplified to the outward radiation, resulting in abnormal noise.

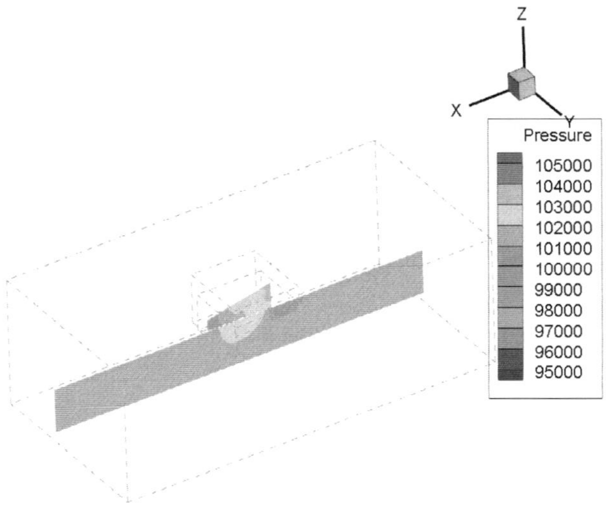

Figure 5. Schematic diagram of the position of the middle longitudinal section

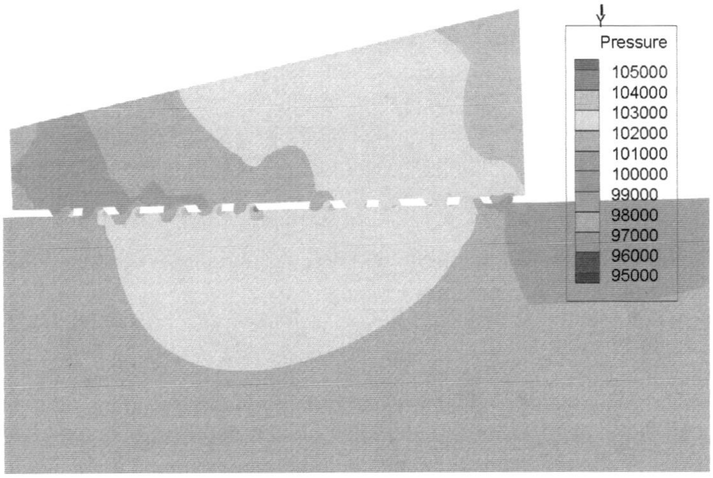

Figure 6. Pressure cloud in the middle longitudinal section at t=0.02s

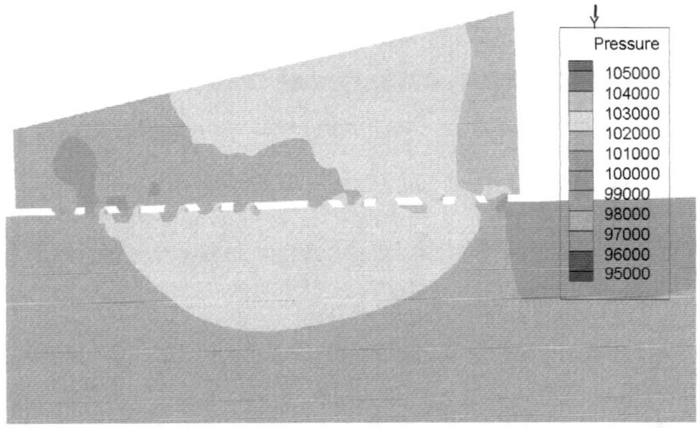

Figure 7. Pressure cloud in the middle longitudinal section at t=0.18s

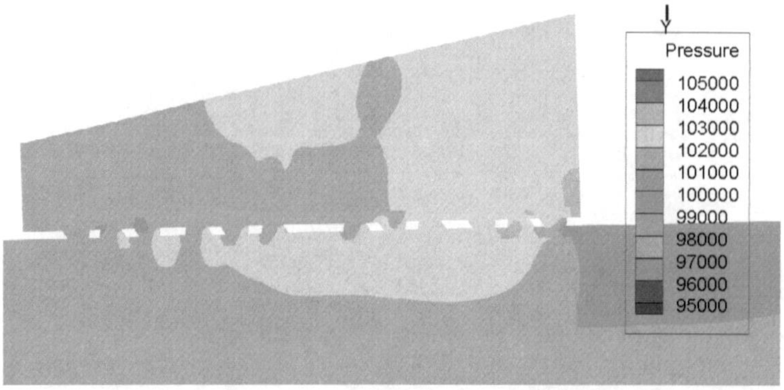

Figure 8. Pressure cloud in the middle longitudinal section at t=0.35s

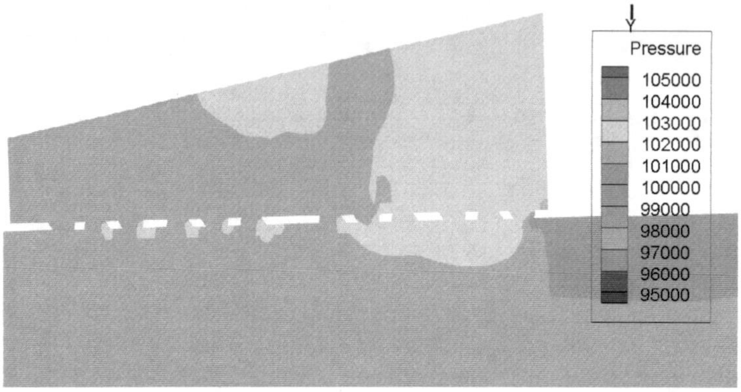

Figure 9. Pressure cloud in the middle longitudinal section at t=0.47s

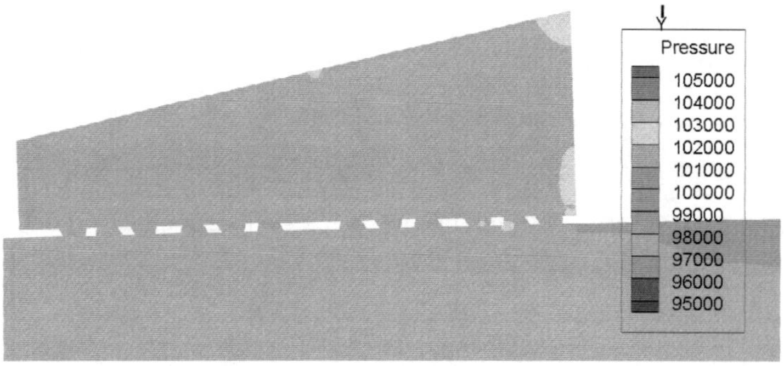

Figure 10. Pressure cloud in the middle longitudinal section at t=0.69s

3. Characterization of flow-excited noise sources

In order to better analyze the spectral characteristics of the flow noise in the shear layer at the opening of the bulkhead, the pressure pulsation on the surface of the bulkhead is imported into Virtual. Lab and Fourier transformed to convert it into a frequency domain signal, where the main order frequencies of the flow noise can be better judged in the frequency domain.

Figure 11. to Figure 18 are the sound pressure level clouds of the flow noise on the surface of the partition at each frequency. It can be seen from the graph that the main energy of the flow noise is concentrated in the low frequency band below 65HZ, and the sound pressure level of the flow noise at the last with the edge is the largest. The maximum sound pressure level of the flow noise is 151 dB at a frequency of 2.213 Hz. As the frequency increases, the sound pressure level of the flow noise decreases.

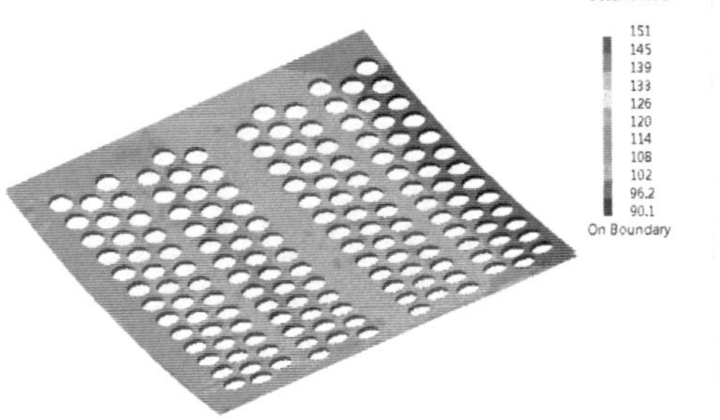

Figure 11. Sound pressure level cloud diagram of diaphragm surface flow noise(2.213Hz)

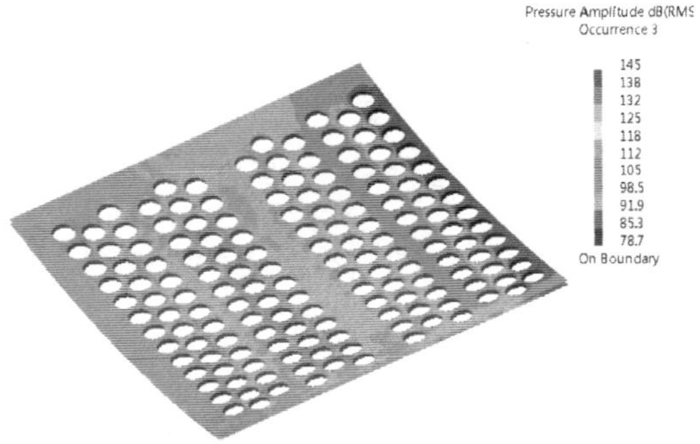

Figure 12. Sound pressure level cloud diagram of diaphragm surface flow noise(3.32Hz)

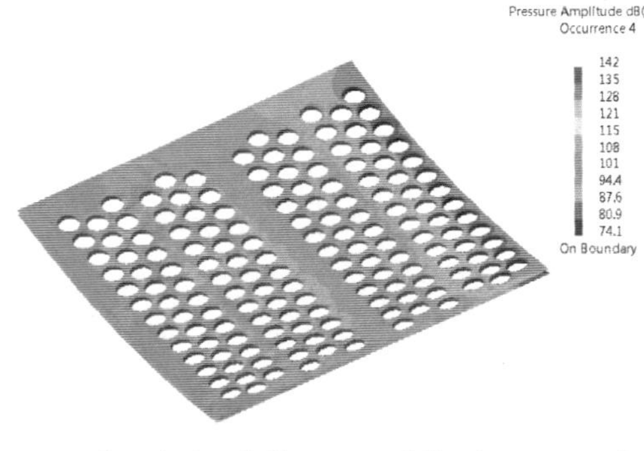

Figure 13. Sound pressure level cloud diagram of diaphragm surface flow noise(4.427Hz)

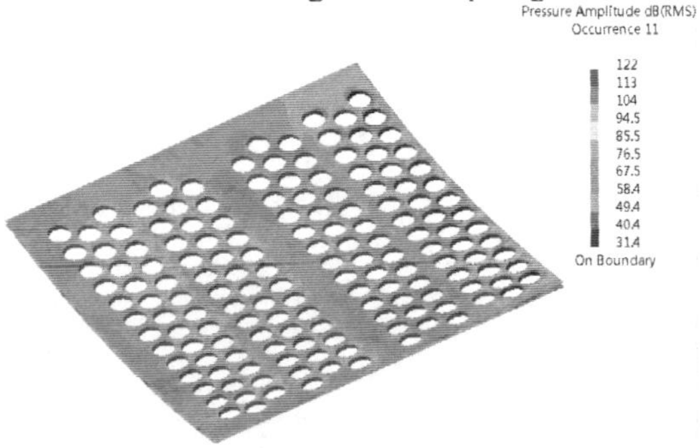

Figure 14. Sound pressure level cloud diagram of diaphragm surface flow noise(12.174Hz)

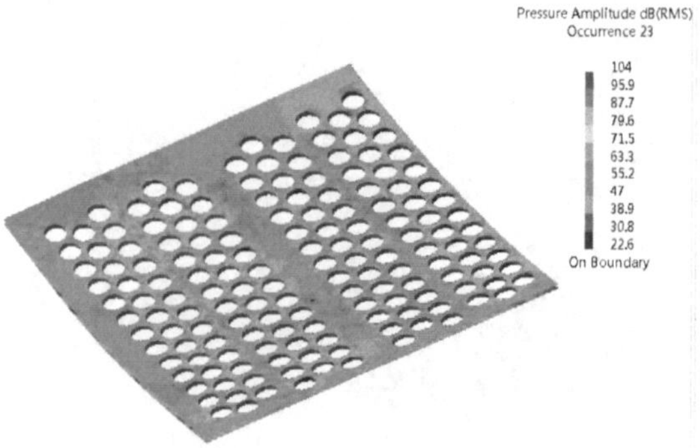

Figure 15. Sound pressure level cloud diagram of diaphragm surface flow noise(25.45Hz)

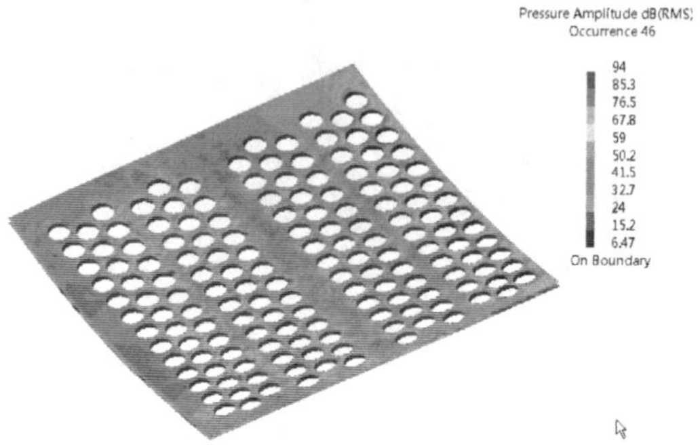

Figure 16. Sound pressure level cloud diagram of diaphragm surface flow noise(50.907Hz)

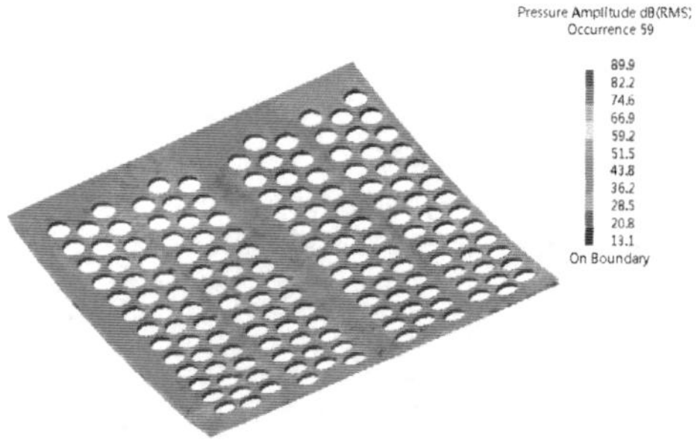

Figure 17. Sound pressure level cloud diagram of diaphragm surface flow noise(65.294Hz)

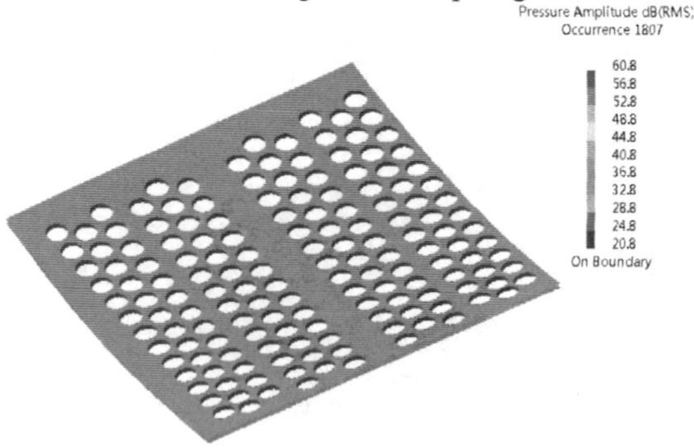

Figure 18. Sound pressure level cloud diagram of diaphragm surface flow noise(1999.779Hz)

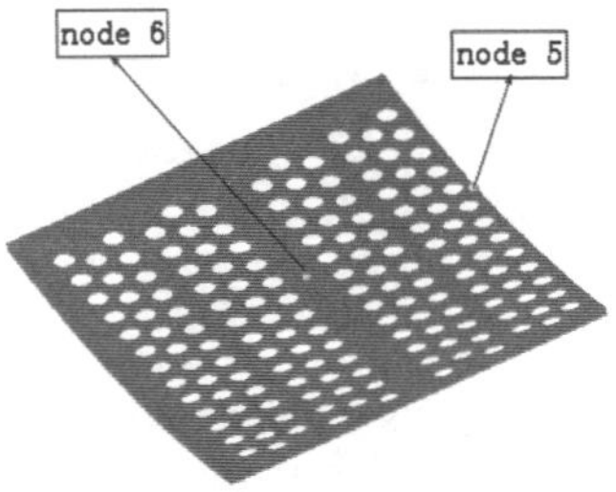

Figure 19. Two typical measurement points on the surface of the partition

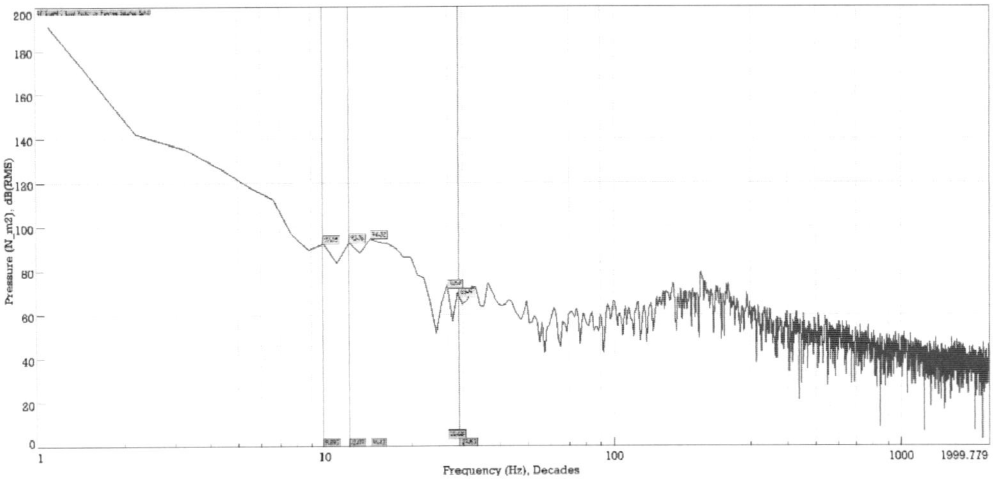

Figure 20. Frequency response curve of sound pressure level at node 5 on the surface of the partition

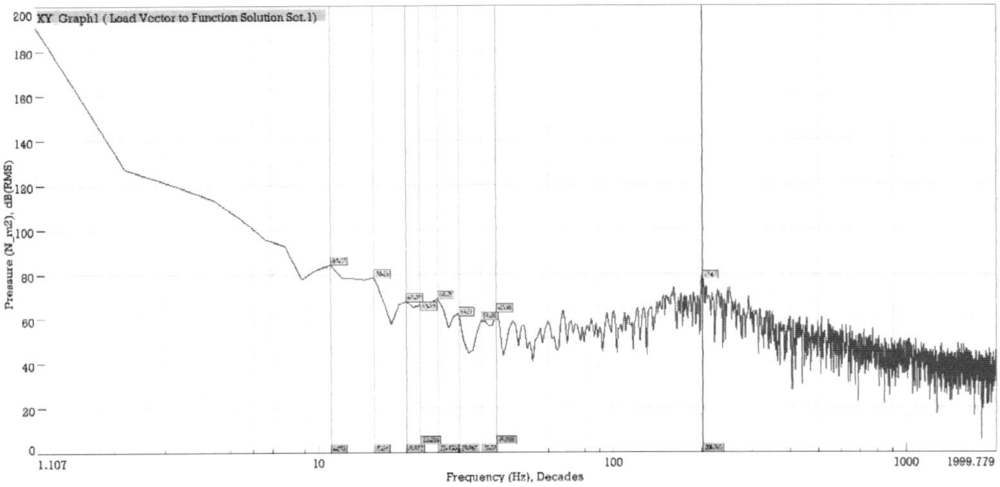

Figure 21. Frequency response curve of sound pressure level at node 6 on the surface of the partition

Table 1. The characteristic frequency of the flow-excited noise source at node 5

Levels	1	2	3	4	5	6	7
Frequency	9.895	12.179	14.43	26.68	29.163	32.15	36.664

Table 2. The characteristic frequency of the flow-excited noise source at node 6

Levels	1	2	3	4	5	6	7
Frequency	11.075	15.54	19.972	22.266	25.455	29.965	35.57

4. Conclusion

In this paper, the flow-excited noise problem at the opening of an underwater unmanned vehicle is analyzed by numerical simulation. The frequency response curves of the stream noise sound pressure level at the two nodes on the surface of the bulkhead given in Figure 20 and Figure 21 can show the characteristics of the stream noise with frequency change more intuitively. From the curve, it can be seen that the flow noise is mainly concentrated in the range of 0~40Hz, and the sound pressure level of the flow noise source decays rapidly to below 90dB in the frequency band above 63Hz. The characteristic frequencies of the flow-excited noise source are compiled in Table 1 and Table 2. If the intrinsic frequency of the structure overlaps with the frequencies in the above table, there is a risk of resonance.

References

[1] Chen Chunji. Discussion on the causes and solutions of cavitation phenomenon in liquefied petroleum gas pump[J]. Chemical Enterprise Management, 2018, (26): 37-38.

[2] PARK S, YANG J, RHEE S H. Parametric study on ship's exhaust-gas behavior using computational fluid dynamics [J]. Engineering Applications of Computational Fluid Mechanics, 2017, 11(1): 159-171.

[3] Zhang Bo,Fu Weibiao,Wu Xianghui.Effect of Hydrogen on Fuel Consumption of Diesel Engine [J]. Journal of Combustion Science and Technology, 2006, (03): 238-242.

[4] MURAI Y. Frictional drag reduction by bubble injection [J]. EXPERIMENTS IN FLUIDS, 2014, 55(7).

Experimental study on Velocity Distribution in Rod Bundle Channels under Flow Fluctuation

Xing Li [1], Hui Wang [1], Qianglong Wang [1, 2], Jinrong Qiu [1], Sichao Tan [3, *]

[1] Wuhan Second Ship Design and Research Institute, Wuhan, Hubei 430205, China
[2] Xi'an Jiaotong University, Xi'an, 710049, China
[3] Harbin Engineering University, Harbin 150001, China

*Corresponding author: tansichao@hrbeu.edu.cn

Abstract. Thermal-hydraulic behaviors within a PWR fuel assembly is very complex under accident conditions and marine conditions, which has a significant impact on reactor safety. It is crucial to carry out studies on thermal-hydraulic properties under transient conditions. A telecentric lens and a pulse signal generator are applied to improve the performance of PIV system, a high temporal and spatial resolution measurement for flow fields in rod bundle channels is achieved, the velocity distribution in rod bundles are obtained under accelerating flows and decelerating flows. The steady-state flow field is measured to use as a reference, the effect of flow fluctuations on the flow fields in the rod bundle is analyzed. The results present, the accelerating flows improve the velocity gradient between adjacent flow layers in rod bundles and weaken the lateral velocity. In the contrary, the decelerating flows weaken the velocity gradient between flow layers in rod bundles and improve lateral velocity. Conversely, the velocity gradient attenuates in decelerating flows between the rod beam layers and increases the lateral velocity. During the accident, the thermal-hydraulic behavior in the reactor has the characteristics of transient complexity and uncertainty. Studying the flow and heat transfer characteristics of the reactor core fuel assembly under accident conditions can improve a understanding for the accident development process.

Keywords: Rod bundle; Flow fluctuation; PIV; Velocity distribution; Turbulence intensity.

1. Introduction

Rod bundle have been widely used in PWR because of its structure characteristics, which can improve the heat exchange efficiency. However, the rod bundle channels induce a complex process of heat and mass transfer compared with those in circular tubes and rectangular channels. Additionally, in nuclear reactor cooling systems, flow fluctuations in fuel assemblies are universal cases due to loss of coolant accident (LOCA) and ocean condition. Therefore, the research on the complex process of heat and mass transfer in rod bundles under flow fluctuations is important for reactor safety.

Quantities of investigations have previously been devoted to identify with the velocity distribution in rod bundle geometries in steady flows. Matozinhos et al. performed a mass of experiments for flow fields measurement, and studied the flow mechanism in rod bundles in steady flows [1]. Zhang et al.

Content from this work may be used under the terms of the Creative Commons Attribution 3.0 licence. Any further distribution of this work must maintain attribution to the author(s) and the title of the work, journal citation and DOI.
Published under licence by IOP Publishing Ltd

used LDV to perform the measurement for velocity and turbulent intensity in steady flows in rod bundles [2]. However, the study on transient flow fields in rod bundles under flow fluctuations is rarely reported.

It is very necessary to carry out research on the instantaneous heat transfer characteristics in the fuel assembly for the safety nuclear reactor. Qi et al. studied the frictional characteristics of a 5×5 rod bundle without spacer girds under pulsating flows, analyzed the discrepancies of time-average friction factor between steady flows and pulsating flows [3]. Zhu et al. carried out the experimental study of the flow resistance characteristics in the rod bundle channel under the condition of natural circulation, analyzed the influence of the pulsation parameters on the flow transition characteristics in the rod bundle channel [4]. However, there is still a lack of detailed research on the flow field in the rod bundle under flow fluctuations, and the study on the transient evolution of flow field should be performed.

There are many studies on the instantaneous resistance characteristics of different flow states in rod bundle under pulsating flows, and the influence of pulsation parameters on the resistance characteristics is basically determined. There are few studies on the resistance characteristics under the flow fluctuation of the beam channel. Therefore, further research is needed to study the instantaneous friction characteristics of rod bundle channels with complex structures and the instantaneous local resistance characteristics of spacer grids. A research of transient evolution of flow fields under accelerating and decelerating flows is performed in a 5×5 rod bundle. The transient evolution of flow fields in rod bundles under flow fluctuations are measured by PIV and MIR (matching index of refraction), the effect of flow fluctuations on velocity distribution and turbulence intensity is analyzed [5].

2. Experimental setup

2.1. Flow Loop

The experimental system is mainly composed of the main flow loop, cooling loop, measurement system, control system and so on. The control system is mainly used to control the temperature of the fluid in the main flow loop and the flow rate in the main flow loop. The measurement system includes a visual measurement system and a system parameter acquisition system. The visual measurement system is used to measure the detailed flow field information in the fuel assembly.

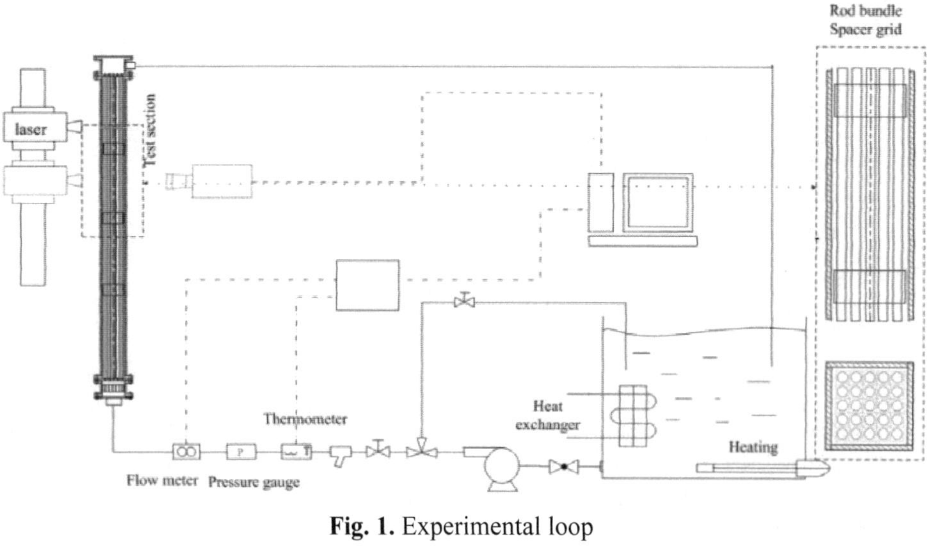

Fig. 1. Experimental loop

2.2. Test section

Fig.2(a) is a 5×5 array of rod bundles, and the rod bundle pitch and rod diameter are 12.6mm and 9.5mm, respectively. The cross section of the test section is shown in Fig.2(b). the thickness of test section is 5 mm. The rods are made of fluorinated ethylene propylene (FEP), which has a refractive index of 1.338.

The inner diameter and outer diameter are 8.8mm and 9.5mm, respectively. The length of test section is 1200mm. In experiments, the rods were filled with water inside. The above measures are used to ensure instability of rods. The high-speed camera is used to monitor the position of rod, and the measures are proved to be effective.

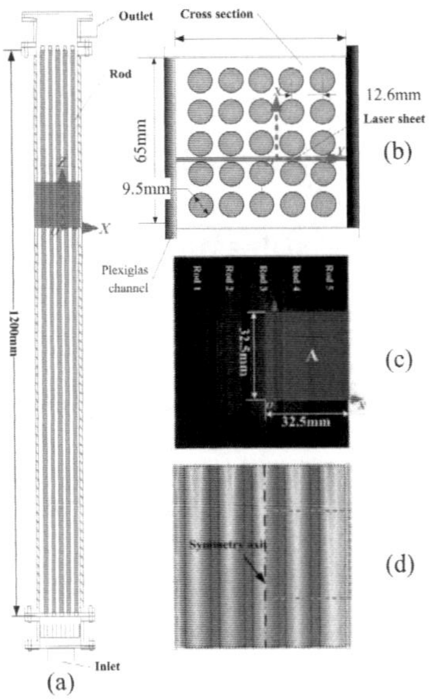

Fig. 2. Test section

2.3. *PIV technique*
The transient flow fields are obtained by using PIV. The average diameter of tracer particles is 10 um, and the specific weight is 1.04 g/cc. The photo-frequency of high-speed camera is from 2000 fps to 4000 fps for ensuring the particle displacement less than 1/4 of the thickness of the laser sheet. The camera is controlled by the pulse signal generator. The output signal frequency of pulse signal generator is 4 Hz. 200 PIV images were taken by the high-speed camera at 4000 fps photo-frequency when a signal is received. Experimental images were computed with the open source MatPIV. The interrogation window size was 16×16 pixels.

2.4. *Data processing*
In order to intuitively display instantaneous flow fields in the whole pulsation period, one whole period is divided into 40 phases, every phase consists of 100 images, thus the data set obtained for each period consists of 40×N instantaneous flow fields. Each instantaneous flow field provides two velocity components u(x, y, Φ, i) and v(x, y, Φ, i) in a spatial point (x, y), at phase Φ of each acoustic oscillation cycle i.

$$u(x,z,\phi)=\frac{1}{N}\sum\nolimits_{i=1}^{n}u_i(x,z,\phi,i) \quad (1)$$

$$v(x,z,\phi)=\frac{1}{N}\sum\nolimits_{i=1}^{n}v_i(x,z,\phi,i) \quad (2)$$

where u (x, y, U) and v (x, y, U) are time-averaged lateral velocity and time-averaged axial velocity at phase U, respectively. N represents the number of images at every phase.

3. Uncertainty analysis
Particle image velocimetry is an optical fluid measurement technology. According to the composition of the measurement system, the measurement uncertainty mainly comes from four aspects: (1) CCD camera background noise; (2) laser energy fluctuation; (3) tracer material tracking performance; (4)

image post-processing error. There are also significant differences in the contribution of different error sources. The analysis is as follows. The tracer particles (10 um diameter, 1.04g/cc density) are used as the tracer particles of PIV. There are many analysis methods for the measurement error and uncertainty of PIV, and this study adopts the analysis method recommended by the Vision Society of Japan (VSJ). The PIV uncertainties were evaluated by the guidelines for (VSJ), the uncertainty analysis of PIV measurement for the flow field is within 6%.

4. Results and discussion

To illustrate the effect of flow fluctuations on the transient flow fields of rod bundles, the velocity distribution and turbulence intensity are investigated under steady flows and flow fluctuations. The flow fields in steady flows are used as a basis for analysis of the transient flow fields.

4.1. Experimental condition

Present experiments are performed at a pressure of 0.1 MPa, and the flow fluctuation is generated by altering the rotational speed of a centrifugal pump controlled by a frequency converter. The curve of flow rate is collected by a flow meter is in Fig.3. From Fig.3, it can be found that the flow rate shows a trapezoidal fluctuation. The curve of flow rate consists of an accelerating flow, a steady-state flows and a decelerating flow. The flow rates in case A1 and case D1 have the same speed value (U=550mm/s) but the reverse acceleration. The case A1 is the accelerating flow, the case D1 is the decelerating flow. The flow fields in case A1 and case D1 are extracted and analyzed.

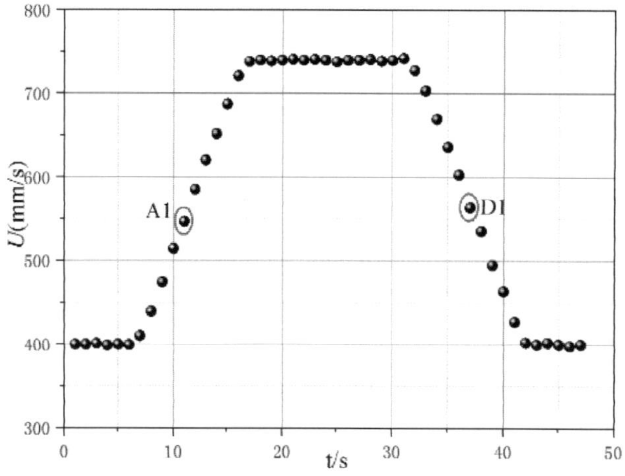

Fig. 3 Flow fluctuations in rod bundle channels

4.2. Effect of flow fluctuations on velocity distribution

Fig.4 shows the distribution of dimensionless lateral velocity (u/U) in case A1 and case D1. A comparison of dimensionless lateral velocity among steady flows, accelerating flows and decelerating flows is performed under the same averaged velocity. It can be found that the dimensionless lateral velocity in decelerating flows is larger than that in accelerating flows and steady flows, the dimensionless lateral velocity in steady flows is larger than that in accelerating flows. The above phenomenon indicates the accelerating flows can suppress the production of lateral velocity in the rod bundles, the decelerating flows can devote the increasing of lateral velocity in the rod bundles. Additionally, the distribution of dimensionless lateral velocity between different flow conditions also presents the significant discrepancies among different sub-channels. For example, in decelerating flows, the axial velocity distribution in sub-channel 5 and sub-channel 6 is larger than that in sub-channel 4. It indicates the wall effect of the test section on lateral velocity distribution is obvious in decelerating flows.

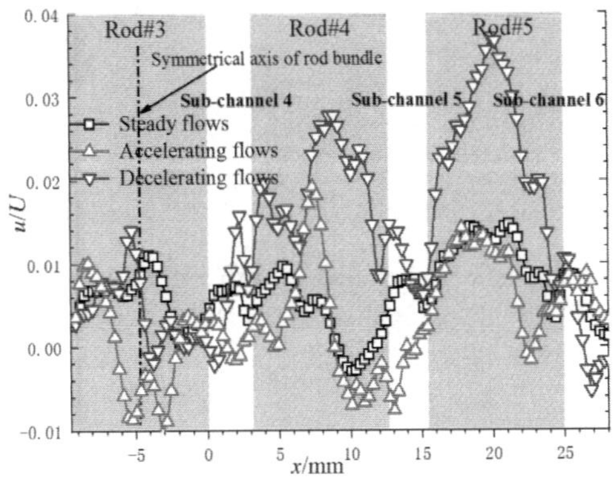

Fig. 4 Lateral velocity distribution in accelerating and deceleration flows

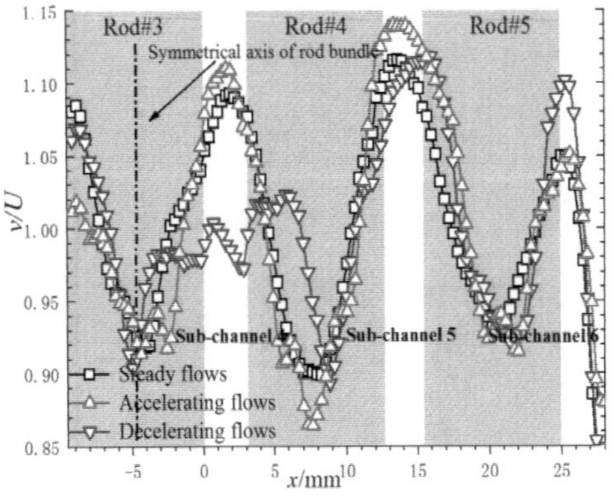

Fig. 5 Axial velocity distribution in accelerating and deceleration flows

Fig.5 shows the distribution of dimensionless axial velocity (u/U) in accelerating flows and decelerating flows. From the Fig.5, the axial velocity distribution under different flow conditions shows significant discrepancies, the speed difference of dimensionless axial velocity in accelerating flows is obvious between the edge of sub-channel and the center of sub-channel. The axial velocity in sub-channel 4 under decelerating flows has a significant drop compared with that under accelerating flows and steady flows. The phenomenon seems to be similar to "annular effect" in a circular tube. It indicates the accelerating flows improve the velocity gradient adjacent flow layers. In previous studies, the accelerating flows can improve the velocity gradient among flow layers, which can be the reason that transient pressure drop in accelerating flows is larger than that in steady-state flows.

5. Conclusion

In paper, the instantaneous flow fields in rod bundle under flow fluctuations were investigated. Experiments in steady flows were conducted in advance verification and datum reference, the effect of flow fluctuations on flow fields is analyzed. The conclusions in paper is as follows.

(1) The accelerating flows can suppress the production of lateral velocity and lateral turbulence intensity in the rod bundles, it induces the lateral velocity in accelerating flows is larger than those in steady flows and decelerating flows; The decelerating flows can improve lateral velocity in the rod bundles, it induces the lateral velocity in decelerating flows is smaller than those in steady flows and accelerating flows.

(2) Accelerating flows improve axial velocity gradient in sub-channel, in other words, viscous shear stresses between adjacent flow layers have an increase; Decelerating flows weaken axial velocity gradient in sub-channel, viscous shear stresses between adjacent flow layers have a decrease.

References

[1] Matozinhos C F, Tomaz G, Nguyen T, et al. Experimental investigation of turbulent flow characteristics in cross-flow planes of a 5×5 rod bundle with a spacer grid [J]. International Journal of Heat and Fluid Flow, 2021, 87: 108-757.

[2] Botao Zhang, Shengjie Gong, Fujun Gan, et al. Flow induced vibration measurement for a 4×4 rod bundle with spacer grids by the Laser Doppler Vibrometer [J]. 2022(166).

[3] Peiyao Qi, Peng Wang, Sijia Hao, et al. Experimental study of flow structures in a large range downstream the spacer grid in a 5×5 rod bundle using TR-PIV [J]. International Journal of Heat and Fluid Flow, 2020(84).

[4] Zhiqiang Zhu, Changqi Yan, Jianjun Wamg, et al. Flow resistance characteristics in rod bundle channel under natural circulation condition [J]. CIESC Journal, 2018.

[5] Aizawa K, Hiyama T, Nishimura M, et al. Velocity distribution in the subchannels of a pin bundle with a wrapping wire [J]. Mechanical Engineering Journal, 2021, 8(4).

Numerical simulation of gas-liquid interface in high-pressure water-blowing chamber

Xionghui Tang [1], **Yiou Liu** [2, *]

[1] Huludao Military Representative Office, Huludao, Liaoning, China
[2] Wuhan Second Ship Design & Research Institute, Wuhan 430064, Hubei, China

*Corresponding author e-mail: best_20_22@163.com

Abstract. The load on the water chamber increases with the pressure level and blowing rate of the system, and the demand for theoretical and experimental studies related to the quantitative parameters of water chamber blowing safety becomes more and more frequent. In this paper, CFD computational fluid dynamics method is used to carry out the dynamics modeling and simulation of three-dimensional non-constant gas-liquid two-phase internal flow field of the test water chamber and the actual water chamber blowdown process to obtain the dynamic distribution of flow parameters in the water chamber under different working conditions. The results found that to ensure the safety of the water tank, as long as to ensure that the high-pressure air into the water tank at the end of the first two stages, the "bubble" movement process always does not interfere with the bulkhead, you can ensure that the high-pressure air in the water tank in the free expansion and gravity floating state, the water tank pressure will be only slightly higher than "gas - water cross-section" of the static pressure, at this moment the water chamber safety.

Keywords: high pressure, blown out water chamber, gas-liquid two-phase flow, numerical simulation.

1. Introduction

With the increase of system pressure grade and blowing rate, the load on the tank becomes more severe, which poses a higher challenge to the safety of the tank. It is necessary to carry out theoretical and experimental studies on quantitative parameters of tank blow-off safety. In this paper, the CFD computational fluid dynamics method is used to carry out the dynamic modeling and simulation of the three-dimensional unsteady gas-liquid two-phase flow field in the experimental tank and the actual tank blowing process, in order to obtain the dynamic distribution of the flow parameters in the tank under different working conditions, such as pressure, mass flow, temperature, gas-liquid interface dynamic changes, etc. The model is modified by the blowing of the test tank under the condition that the virtual simulation results have the same precision as the physical prototype test, it provides the calculation method and data support for evaluating the safety of the tank during the actual blowing process.

Content from this work may be used under the terms of the Creative Commons Attribution 3.0 licence. Any further distribution of this work must maintain attribution to the author(s) and the title of the work, journal citation and DOI.
Published under licence by IOP Publishing Ltd

2. Theoretical basis

The process of high-pressure air blowing is a typical three-dimensional unsteady gas-liquid two-phase flow problem. In this paper, the VOF model is used to simulate the dynamic evolution process of the internal flow field, and the state parameters of the flow field are obtained.

It is not easy to simulate the free surface precisely because of its complex shape and changing spatial distribution. The VOF method can be used to construct and track the free surface according to the function f. When f equal 1, it means that the cell is entirely occupied by the fluid of the specified phase. When f is 0, it means that The unit is a fluid unit without a specified phase. When f is from 0 to1, it means that the unit is an interface unit. The function f (x, y, T) is defined as follows.

$$f(x,y,t)=\begin{cases}1\\0\end{cases} \tag{1}$$

The conservation form of the transport equation is expressed as follows.

$$\frac{\partial f}{\partial t}+\frac{\partial u_f}{\partial x}+\frac{\partial v_f}{\partial y}=0 \tag{2}$$

3. Modeling and simulation of tank blow-off

3.1. System-wide modeling and simulation

3.1.1. 3D modeling. In the process of modeling, the influence of solenoid valve on the dynamic characteristics of the system is neglected. The 15 parallel cylinders are regarded as a single cylinder with a volume of 370L×15 to simplify the complexity of the simulation model under engineering equivalent conditions. The grid is divided as shown in the Figure 1.

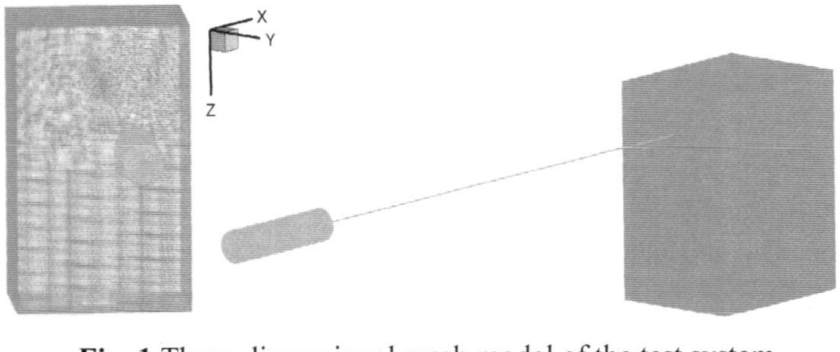

Fig. 1 Three-dimensional mesh model of the test system

3.1.2. Boundary Condition. The initial pressure of high pressure air cylinder is 40 MPa. It is also assumed that the pipeline would have been filled with high-pressure air in its initial state. A water hole is a pressure outlet, whose value is 0.2MPa. Ambient temperature is 24°C. The specific boundary conditions are set as shown in Figure 2.

3.1.3. Analysis results. (1) Distribution of important parameters of flow field at different time

The figures 3-5 show the distribution of the volume fraction of the air in the tank and the pressure distribution of the bulkhead at three different times.

Comparing the results of the three moments, it can be found that over time, “Air bubbles” blow into the water tank at high speed, forming “Bullet” shape in the early days, then the transition to a “Football” shape, the rapid spread to the forward bulkhead. According to the rule of bulkhead pressure change with time, when the high pressure air blow-off begins, in less than 0.13s, the maximum pressure of bulkhead decays rapidly to about 0.32MPA. From the rule of bulkhead pressure change with time, when the high-pressure air blow-off begins, in less than 0.13s, the maximum pressure of the bulkhead decays rapidly to about 0.32MPa, that is about 0.12MPA higher than the external backpressure. As time goes on, the cylinder pressure decreases and the maximum pressure of the bulkhead further decreases.

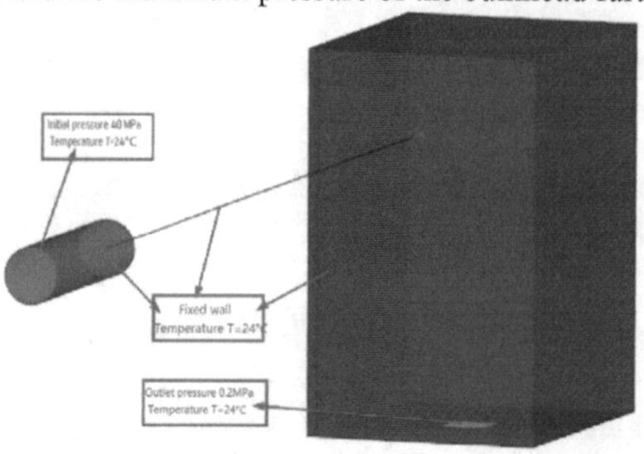

Fig. 2 A sketch of the boundary conditions

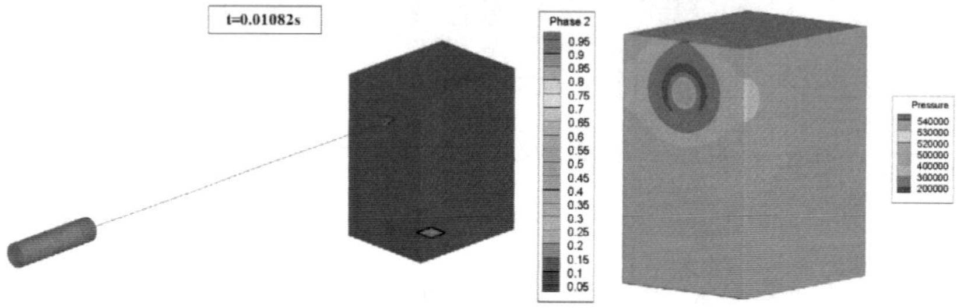

(a) Air volume fraction distribution (b) Pressure distribution of water tank bulkhead

Fig. 3 t=0.01082s

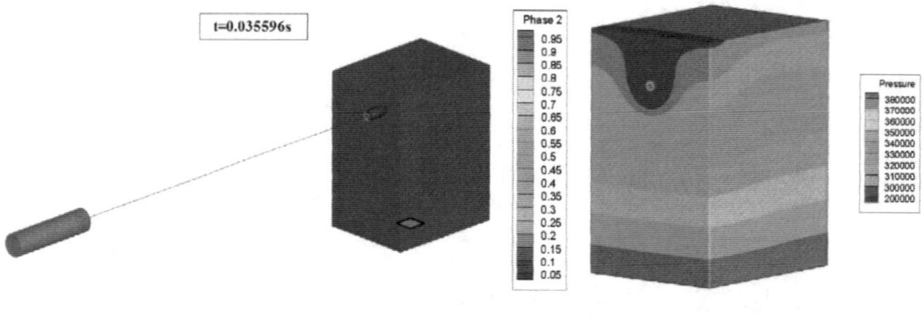

(a) Air volume fraction distribution (b) Pressure distribution of water tank bulkhead

Fig. 4 t=0.035596s

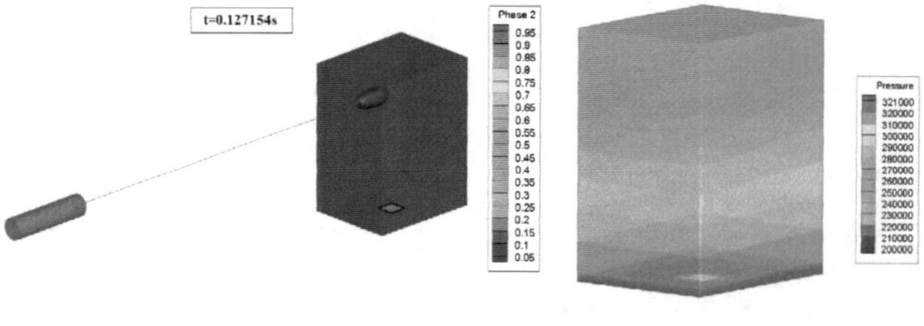

(a) Air volume fraction distribution (b) Pressure distribution of water tank bulkhead

Fig. 5 t=0.127154s

3.1.4. Flow parameter curve. In order to verify the feasibility and validity of CFD simulation of water tank alone, the flow field parameters at 1000mm before the entrance of water tank are monitored, including pressure, mass flow, temperature. Then, the time-domain signals of these parameters are taken as the entrance boundary conditions of the single 3-d CFD simulation of the water tank. The result is shown in figure 6.

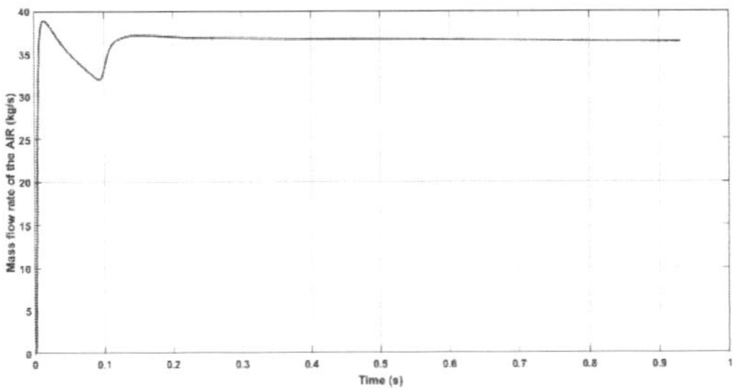

(a) Mass flow curve

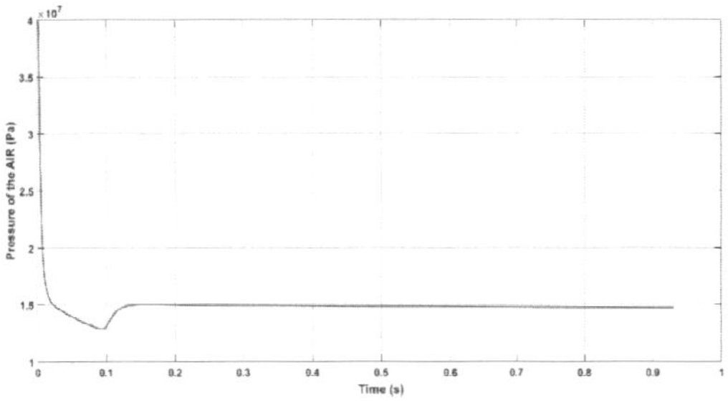

(b) pressure curve

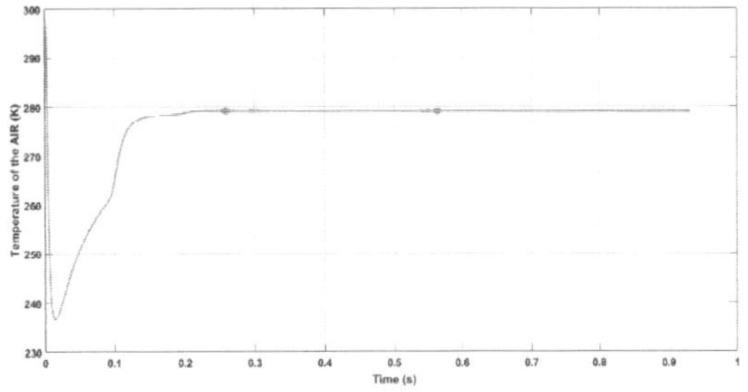

(c) temperature curve

Fig. 6 High pressure air characteristic curve

3.2. CFD analysis of the water tank

3.2.1. 3D modeling. When the actual water tank is blown off, the pipe network system is very complicated. If the above-mentioned three-dimensional unsteady gas-liquid two-phase flow CFD simulation system is used for analysis, it will inevitably lead to a large-scale model, and bring difficulties for solving. Therefore, it is necessary to discuss the three-dimensional CFD modeling and simulation method of single tank blow-out. The discrete case of three-dimensional model and grid inherits the results of the whole system model. The high-pressure air cylinders and most of the piping are cut off. The 1,000mm pipe in front of the tank are remained. The mesh model of the individual tank is shown below.

3.2.2. Boundary Condition. The flow field parameters, including pressure, mass flow rate and temperature as the boundary conditions of the inlet, were obtained by the whole system three-dimensional CFD simulation.

3.3. Results comparison

3.3.1. Air-water interface. Figure 7 shows the isosurface of the volume fraction of high-pressure air at the same time under both algorithms. As can be seen from the graph, the results obtained by the two algorithms are basically the same at the same time.

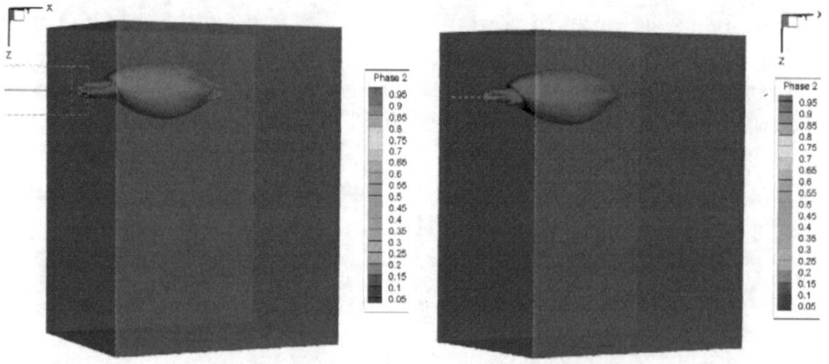

(a) system-wide (b) separate tanks

Fig. 7 Volume fraction of high-pressure air (t=0.239152s)

The gas-liquid cross-sections of the pipelines and tanks are as follows (Figure 8).

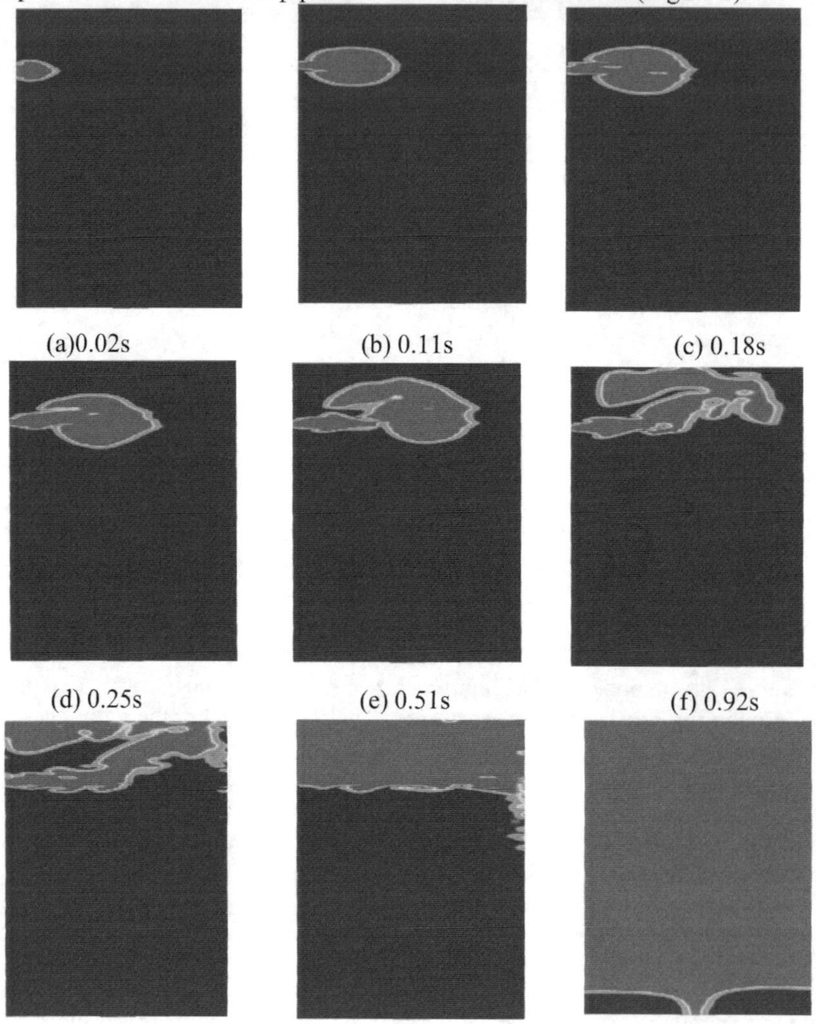

(a)0.02s (b) 0.11s (c) 0.18s

(d) 0.25s (e) 0.51s (f) 0.92s

(g) 2.37s (h) 5.03s (i) 23.03s

Fig. 8 Air volume fraction distribution diagram

It can be found that the process of expansion and diffusion of high pressure air after entering the water tank can be divided into several stages. ① When the high-pressure air enters the cabin, a cone-shaped bubble forms rapidly at the entrance of the bulkhead (Fig. a, 0.02s). Because of the low pressure of the surrounding liquid and the high pressure in the bubble, the bubble diffusion state is mainly "Expansion" (Fig. b, 0.11s). Because the gas velocity in the graph plane perpendicular to each other in the direction of the two axis difference is large, the formation of the bubble in the entrance near the formation of "Ellipsoid". These two processes can be considered as the first stage: "near the entrance bubble formation". ② After the first stage, a more stable "Ellipsoidal bubble" was formed. When the difference between the external pressure and the internal pressure of the bubble is small, the bubble will not expand for a short time. However, because the tail pressure of the "Bubble" is greater than that of the head, the bubble is pushed forward as a whole by the pressure difference (Fig. c, 0.18s), until this pressure difference is not sufficient to overcome the bubble resistance (Fig. e, 0.51s). This stage can be considered as the "Bubble overall forward" stage, recorded as the second stage. ③ When the bubble no longer forward, with the continuous flow of gas mass, back pressure is basically unchanged, the bubble gradually "Expansion" increase. The bubble expands and diffuses toward the top (Fig. g, 2.37s). This is the third stage: the "Inflating up" stage. ④ The fourth stage is the formation of a stable air-water interface, high-pressure air into the cabin, no longer blowing liquid, but only expansion (Fig. h, 5.03s), until almost all the water in the tank was discharged (Fig. i, 23.03s).

At the end of the second stage, the high-pressure air in the tank is in the stage of free expansion and gravity floating, which is close to the steady state. Therefore, the tank safety analysis (maximum pressure in the tank) the main analysis of the first and second stage can be.

3.3.2. Bulkhead pressure distribution. The pressure distribution of the bulkhead under the two algorithms at the same time is shown in figure 9.

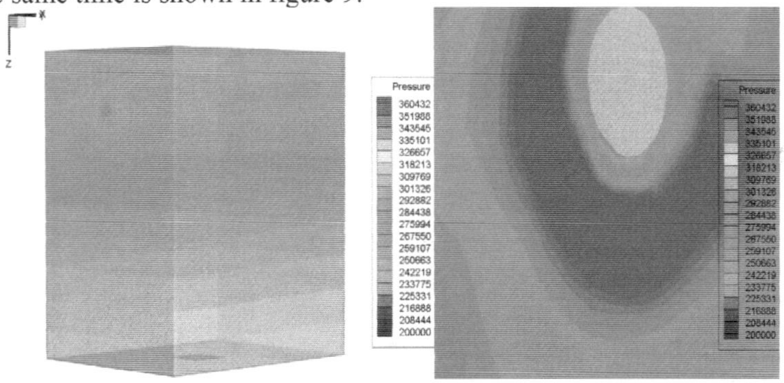

(a) system-wide

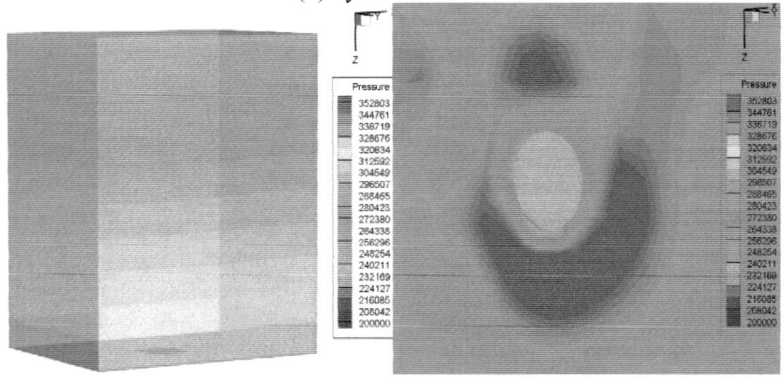

(b) separate tanks

Fig. 9 Cloud map of bulkhead pressure distribution (t=0.239804s)

4. Conclusions

To ensure the safety of the water tank, as long as high-pressure air entering the water tank after the first two stages of the end, "Bubble" movement and the bulkhead has not been interference. This can ensure that the high-pressure air in the tank in free expansion and gravity floating state, tank pressure will only be slightly higher than the "Air-water cross-section" of the static pressure, tank safety. Therefore, in the design of water tank, the distance between the bulkhead and the inlet of high-pressure air pipe should be enough, and the distance of the transverse bulkhead should not be less than the radius of "Ellipsoid bubble".

References

[1] WANG Meng, SUN Nan, WANG Yi, et al. Experimental Study and Numerical Simulation of Gas-Liquid Two Phase Flow Velocity Field Distribution in Aeration Tank[J]. Journal of Hydraulic Engineering, 2016, 47(10): 1322–1331.

[2] ARITOMI M, ZHOU S, NAKAJIMA M, et al. Measurement System of Bubbly Flow Using Ultrasonic Velocity Profile Monitor and Video Data Processing Unit. [J]. Journal of Nuclear Science and Technology, 1996, 33(12): 915–923.

[3] ARITOMI M, ZHOU S, NAKAJIMA M, et al. Measurment System of Bubbly Flow Using Ultrasonic Velocity Profile Monitor and Video Data Processing Unit, (II) [J].Journal of Nuclear Science & Technology, 1997, 33(12):915–923.

[4] Zhou S, Suzuki Y, Aritomi M, et al. Measurement System of Bubbly Flow Using Ultrasonic Velocity Profile Monitor and Video Data Processing Unit (III) [J]. Journal of Nuclear Science & Technology, 1998, 35(5): 335–343.

[5] Murakawa H, Kikura H, Aritomi M. Application of Ultrasonic Multi-Wave Method for Two-Phase Bubbly and Slug Flows [J]. Flow Measurement & Instrumentation, 2008,19(3/4): 205–213.

[6] Wongsaroj W, Hamdani A, Thong-Un N, et al. Extended Short-Time Fourier Transform for Ultrasonic Velocity Profiler on Two-Phase Bubbly Flow Using a Single Resonant Frequency [J]. Applied Sciences, 2018, 9(1): 1-21.

A method for calculating the void fraction of two-phase leakage ports of storage tanks

Jiajia Ma, Min Hua *

Nanjing Tech University, Nanjing, China

*Corresponding author e-mail: 1005260466@njtech.edu.cn

Abstract. The problem of leakage of superheated liquids such as LPG, LNG, LCG and AML is widely valued in the process industry. The key to superheated liquid storage and transportation is to prevent cracks due to unexpected situations such as tank overpressure, pipe corrosion and violent collisions. When the tank wall ruptures, the air pressure inside the tank drops sharply and a violent phase change occurs in the degraded liquid inside the tank to maintain energy balance. The liquid instantly evaporates into a gas, releasing a huge amount of energy to form a flash, and this leakage can lead to violent phase changes and two-phase flash jets containing large amounts of liquid droplets and steam, and further lead to major safety accidents such as fires, explosions and the spread of toxic substances. In this work, the pressure drop in the tank during a leak and the phase change at the leak caused by the pressure change at the orifice are investigated through theoretical analysis and experiments. Based on the ideal gas equation of state, mass conservation, gas-liquid non-slip and thermodynamic equilibrium, the pressure drop inside the tank and the phase change at the orifice during the whole leakage process are modelled and validated, and the results show that the model is in good agreement.

1. Introduction

In the existing tank storage and transportation process, the chemicals in the tank at a certain temperature and pressure conditions to maintain the vapour pressure balance, the tank liquid in the storage and transportation process is generally in a sub-stable state. When the tank wall corrosion, crack collision or overpressure will lead to tank leakage, when the tank wall rupture, the tank air pressure drops sharply, the liquid in the sub-stable state of the tank will undergo a violent phase change to maintain the energy balance, the liquid instantly evaporates into gas, releasing a huge amount of energy to form a flash, which will eventually lead to fire and explosion and other safety accidents. At the leak, as the pressure drops rapidly, the nucleation rate accelerates and the phase change intensifies, resulting in unstable flash jet patterns and jet angles, making it impossible to make an accurate judgement of the hazard range of the jet at the tank leak. The void fraction is one of the most important parameters for assessing the state of the jet, reflecting the proportion of area occupied by gas-liquid two-phase flow bubbles at a given moment. It is also a quantitative parameter that characterises the degree of phase change, and as phase change is the dominant factor affecting the degree of flash jet atomisation, accurate measurement of the bubble void ratio at the leak is important for studying the evolution of two-phase leakage patterns in storage tanks. If the state of change of the void fraction at the leak can be determined, and thus the range

Content from this work may be used under the terms of the Creative Commons Attribution 3.0 licence. Any further distribution of this work must maintain attribution to the author(s) and the title of the work, journal citation and DOI.
Published under licence by IOP Publishing Ltd

of change of the two-phase flow jet pattern and jet angle can be understood, the hazards associated with tank leaks can be effectively controlled.

At present, the measurement of void fraction of two-phase flow at tank leaks has not been extensively studied experimentally in the field of chemical safety. To provide a method for calculating the void fraction of two-phase leakage ports of storage tanks, extensive studies have been performed in the literature. The first method is based on the gas-liquid two-phase slip rate and is often used for the empirical relationship between the slip of the gas-liquid two phases [1]. The second is to develop a model around a uniform void fraction model to predict void fraction by multiplying it with an empirically derived correction element [2]. The third is based on the correlation of drift fluxes and the prediction of the distribution parameters of the void rate and the drift rate based on the drift flux model proposed by Zuber and Findlay [3]. Finally, the fourth one revolves around the relevance of the void fraction miscellaneous terms [4-5], they do not apply to the other groups of empirical relations.

In this work, theoretical analysis and experiments were used to study the pressure drop inside the tank during the leakage process and the phase change at the leakage port caused by the pressure change at the orifice. The pressure drop of the tank during the entire leakage and the resulted phase change at the orifice outlet were modelled based on the ideal gas equation of state, mass conservation, velocity slip between liquid phase and gas phase is ignored and thermodynamic equilibrium.

2. Theory

2.1. Model building

In this paper, the two-phase flow in the horizontal pipe is defined as the steady-state bubble flow in the dispersed flow state, that is, the bubbles are dispersed in the entire liquid, and the gas-liquid two phases are at the same velocity [6]. The void fraction model proposed in this paper is based on a homogeneous model of steady-state flow, i.e. it is assumed that during a leak, the gas and liquid are uniformly distributed in the flow cross-section during the leak, that the mass of the gas phase dissolved in the liquid phase is negligible when the two-phase flow flows in a transparent pipe and that the pressure decay along the pipe is not large, and that the mixed gas-liquid two-phase flow can be regarded as a unidirectional fluid with homogeneous two-phase properties. The model is applicable to the leakage state of different types of gas-liquid two-phase flows in horizontal pipes with different pipe sizes.

In this condition, the gas in the superheated state can be considered as an ideal gas.

$$p_g = \frac{n_g R T_g}{V_g} \tag{1}$$

The differential of p_g can be expressed as:

$$\frac{dp_g}{dt} = -\frac{n_{g,0}RT_{g,0}}{V_g{}^2}\frac{dV_g}{dt} + \frac{RT_{g,0}}{V_{g,0}}\frac{dn_g}{dt} + \frac{Rn_{g,0}}{V_{g,0}}\frac{dT_g}{dt} \tag{2}$$

The water mass in the vessel must be equal to the liquid or water-steam mass released to the ambient and the water mass converted to (from) steam in the vessel. Then, we have the following expression:

$$[\dot{m}_l]_{vessel} = [\dot{m}_l + \dot{m}_g]_{release} + [\dot{m}_g]_{vessel} \tag{3}$$

The water-steam mixture flow released to the ambient is considered as the homogeneous equilibrium flow. Velocity slip between liquid phase and gas phase is ignored. Then, The Eq. (3) can be rewritten by

$$-\rho_l\frac{dV_l}{dt} = \rho_l\frac{dV_g}{dt} = [(\rho_l(1-\alpha) + \rho_g\alpha)\pi(\tfrac{\varphi}{2})^2 u_{mix}] + [M\frac{dn_g}{dt}] \tag{4}$$

In Eq. (4), u_{mix} [7] can be formulated as:

$$u_{mix} = \sqrt{2[\frac{p_v - p_a}{\rho_l} + C_{p,l}(T_l - T_s) - C_{p,g}T_s\, ln(\frac{T_l}{T_s})]} \tag{5}$$

Eq. (4) can be rewritten as:

$$\frac{dn_g}{dt} = \frac{\rho_l}{M}\left(\frac{dV_g}{dt} - \frac{(\rho_l(1-\alpha)+\rho_g\alpha)\pi(\frac{\varphi}{2})^2 u_{mix}}{\rho_l}\right) \tag{6}$$

The differential of gas volume is equal to:

$$\frac{dV_g}{dt} = \pi\left(\frac{D}{2}\right)^2 \frac{dL_g}{dt} \tag{7}$$

Substituting Eq. (6) and (7) into Eq. (2), we have:

$$\frac{dp_g}{dt} = \left(\frac{4\rho_l RT_{g,0}}{M\pi D^2 L_{g,0}} - \frac{16n_{g,0}RT_{g,0}}{\pi^2 D^4 {L_g}^2}\right)\pi\left(\frac{D}{2}\right)^2 \frac{dL_g}{dt} + \frac{4Rn_{g,0}}{\pi D^2 L_{g,0}}\frac{dT_g}{dt} - \frac{RT_{g,0}\rho_l\varphi^2 u_{mix}}{MD^2 L_{g,0}} - \frac{RT_{g,0}(\rho_g-\rho_l)\varphi^2 u_{mix}\alpha}{MD^2 L_{g,0}} \tag{8}$$

Then, the volume fraction of steam in the water-steam mixture flow is determined by Eq. (8) as follows:

$$\alpha(t) = \left[\left(\frac{4\rho_l RT_{g,0}}{M\pi D^2 L_{g,0}} - \frac{16n_{g,0}RT_{g,0}}{\pi^2 D^4 {L_g}^2}\right)\pi\left(\frac{D}{2}\right)^2 \frac{dL_g}{dt} + \frac{4Rn_{g,0}}{\pi D^2 L_{g,0}}\frac{dT_g}{dt} - \frac{dp_g}{dt} - \frac{RT_{g,0}\rho_l\varphi^2 u_{mix}}{MD^2 L_{g,0}}\right]\frac{MD^2 L_{g,0}}{RT_{g,0}(\rho_g-\rho_l)\varphi^2 u_{mix}} \tag{9}$$

Eq. (9) indicates that the volume fraction of steam can be estimated when the state of gas in the vessel (p_g, T_g, and L_g) and geometric parameters of vessel (D and φ) and nozzle are known.

2.2. Model validation

At present, some research methods for measuring void fraction are focused on two-phase flow at room temperature and low flow rate, which cannot meet the dynamic flow conditions at high superheat and high flow rate. To address these problems, a complete system for the detection of void fraction at the gas-liquid two-phase leak is built in this paper, and the void fraction at the leak is measured in real time by the given experimental scheme, through which the void fraction model proposed above can be verified.

The design of the capacitive void rate sensor is shown in Fig.1, with an internal diameter of 10 mm and an overall length of 42 mm, using a polycarbonate suitable for two-phase flow as the dielectric tube material. The capacitive void fraction sensor device for this study is shown in Fig.2. Where (a) is a physical diagram of the capacitive sensor and (b) is a physical diagram of the capacitive data collector.

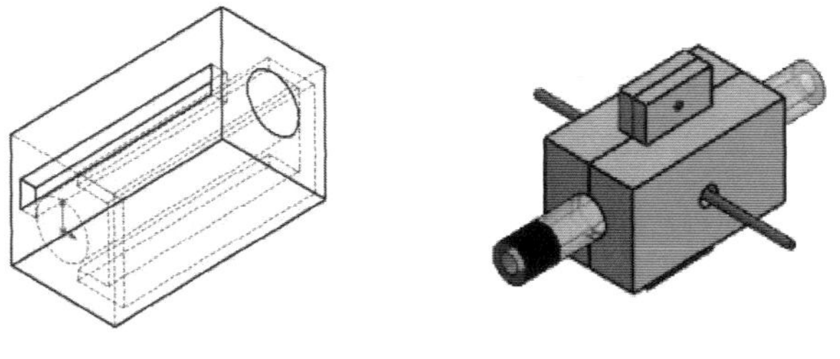

Fig. 1 Capacitive void fraction sensor design diagram.

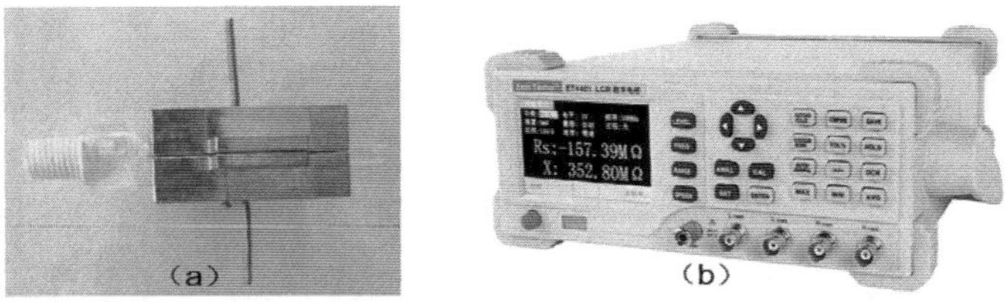

Fig. 2 (a) Physical view of a capacitive sensor, (b) Physical view of a capacitive data collector.

As mentioned above, based on the steady-state leakage conditions, this paper proposes a calculation method for predicting the internal void fraction of the leakage port during two-phase leakage. Fig.1 shows the distribution of void fraction during a two-phase leak with initial leakage conditions: P=11 bar, T=130°C, φ=5mm, between 10s and 40s, as the two-phase leak gradually reaches a steady state leak. The capacitance method does not have such problems, so the model proposed above is validated using the capacitance method. Fig.3 shows the simulated values of the experimental data substituted into the void fraction model for this condition and the capacitively converted void fraction obtained experimentally using a capacitive sensor and fitted to it.

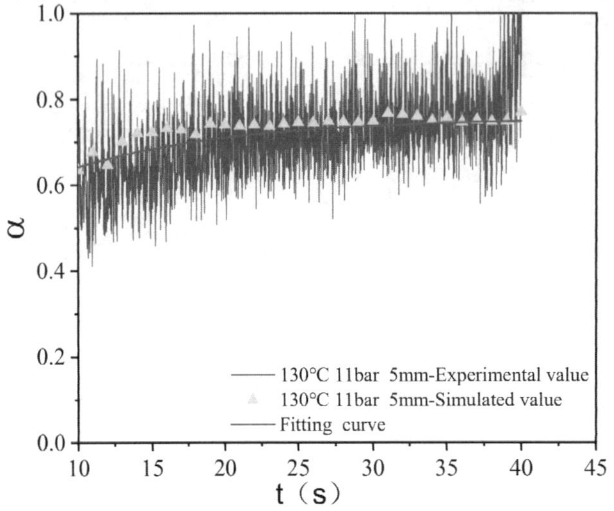

Fig. 3 Comparison of experimental values of void fraction during leakage and the results of equation (10).

At the start of the leak, when the saturation pressure corresponding to the temperature of the liquid in the tank is greater than the external ambient pressure, the liquid overheats, resulting in a violent phase change at the leak, i.e. a flash leak. The violent flash vapour leads to huge fluctuations in the pressure decay, two-phase mixing velocity, two-phase density and other two-phase flow parameters at the leak, i.e., this paper selects between 10s and 40s after the start of the leak, when the leak gradually reaches a steady state, the experimental data of flashing jet in the steady state is substituted into the two-phase leak void fraction model to obtain the distribution of void fraction, and the experimental value of capacitance obtained by the capacitance sensor in the experiment after The results show that the gap rate distribution values were obtained at a leakage time of 10 minutes. The results show that the simulated values are slightly higher than the experimental values for leak times between 10s and 25s, due to the gradual stabilisation of the leakage process. In the leakage time between 25s~40s, the two-phase leakage tends to be stable, the simulated value is basically consistent with the experimental value, and the void fraction tends to be stable at around 70% with an average absolute error of 3.8%, which is a good agreement, i.e. it proves the feasibility of the two-phase leakage void fraction model calculation method.

3. Conclusion

A mathematical calculation model of the internal void fraction of the nozzle during steady-state leakage was established. Based on the fact that there is a certain error in the image processing method under high void fraction operating conditions, a capacitance acquisition device was designed independently in this paper and the capacitance method was used to verify it. The simulated value is slightly higher than the experimental value when the leakage time is between 10s~25s, due to the gradual stabilisation of the leakage process. When the leakage time is between 25s~40s, the two-phase leakage tends to be stable and the simulated value is basically consistent with the experimental value with an average absolute error of 3.8%, which has good agreement, i.e. it proves the feasibility of the two-phase leakage void fraction model calculation method.

Acknowledgments
This work was financially supported by the National Key Research and Development Plan (No.2016YFC0800100) in China.

References

[1] Vijayan P K, Patil A P, Pilkhwal D S, et al. An assessment of pressure drop and void fraction correlations with data from two-phase natural circulation loops[J]. Heat and Mass Transfer, 2000, 36(6): 541-548.
[2] Coddington P, Macian R. A study of the performance of void fraction correlations used in the context of drift-flux two-phase flow models[J]. Nuclear Engineering and Design, 2002, 215(3): 199-216.
[3] Zuber N, Findlay J A. Average volumetric concentration in two-phase flow systems[J]. 1965.
[4] Godbole P V, Tang C C, Ghajar A J. Comparison of void fraction correlations for different flow patterns in upward vertical two-phase flow[J]. Heat Transfer Engineering, 2011, 32(10): 843-860.
[5] Woldesemayat M A, Ghajar A J. Comparison of void fraction correlations for different flow patterns in horizontal and upward inclined pipes[J]. International journal of multiphase flow, 2007, 33(4): 347-370.
[6] Herringe R A, Davis M R. Flow structure and distribution effects in gas-liquid mixture flows[J]. International Journal of Multiphase Flow, 1978, 4(5-6): 461-486.
[7] Polanco G, Holdø A E, Munday G. General review of flashing jet studies[J]. Journal of hazardous materials, 2010, 173(1-3): 2-18.

Simulation of the dynamic behavior of droplet impact on the microstructure surface

Jianhua Yang, Ming Gao*, Hailin Sun, Zhennan Liu and Kaiyuan Zhang

Shanghai Key Laboratory of Multiphase Flow and Heat Transfer in Power Engineering, School of Energy and Power Engineering, the University of Shanghai for Science and Technology, Shanghai 200093, China

*Corresponding author e-mail: gaoming@usst.edu.cn

Abstract. The problem of droplets impacting rough wall has always been a hot spot in the engineering field. In this paper, the single-component multiphase lattice Boltzmann method is used to construct the wall microstructure, and the dynamic behavior characteristics of droplets impacting rough walls are simulated. The results show that the final state of a droplet impacting rectangular microstructure shows that the hydrophilic surface is more hydrophilic and the hydrophobic surface is more hydrophobic, and compared with the smooth surface, the microstructure surface hinders the rebound of the droplet, and the application of microstructure on the wall can promote the spreading of a droplet on the wall.

1. Introduction

Droplet impact has been widely used in engineering and technical fields[1-4]. As one of the most common phenomena in nature, its inertia force, surface tension, and contact surface structure have become sculptors of the final form of droplets. The study of droplet impact behavior first started with Worthington's work [5], which observed the dynamic process of milk and mercury droplets falling on a smoked glass plate. Previous numerical simulation studies on droplet impact mostly assume that the solid wall is smooth [6-8], but due to the limitation of technology, the rough structure of the solid wall can't be ignored in the actual situation, and there are many forms of microstructures. Taking the simplest regular rectangular microstructure as an example, Cheng et al. [9] can be divided into three modes according to the wetting behavior of droplets on its surface: Hemi-Wicking Cassie-Baxter mode [10], Wenzel mode [11], and Cassie-Baxter mode [12]. The droplets in the three modes correspond to different wetting forms respectively. In the Hemi-wicking Cassie-Baxter mode, the droplets will spread outward along the rough microstructure of the solid wall. In Wenzel mode, the droplets will only flow into the rough microstructure of the solid wall below the droplets, and will not spread outward. In Cassie-Baxter mode, the droplets will not flow into the rough microstructure on the solid wall, and at this time, the droplets are completely on the surface microstructure. Table 1 shows a summary of three wetting characteristics of droplets on the microstructure surface.

The previous work, most of them focused on the study of liquid droplets impacting a smooth surface, but the roughness of the actual wall has a great influence on the dynamic characteristics of liquid droplets impacting the wall. In this paper, a two-dimensional rectangular microstructure surface was constructed

Content from this work may be used under the terms of the Creative Commons Attribution 3.0 licence. Any further distribution of this work must maintain attribution to the author(s) and the title of the work, journal citation and DOI.
Published under licence by IOP Publishing Ltd

by using the two-dimensional polyphase lattice Boltzmann method, and the dynamic behavior characteristics of liquid droplets impacting a rough wall were studied.

Table 1. Summary of three wetting modes

Droplet wetting modes	Wetting characteristics
H-W C-B mode	Complete infiltration filling
Wenzel mode	Partial infiltration filling
Cassie-Baxter mode	Suspension (non-infiltration)

2. Methodology

2.1. Numerical method

Lattice Boltzmann method is a mesoscopic numerical simulation method. In the lattice Boltzmann method, the motion of fluid particles is described by the distribution function f of fluid particle clusters. The distribution function F of fluid micro-cluster evolves on the grid nodes, colliding and migrating, thus realizing the numerical simulation of fluid flow. After the evolution, by counting the distribution function F of each grid node, we can get the macroscopic characteristic parameters such as the speed and density of the node at that location.

The evolution equation of the distribution function of LBM particle clusters with a single relaxation time collision operator is:

$$f_i(\mathrm{x} + \mathrm{e}_i\delta_t, t + \delta_t) - f_i(\mathrm{x}, t) = -\frac{1}{\tau}\left(f_i(\mathrm{x}, t) - f_i^{eq}(\mathrm{x}, t)\right) + \Delta f_i(\mathrm{x}, t) \quad (1)$$

where $f_i(\mathrm{x}, t)$is the distribution function of a fluid particle, τ is the dimensionless relaxation time, which is related to the viscosity of the fluid, and is the time step. Generally $\delta_t = 1$, $\Delta f_i(\mathrm{x}, t)$ is the volume force term, $f_i^{\mathrm{eq}}(\mathrm{x}, t)$ is the equilibrium distribution function of fluid particle clusters. Its calculation formula is as follows:

$$f_i^{eq} = \omega_i\rho\left[1 + \frac{\mathrm{e}_i \cdot \mathrm{u}}{c_s^2} + \frac{(\mathrm{e}_i \cdot \mathrm{u})^2}{2c_s^4} - \frac{\mathrm{u}^2}{2c_s^2}\right] \quad (2)$$

where ω_i is the weight coefficient. In the D2Q9 model, the values of the weight coefficient ω_i: $\omega_0 = \frac{4}{9}$, $\omega_{1-4} = \frac{1}{9}$, $\omega_{5-9} = \frac{1}{36}$. C_s is the lattice sound speed, $c_s{}^2 = \frac{c^2}{3}$

In the lattice Boltzmann method, liquid/vapor phase kinematic viscosity is determined by dimensionless relaxation time.

$$\upsilon = c_s^2(\tau - 0.5)\delta_t \quad (3)$$

The form of force greatly influences the accuracy of numerical calculation. At present, there are three kinds of force introduction forms: velocity correction method, discrete force method, and accurate difference method. Gong et al. compared these three force introduction forms with Maxwell's theoretical solutions, and the results show that the accurate difference method is more consistent. In the accurate difference method, the force term is written as:

$$\Delta f_i(\mathrm{x}, t) = f_i^{eq}(\rho(\mathrm{x}, t), \mathrm{u} + \Delta\mathrm{u}) - f_i^{eq}(\rho(\mathrm{x}, t), \mathrm{u}) \quad (4)$$

where $\Delta u = \mathrm{F}\delta_t/\rho$ is the change of speed caused by the resultant force F in unit time step, and the resultant force F is written as:

$$\mathrm{F} = \mathrm{F}_{int} + \mathrm{F}_s + \mathrm{F}_g \quad (5)$$

where F_{int} represents the force between fluid particles, using the formula proposed by Gong et al.:

$$\mathrm{F}_{int} = -\beta\varphi(\mathrm{x})\sum_{x'} G(\mathrm{x}, \mathrm{x'})\varphi(\mathrm{x'})(\mathrm{x'} - \mathrm{x}) - \frac{1-\beta}{2}\sum_{x'} G(\mathrm{x}, \mathrm{x'})\,\varphi^2(\mathrm{x'})(\mathrm{x'} - \mathrm{x}) \quad (6)$$

Where is the weight coefficient, which is related to the selected state equation in lattice Boltzmann method, and $\varphi(\mathrm{x})$ is the "effective mass" of lattice points. The calculation formula is:

$$\varphi(\mathrm{x}) = \sqrt{\frac{2(p - \rho c_s^2)}{c_0 \mathrm{g}}} \tag{7}$$

The P-R equation of state is selected in this paper, which is calculated by the following formula:

$$p = \frac{\rho RT}{1 - b\rho} - \frac{a\rho^2 \alpha(T)}{1 + 2b\rho - b^2\rho^2} \tag{8}$$

$$\alpha(T) = \left[1 + (0.37464 + 1.54226\omega - 0.26882\omega^2)(1 - \sqrt{\frac{T}{T_c}})\right]^2 \tag{9}$$

Where $\omega = 0.344,\ a = \frac{2}{49},\ b = \frac{2}{21},\ R = 1,\ T = 0.8T_{cr},\ T_{cr} = 0.0729$

In formula (5), $\boldsymbol{F}_s$ is the fluid-solid interaction between particles and solid wall, which is obtained from the following calculation formula:

$$\mathrm{F}_s = -G_{ads}\varphi(\mathrm{x}) \sum \omega_i s(\mathrm{x} + \mathrm{e}_i\delta_t)\mathrm{e}_i \tag{10}$$

The Gads is the coefficient of fluid-solid interaction, which determines the hydrophilicity and hydrophobicity of the solid wall. $s(\mathrm{x} + \mathrm{e}_i\delta_t)$ is the indicator function, when its value is 1, it represents a solid lattice point, and when its value is 0, it represents a fluid lattice point.

The $\boldsymbol{F}_g$ in formula (5) is the gravity term, which is given by the following formula:

$$\mathrm{F}_g = (\rho(\mathrm{x}) - \rho_v)\mathrm{g} \tag{11}$$

where $\rho(\mathrm{x})$ is the fluid lattice density and g is the local gravitational acceleration.

After force $\boldsymbol{F}$ is introduced, the macroscopic velocity of the fluid will change. The macroscopic velocity $\boldsymbol{U}$ of the fluid is defined as:

$$\mathrm{U} = \mathrm{u} + \frac{\delta_t}{2\rho}\mathrm{F} \tag{12}$$

2.2. Problem description

Fig.1 shows the setting of the calculation model diagram and regional boundary. Based on the model selected in the previous article, we will study the impact of droplets on rough microstructure surfaces, After grid independence verification, select the grid: $Nx \times Ny = 600 \times 400$. The left and right boundary conditions: Periodic boundary conditions, and the upper and lower boundary conditions: Bounceback boundary conditions.

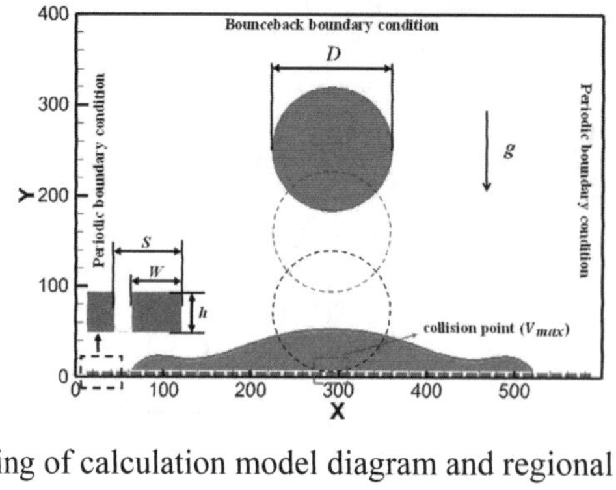

Fig. 1 Setting of calculation model diagram and regional boundary

2.3. Validation of the numerical method

To verify the accuracy of the model in studying the problem of droplet impact, a single droplet impact on a smooth wall is selected for verification. The deformation degree of droplets is measured by the spreading factor. According to the research of Rioboo et al, it is found that the spreading process of droplets impacting the wall has an initial stage, and the spreading factor of droplets in this stage is not related to We number and contact angle, but only to dimensionless time. The relationship between spreading factor ξ and dimensionless time t* can be expressed as:

$$\xi = kt^{*0.5} \tag{13}$$

Where k is the correlation coefficient, Rioboo et al. obtained through experiments: k=2.8

Fig.2 shows the relationship between droplet spreading factor ξ and dimensionless time t^* at the initial stage. It can be found that the droplet spreading factor at this stage does not change with the change of the We number, but only relates to dimensionless time, which is consistent with the results obtained by Rioboo et al.[13] The curve shown in Fig.2 can be obtained by fitting according to formula (13), and the correlation coefficient k=2.769 can be fitted according to the formula (13), which is consistent with the results obtained by Rioboo et al.

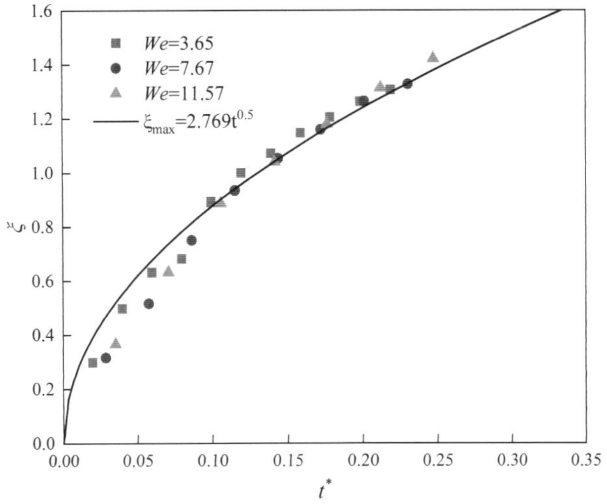

Fig. 2 Comparison of a simulated spreading factor with Rioboo et al.'s correlation equation

3. Results and analysis

3.1. Dynamic behavior of droplet impacting rough microstructure surface

In this paper, the static contact angle of the rough surface is represented by θ *, and that of a smooth solid surface with the same physical properties is represented by θ. Table 1 has given three modes and their characteristics after the droplet impact the microstructure surface. In these three modes, the static contact angle of the rough solid surface and the static contact angle of the smooth solid surface has the following relationship:

$$\cos\theta^* = \phi\cos\theta + (1-\phi), for(\theta < \theta_{c1}) \tag{14}$$

$$\cos\theta^* = Rf\cos\theta, for\ (\theta_{c1} < \theta < \theta_{c2}) \tag{15}$$

$$\cos\theta^* = \phi\cos\theta - (1-\phi), for(\theta > \theta_{c2}) \tag{16}$$

Where θ_{c1} and θ_{c2} are two critical contact angles, corresponding to three different modes. When $\theta<\theta_{c1}$, the droplet is in Hemi-wicking Cassie-Baxter mode; When $\theta_{c1}<\theta<\theta_{c2}$, the droplet is in Wenzel mode, and when $\theta>\theta_{c2}$, the droplet is in Cassie-Baxter mode. The values of θ_{c1} and θ_{c2} are determined by the following formula (17) and (18), which can be determined by the R_f and φ of the microstructure rough surface of the column:

$$\cos\theta_{c1} = (1-\phi)/(Rf-\phi) \tag{17}$$

$$\cos\theta_{c1} = (\phi-1)/(Rf-\phi) \tag{18}$$

According to the setting of roughness coefficient R_f=2.224 and solid fraction φ=0.510 in this paper, two critical contact angles: θ_{c1} = 75.5°and θ_{c2} = 104.5°can be calculated by combining formulas (17) and (18). The static contact angle of the droplet on the rough microstructure surface with the same physical properties (referring to the same fluid-solid interaction coefficient *Gads* in LBM) is different from that on the smooth surface. Fig.3 shows the relationship between the static contact angle of the droplet on the microstructure surface and the fluid-solid interaction coefficient *Gads*.

Fig. 4 shows the dynamic process of the droplet impacting the smooth wall, the hydrophilic rough surface (θ = 80°), and the hydrophobic rough surface (θ = 140°). The impact process of hydrophilic droplets can be divided into four stages: movement stage, spreading stage, retraction stage, and oscillation stage. Different structural surfaces affect the maximum spreading diameter, spreading time, and the final state of the droplet on the solid surface. Through the image of the impact dynamic process, it can be found that the whole process of droplet impact is in Wenzel mode, which is consistent with the theoretical calculation result: 75.5° < 80° < 104.5° (in Wenzel mode). Compared with the hydrophilic surface, the droplet impacting the hydrophobic surface (θ = 140°) bounces off the solid surface after one retraction. In the process of droplet impact, the droplet under the microstructure is in Cassie-Baxter mode.

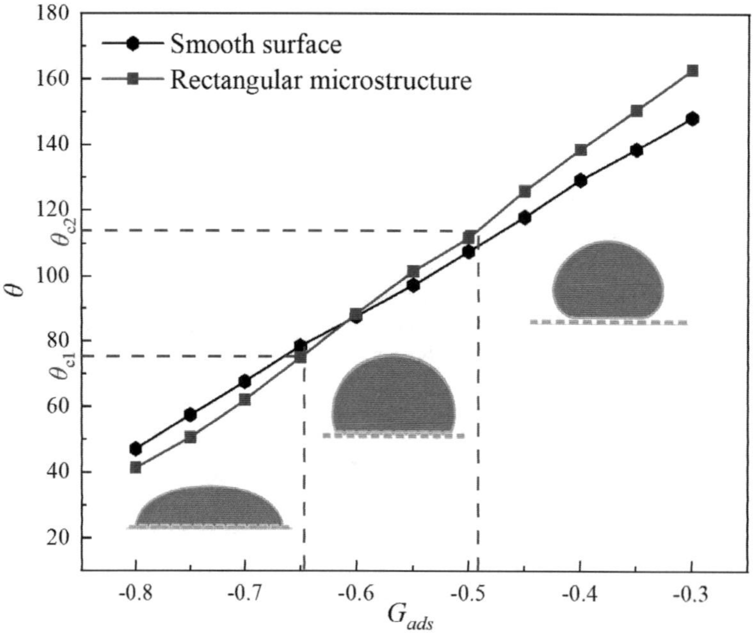

Fig. 3 Relationship between contact angle of microstructure and Gads

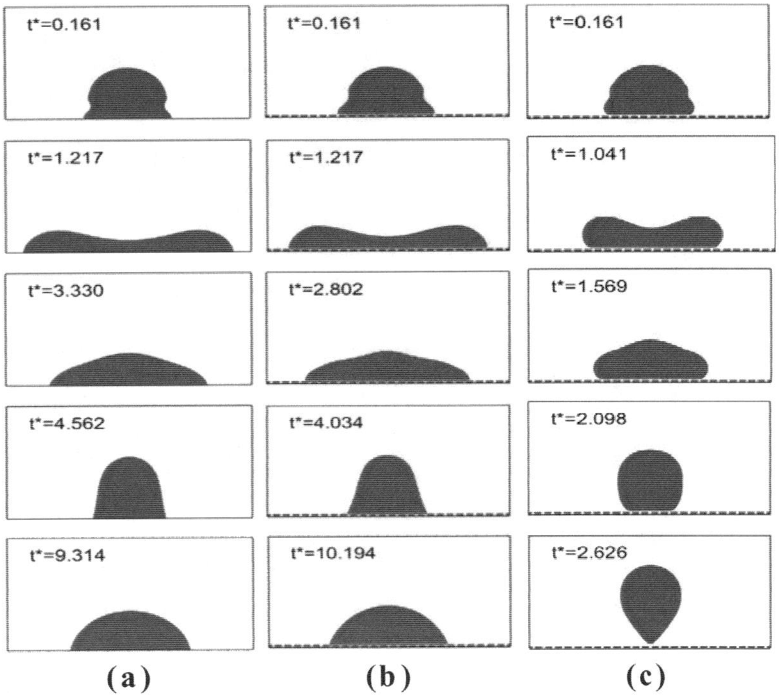

Fig. 4 Dynamic image of a droplet impacting microstructure surface under different conditions (a: smooth hydrophilic surface; b: rough hydrophilic surface; c: rough hydrophobic surface)

3.2. Evolution of droplet impact spreading factor and relative height

Fig. 5 shows the change process of the spreading factor of the droplet on the solid surface after it impacts the hydrophilic-hydrophobic surface of the rectangular microstructure when We=5.08. The symbol fluctuation in Fig 5 shows the oscillation process after the droplet impacts the hydrophilic surface. The area bounded by the black square box in Fig.5 is the initial stage when the droplet just impacts the solid surface. At this stage, the droplet presents a semicircle shape, which is defined by Rioboo et al. [13] as the moving stage ($t^* \ll 1$). After the movement stage, it is the spreading stage of droplet impact, in which the droplets will be flat. Under the hydrophilic condition, the maximum spreading diameter of the microstructure surface is smaller than that of a smooth surface. This is because there is no microstructure on a smooth surface to consume the kinetic energy of droplets, so all the kinetic energy of droplets is used to spread out on the surface. In Fig.5, there is also an impact movement stage when droplets impact the hydrophobic surface, and the droplets bounce off the solid surface when they retract once after reaching the maximum spreading diameter. Under a hydrophobic surface, Whether there is microstructure on the impact surface has little influence on the maximum spreading diameter of droplets after impact. This is because the hydrophobic surface has little adhesion to droplets, the droplets do not have good wettability, and the droplets are not immersed in the microstructure. Therefore, the resistance of microstructure to droplet spreading is small, the kinetic energy consumption of droplets is small, and the droplets all have large spreading diameters.

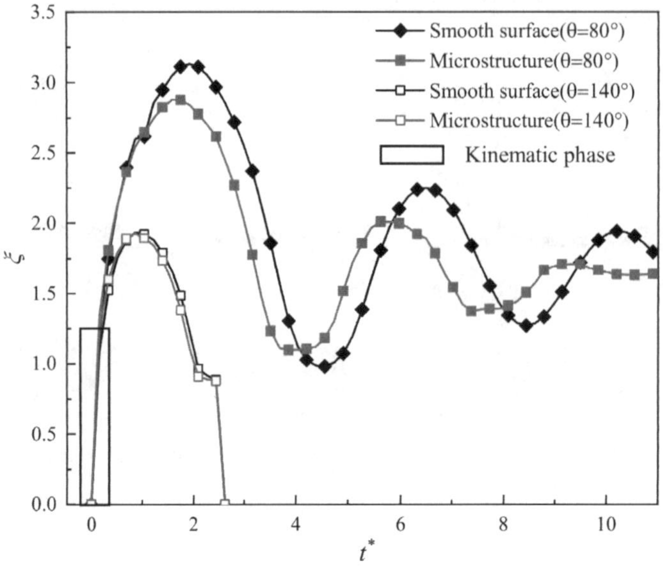

Fig. 5 Relationship between spreading factor and time of droplet impacting hydrophilic and hydrophobic surface

Fig. 6 shows the change in the relative height of the droplet after impacting the hydrophilic-hydrophobic surface with dimensionless time when the droplet *We*=5.08. When the droplet impacts the hydrophilic surface, the microstructure hinders the bounce of the droplet compared with the smooth surface. This is because some kinetic energy is consumed when the liquid drops impact the microstructure surface, and more kinetic energy cannot be converted into surface energy for the liquid drops to bounce. Therefore, the time when the microstructure surface reaches the highest height in the retraction process is somewhat delayed. Compared with the impact on the hydrophilic surface, the process of a droplet impacting the hydrophobic surface is very similar, and the change in relative height is not so great. This is because the droplets on the hydrophobic surface do not have good wettability, and the micro-structure surface has little hindrance to the spreading behavior of the droplets, so the droplets all have large deformation degrees, and the minimum relative height is the same.

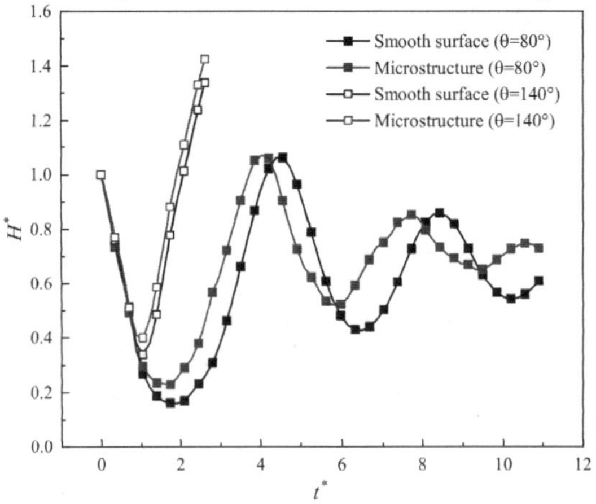

Fig. 6 Relationship between bouncing height and time of droplets impacting hydrophilic and hydrophobic surface

4. Conclusion

In this paper, the dynamic behavior of liquid droplets impinging on a rough microstructure surface is studied by using the single-component multiphase lattice Boltzmann method. The spreading factor and the change of relative height of droplets impacting the hydrophilic-hydrophobic microstructure surface are mainly explored, It is found that the application of micro-structured columns on the hydrophilic wall will make the contact surface more hydrophilic, and the application of micro-structured columns on the hydrophobic wall will make the contact surface more hydrophobic. When the liquid drops impact the hydrophilic surface, the maximum spreading diameter of the liquid drops impacting the micro-structured surface is smaller than that of the smooth surface, and the micro-structured surface hinders the bouncing of the liquid drops compared with the smooth surface. The research in this aspect provides an improved reference for many engineering applications, such as droplet dripping and splashing of 3D printing technology.

Acknowledgments

This work was supported by the National Natural Science Foundation of China through Grant No. 51976127.

References

[1] Henein, H., V. Uhlenwinkel, and U. Fritsching, Metal Sprays and Spray Deposition. 2017: Metal Sprays and Spray Deposition.

[2] Xiong, W., et al., Experimental investigation on the height deviation of bumps printed by solder jet technology. Journal of Materials Processing Technology, 2016. 243: p. 291-298.

[3] Son, Y., et al., Spreading of an inkjet droplet on a solid surface with a controlled contact angle at low Weber and Reynolds numbers. Langmuir the Acs Journal of Surfaces & Colloids, 2008. 24(6): p. 2900-7.

[4] Luo, J., et al., Formation of uniform metal traces using alternate droplet printing. International Journal of Machine Tools & Manufacture, 2017. 122: p. 47-54.

[5] Worthington and M. A., A Second Paper on the Forms Assumed by Drops of Liquids Falling Vertically on a Horizontal Plate. Proceedings of the Royal Society of London, 1876. 25(171-178): p. 261-272.

[6] Xiong, et al., Numerical investigation of air entrapment in a molten droplet impacting and solidifying on a cold smooth substrate by 3D lattice Boltzmann method. INTERNATIONAL JOURNAL OF HEAT AND MASS TRANSFER, 2018.

[7] Amit, G. and K. Ranganathan, Two-Dimensional Lattice Boltzmann Model for Droplet Impingement and Breakup in Low-Density Ratio Liquids. Communications in Computational Physics, 2011. 10(03): p. 767-784.

[8] Ma, Z., W. Xiong, and P. Cheng, 3D Lattice Boltzmann simulations for water droplet's impact and transition from central-pointy icing pattern to central-concave icing pattern on supercooled surfaces. Part I: Smooth surfaces. International Journal of Heat and Mass Transfer, 2021. 171: p. 121097-.

[9] Cheng, P.C., W. C, Similarity solutions for convection of groundwater adjacent to horizontal impermeable surfaces with axisymmetric temperature distribution. Water Resources Research, 1977. 13(4): p. 768-772.

[10] Bico, J., U. Thiele, and D. Quéré, Wetting of textured surfaces. Colloids and Surfaces A: Physicochemical and Engineering Aspects, 2002. 206(1-3): p. 41-46.

[11] Wenzel, R.N., Resistance of solid surfaces to wetting by water. Transactions of The Faraday Society, 1936. 28(8): p. 988-994.

[12] Cassie, A.B.D. and S. Baxter, Wettability of porous surfaces. Transactions of the Faraday Society, 1944. 40: p. 546-551.

[13] Rioboo, R., M. Marengo, and C. Tropea, Time evolution of liquid drop impact onto solid, dry surfaces. Experiments in Fluids, 2002. 33(1): p. 112-124.

Modeling of Micro-downburst accompanied with rain and analysis of its influence on artillery firing accuracy

Jin Wang [1,*], Junwei Tian [1,2], Zhen Zhang [2], Xiaoming Lei [1], Guangyu Sun [2], Xiaodan Guo [2]

[1]Xi'an Technological University School of Ordnance Science and Technology, Xi 'an, Shaanxi Province 710021, China
[2]Xi'an Technological University School of Mechatronic Engineering, Xi 'an, Shaanxi Province 710021, China

*Corresponding author e-mail: 734575032@163.com

Abstract. The characteristics of wet micro burst are more complex, which will cause some deviation on the artillery range, side deviation and other parameters, and have a great influence on the artillery shooting accuracy. Taking 155 mm artillery explosive shell as the research object, based on vortex ring principle and fluid mechanics, the micro-downburst and raindrop motion model was established, which was integrated into the wet micro-downburst model suitable for ballistic analysis and calculation. The wet micro-downburst model was combined with the 6-DOF external ballistic model of artillery. Simulation and analysis of the effect of different initial vortex ring center induced velocity on the firing accuracy of flat firing and curved firing. The experimental results show that for long-range firing, the range of the artillery decreases with the increase of the induction velocity of the initial vortex ring center, which seriously affects the firing accuracy and killing effect of the artillery.

Keywords: Micro-downburst accompanied with rain; Principle of vortex ring; 6-DOF; Artillery exterior trajectory.

1. Introduction

There are various factors that affect the firing accuracy of artillery external ballistic. According to research, among the factors that affect the firing accuracy of artillery, the influence caused by meteorological conditions accounts for more than 70% [1-2], among which the influence of low-altitude wind shear on artillery external ballistic is typical. As one of many types of low-level wind shear, microburst has become a dangerous airflow with serious impact on flight safety due to its small scale, large intensity, short life cycle, and failure to detect and forecast in time [3-5]. In recent years, relevant scholars have carried out a large number of studies on dry micro-downburst, among which Sengupta6 used LES method to simulate and conduct numerical analysis on micro-downburst. Dae [7] established a closed model of non-stationary fluctuation in the case of long pulse duration to quickly evaluate the non-stationary turbulence effect. For wet burst with WDR characteristics, WDR simulation method is mostly used, and Blocken optimizes the particle track model so that it can predict the spatial and temporal distribution of WDR [8]. However, Huang established the WDR Euler multiphase flow model

Content from this work may be used under the terms of the Creative Commons Attribution 3.0 licence. Any further distribution of this work must maintain attribution to the author(s) and the title of the work, journal citation and DOI.
Published under licence by IOP Publishing Ltd

to replace the particle orbit model, and verified the reliability of the method and the simplicity of simulation through experiments [9]. In the military field, Chen combined the dry micro-downburst model with the external ballistic model of rocket, and analyzed its influence on the external ballistic model of rocket, and made certain achievements [10]. But with micro strike critical flow under the different dry, wet strike suddenly and violently under the micro flow phenomenon, there will be a short-term heavy rainfall will increase in the process of the projectile flying air resistance, the impact on the precision of the howitzer is obvious, because of artillery projectile and rocket aerodynamic shape at the same time, the means of transmission, is affected by the strike critical flow under wet micro is different.

Therefore, based on the principle of vortex ring building strike suddenly and violently under the micro flow with fluid mechanics and raindrops fusion for wet micro motion model under critical flow model, and analysis under the wet micro blow air flow characteristics of cascade, and combined the model with the artillery centroid trajectory equation, simulation and analysis of the effect of different initial vortex ring center induced velocity on the firing accuracy of the artillery under the curved firing mode.

2. Airflow wind field modeling

The micro-downburst is driven by the dragging effect of precipitation. In the process of precipitation, the evaporation of rain reduces the temperature of the surrounding air, leading to the collapse of updraft, which is replaced by downdraft, and then spreads around the ground and rises10, thus forming downburst. Fig.1 shows the formation process of micro-downburst. In the area affected by this characteristic air flow, the influence of the air flow at different positions on the artillery's external trajectory is also different.

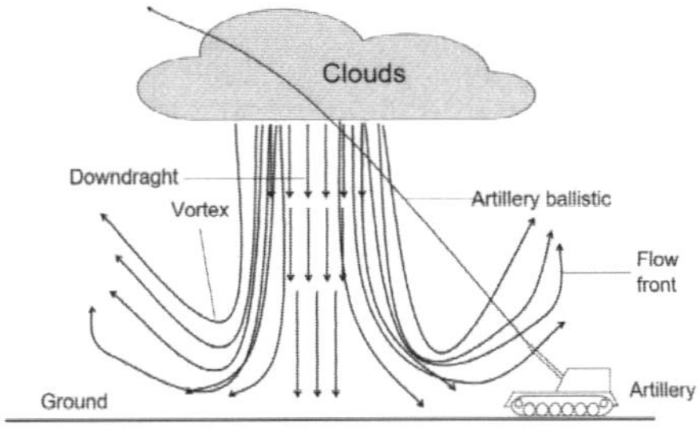

Figure 1. Schematic diagram of microburst formation

2.1. Construction of dry micro-burst model

Firstly, the micro-downburst model is constructed based on the vortex ring principle [12-13]. The ground coordinate system is taken as the reference frame, the center of the main vortex ring is Oa, its coordinate point is (xa, ya, za), the center of the mirror vortex ring is Ob (xb, yb, zb), and the position of the projectile is Oc (xc, yc, zc). When the main vortex ring and the mirror vortex ring are parallel to the ground. The main vortex ring curve equation [10] is as follows

$$\begin{cases} (x-x_a)^2+(y-y_a)^2=R^2 \\ z=z_a \end{cases} \tag{1}$$

In the actual situation, the main vortex ring and the mirror vortex ring respectively have an angle φ on the ground, whose schematic diagram is shown in Fig2. A two-dimensional figure is formed by projecting the three-dimensional inclined vortex ring onto the xoz plane. At this time, the maximum distance between the projectile and the main vortex ring at any position during flight is as follows

$$\begin{cases} r_{\max 1}=\sqrt{(z_c-z_a)^2+(x_c-R)^2} \\ r_{\min 1}=\sqrt{(z_c-z_a)^2+(x_c+R)^2} \end{cases} \tag{2}$$

Similarly, the extreme distance between the projectile and the mirror vortex ring is as follows

$$\begin{cases} r_{\max 2} = \sqrt{(z_c - z_b)^2 + (x_c - R)^2} \\ r_{\min 2} = \sqrt{(z_c - z_b)^2 + (x_c + R)^2} \end{cases} \tag{3}$$

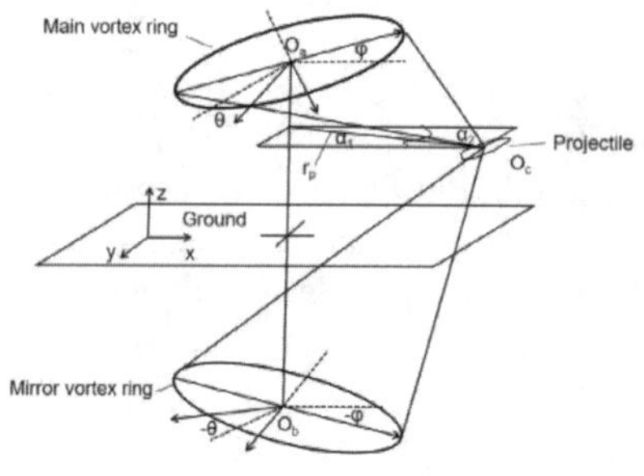

Figure 2. Schematic diagram of tilted vortex ring

The flow function of the main vortex ring at Oc point is as follows [10]

$$\begin{cases} \varphi_a = -\dfrac{2Rv_0}{2\pi}(r_{\max 1} + r_{\min 1})[F_1(k_a) - F_2(k_a)] \\ k_a = |(r_{\max 1} - r_{\min 1})/(r_{\max 1} + r_{\min 1})| \end{cases} \tag{4}$$

$F_1(k_a)$ $F_2(k_a)$ is elliptic integral function, k_a is elliptic integral variable, v_0 is initial airflow velocity. After simplified calculation [10], when $0 \le k_a \le 1$

$$F_1(k_a) - F_2(k_a) = \frac{0.788k_a^2}{0.25 + 0.75\sqrt{1 - k_a^2}} \tag{5}$$

Similarly, the flow function φ_b of the mirrored vortex ring can be obtained, and its expression is as follows

$$\begin{cases} \varphi_b = -\dfrac{\Gamma}{2\pi}(r_{\max 2} + r_{\min 2})\dfrac{0.788k_b^{\;2}}{0.25 + 0.75\sqrt{1 - k_b^2}} \\ k_b = |(r_{\max 2} - r_{\min 2})/(r_{\max 2} + r_{\min 2})| \end{cases} \tag{6}$$

The radial and axial induced velocities of the main vortex ring and the mirror vortex ring can be obtained by taking the partial derivatives of the two vortex circulation line equations, and their expressions are as follows

$$\begin{cases} v_r^a = \dfrac{1}{r_p}\dfrac{\partial \varphi_a}{\partial z_c} & v_r^b = \dfrac{1}{r_p}\dfrac{\partial \varphi_b}{\partial z_c} \\ v_z^a = -\dfrac{1}{r_p}\dfrac{\partial \varphi_a}{\partial r_p} & v_z^b = -\dfrac{1}{r_p}\dfrac{\partial \varphi_b}{\partial r_p} \end{cases} \tag{7}$$

By decompositions the radial velocities of the two vortex rings in the ground coordinate system, the velocities along the Ox axis and Oy axis can be obtained, and their expressions are as follows

$$\begin{cases} v_x^a = v_r^a \cos(\alpha_1) & v_x^b = v_r^b \cos(\alpha_1) \\ v_y^a = v_r^a \sin(\alpha_2) & v_y^b = v_r^b \sin(\alpha_2) \end{cases} \tag{8}$$

By superimposing the equation of the main vortex ring and the mirror vortex ring, and linear superimposing the induced velocity of each axis, the streamline equation and velocity equation of Oc point can be obtained, and their expressions are as follows [7]

$$\begin{cases} \varphi_c = -\varphi_a + \varphi_b \\ v_x^c = v_x^a + v_x^b \\ v_y^c = v_y^a + v_y^b \\ v_z^c = v_z^a + v_z^b \end{cases} \tag{9}$$

According to the velocity equation, when the distance between the Oc point and the vortex core increases, the induced velocity will gradually decrease. However, when the distance between the Oc point and the vortex core is 0, the induced velocity will be $+\infty$, which is not in line with the actual situation. Therefore, the inducible factor ξ is introduced to establish a continuously varying vortex core model, whose expression is as follows

$$\xi = 1 - \exp\left(-\left(\frac{r_{\min 1}}{R}\right)^2 / \left[0.4215\frac{2r}{R} + 0.0822\left(\frac{2r}{R}\right)^2 - 0.0969\left(\frac{2r}{R}\right)^3\right]^2\right) \tag{10}$$

Therefore, the expression of induced velocity after correction is as follow

$$\begin{cases} v_{x1} = \xi v_x^c \\ v_{y1} = \xi v_y^c \\ v_{z1} = \xi v_z^c \end{cases} \tag{11}$$

It can be seen from Equations (6) and (11) that rp in the equation cannot be 0, which is inconsistent with the actual situation. Therefore, in order to solve this problem, the axial induced velocities of the central axis of the main vortex ring and the mirror vortex ring are respectively established based on the turbulent free jet theory, and their expressions are as follows [10]

$$\begin{cases} v_{za} = \Gamma / \left(2R \cdot \left(\left(1 + \left((z_a - z_c)/R\right)^2\right)^{1.5}\right)\right) \\ v_{zb} = \Gamma / \left(2R \cdot \left(\left(1 + \left((z_a + z_c)/R\right)^2\right)^{1.5}\right)\right) \end{cases} \tag{12}$$

Therefore, the axial conjunction velocity at the central axis OaOb is as follows

$$v_{z2} = v_{za} - v_{zb} \tag{13}$$

Since there is an angle between the vortex ring and the ground, and the vector of the angle between the main vortex ring and the ground is (φ θ 0)T, the pitch transformation matrix Lθ and the roll transformation matrix Lφ are introduced to complete the coordinate system transformation of the main vortex ring and the mirror vortex ring [10].

$$L_\theta = \begin{bmatrix} 1 & 0 & 0 \\ 0 & \cos\theta & -\sin\theta \\ 0 & \sin\theta & \cos\theta \end{bmatrix} \tag{14}$$

$$L_\varphi = \begin{bmatrix} \cos\varphi & 0 & \sin\varphi \\ 0 & 1 & 0 \\ -\sin\varphi & 0 & \cos\varphi \end{bmatrix} \tag{15}$$

$$\begin{bmatrix} v_{wx} & v_{wy} & v_{wz} \end{bmatrix}^{\mathrm{T}} = L_\varphi L_\theta \xi \begin{bmatrix} v_x^a & v_y^a & v_z^a \end{bmatrix}^{\mathrm{T}} + L_{-\varphi} L_{-\theta} \xi \begin{bmatrix} v_x^b & v_y^b & v_z^b \end{bmatrix}^{\mathrm{T}} \tag{16}$$

2.2. Construction of wet micro downburst model for the integration of rain and air flow

For the wet microburst, the rain will not be completely evaporated during the landing process, which is bound to have a certain impact on the external ballistic of the artillery. Based on fluid dynamics theory, a raindrop trajectory equation is established.

$$\begin{cases} m_w \dfrac{\mathrm{d}v_x}{\mathrm{d}t} = A_x f_x \\ m_w \dfrac{\mathrm{d}v_y}{\mathrm{d}t} = (m_w - m_a)g - A_y f_y \end{cases} \tag{17}$$

$$\begin{cases} f_x = C_D \rho_a \left(\overline{v}_f - v_x\right)^2/2 \\ f_y = C_D \rho_a v_y^2/2 \end{cases} \tag{18}$$

$$\begin{cases} m_w = \pi\rho_\omega d^3/6 \\ m_a = \pi\rho_a d^3/6 \end{cases} \tag{19}$$

$$C_D = 24/Re + 6/\left(1+\sqrt{Re}\right) + 0.4 \tag{20}$$

Ax and Ay are the projected areas of raindrops on the horizontal and vertical planes. Here, raindrops are abstracted as spheroids with a diameter of 5 mm, mw and ma are raindrop mass and air mass under the same volume. Re is the fluid Reynolds number. Substituting each parameter into Equation (21) and solving it can be obtained:

$$\begin{cases} v_x = \overline{v}_f\left(1 - 1/(1 + 3\rho_a C_D v_0 t/4d\rho_w) + 3\rho_a C_D v_0 t/4d\rho_w\right) \\ v_y = \sqrt{4dg(\rho_w/\rho_a - 1)/3C_D} \cdot \exp\left(2t\sqrt{4dg(\rho_w/\rho_a - 1)/3C_D} - 1\right)/\exp\left(2t\sqrt{4dg(\rho_w/\rho_a - 1)/3C_D} + 1\right) \end{cases} \tag{21}$$

Since the falling time of raindrops cannot be unified with the flying time of artillery projectile, in order to avoid the calculation conflict, the falling time of raindrops is replaced by other quantities by solving the differential equation, and the calculation steps are as follows

$$\int v_y \mathrm{d}t = h_2 = 2\sqrt{3}t/3\mathrm{e}^2\sqrt{dg(\rho_w/\rho_a - 1)/C_D} \tag{22}$$

$$H = h_1 - h_2 = h_1 - 2t\sqrt{3dg(\rho_w/\rho_a - 1)/C_D}/3\mathrm{e}^2 \tag{23}$$

$$t = \mathrm{e}^2(h_1 - H)\sqrt{3C_D/dg(\rho_w/\rho_a - 1)}/2 \tag{24}$$

h_2 is the falling height of raindrops at different times, h1 is the total falling height of raindrops, and H is the height of raindrops from the ground at different times. Substituting equation (28) into equation (25), it can be obtained:

$$\begin{cases} v_x = \overline{v}_f\left(1 - 1/\left(1 + 3\sqrt{3}\rho_a C_D \overline{v}_f \mathrm{e}^2(h_1 - H)\sqrt{C_D/dg(\rho_w/\rho_a - 1)}/8\rho_w\right)\right) \\ v_y = \sqrt{(\rho_W/\rho_a - 1)4dg/3C_D}\exp\left(2\mathrm{e}^2(h_1 - H) - 1\right)/\exp\left(2\mathrm{e}^2(h_1 - H) + 1\right) \end{cases} \tag{25}$$

Due to the influence of airflow, there is an angle between raindrops and the ground, so the falling velocity of raindrops is transformed into a coordinate system, and the velocity equation after transformation is as follows

$$\begin{cases} v_{xs} = \overline{v}_f\left(1 - 1/\left(1 + 3\sqrt{3}\rho_a C_D \overline{v}_f \mathrm{e}^2(h_1 - H)\sqrt{C_D/dg(\rho_w/\rho_a - 1)}/8\rho_w\right)\right)\cos\alpha \\ v_{ys} = \overline{v}_f\left(1 - 1/\left(1 + 3\sqrt{3}\rho_a C_D \overline{v}_f \mathrm{e}^2(h_1 - H)\sqrt{C_D/dg(\rho_w/\rho_a - 1)}/8\rho_w\right)\right)\sin\alpha \\ v_{zs} = \sqrt{4dg(\rho_W/\rho_a - 1)/3C_D}\exp\left(2\mathrm{e}^2(h_1 - H) - 1\right)/\exp\left(2\mathrm{e}^2(h_1 - H) + 1\right) \end{cases} \tag{26}$$

Where α is the forward Angle between the velocity vx and the x axis of the ground coordinate system.

Because of the projectile in flight for high-speed spin of rigid body, the air boundary layer is driven around, make it hard for the rain to by direct contact with the projectile exert forces on its influence on the surface of the projectile flight path, so in order to accurately analyze the comprehensive impact of rainwater and airflow external ballistic, equivalent substitution will flow with the rainwater, the alternative equation is as follows

$$\begin{cases} \bar{v}_x = (v_{xs} + v_{wx})/2 \\ \bar{v}_y = (v_{ys} + v_{wy})/2 \\ \bar{v}_z = (v_{zs} + v_{wz})/2 \end{cases} \tag{27}$$

Since the air state characteristics change after the substitution, the trajectory deviation cannot be accurately calculated by the centroid trajectory equation. In order to make the substitution equation applicable, the state parameters such as near-ground air density and pressure need to be modified, and the modified equation is as follows

$$\begin{cases} \bar{\rho} = (\rho_w V + \rho_a V)/2V \\ p = \bar{\rho}/[R_d(\tau_0 - G_1 z)] \end{cases} \tag{28}$$

Where ρa is the density of wet air near the ground in rainy days, ρw is the density of water, $\bar{\rho}$ is the density of the modified fused gas, Rd is the gas constant of dry air, τ0 is the virtual temperature near the ground, and G1 is taken as 9.83×10^{-3}.

3. External ballistic model of artillery centroid motion

In order to analyze the influence of micro-downburst on the external ballistic deviation of artillery, according to the external ballistic theory of artillery, the 6-DOF external ballistic model of artillery shell is established, and its expression is as follows

$$\begin{cases} \dfrac{dv}{dt} = \dfrac{1}{m} F_{x_2}, \dfrac{d\theta_a}{dt} = \dfrac{1}{mv\cos\varphi_2} F_{y_2}, \dfrac{d\psi_2}{dt} = \dfrac{F_{z_2}}{mv}, \dfrac{d\omega_\xi}{dt} = \dfrac{1}{C} M_\xi \\ \dfrac{d\omega_\eta}{dt} = \dfrac{1}{A} M_\eta - \dfrac{C}{A}\omega_\xi\omega_\zeta + \omega_\zeta^2 \tan\varphi_2 + \dfrac{A-C}{A}\cdot\left(\beta_{D_\eta}\ddot{\gamma} - \beta_{D_\zeta}\dot{\gamma}\right) \\ \dfrac{d\omega_\zeta}{dt} = \dfrac{1}{A} M_\zeta + \dfrac{C}{A}\omega_\xi\omega_\eta - \omega_\eta\omega_\zeta \tan\varphi_2 + \dfrac{A-C}{A}\cdot\left(\beta_{D_\zeta}\ddot{\gamma} + \beta_{D_\eta}\dot{\gamma}\right) \\ \dfrac{d\varphi_a}{dt} = \dfrac{\omega_\zeta}{\cos\varphi_2}, \dfrac{d\varphi_2}{dt} = -\omega_\eta, \dfrac{d\gamma}{dt} = \omega_\xi - \omega_\zeta \tan\varphi_2 \\ \dfrac{dx}{dt} = v\cos\psi_2\cos\theta_a, \dfrac{dy}{dt} = v\cos\psi_2\sin\theta_a, \dfrac{dz}{dt} = v\sin\psi_2 \end{cases} \tag{29}$$

4. Simulation and result analysis

The six degrees of freedom ballistic equation of the gun is combined with the equation of the wind field of the micro-downburst flow. A 155 mm explosive shell is taken as the research object, and its technical data is shown in Table 1. The influence of this kind of wind field on the accuracy of the artillery's external ballistic firing is studied.

Table 1. Technical data of a 155 mm explosive projectile

Parameters	Value
Weight/kg	47
Centroid position/m	0.562
Caliber/m	0.155
Initial velocity/($m\cdot s^{-1}$)	940
Equatorial moment of inertia/ $kg\cdot m^2$	1.814
Polar moment of inertia/ $kg\cdot m^2$	0.163
Direction of artillery fire/°	0

4.1. Simulation of air flow model

According to the mathematical model in section 1.1 and 1.2, the vortex ring parameters are set as follows: the central coordinate O_a of the main vortex ring (3000,3000,800), the radius R is 1000 m, the vortex core R is 100 m, the initial vortex ring center induced velocity v0 is 10 m/s, and the vortex ring is horizontal. As shown in Fig.3 for dry micro strike critical flow under different heights on vertical and horizontal airflow velocity airflow velocity, orientation, right direction is the direction to speed down, because the main vortex ring and parallel to the horizontal ground, thus center symmetric relation with the horizontal airflow velocity, vertical flow in axial symmetry relations, and in the far distances from the center of the vortex ring. The velocity of horizontal and vertical airflow gradually decreases steadily and tends to 0 m/s.

Since there are often raindrops in the wet micro-burst, the rain velocity and air velocity are fused into the wet micro-burst model. The vortex ring and air flow parameters remain unchanged, and the vortex ring tilt angle is set to 0° and 20 °, and the average falling velocity of raindrops $\overline{v}_f$ is 8 m/s. Fig.4 shows the horizontal and vertical flow velocities of wet micro-downburst after the fusion of different angles and upper rain gas at different heights. Since the air pressure and density increase after equivalent, the air viscosity also increases and the fluidity weakens, the air velocity at each point of the equivalent wet micro-burst is slightly smaller than that of the dry micro-burst. Taking the vortex ring with an angle of 0° as an example, the horizontal and vertical air velocity at each point of the equivalent wet micro-burst are slightly smaller than 48.2% and 45.4%, respectively.

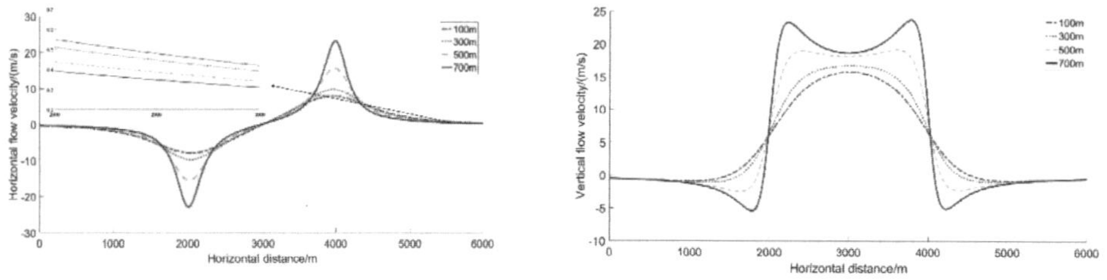

Figure 3. Horizontal and vertical airflow velocities at different heights of dry microbursts

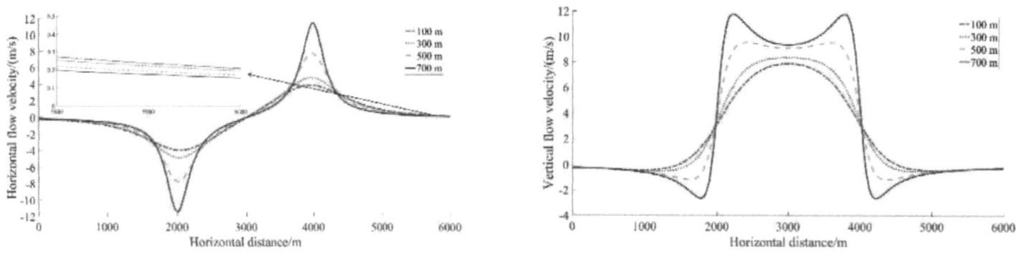

(a). Flow velocity of burst flow at different wet heights at 0° angle

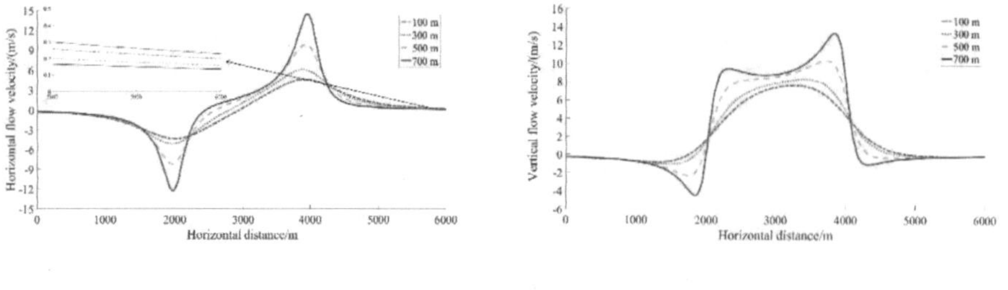

(b). Flow velocity of burst flow at different wet heights at 20° angle

Figure 4. Flow velocity of burst under wet conditions with different dip angles and heights

4.2. Influence of different center induction velocity on long-range firing accuracy of artillery

For the long-range attack target, the projectile is usually launched by the curved firing mode with large firing Angle and high muzzle velocity. Due to the large firing angle, fast muzzle velocity and high maximum firing height of the artillery, the projectile will pass through the vortex ring in a very short time, so it will be affected by wet micro downburst in the rising stage of the projectile. In this section, the firing angle of the artillery is set as 45°, the firing position is set on the right side of the central axis of the vortex ring, the crossing point is near the center of the vortex ring, and the environment outside the vortex ring is a windless environment. The remaining parameters are referred to Section 3.1, as shown in Fig.5. The experimental data are shown in Table2.

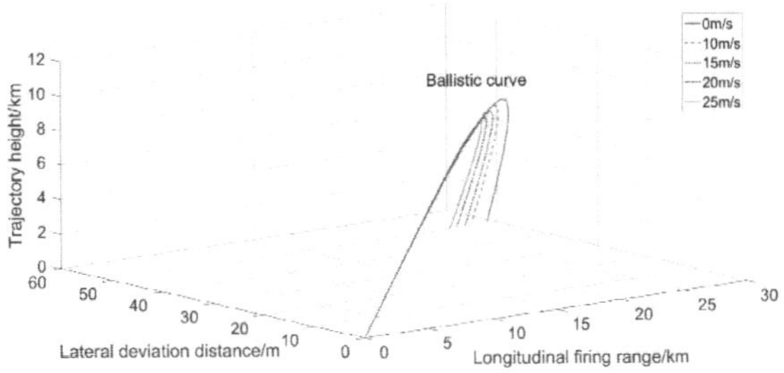

Figure 5. External trajectory diagram of artillery under different induced velocities

Due to the low height of the vortex ring and the high muzzle velocity of the projectile, the passage time through the vortex ring is short and the influence of the micro-downburst is limited. According to the experimental data, compared with the data in no wind environment, the increase of the airflow induced velocity of the central vortex ring has a more obvious impact on the artillery range. Compared with the data in no wind environment, the range decreases by about 2.7% at the induced velocity of 10 m/s, and the range decreases by about 1.3% with the increase of the induced velocity of 5 m/s.

Table 2. Data of influence of different induction speeds on long-range firing accuracy of artillery

Wind speed/(m/s)	Longitudinal firing range/m	Lateral deviation distance/m	Trajectory heighter/m	Placement speed/(m/s)	Yaw Angle of landing/°	Pitch Angle of landing/°
0	29216.55	49.85	10243.88	361.47	0.610	-65.58
10	28417.39	50.18	10170.89	357.65	0.655	-65.82
15	28049.55	50.35	10137.10	357.89	0.715	-65.81
20	27679.07	50.52	10099.60	354.45	0.703	-66.95
25	27308.59	50.68	10062.10	352.91	0.727	-67.53

5. Summarizes

In this paper, based on vortex ring principle and fluid mechanics, the motion model of micro-downburst and raindrop is established, which is fused into a wet micro-downburst model suitable for ballistic analysis and calculation. The influence of the fusion model on the shooting accuracy of a 155 mm artillery is analyzed by introducing the centroid trajectory equation. According to the simulation results, for long-range firing, the range of the artillery is significantly affected by different center induced velocity. At the induced velocity of 10 m/s, the range decreases by about 2.7% compared with that in the wind-free environment, and with the increase of the induced velocity of 5 m/s, the range decreases by about 1.3%, which seriously affects the firing accuracy and killing effect of the artillery.

This research provides a theoretical reference for the artillery firing in the wet micro-burst environment and has positive significance for the application research of the artillery external ballistic. However, in the actual launching process, the meteorological conditions are relatively complex and the coexistence of multiple characteristic airflow is possible. How to integrate this study with other characteristic airflow and analyze its influence on the external ballistic of the artillery, as well as how to correct the deviation caused by the landing point of the artillery in such environment, still needs further research.

References

[1] WU Hanzhou. Low-level wind model building and its influence on trajectory characteristic of projectiles[J]. Journal of Ordance Engineering College, 2015(4): 38-42.

[2] Mukhedkar, R., & Naik, S. Effects of Different Meteorological Standards on Projectile Path. Defence Science Journal, 2013. 63(1), 101-107.

[3] Robinson P A, Reid L D. The modeling of turbulence and downbursts for flight simulators[J]. Journal of Aircraft, 1990, 27(8): 700-707.

[4] FUJIA T T, CARACENA F. An analysis of three weather-related aircraft accidents[J]. Bullentin of the American Meteorological Society, 1977, 58(11): 1164-1181.

[5] FUJIA T T, BYERS R M. Spearhead echo and downbursts in the crash of an airline[J]. Monthly Weather Review, 1977, 105: 129-146.

[6] Anindya Sengupta, Fred L. Haan, Partha P. Sarkar, Vasanth Balaramudu. Transient loads on buildings in microburst and tornado winds[J]. Journal of Wind Engineering & Industrial Aerodynamics, 2008, 96(10).

[7] Dae Kun Kwon, Ahsan Kareem. Towards codification of thunderstorm/downburst using gust front factor: Model-based and data-driven perspectives[J]. Engineering Structures, 2019, 199(C).

[8] Blocken B, Carmeliet J. Spatial and temporal distribution of driving rain on a low-rise building[J]. Wind and Structures, 2002, 5(5): 441-462.

[9] Huang S H, Li Q S. Numerical simulations of wind-driven rain on building envelopes based on Eulerian multiphase mode[J]. Journal of Wind Engineering and Industrial Aerodynamics, 2010, 98(12): 843-857.

[10] Chen Jianwei,Wang Liangming, Li Zijie. Research on Influence of Microburst on Ballistic Characteristics of Rockets[J]. Journal of Physics: Conference Series. 2019(1).

[11] LI Y D, GAO S T, LIU J W. Assessment of Several Moist Adiabatic Processes Associated with Convective Energy Calculation[J]. Advances in Atmospheric Sciences, 2004(06): 941-950.

[12] Nguyen H, Manuel L, Veers P. Simulation of inflow velocity fieldsand wind turbine loads during thunderstorm downbursts[C]||Proc. of the Structural Dynamics and Materials Conference, 2010.

[13] IVAN M. A ring-vortex downburst model for flight simulations[J]. Journal of Aircraft, 2012, 23(3): 232-236.

Thermal fluid lubrication analysis of hydrodynamic journal bearing based on mass conservation boundary conditions

Xinliang Hu*, Kaiping Yu, Junxiong Zhang

Department of Mechanical Engineering, PLA Army Academy of Artillery and Air Defense, Anhui 230000, China.

*Corresponding author e-mail: huxinliang99@yeah.net

Abstract. The thermal-hydrodynamic lubrication analysis of bearing with different bearing speed and eccentricity was carried out under Reynolds boundary conditions and mass conservation boundary conditions, and the variation of performance parameters such as oil film carrying capacity with speed and eccentricity were analyzed in detail for the calculated results, and the calculation results under the two different boundary conditions were compared and analyzed. The results show that the boundary conditions have a greater influence on the lubrication flow calculation results, the Reynolds boundary conditions have a faster calculation process, and the mass conservation boundary conditions are recommended when focusing on the analysis of lubrication flow.

1. Introduction

The hydrodynamic journal bearing has an extremely wide range of applications in various mechanical transmissions, and its working condition is closely related to the lubrication performance. In the literature [1-10], the thermoelastic dynamic lubrication of hydrodynamic journal bearings has been studied in detail, and at present, the analysis of the thermal-hydrodynamic lubrication of journal bearings generally uses Reynolds boundary condition for the joint solution of the generalized Reynolds equation. Compared with the Reynolds boundary condition, the mass conservation boundary condition is closer to the experimental results, and it is more reasonable to use it for lubrication analysis. However, the difference between the results of the two boundary conditions on different working parameters needs to be further studied.

2. Basic equations and formulas

2.1. *Analytical equations for thermal-hydrodynamic lubrication of bearings*

2.1.1. Oil film height equation. T he expression for the height of the oil film for the general case is:

$$h \approx c + e\cos(\theta - \varphi) \tag{1}$$

Where: c is the bearing radius clearance; θ is the angular coordinate measured from the coordinate axis z; ϕ is the eccentricity angle; e is the eccentricity distance.

Content from this work may be used under the terms of the Creative Commons Attribution 3.0 licence. Any further distribution of this work must maintain attribution to the author(s) and the title of the work, journal citation and DOI.
Published under licence by IOP Publishing Ltd

2.1.2. Generalized Reynolds equation with boundary conditions. The generalized Reynolds quation is used for the calculation of lubrication of plain bearings considering thermal effects.

$$\frac{\partial}{\partial\theta}\left(F_2\frac{\partial p}{\partial\theta}\right)+R^2\frac{\partial}{\partial y}\left(F_2\frac{\partial p}{\partial y}\right)=uR\frac{\partial}{\partial\theta}\left(z_m\right) \tag{2}$$

Where: $x = R\theta$, R is the bearing radius, p is the oil film pressure; u is the journal surface linear velocity; $u = R\omega j$, ωj is the journal rotational angular velocity; η is the lubricant viscosity.

$$F_0=\int_0^h\frac{\mathrm{d}z}{\eta};\ F_1=\int_0^h\frac{z\mathrm{d}z}{\eta};\ F_2=\int_0^h\frac{z(z-z_m)}{\eta}\mathrm{d}z;\ z_m=\frac{F_1}{F_0} \tag{3}$$

2.1.3. Lubricant equation of motion. As the journal rotates, it causes a portion of the lubricant to rotate with it, and this portion of the lubricant movement conforms to the following pattern.

$$\frac{\partial p}{\partial\theta}=\frac{\partial}{\partial z}\left(\eta\frac{\partial u}{\partial z}\right);\ \frac{\partial p}{\partial y}=\frac{\partial}{\partial y}\left(\eta\frac{\partial v}{\partial y}\right) \tag{4}$$

Where: u is lubricant circumferential flow rate; v is lubricant axial flow rate.

2.1.4. Energy equation and its boundary condition. If the effect of volume forces and thermal radiation on the bearing operation process, considering the first law of thermodynamics, the common form of the fluid energy equation is :

$$\rho c_f\left(u\frac{\partial T}{R\partial\theta}+v\frac{\partial T}{\partial y}\right)=k_f\frac{\partial^2 T}{\partial z^2}+\eta\left[\left(\frac{\partial u}{\partial z}\right)^2+\left(\frac{\partial v}{\partial z}\right)^2\right] \tag{5}$$

Where: T is oil film temperature; c_f is constant pressure specific heat capacity; k_f is lubricant thermal conductivity; ρ is lubricant density.

2.1.5. Solid heat conduction equations and boundary conditions. On the contact surface between the lubricant and the bearing journal, there is necessarily heat transfer, and the boundary conditions of the energy equation require the heat conduction equation to be determined. The solid heat conduction equation is as follows [11]:

$$\frac{\partial^2 T_b}{\partial r^2}+\frac{1}{r}\frac{\partial T_b}{\partial r}+\frac{\partial^2 T_b}{\partial y^2}+\frac{1}{r^2}\frac{\partial^2 T_b}{\partial\theta^2}=0 \tag{6}$$

Where: T_b is shaft shingle temperature.

2.1.6. Lubricant viscosity temperature relationship equation. Many research results have been published on lubricant viscosity-temperature relationship, and the lubricant used in this paper is CD30 diesel engine oil, and the expression of viscosity-temperature relationship is as follows[12]:

$$\eta=0.05076\times10^3\exp[3434.6/(T-22.29)] \tag{7}$$

Where: T_b is shaft shingle temperature.

2.1.7. Bearing performance parameters calculation formula. Jointly calculate the generalized Reynolds equation, the oil film pressure in the circumferential and axial direction of the bearing will be integrated, the whole oil film bearing capacity is:

$$F=\sqrt{F_x^2+F_y^2} \tag{8}$$

The formula for calculating the bearing end drainage rate is:

$$Q_1=\int_0^{2\pi}\frac{h^3}{12\eta}\frac{\partial p}{\partial y}_{|y=0}R\mathrm{d}\theta;\ Q_2=-\int_0^{2\pi}\frac{h^3}{12\eta}\frac{\partial p}{\partial y}_{|y=L}R\mathrm{d}\theta \tag{9}$$

Adding the drainage flows at both ends to the total flow: $Q=Q_1+Q_2$.

The friction force on the surface of the journal is:

$$F_j=\int_0^L\int_0^{2\pi}\left(\frac{h}{2}\frac{\partial p}{R\partial\theta}+\frac{\eta u}{h}\right)Rd\theta dy \tag{10}$$

2.2. Mass conservation algorithm.

The calculation method proposed by Elrod and other methods are based on the same assumption that the fluid in the cavity area flows in strips and the lubricant is in contact with the upper and lower surfaces of the gap, and that the pressure is constant. The calculation method proposed by Payvar is faster. This method requires the assumption of the variable ϕ and the switching function g. The expression of the equation is:

$$\frac{p-p_c}{p_s-p_c}=g\phi;\ \frac{\rho}{\rho_c}=1+(1-g)\phi;\ g=\begin{cases}1 & (\phi>0)\\ 0 & (\phi\le 0)\end{cases} \tag{11}$$

where p_s represents the reference pressure, p_c represents the cavity pressure, and the rest of the variables are the same as above.

The complete equation in the whole oil film calculation domain is:

$$\frac{\partial}{\partial\theta}\left(F_2\frac{\partial g\phi}{\partial\theta}\right)+R^2\frac{\partial}{\partial y}\left(F_2\frac{\partial g\phi}{\partial y}\right)=\frac{Ru}{p_s-p_c}\frac{\partial}{\partial\theta}\left(z_m\left[1+(1-g)\phi\right]\right) \tag{12}$$

2.3. Calculation process.

The lubrication calculation flow is shown in the figure.

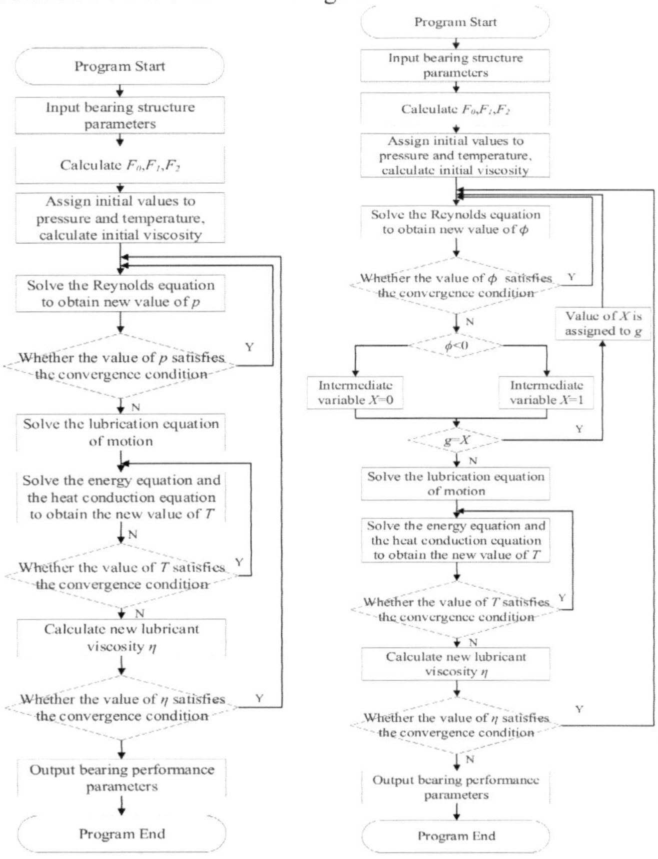

Fig. 1 The lubrication calculation flow

3. Analysis of calculation results

Tab. 1 Calculation parameter table

Bearing width	L(mm)	66
Inner radius of bearing bush	R_2(mm)	30
Outer radius of bearing bush	R_b(mm)	32.5
Radius Gap	c(mm)	0.03
Journal rotation speed	N (r/min)	2000/3000/4000
Lubricant thermal conductivity	k_f(W/m•°C)	0.13
Inlet oil temperature	$T_{in}(K)$	353
Lubricant Density	ρ (kg/m^2)	860
Ambient temperature	T_a(K)	313
Bearing thermal conductivity	k_b(W/m•°C)	250
Journal thermal conductivity	k_j(W/m•°C)	50
Bearing exterior heat transfer coefficient	h_{ab}(W/m•°C)	80

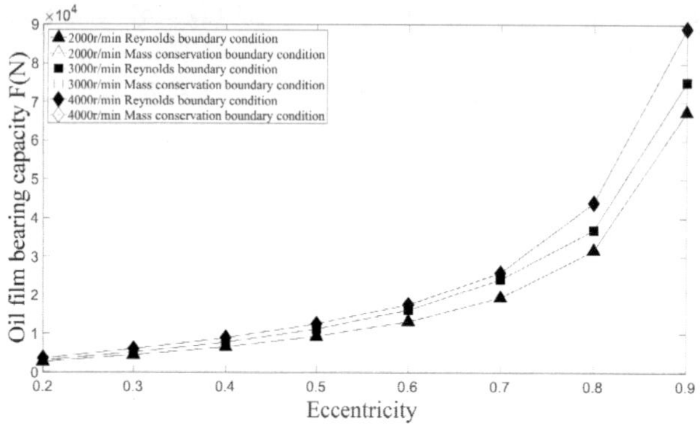

Fig. 2 Bearing capacity of the oil film varying with the bearing speed and eccentricity under different boundary conditions

When the bearing eccentricity is less than 0.6, the oil film bearing capacity calculated by the mass conservation algorithm is larger than the result calculated by the Reynolds boundary condition, and along with the increase of eccentricity, the two calculation results gradually tend to be the same. Since some areas of the oil film are affected by negative pressure when the eccentricity is small, these negative pressure areas will also have a certain influence on the bearing load capacity, and the positive pressure of the oil film will increase with the increase of eccentricity, but the negative pressure will not change significantly. So along with the increase in speed, the gap between the results of the two boundary conditions also gradually decreases. When the eccentricity is equal to 0.9, the difference between the oil film bearing capacity results of the two cases is extremely small.

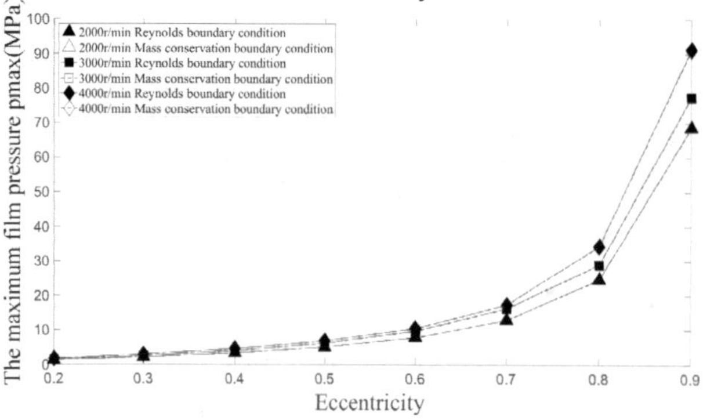

Fig. 3 The maximum film pressure varying with the bearing speed and eccentricity under different boundary conditions

The trends of the oil film bearing capacity and the maximum oil film pressure for the two boundary conditions differ for the same working parameters. When the value of eccentricity is moderate or small, the maximum oil film pressure under the mass conservation boundary condition is smaller than the value under the Reynolds boundary condition, but the bearing load carrying capacity is the opposite. Similarly, the effect of the negative pressure region is the main reason for this result. When the value of eccentricity is large, the effect of the change of boundary conditions on the oil film bearing capacity and maximum oil film pressure is not the same for some working parameters, while the trend is the same for another part of the working conditions.

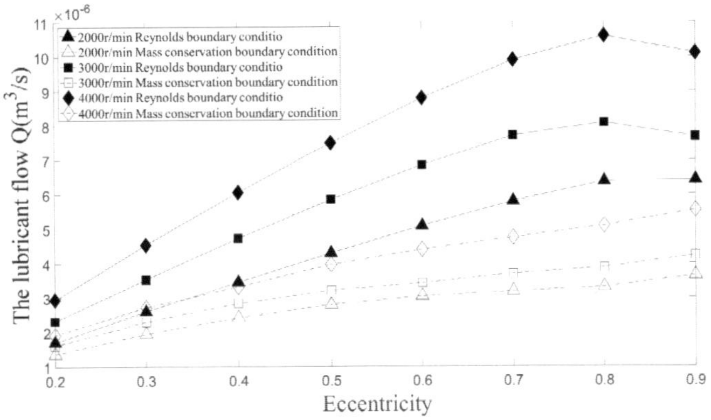

Fig. 4 The lubricant flow varying with the bearing speed and eccentricity under different boundary conditions

When the eccentricity value is below 0.8, the lubricant flow rate at the Reynolds boundary condition will gradually increase as the bearing eccentricity increases; when the bearing eccentricity value is above 0.8, the lubricant flow rate at the Reynolds boundary condition decreases as the eccentricity increases. When the bearing eccentricity value reaches 0.9, the overall range of the main bearing area of the oil film becomes smaller, and the flow rate in all areas of the bearing is affected.

Under the mass conservation boundary condition, the rate of increase is greater after the eccentricity exceeds 0.8. Comparing the corresponding values of Reynolds boundary conditions, the lubricant flows out of both ends of the bearing decreases when the eccentricity value is below 0.8; after the eccentricity value exceeds 0.8, the lubricant flow increases rapidly.

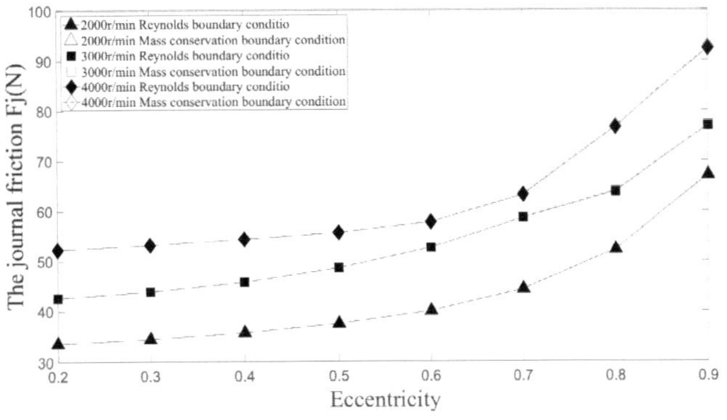

Fig. 5 The journal friction varying with the bearing speed and eccentricity under different boundary conditions

The effect of the change of bearing speed and eccentricity on the journal surface friction in the two cases can be seen in the figure, and there is no significant difference between the calculated results in the different calculation cases. As the speed increases, the journal surface friction increases, and the difference between the calculated and analyzed results under the two boundary conditions increases slightly; as the eccentricity increases, the journal surface friction also increases. When the bearing eccentricity is less than 0.7, the growth rate of journal surface friction is low, while when the bearing eccentricity is greater than 0.7, the speed of journal surface friction becomes faster.

4. Conclusion

(1) Bearing oil film pressure distribution with the increase of eccentricity, the calculation results of the two boundary conditions become closer and closer. In the case of small eccentricity and low speed, the oil film is likely to rupture completely; and due to the increase of bearing eccentricity, the boundary of the formation of the positive pressure region of the oil film will be shifted toward the oil film rupture region.

(2) When the eccentricity value is small, the bearing load capacity under the Reynolds boundary condition is smaller than that under the mass conservation boundary condition, which is due to the influence of the negative pressure region of oil film; when the eccentricity value is larger, the calculation results under the two different boundary conditions are gradually close to each other. And no matter what boundary conditions are used, the increase in bearing speed and eccentricity always leads to an increase in bearing load capacity value.

(3) The influence of the boundary conditions on the maximum oil film pressure of the bearing is different from the bearing load capacity, that is, when the eccentricity is not large, the maximum oil film pressure calculated by the mass conservation boundary conditions is small, and when the eccentricity is large, the results of the two boundary conditions are very close to each other. Similarly, when the bearing speed and eccentricity increase, the maximum oil film pressure under the two boundary conditions are becoming larger.

(4) The calculation results of bearing lubricant flow rate under two boundary conditions are different, and the trend of the results with the speed and eccentricity is also different. When the eccentricity is small, the increase of eccentricity will cause the lubricant flow rate to increase, at this time, the calculation result of the mass conservation boundary condition has a small value and the growth rate is slow. When the eccentricity value exceeds 0.8, the lubricant flow rate of the bearing under the Reynolds boundary condition starts to decrease instead, and the lubricant flow rate under the mass conservation boundary condition starts to increase instead. No matter what boundary conditions are used, the increase in bearing speed always increases the flow of bearing lubricant.

(5) The influence of the two boundary conditions on the friction force of the journal surface is very small, and the difference between the two values is negligible. No matter what boundary conditions are used, the increase in bearing speed and eccentricity will always lead to the increase of journal surface friction.

References

[1] Deng Mei, Sun Jun, Fu Yonghong, et al. Thermoelastohydrodynamic lubrication analysis of bearing considering shaft deformation and surface roughness[J]. Journal of Mechanical Engineering,2010,46(15):95-101.

[2] Dowson D, March C. A thermohydrodynamic analysis of journal bearings[J]. In: Proc. Inst. Mech. Eng, 1967, 181.

[3] Fesanghary M, Khonsari M M. A Modification of the Switch Function in the Elrod Cavitation Algorithm[J]. Journal of Tribology, 2011, 133: 024501(1-4).

[4] Almqvist A, Fabricius J, Larsson R, et al. A New Approach for Studying Cavitation in Lubrication[J]. Journal of Tribology, 2014, 136: 011706(1-6).

[5] Performance of hydrodynamic journal bearing under the combined influence of textured surface and journal misalignment: A numerical survey[J].Comptes Rendus Mécanique Volume 347, Issue 2, February 2019, Pages 141-165.

[6] Iseli Elia, Schiffmann Jürg Experimental and numerical investigation of the unbalance behavior of rigid rotors supported by spiral-grooved gas journal bearings[J] Mechanical Systems and Signal Processing, 2022, 174.

[7] Zhao Chen, Sun Jianliang, Lin Shuilin et al. Rolling mill bearings fault diagnosis based on improved multivariate variational mode decomposition and multivariate composite multiscale weighted permutation entropy[J] Measurement, 2022, 195.
[8] Jiaojiao Ma, Hao Zhang, Shan Lou et al. Analytical and experimental investigation of vibration characteristics induced by tribofilm-asperity interactions in hydrodynamic journal bearings[J] Mechanical Systems and Signal Processing, 2021, 150.
[9] H. Questa, M. Mohammadpour, S. Theodossiades et al. Tribo-dynamic analysis of high-speed roller bearings for electrified vehicle powertrains[J] Tribology International, 2021, 154.
[10] Zhang Chuanbing, Ao Hongrui, Jiang Hongyuan et al. Investigations on start-up performances of novel hybrid metal rubber-bump foil bearings[J] Tribology International, 2021, 154.
[11] Adams D R. Design and analysis: A perspective for the future[J]. In: Proc. 17th Leed-Lyon Symposium on Tribology, 1990: 7-15.
[12] Wen Shizhu, Huang Ping Principles of Tribology (4th Edition) [M] Beijing: Tsinghua University Press, 2012: 9-10

Analysis of Dynamic Performance of Synchronous Double Proportional Valve Based on Digital Simulation

Biaohua Cai [1,*], Maosen Zhang [2], Qiuxu Hu [1], Zheng Zhang [1], Zhaocun Shi [1]

[1] Wuhan No.2 Institute of Ship Design, Wuhan 430061, Hubei, China
[2] Hefei General Machinery Research Institute, Hefei 230031, Anhui, China

[*] Corresponding author e-mail: jxs204204204@163.com

Abstract. The dynamic characteristic parameters of the synchronous dual-proportional pressure integrated valve are of great significance to its research. Among them, the outlet pressure tracking error and the flow resistance characteristics of the internal flow field have a great influence on the dynamic characteristics of the valve. The tracking error of the valve under specific working conditions is analyzed by means of digital simulation, and the flow resistance characteristics of the internal flow field are further obtained.

Key words: Flow resistance characteristics, Synchronous proportional valve, AMEsim, virtual test, Tracking error.

1. Introduction

The dynamic characteristics of proportional pressure integrated valve are very complex. To study its dynamic characteristics, it is necessary to build an expensive and costly test platform, and it is not only a huge economic cost to build a realistic test platform. At the same time, with the huge time cost, it is very difficult to study the dynamic characteristics of the integrated valve. Virtual simulation can save the cost of human and material, and adapt to the uncertainty of engineering because of manufacturing error and the change of using condition1-5. Based on AMEsim, the pressure tracking error of the valve is simulated, and the relationship between the tracking error and the frequency of the input signal is obtained.

2. Virtual test of dynamic characteristics of synchronous dual-ratio pressure integrated valve

2.1. *Multi-area model of valve*

AMESIM is an advanced modeling and simulation platform for engineering systems. It contains components that transmit data in both directions, where all variables are given physical meaning. In order to get the dynamic characteristics of the valve system, the model of the double-proportional pressure integrated valve, as shown in figure 1, was established based on AMEsim, and used for the virtual test of the dynamic characteristics of the valve.

Content from this work may be used under the terms of the Creative Commons Attribution 3.0 licence. Any further distribution of this work must maintain attribution to the author(s) and the title of the work, journal citation and DOI.
Published under licence by IOP Publishing Ltd

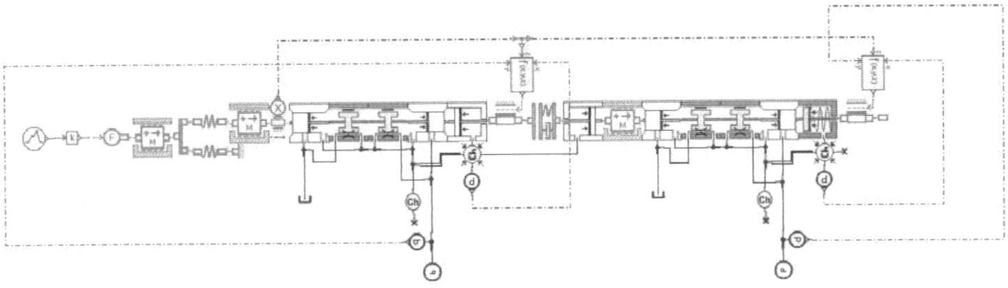

Fig. 1 Multi-area model of synchronous dual-ratio pressure integrated valve

2.2. Virtual test results

According to the actual working conditions of the proportional pressure integrated valve, two typical test conditions are considered, and the synchronization characteristics and tracking error of the proportional pressure integrated valve under the typical test conditions are studied.

Virtual test condition 1: The sine of 6 MPA instruction is input from K port, and the tracking of sinusoids by a1 and a2 ports is observed. When the sinusoidal amplitude of a1 and a2 ports attenuates to 70.7% (-3 db) of the theoretical value, this frequency is the response frequency of the valve. As shown in figure 2, the response frequency of the valve is about 120 Hz (-3 db).

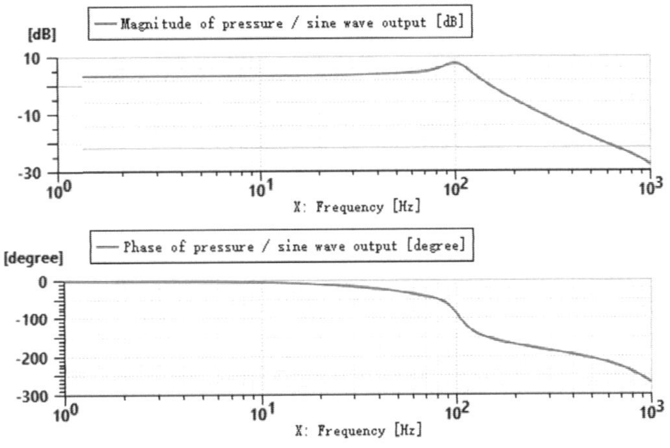

Fig. 2 Pressure response diagram

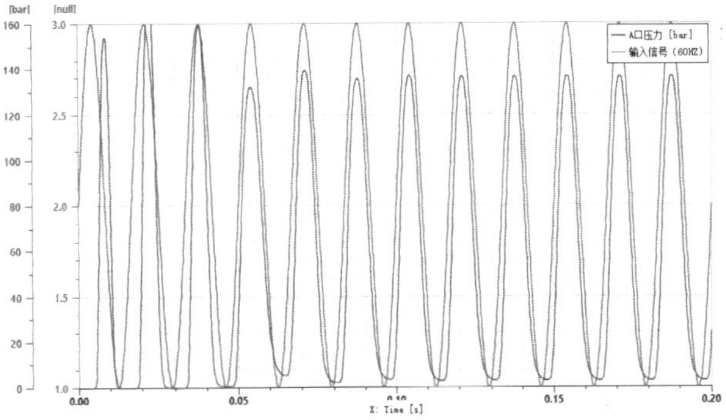

Fig. 3 60HZ command signal amplification and tracking

Journal of Physics: Conference Series **2441** (2023) 012067 doi:10.1088/1742-6596/2441/1/012067

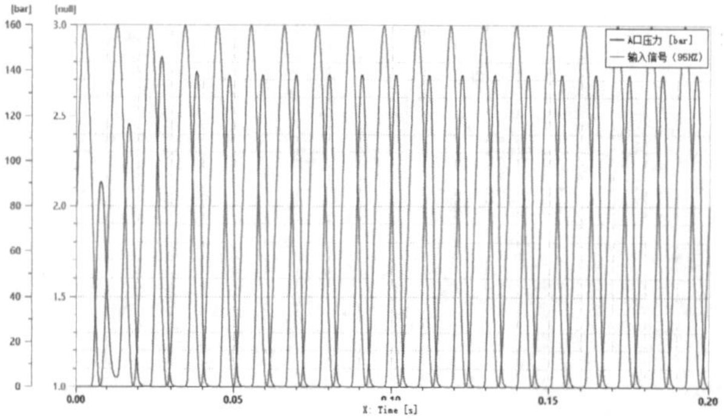

Fig. 4 95HZ command signal amplification and tracking

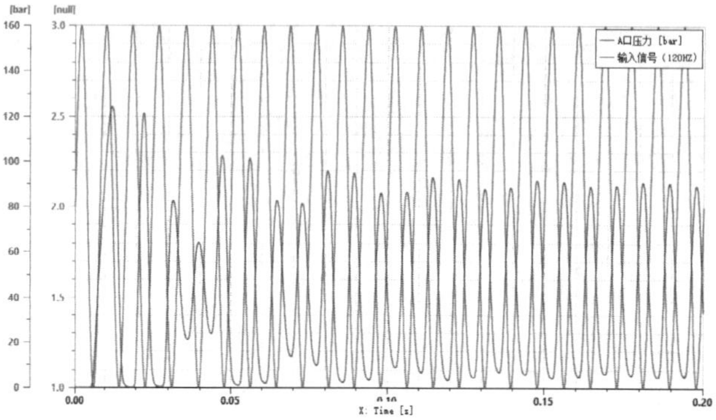

Fig. 5 120HZ command signal amplification and tracking

According to the virtual test results, a-port pressure tracking is better when the input command is at 60 Hz and 95 Hz. But after 120HZ instruction input, a port pressure tracking and instruction input difference of a phase.

Virtual test condition 2: The command pressure of 0→2→4→6→4→2→0 MPa is input from K port. P port pressure is 20MPa. The pressure synchronism error of a1 and a2 was analyzed by observing the pressure tracking of a1 and a2.

The results of virtual experiment show that the pressure tracking of a1 and a2 is good, but there is some tracking error. The pressure synchronization error of a1 and a2 is less than 0.035 MPa.

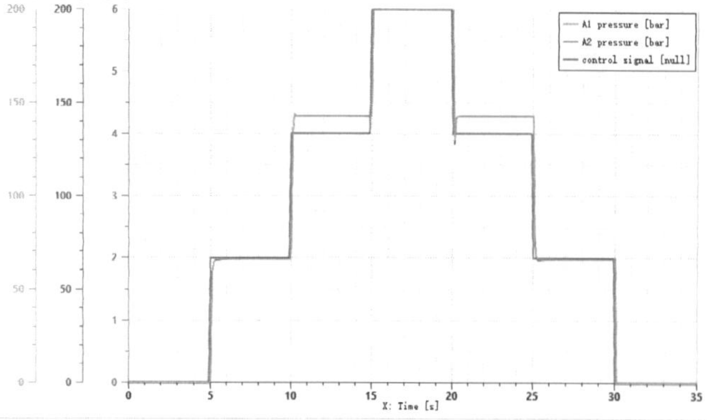

Fig. 6 Step pressure response curve

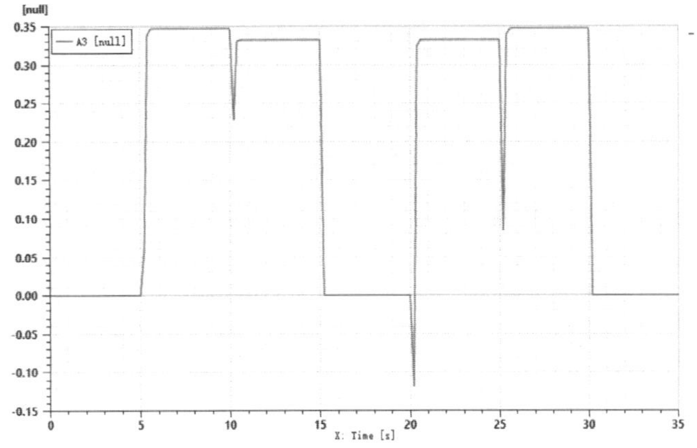

Fig. 7 Pressure synchronization error curve

3. Virtual test for strength/stiffness of main valve construction

3.1. Model

The applied pressure of the virtual test of the pressure resistance of the main valve is shown in figure 8. The static pressure of 35MPa is applied to the internal flow passage of the main valve sleeve and the main valve core.

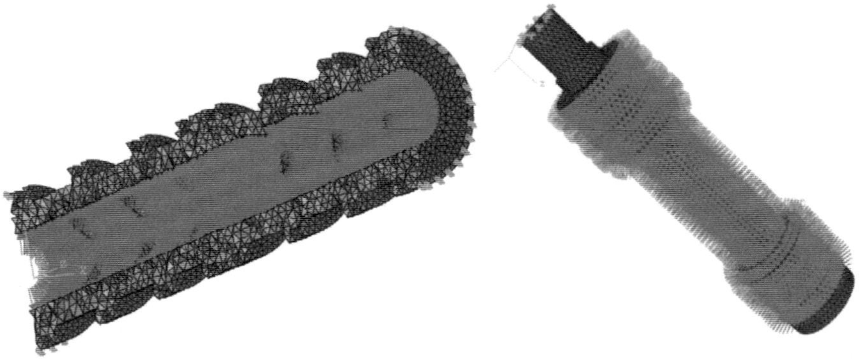

(a) main valve sleeve (b) main spool

Fig. 8 Pressure applied diagram

3.2. Virtual test results

The pressure displacement cloud of the main valve sleeve is shown in figure 9. The maximum deformation is 0.0048 mm, located near the valve port.

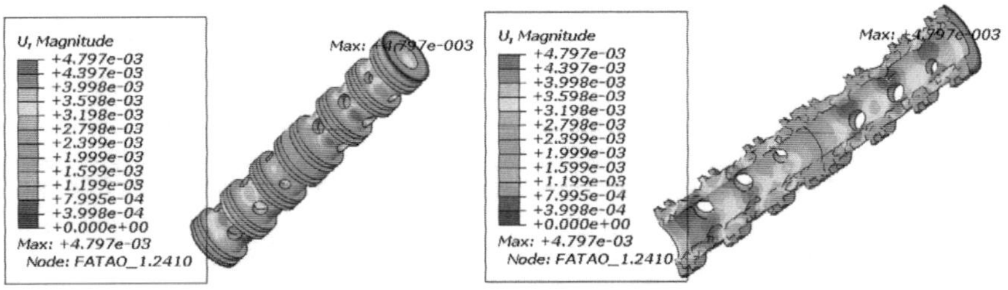

Fig. 9 Main valve sleeve deformation cloud diagram

The stress nephogram of the main valve sleeve is shown in figure 10. The maximum stress is 162.9MPa, and the local maximum stress appears at the valve port.

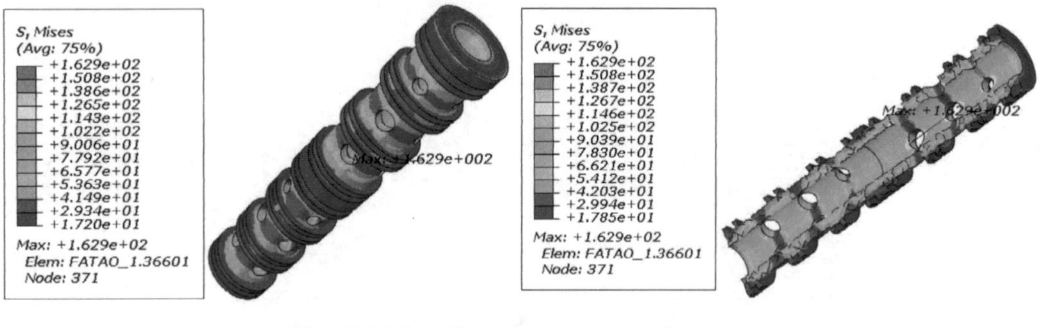

Fig. 10 Main valve sleeve stress nephogram

The stress and deformation cloud diagram of the main valve core is shown in figure 11. The maximum deformation is 0.0178 mm, located near the large port. The maximum stress is 106.6MPa, and the local maximum stress appears at the change of cross section.

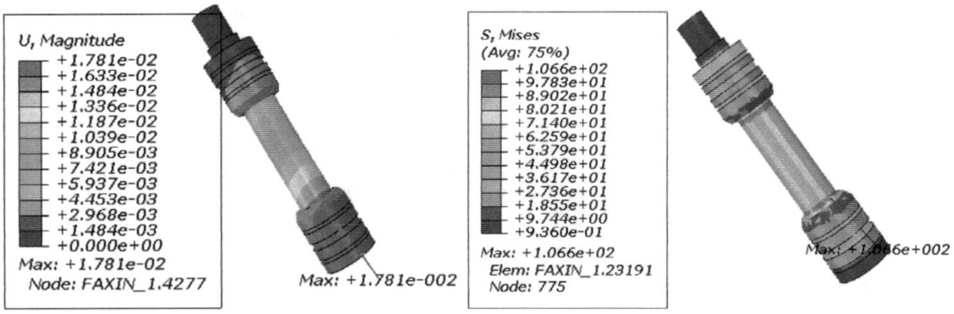

(a) Deformation (b) Stress

Fig. 11 Main valve core deformation and pressure cloud diagram

4. Virtual test of flow field characteristics in valves

The flow resistance and other external characteristics of the valve are determined by the distribution of its internal flow field. Under the given boundary conditions of the virtual test bed, the internal flow field of the valve is analyzed based on the three-dimensional CFD method, and the flow resistance characteristics and other parameters of the valve can be obtained.

4.1. Establishment of internal flow field

The boundary conditions are determined according to the simulation model and operating conditions, which are divided into structural boundary conditions and fluid boundary conditions. P and A ports are connected. P is pressure inlet that is 20MPa. A is the pressure outlet that is 10MPa. B and T ports are connected. B is pressure inlet that is 10MPa. T is the pressure outlet that is 0MPa. The others are fixed wall boundary conditions. The boundary conditions and finite element model are shown in figure 12.

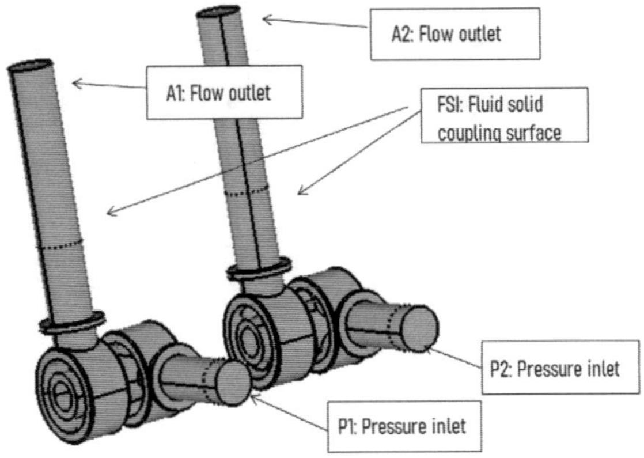

Fig. 12 Boundary Condition

4.2. Virtual test results

The figure 13 shows the simulation results of the physical parameters of the flow field in the valve.

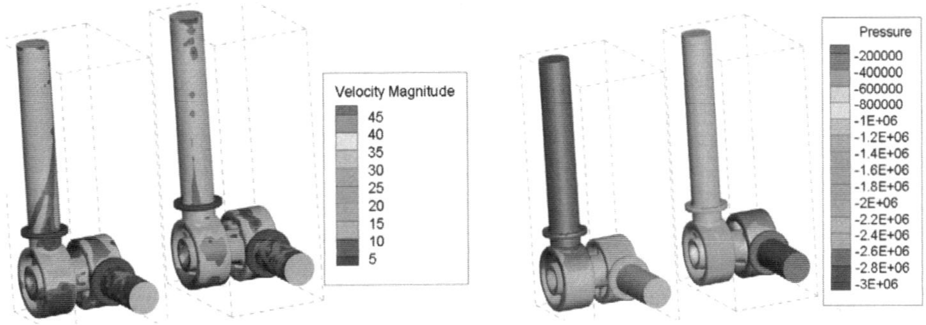

(a) Velocity nephogram (b) Pressure nephogram

Fig. 13 Flow rate and pressure distribution cloud chart

The characteristic curves of valve flow resistance as shown in figure 14 are obtained by virtual test under different conditions.

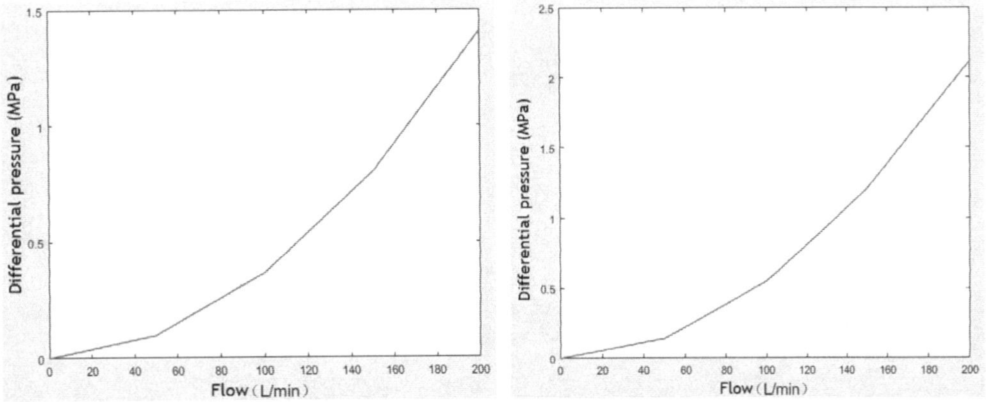

(a) A1 (b) A2

Fig. 14 Flow resistance characteristic curve

5. Conclusion

In this paper, the dynamic characteristics of synchronous dual-ratio pressure integrated valve are studied by means of virtual test method based on AMEsim. The tracking error of the output pressure to the input signal is obtained by simulation. At the same time, the finite element analysis and internal flow field analysis of the main valve are carried out.

References

[1] LI Ruo-yu; WANG Yan-xu; CHEN Juan. Study on pedestrian route choice behavior in multiplayer online virtual evacuation experiments [J]. Journal of Safety and Environment, 2022: 1-10.

[2] Leng Zhenhua. Research on Random Vibration Test Development of Electric Vehicle Power Battery Pack based on Virtual Test Field [J]. Auto Time, 2022, (9): 169-171.

[3] Cai Changpeng;Xue Baiwen;Yang Zhen. Virtual Test Method of Certain Anti-aircraft Gun Weapon [J]. Ordnance Industry Automation, 2022, 41(4): 76-80.

[4] WANG Zhiliang;LIU Zhimin;HUA Liguang, et al. Study on Operating Characteristics of Dry Rolling Granulator Based on Electro-Hydraulic Proportional Valve [J]. Machine Tool & Hydraulics, 2022, 50(13): 162-167.

[5] WANG Cao. Research on Wheel Six-component Force Extraction Method Based on Virtual Proving Ground [D]. Chongqing University of Technology, 2022.

Bubble formation by shearing-to-squeezing transition in a T-junction

Yu Han*

School of Chemical Engineering, Dalian University of Technology, Dalian, China.

*Corresponding author e-mail: 2292135325@dlut.edu.cn

Abstract. A shearing-to-squeezing transition process for bubble formation in a T-junction was reported using microscopy and digital image technology. The investigations were conducted in a polydimethylsiloxane (PDMS) T-junction. According to the aspect ratio of air slug and bubble generation frequency, the bubbles generated in the microchannel were identified to three types, namely dispersed bubble, short-slug bubble and long-slug bubble. The corresponding modes of bubble formation were identified as shearing, transition and squeezing modes, respectively. Additionally, the microbubble size and generation frequency with various gas pressures and velocity of liquid flows in each mode were investigated. The relevant dimensional arguments were performed to characterize the microbubble formation process in the T-junction. A new scaling law, based on the formation mechanisms, was proposed to predict the size of the microbubbles and showed a good agreement with the experimental results.

1. Introduction

The awareness of physical phenomena on microscopic scale in fluid mechanics has been strongly improved due to the development of microfluidic devices [1-3]. Because of the scale miniaturization of the microfluidic devices, the flow process allows the fluids to have a high specific surface area [4], which enhances mass transfer and heat transfer efficiency. Microfluidic devices, therefore, have drawn much attention in biomedicine and chemistry field [5, 6]. Here is more attention paid to the two-phase flow in microfluidic devices for the generation of microbubbles or micro-droplets. The slug flow is quite popular because of its controllability to generate uniform and monodisperse microbubbles or micro-droplets [7, 8]. Therefore, it is of great importance to take the flow patterns and the microbubble hydrodynamics into consideration.

Several microfluidic geometries have been developed to obtain monodisperse microbubbles or micro-droplets, including T-shaped junctions [9, 10], flow-focusing configurations [11] and co-flowing geometries [12]. The underlying bubble or droplet formation mechanisms in these configurations have been investigated in recent studies [13-15]. The squeezing-to-dripping mode was originally proposed to explain the droplet formation process in the T-shaped microchannels [16]. Fu et al. [17] introduced these modes to describe the formation of microbubbles in the T-shaped junction. In the squeezing mode, the shearing force caused by the liquid is much smaller than surface tension force. Therefore, it is not sufficient to make the bubble escape. In the dripping mode, the gas phase thread only occupies small flow area of the channel, and the liquid-induced shearing force is sufficient to pinch off the gas phase when the surface tension and shearing force reach equilibrium. In the transition mode, the gas phase

Content from this work may be used under the terms of the Creative Commons Attribution 3.0 licence. Any further distribution of this work must maintain attribution to the author(s) and the title of the work, journal citation and DOI.
Published under licence by IOP Publishing Ltd

occupies large cross-section of the channel but it does not touch the wall in whole formation process. The latter is controlled simultaneously by both the pressure and shearing force. Until now, both the production of micro-droplets and microbubbles through T-junction microchannels have mainly adopted the feeding mode of continuous phase entering from the main channel and dispersed phase from the branch. The gas-liquid two-phase flow pattern, the formation mechanism and the characteristics of the microbubble in the T-shaped junctions will inevitably be affected by the exchange of inlet positions of liquid and gas phases.

The characteristics of microbubbles have a significant impact on the applications. Actually, gas and liquid velocities, gas and liquid properties, and geometries of microchannels were treated as the main factors to influence the bubble characteristics. Parhizkar et al. [18] investigated the influence of the gas pressure, the velocity of liquid flow, the ratio of velocity of gas-liquid flow and the liquid viscosity on the microbubble size in a T-shaped junction. Dhanaliwala et al. [19] investigated the relationship among the size, the production rate of microbubbles and the velocity of liquid flow to demonstrate the advancement of a novel flow-focusing microreactor. Evangelio et al. [20] demonstrated that the liquid pressure gradient played a critical role in the formation of monodisperse-sized microbubbles and provided the equations to express both the generation frequency and diameter of microbubbles in a flow-focusing junction. Li et al. [21] proposed a device with a convenient method to control bubble sizes. Several researchers have also investigated the bubble characteristics produced in microfluidic junctions and made the prediction of the microbubble characteristics more accurate [22] although the forces involved in the formation process were hardly taken into account.

As described above, although the formation mechanisms of microbubbles in T-shaped junctions with the traditional gas and liquid inlet mode were widely introduced in the previous studies, it was not proper to simply apply these mechanisms to explain the formation process of microbubbles if the feeding positions of gas-liquid phase were exchanged. In this study, an experimental study of the microbubble formation process and the microbubble generation characteristics in the new input mode was carried out in a T-junction microchannel. Based on the force analysis, the effects of gas pressure and velocity of liquid flow on the characteristics of the bubbles generated in the T-junction were studied. Dimensional arguments were proposed to signify the microbubble formation process and a correlation based on the force analysis was formulated to predict the microbubble size.

2. Materials and Methods

The gas phase was air. The liquid phase was the sodium alginate (Sigma Aldrich, U.K.) aqueous solution at the concentration of 0.1%. Tween 20 (Damao Chemical Factory, China) at the concentration of 0.1% was the surfactant. The solution density in the experiments is 999 $kg \cdot m^{-3}$, while the solution viscosity is 1.58 mPa·s and surface tension is 34.17 mN·m.

The diagram of experiments is given in Fig. 1a. It includes three main parts: the microfluidic sample pump, the T-junction microchannel and the microscopy image system. The microfluidic pump (Fluigent Co., France) consists of two pressure-driven pumps, one flow measurement module and two flow reservoirs. The dimensions of the T-shaped microchannel (Wen Hao Technology Co., China) are 150 μm in width and 50 μm in depth. The XSZ microscope (Shangxin Optical Technology Co., China) and FASTCAM SA4 camera (PHOTRON Co., Japan) was applied to capture the gas-liquid interface temporal evolution pictures, and microscope with YM1600 digital camera was used to collect the microbubbles images. The feeding positions of continuous liquid phase and dispersed gas phase in the previous studies [17, 23, 24] and this study are displayed in Fig. 1b. In this study, the gas phase entered from the main channel and the liquid phase entered from the branch channel.

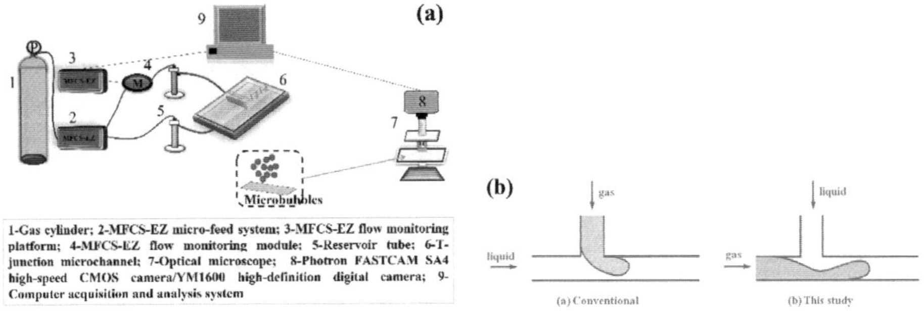

Figure 1. (a) The experimental setup; (b) Positions of gas/liquid input.

3. Results and discussion

3.1. Classification of gas and liquid bubbly flow

Fig. 2 shows the bubble generation procedure. The bubble formation process, as shown in Fig. 2(a.1-a.7), could be divided into two stages: (i) the gas phase penetrated in the liquid phase and a gaseous thread began to expand in the axial direction of the main channel until a neck appeared, and (ii) the liquid flow from the lateral channel pinched off the gaseous thread in the main channel and entrained the bubble downstream. The axial length and the radial length of the gas slug were defined as W_d and W, respectively (Fig. 2(a.7)). Thus, the gas slug aspect ratio in the channel could be expressed as

$$\gamma = Wd/W \tag{1}$$

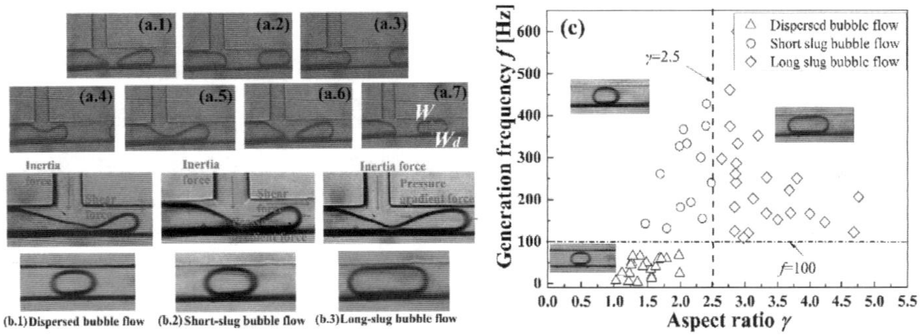

Figure 2. (a) Images of the temporal evolution of the bubble generation process at air pressure P_g= 60 kPa and velocity of liquid flow Q_L=71 μL/min; (b) Patterns of bubbles formed in the T-junction; (c) Bubbly flow under different generation frequency f and aspect ratio γ.

Fig. 2b shows the patterns of gas-slug generated in the T-junction. Fig. 2c gives the classification of bubbly flow generated in the T-junction under different operating conditions according to the bubble generation frequency f and the aspect ratio γ. As shown in Fig. 2(b-c), the bubble generated in the T-junction could be divided into three types based on the bubble generation frequency and aspect ratio, including the dispersed (f<100 Hz, γ <2.5), the short-slug (f>100 Hz, γ <2.5) and the long-slug bubbles (f>100 Hz, γ > 2.5). The upper images in Fig. 2(b) showed the interface of gas-liquid two-phase during the generation of the dispersed bubble, the short-slug bubble and the long-slug bubble, respectively. Fu et al. [17] pointed out that the shearing force and the pressure gradient force of the liquid phase were considered to be the main forces for pinching off the gas slug in the traditional feeding mode. Whereas in the feeding mode of this study, the inertia force also played a crucial role in the bubble formation process. The main forces acting on the gas phase consisted of the inertia force, the shearing force, the pressure gradient existing between up- and downstream, and the surface tension force. Generally, the inertia, pressure and shearing forces acted on the gas phase and drove the gas downstream while the surface tension force resisted the detachment of the bubbles at the junction. The detailed mechanisms were explained as follows:

(i) Dispersed bubbles were generated in the shearing mode at small air pressures or large velocity of liquid flow (Fig. 2(b.1)). The bubbles occupied partial cross-section of the microchannel before the pinching off, and the inertia force and the shearing force acted as the main forces for the bubble breaking-up.

(ii) Long-slug bubbles were obtained in the squeezing mode under large air pressures or small velocity of liquid flow (Fig. 2(b.3)). The front end of the bubbles almost filled the entire channel, and the inertia force and the pressure gradient existing between up- and downstream acted as the main driving force for the bubble squeezing (the effect of shearing force could be ignored).

(iii) Short-slug bubbles were produced in the transition mode under intermediate air pressures and velocity of liquid flow (Fig. 2(b.2)). The bubbles occupied the majority of the cross-section, and the shearing force, the inertia force and the pressure gradient between up- and downstream liquid phases in the bubble formation process shared almost equal importance.

3.2. General scaling law in the shearing-to-squeezing transition

Fig. 3(a-b) illustrates the effect of air pressure on the generation frequency f and average bubble size D at different velocity of liquid flow. As shown in the Fig. 3(a-b), the average generation frequency rose above 100 Hz when the air pressure rose above 35 kPa, and the diameter of the microbubble collected on the slide glass was 140 μm while the aspect ratio γ reached 2.5. No matter in which mode, both the generation frequency and microbubble size increased with increasing air pressure at the same velocity of liquid flow due to the increase of pressure gradient force and shearing force.

Fig. 3(c-d) shows the influence of velocity of liquid flow on the bubble generation frequency f and average bubble size D at different air pressures. As shown in Fig. 3(c-d), the generation frequency increased first and then decreased while the microbubble size decreased with the increase of velocity of liquid flow at the same air pressure in shearing mode (P_g=20 kPa, 30 kPa). The generation frequency increased while the microbubble size decreased with the increase of velocity of liquid flow in the squeezing mode (the rising stage of the curves in Fig. 3(c) for the case of P_g=35 kPa, 38 kPa, 40 kPa, 50 kPa), while both the generation frequency and the microbubble size decreased as the velocity of liquid flow increased in the transition mode (the declining stage of the curves in Fig. 3(c) for the case of P_g=35 kPa, 38 kPa, 40 kPa, 50 kPa). The vertex of the curve in Fig. 3(c) referred to the transition point of squeezing mode and transition mode.

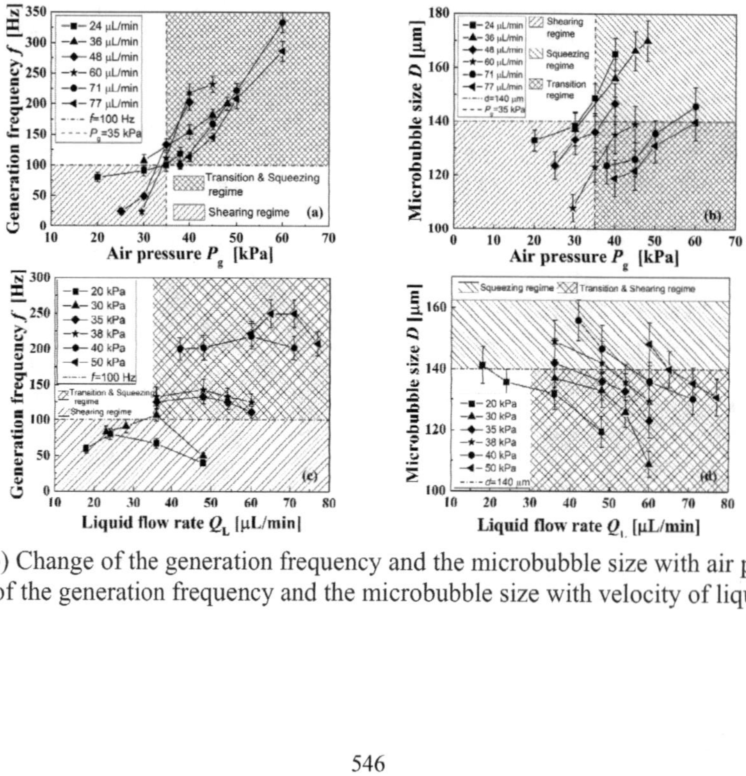

Figure 3. (a-b) Change of the generation frequency and the microbubble size with air pressure; (c-d) Change of the generation frequency and the microbubble size with velocity of liquid flow.

Rodriguez-Rodriguez et al. [24] pointed out that pressure gradients of gas could be ignored and the pressure in the microbubble was approximately uniform during the microbubble formation process. Therefore, the generation process of the bubble in the microchannel could be qualitatively described by the Rayleigh-Plesset equation in the following form:

$$\rho_l\left(RR'' + \frac{3}{2}R'^2\right) = \left(P_g - P_l\right) - 4\mu_l\frac{R'}{R} - \frac{2\sigma}{R} \tag{2}$$

where ρ_l is the liquid density, μ_l is the liquid viscosity, σ is the surface tension. An expression for the computation of bubble diameter D could be written as follows

$$D = \mathrm{F}\left(P_g, P_l, \mu_l, \sigma, \rho_l\right) \tag{3}$$

Since P_l relates to the liquid velocity v_l, eq. (2) could also be modified as

$$D = \mathrm{F}\left(P_g, v_l, \mu_l, \sigma, \rho_l\right) \tag{4}$$

Eq. (4) could become a relation between three dimensionless numbers fully describing the formation mechanism of bubbles using dimensional arguments,

$$\mathrm{Oh} = \Re\left(\mathrm{Ca}, \mathrm{Eu}\right) \tag{5}$$

where $\mathrm{Oh}=\mu_l/(\rho_l\cdot\sigma\cdot D)^{0.5}$, $\mathrm{Ca}=(\mu_l\cdot v_l)/\sigma$, $\mathrm{Eu}=P_g/(\rho_l\cdot v_l^2)$. Referring to the force analysis, dispersed flow, short-slug flow and long-slug flow are dominated by the inertia force, the pressure gradient force, the shearing force and the surface tension force, respectively. The Ca represents the effect of shearing force versus surface tension force, and the Eu is defined as the ratio of pressure gradient force and inertia force. Fig. 4(a) shows the relation of the dimensionless parameters (Ca×Eu) and Oh. The experimental data were closely fitted to the expression

$$\mathrm{Oh} \times 10^2 = 3.39\left(\mathrm{Ca} \times \mathrm{Eu}\right)^{-0.15} \tag{6}$$

This formulation shows the complex relationship between forces in the formation process of bubbles, and indirectly implies the existing relationship between bubble size and dimensionless parameter (Ca×Eu). The relation of bubble diameter and dimensionless parameter (Ca×Eu) is shown in Fig. 4(b). The experimental data were closely fitted to the expression

$$D = 66.53\left(\mathrm{Ca} \times \mathrm{Eu}\right)^{0.27} \tag{7}$$

This formulation presented a new method to predict the microbubble size corresponding to bubble formation mechanism. For (Ca×Eu)⩾15.73, the pressure gradient force acted as the leading role in the formation of long-slug bubbles in squeezing mode (D⩾140 μm). For (Ca×Eu)<15.73, the inertia force and the shearing force of the liquid phase controlled the generation of short-slug bubbles based on shearing mode (D<140 μm) significantly. As the value of dimensionless number (Ca×Eu) increased, the decrease of inertia force caused the increase of D due to the extension of the generation time of the microbubble.

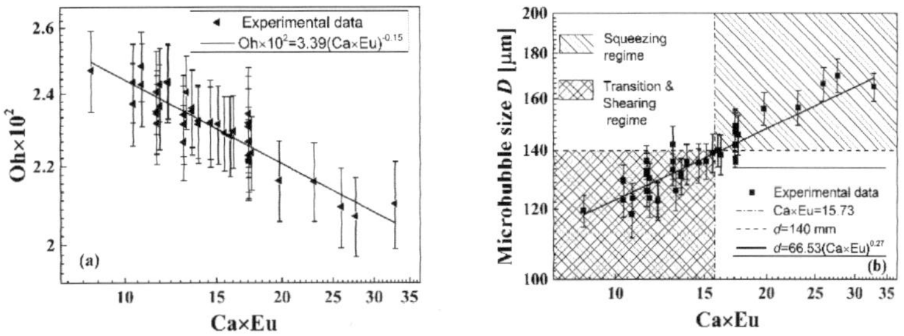

Figure 4. (a) The relation of Oh number and the dimensionless parameter (Ca×Eu); (b) The relation of the microbubble size and the dimensionless parameter (Ca×Eu).

4. Conclusion

The mechanisms of the process of microbubble formation and the effect of air pressure and velocity of liquid flow on the characteristics of microbubbles generated in a microfluidic T-junction were elaborately investigated based on experimental investigation. Three types of bubbles, including dispersed, short-slug and long-slug bubbles, were obtained according to the aspect ratio of the air slug and the microbubble generation frequency. The formation of these three types of bubbles can be accordingly divided into three modes: shearing, transition and squeezing. The air pressure and velocity of liquid played various roles in the bubble formation process and affected the microbubble size and generation frequency under these three modes. No matter in which mode, both the generation frequency and microbubble size increased when air pressure increased at the same velocity of liquid flow. The microbubble size decreased with velocity of liquid flow, however, there appeared a vertex of the variation curve of the generation frequency with velocity of liquid flow. The vertex symbolized the transition point of shearing to squeezing. Further investigation was carried out based on the dimensional analysis, and dimensionless Ca and Eu illustrating the relation of the inertia force, shearing force and pressure gradient force were used to describe the microbubble formation process. The microbubble diameter was well correlated with the dimensionless parameters Ca and Eu.

Acknowledgments

This work was financially supported by the Fundamental Research Funds for the central universities [DUT16QY39].

References

[1] B.K. Moore, D.G. Pahinkar, S. Garimella, Experimental and analytical investigation of displacement flows in microchannels, Int. J. Heat. Mass. Tran. 112 (2017) 745-757.
[2] P. Kluson, P. Stavarek, V. Penkavova, H. Vychodilova, S. Hejda, M. Bendova, M. Došek, Design of the reversible biphasic arrangement in the microfluidic chip reactor for asymmetric hydrogenation reactions, Chem. Eng. Res. Des. 153 (2020) 537-546.
[3] G.T. Vladisavljević, E.E. Ekanem, Z. Zhang, N. Khalid, I. Kobayashi, M. Nakajima, Long-term stability of droplet production by microchannel (step) emulsification in microfluidic silicon chips with large number of terraced microchannels, Chem. Eng. J. 333 (2018) 380-391.
[4] D. Rath, S. Panda, Modeling the transport and capture of analytes in a two-phase heterogeneous microfluidic immunosensor, Chem. Eng. Res. Des. 141 (2019) 272-278.
[5] J. Charmet, P. Arosio, T. Knowles, Microfluidics for Protein Biophysics, J. Mol. Biol. 430 (2018) 565-580.
[6] K. Jähnisch, V. Hessel, H. Löwe, M. Baerns, Chemistry in Microstructured Reactors, Angew. Chem. Edit. 43 (2004) 406-446.
[7] N. Rong, H. Zhou, R. Liu, Y. Wang, Z. Fan, Ultrasound and microbubble mediated plasmid DNA uptake: A fast, global and multi-mechanisms involved process, J. Control. Release. 273 (2018) 40-50.
[8] P. Liu, Y. Zhong, Y. Luo, Preparation of monodisperse biodegradable magnetic microspheres using a T-shaped microchannel reactor, Mater. Lett. 117 (2014) 37-40.
[9] X. Wang, C. Zhu, Y. Wu, T. Fu, Y. Ma, Dynamics of bubble breakup with partly obstruction in a microfluidic T-junction, Chem. Eng. Sci. 132 (2015) 128-138.
[10] A.S. Lobasov, A.V. Minakov, Analyzing mixing quality in a T-shaped micromixer for different fluids properties through numerical simulation, Chem. Eng. Process. 124 (2018) 11-23.
[11] T. Cubaud, M. Tatineni, X. Zhong, C.M. Ho, Bubble dispenser in microfluidic devices, Phys. Rev. E Stat. Nonlin. Soft. Matter. Phys. 72 (2005) 37302.
[12] Y. Hong, F. Wang, Flow rate effect on droplet control in a co-flowing microfluidic device, Microfluid. Nanofluid. 3 (2007) 341-346.
[13] T. Fu, Y. Ma, Bubble formation and breakup dynamics in microfluidic devices: A review, Chem. Eng. Sci. 135 (2015) 343-372.

[14] T. Fu, Y. Ma, D. Funfschilling, H.Z. Li, Bubble formation in non-Newtonian fluids in a microfluidic T-junction, Chem. Eng. Process. 50 (2011) 438-442.
[15] Q. Zhang, C. Zhu, W. Du, C. Liu, T. Fu, Y. Ma, H.Z. Li, Formation dynamics of elastic droplets in a microfluidic T-junction, Chem. Eng. Res. Des. 139 (2018) 188-196.
[16] M. De Menech, P. Garstecki, F. Jousse, H.A. Stone, Transition from squeezing to dripping in a microfluidic T-shaped junction, J. Fluid. Mech. 595 (2008) 141-161.
[17] T. Fu, Y. Ma, D. Funfschilling, C. Zhu, H.Z. Li, Squeezing-to-dripping transition for bubble formation in a microfluidic T-junction, Chem. Eng. Sci. 65 (2010) 3739-3748.
[18] M. Parhizkar, M. Edirisinghe, E. Stride, Effect of operating conditions and liquid physical properties on the size of monodisperse microbubbles produced in a capillary embedded T-junction device, Microfluid. Nanofluid. 14 (2013) 797-808.
[19] A.H. Dhanaliwala, J.L. Chen, S. Wang, J.A. Hossack, Liquid Flooded Flow-Focusing Microfluidic Device for in situ Generation of Monodisperse Microbubbles, Microfluid. Nanofluid. 14 (2013) 457-467.
[20] A. Evangelio, F. Campo-Cortés, J.M. Gordillo, Pressure gradient induced generation of microbubbles, J. Fluid. Mech. 778 (2015) 653-668.
[21] Y.K. Li, K. Wang, J.H. Xu, G.S. Luo, A capillary-assembled micro-device for monodispersed small bubble and droplet generation, Chem. Eng. J. 293 (2016) 182-188.
[22] S. Dicker, M. Mleczko, G. Schmitz, S.P. Wrenn, Size distribution of microbubbles as a function of shell composition, Ultrasonics. 53 (2013) 1363-1367.
[23] J.H. Xu, S.W. Li, J. Tan, G.S. Luo, Correlations of droplet formation in T-junction microfluidic devices: from squeezing to dripping, Microfluid. Nanofluid. 5 (2008) 711-717.
[24] J. Rodríguez-Rodríguez, A. Sevilla, C. Martínez-Bazán, J. M. Gordillo, Generation of Microbubbles with Applications to Industry and Medicine, Annu. Rev. Fluid. Mech. 47 (2015) 405-429.

A CFD-based study of twin-screw expander performance and inlet structure optimization

Shuai Lin

TianGong University, Tianjin, China

2031050621@tiangong.edu.cn

Abstract. Twin-screw expanders are widely used for small to medium-sized organic Rankine cycle (ORC) power generation. In order to improve the performance of twin-screw expanders in ORC, this study aims to improve the inlet efficiency and reduce the inlet pressure loss by changing the geometry of the inlet and shell. The CFD transient simulation is used to study the pressure, gas flow conditions and other parameters in the working chamber that are not easily obtained through experiments. The effects of improved size, speed, pressure ratio and internal volume ratio on the performance of the expander are analyzed based on the results. The results show that for the three expanders with different internal volume ratios, the optimized expander power and filling factor increases by 2%-14%. Isentropic efficiency of the expander reaches a maximum value of 0.73 when the pressure ratio is slightly greater than the internal volume ratio.

Keywords: twin-screw expander; CFD; inlet structural improvements; modeling.

1. Introduction

The Organic Rankine Cycle (ORC) is a promising technology for utilization of low temperature heat sources such as waste heat, solar energy and geothermal energy, by converting these heat sources into mechanical energy that can be used to drive generators to produce electricity.

Expander is key component that affects the overall performance of ORC. Wang et al. [1] made a summary of the application of screw machines and summarized the development history and research methods of screw machines. They pointed out that the development of twin-screw machine covered the research of liquid injection to the screw machine, reduction of synchronous gear consumption and reduction of pressure loss. Dawo et al. [2] conducted experiments on two expanders with different internal volume ratio and verified a mathematical model. For a given case, the model calculated an optimum internal volume ratio of 2.62 and the steam pressure of 6.6 bar. Papes et al. [3] performed 3D transient CFD calculations for twin-screw expander. The method of increasing the air intake by adding additional ports in the radial direction is investigated. The results showed that increasing the intake ports can effectively increase the output power of the expander. But it was accompanied by a decrease of isentropic efficiency and specific power in a certain degree. Byeon et al. [4] investigated three prototype oil-free twin-screw expanders with different intake port areas and rotor lengths by means of p-θ diagrams. Expander operation process was not a simple reversal of the compressor. The inlet pressure loss greatly affects the performance of the expander, especially at high speeds.

Content from this work may be used under the terms of the Creative Commons Attribution 3.0 licence. Any further distribution of this work must maintain attribution to the author(s) and the title of the work, journal citation and DOI.
Published under licence by IOP Publishing Ltd

In this study, a new intake structure is proposed to increase the intake area by increasing the area of the intake adapter plate and removing a part of the shell. This can increase the intake efficiency and also avoid the effect of temperature deformation of the structure in the high pressure area. The working process of twin-screw expander with different improvement sizes is calculated by CFD to find out the suitable improvement size. The performance of the expander before and after improvement is analyzed by changing the internal volume ratio, pressure ratio and rotational speed.

2. Modeling and geometric structure of twin-screw expander

2.1. Selecting a Template

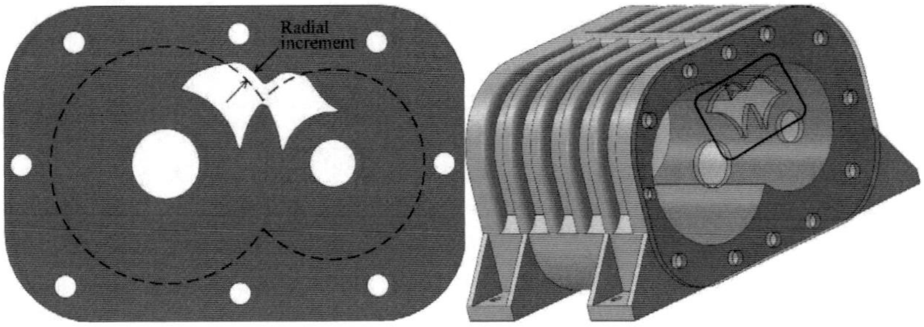

Fig. 1 Optimized adapter plate (above) and shell (below) structure

Fig. 1(above) shows structural improvement for the adapter plate. The two circular arcs are the boundary lines of the inner wall of the shell. The radial diameter of the increased area is defined as radial increment. Fig. 1 (below) indicates the modified structure, where the shell is enlarged by area according to the expansion of the adapter plate. The working mass can enter the working chamber from the enlarged part.

2.2. Theoretical model of twin-screw expander

The performance of the twin-screw expander is based on two indicators: isentropic efficiency and filling factor. For isentropic efficiency, defined as the ratio of the power generated during the expansion of the working fluid to the power generated by the isentropic enthalpy drop of the working fluid, can be expressed by (1).

$$\eta_{is} = \frac{P_{real}}{\dot{m}_{in} \cdot (h_{in} - h_{ex})} \tag{1}$$

Where P_{real} is the shaft power calculated in the software that is the sum of the torque around the shaft applied to the surface of the screw. The filling factor is determined by actual mass flow rate into the working chamber of expander and the theoretical mass flow rate.

$$\Phi = \frac{\dot{m}_{in,chamber}}{\dot{m}_{in,shut,theory} + \dot{m}_{leak}} \tag{2}$$

Where $\dot{m}_{in,shut,theory}$ can be calculated using (3), the density uses the vapor density at the inlet. The filling factor can be used to describe leakage and inlet pressure drop. However, to the best of our knowledge, most of previous studies focus on the leakage and little attention has been paid to the research of the inlet pressure drop. In this study, the actual mass flow rate derived from the density, and thus the filling factor, is more representative of the inlet force loss.

The mass flow rate into the working chamber is calculated using the following equation:

$$\dot{m}_{in,chamber} = N \cdot n \cdot \frac{V_{swept,low}}{r_i} \cdot \rho_{in,chamber} \tag{3}$$

Where N is the number of teeth of the male rotor, r_i is the build-in ratio, $\rho_{in,chamber}$ is density determined according to the chamber pressure $P_{in,chamber}$ and the entropy S_{in} at the inlet. Where $P_{in,chamber}$ is obtained by setting the reference point during the transient simulation analysis.

3. CFD experiment setup

As shown in Fig. 2, the geometry of the twin-screw expander in the current simulation is divided into three main parts: inlet domain, rotor domain, and outlet domain. Since the geometry of the inlet and outlet fluid domains does not move, Cartesian binomial tree grid is adopted. The meshing of the rotor fluid domain is performed using a structured mesh based on the SCORG software, which is meshed by adaptive boundaries [5]. The quality of the grid delineation is critical to the accuracy of the results. Theoretically, the more the number of grids, the more accurate the calculation will be, but the longer the calculation will take. After comparison of the mesh-independent analysis, the number of radial meshes for one chamber part of the male rotor is determined to be 50. Number of cells in each section is shown in Table 1.

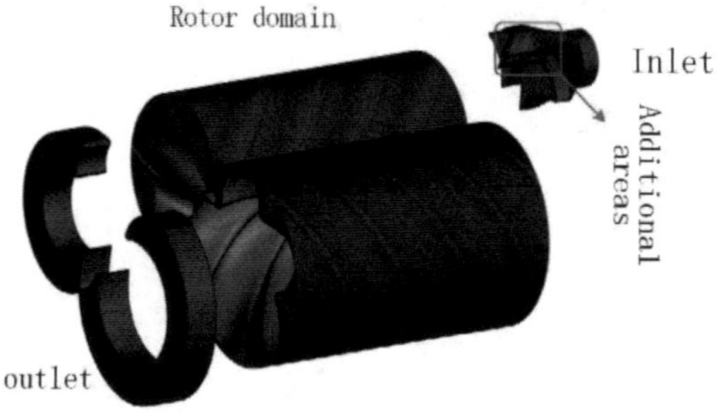

Fig. 2 Meshing of the fluid domain

Table 1. Number of grids in different areas Table Type Styles

	Rotor domain	outlet	4ratio inlet	5ratio inlet	6ratio inlet
Number of cells	506940	62642	68784	45431	39758

The following assumptions are made:

- Leakage of the cross-section is not calculated in the model.
- The mechanical losses during the operation of the expander are not calculated in the model.
- Heat transfer from the machine with the fluid and from the machine with the outside world is not included.
- The deformation of the rotor and shell caused by temperature and pressure during operation is ignored.

The fluid domain calculations of the twin screw expander are performed by PumpLinx software, where the motion of the grid can be set directly in terms of velocity and direction. The equation of state of the gas is set to ideal gas. The energy control equation selects the thermal module. The RNG κ-ε turbulence model is adopted to improve the accuracy. The equation is as follows:

Turbulent pulsation equation (k-equation):

$$\frac{\partial}{\partial t}\int_{\Omega(t)} \rho k d\Omega + \int_{\sigma} \rho((v-v_{\sigma})\cdot n)kd_{\sigma} = \int_{\sigma}\left(\mu + \frac{\mu_t}{\sigma_k}\right)(\nabla k \cdot n)d\sigma + \int_{\Omega}(G_t - \rho\varepsilon)d\Omega$$

Turbulent dissipation rate equation:

$$\frac{\partial}{\partial t}\int_{\Omega(t)} \rho\varepsilon d\Omega + \int_{\sigma} \rho((v-v_\sigma)\cdot n)\varepsilon d_\sigma = \int_{\sigma}\left(\mu+\frac{\mu_t}{\sigma_k}\right)(\nabla\varepsilon\cdot n)d\sigma + \int_{\Omega}(c_1 G_t \frac{\varepsilon}{k} - c_2(RNG)\rho\frac{\varepsilon^2}{k})d\Omega$$

In the formula:

$$c_2(RNG) = c_2 + \frac{c_\mu \eta^3(1-\eta/\eta_0)}{1+\beta\eta^3}, \eta = \frac{k}{\varepsilon}\sqrt{p}$$

The fluid work material in the ORC cycle is selected as R245fa. Their material properties are in TableII. The inlet conditions for the fluid domain of the expander are constant pressure and temperature inlet. The outlet condition is a constant pressure outlet. The fluid domains are connected to each other using interface surfaces. All walls are set to be adiabatic. The output of the CFD results in a positive rotor power and an inlet mass flow rate.

Table 2. Properties of R245fa

Property	Symbol	Value
Molecular weight	M	134.0482 g/mol
Acentric factor	ω	0.3724
Boiling point temperature	T_B	288.29K
Critical pressure	P_C	3.64Mpa
Critical density	ρ_C	517 m^3/kg
Critical temperature	T_C	427.2K

4. Result discussion

Fig.3 shows the pressure distribution in the V-shaped working chamber for one cycle before and after the expander improvement. The internal volume ratio is set to 4.3 and the speed is set to 4000 r/min. It can be seen that the pressure in one working chamber is essentially the same. The moments indicated in the figure are (1) suction process, (2) suction shutdown, (3) expansion process, (4) exhaust after completion of expansion. It can be seen that the pressures in (2), (3) and (4) in the figure below are higher than those in the figure above. This shows that the pressure in the working chamber of the improved expander is higher under the same conditions. This indicates that the improved expander can reduce the effect of inlet pressure loss.

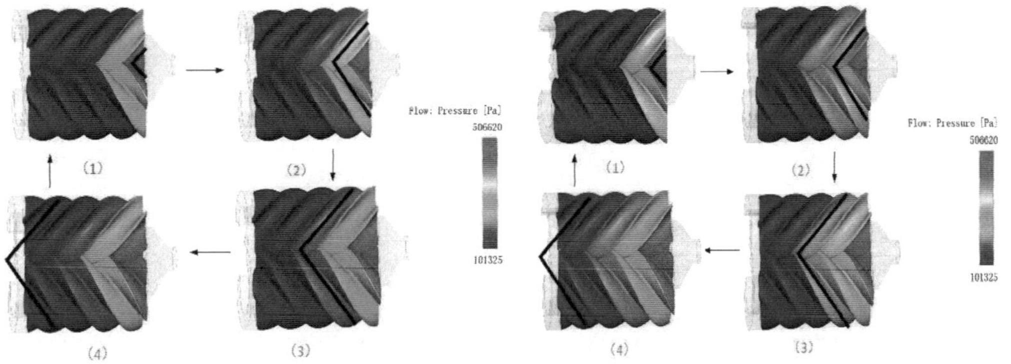

Fig. 3 The pressure distribution befor (above)and after (below) the improvement.

The p-v diagram of the machine can directly reflect the work done by the gas in the machine. In this study, the pressure data in the working chamber is collected by setting up reference points. Fig.4 shows the P-V diagram of the expander with an internal volume ratio of 5 and different radial increments. The inlet pressure is 6 times the atmospheric pressure. It can be seen that the pressure in the chamber increases twice and decreases twice during the inlet process. This is due to the fact that the inlet is a process that grows from 0 to the maximum value and then reduces to 0. It is ideal for the chamber pressure to be equal to the inlet input pressure when the inlet is closed, but the presence of pre-expansion and other phenomena causes the pressure in the working chamber to be lower than the pressure at the inlet when the inlet pressure is closed, a phenomenon also known as inlet pressure loss. The graph shows that the inlet pressure loss ΔP occurs in all three expanders, and ΔP is the largest when the radial increment is not increased, representing the largest pressure loss. When the radial increment is 6 mm, the ΔP value is the smallest. It means that increasing the radial increment can reduce the inlet pressure loss of the expander under this condition. The simulation process also analyzes the case of increasing the radial direction by 4 mm, and the p-v plot is between 2 and 6 mm.

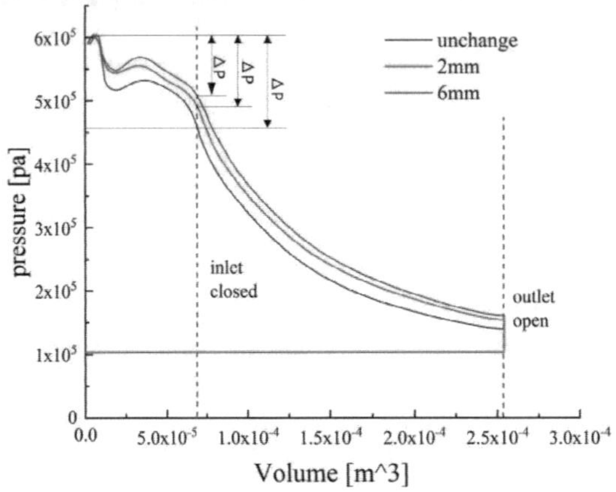

Fig. 4 p-v plot for various radial dimensions

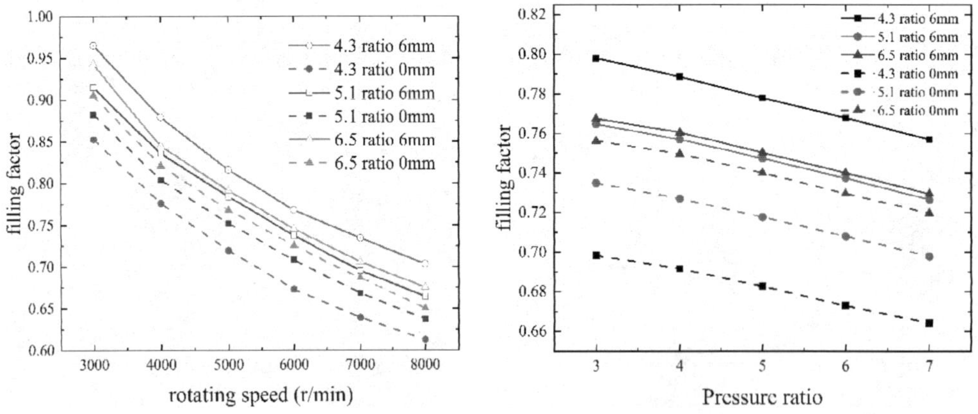

Fig. 5 Variation of filling factor with rotating speed (up)and pressure ratio(down).

Fig. 5 shows the effect of different working conditions of the twin-screw expander on the filling factor. The above figure shows that the filling factor decreases with increasing rotating speed. The rate of increase of the filling factor of the improved expander is related to the internal volume ratio. The greatest increase in the filling coefficient was observed for the expander with an internal volume ratio of 4.3 and the smallest increase in the filling coefficient was observed for the expander with a volume ratio of 6.5. It follows that the smaller the internal volume ratio, the more significant the improvement in fill factor by increasing the radial increment. The graph below shows the effect of pressure ratio on the filling factor, and it can be seen that the larger the pressure ratio the smaller the filling factor. It is due to the fact that an increase in pressure leads to an increase in leakage, so the fill factor decreases.

Fig.6 shows the variation of isentropic efficiency for different operating conditions of the twin-screw expander. It can be seen that the isentropic efficiency first increases and then decreases as the speed and pressure ratio increases. The above graph indicates that at the highest value of isentropic efficiency, the larger the internal volume ratio, the lower the corresponding speed. The expander with a small internal volume ratio has a more significant increase in power after the improvement, but the isentropic efficiency decreases. The below figure indicates that the isentropic efficiency can reach the maximum value when the pressure ratio of the inlet and outlet of the expander is slightly larger than the internal volume ratio. The isentropic efficiency decreases significantly when the pressure ratio is lower than the internal volume ratio, while it decreases slowly when the pressure ratio is higher than the internal volume ratio. This indicates that the effect of over-expansion on the isentropic efficiency of the expander is more than the effect of under-expansion.

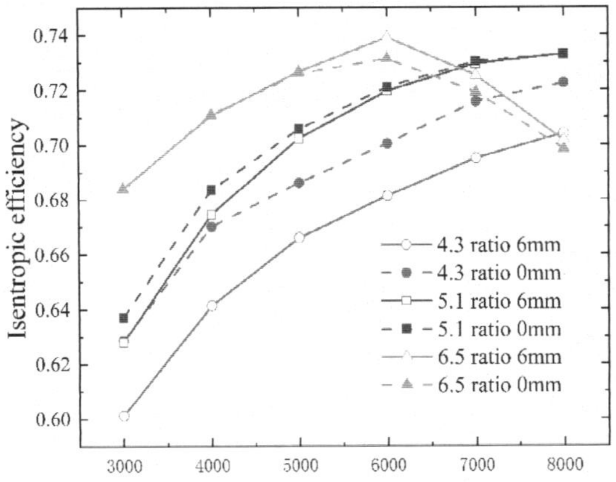

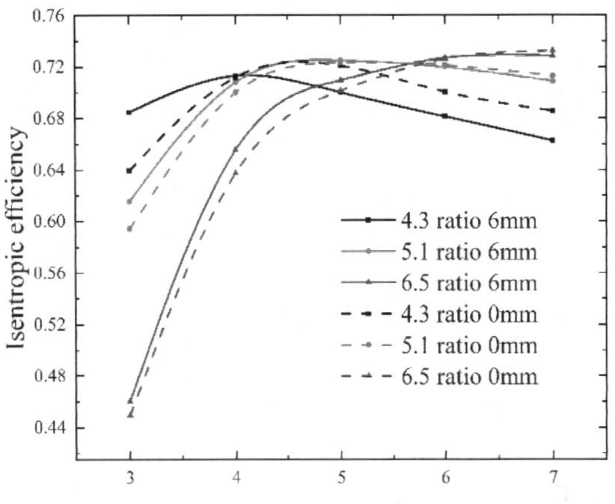

Fig. 6 shows the variation of the filling factor (above) and isentropic efficiency (below) for different cases.

5. Conclusion

The area of the inlet is increased by modifying the structure of the shell and inlet adapter plate. Three-dimensional transient CFD simulation is performed for twin-screw expanders with different internal volume ratios, rotational speed and different degrees of improvement. The mass flow rate and power of the expander increase with increasing pressure ratio and rotational speed for the same exhaust conditions. But the higher the rotational speed, the smaller the filling factor. The improved expanders with internal volume ratios of 4.3, 5.1 and 6.5 can increase the filling factor by 13.9%, 6.7% and 2.1%, respectively. The improved expander reduces the inlet pressure loss. It enables the expander to better avoid the influence of over-expansion phenomenon during the variable working condition with unstable heat source. For expanders with different internal volume ratios, the highest value of isentropic efficiency is achieved when the pressure ratio is slightly greater than the internal volume ratio, with the highest value being about 0.73.

References

[1] Wang C, Xing Z, Sun S, et al. Experimental study on the performance of oil-free twin-screw expanders for recovering energy in fuel cell systems[J]. Applied Thermal Engineering, 2020, 165.

[2] Dawo F, Eyerer S, Pili R, et al. Experimental investigation, model validation and application of twin-screw expanders with different built-in volume ratios[J]. Applied Energy, 2021, 282.

[3] Papes I, Degroote J, Vierendeels J. New insights in twin screw expander performance for small scale ORC systems from 3D CFD analysis[J]. Applied Thermal Engineering, 2015, 91: 535-546.

[4] Byeon S S, Lee J Y, Kim Y J. Performance Characteristics of a 4 × 6 Oil-Free Twin-Screw Compressor[J]. Energies, 2017, 10(7): 945.

[5] Kovacevic A. Boundary adaptation in grid generation for CFD analysis of screw compressors[J]. International Journal for Numerical Methods in Engineering, 2005, 64(3): 401-426.

AUTHOR INDEX